Systematics and Phylogeny of Weevils

Systematics and Phylogeny of Weevils

Volume 1

Special Issue Editors

Rolf Oberprieler
Adriana E. Marvaldi
Chris Lyal

MDPI • Basel • Beijing • Wuhan • Barcelona • Belgrade

Special Issue Editors
Rolf Oberprieler
Australian National Insect Collection
Commonwealth Scientific
and Industrial Research Organisation (CSIRO)
Australia

Adriana E. Marvaldi
The National Scientific and Technical
Research Council (CONICET)
Universidad Nacional de La Plata
Argentina

Chris Lyal
The Natural History Museum
UK

Editorial Office
MDPI
St. Alban-Anlage 66
4052 Basel, Switzerland

This is a reprint of articles from the Special Issue published online in the open access journal *Diversity* (ISSN 1424-2818) from 2018 to 2019 (available at: https://www.mdpi.com/journal/diversity/special_issues/Systematics_Phylogeny_Weevils)

For citation purposes, cite each article independently as indicated on the article page online and as indicated below:

LastName, A.A.; LastName, B.B.; LastName, C.C. Article Title. *Journal Name* **Year**, *Article Number*, Page Range.

Volume 1
ISBN 978-3-03897-656-1 (Pbk)
ISBN 978-3-03897-657-8 (PDF)

Volume 1-2
ISBN 978-3-03897-670-7 (Pbk)
ISBN 978-3-03897-671-4 (PDF)

Cover image courtesy of CSIRO.

Contents

About the Special Issue Editors

Rolf Oberprieler is a senior research scientist at the Australian National Insect Collection of the Commonwealth Scientific and Industrial Research Organisation (CSIRO) in Canberra, Australia. He worked on weevils at the National Insect Collection of the Plant Protection Research Institute in Pretoria, South Africa, for 17 years, mainly on the subfamily Entiminae but also on several other groups. For his Ph.D. degree he studied the systematics and biology of the African cycad-associated weevils, an interest that followed him to Australia in 1997. There he continued taxonomic research on broad-nosed weevils (Entiminae and Cyclominae) but also on some taxa of economic importance. He authored and co-authored a comprehensive catalogue of Australian weevils and several chapters on weevils in the Handbook of Zoology, as well as papers on weevil fossils and phylogeny. With almost 40 years of experience in weevil systematics, he has a broad knowledge of the weevil fauna of the Southern Hemisphere.

Adriana E. Marvaldi is as a research scientist of The National Scientific and Technical Research Council (CONICET) based at the Museum of Natural Sciences in La Plata, Argentina. Insects, especially beetles, have always fascinated her. She studied biology and earned her Doctorate degree in 1995 at the Faculty of Natural Sciences and Museum of the National University of La Plata. Adriana has focused her research on weevil systematics, starting her career at the Argentine Dryland Research Institute (IADIZA) in Mendoza, where she contributed to the foundation of an active entomology group and worked for 20 years. During this period she also spent time at the Museum of Comparative Zoology of Harvard University as a postdoctoral researcher. Her research work initially centred on the comparative morphology of the immature stages of weevils and the use of their characters in phylogenetic reconstruction, but she later also became interested in the morphology of adult weevils and in using molecular data in phylogenetic and evolutionary studies. Since 2015 Adriana has worked at the Entomology Division of the Museum of La Plata, where she continues pursuing her research interests in Coleoptera systematics, in particular of the weevil fauna of South America. Adriana's publications include several articles and book chapters spanning various taxonomic levels, including phylogenetic hypotheses on Curculionoidea, Belidae and curculionid groups such as Entiminae.

Chris Lyal is a scientific associate at the Natural History Museum in London, where he worked as a research scientist for more than 45 years. After research on the morphology and evolution of Heteroptera and Phthiraptera, he moved on to study weevil taxonomy, morphology, nomenclature and systematics. In the 1980s, he spent a year in New Zealand working with Willy Kuschel and preparing a handbook to New Zealand Cryptorhynchinae. He co-authored a global catalogue of weevil genus and family group names and with others is building a global species catalogue. Chris has published on key morphological features, in particular stridulation, sclerolepidia and the rectal system, on weevil seed predators, particularly of the South-East Asian tree family Dipterocarpaceae, and on a variety of other weevil groups. He is currently reviewing all of the tribes and subtribes of the subfamily Molytinae with the intent of creating a firm basis for further systematic research on this subfamily.

Preface to "Systematics and Phylogeny of Weevils"

Weevils (Curculionoidea) are one of the largest superfamilies of animals on Earth, comprising about 62000 described species in 5800 genera, but it has been estimated that about three times as many exist. Their tremendous diversity has been attributed to their close associations and co-radiation with angiosperm plants, but weevils have also evolved intimate relationships with gymnosperms (especially conifers and cycads) and other plant groups. As a consequence of their often highly specialized associations with plants, many weevils are regarded as pests of human agriculture and silviculture, whereas others are used as biological control agents of noxious weeds or as pollinators of crops such as oil palms. Weevils also play critical roles in native ecosystems, from herbivores and seed predators to pollinators to decomposers of dead and dying plants.

Weevil systematics and phylogeny have come a long way since the first comprehensive phylogenetic analysis of the group, published by Willy Kuschel in 1995, and the phylogenetic backbone of the superfamily (its family classification) outlined in that paper has been confirmed several times by later studies and is quite robust. However, intrafamilial relationships and natural groups (subfamilies and tribes) remain much less clear, particularly in the largest family, Curculionidae. Further study is also needed in fields such as comparative morphology, biogeography and patterns of host associations.

Not surprisingly for such a huge and diverse taxonomic group, advances in the systematics and phylogeny of weevils have largely occurred on regional levels and in treatments of genera and other groups scattered across the superfamily, with large-scale studies still needed to address big-picture questions about the evolution of the group effectively. Some collaborative efforts have recently begun to ameliorate this, notably the international cooperation to cover the weevils in the recent Handbook of Zoology and the weevil symposium and follow-up meeting at the 2016 International Congress of Entomology in Orlando, Florida. This Special Issue aims to continue this process and promote collaboration between weevil systematists as well as the dissemination of systematic information on these fascinating beetles. At the same time, it provides an apt forum to recognize and commemorate the significant contributions to the discipline made by the recently deceased Guillermo ("Willy") Kuschel, whose work on especially the phylogeny and higher classification of weevils has shaped our understanding of their evolutionary history like that of no-one else. This Special Issue therefore also serves as a memorial issue for him.

We are thrilled that our call to contribute papers to this Special Issue has been taken up so widely and enthusiastically that it can collate 31 papers spanning over 900 pages, both advancing our knowledge of weevil systematics and phylogeny on a broad front and also paying homage to Kuschel's impact on the field. The papers comprise 24 systematic studies, including seven phylogenetic ones, and five on host associations, diversity, distribution and biocontrol, as well as a summary of the proceedings of the weevil meeting in Orlando and a tribute to Willy Kuschel containing a biography and a summary of his contributions to weevil systematics, including also lists of all his publications and the taxa named after him. We extend our warmest thanks to *Diversity* for inviting this Special Issue, to all the colleagues who contributed their time and research results to this issue, to all the anonymous reviewers who ensured the quality of the papers and to the Editorial Staff of the journal for their sterling efforts in dealing so speedily and efficiently with all the manuscripts, reviewers' comments and various unforeseen problems. We hope that this Special Issue will form another milestone on the road to comprehending and appreciating the evolutionary success of these special beetles.

Rolf Oberprieler, Adriana E. Marvaldi, Chris Lyal
Special Issue Editors

Article

A Tribute to Guillermo (Willy) Kuschel (1918–2017)

Rolf G. Oberprieler [1,*], Christopher H. C. Lyal [2], Kimberi R. Pullen [1], Mario Elgueta [3], Richard A. B. Leschen [4] and Samuel D. J. Brown [5]

1 CSIRO Australian National Insect Collection, G. P. O. Box 1700, Canberra A. C. T. 2601, Australia; Kim.Pullen@csiro.au
2 The Natural History Museum, Cromwell Road, London SW7 5BD, UK; C.lyal@nhm.ac.uk
3 Área de Entomología, Museo Nacional de Historia Natural, Casilla 787, Santiago 8320000, Chile; mario.elgueta@mnhn.cl
4 New Zealand Arthropod Collection, Manaaki Whenua Landcare Research, Private Bag 92170, Auckland Mail Centre, Auckland 1142, New Zealand; leschenr@landcareresearch.co.nz
5 Plant & Food Research, Private Bag 92169, Auckland Mail Centre, Auckland 1142, New Zealand; Samuel.Brown@plantandfood.co.nz
* Correspondence: rolf.oberprieler@csiro.au; Tel.: +612-6246-4271

Received: 10 August 2018; Accepted: 12 September 2018; Published: 14 September 2018

Abstract: This tribute commemorates the life and work of Guillermo (Willy) Kuschel, who made substantial contributions to the understanding of weevil systematics, evolution and biology. Willy was born in Chile in 1918 and studied philosophy, theology and biology. He became fascinated by weevils early on and completed his Ph.D. degree on South American Erirhinini. Subsequent employment by the University of Chile provided him with many opportunities to further his weevil research and undertake numerous collecting expeditions, including to remote and rugged locations such as the Juan Fernandez Islands and southern Chile. In 1963 he accepted a position at the Department of Scientific and Industrial Research in New Zealand, where he became Head of the Systematics Group in the Entomology Division. His emphasis on field work and collections led to the establishment of the New Zealand Arthropod Collection, which he guided through its greatest period of expansion. His retirement in 1983 offered him increased opportunities to pursue his weevil research. In 1988 he presented a new scheme of the higher classification of weevils, which ignited and inspired much subsequent research into weevil systematics. The breadth and quality of his research and his huge collecting efforts have left a legacy that will benefit future entomologists, especially weevil workers, for decades to come. This tribute presents a biography of Willy and accounts of his contributions to, and impact on, the systematics of weevils both regionally and globally. All of his publications and the genera and species named after him are listed in two appendices.

Keywords: biography; obituary; weevils; systematics; publications

---

Guillermo (Willy) Kuschel on his 65th birthday, 13 July 1983.

1. Introduction

Guillermo (Willy) Kuschel was one of the outstanding and most influential weevil systematists of the past century. Over the course of his long life he amassed an immense knowledge of weevils, particularly of those of the Southern Hemisphere, which gave him a unique insight into the diversity, morphology and biology of this huge group of phytophagous beetles. While he made numerous contributions to the taxonomy and phylogeny of a variety of weevil groups, his most influential and enduring achievement is the new classification scheme of weevil families and subfamilies that he first proposed in 1988 at the XVIIIth International Congress of Entomology in Vancouver, BC, Canada. The resulting paper [1] is one of the most widely cited works on weevils of the last quarter of a century and has inspired several generations of subsequent workers to test it, refine it and build on it. Willy's contribution to weevil systematics and entomological science in general is, however, much greater and wider. He was an energetic and thorough field biologist, who organised and participated in expeditions to remote regions and islands throughout the Southern Hemisphere. His collecting and curation efforts, coupled with his distribution of specimens to colleagues around the world, have significantly advanced our understanding of particularly the insect faunas of southern islands and archipelagoes and of specific plant groups, such as conifers and *Nothofagus*.

Willy's contributions to weevil systematics were honoured at a symposium entitled *Phylogeny and Evolution of Weevils (Coleoptera: Curculionoidea): A Symposium in Honor of Dr. Guillermo "Willy" Kuschel*, held in 2016 during the XXVth International Congress of Entomology in Orlando, FL, U.S.A., and at a subsequent International Weevil Meeting that built on the topics and content of the symposium [2]. Due to his age and frail health, Willy was unfortunately unable to attend this symposium and meeting in person, but he sent his thanks and best wishes to the participants. He passed away the following year, shortly after his 99th birthday. As no proceedings of the Orlando weevil symposium and meeting were issued, Willy's colleagues around the world thought it appropriate to commemorate his manifold contributions to weevil systematics with a special journal issue that brings together a number of papers on weevil taxonomy, systematics, biology and evolution.

An obituary of Willy Kuschel was published last year, including an abbreviated list of his scientific publications [3]. In this tribute we pay greater homage to Willy's entomological achievements and the impact he has had on weevil systematics throughout the world. This paper features a more detailed biography of Willy, a complete list of his publications (Appendix A) and a list of all the taxa named after him, which stretches far beyond just weevils (Appendix B).

2. A Biography of Willy Kuschel

Guillermo Kuschel Gerdes was born on 13 July 1918 in Frutillar, southern Chile, where his great-grandfather, Heinrich Kuschel (1823–1873), had settled in 1855 after emigrating from Silesia, then Germany [4]. Willy was the sixth of 11 children born to Germán Pedro Kuschel Kruse (8 June 1887–5 April 1973) (Figure 1a), as the last child of Germán's first wife, Clara Augusta Gerdes Heise (9 August 1891–1918), who died soon after Willy's birth. Germán and Clara were married in 1908 in Puerto Varas. On 14 February 1920 Germán married again, Clara Neumann Wittwer (21 November 1891–19 May 1974), also in Puerto Varas.

Willy grew up on the family farm in a bilingual family, speaking German and Spanish fluently. He left home at the age of eight to attend boarding school, first in Puerto Varas and later in Santiago. After completing his high-school education, Willy entered a long period of continuous tertiary studies. Two years of studying philosophy at the University of Chile in Santiago followed by four years of theology in Buenos Aires led to his ordainment as a priest in the Society of the Divine Word (Sociedad del Verbo Divino, S.V.D.) in 1943. During this time he also developed his interest in science, in particular biology, and in 1945 he began a teaching degree at the University of Chile, while supporting himself by teaching high-school biology at his own Liceo Alemán. Although his initial research interests lay in botany, he quickly became fascinated by weevils through their associations with plants, and in 1947 he took a position at the University of Chile assisting in the Entomology course, which lead to a full research position three years later. At this time he began a doctorate in Biological Sciences, studying the biology and systematics of water weevils in the genus *Lissorhoptrus*. His Ph. D. degree, the first awarded by the University of Chile, was conferred in 1953. This research formed the basis of his lifelong promotion of the study of weevils and their host relationships, an area of research that had been neglected by most workers to that point. Willy was promoted to Head of the Entomology Department in 1956 and remained in that position for six years. During his time at the university, he served as president of the Sociedad Chilena de Entomología twice, from 1950 to 1952 and again in 1956, and he also founded the society's journal, *Revista Chilena de Entomología*, and edited it for six years.

For almost 20 years, from 1944 until his departure from Chile in 1962, Willy untertook collecting expeditions throught the country, from the extreme north to the southern tip, often visiting remote areas that had previously not or only poorly been explored biologically. Between 1951 and 1955 he spent three periods of two months each on the rugged Juan Fernandez Islands, where his determination and physical endurance resulted in the procurement of an enormous and highly important collection of insects. The lengths he went to to obtain specimens included descending into ravines on Masafuera (Alexander Selkirk Island) to collect chironomid midges [5] and scaling El Yunque, the highest point of Masatierra (Robinson Crusoe Island) and a rugged and barely accessible mountain that had only been climbed on seven occasions previously [6], on the summit of which he collected new species of carabid beetles and tipulid flies. Over 40 research papers based on his material were published in the *Revista Chilena de Entomología* between 1952 and 1955, and the value of his efforts was recognised by the Swedish Academy of Sciences awarding him the Linnaeus Medal in 1962 (Figure 1b).

Between March 1953 and March 1954 Willy travelled extensively through Europe, visiting insect collections in twelve different countries to inspect type specimens of weevils. This research resulted in numerous synonymies and other nomenclatural clarifications [7], and even today many specimens in European collections bear his determination and lectotype labels that reflect nomenclatural changes still to be published. This Europe trip was highly significant for Willy as it brought him into personal contact with many of the influential entomologists of the time, including Sir G. A. K. Marshall, Eduard Voss, Fritz van Emden and Willi Hennig. Willy spent three weeks in Berlin with Hennig, who lived in West Berlin but worked in East Berlin. Willy feared that the Russian authorities may have considered him a spy, as his passport showed evidence of his recent extensive travels, and so he left his documentation behind when going across the border with Hennig and friends.

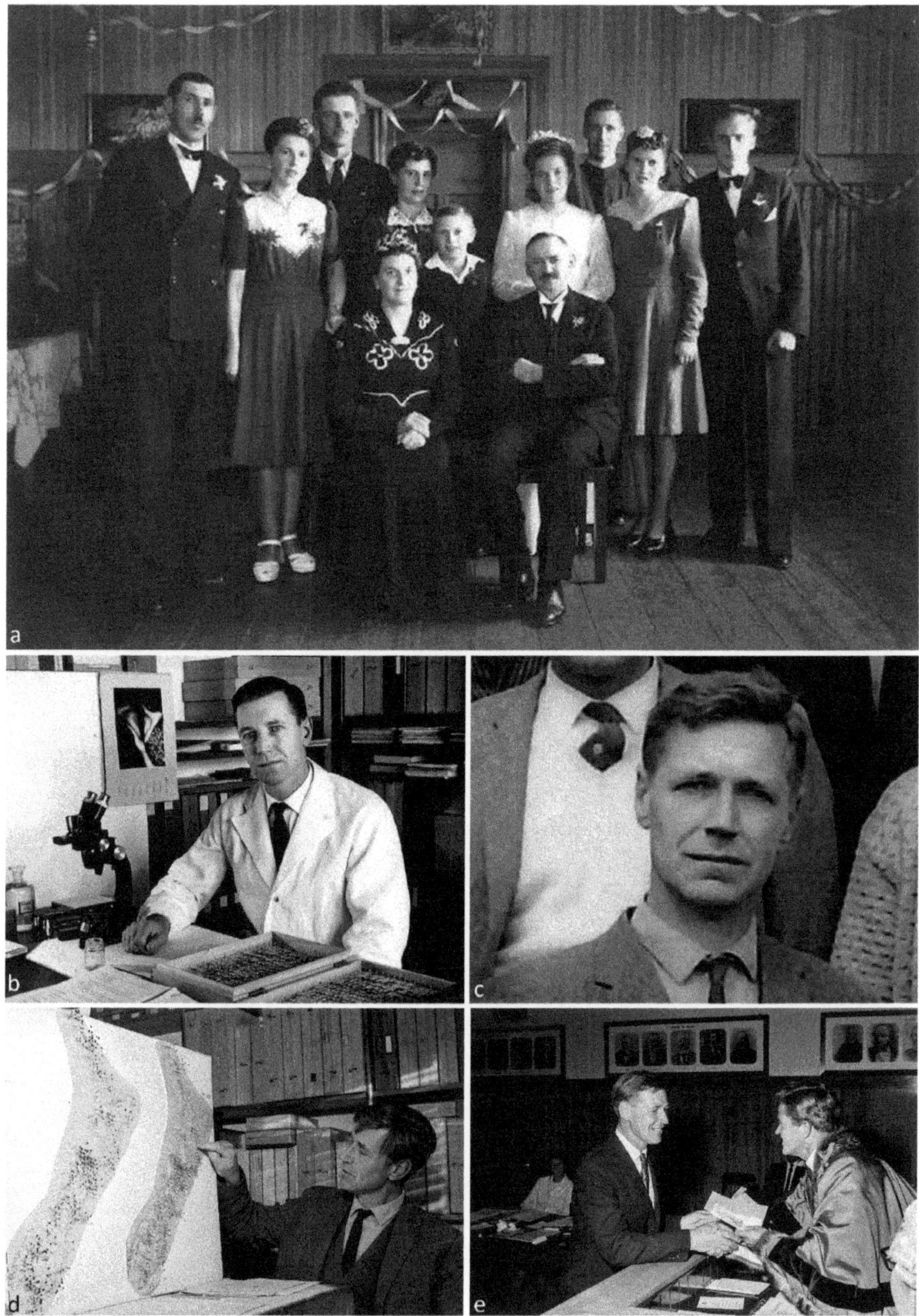

Figure 1. A portrait of Willy Kuschel: (**a**) with parents (Clara Neumann Wittwer and Germán Pedro Kuschel Kruse; seated) and siblings (f.l.t.r. Alberto, Clara, Arnoldo, María, Oscar, Olga, Guillermo, Adela, Evaldo) at parents' Silver Wedding anniversary, Frutillar, Chile, February 1945; (**b**) at time of reception of Linnaeus medal, 1962; (**c**) on D.S.I.R. staff photo, Entomology Division, Nelson, 1967; (**d**) indicating areas on New Zealand's South Island for further sampling, June 1969 (© NPN); (**e**) receiving his New Zealand citizenship papers from the Mayor of Nelson, April 1969 (© NPN).

In 1958/1959 Willy represented the University of Chile on an expedition to southern Chile, which was arranged by the Royal Society of London and also included three New Zealand scientists, among them the botanist Eric Godley, with whom Willy struck up a long-lasting friendship. In 1961 he was invited by the Royal Society of London to visit New Zealand and Australia. During this period of nine months he attended the 1961 Pacific Science Congress in Honolulu, Hawaii, where he presented papers on insect biogeography of southern South America and on his work on the insect faunas of the islands of the Eastern Pacific. After the conference he travelled to Australia and then worked in New Zealand for three months on the invitation of Eric Godley. During this time he met many people who were to become important associates, including John Townsend and Beverley Holloway.

In 1962 Dr. W. Cottier invited Willy to join the Department of Scientific and Industrial Research (D.S.I.R.) in New Zealand. This invitation gave Willy the opportunity to continue the westward research focus that he had already embarked upon, and he accepted. He applied for a year's unpaid leave from the University of Chile in early November and left for New Zealand the following month. His departure from Chile was necessarily abrupt, precipitated by political and personal differences with the director of the Centro de Investigaciones Zoológicas. Subsequent events in Chile proved the wisdom in his move, and Willy embraced his new life in New Zealand, becoming a New Zealand citizen in 1969 (Figure 1e). However, he never forsook his country of origin. He filled his garden with South American plants, and he was able to return to Chile on several occasions between 1983 and 2003, sometimes with his family. These trips usually combined visits to family with continuing research on the weevils of Chile (Figure 2c).

Willy's arrival at the D.S.I.R. spawned the establishment, in 1963, of a Systematics Group in the Entomology Division, which was initially located in Nelson. Under Willy's leadership (Figure 1c), the Systematics Group placed priority on comprehensive collecting in New Zealand (Figure 1d), initially focusing on previously unexplored habitats, such as alpine environments, but ultimately covering most of the country. Over the period 1965 to 1973, major expeditions were mounted (Figure 2a), with most available habitats thoroughly sampled and over 500 litter samples processed annually. Willy personally accompanied many of these collecting expeditions, often in association with Charles Watt, John Dugdale and John Townsend. Willy had an instinctive knack for collecting and an extraordinary ability to predict localities of significant diversity and abundance. These expeditions also resulted in legendary stories, such as his using a scalpel to butcher a sheep on the Chatham Islands. Collecting expeditions were also undertaken to the Galapagos Islands in 1964, Norfolk Island in 1967, Niue in 1975, Fiji in 1977 and New Caledonia in 1978 (Figure 2b).

In 1963 Willy and Beverley were married in a low-key ceremony and spent their honeymoon in Karamea. Shortly afterwards they began their family. Willy was not heavily involved in raising the children, especially during their early years, and Beverley shouldered the bulk of the domestic duties, particularly during Willy's frequent absences for collecting and research. His children remember him during their early years as being loving and kind, though they often felt they had to compete with insects for his attention. Family holidays were organised with insect collecting in mind, and the house was filled with entomological paraphernalia. As the children grew up, Willy's relationship with them became stronger, and he was proud of their achievements. Beverley was an excellent systematic entomologist in her own right and strongly influenced Willy's thoughts and ideas about character systems. Although they did not formally collaborate on any publications, Willy freely acknowledged his debt to her knowledge and insight. His achievements in New Zealand, especially his productivity during his retirement, were made possible through the love, support and patience of his family.

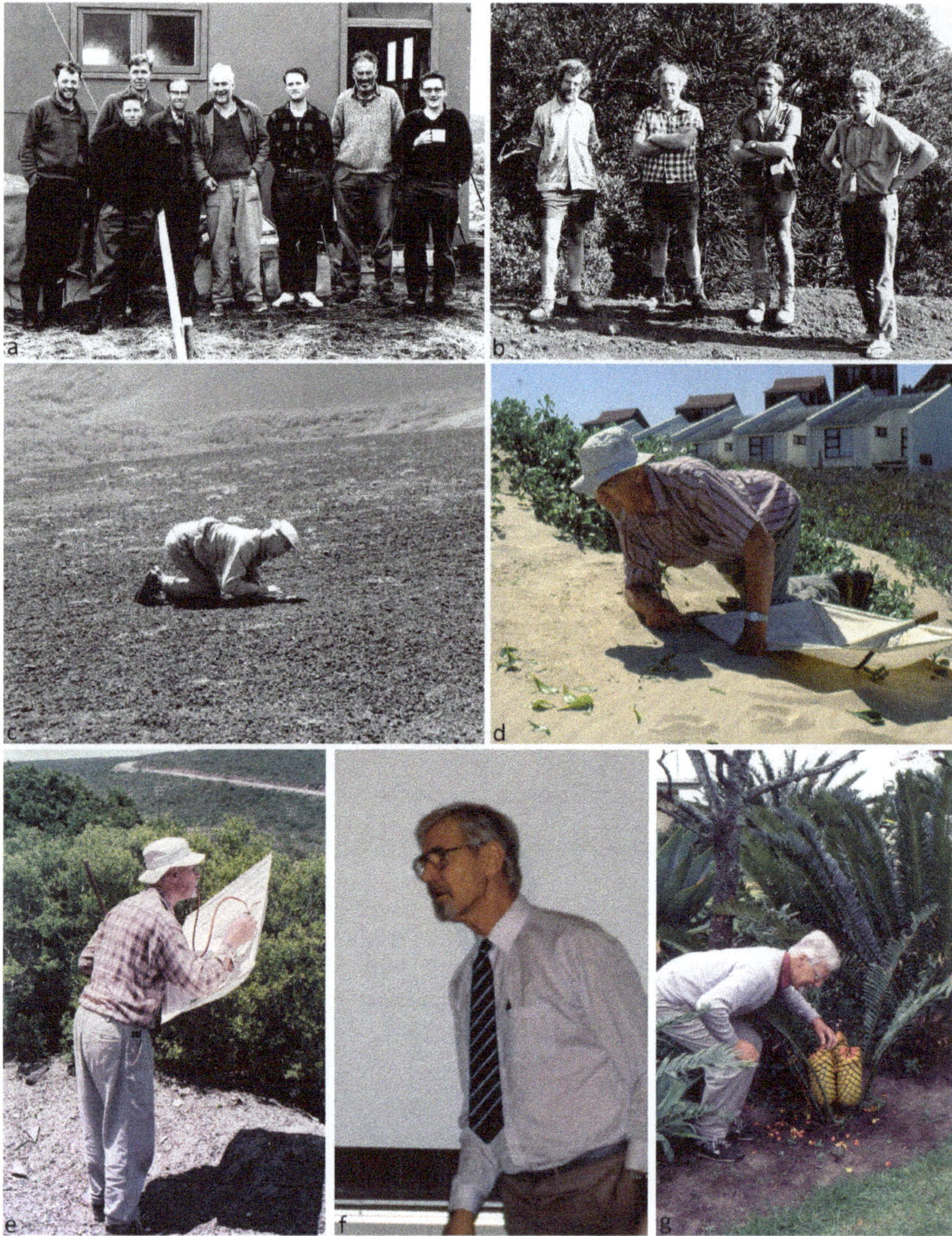

Figure 2. Willy Kuschel in action: (**a**) University of Canterbury Antipodes Expedition, 1969 (f.l.t.r. Rowley H. Taylor, Brian Bell, Guillermo Kuschel, John Warham, Eric Godley, Ian Mannering, Robert Stanley, Peter M. Johns) (photo: John Warham); (**b**) New Caledonia, 1978 (f.l.t.r. Charles Watt, John Dugdale, Peter Johnson, Guillermo Kuschel) (photo: Ken Fox); (**c**) investigating alpine plants, Antillanca, southern Chile, February 1997 (photo: Gerda Kuschel); (**d**) Port Alfred, South Africa, November 1992 (photo: RGO); (**e**) collecting the rare *Hispodes spicatus*, Ecca Pass, South Africa, November 1992 (photo: RGO); (**f**) lecturing at the I.C.E. weevil symposium in Vancouver, July 1988 (photo: RGO); (**g**) inspecting a cycad cone, Komga, South Africa, November 1992 (photo: RGO).

In 1973 the Systematics Group was moved from Nelson to Auckland. The disruption caused by this unpopular decision placed much strain on Willy. However, the event encouraged him to invest substantial time in the curation of the collection of New Zealand weevils. This massive effort of identifying and sorting specimens has resulted in the single-most useful resource currently available for weevils in the country. His comparison of specimens with the Broun types held by the Natural History Museum in London has allowed Broun's names to be used with a high degree of confidence, despite the lack of recent revisionary work.

Soon after arriving in Auckland, Willy started collecting insects in a small area of native bush close to his home in Lynfield. Before long this turned into a major study of the diversity of Coleoptera in an urban setting. It culminated in the publication of *Beetles in a suburban environment: a New Zealand case study* [8], usually termed the "Lynfield Catalogue", in which were provided details of the abundance, provenance and biology of 932 beetle species. On a personal level, the Lynfield Catalogue provided a useful memory aid for Willy. A copy was kept by the dining table and was frequently consulted when he needed to remind himself about names, host plants or abundance of beetles that came up in conversation. In May 1983 he participated in the retrieval of a rare deposit of subfossil beetles from the famous Waitomo Caves (Figure 3a), which included fragments of a large extinct molytine weevil he subsequently described as *Tymbopiptus valeas* [9].

Willy formally retired from the D.S.I.R. in 1983 but remained a research fellow with the Department (Figure 3b). His contributions to New Zealand entomology were recognised by his election as the inaugural Fellow of the Entomological Society of New Zealand in 1988. The Lynfield project and the extensive collections made by the D.S.I.R. Systematics Group, as well as his work on weevil systematics, were cited as his crowning achievements. Unfortunately, disagreements and personality clashes led to Willy's disillusionment with the D.S.I.R., with the result that he turned his research focus to the weevils of the Pacific, particularly those of New Caledonia. However, he retained a working relationship with staff at the New Zealand Arthropod Collection and periodically visited the collection until only a few months before his death.

In 1992 Willy had a chance to visit the only continent he had not yet been to: Africa. On his way to visit relatives in Chile he stopped over in South Africa, where he was hosted by Rolf Oberprieler and Schalk Louw. He spent a week with Rolf in Pretoria, looking at various wondrous African weevils in the National Collection of Insects and exploring the surrounding hills, then travelled to Bloemfontein to be impressed by the huge *Brachycerus* and other terricolous weevils at Schalk's breeding site and on to the Eastern Cape province (Figure 2d), where he encountered South African rarities such as *Somatodes* and *Hispodes* (Figure 2e) and various cycad weevils (Figure 2g) [10]. Back in Pretoria he studied several Cretaceous weevil fossils from Orapa, Botswana, with Rolf.

Willy was not only an entomologist but also an accomplished linguist. He grew up bilingual, speaking German and Spanish at home. During his studies in Chile he learned French and Italian, and classical Greek, Latin and Hebrew as part of his theological training. Only later in life did he add English to his linguistic repertoire, while assisting two English-speaking entomologists with their fieldwork in Chile. Over the period of this expedition, Willy taught himself English with the aid of an issue of *Time* magazine. He remained a subscriber to this magazine to the end of his life. Willy was passionate about the correct usage of language and terminology and enjoyed lengthy discussions about the origins, meanings and pronunciations of words. The numerous names he gave to new genera and species are not only etymologically correct but also commendably euphonic.

In his later years (Figure 3e), Willy found a lot of enjoyment in his garden (Figure 3c,d) and managed to pack an impressive number of plants into his backyard. These included flora from his native Chile (especially bromeliads), fruit and vegetables as well as several host plants for weevils. His garden provided many fascinating biological observations, including of *Nephila* golden orb-web spiders blown over from Australia and the first New Zealand record for several species of beetles. He enjoyed spending time in the backyard pool, often late at night, despite being unable to swim.

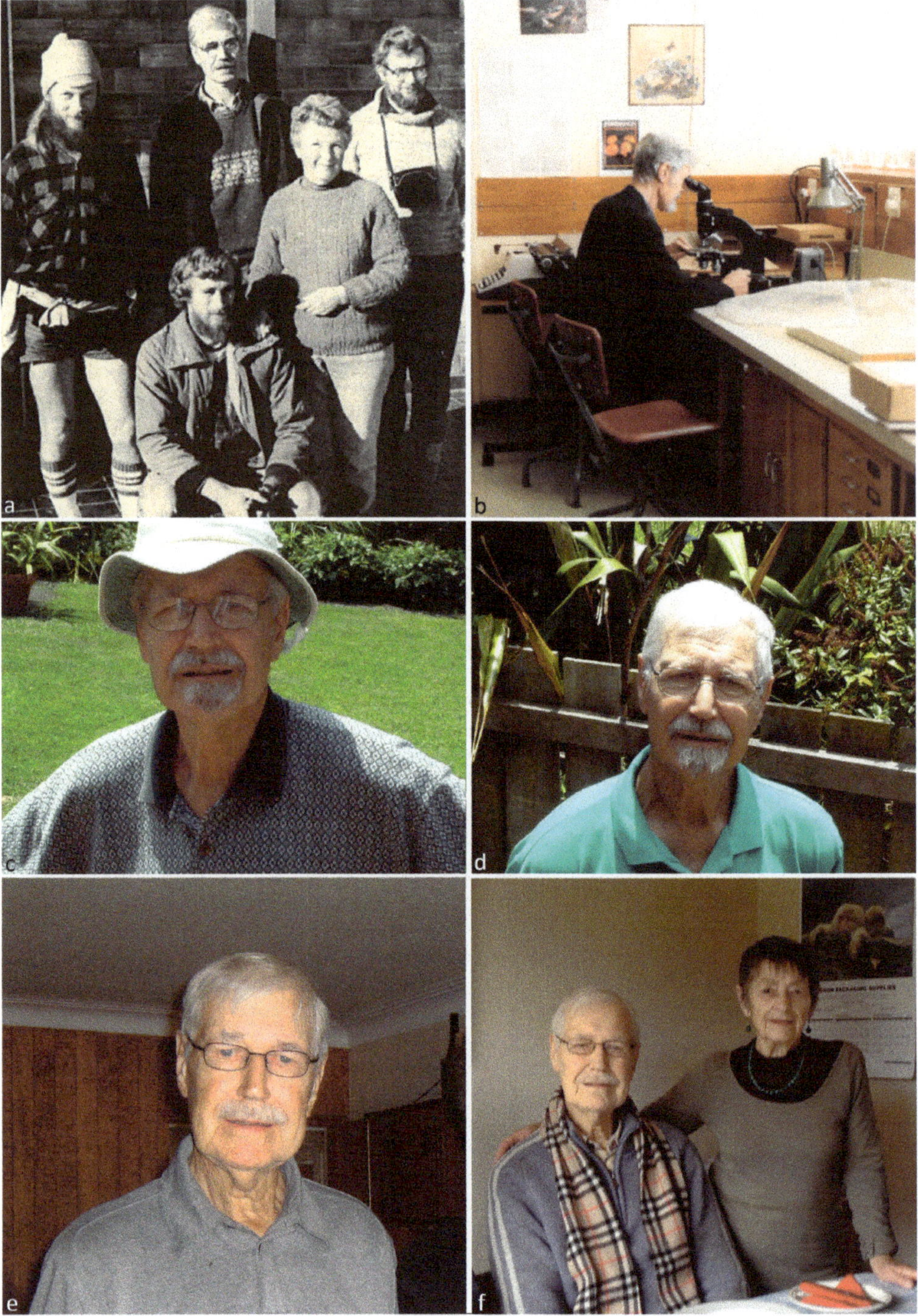

Figure 3. Willy Kuschel in later years: (**a**) with the beetle hunting team of the Waitomo Caves, New Zealand, May 1983 (f.l.t.r. Trevor Crosby, Charles Watt, Willy Kuschel, Brenda May, Trevor Worthy) (© Waitomo News); (**b**) working in his office, 1985 (photo: CHCL); (**c**) in his garden, January 2013; (**d**) in his garden, February 2013; (**e**) at home in Auckland, January 2011 (photo: SDJB); (**f**) at dinner at home with Beverley, September 2013.

Willy's emphasis on collecting, his broad taxon focus and his generosity with the specimens resulting from it have led to his being honoured by one tribe, 28 genera and 212 species from 23 orders named after him (Appendix B). They are also are a testament of the high esteem in which his scientific colleagues have held him throughout his life.

Despite a critical and sometimes adversarial manner, for Willy scientific research was very much about people. Despite his extensive fieldwork, he rarely spoke of it, instead discussing the work and ideas of others. He maintained extensive correspondence with scientists throughout the world. Praise did not come easily to him, but those he critiqued he generally held in high regard. His relationship with Elwood Zimmerman in Australia exemplified this characteristic of Willy's. Despite many disagreements between these two influential scientists, they kept in close contact and Willy felt Zimmie's death strongly.

Willy passed away in his sleep on 1 August 2017, three weeks after his 99th birthday. He is survived by his wife, Beverley (Figure 3f), their three children Gerda, Carl and Erika and their four grandchildren Alex, Oliver, Abigail and Elizabeth.

3. Willy Kuschel's Contributions to, and Impact on, the Systematics of Weevils

Willy Kuschel's contributions to weevil systematics extended over three quarters of a century, starting in 1943, when he was only 25 years of age, and ending in 2017, when he was 99. Three epochs can be identified in his work, the first of 20 years in Chile, another of two decades in New Zealand until his retirement, in 1983, and then another of almost 35 years in retirement, when he was relieved of administrative burdens and could direct his research interests more freely.

In Chile his work was largely concerned with collecting, taxonomic descriptions and revisions and some faunistics, but as it was published in Spanish and German, it reached mainly a regional audience. Towards the end of his time in Chile, his biogeographical publications, especially those about the eastern Pacific islands, brought him into contact with the broader scientific community and ultimately paved the way for his migration to New Zealand. In New Zealand he placed more emphasis on the exploration and study of island faunas, both of New Zealand and of other southern continents, as well as on a long-term study of the suburban beetle fauna of Lynfield in Auckland, near where he lived. In his retirement he published almost as much as he did during his employment years, and also his most significant works, in particular those on the world fauna of Nemonychidae, the new chrysomelid subfamily Palophaginae, parts of the New Caledonian fauna and, most importantly, those on weevil phylogeny and fossils. He made use of characters that had been largely neglected, even though, like some of his taxonomic changes, he mentioned them almost in passing in a paper apparently about something different. The breadth and depth of his work means that in almost any group of weevils, in any area of the world, there will be some contribution of Willy's that is relevant, sometimes crucially so.

3.1. South America

Willy published widely on the weevils of South America, particularly on the Chilean fauna (Appendix A), covering numerous groups in larger or smaller detail. Of particular significance is his work on the primitive families Nemonychidae and Belidae [11,12], the Erirhininae [13], Aterpini [14] and Listroderini and the entimine tribes Cylydrorhinini [15], Epistrophini [16] and Premnotrypini [17]. He also made significant contributions to the knowledge of the weevils associated with *Araucaria* and *Nothofagus*, partly scattered through his publications but the former associations later summarised more comprehensively [18]. Having also spent a large amount of time on collecting expeditions and departing from Chile rather abruptly in 1962, after 13 years of work at the Instituto de Zoología of the University of Chile, Willy necessarily had to leave quite a number of projects on South American weevils unfinished. Among them was his study of the Erirhininae, which he had expanded from his early work on *Lissorhoptrus* to all the South American genera, and even though he later translated his key to these genera into English and enlarged it to cover the world genera of the group, it has remained unpublished. One of his particular regrets (and sources of annoyance) also was that work he had commenced on the Cossoninae in South America had to be abandoned when he left Chile, together with the collection. Although he returned to Cossoninae several times (and was working on a paper on them when he died), he never managed to treat the group in the depth that he had planned.

Apart from his taxonomic studies of the weevils of South America and specifically of Chile, Willy's work in the region had a huge impact though his manifold contributions to the exploration of the entomofauna of especially Chile, both continental and insular. On his numerous expeditions he collected not only weevils but also other beetles, insects and invertebrates and even plants, and he was not content with having collected them but also went to great lengths to make the material available to specialists for study.

Between 1946 and 1949 Willy participated in three expeditions to the extreme north of Chile, the initial one being the first visit of any entomologist to this region, and besides insects he also collected plants for the herbarium of the Museo Nacional de Historia Natural in Chile. From 1946 on he went on several expeditions to the south of Chile, from Biobío to Llanquihue, the first together with his late Chilean friends Luis E. Peña (specialist on Tenebrionidae) and Ramón Gutiérrez (expert on Scarabaeidae) and later ones to Aisén, Magallanes, Tierra del Fuego and Navarino Island. He spent two months collecting on the Juan Fernandez archipelago on three occasions, in February/March 1951, from December 1951 to February 1952 and from December 1954 to February 1955, the last visit together with Prof. Carl Skottsberg. These expeditions yielded large and important collections of insects, which were studied by entomologists around the world and published on in volumes 1–5 of the *Revista Chilena de Entomología.* Of special interest are the weevils Willy collected on these islands because they are accompanied by important data of their host plants; this collection, presently at Landcare Research, New Zealand, still awaits study.

From September 1958 to March 1959, Willy participated in the expedition to southern Chile organised by the Royal Society of London, which was led by Martin Holdgate (University of Durham, England) and also included the New Zealanders Eric Godley (botanist at D.S.I.R.), George Knox (marine biologist at Canterbury University) and William Watters (geologist with the New Zealand Geological Survey). The expedition explored the region from the Chiloe Archipelago and Wellington Island southwards to Navarino Island in the Beagle Channel and Cape Horn.

In November 1960 Willy spent a month on the isolated volcanic island of San Ambrosio, one of the larger islands of the Desventuradas, studying its topography, naming ravines and plains and preparing a synoptic map as well as describing its vegetation and bird fauna and collecting plants and invertebrates [19]. He sent a sample of the insects to the British Museum of Natural History in London, and from his plant samples Carl Skottsberg described a new genus of Cariophyllaceae and a new species of *Eragrostis* (Poaceae). Afterwards Willy spent 15 days on the small continental Mocha Island, near the coast of Arauco in the Biobío Region of southern Chile, which is of interest because of the absence of *Nothofagus* species on it, despite their presence at the same latitude on the nearby continent.

Willy also explored the insect faunas of several other South American countries. He made collections in Argentina in 1943, 1944, 1948, 1956 and 1957, in the Paraná Delta, Uspallata, Mendoza, Buenos Aires and Tucumán. He collected on the altiplano of Bolivia and Peru on three occasions. At the end of 1946 he visited Lima, Junín and Tingo María in Peru, from December 1948 to March 1949 he explored the Yungas on the eastern slopes of the Andean Range and the basins of the rivers Beni and Mamoré, visiting places such as Titicaca, Oruro, Cochabamba and Trinidad in Bolivia and Puno, Marcapata and Cuzco in Peru, and in July 1957 he again visited Lake Titicaca and Rurrenabaque in Bolivia. In 1964, after he had left Chile, he was one of the nearly fifty researchers invited by the California Academy of Sciences, University of California at Berkeley and the Bernice P. Bishop Museum to visit the Galápagos Islands of Ecuador.

3.2. *New Zealand*

Willy's arrival in New Zealand heralded a substantial change in focus from his previous work. His appointment as head of the Systematics Group initiated the establishment of the New Zealand Arthropod Collection, which employed his talent and passion for collecting and curation as well as his international connections, charisma, charm and ability to inspire others. It did, however, take time away from research and the preparation of publications. Although Willy began much research on many groups of weevils during his early years in New Zealand, a lot of the work he did was not published in his lifetime, and his archives contain a wealth of research results. He compiled a set of index cards for most if not all New Zealand weevils, and on these he recorded unpublished synonymies and new combinations as well as other observations.

Soon after his arrival, Willy thoroughly revised the weevils of New Zealand's subantarctic islands. This work culminated in two papers [20,21], which still provide the most detailed study of members for many New Zealand weevil groups. In these papers Willy also started to develop his ideas of weevil classification, in particular recognising the basal position of the Erirhininae. Another hallmark of the subantarctic papers was his restoration of the names of weevils published before, and overlooked by, Thomas Broun's seminal work on New Zealand beetles, including by Fabricius [22] and Schoenherr [23].

Willy's side project on the beetle fauna of the parks and reserves around his home in Lynfield again demonstrated his ability to distribute specimens to the right people, who were to provide identifications and descriptions of new species. Some of the beetle species described from the Lynfield material include scydmaenine rove beetles [24], scirtid beetles [25], ptiliid beetles [26] and weevils [8,27]. The Lynfield work also highlighted the diversity of beetles surviving in urban settings and the importance of forest fragments. It was an early and influential work in urban ecology, especially in the New Zealand and invertebrate contexts. Finally, Willy's attention to the biological information is apparent in the Lynfield Catalogue, with at least a modicum of biological information available for every species included in it. In many cases, these are the only biological data available for these species. Willy's attempts to understand the plant associations of weevils included detailed records of all plants growing around leaf litter sampling sites, to a level probably unmatched in other collecting regimes.

3.3. *New Caledonia*

Willy first visited New Caledonia in 1963 with colleagues from the Bishop Museum. His second visit took place from 3 October to 3 November 1978, with John Dugdale, Charles Watt, Ken Fox and Peter Johnson (Figure 2b). This expedition amassed a vast amount of material, which was to form the basis of much future work. After Willy's retirement, there was a time when his relationship with the D.S.I.R. became strained, and he began working in earnest on the weevil fauna of New Caledonia. Between 1990 and 2017 he published 10 papers on the weevils of New Caledonia, covering the Nemonychidae, Anthribidae, Curculioninae, Entiminae, Aterpini, Gonipterini and Myrtonymini. Three of these papers he contributed to the *Zoologia Neocaledonica* series, even though it meant quite lengthy delays in publication. Willy's body of work on the New Caledonian weevils is the most comprehensive coverage of the fauna of this island by a single author since Karl Heller [28].

3.4. *Australia*

Although Willy was also keenly interested in the Australian weevil fauna as it shares numerous elements with New Zealand and the wider Pacific region, he only got involved in its taxonomic study to a limited extent. This was partly because his focus lay on the New Zealand fauna and partly because in 1972, not too long after he arrived in New Zealand, Elwood Zimmerman ('Zimmie') migrated to Australia and embarked on an ambitious study of the fauna of this island continent. Willy had collected some weevils in eastern Australia in 1961 (and also during a later visit, in October 1979), and in the late 1960s he started taxonomic work on the Australian Phrynixini, Cossoninae and Erirhininae.

He published his study of the Phrynixini in 1972 [29] but handed over to Zimmie his work and specimens of the Cossoninae and Erirhininae, among which he had identified numerous new genera and species. Zimmie in turn invited Willy to study the Australian Nemonychidae for inclusion in his Australian Weevils monograph series [30]. Willy reciprocated by including Zimmie in his 2000 study of the Platypodinae [31], although Zimmie only agreed to this with hesitation as he felt that he had not contributed much and was not comfortable with phylogenetic analyses such as included in the study. In a number of later studies of weevils of the Pacific region, Willy included relevant Australian taxa, i.e. of Orthorhinini in 2008 [32], Cranopoeini in 2009 [33] and Myrtonymini in 2014 [34], and he also included the Australian genera of Belinae and Nemonychidae in phylogenetic analyses (with Rich Leschen), respectively in 2003 [35] and 2011 [36]. Numerous Australian weevil taxa were thus described by Willy Kuschel.

Willy visited Australia for a last time in December 1999, when he was invited to attend the John Lawrence Celebration Symposium in Canberra. The main drawing card for him was the attendance of Vladimir Zherikhin, of the Palaeontological Institute in Moscow, Russia, of the same symposium and the chance to discuss weevil fossils with him. Zherikhin had published some major papers on weevil fossils, but Willy did not agree with some of the interpretations and conclusions and was keen to debate these with Zherikhin in person. Zherikhin had brought a number of critical fossils with him from Moscow, in particular some Obrieniidae (one genus of which he had named after Willy), and before long an in-depth and lengthy discussion ensued between the two, evidently to mutual benefit as Zherikhin subsequently also excluded the obrieniids from Curculionoidea.

Willy could not meet Zimmie during this visit as the latter did not attend the symposium, but he kept in regular contact with Zimmie by phone. They discussed various weevil issues, mainly their differences of opinion on weevil classification, and struck up a strange but amicable relationship in this way, reminiscent of two old warhorses grazing together on the same paddock in their old days. On Zimmie's 91st birthday, in 2003, Willy sent him a congratulatory poem that he had composed in Latin. Zimmie treasured this as one of his most valuable birthday presents ever and lamented: "If only I could reply to him in kind!".

3.5. The World

The uniqueness of Willy Kuschel's contributions to global weevil systematics was probably his integration of the fauna of the Southern Hemisphere into the mainstream understanding of weevil classification and biology, which had evolved in Europe and North America and was centred on the fauna of the Northern Hemisphere. Other weevil taxonomists had of course studied the southern fauna before him, such as Fiedler, Hustache, Voss and others in South America, Broun in New Zealand, Blackburn and Lea in Australia and Marshall in Africa, but they generally tried to slot the faunas of these continents into the European framework of classification. Willy, in contrast, grew up and studied entomology in the Southern Hemisphere, learning about its weevils and their hostplants in the field and increasingly realising that they did not properly fit into the Lacordairean system. He was among the first to recognise the crucial differences in the male genitalia and accordingly redefined the Erirhininae, he proposed a new concept of Molytinae and he thoroughly revised the world fauna of Nemonychidae, in a number of papers. He studied poorly known southern groups of Curculionidae, such as Aterpini, Cranopoeini, Cylydrorhinini, Ectemnorhinini, Listroderini, Myrtonymini, Orthorhinini, Phrynixini and Premnotrypini. He was also well acquainted with the phylogenetically basal families Nemonychidae, Anthribidae, Belidae and Caridae and their characters, and when the method of cladistic analysis came of age in the 1980s, he had a character set available for all weevils to try it out. The analysis took several iterations, but by the time of the XVIIIth International Congress of Entomology, held in Vancouver in 1988 and for which a special weevil symposium was being organised, he had a revolutionary new classification in hand (Figure 2f). The abstract of his talk was innocuously titled 'Thoughts on past classifications of the weevils—how a new scheme may be attempted', but it was much more than an attempt, it was a well thought-through system of families and subfamilies that, published in 1995 [1], has stood the test of time and become synonymous with Willy's name.

Willy had perhaps the widest grasp of weevil morphology and diversity of any worker of his day. This enabled him to make connections and see patterns with great clarity, and it underpinned the systematic changes he proposed. His profound knowledge of weevil characters and higher taxa also allowed him to assess the weevil fossils that were described from Russia in the 1970s and 1980s. He concluded early (in 1983) that Arnoldi's Eobelidae were in fact extinct representatives of Nemonychidae [37], and he assessed these and other fossils (including the contentious Obrieniidae) in more detail in a later study of the Nemonychidae, Belidae and Brentidae of New Zealand [38]. He also described a few Cretaceous fossils from Chile, Botswana and Lebanon and reassessed the Baltic amber weevils described by Eduard Voss, which resulted in the recognition of a new subfamily of Brentidae, the Carinae (now the family Caridae) [39].

Willy Kuschel has had an outstanding impact on the development of weevil taxonomy and systematics. The breadth of his knowledge and publications, the challenging insights he developed and the freshness of his views make him one of the key workers in the taxonomic history of the group. Much more than that, he was a unique and powerful character, and anyone who met him will recall intense and lengthy discussions on topics of interest—weevils of course, but also linguistics, terminology and all the things that interested him. His contacts with researchers worldwide, his generosity in sharing his knowledge, his friendship and continued intellectual vitality have left an indelible mark on several generations, and we miss him.

Author Contributions: R.G.O. conceived the layout and wrote the first draft of the paper, and all others authors contributed to the final text. R.G.O. and S.D.J.B. compiled Appendix A, and K.R.P. and S.D.J.B. composed Appendix B.

Funding: This research received no specific external funding.

Acknowledgments: We thank Peter Atkins, John Dugdale, Beverley Holloway, Peter Johns, Gerda Kuschel and Annette Walker for sharing their memories and experiences with Willy. We are grateful to Landcare Research, New Zealand, for the use of Figures 1a–c and 3a,b from the Kuschel archives and to the Kuschel family for permission to publish them. For the use of Figure 1d,e we acknowledge the Nelson Provincial Museum in Nelson, New Zealand, as follows: Figure 1d–Guillermo Kuschel, Head of the Systematics Section, 1969. *Nelson Photo News*, Issue 104, June 1969, p. 50. Nelson Provincial Museum, Barry Simpson Collection: 35mm 1507_fr10; and Figure 1e–Guillermo Kuschel receives his papers from the Mayor, 1969. *Nelson Photo News*, Issue 101, April 1969, p. 78. Nelson Provincial Museum, Barry Simpson Collection: 35mm 1427A_fr12. Figure 2a is reproduced courtesy of the *Waitomo News* and was originally published on 31 May 1983. Other photo credits are: CHCL–Chris Lyal, RGO–R. Oberprieler, SDJB–Sam Brown. We sincerely thank Debbie Jennings (ANIC) for compiling the figures.

Conflicts of Interest: The authors declare no conflict of interest.

Appendix A

Publications by Guillermo Kuschel

1. KUSCHEL, G. (1943) Un gorgojo acuático del arroz" argentino, *Lissorhoptrus bosqi* n. sp. (Col. Curculionidae). *Notas del Museo de La Plata*, 8, 305–315.
2. KUSCHEL, G. (1945) Aportes entomológicos I (Curculionidae). *Anales de la Sociedad Científica Argentina*, 139, 120–136.
3. KUSCHEL, G. (1945) Aportes entomológicos (II) (Coleop. Curculionidae). *Revista de la Sociedad Entomológica Argentina*, 12 (5), 359–381.
4. KUSCHEL, G. (1946) Comentario a los tipos más antiguos de *Listroderes* de la obra de Schönherr (Aporte 4 de Col. Curculionidae). *Agricultura Técnica*, 6 (2), 135–140.
5. KUSCHEL, G. (1949) Los Curculionidae del extremo norte de Chile (Coleoptera, Curcul. Ap. 6). *Acta Zoologica Lilloana*, 8, 5–54.
6. KUSCHEL, G. (1950) I. Nuevos Curculionidae de Bolivia y Perú. II. Notas a algunas especies de Brèthes (Ap. 7 de Col. Curcul.). *Revista del Museo de La Plata*, 6, 69–116.
7. KUSCHEL, G. (1950) Nuevas sinonimias, revalidaciones y combinaciones (9 aporte a Col. Curculionidae). *Agricultura Técnica*, 10 (1), 10–21.
8. KUSCHEL, G. (1950) Nuevos Brachyderinae y Magdalinae chilenos (Coleoptera Curculionidae) (Aporte 5). *Arthropoda*, 1 (2/4), 181–195.
9. KUSCHEL, G. (1950) Die Gattung *Priocyphus* Hust. 1939 (10. Beitrag zu Col. Curculionidae). *Revista de Entomología*, 21 (3), 545–550.
10. KUSCHEL, G. (1950). Los Curculionidae de Tarapacá y Antofagasta (Insecta, Coleoptera). *Investigaciones Zoológicas Chilenas*, 1, 13–14.
11. KUSCHEL, G. (1952) Cylindrorhininae aus dem Britischen Museum (Col. Curculionidae, 8. Beitrag). *Annals and Magazine of Natural History*, (12), 5, 121–137.
12. KUSCHEL, G. (1952 ("1951")) Revisión de *Lissorhoptrus* LeConte y géneros vecinos de América (Ap. 11 de Coleoptera Curculionidae). *Revista Chilena de Entomología*, 1, 23–74.
13. KUSCHEL, G. (1952 ("1951")) Entomologische Arbeiten, Museum G. Frey, München. *Revista Chilena de Entomología*, 1, 128.
14. KUSCHEL, G. (1952 ("1951")) Las palabras compuestas de "tipo" son graves o esdrújulas en castellano? *Revista Chilena de Entomología*, 1, 146.
15. KUSCHEL, G. (1952 ("1951")) Conspice naturam; inspice structuram. *Revista Chilena de Entomología*, 1, 174.
16. KUSCHEL, G. (1952 ("1951")) IX. Congreso internacional de entomología. *Revista Chilena de Entomología*, 1, 204.
17. KUSCHEL, G. (1952 ("1951")) La subfamilia Aterpinae en América (Ap. 12 de Coleoptera Curculionidae). *Revista Chilena de Entomología*, 1, 205–244.
18. KUSCHEL, G. (1952) Los insectos de las Islas Juan Fernández. Introducción. *Revista Chilena de Entomología*, 2, 3–6.
19. KUSCHEL, G. (1952) Los Curculionidae de la cordillera chileno-argentina (I. parte) (Aporte 13 de Coleoptera Curculionidae). *Revista Chilena de Entomología*, 2, 229–279.
20. KUSCHEL, G. (1952) Dr. Herman Lent. *Revista Chilena de Entomología*, 2, 314.
21. KUSCHEL, G. (1952) Sr. Walter Wittmer. *Revista Chilena de Entomología*, 2, 314.
22. KUSCHEL, G. (1952) Prof. Dr. Kurt Wolfgang Wolffhügel (1969–1851). *Revista Chilena de Entomología*, 2, 314–315.
23. KUSCHEL, G. (1952) Willi Hennig, Die Larvenformen der Dipteren, Akademie-Verlag, Berlin. Tomo I (1948): 185 pp., 63 figs., 3 láminas; Tomo II (1950): 458 pp., 236 figs., 10 láminas; Tomo III (1952): 628 pp., 338 figs., 21 láminas. *Revista Chilena de Entomología*, 2, 319.

24. KUSCHEL, G. (1954) La familia Nemonychidae en la Región Neotropical (Aporte 15 de Col. Curculionidae) *Revista Chilena de Historia Natural*, 54 (9), 97–126.
25. KUSCHEL, G. (1954) Un gorgojo ciego de Otiorhynchinae de Madagascar (Aporte 14 de Col. Curculionidae). *Revue française d'Entomologie*, 21, 286–289.
26. KUSCHEL, G. (1955) A propos du *Typhlorhinus jeanneli* Kuschel (1954, *Rev. Fr. d'Ent.* XXI, p. 288). *Revue Française d'Entomologie*, 22 (1), 74.
27. KUSCHEL, G. (1955) Una nueva especie de *Cheloderus* Castelnau (Coleoptera Cerambycidae). *Revista Chilena de Entomología*, 4, 251–254.
28. KUSCHEL, G. (1955) Nuevas sinonimias y anotaciones sobre Curculionoidea (1) (Coleoptera). *Revista Chilena de Entomología*, 4, 261–312.
29. KUSCHEL, G. (1955) *Compsus serrans* n. sp., gorgojo dañino de la caña de azúcar en Venezuela (Aporte 20 de Coleoptera, Curculionidae). *Boletín de Entomología Venezolana*, 11 (3/4), 133–140.
30. KUSCHEL, G. (1956) Attelabidae und Curculionidae aus El Salvador (Ins. Col. Curculionidae, 21. Beitrag). *Senckenbergiana biologica*, 37 (3/4), 319–339.
31. KUSCHEL, G. (1956) Revisión de los Premnotrypini y adiciones a los Bagoini (Aporte 17 sobre Coleoptera Curculionoidea). *Boletín Museo Nacional de Historia Natural*, 26, 187–235.
32. KUSCHEL, G. (1957) Las especies sudamericanas de *Grypidiopsis* Champion (Aporte 22 de Col. Curculionoidea). *Revista Brasileira de Biologia*, 17 (1), 65–72.
33. KUSCHEL, G. (1957) Revisión de la subtribe Epistrophina (Aporte 19 de Col. Curculionoidea). *Revista Chilena de Entomología*, 5, 251–364.
34. KUSCHEL, G. (1958) Nuevo gorgojo de Costa Rica dañino al café (Col. Curculionoidea, aporte 24). *Investigaciones Zoológicas Chilenas*, 4, 135–137.
35. KUSCHEL, G. (1958) Nuevos Cylydrorhininae de la Patagonia. (Col. Curculionoidea, aporte 18). *Investigaciones Zoológicas Chilenas*, 4, 231–252.
36. KUSCHEL, G. (1958) Neotropische Rüsselkäfer aus dem Museum G. Frey (Col. Curcul.). 23. Beitrag. *Entomologische Arbeiten aus dem Museum G. Frey*, 9 (3), 750–798.
37. KUSCHEL, G. (1959) Un curculiónido del cretáceo superior, primer insecto fósil de Chile. *Investigaciones Zoológicas Chilenas*, 5, 49–54.
38. KUSCHEL, G. (1959) Nemonychidae, Belidae y Oxycorynidae de la fauna chilena, con algunas consideraciónes biogeográficas (Coleoptera Curculionoidea, aporte 28). *Investigaciones Zoológicas Chilenas*, 5, 229–271.
39. KUSCHEL, G. (1959) Reforestación e insectos. *Noticiario Mensual Museo Nacional de Historia Natural*, 40, 3.
40. KUSCHEL, G. (1959) Beiträge zur Kenntnis der Curculioniden von Venezuela und Trinidad-Insel (1. Lieferung). (Col. Curculionidea [*sic*], 25. Beitrag). *Entomologische Arbeiten aus dem Museum G. Frey*, 10 (2), 478–514.
41. KUSCHEL, G. (1959) Beiträge zur Kenntnis der Insektenfauna Boliviens. Teil XII. Coleoptera XI. Curculionidae (1. Teil). Cossoninae, Amalactinae, Ithaurinae. (Col. Curculionoidea, 26. Beitrag). *Veröffentlichungen der Zoologischen Staatssammlung*, 6, 29–80.
42. KUSCHEL, G. (1960) Terrestrial zoology in southern Chile. *Proceedings of the Royal Society*, (B), 152, 540–550.
43. KUSCHEL, G. (1961) On problems of synonymy in the *Sitophilus oryzae* complex (30th contribution, Col. Curculionoidea). *Annals and Magazine of Natural History*, (13), 4, 241–244.
44. KUSCHEL, G. (1961) Composition and origin of the insect fauna of southern South America. *Abstracts, 10th Pacific Science Congress*, p. 231.
45. KUSCHEL, G. (1961) Composition and origin of the insect fauna off the west coast of South America. *Abstracts, 10th Pacific Science Congress*, pp. 466–467.

46. KUSCHEL, G. (1962) Some notes on the genus *Caulophilus* Wollaston with a key to the species (Coleoptera: Curculionidae) (29th contribution, Col. Curculionoidea). *The Coleopterists' Bulletin*, 16 (1), 1–4.
47. KUSCHEL, G. (1962) The Curculionidae of Gough Island and the relationships of the weevil fauna of the Tristan da Cunha group. *Proceedings of the Linnean Society of London*, 173 (2), 69–78.
48. KUSCHEL, G. (1962) Zur Naturgeschichte der Insel San Ambrosio (Islas Desventuradas, Chile). 1. Reisebericht, geographische Verhältnisse und Pflanzenverbreitung. *Arkiv for Botanik*, (2), 4 (12), 413–419.
49. KUSCHEL, G. (1963) Composition and relationships of the terrestrial faunas of Easter, Juan Fernández, Desventuradas and Galápagos Islands. *Occasional Papers of the California Academy of Sciences*, 44, 79–95.
50. KUSCHEL, G. (1964 ("1963")) Problems concerning an Austral Region. Pp. 443–449. *In*: GRESSITT, J. L. (Ed.) *Pacific Basin Biogeography: A Symposium*. Bishop Museum Press, Honolulu, Hawaii, p. 563 [issued 20 February 1964].
51. KUSCHEL, G. (1964) Insects of Campbell Island. Coleoptera: Curculionidae of the Subantarctic Islands of New Zealand. *Pacific Insects Monograph*, 7, 416–493.
52. KUSCHEL, G. (1966) A cossonine genus with bark-beetle habits, with remarks on relationships ad biogeography (Coleoptera Curculionidae). *New Zealand Journal of Science*, 9 (1), 3–29.
53. KUSCHEL, G. (1967) New synonymies in the genus *Promecops* Sahlberg (Coleoptera Curculionidae). *New Zealand Journal of Science*, 10 (3), 841–842.
54. KUSCHEL, G. (1969) Biogeography and ecology of South American Coleoptera. In: FITTKAU, E. J., ILLIES, J., KLINGE, H., SCHWABE, G. H., & SIOLI, H. (Eds.), *Biogeography and Ecology in South America. Monographiae Biologicae*, 19, 709–722.
55. KUSCHEL, G. (1969) The genus *Catoptes* Schönherr and two *species oblitae* of Fabricius from New Zealand (Coleoptera Curculionidae). *New Zealand Journal of Science*, 12, 789–810.
56. KUSCHEL, G. (1970) New Zealand Curculionoidea from Captain Cook's voyages (Coleoptera). *New Zealand Journal of Science*, 13 (2), 191–205.
57. KUSCHEL, G. (1970) Coleoptera: Curculionidae of Heard Island. *Pacific Insects Monograph*, 23, 255–260.
58. KUSCHEL, G. (1971) Entomology of the Aucklands and other islands south of New Zealand: Coleoptera: Curculionidae. *Pacific Insects Monograph*, 27, 225–259.
59. KUSCHEL, G. (1971) Chapter Twenty-seven. Curculionidae. Pp. 355–359. *In*: VAN ZINDEREN BAKKER, E. M., WINTERBOTTOM, J. M., DYER, R. A. (Eds.) *Marion and Prince Edwards Islands: Report on the South African Biological and Geological Expedition 1965–1966*. A. A. Balkema, Cape Town.
60. KUSCHEL, G. (1972) The Australian Phrynixinae (Coleoptera: Curculionidae). *New Zealand Journal of Science*, 15 (2), 209–231.
61. KUSCHEL, G. (1972) The biogeographical elements of New Zealand. *Abstracts, XIVth International Congress of Entomology*, p. 97.
62. KUSCHEL, G. (1972) The foreign Curculionoidea established in New Zealand (Insecta: Coleoptera). *New Zealand Journal of Science*, 15 (3), 273–289.
63. KUSCHEL, G. (1975) Introduction. Pp. xv–xvi. *In*: KUSCHEL, G. (Ed.). Biogeography and Ecology in New Zealand. *Monographiae Biologicae*, 27, xvi + 689 pp.
64. KUSCHEL, G. (1978) Notes on the identity of *Sitophilus zeamais* Motschulsky based on type material examination (Coleoptera). *Journal of Natural History*, 12, 231.
65. KUSCHEL, G. (1979) The genera *Monotoma* Herbst (Rhizophagidae) and *Anommatus* Wesmael (Cerylidae) in New Zealand (Coleoptera). *New Zealand Entomologist*, 7, 44–48.
66. KUSCHEL, G. (1982) Apionidae and Curculionidae (Coleoptera) from the Poor Knights Islands, New Zealand. *Journal of the Royal Society of New Zealand*, 12 (3), 273–282.

67. KUSCHEL, G. (1983) New synonymies and combinations of Baridinae from the Neotropic and Nearctic regions (Coleoptera: Curculionidae). *The Coleopterists' Bulletin*, 37 (1), 34–44.
68. KUSCHEL, G. (1983) Past and present of the relict family Nemonychidae (Coleoptera: Curculionidae). *GeoJournal*, 7.6, 499–504.
69. KUSCHEL, G. (1983) Distribution patterns, host plant associations and feeding habits in the relict family Nemonychidae (Coleoptera: Curculionidae). *Programme and Abstracts, 15th Pacific Science Congress*, 1, 137.
70. KUSCHEL, G. (1986) [Replacement names, transfers, new synonymies and combinations]. *In*: WIBMER, G. J., & O'BRIEN, C. W. Annotated checklist of the weevils (Curculionidae sensu lato) of South America (Coleoptera: Curculionoidea). *Memoirs of the American Entomological Institute*, 39, i–xvi, 1–563.
71. KUSCHEL, G. (1987) The subfamily Molytinae (Coleoptera: Curculionidae): general notes and description of new taxa from New Zealand and Chile. *New Zealand Entomologist*, 9, 11–29.
72. KUSCHEL, G. (1987) A New Zealand histerid beetle of Fabricius mistakenly described from Australia (Coleoptera: Histeridae). *New Zealand Entomologist*, 9, 56–57.
73. KUSCHEL, G. (1988) Thoughts on past classifications of the weevils – how a new scheme may be attempted. P. 40. *Proceedings of the XVIII. International Congress of Entomology, Vancouver, Canada, 3–9 July*.
74. KUSCHEL, G. (1989) Terminology affecting the spermatheca. *Curculio*, 27, 4.
75. KUSCHEL, G. (1989) The Nearctic Nemonychidae (Coleoptera: Curculionidae). *Entomologica scandinavica*, 20 (2), 121–171.
76. KUSCHEL, G. (1990) Some weevils from Winteraceae and other hosts from New Caledonia. *Tulane Studies in Zoology and Botany*, 27 (2), 29–47.
77. KUSCHEL, G. (1990) Beetles in a suburban environment: A New Zealand case study. The identity and status of Coleoptera in the natural; and modified habitats of Lynfield, Auckland (1974–1989). *DSIR Plant Protection Report*, 3, 1–118.
78. KUSCHEL, G., & MAY, B. M. (1990) Palophaginae, a new subfamily for leaf-beetles, feeding as adult and larva on araucarian pollen in Australia (Coleoptera: Megalopodidae). *Invertebrate Taxonomy*, 3, 697–719.
79. KUSCHEL, G. (1991) Degenerate trend engendered in gender endings. *Curculio*, 30, 5–6.
80. KUSCHEL, G. (1991) Biogeographic aspects of the subantarctic islands. *In*: International Symposium on Biogeographical Aspects of Insularity, Rome, 18–22 May 1987. *Atti dei Convegni Lincei*, 85, 575–591.
81. KUSCHEL, G. (1991) A trap for hypogean fauna. *Curculio*, 31, 5.
82. KUSCHEL, G. (1992) Reappraisal of the Baltic Amber Curculionoidea described by E. Voss. *Mitteilungen aus dem Geologisch-Paläontologischen Institut der Universität Hamburg*, 73, 191–215.
83. KUSCHEL, G. (1993) The Palaearctic Nemonychidae (Coleoptera: Curculionoidea). *Annales de la Société entomologique de France (N. S.)*, 29 (1), 23–46.
84. KUSCHEL, G., & POINAR, G. O. (1993) *Libanorhinus succinus* gen. & sp. n. (Coleoptera: Nemonychidae) from Lebanese amber. *Entomologica scandinavica*, 24, 143–146.
85. KUSCHEL, G. (1994) Nemonychidae of Australia, New Guinea and New Caledonia. Pp. 563–637. *In*: ZIMMERMAN, E. C. *Australian Weevils (Coleoptera: Curculionoidea). Vol. 1. Orthoceri: Anthribidae to Attelabidae: The Primitive Weevils.* Melbourne, CSIRO, xxxii + 741 pp.
86. KUSCHEL, G., OBERPRIELER, R. G., & RAYNER, R. J. (1994) Cretaceous weevils from southern Africa, with description of a new genus and species and phylogenetic and zoogeographical comments (Coleoptera: Curculionoidea). *Entomologica Scandinavica*, 25, 137–149.
87. CHOWN, S. L., & KUSCHEL, G. (1994) New *Bothrometopus* species from Possession Island, Crozet Archipelago, with nomenclatural amendments and a key to its weevil fauna (Coleoptera: Curculionidae: Brachycerinae). *African Entomology*, 2 (2), 149–154.

88. KUSCHEL, G. (1995) A phylogenetic classification of Curculionoidea to families and subfamilies. *Memoirs of the Entomological Society of Washington*, 14, 5–33.
89. KUSCHEL, G. (1995) *Oxycorynus missionis* spec. nov. from NE Argentina, with key to the South American species of Oxycoryninae (Coleoptera Belidae). *Acta Zoológica Lilloana*, 43 (1), 45–48.
90. KUSCHEL, G., & CHOWN, S. L. (1995) Phylogeny and systematics of the *Ectemnorhinus*-group of genera (Insecta: Coleoptera). *Invertebrate Taxonomy*, 9, 841–863.
91. KUSCHEL, G., & MAY, B. M. (1996) Discovery of Palophaginae (Coleoptera: Megalopodidae) on *Araucaria araucana* in Chile and Argentina. *New Zealand Entomologist*, 19, 1–13.
92. KUSCHEL, G., & WORTHY, T. H. (1996) Past distribution of large weevils (Coleoptera: Curculionidae) in the South Island, New Zealand, based on Holocene fossil remains. *New Zealand Entomologist*, 19, 15–22.
93. KLIMASZEWSKI, J., & KUSCHEL, G. (1996) Annual variation in the beetle fauna associated with the Hard Beech (*Nothofagus truncata*) litter of the Orongorongo Valley, New Zealand. *Giornale italiano di Entomología*, 8 (44), 157–166.
94. KUSCHEL, G., & MAY, B. M. (1996) Palophaginae, their systematic position and biology. Pp. 173–185. *In*: JOLIVET, P. H. A., & COX, M. L. (Eds.) *Chrysomelidae Biology, Vol. 3: General Studies*. SPB Academic Publishing, Amsterdam.
95. BARRATT, B. I. P., & KUSCHEL, G. (1996) Broad-nosed weevils (Curculionidae: Brachycerinae: Entimini) of the Lammermoor and Rock and Pillar Ranges in Otago, with descriptions of four new species of *Irenimus*. *New Zealand Journal of Zoology*, 23, 359–374.
96. KUSCHEL, G., & MAY, B. M. (1997) A new genus and species of Nemonychidae (Coleoptera) associated with *Araucaria angustifolia* in Brazil. *New Zealand Entomologist*, 20, 15–22.
97. KUSCHEL, G. (1997) Description of two new *Microcryptorhynchus* species from Lynfield, Auckland City, New Zealand (Coleoptera: Curculionidae). *New Zealand Entomologist*, 20, 23–27.
98. KUSCHEL, G. (1998) Comments on some Anthribidae described by MONTROUZIER (Coleoptera). *Bulletin de l'Institut Royal des Sciences Naturelles de Belgique, Entomologie*, 68, 193–195.
99. KUSCHEL, G. (1998) The subfamily Anthribidae in New Caledonia and Vanuatu (Coleoptera: Anthribidae). *New Zealand Journal of Zoology*, 25, 335–408.
100. KUSCHEL, G. (1998) Brenda May (1917–1998). *Curculio*, 43, 14–15.
101. KUSCHEL, G. (1999) New generic descriptions. Genus *Pacindonus*. Kuschel, h. o., gen. n. P. 265. In: ALONSO-ZARAZAGA, M. A., & LYAL, C. H. C., *A World Catalogue of Families and Genera of Curculionoidea (Insecta: Coleoptera) (excepting Scolytidae and Platypodidae)*. Entomopraxis, S.C.P., Barcelona, p. 315
102. KUSCHEL, G., LESCHEN, R. A. B., & ZIMMERMAN, E. C. (2000) Platypodidae under scrutiny. *Invertebrate Taxomony*, 14, 771–805.
103. KUSCHEL, G. (2001) La fauna curculiónica (Coleoptera: Curculionoidea) de la *Araucaria araucana*. *Revista Chilena de Entomología*, 27, 41–51.
104. KUSCHEL, G. (2001) Book Review. A world catalogue of families and genera of Curculionoidea (Insecta: Coleoptera) (excepting Scolytidae and Platypodidae). M. A. Alonso-Zarazaga & C. H. C. Lyal, Entomopraxis, S. C. P., Apartado 36164, 08080 Barcelona, Spain. *New Zealand Journal of Zoology*, 28, 245.
105. ELGUETA, M., & KUSCHEL, G. (2002) *Aegorhinus* Erichson, 1834 (Insecta, Coleoptera): proposed precedence over *Psuchocephalus* Latreille, 1828. *Bulletin of Zoological Nomenclature*, 59 (4), 253–255.
106. KUSCHEL, G. (2003) Nemonychidae, Belidae, Brentidae (Insecta: Coleoptera: Curculionoidea). *Fauna of New Zealand*, 45, 1–100.
107. KUSCHEL, G., & LESCHEN, R. A. B. (2003) Appendix 1. Phylogenetic relationships of the genera of Belinae. Pp. 48–55. *In*: KUSCHEL, G., Nemonychidae, Belidae, Brentidae (Insecta: Coleoptera: Curculionoidea)'. *Fauna of New Zealand*, 45, 1–97.

108. LESCHEN, R. A. B., LAWRENCE, J. F., KUSCHEL, G., THORPE, S., & WANG, Q. (2003) Coleoptera genera of New Zealand. *New Zealand Entomologist*, 26, 15–28.
109. ASHWORTH, A. C., & KUSCHEL, G. (2003) Fossil weevils (Coleoptera: Curculionidae) from latitude 85° S Antarctica. *Palaeogeography, Palaeoclimatology, Palaeoecology*, 191, 191–202.
110. KUSCHEL, G. (2003) A ball-forming weevil from young *Nothofagus* leaves in Chile (Coleoptera: Curculionidae: Curculioninae: Sphaeriopoeini). *Revista Chilena de Entomología*, 29, 59–65.
111. KUSCHEL, G., & EMBERSON, R. M. (2008) Notes on *Hybolasius trigonellaris* Hutton from the Chatham Islands (Coleoptera: Cerambycidae: Lamiinae). *New Zealand Entomologist*, 31, 89–92.
112. KUSCHEL, G. (2008) Curculionoidea (weevils) of New Caledonia and Vanuatu: basal families and some Curculionidae. *In*: GRANDCOLAS, P. (Ed.). *Zoologia Neocaledonica 6. Biodiversity Studies in New Caledonia. Mémoires du Muséum National d'Histoire Naturelle*, 197, 99–249.
113. KUSCHEL, G. (2009) New tribe, new genus and species for an Australasian weevil group with notes and keys [Coleoptera, Curculionoidea]. *Revue française d'Entomologie (N. S.)*, 30 (2–4), 41–66.
114. KUSCHEL, G., & LESCHEN, R. A. B. (2011) Phylogeny and taxonomy of the Rhinorhynchinae (Coleoptera: Nemonychidae). *Invertebrate Systematics*, 24, 573–615.
115. KUSCHEL, G. (2014) The New Caledonian and Fijian species of Aterpini and Gonipterini (Coleoptera: Curculionidae). *In*: GUILBERT, É., ROBILLARD, T., JOURDAN, H., & GRANDCOLAS, P. (Eds.). *Zoologia Neocaledonica 8. Biodiversity Studies in New Caledonia. Mémoires du Muséum National d'Histoire Naturelle*, 206, 133–163.
116. KUSCHEL, G. (2014) The blind weevils of Myrtonymina in New Caledonia and Australia (Curculionidae: Curculioninae: Erirhinini: Myrtonymina). *In*: GUILBERT, É., ROBILLARD, T., JOURDAN, H., & GRANDCOLAS, P. (Eds.). *Zoologia Neocaledonica 8. Biodiversity Studies in New Caledonia. Mémoires du Muséum National d'Histoire Naturelle*, 206, 165–180.
117. KUSCHEL, G. (2017) First zygopine weevil from Chile (Coleoptera: Curculionoidea). *Revista Chilena de Entomología*, 43, 19–23.
118. MCKENNA, D. D., CLARKE, D. J., ANDERSON, R. [S.], ASTRIN, J. J, BROWN, S., CHAMORRO, L., DAVIS, S. R., DE MEDEIROS, B., DEL RIO, M. G., HARAN, J., KUSCHEL, G.†, FRANZ, N., JORDAL, B., LANTERI, A., LESCHEN, R. A. B., LETSCH, H., LYAL, C. [H. C.], MARVALDI, A. [E.], MERMUDES, J. R., OBERPRIELER, R. G., SCHÜTTE, A., SEQUEIRA, A., SHIN, S., VAN DAM, M. H., & ZHANG, G. 2018 (18 July online). Morphological and molecular perspectives on the phylogeny, evolution and classification of weevils (Coleoptera: Curculionoidea): Proceedings from the 2016 International Weevil Meeting. *Diversity*, 10 (3), 64, 1–33. (https://doi.org/10.3390/d10030064).

Appendix B

Taxa Named after Guillermo Kuschel (* Extinct)

I. Tribes and Genera

Name	Reference	Order: Family	Current Status
Kuschelomacrini * Riedel, 2010	*Insect Systematics and Evolution*, 41, 31	Coleoptera: Nemonychidae	**Note**: published as Kuschelomacerini, based on incorrect stem formation of *Kuschelomacer*
Kuschelia Malaise, 1949	*Arkiv för Zoologi*, 42A (9), 21	Hymenoptera: Tenthredinidae	syn. of *Trichotaxonus* Rohwer
Kuschelenia Hylton Scott, 1951	*Acta Zoologica Lilloana*, 12, 539	Panpulmonata: Bulimulidae	
Kuschelina Bechyné, 1952	*Revista Chilena de Entomología*, 1, 110	Coleoptera: Chrysomelidae	
Kuschelochilis Wygodzinsky, 1952	*Revista Chilena de Entomología*, 1, 199	Archaeognatha: Machilidae	syn. of *Allomachilis* Silvestri
Kuscheliana Carvalho, 1952	*Revista Chilena de Entomología*, 2, 21	Hemiptera: Miridae	
Kuschelachertus De Santis, 1955	*Revista Chilena de Entomología*, 4, 172	Hymenoptera: Elachertidae	
Kuscheliola Evans, 1957	*Revista Chilena de Entomología*, 5, 372	Hemiptera: Cicadellidae	
Kuschelomyia Souza Lopes, 1961	*Revista Brasileira de Biologia*, 21, 455	Diptera: Calliphoridae	syn. of *Toxotarsus* Macquart
Kuscheloniscus Strouhal, 1961	*Annalen des Naturhistorischen Museums in Wien*, 64, 217	Isopoda: Styloniscidae	
Kuschelia China, 1962	*Transactions of the Royal Entomological Society of London*, 114, 153	Hemiptera: Peloridiidae	junior homonym (repl. name *Kuscheloides* Evans, 1982)
Kuscheliotes Jeannel, 1962	*Biologie de l'Amérique australe. Vol. 1. Études sur la faune du Sol*, 1, 321	Coleoptera: Staphylinidae	
Kuschelinus Straneo, 1963	*Revue Française d'Entomologie*, 30, 124	Coleoptera: Carabidae	
Kuschelita Climo, 1974	*New Zealand Journal of Zoology*, 1 (3), 265	Littorinimorpha: Tateidae	
Kuschelydrus Ordish, 1976	*New Zealand Journal of Zoology*, 3, 6	Coleoptera: Dytiscidae	
Kuschelodesmus Hoffman, 1979	*Revue Suisse de Zoologie*, 86 (3), 629	Polydesmida: Dalodesmidae	
Kuschelius Sublette & Wirth, 1980	*New Zealand Journal of Zoology*, 7 (3), 314	Diptera: Chironomidae	
Kuschelidium Johnson, 1982	*New Zealand Journal of Zoology*, 9, 337	Coleoptera: Ptiliidae	
Kuscheloides Evans, 1982	*Records of the Australian Museum*, 34 (5), 384	Hemiptera: Peloridiidae	repl. name for *Kuschelia* China (*non Kuschelia* Malaise, 1949)
Kuschelus Kaszab, 1982	*Folia Entomologica Hungarica*, 43, 112	Coleoptera: Tenebrionidae	
Kuschelacarus Cook, 1992	*Stygologia*, 7, 58	Trombidiformes: Mideopsidae	
Kuschelaxius Howden, 1992	*Memoirs of the Entomological Society of Canada*, 124, 43	Coleoptera: Curculionidae	
Guillermia * Zherikhin & Gratshev, 1994	*Paleontological Journal*, 27, 57	Coleoptera: Obrieniidae	
Kuschelanthus Alonso-Zarazaga & Lyal, 1999	*World Catalogue of Families and Genera of Curculionoidea*, 42	Coleoptera: Curculionidae	
Kuschelengis Skelley & Leschen, 2007	*Fauna of New Zealand*, 59, 14	Coleoptera: Erotylidae	
Kuschelomacer * Riedel, 2010	*Insect Systematics and Evolution*, 41, 31	Coleoptera: Nemonychidae	
Kuschelorhynchus Jennings & Oberprieler, 2018	*Diversity*, 10 (3), 71, 25	Coleoptera: Curculionidae	
Kuschelysius Brown & Leschen, 2018	*Diversity*, 10 (3), 75, 2	Coleoptera: Curculionidae	
Kuschelorhinus Anderson & Setliff, 2018	*Diversity*, 10 (3), 83, 2	Coleoptera: Curculionidae	

II. Species

Name	Reference	Order: Family	Origin	Current Status
Dasytes kuscheli Wittmer, 1942	*Revue d'Entomologie*, 12, 513	Coleoptera: Melyridae	Chile	syn. of *Hylodanacaea derbesii* (Solier)
Borgmeierus kuscheli Bondar, 1945	*Revista de Entomología, Rio de Janeiro*, 16 (1–2), 110	Coleoptera: Curculionidae	Argentina	*Demoda*
Plectonotum kuscheli Wittmer, 1945	*Revista de la Sociedad Entomológica Argentina*, 12 (4), 322	Coleoptera: Cantharidae	Chile	syn. of *Hyponotum philippii* Gemminger
Thaliabaris kuscheli Bondar, 1945	*Revista de Entomología, Rio de Janeiro*, 16 (1–2), 105	Coleoptera: Curculionidae	Argentina	*Odontobaris*
Teriocolias atinas kuscheli Ureta, 1947	*Boletín del Museo Nacional de Historia Natural, Santiago de Chile*, 23, 49	Lepidoptera: Pieridae	Chile	subsp. of *Eurema (Teriocoleus) riojana*
Megavallius kuscheli Bondar, 1948	*Revista de Entomología, Rio de Janeiro*, 19 (1–2), 28	Coleoptera: Curculionidae	Brazil	
Chuquiraga kuschelii Acevedo de Vargas, 1949	*Boletín del Museo Nacional de Historia Natural, Chile*, 24, 86	Asterales: Asteraceae	Chile	
Oogenius kuscheli Gutiérrez, 1949	*Anales de la Sociedad Científica Argentina*, 148 (1), 29	Coleoptera: Scarabaeidae	Chile	
Thecla kuscheli Ureta, 1949	*Boletín del Museo Nacional de Historia Natural, Santiago de Chile*, 24, 98	Lepidoptera: Lycaenidae	Chile	*Chlorostrymon*
Antitypona kuscheli Bechyné, 1950	*Entomologische Arbeiten aus dem Museum Georg Frey*, 1, 210	Coleoptera: Chrysomelidae	Bolivia	
Bergemesa kuscheli Wygodzinsky, 1950	*Anales de la Sociedad Científica Argentina*, 150, 44	Hemiptera: Reduviidae	Peru	
Cryptotarsus kuscheli Wittmer, 1950	*Revista de Entomología, Rio de Janeiro*, 21, 256	Coleoptera: Melyridae	Peru	*Engilemphus?*
Caryonoda kuscheli Bechyné, 1951	*Entomologische Arbeiten aus dem Museum Georg Frey*, 2, 265	Coleoptera: Chrysomelidae	Bolivia	
Chalcophana kuscheli Bechyné, 1951	*Entomologische Arbeiten aus dem Museum Georg Frey*, 2, 332	Coleoptera: Chrysomelidae	Bolivia	
Coelioxys kuscheli Moure, 1951	*Dusenia*, 2 (6), 406	Hymenoptera: Megachilidae	Chile	
Dachrys kuscheli Monros, 1951	*Revista de la Sociedad Entomológica Argentina*, 15, 154	Coleoptera: Chrysomelidae		
Maecolaspis kuscheli Bechyné, 1951	*Entomologische Arbeiten aus dem Museum Georg Frey*, 2, 313	Coleoptera: Chrysomelidae	Bolivia	*Syphraea*
Midacritus kuscheli Séguy, 1951	*Revue française d'Entomologie*, 18, 12	Diptera: Mydidae	Chile	

Name	Reference	Order: Family	Origin	Current Status
Typophorus kuscheli Bechyné, 1951	*Entomologische Arbeiten aus dem Museum Georg Frey*, 2, 343	Coleoptera: Chrysomelidae	Bolivia	
Lactica kuscheli Bechyné, 1952	*Revista Chilena de Entomología*, 1, 100	Coleoptera: Chrysomelidae	Peru	
Metapterus kuscheli Wygodzinsky, 1952	*Revista Chilena de Entomología*, 1, 126	Hemiptera: Reduviidae	Juan Fernandez Islands	*Pseudometapterus*
Ogcodes kuscheli Sabrosky, 1952	*Revista Chilena de Entomología*, 1, 189	Diptera: Acroceridae	Juan Fernandez Islands	
Pityophthorus kuscheli Schedl, 1952	*Revista Chilena de Entomología*, 1, 19	Coleoptera: Curculionidae	Chile	
Pnigomenus kuscheli Bosq, 1952	*Revista Chilena de Entomología*, 1, 196	Coleoptera: Cerambycidae	Chile	
Rhynchitomacer (*Rhynchitomace-rinus*) *kuscheli* Voss, 1952	*Revista Chilena de Entomología*, 1, 179	Coleoptera: Nemonychidae	Chile	
Gigantodax kuscheli Wygodzinsky, 1952	*Revista Chilena de Entomología*, 2, 81	Diptera: Simuliidae	Juan Fernandez Islands	
Limonia (*Dicranomyia*) *kuscheliana* Alexander, 1952	*Revista Chilena de Entomología*, 2, 47	Diptera: Tipulidae	Juan Fernandez Islands	
Micrymenus kuscheli Kormilev, 1952	*Revista Chilena de Entomología*, 2, 12	Hemiptera: Rhyparochromidae	Juan Fernandez Islands	
Minotula kuscheli Bechyné, 1952	*Revista Chilena de Entomología*, 2, 117	Coleoptera: Chrysomelidae	Juan Fernandez Islands	
Phantasiosiphona kuscheli Cortes, 1952	*Revista Chilena de Entomología*, 2, 110	Diptera: Tachinidae	Juan Fernandez Islands	*Siphona*
Podonomus kuscheli Wirth, 1952	*Revista Chilena de Entomología*, 2, 95	Diptera: Chironomidae	Juan Fernandez Islands	
Rhantus signatus kuscheli Guignot, 1952	*Revista Chilena de Entomología*, 2, 114	Coleoptera: Dytiscidae	Juan Fernandez Islands	
Scolopsopteron kuscheli Ogloblin, 1952	*Revista Chilena de Entomología*, 2, 128	Hymenoptera: Mymaridae	Juan Fernandez Islands	*Cremnomymar*
Shannonomyia kuscheli Alexander, 1952	*Revista Chilena de Entomología*, 2, 57	Diptera: Tipulidae	Juan Fernandez Islands	
Lancetes kuscheli Guignot, 1953	*Revue française d'Entomologie*, 20, 114	Coleoptera: Dytiscidae	Chile	syn. of *Lancetes nigriceps* (Erichson)
Lepidosternopsis kuscheliana Ogloblin, 1953	*Revista Chilena de Entomología*, 3, 102	Hymenoptera: Bethylidae	Juan Fernandez Islands	*Sclerodermus*
Merianina kuscheli Freeman, 1954	*Revista Chilena de Entomología*, 3, 33	Diptera: Sciaridae	Juan Fernandez Islands	
Psectrascelis kuscheli Kulzer, 1954	*Entomologische Arbeiten aus dem Museum Georg Frey*, 5, 178	Coleoptera: Tenebrionidae	Chile	
Tachygonus kuscheli Viana, 1954	*Comunicaciones del Instituto Nacional de Investigación de las Ciencias Naturales, Ciencias Zoológicas*, 2 (11), 151	Coleoptera: Curculionidae	Bolivia	
Trechisibus kuscheli Jeannel, 1954	*Revue française d'Entomologie*, 21, 92	Coleoptera: Carabidae	Juan Fernandez Islands	

Name	Reference	Order: Family	Origin	Current Status
Chelanops kuscheli Beier, 1955	*Revista Chilena de Entomología*, 4, 212	Pseudoscorpiones: Chernetidae	Juan Fernandez Islands	
Conchopterella kuscheli Handschin, 1955	*Revista Chilena de Entomología*, 4, 10	Neuroptera: Hemerobiidae	Juan Fernandez Islands	
Delphacodes kuscheli Fennah, 1955	*Proceedings of the Royal Entomological Society of London*, 24, 137	Hemiptera: Delphacidae	Juan Fernandez Islands	
Evansiella kuscheli China, 1955	*Revista Chilena de Entomología*, 4, 200	Hemiptera: Cicadellidae	Juan Fernandez Islands	
Hemencyrtus kuscheli De Santis, 1955	*Revista Chilena de Entomología*, 4, 193	Hymenoptera: Encyrtidae	Juan Fernandez Islands	*Deloencyrtus*
Hydrophorus kuscheli Harmston, 1955	*Revista Chilena de Entomología*, 4, 35	Diptera: Dolichopodidae	Juan Fernandez Islands	
Magellomyia kuscheli Schmid, 1955	*Mémoires de la Société vaudoise des Sciences Naturelles*, 11, 138	Trichoptera: Limnephilidae	Chile	*Verger*
Metius kuscheli Straneo *in* Straneo & Jeannel, 1955	*Revista Chilena de Entomología*, 4, 137	Coleoptera: Carabidae	Juan Fernandez Islands	
Notoschoenomyza kuscheli Hennig, 1955	*Revista Chilena de Entomología*, 4, 27	Diptera: Muscidae	Juan Fernandez Islands	
Opius kuscheli Nixon, 1955	*Revista Chilena de Entomología*, 4, 159	Hymenoptera: Braconidae	Juan Fernandez Islands	
Parachernes (*Argentochernes*) *kuscheli* Beier, 1955	*Revista Chilena de Entomología*, 4, 208	Pseudoscorpiones: Chernetidae	Juan Fernandez Islands	*Parachernes (P.)*
Peloridora kuscheli China, 1955	*Entomologist's Monthly Magazine*, 91, 82	Hemiptera: Peloridiidae	Chile	
Pterostichus kuscheli Straneo *in* Straneo & Jeannel, 1955	*Revista Chilena de Entomología*, 4, 131	Coleoptera: Carabidae	Juan Fernandez Islands	junior homonym in *Metius*; repl. name *M. guillermoi* Will
Quadraceps kuscheli Timmermann, 1955	*Annals and Magazine of Natural History*, Series 12, 8 (91), 521	Psocodea: Philopteridae	South America	
Scatella kuscheli Wirth, 1955	*Revista Chilena de Entomología*, 4, 61	Diptera: Ephydridae	Juan Fernandez Islands	
Trachysarus kuscheli Straneo & Jeannel, 1955	*Revista Chilena de Entomología*, 4, 141	Coleoptera: Carabidae	Juan Fernandez Islands	
Mastinocerus kuscheli Wittmer, 1956	*Entomologische Arbeiten aus dem Museum Georg Frey*, 7, 225	Coleoptera: Phengodidae	Chile	
Megachile (*Dasymegachile*) *kuscheli* Moure, 1956	*Dusenia*, 7, 105	Hymenoptera: Megachilidae	Bolivia	
Philorea kuscheli Kulzer, 1956	*Entomologische Arbeiten aus dem Museum Georg Frey*, 7, 927	Coleoptera: Tenebrionidae	Chile	
Thinobatis kuscheli Kulzer, 1956	*Entomologische Arbeiten aus dem Museum Georg Frey*, 7, 903	Coleoptera: Tenebrionidae	Chile	
Adalia kuscheli Mader, 1957	*Revista Chilena de Entomología*, 5, 91	Coleoptera: Coccinellidae	Chile	
Anthidium kuscheli Moure, 1957	*Revista Chilena de Entomología*, 5, 213	Hymenoptera: Megachilidae	Chile	syn. of *Anthidium rubripes* Friese

Name	Reference	Order: Family	Origin	Current Status
Auletobius kuscheli Voss, 1957	*Revista Chilena de Entomología*, 5, 98	Coleoptera: Attelabidae	Bolivia	*Gymnauletes*
Baeus kuscheli Ogloblin, 1957	*Revista Chilena de Entomología*, 5, 438	Hymenoptera: Scelionidae	Juan Fernandez Islands	
Dasypelates kuscheli Jeannel, 1957	*Revista Chilena de Entomología*, 5, 48	Coleoptera: Leiodidae	Chile	*Chiliopelates*
Drosophila kuscheli Brncic, 1957	*Revista Chilena de Entomología*, 5, 394	Diptera: Drosophilidae	Juan Fernandez Islands	*Hirtodrosophila*
Edrabius kuscheli Scheerpeltz, 1957	*Revista Chilena de Entomología*, 5, 220	Coleoptera: Staphylinidae	Chile	
Hemidianeura kuscheli Malaise, 1957	*Entomologisk Tidskrift*, 78, 12	Hymenoptera: Argidae	Bolivia	*Ptenos*
Lechytia kuscheli Beier, 1957	*Revista Chilena de Entomología*, 5, 453	Pseudoscorpiones: Lechytiidae	Juan Fernandez Islands	
Neocamiarus kuscheli Jeannel, 1957	*Revista Chilena de Entomología*, 5, 61	Coleoptera: Leiodidae	Chile	
Radiodiscus kuscheli Hylton Scott, 1957	*Neotropica*, 3 (10), 7–16	Panpulmonata: Helicodiscidae	Chile	*Glabrogyra?*
Rhizobius kuscheli Mader, 1957	*Revista Chilena de Entomología*, 5, 73	Coleoptera: Coccinellidae	Chile	
Plathesthes kuscheli Kulzer, 1958	*Entomologische Arbeiten aus dem Museum Georg Frey*, 9, 10	Coleoptera: Tenebrionidae	Chile	
Praocis (*Praocida*) *kuscheli* Kulzer, 1958	*Entomologische Arbeiten aus dem Museum Georg Frey*, 9, 91	Coleoptera: Tenebrionidae	Bolivia	
Capraita kuscheli Bechyné, 1959	*Beiträge zur Kenntnis der Alticidenfauna Boliviens (Coleopt. Phytoph.). Beiträge zur Neotropischen Fauna*, Jena, 1, 364	Coleoptera: Chrysomelidae	Bolivia	
Diosyphraea kuscheli Bechyné, 1959	*Beiträge zur Kenntnis der Alticidenfauna Boliviens (Coleopt. Phytoph.). Beiträge zur Neotropischen Fauna*, Jena, 1, 307	Coleoptera: Chrysomelidae	Peru	
Huarinillasa kuscheli Bechyné, 1959	*Beiträge zur Kenntnis der Alticidenfauna Boliviens (Coleopt. Phytoph.). Beiträge zur Neotropischen Fauna*, Jena, 1, 367	Coleoptera: Chrysomelidae	Bolivia	
Ocnoscelis kuscheli Bechyné, 1959	*Beiträge zur Kenntnis der Alticidenfauna Boliviens (Coleopt. Phytoph.). Beiträge zur Neotropischen Fauna*, Jena, 1, 301	Coleoptera: Chrysomelidae	Bolivia	
Heliofugus (*Inscutoheliofugus*) *kuscheli* Freude, 1960	*Proceedings of the California Academy of Sciences*, 31 (6), 130	Coleoptera: Tenebrionidae	Chile	
Klapopteryx kuscheli Illies, 1960	*Zoologischer Anzeiger*, 164 (1–2), 35	Plecoptera: Austroperlidae	Chile	
Megandiperla kuscheli Illies, 1960	*Mitteilungen der Schweizerischen entomolo-gischen Gesellschaft*, 33 (3), 162	Plecoptera: Gripopterygidae	Chile	
Frutillaria kuscheli Richards, 1961	*Proceedings of the Royal Entomological Society of London B*, 30, 66	Diptera: Sphaeroceridae	Chile	

Name	Reference	Order: Family	Origin	Current Status
Oniscophiloscia kuscheli Strouhal, 1961	*Annalen des Naturhistorischen Museums in Wien*, 64, 238	Isopoda: Philosciidae	Juan Fernandez Islands	
Achilia kuscheli Jeannel, 1962	*Biologie de l'Amérique australe. Vol. 1, Études sur la faune du sol*, 425	Coleoptera: Staphylinidae	Chile	
Dalminiastes kuscheli Jeannel, 1962	*Biologie de l'Amérique australe. Vol. 1, Études sur la faune du sol*, 375	Coleoptera: Staphylinidae	Chile	
Frutillariotes kuscheli Jeannel, 1962	*Biologie de l'Amérique australe. Vol. 1, Études sur la faune du sol*, 382	Coleoptera: Staphylinidae	Chile	
Golasa kuscheli Jeannel, 1962	*Biologie de l'Amérique australe. Vol. 1, Études sur la faune du sol*, 308	Coleoptera: Staphylinidae	Chile	
Nemadiolus kuscheli Jeannel, 1962	*Biologie de l'Amérique australe. Vol. 1, Études sur la faune du sol*, 525	Coleoptera: Leiodidae	Chile	
Notaphus (*Austronotaphus*) *kuscheli* Jeannel, 1962	*Biologie de l'Amérique australe. Vol. 1, Études sur la faune du sol*, 620	Coleoptera: Carabidae	Chile	*Bembidion (Notaphus)*
Notholopha (*Pacmophena*) *kuscheli* Jeannel, 1962	*Biologie de l'Amérique australe. Vol. 1, Études sur la faune du sol*, 637	Coleoptera: Carabidae	Chile	junior homonym in *Bembidion;* repl. name *B. (Pacmophena) penai* Toledano
Omalodera dentimaculata kuscheli Jeannel, 1962	*Biologie de l'Amérique australe. Vol. 1, Études sur la faune du sol*, 546	Coleoptera: Carabidae	Chile	
Paractium kuscheli Jeannel, 1962	*Biologie de l'Amérique australe. Vol. 1, Études sur la faune du sol*, 335	Coleoptera: Staphylinidae	Chile	
Parapteracmes kuscheli Jeannel, 1962	*Biologie de l'Amérique australe. Vol. 1, Études sur la faune du sol*, 362	Coleoptera: Staphylinidae	Chile	
Rybaxidia kuscheli Jeannel, 1962	*Biologie de l'Amérique australe. Vol. 1, Études sur la faune du sol*, 393	Coleoptera: Staphylinidae	Chile	
Salagosa kuscheli Jeannel, 1962	*Biologie de l'Amérique australe. Vol. 1, Études sur la faune du sol*, 316	Coleoptera: Staphylinidae	Chile	
Ambrosiella kuscheli Odhner, 1963	*Proceedings of the Malacological Society of London*, 35, 208	Panpulmonata: Tornatellinidae	Chile	
Aubertoperla kuscheli Illies, 1963	*Mitteilungen der Schweizerischen entomolo-gischen Gesellschaft*, 36, 190	Plecoptera: Gripopterygidae	Chile	
Eragrostis kuschelii Skottsberg, 1963	*Arquivos de Botânica do Estado de São Paulo n.s., f.m.*, 4, 485	Poales: Poaceae	Desventuradas Islands	
Katianna kuscheli Delamare Deboutteville & Massoud, 1963	*Collemboles Symphypléones, in: Delamare Deboutteville, C. & Rapoport, E. (eds), Biol. Amer. Australe*, Paris, 2, 228	Collembola: Katiannidae	Argentina?	

Name	Reference	Order: Family	Origin	Current Status
Parahelops kuscheli Kulzer, 1963	*Entomologische Arbeiten aus dem Museum Georg Frey*, 14, 611	Coleoptera: Promecheilidae	Chile	
Asterochernes kuscheli Beier, 1964	*Annalen des Naturhistorischen Museums in Wien*, 67, 352	Pseudoscorpiones: Chernetidae	Chile	
Bubekiana kuscheli De Santis, 1964	*Revista del Museo de La Plata, Sección Zoología*, 8 (57), 9, 22	Hymenoptera: Pteromalidae	Juan Fernandez Islands	
Liriomyza kuscheli Spencer, 1964	*Pacific Insects*, 6 (2), 253	Diptera: Agromyzidae	Juan Fernandez Islands	
Parazaona kuscheli Beier, 1964	*Annalen des Naturhistorischen Museums in Wien*, 67, 356	Pseudoscorpiones: Chernetidae	Chile	
Pseudopilanus kuscheli Beier, 1964	*Annalen des Naturhistorischen Museums in Wien*, 67, 359	Pseudoscorpiones: Chernetidae	Chile	
Thaumatolpium kuscheli Beier, 1964	*Annalen des Naturhistorischen Museums in Wien*, 67, 330	Pseudoscorpiones: Garypinidae	Chile	
Fernandocrambus kuscheli Clarke, 1965	*Proceedings of the United States National Museum, Washington*, 117, 24	Lepidoptera: Crambidae	Chile	
Macrurohelea kuscheli Wirth, 1965	*Pan-Pacific Entomologist*, 41, 49	Diptera: Ceratopogonidae	Chile	
Nesoeme kuscheli Linsley & Chemsak, 1966	*Proceedings of the California Academy of Sciences*, Series 4, 33 (8), 211	Coleoptera: Cerambycidae	Galapagos Islands	
Urgleptes kuscheli Linsley & Chemsak, 1966	*Proceedings of the California Academy of Sciences*, Series 4, 33 (9), 245	Coleoptera: Cerambycidae	Costa Rica	
Tipula (*Eumicrotipula*) *kuscheli* Alexander, 1967	*Studia Entomologica*, 10, 463	Diptera: Tipulidae	Chile	
Andrewesella kuscheli Straneo, 1969	*Annales de la Société entomologique de France*, 5 (1), 970	Coleoptera: Carabidae	Chile	*Euproctinus*, syn. of *E. fasciatus* (Solier)
Genaphthona kuscheli Bechyné, 1959	*Beiträge zur Kenntnis der Alticidenfauna Boliviens (Coleopt. Phytoph.). Beiträge zur Neotropischen Fauna. Jena (privately published)*, 1, 278	Coleoptera: Chrysomelidae	Bolivia	
*Ocenebra kuscheli** Fleming, 1972	*Philosophical Transactions of the Royal Society B, Biological Sciences*, 263 (853), 395	Neogastropoda: Muricidae	Chile	*Muregina*
Orynipus kuscheli Hofmann, 1972	*Mitteilungen der Münchner entomologischen Gesellschaft, München*, 62, 81	Coleoptera: Coccinellidae	Chile	
Pachycotes kuscheli Schedl, 1972	*New Zealand Journal of Science*, 15 (3), 266	Coleoptera: Curculionidae	Norfolk Island	

Name	Reference	Order: Family	Origin	Current Status
Tiphobiosis kuscheli Wise, 1972	*Records of the Auckland Institute and Museum*, 9, 258	Trichoptera: Hydrobiosidae	Auckland Islands	
Homalinotus kuscheli Vaurie, 1973	*Bulletin of the American Museum of Natural History*, 152, 35	Coleoptera: Curculionidae	Brazil	
Novolopa kuscheli Knight, 1973	*New Zealand Journal of Science*, 16, 978	Hemiptera: Cicadellidae	New Zealand	
Aegorhinus kuscheli Elgueta, 1974	*Revista Chilena de Entomología*, 8, 133	Coleoptera: Curculionidae	Chile	
Apteryoperla kuscheli Illies, 1974	*New Zealand Journal of Zoology*, 1, 288	Plecoptera: Gripopterygidae	Auckland Islands	*Aucklandobius*
Opacuincola kuscheli Climo, 1974	*New Zealand Journal of Zoology*, 1, 269	Littorinimorpha: Tateidae	New Zealand	
Euconnus (*Allomaoria*) *kuschelianus* Franz, 1975	*Revision der Scydmaeniden von Australien, Neuseeland und den benachbarten Inseln*, 70	Coleoptera: Scydmaenidae	New Caledonia	
Euconnus (*Maoria*) *kuscheli* Franz, 1975	*Revision der Scydmaeniden von Australien, Neuseeland und den benachbarten Inseln*, 48	Coleoptera: Scydmaenidae	New Zealand	
Apatochernes kuscheli Beier, 1976	*New Zealand Journal of Zoology*, 3, 235	Pseudoscorpiones: Chernetidae	New Zealand	
Nesidiochernes kuscheli Beier, 1976	*New Zealand Journal of Zoology*, 3, 223	Pseudoscorpiones: Chernetidae	New Zealand	
Phaulochernes kuscheli Beier, 1976	*New Zealand Journal of Zoology*, 3, 244	Pseudoscorpiones: Chernetidae	New Zealand	
Tomogenius kuscheli Dahlgren, 1976	*Journal of the Royal Society of New Zealand*, 6, 409	Coleoptera: Histeridae	New Zealand	
Eutricimba kuscheli Spencer, 1977	*Journal of the Royal Society of New Zealand*, 7 (4), 456	Diptera: Chloropidae	New Zealand	*Tricimba*
Hyphalus kuscheli Britton, 1977	*Records of the Auckland Museum*, 14, 82	Coleoptera: Limnichidae	New Zealand	
Neuraphoconnus kuscheli Franz, 1977	*Koleopterologische Rundschau*, 53, 18	Coleoptera: Scydmaenidae	New Zealand	
Stenichnus (*Austrostenichnus*) *kuschelianus* Franz, 1977	*Koleopterologische Rundschau*, 53, 19	Coleoptera: Scydmaenidae	New Zealand	
Vesicaperla kuscheli McLellan, 1977	*New Zealand Journal of Zoology*, 4, 140	Plecoptera: Gripopterygidae	New Zealand	
Zelandopsocus kuscheli Thornton, Wong & Smithers, 1977	*Pacific Insects*, 17 (2–3), 209	Psocodea: Philotarsidae	New Zealand	
Culicoides kuscheli Wirth & Blanton, 1978	*Pan-Pacific Entomologist*, 54, 236	Diptera: Ceratopogonidae	Chile	
Asilis kuscheli Wittmer, 1979	*Entomologica Basiliensia*, 4, 298	Coleoptera: Cantharidae	New Zealand	
Coccotrypes kuscheli Schedl, 1979	*New Zealand Entomologist*, 7, 104	Coleoptera: Curculionidae	Fiji	syn. of *Ozopemon augustae* Eggers
Baeocera kuscheli Löbl, 1980	*New Zealand Journal of Zoology*, 7, 384	Coleoptera: Staphylinidae	Fiji	

Name	Reference	Order: Family	Origin	Current Status
Baeocera kuscheliana Löbl, 1980	*New Zealand Journal of Zoology*, 7, 386	Coleoptera: Staphylinidae	Fiji	
Forcipomyia (Trichohelea) kuscheli Sublette & Wirth, 1980	*New Zealand Journal of Zoology*, 7, 330	Diptera: Ceratopogonidae	Auckland Islands	
Hygranillus kuscheli Moore, 1980	*New Zealand Journal of Zoology*, 7, 404	Coleoptera: Carabidae	New Zealand	
Mniovelia kuscheli Andersen & Polhemus, 1980	*Entomologica Scandinavica*, 11, 377	Hemiptera: Mesoveliidae	New Zealand	
Proeulia kuscheli Clarke, 1980	*Journal of the Lepidopterists' Society*, 34, 184	Lepidoptera: Tortricidae	Chile	
Scaphisoma kuscheli Löbl, 1980	*New Zealand Journal of Zoology*, 7, 389	Coleoptera: Staphylinidae	Fiji	
Semiocladius kuscheli Sublette & Wirth, 1980	*New Zealand Journal of Zoology*, 7, 319	Diptera: Chironomidae	Auckland Islands	
Australomalus kuscheli Mazur, 1981	*New Zealand Journal of Zoology*, 8, 381	Coleoptera: Histeridae	Fiji	*Cryptomalus*
Insulahypnus kuscheli Stibick, 1981	*Eos*, 55–56, 234	Coleoptera: Elateridae	New Zealand	
Scaphisoma kuschelianum Löbl, 1981	*Revue Suisse de Zoologie*, 88, 376	Coleoptera: Staphylinidae	New Caledonia	
Callismilax kuscheli Kaszab, 1982	*Folia Entomologica Hungarica*, 43 (2), 198	Coleoptera: Tenebrionidae	New Caledonia	
Cymbeba kuscheli Kaszab, 1982	*Folia Entomologica Hungarica*, 43 (2), 260	Coleoptera: Tenebrionidae	New Caledonia	
Cymbeba kuscheliana Kaszab, 1982	*Folia Entomologica Hungarica*, 43 (2), 265	Coleoptera: Tenebrionidae	New Caledonia	
Isopus kuscheli Kaszab, 1982	*Folia Entomologica Hungarica*, 43 (2), 142	Coleoptera: Tenebrionidae	New Caledonia	
Menimus kuscheli Kaszab, 1982	*Folia Entomologica Hungarica*, 43 (2), 61	Coleoptera: Tenebrionidae	New Caledonia	
Notoptenidium kuscheli Johnson, 1982	*New Zealand Journal of Zoology*, 9, 346	Coleoptera: Ptiliidae	New Zealand	
Tagalinus kuscheli Kaszab, 1982	*Folia Entomologica Hungarica*, 43 (2), 72	Coleoptera: Tenebrionidae	New Caledonia	
Uloma kuscheli Kaszab, 1982	*Folia Entomologica Hungarica*, 43 (2), 91	Coleoptera: Tenebrionidae	New Caledonia	
Xyletobius kuscheli Español, 1982	*Miscelánea Zoológica*, 6, 64	Coleoptera: Anobiidae	New Zealand	
Carphurus kuscheli Wittmer, 1983	*New Zealand Journal of Zoology*, 10, 336	Coleoptera: Melyridae	Fiji	
Acritus kuscheli Gomy, 1984	*Annales de la Société entomologique de France*, 20 (2), 196	Coleoptera: Histeridae	Tasmania	
Gomyopsis kuscheli Dégallier, 1984	*Nouvelle Revue d'Entomologie (N.S.)*, 1 (1), 57	Coleoptera: Histeridae	Fiji	
Podaena kuscheli Ordish, 1984	*Fauna of New Zealand*, 6, 15	Coleoptera: Hydraenidae	New Zealand	
Onthobium kuscheli Paulian & Pluot-Sigwalt, 1985	*Bulletin du Muséum d'histoire naturelle, Paris*, (4) 4, 1106	Coleoptera: Scarabaeidae	New Caledonia	
Platolenes kuscheli Kaszab, 1985	*Folia Entomologica Hungarica*, 46, 52	Coleoptera: Tenebrionidae	Fiji	*Amarygmus*
Pounamuella kuscheli Forster & Platnick, 1985	*Bulletin of the American Museum of Natural History*, 181 (1), 203	Araneae: Orsolobidae	New Zealand	

Name	Reference	Order: Family	Origin	Current Status
Heptathrips kuscheli Mound & Walker, 1986	*Fauna of New Zealand*, 10, 26	Thysanoptera: Phlaeothripidae	New Zealand	
Hiotus kuscheli Wibmer & O'Brien, 1986	*Memoirs of the Americam Entomological Institute*, 39, 287	Coleoptera: Curculionidae	French Guiana	repl. name for *Curculio gagates* Fabricius, 1792
Austrotoxeuma kuscheli Bouček, 1988	*Australasian Chalcidoidea (Hymenoptera). A Biosystematic Revision of Genera of Fourteen Families, with a Reclassification of Species. CABI, Wallingford, U.K.*, p. 504	Hymenoptera: Perilampidae	New Zealand	
Diphoropria kuscheli Naumann, 1988	*Fauna of New Zealand*, 15, 36	Hymenoptera: Diapriidae	New Zealand	
Anomobrenthus kuscheli Damoiseau, 1989	*Bulletin de l'Institut Royal des Sciences Naturelles de Belgique, Entomologie*, 59, 55	Coleoptera: Brentidae	Fiji	
Anthonomus kuscheli Clark, 1989	*Proceedings of the Entomological Society of Washington*, 91, 101	Coleoptera: Curculionidae	Chile	
Argentinorhynchus kuscheli O'Brien & Wibmer, 1989	*Southwestern Entomologist*, 14, 215	Coleoptera: Curculionidae	Bolivia	
Zealanapis kuscheli Platnick & Forster, 1989	*Bulletin of the American Museum of Natural History*, 190, 54	Araneae: Anapidae	New Zealand	
Priocyphus kuscheli Lanteri, 1990	*Revista Brasileira de Entomología*, 34 (2), 412	Coleoptera: Curculionidae	Paraguay	
Sphaerothorax kuscheli Endrödy-Younga, 1990	*New Zealand Journal of Zoology*, 17, 123	Coleoptera: Clambidae	New Zealand	
Edaphus kuscheli Puthz, 1991	*Deutsche Entomologische Zeitschrift*, 38, 271	Coleoptera: Staphylinidae	Fiji	
Edaphus kuschelianus Puthz, 1991	*Deutsche Entomologische Zeitschrift*, 38, 270	Coleoptera: Staphylinidae	Fiji	
Falklandius kuscheli Morrone, 1992	*Acta Entomológica Chilena*, 17, 165	Coleoptera: Curculionidae	Falkland Islands	
Nudomideopsis kuscheli Cook, 1992	*Stygologia*, 7, 54	Trombidiformes: Nudomideopsidae	New Zealand	
Carpophilus kuscheli Dobson, 1993	*Storkia*, 2, 2	Coleoptera: Nitidulidae	Norfolk Island	
Zelandobius kuscheli McLellan, 1993	*Fauna of New Zealand*, 27, 25	Plecoptera: Gripopterygidae	New Zealand	
Obrienia kuscheli Zherikhin & Gratshev, 1994	*Paleontological Journal*, 27 (1A), 53	Coleoptera: Obrieniidae	Kyrgyzstan	

Name	Reference	Order: Family	Origin	Current Status
Lissorhoptrus kuscheli O'Brien, 1996	*Transactions of the American Entomological Society*, 122 (2–3), 119	Coleoptera: Curculionidae	Venezuela	
Listrocerus kuscheli Majer, 1997	*Entomologische Abhandlungen des Staatlichen Museums für Tierkunde Dresden*, 58 (1), 70	Coleoptera: Melyridae	Chile	
Todimopsis kuscheli Ślipiński & Lawrence, 1997	*Annales Zoologici*, 47, 421	Coleoptera: Zopheridae	New Caledonia	
Lutzomyia (*Pintomyia*) *kuscheli* Le Pont, Mar- tinez, Torrez-Espejo & Dujardin, 1998	*Bulletin de la Société Entomologique de France*, 103 (2), 163	Diptera: Psychodidae	Bolivia	*Pintomyia (P.)*
Megarthrus kuscheli Cuccodoro, 1998	*Tropical Zoology*, 11, 108	Coleoptera: Staphylinidae	New Caledonia	
Araucarius kuscheli Mecke, 2000	*Studies in Neotropical Fauna and Environment*, 35, 196	Coleoptera: Curculionidae	Brazil	
Coconotus kuscheli Anderson & Lanteri, 2000	*American Museum Novitates*, 3299, 12	Coleoptera: Curculionidae	Cocos Island	
Zelodes kuscheli Leschen, 2000	*New Zealand Entomologist*, 22, 41	Coleoptera: Leiodidae	New Zealand	
Pogonapion kuscheli Wanat, 2001	*Genera of the Australo-Pacific Rhadinocybinae and Myrmacicelinae, with biogeography of the Apionidae (Coleoptera: Curculionoidea) and phylogeny of the Brentidae (s. lato)*. Olsztyn, Poland: Mantis, p. 129	Coleoptera: Brentidae	New Caledonia	
Noterapion kuscheli Kissinger, 2002	*Insecta Mundi*, 16 (4), 231	Coleoptera: Brentidae	Chile	
Estola kuscheli Barriga, Moore & Cepeda, 2005	*Gayana*, 69 (2), 398	Coleoptera: Cerambycidae	Chile	
Metius guillermoi Will, 2005	*Pan-Pacific Entomologist*, 81, 69	Coleoptera: Carabidae	Juan Fernandez Islands	repl. name for *Ptero-stichus kuscheli* Straneo
Ectemnorhinus kuscheli Grobler, van Rensburg, Bastos, Chimimba & Chown, 2006	*Journal of Zoological Systematics and Evolutionary Research*, 44 (3), 210	Coleoptera: Curculionidae	Prince Edward Island	
Anischia kuscheli Lawrence, 2007	*Insect Systematics and Evolution*, 38 (2), 217	Coleoptera: Eucnemidae	New Caledonia	
Kiwiaesthetus kuscheli Puthz, 2008	*Zeitschrift der Arbeitgemeinschaft Österreichischer Entomologen*, 60, 62	Coleoptera: Staphylinidae	New Zealand	

Name	Reference	Order: Family	Origin	Current Status
Proxylastodoris kuscheli van Doesburg, Cassis & Monteith, 2010	*Zoologische Mededelingen*, 84 (6), 97	Hemiptera: Thaumastocoridae	New Caledonia	
Psalitrus kuscheli Fikáček, 2010	*Koleopterologische Rundschau*, 80, 351	Coleoptera: Hydrophilidae	New Caledonia	
Gromilus kuschelii Morrone, 2011	*Zootaxa*, 3119, 23	Coleoptera: Curculionidae	New Zealand	
Microconilon kuscheli Besuchet & Hlaváč, 2011	*Acta Entomologica Musei Nationalis Pragae*, 51 (2), 523	Coleoptera: Staphylinidae	Fiji	
Archicorynus kuscheli Anderson & Marvaldi, 2013	*Coleopterists Bulletin*, 67, 70	Coleoptera: Belidae	Nicaragua	
Eocaenonemonyx kuscheli * Legalov, 2013	*Paleontological Journal*, 47, 411	Coleoptera: Nemonychidae	U.S.A.	
Palaeophelypera kuscheli * Legalov, 2013	*Historical Biology*, 25, 75	Coleoptera: Curculionidae	Russia (Baltic amber)	
Trigonopterus kuscheli Rheinheimer, 2013	*Koleopterologische Rundschau*, 83, 219	Coleoptera: Curculionidae	New Caledonia	
Sagola kuscheli Park & Carlton, 2014	*Coleopterists Society Monograph*, 13, 14	Coleoptera: Staphylinidae	New Zealand	
Nunnea kuscheli Park & Carlton, 2015	*Florida Entomologist*, 98, 591	Coleoptera: Staphylinidae	New Zealand	
Pactola kuscheli Mazur, 2016	*Austral Entomology*, 56 (3), 273	Coleoptera: Curculionidae	New Caledonia	
Auletanus (*Neauletes*) *kuscheli* Legalov, 2018	*Ukrainian Journal of Ecology*, 8 (1), 798	Coleoptera: Attelabidae (as Rhynchitidae)	Philippines	
Sclerocardius kuscheli Lyal, 2018	*Diversity*, 10 (3), 74, 18	Coleoptera: Curculionidae	Angola	
Philenis kuscheli Hespenheide, 2018	*Diversity*, 10 (3), 84, 21	Coleoptera: Curculionidae	Ecuador	
Afroryzophilus kuscheli Caldara & Košťál, 2018	*Diversity*, 10 (3), 86, 4	Coleoptera: Curculionidae	Senegal	
Aphanommata kuscheli Skuhrovec, Hlaváč & Batelka, 2018	*Diversity*, 10 (3), 87, 2	Coleoptera: Curculionidae	Cape Verde Islands	

References

1. Kuschel, G. A phylogenetic classification of Curculionoidea to families and subfamilies. *Mem. Entomol. Soc. Wash.* **1995**, *14*, 5–33.
2. McKenna, D.D.; Clarke, D.J.; Anderson, R.S.; Astrin, J.; Brown, S.; Chamorro, L.; Davis, S.R.; del Rio, M.; Haran, J.; Kuschel, G.; et al. Proceedings from the 2016 International Weevil Meeting: Morphological and molecular perspectives on the phylogeny, evolution and classification of weevils (Coleoptera: Curculionoidea). *Diversity* **2018**, *10*, 64. [CrossRef]
3. Brown, S.; Oberprieler, R.G.; Leschen, R.; Crosby, T. Guillermo (Willy) Kuschel (13 July 1918–1 August 2017). *N. Z. Entomol.* **2017**, *40*, 92–97. [CrossRef]
4. Cepeda, M.P. Genealogía de la Familia Kuschel. 2018. Available online: http://www.genealog.cl/-Alemanes/K/Kuschel/ (accessed on 27 July 2018).
5. Wirth, W.W. Los insectos de las Islas Juan Fernandez 7. Heleidae and Tendipedidae (Diptera). *Rev. Chil. Entomol.* **1952**, *2*, 87–103.
6. Alexander, C.P. Los insectos de las Islas Juan Fernandez 5. Tipulidae (Diptera). *Rev. Chil. Entomol.* **1952**, *2*, 35–80.
7. Kuschel, G. Nuevas sinonimias y anotaciones sobre Curculionoidea (Coleoptera). *Rev. Chil. Entomol.* **1955**, *4*, 261–312.
8. Kuschel, G. Beetles in a suburban environment: A New Zealand case study. *DSIR Plant Prot. Rep.* **1990**, *3*, 1–118.
9. Kuschel, G. The subfamily Molytinae (Coleoptera: Curculionidae): General notes and description of new taxa from New Zealand and Chile. *N. Z. Entomol.* **1987**, *9*, 11–29. [CrossRef]
10. Louw, S.; Oberprieler, R.G. Willy Kuschel visits South Africa. *Curculio* **1993**, *34*, 5–6.
11. Kuschel, G. La familia Nemonychidae en la Región Neotropical (Aporte 15 de Col. Curculionidae). *Rev. Chil. Hist. Nat.* **1954**, *54*, 97–126.
12. Kuschel, G. Nemonychidae, Belidae y Oxycorynidae de la fauna chilena, con algunas consideraciónes biogeográficas (Coleoptera Curculionoidea, aporte 28). *Investig. Zool. Chil.* **1959**, *5*, 229–271.
13. Kuschel, G. Revisión de *Lissorhoptrus* LeConte y géneros vecinos de América. *Rev. Chil. Entomol.* **1952**, *1*, 23–74.
14. Kuschel, G. La subfamilia Aterpinae en América (Ap. 12 de Coleoptera Curculionidae). *Rev. Chil. Entomol.* **1952**, *1*, 205–244.
15. Kuschel, G. Cylindrorhininae aus dem Britischen Museum (Col. Curculionidae, 8. Beitrag). *Ann. Mag. Nat. Hist.* **1952**, *5*, 121–137. [CrossRef]
16. Kuschel, G. Revisión de la subtribe Epistrophina (Aporte 19 de Col. Curculionoidea). *Rev. Chil. Entomol.* **1957**, *5*, 251–364.
17. Kuschel, G. Revisión de los Premnotrypini y adiciones a los Bagoini (Aporte 17 sobre Coleoptera Curculionoidea). *Bolet. Museo Nacional Hist. Nat.* **1956**, *26*, 187–235.
18. Kuschel, G. La fauna curculiónica (Coleoptera: Curculionoidea) de la *Araucaria araucana*. *Rev. Chil. Entomol.* **2001**, *27*, 41–51.
19. Kuschel, G. Zur Naturgeschichte der Insel San Ambrosio (Islas Desventuradas, Chile). 1. Reisebericht, geographische Verhältnisse und Pflanzenverbreitung. *Arkiv Botanik* **1962**, *4*, 413–419.
20. Kuschel, G. Insects of Campbell Island. Coleoptera: Curculionidae of the Subantarctic Islands of New Zealand. *Pac. Islands Monogr.* **1964**, *7*, 415–493.
21. Kuschel, G. Entomology of the Aucklands and other islands south of New Zealand: Coleoptera: Curculionidae. *Pac. Islands Monogr.* **1971**, *27*, 225–259.
22. Kuschel, G. The genus *Catoptes* Schönherr and two *Species oblitae* of Fabricius from New Zealand (Coleoptera Curculionidae). *N. Z. J. Sci.* **1969**, *12*, 789–810.
23. Kuschel, G. New Zealand Curculionoidea from Captain Cook's voyages (Coleoptera). *N. Z. J. Sci.* **1970**, *13*, 191–205.
24. Franz, H. Neue Scydmaeniden (Coleoptera) aus Neuseeland, von Samoa, den Tonga-Inseln und Cook-Inseln. *Koleopterol. Rund.* **1977**, *53*, 15–25.
25. Nyholm, T. New species, taxonomic notes and genitalia of New Zealand *Cyphon* (Coleoptera: Scirtidae). *N. Z. Entomol.* **2000**, *22*, 45–67. [CrossRef]

26. Johnson, C. An introduction to the Ptiliidae (Coleoptera) of New Zealand. *N. Z. J. Zool.* **1982**, *9*, 333–376. [CrossRef]
27. Kuschel, G. Description of two new *Microcryptorhynchus* species from Lynfield, Auckland City, New Zealand (Coleoptera: Curculionidae). *N. Z. Entomol.* **1997**, *20*, 23–27. [CrossRef]
28. Heller, K.M. Die Käfer von Neu Caledonien und den benachbarten Inselgruppen. *Nova Caled. A Zool.* **1916**, *2*, 229–365.
29. Kuschel, G. The Australian Phrynixinae (Coleoptera: Curculionidae). *N. Z. J. Sci.* **1972**, *15*, 209–231.
30. Kuschel, G. Nemonychidae of Australia, New Guinea and New Caledonia. In *Australian Weevils (Coleoptera: Curculionoidea). Volume 1. Orthoceri: Anthribidae to Attelabidae: The Primitive Weevils*; Zimmerman, E.C., Ed.; CSIRO: Melbourne, Australia, 1994; pp. 563–637.
31. Kuschel, G.; Leschen, R.A.B.; Zimmerman, E.C. Platypodidae under scrutiny. *Invertebr. Taxomony* **2000**, *14*, 771–805. [CrossRef]
32. Kuschel, G. Curculionoidea (weevils) of New Caledonia and Vanuatu: Basal families and some Curculionidae. In *Zoologia Neocaledonica 6. Biodiversity Studies in New Caledonia*, Grandcolas, P., Ed. *Mém. Mus. Hist. Nat.* **2008**, *197*, 99–249.
33. Kuschel, G. New tribe, new genus and species for an Australasian weevil group with notes and keys [Coleoptera, Curculionoidea]. *Rev. Fr. d'Entomol.* **2009**, *30*, 41–66.
34. Kuschel, G. The blind weevils of Myrtonymina in New Caledonia and Australia (Curculionidae: Curculioninae: Erirhinini: Myrtonymina. In *Zoologia Neocaledonica 8. Biodiversity Studies in New Caledonia*, Guilbert, É., Robillard, T., Jourdan, H., Grandcolas, P., Eds. *Mém. Mus. Hist. Nat.* **2014**, *206*, 165–180.
35. Kuschel, G.; Leschen, R.A.B. Appendix 1. Phylogenetic relationships of the genera of Belinae. In *Nemonychidae, Belidae, Brentidae (Insecta: Coleoptera: Curculionoidea)*; Kuschel, G., Ed.; Fauna of New Zealand: Canterbury, New Zealand, 2003; Volume 45, pp. 48–55.
36. Kuschel, G.; Leschen, R.A.B. Phylogeny and taxonomy of the Rhinorhynchinae (Coleoptera: Nemonychidae). *Invertebr. Syst.* **2011**, *24*, 573–615. [CrossRef]
37. Kuschel, G. Past and present of the relict family Nemonychidae (Coleoptera: Curculionidae). *GeoJournal* **1983**, *7*, 499–504.
38. Kuschel, G. *Nemonychidae, Belidae, Brentidae (Insecta: Coleoptera: Curculionoidea)*; Fauna of New Zealand: Canterbury, New Zealand, 2003; Volume 45, pp. 1–100.
39. Kuschel, G. Reappraisal of the Baltic Amber Curculionoidea described by E. Voss. *Mitteilungen aus dem Geologisch-Paläontologischen Institut der Universität Hamburg* **1992**, *73*, 191–215.

Conference Report

Morphological and Molecular Perspectives on the Phylogeny, Evolution, and Classification of Weevils (Coleoptera: Curculionoidea): Proceedings from the 2016 International Weevil Meeting

Duane D. McKenna [1,*], Dave J. Clarke [1], Robert Anderson [2], Jonas J. Astrin [3], Samuel Brown [4], Lourdes Chamorro [5], Steven R. Davis [6], Bruno de Medeiros [7], M. Guadalupe del Rio [8], Julien Haran [9], Guillermo Kuschel [†], Nico Franz [10], Bjarte Jordal [11], Analia Lanteri [8], Richard A. B. Leschen [12], Harald Letsch [13], Chris Lyal [14], Adriana Marvaldi [8], Jose Ricardo Mermudes [15], Rolf G. Oberprieler [16], André Schütte [3], Andrea Sequeira [17], Seunggwan Shin [1], Matthew H. Van Dam [18] and Guanyang Zhang [19]

1. Department of Biological Sciences, University of Memphis, 3700 Walker Avenue, Memphis, TN 38152, USA; dclarke@fieldmuseum.org (D.J.C.), sshin@memphis.edu (S.S.)
2. Research and Collection Division, Canadian Museum of Nature, P.O. Box 3443, Station D, Ottawa, ON K1P 6P4, Canada; randerson@nature.ca
3. Zoologisches Forschungsmuseum Alexander Koenig, Adenauerallee 160, Bonn 53113, Germany; J.Astrin.ZFMK@uni-bonn.de (J.J.A.), schuette@uni-bonn.de (ZFMK) (A.S.)
4. Bio-Protection Research Centre, P.O. Box 85084, Lincoln University, Lincoln 7647, New Zealand; xsdjbx@gmail.com
5. Systematic Entomology Laboratory, Agricultural Research Service, U.S. Department of Agriculture, c/o National Museum of Natural History, Smithsonian Institution, P.O. Box 37012, MRC-168, Washington, DC 20013-7012, USA; lourdes.chamorro@ars.usda.gov
6. Division of Invertebrate Zoology, American Museum of Natural History, Central Park West at 79th Street, New York, NY 10024, USA; sdavis@amnh.org
7. Museum of Comparative Zoology, Department of Organismic & Evolutionary Biology, Harvard University, 26 Oxford Street, Cambridge, MA 02138, USA; souzademedeiros@fas.harvard.edu
8. División Entomología, Facultad de Ciencias Naturales y Museo, Universidad Nacional de La Plata, CONICET, Paseo del Bosque s/n, La Plata, Buenos Aires B1900FWA, Argentina; gdelrio@fcnym.unlp.edu.ar (M.G.d.R.); alanteri@fcnym.unlp.edu.ar (A.L.); marvaldi@fcnym.unlp.edu.ar (A.M.)
9. CIRAD, CBGP, Montpellier, France/CBGP, CIRAD, INRA, IRD, Montpellier SupAgro, University Montpellier, 34000 Montpellier, France; julien.haran@cirad.fr
10. School of Life Sciences, Arizona State University, P.O. Box 874501, Tempe, AZ 85287-4501, USA; nico.franz@asu.edu
11. Natural History Museum, The University Museum, University of Bergen, NO-5007 Bergen, Norway; Bjarte.Jordal@uib.no
12. Manaaki Whenua, New Zealand Arthropod Collection, Private Bag 92170, Auckland 1142, New Zealand; leschenr@landcareresearch.co.nz
13. Department für Botanik und Biodiversitätsforschung, Universität Wien, Rennweg 14, 1030 Wien, Austria; harald.letsch@univie.ac.at
14. Department of Entomology, The Natural History Museum, Cromwell Road, London SW 7 5BD, UK; C.lyal@nhm.ac.uk
15. Laboratório de Entomologia, Departamento de Zoologia, Instituto de Biologia, Universidade Federal do Rio de Janeiro, A1–107, Bloco A, Av. Carlos Chagas Filho, 373, Cidade Universitária, Ilha do Fundão, Rio de Janeiro A1–107, Brazil; jrmermudes@gmail.com
16. CSIRO, Australian National Insect Collection, GPO Box 1700, Canberra, ACT 2601, Australia; Rolf.Oberprieler@csiro.au
17. Department of Biological Sciences, Wellesley College, 106 Central Street Wellesley, MA 02481, USA; asequeir@wellesley.edu

[18] Entomology Department, California Academy of Sciences, 55 Music Concourse Drive, San Francisco, CA 94118, USA, matthewhvandam@gmail.com
[19] Florida Museum of Natural History, University of Florida, 1659 Museum Road, P.O. Box 117800, Gainesville, FL 32611, USA; gyz151@gmail.com
* Correspondence: dmckenna@memphis.edu; Tel.: +1-901-678-1386
† Deceased.

Received: 30 June 2018; Accepted: 5 July 2018; Published: 18 July 2018

Abstract: The 2016 International Weevil Meeting was held immediately after the International Congress of Entomology (ICE). It built on the topics and content of the 2016 ICE weevil symposium *Phylogeny and Evolution of Weevils (Coleoptera: Curculionoidea): A Symposium in Honor of Dr. Guillermo "Willy" Kuschel*. Beyond catalyzing research and collaboration, the meeting was intended to serve as a forum for identifying priorities and goals for those who study weevils. The meeting consisted of 46 invited and contributed lectures, discussion sessions and introductory remarks presented by 23 speakers along with eight contributed research posters. These were organized into three convened sessions, each lasting one day: (1) weevil morphology; (2) weevil fossils, biogeography and host/habitat associations; and (3) molecular phylogenetics and classification of weevils. Some of the topics covered included the 1K Weevils Project, major morphological character systems of adult and larval weevils, weevil morphological terminology, prospects for future morphological character discovery, phylogenetic analysis of morphological character data, the current status of weevil molecular phylogenetics and evolution, resources available for phylogenetic and comparative genomic studies of weevils, the weevil fossil record, weevil biogeography and evolution, weevil host plants, evolutionary development of the weevil rostrum, resources available for weevil identification and the current status of and challenges in weevil classification.

Keywords: 1K Weevils Project; biogeography; classification; Curculionidae; Curculionoidea; fossils; Guillermo Kuschel; morphology; molecular phylogenetics; DNA barcoding; phylogeny; phytophagy; weevils

---

1. Introduction

The 2016 International Weevil Meeting was held from 1 to 3 October 2016 at the Rosen Centre Hotel (Orlando, FL, U.S.A.) immediately after the International Congress of Entomology (ICE). It built on the topics and content of the 2016 ICE weevil symposium *Phylogeny and Evolution of Weevils (Coleoptera: Curculionoidea): A Symposium in Honor of Dr. Guillermo "Willy" Kuschel* (Figure 1). The meeting was convened by researchers from the 1K Weevils Project (funded by the U.S. National Science Foundation; Figure 2) but was open to attendance by all interested parties. Thirty-two people attended representing 13 countries on five continents (Figure 3; Table 1). Beyond catalyzing research and collaboration, the meeting was intended to serve as a forum for identifying priorities and goals for those who study weevils or otherwise have an interest in them. The meeting consisted of 46 invited and contributed lectures, pre-arranged discussion sessions and introductory remarks presented by 23 speakers along with eight contributed research posters (three of these are summarized herein). These were organized into three convened sessions, each lasting one day: (1) weevil morphology; (2) weevil fossils, biogeography and host/habitat associations; and (3) molecular phylogenetics and classification of weevils. At the close of the meeting, the attendees discussed ideas for future research and presented a leather-bound journal for Dr. Kuschel (delivered to him after the meeting by Samuel Brown) with their personal notes and thanks in recognition of his friendship, collaboration and many important contributions to the study of weevils and beyond. This paper reports on the meeting, including a summary of the scientific content presented.

Figure 1. Guillermo "Willy" Kuschel at his home in Lynnfield, AK, New Zealand (5 January 2017). Courtesy, D. Clarke.

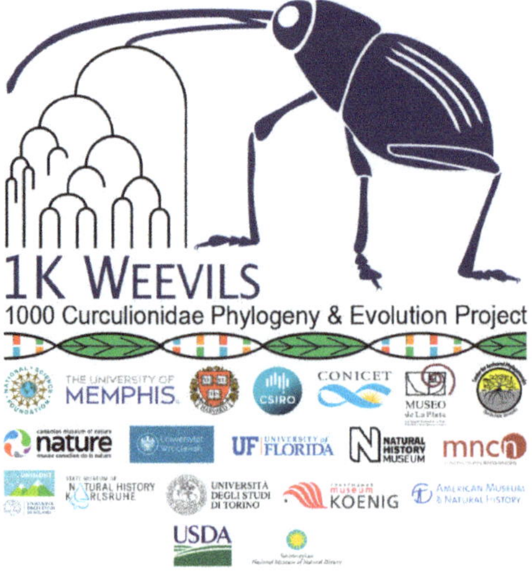

Figure 2. 1000 Curculionidae Phylogeny and Evolution Project (1K Weevils) logo.

Figure 3. Meeting Attendees (from left to right): B. de Medeiros (Harvard University Museum of Comparative Zoology, Cambridge, MA, USA), S. Davis (American Museum of Natural History, New York, NY, USA), S. Shin (University of Memphis, Memphis, TN, USA), D. Clarke (University of Memphis, Memphis, TN, USA), H. Letsch (University of Vienna, Vienna, Austria), (D. McKenna University of Memphis, Memphis, TN, USA), N. Franz (Arizona State University, Tempe, AZ, USA), M. Van Dam (Bavarian State Collection of Zoology, München, Germany), R. Anderson (Canadian Museum of Nature, Ontario, Canada), A. Marvaldi (CONICET, Universidad Nacional de La Plata, Argentina), S. Brown (Lincoln University, Christchurch, New Zealand), A. Lanteri (CONICET, Universidad Nacional de La Plata, Argentina), C. Lyal (The Natural History Museum, London, UK), L. Chamorro (Systematic Entomology Laboratory - ARS, USDA, Beltsville, MD, USA), R. Oberprieler (CSIRO, Australian National Insect Collection, Canberra, Australia), G. Setliff (Kutztown University of Pennsylvania, Kutztown, PA, USA), G. del Rio (CONICET, Universidad Nacional de La Plata, La Plata, Argen), B. Jordal (University of Bergen, Bergen, Norway), S. Il Kim (Harvard University Museum of Comparative Zoology, Cambridge, MA, USA), R. Leschen (New Zealand Arthropod Collection, Auckland, New Zealand), G. Zhang (Arizona State University, Tempe, AZ, USA), R. Whitehouse (Mississippi State University Entomological Museum, Starkville, MS, USA), J. Haran (Centre for Biology and Mgmt. of Populations (CIRAD), UMR CBGP, Montpellier, France), and M. Barrios (Centro Universitario de Zacapa, Universidad de San Carlos de Guatemala). Not shown: S. Anzaldo, C. Beza, P. Biedermann, D. Furth, J. Morillo, R. Mueller and A. Riedel.

Table 1. Affiliations of Attendees *.

Name	Primary Affiliation	City-Country
Robert Anderson	Canadian Museum of Nature	Ottawa, Canada
Salvatore Anzaldo	Arizona State University	Tempe, AZ, USA
Manuel Barrios	Centro Universitario de Zacapa Universidad de San Carlos de Guatemala	Zacapa, Guatemala
Cristian Beza	University of Memphis	Memphis, TN, USA
Peter Biedermann	Max-Planck-Institut for Chemical Ecology	Jena, Germany
Samuel Brown	Lincoln University	Christchurch, New Zealand
Lourdes Chamorro	Systematic Entomology Laboratory, ARS, USDA	Beltsville, MD, USA
Dave Clarke	University of Memphis	Memphis, TN, USA
Steve Davis	American Museum of Natural History	New York, NY, USA
Bruno de Medeiros	Harvard University Museum of Comparative Zoology	Cambridge, MA, USA
Guadalupe del Rio	CONICET, Universidad Nacional de La Plata	La Plata, Argentina
Nico Franz	Arizona State University	Tempe, AZ, USA

Table 1. *Cont.*

Name	Primary Affiliation	City-Country
David Furth	Smithsonian Institution	Washington D.C., USA
Julien Haran	Centre for Biology & Mgmt. of Populations (CBGP)/CIRAD	Montpellier, France
Bjarte Jordal	University of Bergen	Bergen, Norway
Sang Il Kim	Harvard University Museum of Comparative Zoology	Cambridge, MA, USA
Analia Lanteri	CONICET, Universidad Nacional de La Plata	La Plata, Argentina
Richard Leschen	Manaaki Whenua, New Zealand Arthropod Collection	Auckland, New Zealand
Harald Letsch	University of Vienna	Vienna, Austria
Chris Lyal	The Natural History Museum	London, UK
Adriana Marvaldi	CONICET, Universidad Nacional de La Plata	La Plata, Argentina
Duane McKenna	University of Memphis	Memphis, TN, USA
Jose Ricardo Mermudes ^	Universidade Federal do Rio de Janeiro	Rio de Janeiro, Brazil
Jhunior Morillo	The City College of New York	New York, NY, USA
Robert Mueller	Western Sydney University	Sydney, Australia
Rolf Oberprieler	CSIRO, Australian National Insect Collection	Canberra, Australia
Alex Riedel	Staatliches Museum für Naturkunde	Karlsruhe, Germany
Gregory Setliff	Kutztown University of Pennsylvania	Kutztown, PA, USA
Seunggwan Shin	University of Memphis	Memphis, TN, USA
Matthew Van Dam	Bavarian State Collection of Zoology	München, Germany
Ryan Whitehouse	Mississippi State University Entomological Museum	Starkville, MS, USA
Guanyang Zhang	Arizona State University	Tempe, AZ, USA

* Several colleagues who originally intended to take part in the meeting were ultimately unable to attend. These included Miguel Alonso-Zarazaga (several talks had been proposed), Anthony Cognato (had intended to present a talk on Scolytinae), Charlie O'Brien (had intended to present a talk on New World Stenopelmini) and Marek Wanat (had intended to present a talk on Apioninae). ^ Unable to attend, but presented a recorded talk.

2. Scientific Program

The 2016 International Weevil Meeting was organized by Duane McKenna and Dave Clarke (University of Memphis) as part of the U.S. National Science Foundation-funded 1K Weevils Project. Meeting conveners included (in addition to D. McKenna and D. Clarke): Robert Anderson (Canadian Museum of Nature, Ottawa, ON, Canada), Chris Lyal (Natural History Museum, London, UK), Adriana Marvaldi (CONICET, Universidad Nacional de La Plata, La Plata, Argentina) and Rolf Oberprieler (CSIRO, Australian National Insect Collection, Canberra, Australia).

2.1. DAY 1: *Weevil Morphology*

2.1.1. Introduction to Cucujiform Systematics and Morphology (Richard Leschen (Speaker); John Lawrence)

Richard Leschen started the morning session with a presentation on the morphology and classification of the series Cucujiformia, with a focus on the superfamily Cucujoidea and the Phytophaga (comprised of the sister superfamilies Chrysomeloidea and Curculionoidea) [1,2]. Among other things, he commented on the characteristic double tegmen in the superfamily Cleroidea (and a few other clear morphological apomorphies of this superfamily), the difficulties of resolving deep relationships in the superfamily Tenebrionoidea and the non-monophyly of Cucujoidea, which is now thought to comprise two groups: Coccinelloidea (the cerylonid group) and the core Cucujoidea (see [3–5]). He went on to discuss general features of the adult and larval morphology of Phytophaga.

2.1.2. Characters, Homology Assessment and 1K Weevils Morphological Phylogenetic Analyses (Dave Clarke (Speaker); Adriana Marvaldi; Duane McKenna)

Dave Clarke provided an overview of the 1K Weevils morphology project and discussed theoretical and practical aspects of large-scale morphological dataset construction. He began by outlining the three main goals of the project: (1) Compile a comprehensive dataset sampling ~750 genera of Curculionoidea and representing all major lineages of Curculionidae; (2) provide an independent morphological test of weevil relationships by integrating >700 adult and immature characters into one framework; and (3) improve the higher classification of Curculionidae (subfamilies/tribes) using subsets of this information for diagnosing higher taxa and by comprehensively illustrating and integrating this information into a Lucid (or similar) platform. He introduced his literature-based synthetic compilation of character data from throughout Curculionoidea by describing the relationship between the numbers of new characters introduced to the 'global' character database as new studies of weevils appear over time. From this work, he has developed a consensus character list and database comprising both internal and external characters sampled across Curculionoidea, which tracks the usage of individual characters by various studies and thus widely across weevil phylogeny. He advocated for the use of a structured language for character construction (distinct from anatomical/morphological terminology as discussed by Chris Lyal below), such as that described by [6] and for organizing characters in an anatomically logical way, e.g., by body region. The character database is dominated by adult external characters, with internal and larval characters forming a smaller proportion. Adult characters are comparatively evenly distributed among the body regions/primary anatomical divisions. He noted that many characters are probably not independent and discussed the needed balance between quality and quantity of characters. He outlined aspects of what is meant by 'quality', including the importance of understanding the 'classification of characters' and what this means for character construction and phylogeny inference. For example, he emphasized a distinction between 'neomorphic' characters, e.g., 'new' setae or appearance of novel structures (presence/absence, etc.) and transformational characters (similar to [6]). Neomorphic characters are largely presence/absence or meristic characters, whereas transformational characters can be divided into 10 or more categories. He also discussed the use of explicit criteria for representing homology, e.g., position, fine structure, and connectivity with other structures, as well as the implications of primary homology assessment when this step of a phylogenetic analysis accounts for the various categories and types of morphological characters. In addition to the more obvious positive relationship between increasing taxon and character sampling, an emergent property of the synthesis of morphological character analysis that he is conducting is the notion of character scope (e.g., [7,8]). Characters originally circumscribed for a narrow set of taxa (local area of weevil phylogeny) may require substantial re-evaluation with an expanded taxon sample. Dehiscent mandibular cusps are a potential example of this conceptual problem that was discussed in relation to applying explicit homology criteria to the formulation of character statements (e.g., position, fine structure, connectivity). These cusps are typically associated with adult Entiminae (though cusps, or the scars indicating their dehiscence, are not known from all Entiminae) though Steve Davis noted that some Rhynchitinae and Baridini (Conoderinae) also have such cusps. Their apparent scattered appearance may therefore indicate a kind of developmental homology deeper in the tree, reflecting genes being switched on independently in different groups; that is, reflected as homoplasy in terms of character appearance. It may also reflect an incomplete understanding of 'cusps' at the comparative morphology level. Other examples of this problem include the corbels of Entiminae, which was clarified by Rolf Oberprieler's talk. This methodological problem is therefore of key importance to the 1K Weevils morphological study as it incorporates a large and diverse taxon sample involving many characters prone to these and other comparative morphological problems.

2.1.3. Internal Character Systems (Adults) (Steve Davis)

Steve Davis spoke about internal character systems of adult weevils. Some of his reported work was pursued using confocal microscopy (with staining) or micro computed tomography (CT) scanning. He noted that there are an abundance of potentially important morphological characters in the mouthparts, including mandibular structure and articulation, and the form and articulation of the pharyngeal plate. He also noted that the pharyngeal structure of Platypodinae shows similarities to Dryophthorinae consistent with recent papers that recover a close relationship between these subfamilies (e.g., [9–12]). The mesonotum was found to exhibit differences useful in separating genera of Curculionoidea and the metanotum provides many useful family-level characters, such as the shapes of the scutellar groove and the metascutellum. The metendosternite is complex across weevils, differing across subfamilies and sometimes within them, but has potentially good characters, including from musculature. Other character systems were discussed, including characters of the wings (e.g., general venation; a broad lobe at the base of the costal vein; an acute lobe or spine on the 3Ax sclerite; and in Baridini, variation particularly in the cubitus and radial areas), elytra (which have characteristic wing binding patches, submarginal ridges that may be diagnostic for subfamilies and potentially informative patterns of tracheation), legs (with potentially important characters at the base of the coxa (trochantin/pleurotrochantin), the propleurotrochantin (with different elements differentially elongated in different groups, suggesting independent elongation, e.g., Anthribidae versus Scolytinae)), the abdominal tergum (sometimes strongly sclerotized, sometimes subdivided; with locking mechanisms present), new aedeagal structures (e.g., ventral struts from its posterior side in Dryophthorinae (separate from the larger and well-characterized dorsal struts)) and scale ultrastructure (a loose internal structure of cuticular spheres and webbing in primitive weevils becoming more packed and organized in derived weevils, with photonic crystal lattices in scales of the Cyclominae, Entiminae, Gonipterini and Hyperini (CEGH clade)). Cuticular structure was observed to vary across weevils. Embryological characters are informative, e.g., the ephemeral presence of thoracic leg tissues (which degenerate before egg hatching) in families other than Nemonychidae and Anthribidae (i.e., those families in which the larvae have lost thoracic legs). Additionally, the tracheal system contains potentially useful characters mostly in the thorax and elytra.

2.1.4. Larval Character Systems (Adriana Marvaldi)

Adriana Marvaldi discussed the phylogenetic value of larval and pupal character systems. She noted that larval characters, particularly those of the head-capsule and mouthparts, provide useful family- and subfamily-level information [13,14]. Examples include the fronto-epicranial bracon, endocarinal line, shape and setae of the maxilliary mala, palpomere numbers, and the shape of the labral sclerite. Antennal characters are similarly informative, for example, the shape of the sensorium, but this is sometimes difficult to compare across groups. The larvae of Belidae and Attelabidae, for example, have 2-segmented antennae; however, these may not be homologous: Belidae appear to have lost the 1st segment, while Attelabidae may have lost the 2nd segment. The thorax also has useful characters, including the prothoracic shield, which has numerous setae that may be numbered. However, setal numbering may be misleading; the relative position of setae is also significant (in order to ensure comparison of homologous structures). Larval legs are present in Caridae, Nemonychidae, Anthribidae and Brentidae and a pretarsal claw is present only in *Nemonyx* Redtenbacher and Caridae. The sternellum is only present in larvae of Curculionidae. The thoracic spiracle is located on the prothorax in the larvae of Curculionidae, but is intersegmental or placed on the mesothorax in the larvae of the other weevil families. Regarding abdominal features, typical abdominal segments (usually second to seventh) have three or four transverse dorsal folds in Curculionidae, while other families have only two folds. Within Curculionidae, larval characters can define large groups, such as Entiminae (antennal sensorium cushion-like and elliptical in apical view) and the hypothesized clade comprising Dryophthorinae and Platypodinae (abdominal pleural division). Some characters of weevil larvae are clearly linked to habit [15], for example, the head capsule posteriorly emarginated and labrum with reduced setae in leaf miners, spiracles externally tipped as plant-piercing structures in larvae living in aquatic habitats, the concave labrum shape and abdominal ambulatory lobes or pygopods in aerial plant feeders, and the various adaptations of some larvae to wood boring, such as mandibles with the cutting edge raised and with a grinding surface. Many pupal characters foreshadow adult characters and are therefore not very useful; however, pupa-specific chaetotaxi and the shape of pupal urogomphi can be diagnostic at the genus or species levels, while relatively few pupal features can characterize large groups (e.g., most Entiminae have one to two setae on the mandibular thecae). There are no known distinctions between the pupae of Curculionoidea and those of other Phytophaga.

2.1.5. Weevil Morphological Terminology (Chris Lyal)

Chris Lyal spoke on weevil morphological terminology. Currently, many terms are in use for structures across the weevils, but different terms are often used for the same structures. These differences may be linguistic: whether the term is latinized, in English, or another language; applicable only within the weevils or with a wider homology, such as the antennal nomenclature (scape + pedicel + flagellum c.f. scape + funicle + club); they may reflect a functional rather than morphological terminological basis (penis as a morphological structure c.f. the intromittent organ); or they may relate to nomenclatural precision (tarsomere rather than tarsal segment). Some structures, such as the spermatheca, lack a sufficient terminology, in this case because of the lack of landmarks or, as in the case of the antennal club, because morphological problems are unresolved. Authors may also use their own idiosyncratic terminology. The current situation is confusing and limits engagement; it needs improvement. Agreement on terms is necessary, including clarity on their applicability and whether they represent homology or not, which will particularly assist the reuse of characters in phylogenetic analysis. In that context, it is worth considering whether characters particularly useful for descriptive work and those used in reconstructing phylogeny should be noted as such and authors might consider identifying diagnostic and phylogenetic sections separately in descriptions of new taxa. Reaching for some agreement in terminology requires both open discussion and a means of presenting any conclusions. However, although consensus is important, it is also important to facilitate cross-linking terminological synonyms used by different people; this would avoid such an approach appearing as exclusionary and authoritarian. An output could be an atlas of weevil morphology of all stages, from egg to adult. Certainly, in any development of a resource, images illustrating morphological terminology would be a necessary component. The weevil community could build on the glossary in the Handbook of Zoology [16] and the glossary in the International Weevil Community Website (http://weevil.info/). When possible, we should seek congruence with other terminology used in studies of Coleoptera with an additional set of terms for weevil-specific characters. The platform on http://weevil.info/ is open to all to contribute and might be used both as a discussion forum and the basis of a joint publication by all contributors.

2.1.6. Evolutionary Development of the Weevil Rostrum (Steve Davis)

Steve Davis opened the afternoon session with a presentation on the weevil rostrum. He noted that certain Salpingidae (Tenebrionoidea) also have a rostrum (like Curculionoidea), but that the salpingid rostrum differs in having free tendons. In Curculionoidea, the tendons are supported by internal apodemes that extend from the head sulci. He used micro CT scanning to see more detail. He also obtained histological sections across the length of the rostrum from exemplar weevils from different lineages of Curculionoidea and found that apodemes sometimes fuse internally, especially posteriorly. Anteriorly, the pharyngeal plate is supported by apodemes. The weevil phylogeny reconstructed on the basis of rostrum anatomy [9] is similar to phylogenies reconstructed from molecular data (e.g., [10]). He sequenced transcriptomes from the heads of developing weevils and functionally tested (by RNAi) genes for a role in head (and particularly rostrum) morphology and development. His results indicate a wide range of involvement for his previously identified candidate genes in development [17]. Notably, when the gene *sex combs reduced* (Scr) is knocked out, the posterior tentorial pits appear as paired (they are fused into a single pit in the wild-type), the gula appears (this also reverses the fusion of the ventral head sulci (the subgenal sulci reappear)) and the pleurostomal sulcus also reappears. The pharyngeal plate appears to be a novel structure in weevils, although it may share homologies with the sitophore sclerite in other insects or with the hypopharyngeal sclerite and connecting apodemes in other Coleoptera; the apodemes supporting the plate show a general trend from weak/reduced in basal weevil families to robust and rigid in Curculionidae, possibly serving as an adaptive response to changes in feeding strategies and oviposition behavior. Apodemes (when present) are in two pairs arising from the postmentum (the posterior arms) and the postcoila of the anterior mandibular articulations (the anterior arms). The pharyngeal plate is present in Platypodinae but appears to have been lost in Scolytinae (although apodemes remain present).

2.1.7. Morphological Character Evolution in Scolytinae (Bjarte Jordal)

Bjarte Jordal introduced his talk on morphological character evolution in Scolytinae by noting that similar adaptations to subcortical life are present in several other groups of Curculionidae. For example, *Homoeometamelus* Hustache (Conoderinae) produces galleries similar to those produced by Cossoninae (and Scolytinae). Morphological characters have not been well-developed for higher-level studies of Scolytinae, with a strong focus on head, pronotum and elytral declivity. Characters in these body regions are most useful in species identification and were therefore early established as the prime characters for classification. He noted that these body parts are under strong selective pressure in Scolytinae due to specialized courtship behavior in the later arriving, courting sex. Hidden or less extravagant characters have therefore a greater potential for phylogenetic resolution. Examples of such characters include the intersegmental transverse suture of the metanotum, a longitudinal scutoscutellar groove, the elytral locking system anteriorly on the metanepisternum and nodules and pits along the elytral suture. There may also be useful characters in the mouthparts of Scolytinae, as well as along the internal digestive tract, such as the proventriculus [18]. Male genitalia are occasionally used in taxonomy, but exhibit informative characters of greater use at higher taxonomic levels. Care is needed when using generalized characters from the literature and he recommended looking at the actual specimen for studies that rely on these data. Regarding taxon sampling, *Scolytus* Geoffroy is not representative for the subfamily Scolytinae (e.g., [19]) (also noted by several other speakers). More characters resulting from ongoing work in the Jordal lab will be described and used in the future. Emerging patterns of relationships on the basis of morphological characters are largely congruent with relationships reconstructed from analysis of molecular data (e.g., see [20–22]). However, the deep nodes in the phylogeny of Scolytinae are not supported in the published few-gene molecular data sets: more data are apparently needed to resolve these relationships (see [23]).

2.1.8. Systematics of Neotropical Entiminae: Eustylini and Geonemini (Guanyang Zhang (Speaker); Sara Tanveer; Nico Franz)

Guanyang Zhang presented on the systematics of Neotropical Entiminae: Eustylini and Geonemini. He and his coauthors are also building a larger phylogeny of all groups of Neotropical Entiminae. An analysis of biogeographic trends in the Caribbean *Exophthalmus* Schoenherr genus complex is now published [24]. In a larger, still unpublished phylogeny that samples more than 200 terminals, Eustylini are recovered in four places, but they think this may be due to limited taxon sampling. They also studied host plant associations in Eustylini, particularly among taxa from Cuba and Central America. They did so by extracting plant DNA from weevil guts. Across various taxonomic ranks, there were variable oligo to polyphagous patterns of host plant usage not consistent with close and well-conserved cocladogenetic or coevolutionary interactions. Adult Eustylini are generally found on fresh shoots of their hosts, and some species can consume and retain in their guts leaves from up to five host plant species at the same time. There is a likely association between weevil and endosymbiont taxa, with unknown connections to host plant usage. Buchner [25] cites a large number of symbionts in weevils. The present study identified 947 bacterial operational taxonomic units, 44.5% of which were not assigned to genus and many were not assigned at all. Thirty six percent of weevil samples contained bacteria in the genus *Nardonella*. There was some evidence for coevolution between weevils and their associated bacterial faunas. Results of this broad and exploratory survey are published in Zhang [26]. The talk closed with an introduction to the Symbiota Collections of Arthropods Network (SCAN), which is available to facilitate collaboration at the specimen level (see [27]).

2.1.9. Dryophthorinae Larval Morphology (Lourdes Chamorro)

Lourdes Chamorro presented on Dryophthorinae larval morphology. Of the almost 1200 species and 152 genera classified under Dryophthorinae, approximately 70 species in 29 genera are known by their adult and immature forms. She is currently describing the larvae of 37 genera (representing all currently known taxa), including 7 previously unknown genera. Generally, larvae and adults are found simultaneously within their host plant, which include economically important herbaceous plants, such as banana, sugar cane, bamboo, orchids, and a few in seeds or rotting wood. With a few exceptions prior to pupation, the larvae construct cocoons using plant fibers. She reported on new discoveries of the immature stages of Dryophthorinae, including the possible association of *Cryptoderma* Ritsema with ferns, as well as clarification and addition of new characters that will aid in the identification and phylogenetic inference of Dryophthorinae.

Among the characters discussed were the terminal abdominal processes, which differ between taxa and may be entire, or with two (*Scyphophorus* Schoenherr) or six (*Sipalinus* Marshall and *Nephius* Pascoe (=*Anius* Pascoe) digitate processes. In larval *Temnoschoita* Chevrolat, the anterior and posterior stemmata are visible but difficult to see once the specimen is cleared and the posterior stemmata may only be present in early instars. Another set of very useful characters consist of the number, shape, arrangement and relative size of the malar setae, which, among other states, may be branched/unbranched, may vary in number from 8 to more than 20, may or may not possess mircrosetae or denticles basally and may be arranged in a row or in general clumps. The number of setae, particularly the ventral malar setae, may be variable, but perhaps within limits and can be challenging to count when reduction in size has occurred for one or more of the setae. Malar shape is distinct for *Rhynchophorus* Herbst and may be an important character to distinguish the larvae of the various subtribes within Rhynchophorini. Setae and the arrangement of these setae and pores on the epipharynx and labrum appear to be among the most important larval features for the subfamily Dryophthorinae; however, terminology and homology of pores, for example epipharyngeal sensory pores, cf. accessory sensory pores and pores of the anterolateral and anteromedian epipharyngeal setae [28], need to be addressed. Mandibles show differences as well as the shape and number of spiracles. vSysLab is used to code characters http://www.vsyslab.osu.edu/. Each image has a unique identifier to link to the specimen, including an annotation of the provenance of each character. This approach allows one to enter characters, view the character state and code each state collaboratively if necessary. It is also possible to export these data to produce a Lucid key (with images) and also create a natural language description for each taxon. The data can also be linked to Morphbank and Zoobank and can be exported as a nexus file. The complete work is published in this volume.

2.1.10. Logically Reconciling Conflicting Belid Weevil Classifications and Phylogenies (Nico Franz)

Nico Franz's talk focused mainly on collaboration, data culture and specimen management. In particular, given that weevil phylogenies and classifications are expected to change and conflict with each other for decades to come, he asked how (well) are we keeping up with this change, including in our data repositories? He showed logic-reasoned examples of multi-phylogeny and -classification alignments of different succeeding or simultaneously endorsed arrangements for Belidae (cf., [29,30]). Regarding specimen management, he further introduced the Symbiota Collections of Arthropods (SCAN) as a portal for collaborations involving specimens. As of April 2018, this portal holds >180,000 records of >12,000 therein recognized species of Curculionoidea. He then talked about the Open Tree of Life Project, including the possibility of using the Pensoft Arpha Writing Tool for collaborative studies of weevil morphology. He proposed a joint paper focused on existing phylogenetic trees for weevils (and other Phytophaga).

2.1.11. The 1K Weevils Project (Duane McKenna, on Behalf of the 1K Weevils Project Consortium)

Duane McKenna presented an overview of the 1K Weevils Project's goals on behalf of the 1K Weevils Project Consortium. The weevil family Curculionidae, with approximately 51,000 described species in more than 4600 genera [31], is the largest family of weevils and the second largest family of metazoans [32]. Subfamily concepts and interrelationships in Curculionidae remain controversial, though some natural groupings (e.g., the "higher Curculionidae"; those species with the derived pedal type of male genitalia) have come to light through studies of adult morphology [33–38] and phylogenetic studies employing morphological and/or molecular data [10,12,14,39–44]. Nonetheless, outside of the morphologically distinct early-divergent subfamilies Dryophthorinae and Platypodinae, the identity and interrelationships of major lineages corresponding to other subfamilies in Curculionidae remain tentative. This is especially true among the higher Curculionidae. The 1K Weevils Project seeks to infer a molecular phylogeny and chronogram (timetree) for the family Curculionidae using phylogenomic data (1000 species, ~500 genes [45]), thereby further establishing Curculionoidea as a model system for testing predictions and refining general theories about the evolution of insect–plant interactions. A corresponding morphological study (detailed in a talk by Dave Clarke) will permit independent reconstruction of curculionid phylogeny and assessment of curculionid morphology in light of the results from analysis of these and the molecular data. Both studies are focused on sampling type genera and species when possible. The resulting phylogenetic and temporal hypotheses will provide a framework for investigation of curculionid relationships and evolution and will contribute to resolving the taxonomic problems that pervade curculionid internal higher (subfamilial, tribal) classification. Anticipated direct outcomes from this project include: (1) a deeply gene- and taxon-sampled molecular phylogeny of Curculionidae and relatives (other families of Curculionoidea and select outgroups from superfamily Chrysomeloidea); (2) a corresponding molecular chronogram (timetree); (3) a reconstructed evolutionary history of diversification in Curculionidae, particularly in relation to the rise of angiosperms to floristic dominance and the evolution of weevil–angiosperm interactions; (4) an extensive morphological data matrix and associated phylogeny for Curculionidae and relatives and (5) clarification of group concepts and relationships in Curculionidae based on molecular and morphological data. Additional outcomes include undergraduate, graduate and postdoctoral training and mentoring, teacher training and extensive youth and adult outreach and education.

2.2. *DAY 2: Weevil Fossils, Biogeography, and Host/Habitat Associations*

2.2.1. Fossil Overview and the Daohugou and Karatau Weevil Faunas (Rolf Oberprieler)

The day started with a talk by Rolf Oberprieler on the oldest known weevil fossils, those preserved in the deposits of Daohugou in China (ca. 164 Ma) and Karatau in Kazakhstan (163–152 Ma). Whereas the few known Daohugou fossils remain undescribed, those from Karatau have been described as comprising 36 genera and 70 species. Rolf revealed a curious anomaly in the classification of the Karatau fossils following Arnoldi [46] and Gratshev and Zherikhin [47,48] in that the lateral impressions are nearly always classified as Belidae (subfamilies Eobelinae and Oxycorynoidinae) but the dorsoventral impressions as Nemonychidae (in a subfamily Brenthorrhininae). He showed that, if one sorts both types of impression by size and caters for missing features, inaccurate or incorrect depiction of the specimens in the literature and sexual variation as exhibited by extant weevils, the diversity of the Karatau weevils appears considerably smaller. Four size classes are thus identifiable, corresponding to the genera *Eobelus* Arnoldi, *Archaeorrhynchus* Martynov, *Scelocamptus* Arnoldi and *Oxycorynoides* Arnoldi, each comprising about half a dozen recognizable species. All of them are readily comparable with extant Nemonychidae, although assignment to the extant nemonychid subfamilies Cimberidinae or Rhinorhynchinae is difficult and compromised by uncertainties surrounding critical characters in the fossils (e.g., elytral striae, condition of tarsal segments and claws). He further showed that the Karatau fossils recently described or classified as Anthribidae (including the subfamily

Protoscelidinae) exhibit no convincing characters of extant members of this family, being either indistinguishable from "brentorrhinines" or not ostensibly representing weevils. There are likewise no fossils known from Karatau that can be assigned to the extant family Caridae. He concluded that the Karatau fauna is in dire need of comprehensive revision and taxonomic reassessment based on the actual specimens, and he acknowledged the invaluable contributions of Willy Kuschel in this evaluation of weevil fossils.

2.2.2. Yixian Formation Fossils (Steve Davis)

Steve Davis talked about Yixian formation weevil fossils (Early Cretaceous, Barremian–Aptian, ~129.7–122.1 Ma), as well as primitive possible carid taxa being studied from Burmese amber (Early Cretaceous, ca. 99 Ma) and various basal curculionoids from the Crato Formation (Early Cretaceous, ca. 108 Ma). Regarding the Yixian weevil fauna, it has been found to thus far largely comprise nemonychids, as well as belids, anthribids and carids that were originally placed in the subtribe Baissorhynchina (Baissorhynchini) [49]. The other subtribe circumscribed within this tribe, Abrocarina, comprises nemonychids. Preservation of this material is quite good in many cases, with structures such as head sutures/sulci, fine setae and details of the meso-and metanota visible, allowing appropriate morphological comparisons with extant taxa and proper taxonomic placement. In regards to a few new Burmese carid taxa, confocal microscopy and micro-CT scanning was undertaken to examine preserved internal features such as the meso- and metanota and metendosternite, features shared with extant Caridae. A review of carid fossils, included in a phylogeny of the family, was presented. Micro-CT scanning also was implemented for several Crato Formation compressions (AMNH material), most representing Nemonychidae or Belidae. Scans of this material revealed finely detailed preservation of external and internal anatomy, including internal apodemes, tendons and musculature. Work on some new Nemonychidae and Belidae from Daohugou (Jiulongshan Formation; ca. 164 Ma), China, also was presented.

2.2.3. Santana Formation Fossils (Jose Ricardo Mermudes (Recorded Presentation))

Jose Ricardo Mermudes talked about Santana formation fossils by way of a recorded presentation (he did not attend the meeting). He discussed the registry of terrestrial Coleoptera, which was produced as part of the master's degree of his student Márcia F. de Aquino dos Santos. The Santana Formation is from the Araripe Basin and is comprised of sequences of carbonates from the Aptian-Albian, at least 125–112 Ma (Lower Cretaceous), with 300 described species of insects comprising 100 families and 18 orders. Among beetles, the specimens already studied include Carabidae, Buprestidae, Belidae, and Curculionidae. With respect to the species of Curculionoidea described, he noted that *Preclarusbelus vanini* Santos, Mermudes and Fonseca and *Arariperhinus monnei* Santos, Mermudes and Fonseca, previously described as Belidae [50] and Curculioninae [51], are now interpreted as Nemonychidae [52] and Brachycerinae [53], respectively.

2.2.4. The Burmese Amber Weevil Fauna (Rolf Oberprieler)

Rolf Oberprieler presented an overview of the weevil fauna preserved in Burmese amber. The amber originates from a mine in the Hukawng Valley in Mynamar, is dated as 99 Ma in age and is apparently derived from araucariaceous trees. Numerous insects have been described from Burmese amber, including five weevils (four more specimens have been described since, all mesophyletines). Three of these are classified in a subfamily Mesophyletinae, originally placed in Brentidae but as yet of uncertain affinity, whereas two specimens have been assigned to Curculionidae, one to Molytinae (later to Erirhininae, but unconvincingly so) and the other to Scolytinae. Rolf showed a selection of new weevil fossils in Burmese amber that he and Dave Clarke are studying, comparing their salient observable features with diagnostic characters of extant Nemonychidae, Attelabidae and Caridae. The specimens shown included a clear nemonychid (with a free labrum, falcate mandibles, non-geniculate antennae and punctostriate elytra), a specimen with non-geniculate, subbasal antennae and open notosternal sutures (resembling Rhynchitinae but with dentate tarsal claws) and a dozen mesophyletines representing the variety of specimens under study. From these fossils, he characterized Mesophyletinae as having a densely to sparsely setose (never squamose) body, large protruding eyes, no labrum, geniculate antennae with a 7-segmented funicle, exodont mandibles, closed notosternal sutures, punctostriate elytra without a scutellary striole, short trochanters, unarmed femora, often crenulate or serrulate tibiae with two apical spurs and long, deeply lobed tarsi with divaricate, dentate (rarely simple) claws. This combination of characters indicates an affinity with Attelabidae but does not fit in the current concept of this family. Rolf suggested that similar fossils described from other Cretaceous ambers, such as *Albicar* Peris et al. (Spanish amber), *Gobicar* Gratshev and Zherikhin (New Jersey amber), *Gratshevbelus* Soriano (French amber) and *Antiquis* Peris et al. (French amber), which also show geniculate antennae and a similar body shape, may actually belong to Mesophyletinae and that the group may have been more diverse and widespread in the Cretaceous than is currently recognized. Among almost 100 weevils known from Burmese amber to date, and many thousands of Coleoptera, no additional specimens of Curculionidae have come to light, throwing further doubt on the origin of the two members of Curculionidae described thus far from Burmese amber.

2.2.5. Effects of Insect–Host Interactions on the Diversification of Palm-Associated Weevils (Bruno de Medeiros (Speaker); Brian Farrell)

Bruno de Medeiros presented his PhD thesis work on weevils associated with *Syagrus* Mart. palms in Brazil. Multiple species of weevils (mostly Derelomini and Bariditae) are associated with flowers of *Syagrus* [54–61], a genus of palms native to South America and closely related to the coconut [62,63]. Based on a review of the literature and extensive fieldwork, he found that the most abundant weevils visiting flowers of *Syagrus coronata* (Mart.) Becc. and *Syagrus botryophora* (Mart.) Mart. can be classified into three ecological groups: brood pollinators [64,65], non-pollinators feeding on living tissues, and non-pollinators feeding on decaying palm floral tissues. Among weevils, species in the genus *Anchylorhynchus* Schoenherr are important pollinators, consistent with congeneric species in other palms [58,61,66–68], while other Derelomini are typically non-pollinators feeding on decaying floral tissues and Bariditae typically feed on living floral tissues. By using double-digest RAD-seq [69] to generate genetic data for both weevils and plants, Bruno is comparing the degree to which beetle population structure is associated with the population structure in their host plants and whether this varies between weevils with different modes of interaction. He also presented his ongoing work to develop PCR-generated target enrichment probes [70] and RAD-seq [71] for reconstructing lower-level phylogenies among the species of weevils associated with palms and also briefly shared highlights from his work mining the literature for information on egg morphospace. Bruno and collaborators developed a database of egg size measurements gleaned from thousands of references and linked this data to phylogenetic information. This could provide a model on how to gather data from the literature to study weevil biology once a comprehensive phylogenetic tree is produced under the 1K Weevils project.

2.2.6. Weevil Habitat Associations and Host Evolution/Coevolution (Robert Anderson)

Robert Anderson initiated the afternoon session with a talk on weevil habitat associations and host evolution/coevolution. He noted that there is generally sexual dimorphism in rostrum length for most weevils with a strong correlation between female rostrum length and ovipositor length in (at least) species of the genus *Curculio* Linnaeus [72]. This oviposition deep into host substrates facilitated by the elongate rostrum likely ensures enhanced protection for eggs and larvae from desiccation, parasitism and predation; this also allows weevils to exploit novel food sources not available to other beetles. He also noted the differential taxonomic diversity between angiosperm-associated and gymnosperm-associated groups of weevils. The issue of directionality of host shifts was raised, e.g., is it most typically from gymnosperms to angiosperms or the other way around? Clearly, this goes both ways. *Blepharida* Chevrolat (Chrysomelidae: Chrysomelinae) was given as an example of when host associations are better explained by host plant allelochemicals than taxonomy [73]. Species of various semiaquatic weevils (e.g., *Bagous* Germar, *Notiodes* Schoenherr, *Listronotus* Jekel) are generally associated with unrelated plants likely because these groups of plants live in the same habitat as the weevils [74]. Weevil associations with plants in the family Cyclanthaceae have a recently elucidated complex history with no simple cophyletic basis or host range pattern [75]. Robert introduced his leaf litter project in Central America (LLAMA; Leaf Litter Arthopods of Mesoamerica; https://sites.google.com/site/longinollama/), where he and collaborators sampled at low, middle and high elevations, obtaining distinct faunas at each elevation for each locality sampled. They found that low-elevation sites have more in common with one another than high-elevation sites and there was no overlap between high- and low-elevation sites in their litter weevil faunas. He encountered massive undescribed diversity in these samples and briefly discussed the implications for global weevil biodiversity. Robert gave particular attention to Belidae and Nemonychidae in his talk, including discussion of the new belid genus and species *Archicorynus kuscheli* Anderson and Marvaldi, which was unexpectedly discovered through his work with LLAMA [76]. The nemonychid genus *Atopomacer* Kuschel from Central America, which is associated with Podocarpaceae, was also mentioned.

2.2.7. A Biogeographic Overview of the New Zealand Weevil Fauna (Samuel Brown)

Samuel Brown reported that the New Zealand weevil fauna includes approximately 1300 described species in 290 genera. In common with many other New Zealand taxa [77,78], the fauna shows evidence of radiations forming from a few lineages that dispersed to New Zealand or persisted through Oligocene inundations [79]. This is demonstrated by the absence of endemic species of several widely distributed clades, such as the Attelabidae, Dryophthorinae, Conoderinae, Lixini, Hyperini, Rhamphini, Bagoini, Nanophyinae and Mesoptiliinae, and with depauperate faunas of Apioninae, Brentinae and Scolytinae. The major clades in New Zealand are the Anthribidae (29 genera), Cryptorhynchini (43 genera), Phrynixini (31 genera), Cossoninae (38 genera), Eugnomini (17 genera), Storeini (18 genera) and Entiminae (40 genera).

The families of Curculionoidea outside of Curculionidae have been well-revised [80,81]. However, substantial work remains to be done on the Curculionidae at both the genus and species levels. The genera of Cryptorhynchini have been comprehensively revised [82] and recent work on the Entiminae is starting to clarify formerly uncertain generic limits [83]. A genus checklist [84] and species checklist [85] have been relatively recently published. Although the classifications used by these checklists do not reflect the current understanding of weevil evolution, Leschen et al. [84] is a useful classification in the New Zealand context in that the major groups within the fauna are clearly delineated.

A hallmark of the New Zealand weevil fauna is high endemism, with 184 endemic genera. Those genera found elsewhere are primarily shared with New Caledonia, Australia, and the South Pacific; with fewer shared with South America and elsewhere. The Cossoninae are the most cosmopolitan of the major groups, with 13 (35%) non-endemic genera. This proportion of endemism is likely to be slightly exaggerated due to poor understanding of the weevil fauna of the South Pacific. Recent work suggests that greater linkages between New Zealand and the South Pacific will be uncovered as the fauna becomes better known [86,87]. Recent fossil evidence from Miocene sites in Otago indicate a greater diversity of non-curculionid Curculionoidea in the past and a fauna that appears to have been similar to the contemporary New Caledonian fauna [88]. Research on New Zealand weevil taxa has usually been done in isolation. Thus, the monophyly of endemic clades remains untested and their sister taxa elsewhere are unknown. Future research should contextualise the New Zealand weevil fauna to a greater extent by attempting to identify their relationships with taxa from Australia, New Caledonia and the South Pacific.

A number of New Zealand weevil taxa show unusual life history traits or character combinations with potential to inform hypotheses of weevil classification and evolution on a global scale. These include genera, such as *Inosomus* Broun, *Xenocnema* Wollaston, *Novitas* Broun (Cossoninae), *Bantiades* Broun, *Etheophanus* Broun, *Phronira* Broun (Molytinae), *Philacta* Broun (Eugnomini), *Abantiadinus* Schenkling (Storeini), and *Myrtonymus* Kuschel (Myrtonomini) [89]. Several other endemic weevils have been the focus of ecological research, including sexual selection [90,91] and population dynamics [92]. A sound taxonomic and systematic foundation will unlock the potential of this fascinating fauna for informing more biogeographic, evolutionary and ecological hypotheses.

2.2.8. Evolution of Parthenogenesis in South American Naupactini: Insights into Its Origin and Consequences (Analia Lanteri (Speaker); Marcela Rodriguero; Viviana Confalonieri; Noelia Guzmán)

Analia Lanteri reported that there are repeated independent origins of parthenogenesis in South American Naupactini [93]. Some taxa are infected by different strains of *Wolbachia* bacteria of supergroup B, which are transmitted through females (infection cannot be transmitted via males). These bacteria may cause male mortality, cytoplasmic incompatibility, or the development of haploid eggs into females [94]. She noted that parasitoids could play an important role in the horizontal transmission of the bacteria [95]. Members of her research group are developing the following projects: (1) a curing experiment using antibiotics to compare infected hosts with artificially cured hosts to test the effects of the bacterial infection on weevil females; (2) studies of ploidy using confocal microscopy in ovocites taken from ovarioles and in embryos; (3) transcriptomic studies on *Naupactus cervinus* Boheman and *Naupactus leucoloma* Boheman with the goal of identifying genes related to their colonization capacities (in collaboration with A. Sequeira); (4) hybridization as a possible mechanism for the origin of parthenogenetic reproduction in weevils through the study of intra-individual variation in ribosomal sequences [96].

2.2.9. The Contribution of Mitogenome Sequences to the Reconstruction of the Phylogeny of Weevils (Julien Haran (Speaker); Martijn Timmermans; Alfried P. Vogler)

Julien Haran reported on his collaborative works that used mitochondrial genome (mitogenome) sequences in reconstructing the phylogeny of weevils. The development of high-throughput sequencing of full mitochondrial genomes in beetles [97] has made possible the exploration of weevil phylogenetic relationships based on this relatively large and easy-to-handle sequence dataset (12 protein coding genes, approximately 10,000 base pairs). Phylogenetic analyses were first conducted on 27 taxa, including major basal groups and subfamilies, representing the main lineages of Curculionidae [43]. These data were later included in a dataset of 245 mitogenome sequences of Coleoptera to assess the position of weevils in the order Coleoptera [98]. In parallel, a protocol to assemble new mitogenomes was developed and taxon sampling was increased to 122 species of Curculionoidea [44].

All trees reconstructed from the analysis of mitogenome sequences were consistent with previous molecular phylogenetic reconstructions [10,14]. The basal position of families considered as primitive (Anthribidae, Nemonychidae, Attelabidae, Rhynchitidae, Brentidae incl. Nanophyinae, and Apioninae) was supported. The results also supported the existence of an intermediate clade containing the Dryophthorinae, the Brachycerinae, and the Platypodinae at the base of a large clade containing all true weevil lineages (Curculionidae and Scolytinae). Contrary to some previous studies (e.g., [20]), the wood-boring Scolytinae, Cossoninae, and Platypodinae did not form a monophyletic clade [43,44], suggesting strong morphological and behavioral convergence among these groups. In Curculionidae, the largely ectophagous broad-nosed taxa (Entiminae, Cyclominae and Hyperini) formed a monophyletic clade separate from the largely endophagous lineages, highlighting the importance of larval feeding strategy in the early diversification of weevils [43].

Mitogenome sequences robustly recovered the earliest nodes in weevils, thereby "stabilizing" the phylogeny of the superfamily; however, the more recent splits within Curculionidae had lower statistical measures of nodal support. Interestingly, translocations changing the order of some mitochondrial genes (t-RNA) were observed among Curculionidae. These translocations are rare in beetles [99] and seem to be more abundant than expected in this family. As they are evidently specific to certain clades, they are potentially useful for identifying the members of those clades (e.g., Entiminae, Hyperini, Sitonini).

The molecular rate of the mitochondrial genome was found to vary substantially between species. Wood-boring lineages, for instance, show a higher mutation rate than the average rate in all other weevil lineages [43]. The causes remain unclear, but this should be taken into account to avoid long-branch attraction artefacts in phylogeny reconstruction.

2.2.10. Timing and Host Plant Associations in the Evolution of the Weevil Tribe Apionini (Apioninae, Brentidae) Indicate an Ancient Co-Diversification Pattern of Beetles and Flowering Plants (Sven Winter; Ariel Friedman; Jonas Astrin; Brigitte Gottsberger; Harald Letsch (Speaker))

Harald Letsch presented results from some of his collaborative work reconstructing timing and patterns of host plant associations in the weevil tribe Apionini (Apioninae, Brentidae). His work indicates a pattern of ancient co-diversification between Apionini and flowering plants. Most species are monophagous. Host plant use in Apionini is generally conserved, meaning that closely related species feed on closely related plants. Their analysis showed Nanophyinae as the sister group of Brentinae + *Eurhynchus* Kirby + Apioninae and *Antliarhis* Billberg as the sister group of Apioninae. Most subtribes were monophyletic and only a few were paraphyletic. Apionini emerged in the Cretaceous 80 Ma. The supertribes defined by Alonso-Zarazaga [100] were found not to be monophyletic. Wanat's [101] suggestion of Aplemonini forming the sister group of the rest of Apioninae was supported. Some tribes appear largely restricted to specific host families and host family shifts are rare without any shifts to previous hosts: members of the subtribes Piezotrachelina, Oxystomatina, Trichapiina, and Exapiina are all associated with plants in the family Fabaceae, Ceratapiina with Asteraceae and Malvapiina and Aspidapina with Malvaceae. In contrast, weevils of the subtribe Aplemonina are associated with several different plant families: Tamaricaceae, Polygonaceae, Cistaceae, Hypericaceae, Plumbaginaceae and Crassulaceae. Members of the Kalcapiina feed on Euphorbiaceae, Lamiaceae and Urticaceae. The ancestral host for all Apionini remains ambiguous, with either Fabaceae or "basal Caryophyllales" as potential ancestral host plant groups. The comparison of weevil divergence times with the appearance of their host families indicated a simultaneous occurrence of several families of flowering plants and their occupation by Apionini. The analyses further supported the suggestions by Wanat [101] that Apionini originated in Africa and that only dry-adapted Apionini were able to cross the arid northern African zones facilitated by the distribution of dry-adapted plant hosts. However, the authors' focus on the European fauna meant omission of some African and Asian plant hosts and a potential underrepresentation of host use plasticity. Thus, further work is needed to complete the picture of apionine–host coevolution. He also mentioned ongoing work intended to reconstruct the phylogeny and evolution of Ceutorhynchinae [102].

2.2.11. The Vexing Corbels of Entiminae (Rolf Oberprieler)

Rolf Oberprieler gave a short unscheduled talk about the corbels in the subfamily Entiminae. Beginning with pictures of an unmodified tibial apex, which features a circle of fringing setae interrupted on the inside by a pair of socketed spurs (sometimes only one or none) and usually a fixed, perpendicular mucro, he showed that there are three types of corbels that can be derived from this condition: the setose/squamose corbel, the bare corbel and the false corbel. In the first type, the outer tibial edge is bent inwards, first forming a simple bevel without any secondary outer setae, but later such setae develop along the edge of the bend to form a second, outer row of setae that meets the outer row of fringing setae to form a complete circle of setae. In this type, the inner surface of the bevel or corbel is covered with the same setae or scales as occur on the outside of the tibia. This type has been referred to as a "closed" or "enclosed" corbel in the past. In the second type, the narrow strip of integument just above the outer row of fringing setae widens to form a narrow, lenticular surface, whose outer edge is demarcated by either a slight rim or a strong carina or a secondary row of setae. The inner surface of this corbel is always bare (devoid of setae or scales). This type has apparently not been recognised in the literature before. In the third type, a flange develops on the inside of the outer row of fringing setae next to the tarsal socket. This flange is almost always bare and never has an inner row of setae. It has been called a "semi-enclosed" corbel in the literature, but it is not a true corbel and has, correctly, been termed a false corbel in the old German literature (e.g., [103]). An "open" corbel, as often named in the literature, refers to the unmodified condition of the tibial apex and in fact signifies the absence of any type of corbel. The distribution of these corbel types among Entiminae requires further investigation, but tibiae without corbels are found mainly in the tribes Brachyderini, Cyphicerini, Ectemnorhinini, Oosomini, Otiorhynchini, Phyllobiini, Sitonini and Tanyrhynchini, setose/squamose corbels in Embrithini, Leptopiini and Naupactini, bare corbels in Leptopiini and Tanymecini and false corbels in Celeuthetini, Ottistirini and Pachyrhynchini. Some tribes appear to have only one type of corbel, whereas in others, such as Leptopiini and Naupactini, several types occur. Outside of Entiminae, setose bevels and corbels are only known to occur in Brachycerinae, whereas bare corbels occur seemingly nowhere else. False corbels, however, are found sporadically throughout Curculionidae. It is imperative both in descriptions and in phylogenetic analyses to accurately identify and code the type of corbels that may be present.

2.3. DAY 3: Molecular Phylogenetics and Classification of Weevils

2.3.1. Overview of Genomic Resources for Studying Weevils (Duane McKenna (Speaker); Seunggwan Shin; Asela Wijeratne)

The day started with a talk by Duane McKenna. Beyond discussion of development of the present paper, he discussed the genomic resources available for weevils, stressing that this is a community effort. Resources include transcriptomes and genomes. The 1KITE project is adding many more transcriptomes for beetles, including 10 weevils and is contributing substantially to the development of analytical approaches for dealing with large phylogenomic data sets (e.g., [104–106]). Two scolytine genomes have been published (see [107]), but many more have now been sequenced by the McKenna lab from across the Phytophaga, including another 22 genomes from exemplars of Curculionoidea. He discussed collaborative studies on the glycoside hydrolase family of genes in the Phytophaga, some of which function in the digestion of wood (e.g., [108,109]). He talked briefly about the evolution of genes underlying specialized phytophagy in beetles and how this is related to the macroevolution of specialized plant feeding in weevils and other phytophagous beetles [110]. He also briefly mentioned collaborative studies to reconstruct the evolution of genes involved in color vision in beetles (e.g., [111]) and collaborative studies of olfaction in beetles (e.g., [112]). On a different note, he talked about the biodiversity crisis and specifically the loss of tropical biodiversity, with particular reference to the current status of forests in the Greater Congo Basin (http://tropicalbiology.org/congo-resolution/). He is currently collaborating with researchers in the Democratic Republic of the Congo (DRC) on studies involving the regional beetle fauna, with a focus on forested areas in the eastern DRC. A brief summary was given of the 1KITE beetle project, including the higher-level relationships of beetles, with special reference to the phylogeny of Cucujiformia and Phytophaga. Notably, he shared the observation that Cucujoidea is not monophyletic and forms a paraphyletic grade subtending the Phytophaga. Additionally, he discussed the kinds of molecular data (e.g., genomes, transcriptomes, anchored hybrid enrichment, ultraconserved elements) that his lab is currently using to reconstruct beetle phylogeny and he shared information about several different target enrichment probe sets under development in his lab for studying beetle phylogeny and evolution. These included the probes currently being used across the Phytophaga (including the 1K Weevils project [12] and studies of

Scolytinae [23] as well as ongoing studies of longhorned beetles [45] and Buprestidae) and new probe sets under development specifically for use in the Phytophaga.

2.3.2. 1K Weevils Phylogenomic Data and Analyses (Seunggwan Shin (Speaker); Duane McKenna)

Seunggwan Shin presented a talk regarding phylogenomic data analyses for the 1K Weevils Project. This presentation included a general overview of phylogenetics and next-generation sequencing (NGS)-based research for phylogenomics. The goals of the 1K Weevils molecular project were described and the data being used to generate a molecular phylogeny for Curculionoidea were discussed. This project was designed to use anchored hybrid enrichment (AHE) to gather phylogenomic data for more than 1000 weevil species (focused on Curculionidae). AHE is a method of sequencing known targeted genes (here, known 1:1 orthologs) using probes, the design of which is taxon-specific. In this presentation, the workflow and general concept of AHE were presented (see [12,45] for more information on the probes and our evolving analytical pipeline). Using our AHE probes we can obtain DNA sequence data from up to approximately 500 genes for each taxon of interest. Long-branched taxa lacking models in the probe set (e.g., Platypodinae) often produce data from somewhat fewer than 500 genes (but rarely fewer than half of the genes). We have been successful in generating AHE data from specimens preserved in EtOH, RNAlater, and dry (pinned) specimens, including some that are more than 30 years old. Additionally, we have sequenced low-coverage (incomplete) draft genomes for nearly 50 beetles, including more than 30 weevils and other Phytophaga, for use in assessing the position of intron–exon boundaries and otherwise refining the selection of target genes/exons and other features of our probe sets.

2.3.3. Unraveling an Adaptive Radiation: Exploring Genomic Data and Phylogenomic Methods in the Eupholini Weevils of New Guinea (Matthew Van Dam (Speaker); Athena Lam; Alex Riedel; Michael Balke)

Matthew Van Dam spoke about the use of phylogenomic data (ultraconserved elements) for inferring the phylogeny of the Eupholini of New Guinea. Ultraconserved genomic elements [113] have been used in a wide variety of organisms to help resolve both young and old evolutionary radiations. His study aimed to test their use in a tribe of weevils, the Eupholini, found throughout Australasia [114]. Eupholini represent an ecologically diverse clade of weevils found from sea level to the alpine grasslands of New Guinea. The uplift of New Guinea's Central Highlands may have created ecological opportunities for the diversification of Eupholini, resulting in many novel ecomorphologies, such as in the subgenus *Symbiopholus* Gressitt, which has pitted elytra and specialized setae that promote the growth of epizoic symbiosis for camouflage [115]. He and his colleagues examined whether the diversification of the New Guinea Eupholini coincided with the uplift of the Central Highlands. In addition to the collection of fresh specimens, they also utilized museum specimens as a genomic resource. They used both concatenated (RAxML [116]) and species-tree (ASTRAL-III and SVDquartets [117,118]) analyses to examine the relationships and taxonomy of this group. Their findings demonstrated that the current taxonomy renders polyphyletic many of the clades recovered in the analysis, largely grouping species on the basis of similar coloration. In addition, they demonstrated that the majority of loci require multiple partitioning strategies for nucleotide rate substitution. They were successful in gathering hundreds of loci from the nine museum specimens used. Lastly, an elevated speciation rate did coincide with the uplift of the Central Highlands. Highlights from this research can be found in [119].

2.3.4. Bark and Ambrosia Beetle Phylogeny and Diversification (Bjarte Jordal)

Bjarte Jordal presented a talk on bark and ambrosia beetle phylogeny and diversification. A range of interesting behaviors in the subfamily makes the group very useful for studies on evolution. Different reproductive systems have evolved multiple times, such as monogyny, bigyny, harem polygyny, parthenogenesis and permanent inbreeding, as has fungus farming and intricate associations with host plants. Previous attempts to establish a robust phylogeny of the group have failed, particularly due to the lack of resolution at basal nodes. A recent study focused on PCR screening of 100 genes, resulting in 16 promising markers and 13 that were ultimately used for phylogenetics [22]. Some improvement was obtained in the present study, with increased, albeit still limited, resolution of basal nodes. Comparative analyses of diversification indicated a very rapid adaptive radiation by the permanently inbreeding and fungus-cultivating lineage Xyleborini beginning around 20 Ma: the highest speciation rate for any lineage of Scolytinae. More detailed analyses using the software SLOUCH revealed significant effects of inbreeding and deep host shifts for the diversification of the subfamily [120]. High diversification rates were further emphasized by the likely high levels of cryptic species in some inbreeding lineages, e.g., *Hypothenemus eruditus* (Westwood) and allied species groups [121]. Over larger evolutionary time spans, it is also clear that radical host shifts have facilitated high diversification rates given the radically new host opportunities in these lineages.

2.3.5. Phylogeny of Dryophthorinae (Lourdes Chamorro)

Lourdes Chamorro presented a preliminary Bayesian phylogeny of Dryophthorinae based on two molecular markers, 18S and 28S rRNA, aligned using primary and secondary structure. The analysis included a broad sampling of 67 taxa, 2 platypodine outgroup taxa and 65 ingroup taxa representing all 5 tribes of Dryophthorinae and all but one subtribe, Ommatolampina, and a putative African lineage that includes *Ichthyopisthen* Aurivillius, *Korotayeavius* Alonso-Zarazaga and Lyal, etc. Results support a monophyletic Dryophthorinae; however, a long branch subtended the clade suggesting a need for broader outgroup sampling. The small, generally detritovorous dryophthorine tribes Dryophthorini and Stromboscerini were each monophyletic and sister taxa. This clade was in turn sister to a monophyletic Orthognathini; however, this was not strongly supported (posterior probability (PP) was below 95%). The Rhynchophorini was also monophyletic, albeit also with less than 95% PP support, with the possible inclusion of *Cryptoderma* Ritsema, which is currently classified in its own tribe as the Cryptodermatini within this clade. She provided an overview of known host plant preferences and possible evolutionary shifts of these preferences among major groups within dryophthorines. She is currently working to add additional molecular markers as well as morphological data of the adults and immatures. She concluded with an overview of ongoing efforts to generate two-dimensional (2D), and possibly three-dimensional (3D), images of all North American weevil types housed at the United States National Museum (USNM). These images and corresponding occurrence data are being made available through the Symbiota Collections of Arthropods Network (SCAN) as part of the Weevils of North America (WoNA) project http://scan-bugs.org/portal/checklists/checklist.php?cl=1&proj=&dynclid=0.

2.3.6. Structurally Aligned rRNA Sequences and Weevil Phylogeny (Adriana Marvaldi)

Adriana Marvaldi presented a talk on structural alignment of sequences of the ribosomal markers 18S (complete), 28S (D2 and D3), and 16S (IV and V) performed for phylogenetic analysis of weevils. She constructed an annotated alignment for about 270 weevils plus 30 outgroup taxa using as reference the structural model of rRNA as currently predicted for Arthropoda, including beetles [122,123]. A well-resolved tree was produced, but resolution was least in Curculionidae, especially in the CCCMS (Curculioninae, Conoderinae, Cossoninae, Molytinae, Scolytinae) clade. Addition of the protein-coding mitochondrial cytochrome oxidase I (COI) gene improved resolution. The main higher-level relationships found involve: the paraphyly of Nemonychidae, whereby the bulk of included taxa formed a basal weevil clade but part was resolved as closer to Anthribidae: (*Nemonyx* (Urodontinae (Anthribinae))); the monophyly of Belidae and Attelabidae as well as of their two respective subfamilies; and the placement of Caridae as sister group of the clade Brentidae + Curculionidae. The main relationships in Curculionidae worth mentioning were: non-monophyly (forming a basal grade) of Brachycerinae (incl. Erirhininae); Dryophthorinae and Platypodinae were sister taxa; recovery of a clade of "higher" curculionids (with a pedal type of male aedeagus) divided into two sister clades, the CEGH clade (Cyclominae, Entiminae, and allies) and the CCCMS clade. She proposed that it could be helpful to develop a matrix or resource of annotated secondary structural alignments to be available for updates and for use as a template for further phylogenetic analyses of weevils or beetles. Through discussion, Bjarte said that he pursued something similar for 28S (D2–D3) in scolytines but did not find it of great use in providing much more resolution. Duane commented that he and collaborators did this for 28S, 18S, and 16S in Curculionoidea but ultimately decided not to publish the results because they did not noticeably improve resolution in the resulting phylogeny. Adriana insisted that it is still worth the effort to align by structure because it uses a biological criterion to objectively identify homologous positions within length-heterogeneous alignments and to recognize regions of ambiguous alignment.

2.3.7. Some Outstanding Phylogenetic Problems in Broad-Nosed Weevils: The Entiminae, Cyclominae and Allies (CEGH clade) (Adriana Marvaldi; Rolf Oberprieler)

Adriana Marvaldi and Rolf Oberprieler gave a combined talk on the current composition and classification of the CEGH clade, the so-called "broad-nosed weevils", comprising the current subfamilies Cyclominae and Entiminae and the tribes Hyperini, Gonipterini, Viticiini and Phrynixini. Adriana spoke about the CEGH clade and the subfamily Entiminae (e.g., [124]), and this talk was followed by Rolf's exposition on the Cyclominae, Gonipterini and Hyperini (e.g., [125]). The CEGH clade is characterized by usually possessing a short, broad rostrum that is not used in oviposition and by larvae feeding ectophytically on roots in the soil or on aerial plant parts, although the larvae of a number of taxa, especially Cyclominae, bore in stems, trunks or roots (are endophytic). Common characters (putative synapomorphies) of the group include the largely sclerotized, bilobed basal part of sternite IX (c.f., y-shaped with only the arms sclerotized) in the male genitalia, the unarmed or mucronate (not uncinate) meso- and metatibiae and the three-dimensional photonic crystals embedded in a nano-scale chitin lattice in their iridescent scales (in Entiminae, Cyclominae, Hyperini and Viticiini). Although the latter character is proposed as a putative synapomorphy of the CEGH group (to be further investigated), such scales also occur in some Australian genera currently placed in Storeini (though it is likely that these are misclassified and actually belong in the CEGH clade), some Tychiini (both tribes currently in the CCCMS clade of Curculionidae) and in Cerambycidae. The subfamily Entiminae (e.g., [124]), comprising ca. 12,000 described species placed in 1370 genera, includes mainly taxa that were grouped in the section Adelognatha of old classifications. Most have soil-dwelling larvae (few exceptions) and many are polyphagous, some parthenogenetic. The subfamily is indicated to be a monophyletic group based on a number of synapomorphic characters in the adult (e.g., the mandibles typically have deciduous mandibular processes or their scar when broken off) as well as in the larva (two larval characters have been proposed as supporting the monophyly of entimines), but it has not yet been shown to be monophyletic in molecular analyses [12,124]. Its classification into tribes (currently 55) is highly artificial and unsatisfactory, with most indicated to be para- or polyphyletic. Adriana also emphasized that members of tribe Thecesternini lack those above-mentioned entimine features as well as the CEGH apomorphies in spite of the fact that they have a short rostrum and subterranean larvae. She suggested that they belong instead in the CCCMS clade based on the larvae having the *des3* setae on the epicranium and the adults having strikingly uneven elytral sutural flanges. Rolf then provided a synthesis of the Cyclominae. This largely southern-hemisphere subfamily, including ca. 1550 species in 148 genera (currently in 8 tribes), also generally has larvae that live in soil, feeding on or in roots and underground stems. The group is probably not monophyletic in its present composition, lacking any clear synapomorphies. As those of Entiminae, cyclomine larvae have the *des3* setae on the frontal line or on the frons (but this seems to be symplesiomorphic). Of the eight current tribes, Cyclomini, Amycterini and Hipporhinini are indicated to be monophyletic but the others not [125,126]. The tribe Hyperini, a cosmopolitan group comprising ca. 500 species in 44 genera, is characterized by ectophytic and ectophagous larvae (though some mine in leaves, e.g., *Gerynassa* Pascoe) and a meshed cocoon made from fibers secreted by the Malphigian tubules (a putative synapomorphy), but its concept and definition are unclear (no morphological characters indicating monophyly have been found) and the northern- and southern-hemisphere faunas may not be closely related. The small Australo-Pacific tribe Gonipterini (ca. 130 species in 9 genera) is monophyletic based on its peculiar internal proventricular projections. Adults and larvae generally feed on leaves and known larvae are ectophytic, but the larvae of the longer-snouted genera are unknown and probably endophytic. Other tribes and genera indicated to belong in the CEGH clade are the also small tribes Viticiini and Phrynixini (except the *Syagrius* Pascoe group, likely belonging in Molytinae). The Australo-Pacific tribe Viticiini (18 species in 8 genera, forming two groups) is probably monophyletic but no autapomorphies have been identified for it as yet. Its known larvae are all leaf miners. The Australo-Pacific tribe Phrynixini is not monophyletic in its current concept in that the *Syagrius* group belongs in or near Molytinae. Phrynixine larvae probably develop in leaf litter (as known for *Geochus* Broun). Although the CEGH clade is well-supported in all recent molecular analyses, relationships within the group are unresolved and will remain elusive until much larger taxon sets can be analyzed. It is likely that more subfamilies need to be recognized in it but fewer tribes in Entiminae. Adriana and Rolf discussed a number of topics mainly related to the fact that monophyly of these diverse groups has not been properly tested yet. For those groups that have been studied (and their monophyly supported/suggested), their position in the subfamily or family Curculionidae nevertheless remains unclear. At lower levels, several tribal concepts are controversial, e.g., that of Tropiphorini in Entiminae.

2.3.8. Big Blocks and Little Blocks: Phylogeny of the CCCMS Clade (Curculioninae, Conoderinae, Cossoninae, Molytinae, Scolytinae) (Chris Lyal)

Chris Lyal provided an overview of the members of the CCCMS clade and some of the problems that need addressing. This clade, comprising 12 subfamilies recognized in [127] (Baridinae, Ceutorhynchinae, Conoderinae, Cossoninae, Curculioninae, Lixinae, Mesoptiliinae, Molytinae, Cryptorhynchinae, Orobitidinae, Scolytinae, and probably Xiphaspidinae) has been recovered in several molecular analyses as one of the two major clades of the higher weevils [10,12,43,44,128]. Its sister group, the CEGH clade, is discussed by Marvaldi and Oberprieler in this paper (see above). There is some support for the CCCMS clade from morphological characters: elytro-tergal stridulation [129] may be apomorphic for the clade and sclerolepidia [130] may be apomorphic for a clade excluding *Scolytus* Geoffroy and its close relatives, although no analysis has yet been carried out to test these propositions and more work is needed. The ranks given to the family-group taxa vary [127,131,132], but whatever the rank there are some 244 family-group taxa currently included. Twenty-nine of the 244 contain only a single genus and may be assumed to be monophyletic, but fewer than 50% of the others have had any hypothesis of monophyly presented, including only five of the subfamilies listed above. The others may be paraphyletic or even polyphyletic. The large subfamilies Molytinae and Curculioninae lack any known autapomorphies; Cryptorhynchinae may lie inside Molytinae; Cossoninae have many character states uniting them but none appear to be unequivocally autapomorphic; and Baridinae and Conoderinae are not demonstrably monophyletic. Lixinae may be monophyletic but their sister-group is unknown. Scolytinae increasingly appear to be paraphyletic with respect to the rest of the CCCMS clade.

The current situation is not conducive to successful studies of higher-level relationships; genera cannot easily be taken as representing subtribes or tribes if these are not demonstrably monophyletic and thus applying family-group names to tree branches lacks clear meaning. This may be addressed by developing and testing hypotheses of monophyly for extant higher taxa. The option of sampling thousands of genera is probably unrealistic and so reducing this number to a few hundred units, the monophyly of each of which is testable, seems a sensible move. Undoubtedly, other clades will become apparent whether there is a formal name available for them or not. From this, it is more feasible to develop trees at the higher level. It is also important to understand the sampling intensity necessary to deliver a robust tree; e.g. the tree in [79] is far more coherent for Old-World Cryptorhynchini than other cryptorhynchines and Molytinae with a lower sampling density. Establishing phylogenetically reliable bricks may help us build the larger construction that is the phylogeny of the CCCMS clade.

2.3.9. Regional Databases in Curculionidae: An Example for the Naupactini (Curculionidae: Entiminae) from Argentina and Uruguay (M. Guadalupe del Río)

Guadalupe del Río talked about cybertaxonomy (e-taxonomy = web-based taxonomy) and how systematics has evolved, integrating standardized electronic tools, cyber-infrastructure, computer science and computer engineering. Through user-friendly interfaces, these tools make it easier for taxonomists and the general public to identify species and access the world´s total knowledge of biodiversity [133]. She gave a screenshot of the main biodiversity portals, e.g., taxonomic data (Antweb, Avibase, FishBase, species file OSF); biodiversity data (GenBank, iBol, BOLD); morphological data (Morphbank); phylogenetic data (Tree of Life); biodiversity initiatives (Species 2000, GBIF, CoL, EOL); bibliographic information (Biodiversity Heritage Library); and tools for registration and recovery of taxonomic collections (life science identifiers [LSIDs], Darwin Core, digital object identifiers [DOIs]). In the case of weevils, there is an International Weevil Community Website [134] that facilitates cooperation and information sharing worldwide and that also includes a very useful glossary of weevil structures [135]. There are several other websites, for example Wtaxa: Electronic Catalogue of weevil names [136], New Zealand weevil Images [137], Potential invasive weevils of the world [138], Coleoptera Neotropical [139], Symbiota Collections of Arthropods Network (SCAN) [140] and Beetles and Coleopterists [141] among others.

Analía Lanteri and M. Guadalupe del Río generated a regional database of "Naupactini (Curculionidae: Entiminae) from Argentina and Uruguay" that includes more than 120 species, some of which are also distributed in other South American countries and have been introduced in other continents. For each species, the authors give the synonyms, a short diagnosis in English and Spanish, complete data of distribution, host plants and biological observations. This information is complemented with pictures of adult females and males and maps of their distribution in Argentina and Uruguay. The database will be available through the "Portal of Biodiversity of Insects from Argentina" hosted at CEPAVE (Centro de Estudios de Parásitos y Vectores, CONICET-UNLP) and will facilitate the rapid identification of naupactines harmful to agriculture.

2.3.10. Comments on Weevil Classification (Contributed by Guillermo Kuschel; Read by Richard Leschen)

Guillermo (Willy) Kuschel dictated various topics to Richard Leschen who presented these after a brief introduction to Willy's career [36] that highlighted weevil specialists who had studied in or visited New Zealand. Ventral structures of beetles: The current sternal structures of beetles are based on the invagination of what are thought to be primitively external structures and referred to as meso-, meta-, or abdominal ventrites. This is not acceptable to Willy, who maintains that the structural homologies as presented by Snodgrass and others are more accurate. Terminology: Willy was a linguistic scholar and adamant about proper syntax and names and he had disdain for inadequate use of Greek and Latin with respect to taxonomy and morphological terminology. Morphology: The embryonic origin of the dermal layers of the fore- and hind-gut informs the origin and naming of the male genitalia, and Willy believes that David Sharp's work in this regard was exemplary. Classification: The most vexing problem in weevil classification to Willy is the placement of Cossoninae and that host plants should be used as phylogenetic characters to help resolve systematic problems in weevil phylogeny.

2.3.11. Curculionidae Classification: Viewpoints and Vision (Chris Lyal)

Chris Lyal discussed his work on the classification of Curculionidae. An overarching vision is of a time when we know everything about weevil phylogeny, when fitting new taxa into the model is simple, when phylogeny illuminates hypotheses of biology and evolution and when we can address novel questions based on a clear phylogenetic framework supported by a digital databank of all known information about weevils and their associations. Until then ... we are currently in a key position. We have a working community that is building on a long-term engagement around the goal of a functioning classification based on phylogeny and with a recent record of successful interactions in creating the Handbook [142]. We are using multiple methodologies, but these are increasingly interlocking. Our observational precision is improving: the reports by Oberprieler on weevil fossils and of Davis on the weevil rostrum above are outstanding examples of this. We are already sharing information well and have the opportunity to design and build an information system to support our studies. Data management and analysis systems are improving greatly; genomic data and new analytical tools are opening doors to novel insights. Much of what has been discussed in the workshop has been academic and the academic progress is considerable. However, we are living through a biodiversity crisis and we need to develop the use of academic research to combat this. As we progress we should consider the ways in which our work can be at the forefront of effective biodiversity information provision. How do we make host information available? How do we inform Invasive Alien Species risk analysis? How do we develop and disseminate identification tools suitable for use by non-specialists around the world? We need to consider how our outputs can be made multi-purpose, so that what we discover is valuable for non-specialists and made available to them to solve real-world problems. My vision, therefore, is that we carry on with intellectually challenging and rewarding research and use it to safeguard biodiversity and human livelihood.

3. Posters

3.1. Introductions and the Potential for Interspecies Gene Flow in Endemic Galápagos Weevils (Entiminae: Naupactini) (Andrea Sequeira (lead author); Flavia Mendonca de Sousa; Sarah Pangburn; Sara Eslami; Mary Kate Dornon; Anna Hakes)

We aimed to study the potential for interspecies gene flow between introduced and endemic Galápagos weevils (Entiminae: Naupactini). Introduced species can threaten endemic species through competitive exclusion, niche displacement, introgression, predation and hybridization [143]. As a result of hybridization, endemic species can lose important adaptations. Small-island populations of endemic species may be especially threatened due to hybridization with introduced close relatives [144]. The weevil genus *Galapaganus* Lanteri contains 15 species [145]. Thirteen of these species are flightless and 10 are endemic to the Galápagos Archipelago. Phylogenetic reconstructions propose that a *Galapaganus* ancestor colonized the archipelago between 8.6 and 11.5 million years ago from continental Ecuador [146]. The proposed scenario is that original colonizers reached now submerged seamounts, later colonizing the younger islands that exist today. One member of the *Galapaganus* genus, *G. h. howdenae* Lanteri, has a very different colonization history into the islands: it was accidentally introduced into Santa Cruz from mainland Ecuador with the aid of humans. Genetic estimates indicate that this occurred during the colonization period (1832–1959) [147,148] prior to the spurt in human population growth on the island. The six *Galapaganus* species we focused on included five endemics and one introduced species in two islands: *G. h. howdenae, G. conwayensis* Lanteri, and *G. ashlocki* Lanteri (in Santa Cruz Island) and *G. galapagoensis* Lanteri, *G. collaris* Lanteri, and *G. vandykei* Lanteri (in San Cristóbal). Range expansion of *G. h. howdenae* into the highlands prompts us to ask if the introduced population could be hybridizing with the highland endemic *G. ashlocki*, which is also a single-island endemic. We predicted that if hybridization is occurring and is recent, larger estimates of genetic exchange should be found between the introduced *G. h. howdenae* and the Santa Cruz endemics *G. conwayensis* and *G. ashlocki*. Moreover, the effects of hybridization should be greater when analyzing patterns with mitochondrial DNA than with nuclear DNA due to the maternal inheritance of mtDNA.

We estimated six parameters for six pairwise comparisons using IMa2 [149,150] using mitochondrial and nuclear DNA sequence datasets. The estimated parameters included migration rates in both directions (m); a time estimate for when the ancestral population diverged into the two populations (t); population sizes dependent upon the mutation rate (4mu); and the overall effect of migration (2Nm). IMFig [151] was used to obtain significance values for 2Nm. The largest migration rate estimate (m) using mitochondrial DNA was found to be between Santa Cruz species from *G. howdenae* to *G. ashlocki*. The mitochondrial 2Nm values, which give an overall estimate of the effect of migration, show the highest, significant results from the endemic *G. ashlocki* population to the introduced *G. h. howdenae* population. These results indicate recent gene exchange between the endemic and introduced populations, but because population size estimates for *G. ashlocki* were larger than those of *G. h. howdenae*, the effective gene exchange appears to be in the opposite direction as expected. In any case, our results suggest mitochondrial gene exchange between endemic and introduced *Galapaganus*.

The "El Niño" events and subsequent "La Niña" droughts, which affect the islands approximately every ten years, most often cause natural range expansions and provide opportunities for range overlap between previously allopatric populations and species. The expansion of the range of the introduced *G. h. howdenae* and the endemic *G. conwayensis* into the highlands is an example of one of those natural mixtures. Our results provide genetic evidence of the effects of the recent introduced population expansion into the highlands and suggest that an introduced population could have an impact despite low population size or low genetic variation and possibly independently from their competitive abilities. This gene exchange between endemic and introduced species has possible genetic impacts: possible loss or gain of variation and loss or gain of adaptation in the endemic species as well as blurred species boundaries [152], which could lead to the complete loss of the genetic identity of the endemic species. Gene exchange between introduced and wild populations has been documented in species of *Viola* L. [143], smooth cordgrass [153], and tiger salamanders [154]. In the case of the salamander, the speed of spread of introduced alleles into an endangered species underscores the importance of the genetic impact of an introduction even in the absence of ecological dominance. Within the Galapágos, *G. h. howdenae* has been introduced only in Santa Cruz to date. Even though the existing barriers to its dispersal to other islands are effective, they are not insurmountable. As a result, *G. h. howdenae* adults could disperse, via flight or transportation on plants that are exported from Santa Cruz, to other islands, such as Isabela, and potentially encounter and impact populations of other endemic *Galapaganus*.

3.2. The Molecular Weevil Identification (MWI) Project (André Schütte; Peter Stüben; Jonas Astrin (Correspondence))

Since 2011, the Molecular Weevil Identification (MWI) project has strived to build a starting infrastructure for DNA-based research on European and Macaronesian species of Curculionoidea and to foster research on this group of beetles. MWI aggregates CO1 barcodes and centrally archives tissue vouchers, DNA vouchers and a morphological collection. The project involves the Zoological Research Museum Alexander Koenig (ZFMK) in Bonn, Germany, in conjunction with the pan-European entomologists' association Curculio-Institute (CURCI), which is based in Mönchengladbach, Germany.

The early steps in the MWI workflow center on the robust morphological identification of the reference specimens performed by taxonomic and faunistic experts within CURCI. Using these same (subsequently pinned) specimens, we then generate DNA barcodes to validate the morphological identifications and to serve, in combination with these, as the scaffold for ensuing integrative taxonomic research. Several 'cryptic' species and synonyms have already been identified within MWI [155–159], with a strong focus on the subfamily Cryptorhynchinae. In its last project phase, MWI now investigates, among other aspects, the feasibility of using 'barcode gaps' in a quick approximation to pre-assign unknown specimens to species or to get a first heuristic hypothesis on species limits that will be subsequently tested by additional evidence (morphology, ecology). While a general barcode gap for such a diverse group of organisms would not hold, adapted gap values can still be conveniently applied for the various taxonomic and ecological groups.

Currently, the MWI reference database contains roughly 1400 species, with usually several specimens per species, from 330 genera [160,161]. Using synergies with the ZFMK-coordinated project German Barcode of Life (www.bolgermany.de), a regional sampling focus of MWI lies on the weevil species occurring in Germany. In summary, MWI compiles a validated reference database with DNA barcodes of western Palearctic weevils (to be released soon) and offers cross-linked morphological and molecular collections that we encourage entomologists to use.

3.3. Taxonomy and Evolution of New Zealand Broad-Nosed Weevils (Coleoptera: Curculionidae: Entiminae) (Samuel D. J. Brown, Karen F. Armstrong, Barbara I. P. Barratt, Rob Cruickshank, Craig Phillips).

New Zealand has a diverse broad-nosed weevil (Coleoptera: Curculionidae: Entiminae) fauna, which is particularly speciose in alpine regions of the southern South Island [83,162]. A number of species show remarkable sexual dimorphism, where females possess exaggerated structures on the elytra and abdominal ventrites [83]. These structures can be classified into five forms: (1) tubercles at the top of the elytral declivity; (2) prolongation of the apex of the elytra; (3) swelling of the disc of ventrite V; (4) emargination of the apex of ventrite V; and (5) the posterior margin of ventrite IV produced into a lamina. The evolution of these structures is of interest from systematic and functional morphological viewpoints. A phylogeny was inferred from four gene regions (28S, COI, ArgK, and CAD), sequenced from 316 individuals representing 106 species and a species tree inferred using *BEAST. Sexually dimorphic traits were scored and mapped onto the tree. Competing hypotheses of trait evolution were evaluated using BiSSE and Markov models. Sexually dimorphic species had a greater speciation rate, indicating that some form of sexual selection is in operation. It is hypothesized that this takes the form of sexual conflict through energetic costs incurred by females carrying males during prolonged copulation. The evolution of the number of dimorphic traits was best modeled with a progression model, where transitions between n -> n + 1 traits had greater rates. Laminae on ventrite IV are labile and plesiomorphic within NZ Entiminae; however, these structures may prove to be apomorphic at higher levels of entimine weevil systematics.

4. Conclusions

The 2016 International Weevil Meeting, convened by researchers from the 1K Weevils Project (funded by the U.S. National Science Foundation), was a success in catalyzing research and collaboration (particularly involving the 1K Weevils Project) and serving as a forum for identifying priorities and goals for those who study weevils. The meeting collectively hosted 46 invited and contributed lectures, pre-arranged discussion sessions and introductory remarks presented by 23 speakers along with eight contributed research posters. This report presents a summary of the meeting, with a focus on invited and contributed lectures and select posters, including new research findings and ideas contributed via these contributions. We hope that this report will be followed at regular intervals with others reporting on future such meetings.

Author Contributions: Conceptualization, D.D.M., D.J.C., A.M., R.G.O.; Writing-Original Draft Preparation, D.D.M., D.J.C., R.A., J.J.A., S.B., L.C., S.R.D., B.D., B.d.M., J.H., G.K., N.F., B.J., A.L., R.A.B.L., H.L., C.L., A.M., J.M., R.O., A.S., A.S., S.S., M.V., G.Z.; Writing-Review & Editing, D.M., D.C., R.A., J.J.A., S.B., L.C., S.D., B.D., M.G.d.R., J.H., N.F., B.J., A.L., R.A.B.L., H.L., C.L., A.M., J.R.M., R.G.O., A.S. (André Schütte), A.S. (Andrea Sequeira), S.S., M.H.V.D., G.Z.; Project Administration, D.D.M.; Funding Acquisition, D.D.M.

Funding: This research was funded in part by the United States National Science Foundation grant number DEB1355169 to DDM, the University of Memphis and the Entomological Society of America.

Acknowledgments: Thanks are extended to Rosina Romano and Becky Anthony (Entomological Society of America) for assistance with logistics at the meeting venue. This meeting would not have been possible without the substantial time and energy contributed by the program conveners. We would also like to thank the coauthors of presentations: John Lawrence, CSIRO, Australian National Insect Collection, Australia (Leschen); Sara Tanveer, Arizona State University, Tempe, AZ, USA (Zhang); Brian Farrell, Museum of Comparative Zoology, Harvard University, Cambridge, MA, USA (de Medeiros); Marcela Rodriguero, Viviana Confalonieri, and Noelia Guzmán, CONICET, Universidad de Buenos Aires, Argentina (Lanteri); Martijn Timmermans (Department of Life Sciences, Natural History Museum, London, United Kingdom; and Department of Natural Sciences, Middlesex University, Hendon Campus, London, United Kingdom), Alfried Vogler (Department of Life Sciences, Natural History Museum, London, United Kingdom; and Department of Life Sciences, Imperial College London—Silwood Park Campus, Ascot, United Kingdom) (Haran); Sven Winter (University of Vienna, Vienna, Austria), Ariel Friedman (Tel Aviv University, Tel Aviv, Israel), Brigitte Gottsberger (University of Vienna, Vienna, Austria)

(Letsch); Asela Wijeratne (Arkansas State University, Jonesboro, AR, USA) (McKenna); Athena Lam (Bavarian State Collection of Zoology, Munich, Germany), Alex Riedel (State Museum of Natural History Karlsruhe, Germany), Michael Balke (Bavarian State Collection of Zoology, Munich, Germany); Flavia Mendonca de Sousa, Sarah Pangburn, Sara Eslami, Mary Kate Dornon, and Anna Hakes (all Wellesley College, Wellesley, MA, USA) (Andrea Sequeira); Peter Stüben (Curculio-Institute, Mönchengladbach, Germany) (André Schütte); Karen F. Armstrong (Lincoln University, New Zealand), Barbara I. P. Barratt (AgResearch, NZ), Rob Cruickshank (Christchurch City Libraries, NZ) and Craig Phillips (AgResearch, NZ) (Samuel D. J. Brown). USDA is an equal opportunity employer.

Conflicts of Interest: The authors declare no conflict of interest. The funding sponsors had no role in the design of the study; in the collection, analyses, or interpretation of data; in the writing of the manuscript and in the decision to publish the results.

References

1. McKenna, D.D. Molecular phylogenetics and evolution of Coleoptera. In *Handbook of Zoology, Volume IV: Arthropoda: Insecta. Part 38 Coleoptera, Beetles Volume 3: Morphology and Systematics (Phytophaga)*; Leschen, R.A.B., Beutel, R.G., Eds.; Walter de Gruyter: Berlin, Germany, 2014; pp. 1–10, ISBN 978-3-11-027370-0.
2. Haddad, S.; McKenna, D.D. Phylogeny and evolution of superfamily Chrysomeloidea (Coleoptera: Cucujiformia). *Syst. Entomol.* **2016**, *41*, 697–716. [CrossRef]
3. Robertson, J.A.; Slipinski, A.; Moulton, M.; Shockley, F.W.; Giorgi, A.; Lord, N.P.; McKenna, D.D.; Tomaszewska, W.; Forrester, J.; Miller, K.B.; et al. Phylogeny and classification of Cucujoidea and the recognition of a new superfamily Coccinelloidea (Coleoptera: Cucujiformia). *Syst. Entomol.* **2015**, *40*, 745–778. [CrossRef]
4. McKenna, D.D.; Wild, A.L.; Kanda, K.; Bellamy, C.L.; Beutel, R.G.; Caterino, M.S.; Farnum, C.W.; Hawks, D.C.; Ivie, M.A.; Jameson, M.L.; et al. The beetle tree of life reveals that Coleoptera survived end-Permian mass extinction to diversify during the Cretaceous terrestrial revolution. *Syst. Entomol.* **2015**, *40*, 835–880. [CrossRef]
5. McKenna, D.D. *Molecular Systematics of Coleoptera*; Leschen, R.A.B., Beutel, R.G., Eds.; Walter de Gruyter: Berlin, Germany, 2016; Volume 1, pp. 23–34.
6. Sereno, P.C. Logical basis for morphological characters in phylogenetics. *Cladistics* **2007**, *23*, 565–587. [CrossRef]
7. Franz, N.M. Anatomy of a cladistic analysis. *Cladistics* **2014**, *30*, 294–321. [CrossRef]
8. Rieppel, O. The performance of morphological characters in broad-scale phylogenetic analyses. *Biol. J. Linn. Soc.* **2007**, *92*, 297–308. [CrossRef]
9. Davis, S.R. The weevil rostrum (Coleoptera: Curculionoidea): Internal structure and evolutionary trends. *Bull. Am. Mus. Natl. Hist.* **2017**, *416*, 1–76. [CrossRef]
10. McKenna, D.D.; Sequeira, A.S.; Marvaldi, A.E.; Farrell, B.D. Temporal lags and overlap in the diversification of weevils and flowering plants. *Proc. Natl. Acad. Sci. USA* **2009**, *106*, 7083–7088. [CrossRef] [PubMed]
11. McKenna, D.D. Temporal lags and overlap in the diversification of weevils and flowering plants: Recent advances and prospects for additional resolution. *Am. Entomol.* **2011**, *57*, 54–55. [CrossRef]
12. Shin, S.; Clarke, D.J.; Lemmon, A.R.; Lemmon, E.M.; Aitken, A.L.; Haddad, S.; Farrell, B.D.; Marvaldi, A.E.; Oberprieler, R.G.; McKenna, D.D. Phylogenomic data yield new and robust insights into the phylogeny and evolution of weevils. *Mol. Biol. Evol.* **2018**, *35*, 823–836. [CrossRef] [PubMed]
13. Marvaldi, A.E. Higher level phylogeny of Curculionidae (Coleoptera: Curculionoidea) based mainly on larval characters, with special reference to broad-nosed weevils. *Cladistics* **1997**, *13*, 285–312. [CrossRef]
14. Marvaldi, A.E.; Sequeira, A.S.; O'Brien, C.W.; Farrell, B.D. Molecular and morphological phylogenetics of weevils (Coleoptera, Curculionoidea): Do niche shifts accompany diversification? *Syst. Biol.* **2002**, *51*, 761–785. [CrossRef] [PubMed]
15. May, B.M. *Fauna of New Zealand, 28. Larvae of Curculionoidea (Insecta: Coleoptera): A Systematic Overview*; Manaaki Whenua Press: Lincoln, New Zealand, 1993; 226p.
16. Lawrence, J.F.; Beutel, R.G.; Leschen, R.A.B.; Slipinski, A. Arthropoda: Insecta. Part 38 Coleoptera, Beetles. Morphology and Systematics (Phytophaga). In *Handbook of Zoology Glossary of Morphological Terms*; Leschen, R.A.B., Beutel, R.G., Eds.; Walter de Gruyter: Berlin, Germany, 2010; Volume 2.4, pp. 9–20.

17. Davis, S.R. Developmental Genetics in a Complex Adaptive Structure, the Weevil Rostrum. *bioRxiv* **2018**. [CrossRef]
18. Nobuchi, A. Studies on Scolytidae (Coleoptera) XXI. Three new genera and species from Japan. *Kontyû* **1981**, *49*, 12–18.
19. Hulcr, J.; Atkinson, T.H.; Cognato, A.I.; Jordal, B.H.; McKenna, D.D. Morphology, Taxonomy, and Phylogenetics of Bark Beetles. In *Bark Beetles: Biology and Ecology of Native and Invasive Species*; Vega, F., Hofstetter, R., Eds.; Elsevier: Orlando, FL, USA, 2014; pp. 41–81.
20. Jordal, B.H.; Sequeira, A.S.; Cognato, A.I. The age and phylogeny of wood boring weevils and the origin of subsociality. *Mol. Phylogenet. Evol.* **2011**, *59*, 708–724. [CrossRef] [PubMed]
21. Jordal, B.H.; Kaidel, J. Phylogenetic analysis of Micracidini bark beetles (Coleoptera: Curculionidae) demonstrates a single trans-Atlantic disjunction and inclusion of *Cactopinus* in the New World clade. *Can. Entomol.* **2017**, *149*, 8–25. [CrossRef]
22. Pistone, D.; Gohli, J.; Jordal, B.H. Molecular phylogeny of bark and ambrosia beetles (Curculionidae: Scolytinae) based on 18 molecular markers. *Syst. Entomol.* **2018**, *43*, 387–406. [CrossRef]
23. Johnson, A.J.; McKenna, D.D.; Jordal, B.H.; Cognato, A.I.; Smith-Cognato, S.M.; Lemmon, A.R.; Lemmon, E.L.; Hulcr, J. Phylogenomics reveals repeated evolutionary origins of mating systems and fungus farming in bark beetles. *Mol. Phylogenet. Evol.* **2018**, in press. [CrossRef] [PubMed]
24. Zhang, G.; Basharat, U.; Matzke, N.; Franz, N.M. Model selection in statistical historical biogeography of Neotropical insects—The *Exophthalmus* genus complex (Curculionidae: Entiminae). *Mol. Phyologenet. Evol.* **2017**, *109*, 226–239. [CrossRef] [PubMed]
25. Buchner, P. *Endosymbiosis of Animals with Plant Micro-Organisms*; Interscience Publishers: New York, NY, USA, 1965; 160p.
26. Zhang, G.; Browne, P.; Zhen, G.; Johnston, A.; Cadillo-Quiroz, H.; Franz, N. Endosymbiont diversity and evolution across the weevil tree of life. *bioRxiv* **2017**. [CrossRef]
27. Gries, C.; Gilbert, E.E.; Franz, N.M. Symbiota—A virtual platform for creating voucher-based biodiversity information communities. *Biodivers. Data J.* **2014**, *2*, e1114. [CrossRef] [PubMed]
28. Anderson, W.H. Larvae of some genera of Calendrinae (=Rhynchophorinae) and Stromboscerinae. *Ann. Entomol. Soc. Am.* **1948**, *41*, 413–437. [CrossRef]
29. Franz, N.M.; Pier, N.M.; Reeder, D.M.; Chen, M.; Yu, S.; Kianmajd, P.; Bowers, S.; Ludäscher, B. Two influential primate classifications logically aligned. *Syst. Biol.* **2016**, *6*, 561–582. [CrossRef] [PubMed]
30. Franz, N.M.; Musher, L.J.; Brown, J.W.; Yu, S.; Ludäscher, B. Verbalizing phylogenomic conflict: Representation of node congruence across competing reconstructions of the neoavian explosion. *bioRxiv* **2017**. [CrossRef]
31. Oberprieler, R.G.; Anderson, R.S.; Marvaldi, A.E. 3. Curculionoidea Latreille, 1802: Introduction, Phylogeny. In *Handbook of Zoology: Arthropoda: Insecta. Part 38. Coleoptera, Beetles Volume 3: Morphology and Systematics (Phytophaga)*; Leschen, R.A.B., Beutel, R.G., Eds.; Walter de Gruyter: Berlin, Germany, 2014; Volume 3, pp. 285–300, ISBN 978-3-11-027370-0.
32. McKenna, D.D.; Farrell, B.D.; Caterino, M.S.; Farnum, C.W.; Hawks, D.C.; Maddison, D.R.; Seago, A.E.; Short, A.E.Z.; Newton, A.F.; Thayer, M.K. Phylogeny and evolution of Staphyliniformia and Scarabaeiformia (rove and scarab beetles): Forest litter as a stepping-stone for diversification of non-phytophagous beetles. *Syst. Entom.* **2014**, *40*, 35–60. [CrossRef]
33. Morimoto, K. Comparative morphology and phylogeny of the superfamily Curculionoidea of Japan. *J. Fac. Agric. Kyushu Univ.* **1962**, *11*, 331–373.
34. Kuschel, G. Entomology of the Aucklands and other islands south of New Zealand: Coleoptera: Curculionidae. *Pac. Insects Monogr.* **1971**, *27*, 225–259.
35. Thompson, RT. Observations on the morphology and classification of weevils (Coleoptera, Curculionoidea) with a key to major groups. *J. Nat. Hist.* **1992**, *26*, 835–891. [CrossRef]
36. Zimmerman, E.C. *Australian Weevils (Coleoptera: Curculionoidea). III. Nanophyidae, Rhynchophoridae, Erirhinidae, Curculionidae: Amycterinae, Literature Consulted*; CSIRO: Melbourne, Australia, 1994.
37. Zimmerman, E.C. *Australian Weevils (Coleoptera: Curculionoidea). I. Orthoceri, Anthribidae to Attelabidae*; CSIRO: Melbourne, Australia, 1994.
38. Alonso-Zarazaga, M.A. On terminology in Curculionoidea (Coleoptera). *Bol. Soc. Entomól. Aragon.* **2007**, *40*, 210.

39. Kuschel, G. A phylogenetic classification of Curculionoidea to families and subfamilies. *Mem. Entomol. Soc. Wash.* **1995**, *14*, 5–33.
40. Marvaldi, A.E.; Morrone, J.J. Phylogenetic systematics of weevils (Coleoptera: Curculionoidea): A reappraisal based on larval and adult morphology. *Insect Syst. Evol.* **2000**, *31*, 43–58. [CrossRef]
41. Morimoto, K.; Kojima, H. New taxa. Curculionoidea: General introduction and Curculionidae: Entiminae (part 1) Phyllobiini, Polydrusini and Cyphicerini (Coleoptera). In *The Insects of Japan*; Morimoto, K., Kojima, H., Eds.; Fukuoka: Touka Shobo, Japan, 2006; Volume 3, pp. 1–406.
42. Hundsdoerfer, A.K.; Rheinheimer, J.; Wink, M. Towards the phylogeny of the Curculionoidea (Coleoptera): Reconstructions from mitochondrial and nuclear ribosomal DNA sequences. *Zool. Anz. J. Comp. Zool.* **2009**, *248*, 9–31. [CrossRef]
43. Haran, J.; Timmermans, M.J.T.N.; Vogler, A.P. Mitogenome sequences stabilize the phylogenetics of weevils (Curculionoidea) and establish the monophyly of larval ectophagy. *Mol. Phylogenet. Evol.* **2013**, *67*, 15–166. [CrossRef] [PubMed]
44. Gillett, C.P.D.T.; Crampton-Platt, A.; Timmermans, M.J.T.N.; Jordal, B.H.; Emerson, B.C.; Vogler, A.P. Bulk de novo mitogenome assembly from pooled total DNA elucidates the phylogeny of weevils (Coleoptera: Curculionoidea). *Mol. Biol. Evol.* **2014**, *31*, 2223–2237. [CrossRef] [PubMed]
45. Haddad, S.; Shin, S.; Lemmon, A.R.; Moriarty Lemmon, E.; Svacha, P.; Farrell, B.D.; Ślipiński, A.; Windsor, D.; McKenna, D.D. Anchored hybrid enrichment provides new insights into the phylogeny and evolution of longhorned beetles (Cerambycidae). *Syst. Entomol.* **2018**, *43*, 68–89. [CrossRef]
46. Arnoldi, L.V. *Mesozoic Coleoptera. Eobelidae*; Arnoldi, L.V., Zherikhin, V.V., Eds.; Nauka: Moscow, Russia, 1977; pp. 144–176.
47. Gratshev, V.G.; Zherikhin, V.V. A revision of the Late Jurassic nemonychid weevil genera *Distenorrhinus* and *Procurculio* (Insecta, Coleoptera, Nemonychidae). *Paleontol. Zhurnal* **1995**, *2*, 83–84.
48. Gratshev, V.G.; Zherikhin, V.V. A revision of the nemonychid weevil subfamily Brethorrhininae [sic] (Insecta, Coleoptera, Nemonychidae). *Paleontol. J.* **1995**, *29*, 112–127.
49. Davis, S.R.; Engel, M.S.; Legalov, A.; Ren, D. Weevils of the Yixian Formation, China (Coleoptera: Curculionoidea): Phylogenetic considerations and comparison with other Mesozoic faunas. *J. Syst. Palaeontol.* **2013**, *11*, 399–429. [CrossRef]
50. Santos, M.F.A.; Mermudes, J.R.M.; Fonseca, V.M.M. Description of a new genus and species of Belinae (Belidae, Curculionoidea, Coleoptera) from the Santana Formation (Crato member, Lower Cretaceous) of the Araripe basin, northeastern Brazil. In *Paleontology: Life Scenarios, v.1*; Carvalho, I.S., Cassab, R.C.T., Schwanke, C., Eds.; Interscience: Rio de Janeiro, Brazil, 2007; pp. 449–455.
51. Santos, M.F.A.; Mermudes, J.R.M.; Fonseca, V.M.M. A specimen of Curculioninae (Curculionidae, Coleoptera) from the Lower Cretaceous, Araripe Basin, north-eastern Brazil. *Palaeontology* **2011**, *54*, 807–814. [CrossRef]
52. Anderson, R.S.; Oberprieler, R.G.; Marvaldi, A.E. 3.1 Nemonychidae Bedel, 1882. In *Handbook of Zoology, Volume IV: Arthropoda: Insecta. Part 38 Coleoptera, Beetles Volume 3: Morphology and Systematics (Phytophaga)*; Leschen, R.A.B., Beutel, R.G., Eds.; Walter de Gruyter: Berlin, Germany, 2014; Volume 3.4, pp. 301–309.
53. Caldara, R.; Franz, N.M.; Oberprieler, R.G. Curculionidae Latreille, 1802. 3.7.19 Curculionidae Latreille, 1802. In *Handbook of Zoology, Volume IV: Arthropoda: Insecta. Part 38 Coleoptera, Beetles Volume 3: Morphology and Systematics (Phytophaga)*; Leschen, R.A.B., Beutel, R.G., Eds.; Walter de Gruyter: Berlin, Germany, 2014; pp. 589–628.
54. Bondar, G.G. Notas Entomológicas da Bahia. XIII. *Rev. Entomol.* **1943**, *14*, 337–388.
55. Franz, N.M.; Valente, R.M. Evolutionary trends in derelomine flower weevils (Coleoptera: Curculionidae): From associations to homology. *Invertebr. Syst.* **2005**, *19*, 499–530. [CrossRef]
56. De Medeiros, B.A.S.; Núñez-Avellaneda, L.A. Three new species of *Anchylorhynchus* Schoenherr, 1836 from Colombia (Coleoptera: Curculionidae; Curculioninae; Acalyptini). *Zootaxa* **2013**, *3636*, 394–400. [CrossRef] [PubMed]
57. Valente, R.M.; de Medeiros, B.A.S. A new species of *Anchylorhynchus* Schoenherr (Coleoptera: Curculionidae) from the Amazon, with a record of a new host palm for the genus. *Zootaxa* **2013**, *3709*, 394–400. [CrossRef]
58. Silberbauer-Gottsberger, I.; Vanin, S.A.; Gottsberger, G. Interactions of the Cerrado Palms *Butia paraguayensis* and *Syagrus petraea* with Parasitic and Pollinating Insects. *Sociobiology* **2013**, *60*, 306–316. [CrossRef]

59. Valente, R.M.; da Silva, P.A.L. The first Amazonian species of *Andranthobius* Kuschel (Coleoptera: Curculionidae), with records of new host palms for the genus. *Zootaxa* **2014**, *3786*, 458–468. [CrossRef] [PubMed]
60. Guerrero-Olaya, N.Y.; Núñez-Avellaneda, L.A. Ecología de la polinización de *Syagrus smithii* (Arecaceae), una palma cantarofila de la Amazonia Colombiana. *Rev. Peru. Biol.* **2017**, *24*, 43–54. [CrossRef]
61. Nuñez Avellaneda, L.A.; Carreño, J.I. Polinización por abejas en Syagrus orinocensis (Arecaceae) en la Orinoquia colombiana. *Acta Biol. Colomb.* **2017**, *22*, 221–233. [CrossRef]
62. Meerow, A.W.; Noblick, L.R.; Salas-Leiva, D.E.; Sanchez, V.; Francisco-Ortega, J.; Jestrow, B.; Nakamura, K. Phylogeny and historical biogeography of the cocosoid palms (Areaceae, Arecoideae, Cocoseae) inferred from sequences of six WRKY gene family loci. *Cladistics* **2015**, *31*, 509–534. [CrossRef]
63. Noblick, L.R. A revision of the genus *Syagrus* (Arecaceae). *Phytotaxa* **2017**, *294*, 1. [CrossRef]
64. Sakai, S. A review of brood-site pollination mutualism: Plants providing breeding sites for their pollinators. *J. Plant Res.* **2002**, *115*, 161–168. [CrossRef] [PubMed]
65. Hembry, D.H.; Althoff, D.M. Diversification and coevolution in brood pollination mutualisms: Windows into the role of biotic interactions in generating biological diversity. *Am. J. Bot.* **2016**, *103*, 1783–1792. [CrossRef] [PubMed]
66. Núñez-Avellaneda, L.A.; Rojas-Robles, R. Biología Reproductiva y Ecología de la Polinización de la Palma Milpesos *Oenocarpus bataua* en los Andes Colombianos. *Caldasia* **2008**, *30*, 101–125.
67. De Medeiros, B.A.S.; Bená, D.C.; Vanin, S.A. Curculio Curculis lupus: Biology, behavior and morphology of immatures of the cannibal weevil *Anchylorhynchus eriospathae* G. G. Bondar. *PeerJ* **1943**, *2*, 1–26.
68. Núñez-Avellaneda, L.A.; Isaza, C.; Galeano, G. Ecología de la polimización de tres especies de Oenocarpus (Arecaceae) simpátricas en la Amazonia colombiana. *Rev. Biol. Trop.* **2015**, *63*, 35–55. [CrossRef]
69. Peterson, B.K.; Weber, J.N.; Kay, E.H.; Fisher, H.S.; Hoekstra, H.E. Double Digest RADseq: An Inexpensive Method for De Novo SNP Discovery and Genotyping in Model and Non-Model Species. *PLoS ONE* **2012**, *7*, e37135. [CrossRef] [PubMed]
70. Peñalba, J.V.; Smith, L.L.; Tonione, M.A.; Sass, C.; Hykin, S.M.; Skipwith, P.L.; McGuire, J.A.; Bowie, R.C.K.; Moritz, C. Sequence capture using PCR-generated probes: A cost-effective method of targeted high-throughput sequencing for nonmodel organisms. *Mol. Ecol. Res.* **2014**, *14*, 1000–1010. [CrossRef] [PubMed]
71. De Medeiros, B.A.S.; Farrell, B.D. Whole-genome amplification in double-digest RAD-seq results in adequate libraries but fewer sequenced loci. *PeerJ* **2018**, *6*, e5089. [CrossRef]
72. Anderson, R.S. An evolutionary perspective on diversity in Curculionoidea. In *Biology and Phylogeny of Curculionoidea, Proceedings of the XVIII International Congress of Entomology, Vancouver, BC, Canada, 3–9 July 1988*; Anderson, R.S., Lyal, C.H.C., Eds.; Entomological Society of Washington: Washington, DC, USA, 1988; Volume 14, pp. 103–114.
73. Becerra, J.X.; Venable, D.L. Macroevolution of insect–plant associations: The relevance of host biogeography to host affiliation. *Proc. Natl. Acad. Sci. USA* **1999**, *96*, 12626–12631. [CrossRef] [PubMed]
74. Anderson, R.S. Weevils and plants: Phylogenetic versus ecological mediation of evolution of host plant associations in Curculionidae (Curculioninae). *Mem. Entomol. Soc. Can.* **1993**, *165*, 197–232. [CrossRef]
75. Franz, N. Analysing the history of the derelomine flower weevil–Carludovica association (Coleoptera: Curculionidae; Cyclanthaceae). *Biol. J. Linn. Soc.* **2004**, *81*, 483–517. [CrossRef]
76. Anderson, R.S.; Marvaldi, A.E. Finding unexpected beetles in odd places: *Archicorynus kuscheli* Anderson and Marvaldi, a new genus and species representing a new tribe, Archicorynini, of Oxycoryninae (Coleoptera: Belidae) from Nicaragua. *Coleopt. Bull.* **2013**, *67*, 61–71. [CrossRef]
77. Arensburger, P.; Buckley, T.R.; Simon, C.; Moulds, M.; Holsinger, K. Biogeography and phylogeny of the New Zealand cicada genera (Hemiptera: Cicadidae) based on nuclear and mitochondrial DNA data. *J. Biogeogr.* **2004**, *31*, 557–569. [CrossRef]
78. Boyer, S.L.; Giribet, G. Welcome back New Zealand: Regional biogeography and Gondwanan origin of three endemic genera of mite harvestmen (Arachnida, Opiliones, Cyphophthalmi). *J. Biogeogr.* **2009**, *36*, 1084–1099. [CrossRef]
79. Riedel, A.; Tänzler, R.; Pons, J.; Suhardjono, Y.R.; Balke, M. Large-scale molecular phylogeny of Cryptorhynchinae (Coleoptera, Curculionidae) from multiple genes suggests American origin and later Australian radiation. *Syst. Entomol.* **2016**, *41*, 492–503. [CrossRef]

80. Holloway, B.A. *Anthribidae (Insecta: Coleoptera)*; Department of Science, Industrial Research: Lincoln, NE, USA, 1984; Volume 3, p. 269.
81. Kuschel, G. *Nemonychidae, Belidae, Brentidae (Insecta: Coleoptera: Curculionoidea)*; Manaaki Whenua Press: Lincoln, NE, USA, 2003; 100p.
82. Lyal, C.H.C. *Cryptorhynchinae (Insecta: Coleoptera: Curculionidae)*; Landcare Research: Lincoln, NE, USA, 1993; Volume 29, 305p.
83. Brown, S.D.J. *Austromonticola*, a new genus of broad-nosed weevil (Coleoptera: Curculionidae: Entiminae) from montane areas of New Zealand. *ZooKeys* **2017**, *707*, 73–130. [CrossRef] [PubMed]
84. Leschen, R.A.B.; Lawrence, J.F.; Kuschel, G.; Thorpe, S.; Wang, Q. Coleoptera genera of New Zealand. *N. Z. Entomol.* **2003**, *26*, 15–28. [CrossRef]
85. Macfarlane, R.P.; Maddison, P.A.; Andrew, I.G.; Berry, J.A.; Johns, P.M.; Hoare, R.J.B.; Lariviere, M.C.; Greenslade, P.; Henderson, R.C.; Smithers, C.; et al. Chapter Nine. Phylum Arthropoda; subphylum Hexapoda; Protura, springtails, Diplura and insects. In *New Zealand Inventory of Biodiversity, Kingdom Animalia, Chaetognatha, Ecdysozoa, Ichnofossils*; Gordon, D.P., Ed.; Canterbury University Press: Canterbury, Australia, 2011; Volume 2, pp. 233–467.
86. Mazur, M.A. First record of the tribe Eugnomini Lacordaire, 1863 (Coleoptera: Curculionidae) from Fiji with description of *Pactola fiji* sp. n. *Zootaxa* **2012**, *3517*, 63–70.
87. Mazur, M.A. Review of the New Caledonian species of the genus *Pactola* Pascoe, 1876 (Coleoptera: Curculionidae: Eugnomini), with description of two new species. *Zootaxa* **2014**, *3814*, 202–220. [CrossRef] [PubMed]
88. Kaulfuss, U.; Brown, S.D.J.; Henderson, I.; Szwedo, J.; Lee, D. First insects from the Manuherikia Group, early Miocene, New Zealand. *J. R. Soc. N. Z.* **2018**, 1–14. [CrossRef]
89. Grebennikov, V.V. First *Alaocybites* weevil (Insecta: Coleoptera: Curculionoidea) from the Eastern Palaearctic: A new micropthalmic species and generic relationships. *Arthropod Syst. Phyl.* **2010**, *68*, 331–365.
90. Painting, C.J.; Holwell, G.I. Exaggerated trait allometry, compensation and trade-offs in the New Zealand giraffe weevil (*Lasiorhynchus barbicornis*). *PLoS ONE* **2013**, *8*, e82467. [CrossRef] [PubMed]
91. Painting, C.J.; Holwell, G.I. Temporal variation in body size and weapon allometry in the New Zealand giraffe weevil. *Ecol. Entomol.* **2015**, *40*, 486–489. [CrossRef]
92. Johst, K.; Schöps, K. Persistence and conservation of a consumer-resource metapopulation with local overexploitation of resources. *Biol. Conserv.* **2003**, *109*, 57–65. [CrossRef]
93. Lanteri, A.A.; Normark, B.B. Parthenogenesis in the tribe Naupactini (Coleoptera: Curculionidae). *Ann. Entomol. Soc. Am.* **1995**, *88*, 722–731. [CrossRef]
94. Rodriguero, M.S.; Confalonieri, V.A.; Guedes, J.V.C.; Lanteri, A.A. *Wolbachia* infection in the tribe Naupactini (Coleoptera, Curculionidae): Association between thelytokous parthenogenesis and infection status. *Insect Mol. Biol.* **2010**, *19*, 599–705.
95. Rodriguero, M.S.; Aquino, D.A.; Loiácono, M.S.; Elías Costa, A.J.; Confalonieri, V.A.; Lanteri, A.A. Parasitoidism of the "Fuller's rose weevil" Naupactus cervinus by Microctonus sp. larvae. *BioControl* **2014**, *59*, 547–556. [CrossRef]
96. Rodriguero, M.S.; Wirth, S.A.; Alberghina, J.S.; Lanteri, A.A.; Confalonieri, V.A. A tale of swinger insects: Signatures of past sexuality between divergent lineages of a parthenogenetic weevil revealed by ribosomal intraindividual variation. *PLoS ONE* **2018**, *13*, e0195551. [CrossRef] [PubMed]
97. Timmermans, M.J.T.N.; Dodsworth, S.; Culverwell, C.L.; Bocak, L.; Ahrens, D.; Littlewood, D.T.J.; Pons, J.; Vogler, A.P. Why barcode? Highthroughput multiplex sequencing of mitochondrial genomes for molecular systematics. *Nucleic Acids Res.* **2010**, *38*, 1–14. [CrossRef] [PubMed]
98. Timmermans, M.J.T.N.; Barton, C.; Haran, J.; Ahrens, D.; Culverwell, L.; Ollikainen, A.; Dodsworth, S.; Foster, P.G.; Bocak, L.; Vogler, A.P. Family-level sampling of mitochondrial genomes in Coleoptera: Compositional heterogeneity and phylogenetics. *Gen. Biol. Evol.* **2016**, *8*, 161–175. [CrossRef] [PubMed]
99. Timmermans, M.J.T.N.; Vogler, A.P. Phylogenetically informative rearrangements in mitochondrial genomes of Coleoptera, and monophyly of aquatic elateriform beetles (Dryopoidea). *Mol. Phylogenet. Evol.* **2012**, *63*, 299–304. [CrossRef] [PubMed]
100. Alonso-Zarazaga, M.A. Revision of the supraspecific taxa in the Palaearctic Apionidae Schoenherr, 1823 (Coleoptera, Curculionoidea). 2. Subfamily Apioninae Schoenherr, 1823: Introduction, keys and descriptions. *Graellsia* **1990**, *46*, 19–156.

101. Wanat, M. *Genera of Australo-Pacific Rhadinocybinae and Myrmacicelinae: With Biogeography of the Apionidae (Coleoptera: Curculionoidea) and Phylogeny of the Brentidae (s. lato)*; Mantis: Oslztyn, Poland, 2001; 432p, ISBN 978-8391433614.
102. Letsch, H.; Gottsberger, B.; Metzl, C.; Astrin, J.J.; Friedman, A.L.L.; McKenna, D.D.; Fiedler, K. Climate and host plant associations shaped the evolution of weevils (Curculionidae: Ceutorhynchinae) throughout the Cenozoic. *Evolution* **2018**, in press. [CrossRef]
103. Faust, J. Stellung und neue Arten der asiatischen Rüsselkäfergattung *Catapionus*. *Deutsch. Entomol. Z.* **1883**, *27*, 81–98.
104. Niehuis, O.; Hartig, G.; Grath, S.; Pohl, H.; Lehmann, J.; Tafer, H.; Donath, A.; Krauss, V.; Eisenhardt, C.; Hertel, J.; et al. Genomic and morphological evidence converge to resolve the enigma of Strepsiptera. *Curr. Biol.* **2012**, *23*, 1388. [CrossRef]
105. Misof, B.; Liu, S.; Meusemann, K.; Peters, R.S.; Donath, A.; Mayer, C.; Frandsen, R.B.; Ware, J.; Flouri, T.; Beutel, R.G.; et al. Phylogenomics Resolves the Timing and Pattern of Insect Evolution. *Science* **2014**, *346*, 763–767. [CrossRef] [PubMed]
106. Kjer, K.H.; Aspock, U.; Aspock, R.G.; Beutel, A.; Blanke, A.; Donath, T.; Flouri, P.; Frandsen, L.; Jermiin, P.; Kapli, A.; et al. Response to comment on 'Phylogenomics resolves the timing and pattern of insect evolution'. *Science* **2015**, *349*, 487–488. [CrossRef] [PubMed]
107. McKenna, D.D. Beetle systematics in the 21st Century: Prospects and progress from studies of genes and genomes. *Curr. Opin. Insect Sci.* **2018**, *24*, 76–82. [CrossRef] [PubMed]
108. Scully, E.D.; Geib, S.M.; Carlson, J.E.; Tien, M.; McKenna, D.D.; Hoover, K. Functional genomics and microbiome profiling of the Asian longhorned beetle (*Anoplophora glabripennis*) reveal new insights into the digestive physiology and nutritional ecology of wood-feeding beetles. *BMC Genom.* **2014**, *15*, 1096. [CrossRef] [PubMed]
109. McKenna, D.D.; Scully, E.D.; Pauchet, Y.; Hoover, K.; Kirsch, R.; Geib, S.M.; Mitchell, R.F.; Waterhouse, R.M.; Ahn, S.; Arsala, D.; et al. Genome of the Asian longhorned beetle (*Anoplophora glabripennis*), a globally significant invasive species, reveals key functional and evolutionary innovations at the beetle-plant interface. *Genome Biol.* **2016**, *17*, 227. [CrossRef] [PubMed]
110. McKenna, D.D. Towards a temporal framework for "Inordinate Fondness": Reconstructing the macroevolutionary history of beetles (Coleoptera). *Entomol. Am.* **2011**, *117*, 28–36. [CrossRef]
111. Sharkey, C.R.; Fujimoto, M.S.; Lord, N.P.; Shin, S.; McKenna, D.D.; Suvorov, A.; Martin, G.J.; Bybee, S.M. Beetle UV opsin gene duplications restore the lost short wave insect opsin class. *Sci. Rep.* **2017**, *7*, 8. [CrossRef] [PubMed]
112. Mitchell, R.F.; Hall, L.P.; Reagel, P.F.; McKenna, D.D.; Baker, T.C.; Hildebrand, J.G. Odorant receptors and antennal lobe morphology offer a new approach to understanding the olfactory biology of the Asian longhorned beetle (*Anoplophora glabripennis*). *J. Comp. Physiol. A* **2017**, *203*, 99–109. [CrossRef] [PubMed]
113. Faircloth, B.C.; McCormack, J.E.; Crawford, N.G.; Harvey, M.G.; Brumfield, R.T.; Glenn, T.C. Ultraconserved Elements Anchor Thousands of Genetic Markers Spanning Multiple Evolutionary Timescales. *Syst. Biol.* **2012**, *61*, 717–726. [CrossRef] [PubMed]
114. Riedel, A. Revision of the genus Penthoscapha Heller (Coleoptera, Curculionoidea, Entiminae, Eupholini) with notes on the genera of Eupholini from New Guinea. *Zootaxa* **2009**, *2224*, 1–29. [CrossRef]
115. Gressitt, J.L.; Sedlacek, J.; Szent-Ivany, J.J.H. Flora and Fauna on Backs of Large Papuan Moss-Forest Weevils. *Science* **1965**, *80*, 150. [CrossRef] [PubMed]
116. Stamatakis, A. RAxML version 8: A tool for phylogenetic analysis and post-analysis of large phylogenies. *Bioinformatics* **2014**, *30*, 1312–1313. [CrossRef] [PubMed]
117. Chifman, J.; Kubatko, L. Quartet Inference from SNP Data under the Coalescent Model. *Bioinformatics* **2014**, *30*, 3317–3324. [CrossRef] [PubMed]
118. Zhang, C.; Sayyari, E.; Mirarab, S. ASTRAL-III: Increased Scalability and Impacts of Contracting Low Support Branches. In *Lecture Notes in Computer Science. Comparative Genomics, RECOMB-CG*; Meidanis, J., Nakhleh, L., Eds.; Springer: Cham, Switzerland, 2017; Volume 1, pp. 53–75.
119. Van Dam, M.H.; Lam, A.W.; Sagata, K.; Gewa, B.; Laufa, R.; Balke, A. Ultraconserved elements (UCEs) resolve the phylogeny of Australasian smurf-weevils. *PLoS ONE* **2017**, *12*, e0188044. [CrossRef] [PubMed]
120. Gohli, J.; Kirkendall, L.R.; Smith, S.M.; Cognato, A.I.; Hulcr, J.; Jordal, B.H. Biological factors contributing to bark and ambrosia beetle species diversification. *Evolution* **2017**, *71*, 1258–1272. [CrossRef] [PubMed]

121. Kambestad, M.; Kirkendall, L.R.; Knutsen, I.L.; Jordal, B.H. Cryptic and pseudo-cryptic diversity in the world's most common bark beetle—*Hypothenemus eruditus*. *Org. Divers. Evol.* **2017**, *17*, 633–652. [CrossRef]
122. Gillespie, J.J.; Johnston, J.S.; Cannone, J.J.; Gutell, R.R. Characteristics of the Nuclear (18S, 5.8S, 28S and 5S) and mitochondrial (12S and 16S) rRNA genes of *Apis mellifera* (Insecta: Hymenoptera): Structure, organization, and retrotransposable elements. *Insect Mol. Biol.* **2006**, *15*, 657–686. [CrossRef] [PubMed]
123. Marvaldi, A.E.; Duckett, C.N.; Kjer, K.; Gillespie, J. Structural alignment of 18S and 28S rDNA sequences provides insights into phylogeny of Phytophaga (Coleoptera: Curculionoidea and Chrysomeloidea). *Zool. Scr.* **2009**, *38*, 63–77. [CrossRef]
124. Marvaldi, A.E.; Lanteri, A.A.; del Río, M.G.; Oberprieler, R.G. 3.7.5 Entiminae Schoenherr, 1823. In *Handbook of Zoology, Volume IV: Arthropoda: Insecta. Part 38 Coleoptera, Beetles Volume 3: Morphology and Systematics (Phytophaga)*; Leschen, R.A.B., Beutel, R.G., Eds.; Walter de Gruyter: Berlin, Germany, 2014; pp. 503–522.
125. Oberprieler, R.G. A reclassification of the weevil subfamily Cyclominae (Coleoptera: Curculionidae). *Zootaxa* **2010**, *2515*, 1–35.
126. Oberprieler, R.G. 3.7.4 Cyclominae Schoenherr, 1826. In *Handbook of Zoology, Volume IV: Arthropoda: Insecta. Part 38 Coleoptera, Beetles Volume 3: Morphology and Systematics (Phytophaga)*; Leschen, R.A.B., Beutel, R.G., Eds.; Walter de Gruyter: Berlin, Germany, 2014; pp. 483–502, ISBN 978-3-11-027370-0.
127. Bouchard, P.; Bousquet, Y.; Davies, A.E.; Alonso-Zarazaga, M.A.; Lawrence, J.F.; Lyal, C.H.C.; Newton, A.F.; Reid, C.A.M.; Schmitt, M.; Ślipiński, S.A.; et al. Family-group names in Coleoptera (Insecta). *ZooKeys* **2011**, *88*, 972. [CrossRef] [PubMed]
128. Gunter, N.L.; Oberprieler, R.G.; Cameron, S.L. Molecular phylogenetics of Australian weevils (Coleoptera: Curculionoidea): Exploring relationships in a hyperdiverse lineage through comparison of independent analyses. *Austral Entomol.* **2015**, *55*, 217–233. [CrossRef]
129. Lyal, C.H.C.; King, T. Elytro-tergal stridulation in weevils (Insecta: Coleoptera: Curculionoidea). *J. Nat. Hist.* **1996**, *30*, 703–773. [CrossRef]
130. Lyal, C.H.C.; Douglas, D.; Hine, S.J. Morphology and systematic significance of sclerolepidia in the weevils (Coleoptera: Curculionoidea). *Syst. Biodivers.* **2006**, *4*, 203–241. [CrossRef]
131. Alonso-Zarazaga, M.A.; Lyal, C.H.C. *A World Catalogue of Families and Genera of Curculionoidea (Insecta: Coleoptera) (Excepting Scolytidae and Platypodidae)*; Entomopraxis: Barcelona, Spain, 1999; 315p.
132. Oberprieler, R.G.; Marvaldi, A.E.; Anderson, R.S. Weevils, Weevils, Weevils Everywhere. *Zootaxa* **2007**, *1668*, 491–520.
133. Cigliano, M.M.; Pocco, M.E.; Pereira, H.L. Avances tecnológicos y sus aplicaciones en la cibertaxonomía. *Rev. Soc. Entomol. Argic.* **2014**, *73*, 3–15.
134. International Weevil Community Website. Available online: http://weevil.info/ (accessed on 20 April 2018).
135. Glossary of Weevil Characters. Available online: http://weevil.info/glossary-weevil-characters/ (accessed on 20 April 2018).
136. WTaxa: Electronic Catalogue of Weevil Names (Curculionoidea). Web Version 3.0, Database Version 18. Available online: http://wtaxa.csic.es/ (accessed on 20 April 2018).
137. New Zealand Weevil Images. Available online: http://weevils.landcareresearch.co.nz/ (accessed on 20 April 2018).
138. Potential Invasive Weeds of the World. Available online: http://www.piweevils.com/ (accessed on 20 April 2018).
139. Coleoptera Neotropical. Available online: http://www.coleoptera-neotropical.org (accessed on 20 April 2018).
140. Symbiota Collections of Arthropods Network (SCAN). Available online: http://scan-bugs.org/portal/ (accessed on 20 April 2018).
141. Beetles and Coleopterists. Available online: http://www.zin.ru/ANIMALIA/Coleoptera/ (accessed on 20 April 2018).
142. Leschen, R.A.B.; Beutel, R.G. *Handbook of Zoology, Vol. IV: Arthropoda: Insecta. Part 38 Coleoptera, Beetles, Volume 3: Morphology and Systematics (Phytophaga)*; Walter de Gruyter: Berlin, Germany, 2014; 675p, ISBN 978-3-11-027370-0.
143. Krahulcova, A.; Krahulec, F.; Kirschner, J. Introgressive hybridization between a native and an introduced species: *Viola lutea* subsp *sudetica* versus. *V. tricolor*. *Folia Geobot.* **1996**, *31*, 219. [CrossRef]
144. Engilis, A., Jr.; Pratt, T. Status and population trends of Hawaii's native waterbirds. *Wilson Bull.* **1993**, *105*, 142–158.

145. Lanteri, A.A. Systematics, cladistics and biogeography of a new weevil genus, *Galapaganus* (Coleoptera: Curculionidae) from the Galápagos Islands, and Coasts of Ecuador and Perú. *Trans. Am. Entomol. Soc.* **1992**, *118*, 227–267.
146. Sequeira, A.S.; Lanteri, A.A.; Albelo, L.R.; Bhattacharya, S.; Sijapati, M. Colonization history, ecological shifts and diversification in the evolution of endemic Galapagos weevils. *Mol. Ecol.* **2008**, *17*, 1089–1107. [CrossRef] [PubMed]
147. Mok, H.F.; Stepien, C.C.; Kaczmarek, M.; Albelo, L.R.; Sequeira, A.S. Genetic status and timing of a weevil introduction to Santa Cruz Island, Galapagos. *J. Hered.* **2014**, *105*, 365–380. [CrossRef] [PubMed]
148. Sequeira, A.S.; Cheng, A.; Pangburn, S.; Troya, A. Where can introduced populations learn their tricks? Searching for the geographical source of a species introduction to the Galápagos archipelago. *Conserv. Genet.* **2017**, *18*, 1403–1422. [CrossRef]
149. Hey, J.; Nielsen, R. Multilocus methods for estimating population sizes, migration rates and divergence time, with applications to the divergence of *Drosophila pseudoobscura* and *D. persimilis*. *Genetics* **2004**, *167*, 747–760. [CrossRef] [PubMed]
150. Hey, J. *Documentation for IMa2*; Department of Genetics, Rutgers University: Piscataway, NJ, USA, 2011.
151. Hey, J. *Using the IMfig Program*; Department of Genetics, Rutgers University: Piscataway, NJ, USA, 2011.
152. Hedrick, P.W. Adaptive introgression in animals: Examples and comparison to new mutation and standing variation as sources of adaptive variation. *Mol. Ecol.* **2013**, *22*, 4606–4618. [CrossRef] [PubMed]
153. Daehler, C.C.; Strong, D.R. Hybridization between introduced smooth cordgrass (*Spartina alterniflora;* Poaceae) and native California cordgrass (*S. foliosa*) in San Francisco Bay, California, USA. *Am. J. Bot.* **1997**, *84*, 607–611. [CrossRef] [PubMed]
154. Fitzpatrick, B.M.; Johnson, J.R.; Kump, D.K.; Smith, J.J.; Voss, S.R.; Shaffer, H.B. Rapid spread of invasive genes into a threatened native species. *Proc. Natl. Acad. Sci. USA* **2010**, *107*, 3606–3610. [CrossRef] [PubMed]
155. Stüben, P.E.; Schütte, A. *Silvacalles* (s. str.) *carlinavorus* sp.n. from La Gomera (Canary Islands) (Coleoptera: Curculionidae: Cryptorhynchinae). *Snudebiller* **2014**, *15*, 1–8.
156. Stüben, P.E.; Schütte, A.; Astrin, J.J. Molecular phylogeny of the weevil genus *Dichromacalles* Stüben (Curculionidae: Cryptorhynchinae) and description of a new species. *Zootaxa* **2013**, *3718*, 101–127. [CrossRef] [PubMed]
157. Schütte, A.; Stüben, P.E. Molecular systematics and morphological identification of the cryptic species of the genus *Acalles* Schoenherr, 1825, with descriptions of new species (Coleoptera: Curculionidae: Cryptorhynchinae). *Zootaxa* **2015**, *3915*, 1–51. [CrossRef] [PubMed]
158. Germann, C.; Wolf, I.; Schütte, A. *Echinodera (Ruteria) soumasi* sp. n. from Greece (Coleoptera, Curculionidae). *Mitteilungen Schweizerischen Entomologischen Gesellschaft* **2015**, *88*, 285–293.
159. Haran, J.; Schütte, A.; Friedman, A.L. A review of *Smicronyx* Schoenherr (Coleoptera, Curculionidae) of Israel, with description of two new species. *Zootaxa* **2017**, *4237*, 17–40. [CrossRef] [PubMed]
160. Schütte, A.; Stüben, P.; Sprick, P. The Molecular Weevil Identification Project (Coleoptera: Curculionoidea), Part I—A contribution to Integrative Taxonomy and Phylogenetic Systematics. *Snudebiller* **2013**, *14*, 1–77.
161. Stüben, P.E.; Schütte, A.; Bayer, C.; Astrin, J.J. The Molecular Weevil Identification Project (Coleoptera: Curculionoidea), Part II—Towards an Integrative Taxonomy. *Snudebiller* **2015**, *16*, 1–294.
162. Brown, S.D.J. A revision of the New Zealand weevil genus *Irenimus* Pascoe, 1876 (Coleoptera: Curculionidae: Entiminae). *Zootaxa* **2017**, *4263*, 1–42. [CrossRef] [PubMed]

Article

The Genus *Urodontidius* Louw (Anthribidae: Urodontinae) Rediscovered and Its Biological Secrets Revealed: A Tribute to Schalk Louw (1952–2018)

Rolf G. Oberprieler [1,*] and Clarke H. Scholtz [2]

[1] CSIRO Australian National Insect Collection, G.P.O. Box 1700, Canberra, A.C.T. 2601, Australia
[2] Department of Zoology & Entomology, University of Pretoria, Pretoria 0002, South Africa; chscholtz@zoology.up.ac.za
* Correspondence: rolf.oberprieler@csiro.au; Tel.: +61-2-62464271

Received: 9 July 2018; Accepted: 11 August 2018; Published: 14 August 2018

Abstract: The paper records the rediscovery of the rare *Urodontidius enigmaticus* Louw, 1993 in South Africa, based on specimens reared from galls in the succulent leaves of *Ruschia versicolor*. The original account of some of the morphological characters of the species is corrected, and its habitus, antennae, pygidium and genitalia are illustrated. Its life history and galling habit on its host plant are described and illustrated, and its larva is compared with those of the genera *Urodontellus* Louw and *Urodontus* Louw, which represent different larval types with different life histories. The silk-spinning habits of the *Urodontellus* larva are briefly described. A tribute to the late Schalk Louw is presented, together with a list of his publications on weevils.

Keywords: Urodontinae; *Urodontidius*; genitalia; larva; life history; galling habit; silk production

---

1. Introduction

The anthribid subfamily Urodontinae is remarkable in many ways. It is morphologically quite different from Anthribinae (including Choraginae), i.a., by not having a bracteate pronotal carina, the mandibles without a mycetangial pocket, the hind wings with only three veins, the hind gut with a rectal ring, tergite VIII exposed in the males and the gonocoxites of the ovipositor not apically sclerotised and dentate [1]. Biologically, it differs from Anthribinae in that its larvae do not feed on fungus-infected dead wood but on living plant tissues, developing in soft stems and seeds of particular plant families. The subfamily is also unusual in its distribution, being restricted to the Afrotropical and western Palaearctic regions but with a distinct centre of diversity in southern Africa and another in the western Palaearctic, the former though with a much higher generic diversity [2,3]. In contrast to the Palaearctic fauna, the southern African one remained very poorly known until Schalk Louw thoroughly revised it, describing several new genera and many new species, summarising the biological information available for it and attempting a first analysis of its phylogenetic relationships [2]. Little work has been done on this fauna since then, and many aspects of the taxonomy and biology of particular genera remain unknown.

During 2015, one of us (CHS) reared a species of Urodontinae from galls in the succulent leaves of *Ruschia versicolor* L. Bolus (Aizoaceae) in west-coastal Namaqualand, South Africa, and recorded its life history. He handed the specimens to Schalk Louw for identification, who surmised it to be a new species of *Urodontus* Louw and was busy studying it at the time of his sudden and untimely death, in April 2018. The study would have formed part of a larger treatment by him and CHS of gall formation in Urodontinae and its phylogenetic and evolutionary implications. Unfortunately this did not come to fruition. After Schalk's death, the first author of this paper, also an old friend and colleague of Schalk, joined CHS to complete the study of this species, which, on closer inspection and comparison with type specimens, turned out to be the arcane *Urodontidius enigmaticus* Louw.

Urodontidius enigmaticus is the sole known species of the genus *Urodontidius*, which was described by Schalk Louw in his revision of the Urodontinae of southern Africa [2]. The species is remarkable for its extraordinary antennal structure in the male. Its description was based on only four specimens, one pair collected in 1985 in the Namaqualand region of the western Northern Cape province of South Africa and two males taken much earlier at Willowmore in the Eastern Cape province of the same country. No further specimens appear to have been collected since. The hosts and life history of the species also remained unknown, aside from the fact that the pair from Namaqualand had been found on flowers of a species of *Eberlanzia* (Aizoaceae) [2].

In this paper, we report the rediscovery of this rare species, describe and illustrate additional morphological aspects of it and record the extraordinary galling habit and life history of its larva. We also take this opportunity to pay tribute to the late Schalk Louw and his contributions to the study of weevils, particularly in southern Africa.

2. Materials and Methods

2.1. Specimens

The study is based on 1 male and 28 adult females and 4 larvae of *Urodontidius enigmaticus*, as sent to RGO from the University of the Free State in Bloemfontein, South Africa, where they were retrieved from Schalk Louw's office by his colleague Charles Haddad. All these specimens had been reared from succulent leaf galls on *Ruschia versicolor* at Kommandokraal (31.50°S, 18.21°E) in the Western Cape province of South Africa in November 2015 by CHS. We also studied photographs of the holotype and only female paratype of *U. enigmaticus*, housed in the National Museum in Bloemfontein, kindly provided to us by Burgert Muller. Specimens of other genera and species of Urodontinae in the Australian National Insect Collection (ANIC) were studied for comparison.

2.2. Illustrations

Photographs of specimens and structures were taken using a Leica DFC500 camera mounted on a Leica M205C microscope. Photographs taken at different focus levels were combined into single images using the software program Leica Application Suite V4.9, and these images were enhanced as necessary using the Adobe Photoshop CS6 software.

3. Diagnosis, Distribution and Life History

Genus ***Urodontidius*** Louw, 1993

Urodontidius Louw, 1993: 11 [2].

Type species, by original designation: *Urodontidius enigmaticus* Louw, 1993.

Urodontidius enigmaticus Louw, 1993

Urodontidius enigmaticus Louw, 1993: 12, Figures 7, 40 and 52 [2].

Diagnosis. *Urodontidius enigmaticus* is readily distinguishable from all other Urodontinae by its antennae, especially the extraordinary clubs of the male (Figure 1e). Its large size, dark-brown colour, very sparse vestiture and shape of the elytra and pygidium (Figure 1a–d) also set it apart from all other urodontines. Louw's description of the genus and the species [2] is accurate except for that of the antennae. He described these as being 10-segmented, but this is incorrect. The antennae of the female comprise 11 segments: a short, medially constricted scape, a 7-segmented funicle (the 1st segment enlarged, the 7th broadened) and a 3-segmented club (Figure 1f). The segmentation of the club is astonishingly variable, ranging from two loose segments and one fused (Figure 1g) to three fused ones (Figure 1f, h) to two fused ones (Figure 1i) to a large single one (Figure 1j). The fusion line between the last two segments is fairly distinct in some specimens (in lateral outline as well as on the surface) but almost obsolete in others. The apical edge of the terminal (third) club segment also varies, from symmetrically subtruncate to slightly emarginate to asymmetrically excised (Figure 1g–j). In the male, the club is modified into a single, short but very broad, asymmetrical segment (Figure 1e). Louw misinterpreted the funicle as being 6-segmented and the 7th, broader funicle segment as belonging to the club, but the club of the female is distinctly 3-segmented in most specimens (including in the female paratype), without the 7th funicle segment. Louw's drawing of the club of the male as being 2-segmented, consisting of a smaller penultimate segment in addition to the large terminal one (Figure 7 [2]), is also incorrect, as we could ascertain from a photo of the antenna of the holotype.

As in other urodontines, the pygidium is sexually dimorphic in *Urodontidius*, in the female formed by tergite VII (Figure 2c) and in the male by tergite VII as well as a small tergite VIII (Figure 2a,b). The latter was referred to as a supplementary sclerite by Louw [2]. In the female, this last tergite is concealed beneath tergite VII.

In the male genitalia, the penis consists of a flat, slightly curved, elongate, parallel-sided pedon and a narrowly triangular tectum extending only over the basal half of the pedon (Figure 2d,e). The temones (apodemes) are as long as the pedon, very broad (deep) at the base (with the tectal and pedal arms discernible but connected) but narrower in the distal half, and they are only membranously connected to the pedon and tectum (Figure 2e). The endophallus is equipped with two long, narrow, parallel, medially connected sclerites, possibly a flattened flagellum, of about half the length of the pedon (Figure 2d,e). The tegmen is quite reduced, the sides of the ring broad but weakly sclerotised, not articulated in the middle (Figure 2e), and the parameral sector narrow, very weakly sclerotised, with a pair of apical clusters of few strong, long setae (Figure 2d). In the female, the ovipositor is a strong, flat, apically truncate tube with large, broad, flattened lateral rods, without a median tranverse bar and without styli and apical teeth, and the median rod is long (protruding beyond the apices of the lateral rods) and thick, covered with large, sharp denticles directed caudad (especially in the apical third) and anteriorly flanked by two pairs of narrow lateral rods (Figure 2f). The spermatheca (Figure 2g) is strongly S-shaped, weakly sclerotised and undifferentiated, with an apparently small gland and the duct entering the bursa copulatrix in an apical position. The densely, coarsely dentate internal rod of the ovipositor appears to function like a round grater or coarse file that can be pushed in and out of plant tissues to create a hole for oviposition.

Distribution. *Urodontidius enigmaticus* occurs in the Namaqualand area of the Northern and Western Cape provinces of South Africa, less certainly also in the Eastern Cape (Figure 3). It was previously only known from the four specimens comprising its type series [2], a pair collected at Steenbok in the Northern Cape province (close to the Atlantic coast, between Port Nolloth and Nigramoep) and two males from Willowmore in the Eastern Cape province, collected by Dr. J. H. J. C. Brauns (1857–1929) about a century ago. The locality of our specimens, Kommandokraal, is about 100 km south of Steenbok. The occurrence of the species at or near Willowmore, very distant from the other two known localities, requires confirmation.

Host plants. Two hosts are recorded for *U. enigmaticus*, a species of *Eberlanzia*, on whose flowers Schalk Louw found a pair of specimens in 1985 [2], and *Ruschia versicolor* (Figure 4a,b), from whose leaf galls our specimens were reared. Both genera belong to the family Aizoaceae and are some of the many "mesembs" for which Namaqualand is famous. *Ruschia versicolor* is a perennial prostrate shrub that can grow up to 400 mm high. Its fleshy, succulent leaves are cylindrical and may grow to 60 mm in length. Some leaves form clusters. Leaves are initially green but turn pink when they get older. According to the Botanical Database of Southern Africa (BODATSA), the species has a fairly limited distribution (Figure 3) in coastal Namaqualand Sandveld [4], but it is apparently not threatened. *Eberlanzia* is closely related to *Ruschia* (also placed in the tribe Ruschiae of the subfamily Ruschioideae) but much smaller, comprising only eight species in southern Namibia and south-western South Africa [5]. Its leaves are also succulent but much shorter, sometimes spinose. While the identification underlying this host record cannot be verified, it seems likely that *U. enigmaticus* may also form galls in the succulent leaves of *Eberlanzia*.

Figure 1. Morphological aspects of *Urodontidius enigmaticus* Louw: (**a**) habitus of male, dorsal view; (**b**) habitus of female, dorsal view; (**c**) habitus of male, lateral view (antennal club missing); (**d**) habitus of female, lateral view; (**e**) right antenna of male, lateral view; (**f**) left antenna of female, lateral view; (**g–j**) variation in shape of antennal club of female, from (**g**) 3-segmented to (**j**) 1-segmented.

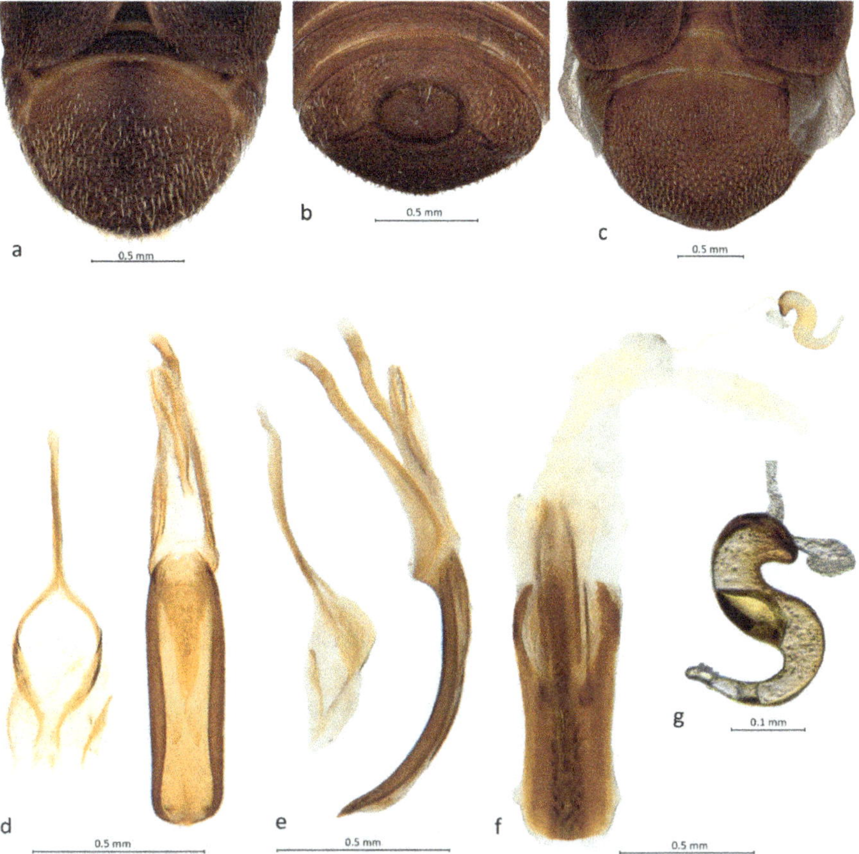

Figure 2. Abdomen and genitalia of *Urodontidius enigmaticus* Louw: (**a**) pygidium of male, caudal view; (**b**) pygidium of male, ventral view; (**c**) pygidium of female, caudal view; (**d**) male genitalia, dorsal view; (**e**) male genitalia, lateral view; (**f**) female genitalia, dorsal view; (**g**) spermatheca.

Life history and gall development. The larva of *U. enigmaticus* is a fairly flaccid white grub without appendages (Figure 4e,f,h) and totally lacking in areas of sclerotisation except for the sharp mandibles (Figure 4g). It incites a gall in the fleshy leaves of its host, in which it evidently feeds on the jelly-like cells lining the inside of the galls. The small quantity of frass present in the gall suggests that it feeds on low-fibre, highly nutritious food. Soon after the larva has established in the soft, succulent tissue, an elongate, hard woody capsule starts to develop around it (Figure 4e), while the soft tissue surrounding the capsule swells noticeably to two to three times its size. The gall is often bilobed (Figure 4b) as a result of swollen development of two deformed leaves growing from the galled leaf bud. The internal capsule quickly develops to its final size and initially dwarfs the young larva, but as the larva grows it fills more and more of the available space until eventually the enclosed adult fills it almost completely (Figure 4d). The outer cell layers of the capsule are woody from the beginning, whereas initially it is lined internally by layers of spongy tissue. As these are consumed by the larva and it grows, the capsule sides become smoother. Duration of the larval development is about five months, and the pupal stage lasts about two weeks. *Urodontidius enigmaticus* has two generations per year on *Ruschia versicolor* at Kommandokraal. First-generation adults (Figure 4c,d) emerge in winter (June), mate and lay eggs on developing leaf buds; those of the second generation emerge during November. The relatively short development period of the larva supports the assumption of two generations existing per year, as compared to the much longer larval development and single generation per year of *Urodontus scholtzi* Louw, which induces woody stem galls in *Galenia* (also Aizoaceae) [2,6].

Figure 3. Distribution of *Urodontidius enigmaticus* and *Ruschia versicolor*.

4. Discussion

In his phylogenetic reconstruction of the genera of Urodontinae, Louw [2] related *Urodontidius* most closely to *Breviurodon* Strejcek, based on the fusion of the club segments in both. However, this fusion as well as the segmentation of the antennae are different in the two genera. In *Urodontidius* the antennae are 11-segmented, comprising a 7-segmented funicle and a variably fused club (3-segmented in the female, 1-segmented enlarged in the male), whereas in *Breviurodon* the antennae are 10-segmented, the funicle being 6-segmented and the club spindle-shaped with three fused but distinct segments in both sexes [7,8]. Louw ([8], Figure 4) drew the funicle of *Breviurodon decellei* Louw as being 7-segmented, but this is evidently an error. *Urodontidius* is thus unlikely to be as closely related to *Breviurodon* as Louw concluded from his phylogenetic analysis [2]. Although its 11-segmented antennae may instead suggest a closer relationship to *Bruchela* Dejean and *Urodontellus* Louw, its larval development in aizoaceous hosts indicates that it may indeed belong in the clade of genera with 10-segmented antennae. Further phylogenetic study is necessary to resolve its relationships and elucidate the origin of its unusual galling habit.

Figure 4. Biological aspects of *Urodontidius enigmaticus* Louw: (**a**) host plant (*Ruschia versicolor*, Aizoaceae) in habitat; (**b**) leaf gall on *Ruschia versicolor* incited by larva; (**c**) female emerged from gall; (**d**) female next to larval feeding chamber in gall; (**e**) larva in feeding chamber in gall; (**f**) larva, ventral view; (**g**) larva, head with mandibles; (**h**) larvae of *U. enigmaticus*, *Urodontellus lilii* (Fåhraeus) and *Urodontus scholtzi* Louw in comparison, lateral view. (Photographs c, d, e by Hennie de Klerk).

In contrast to other Anthribidae, all species of Urodontinae as known develop in living plant tissues. The life styles and hosts of the subfamily fall in two groups: the larvae of the Palaearctic genus *Bruchela* and of the southern African genus *Urodontellus* develop in seed capsules of, respectively, the brassicalean families Resedaceae and Brassicaceae and the monocotyledonous families Iridaceae and Asphodelaceae, whereas the larvae of the southern African genera *Urodontidius* and *Urodontus* Louw (and probably of *Urodoplatus* Motschulsky as well) develop predominantly in the stems and flower heads of Aizoaceae [2]. The larvae of these two groups also differ markedly (Figure 4h). The larvae of *Urodontellus* (*U. lilii* (Fåhraeus)) have a prognathous, sclerotised head; large antennae; broad, bluntly bidentate mandibles; a small clypeus and labrum; a semilunar, anteriorly evenly thickly cylindrical body rapidly thinning posteriad from abdominal segment V (A-V), with A-VIII and A-IX small, weakly sclerotised and A-X with narrowly sclerotised lateral anal lobes; a dorsal pair of conical ambulatory ampullae on each of segments T-III and A-1 to A-VII; distinct clusters of long fine setae ventrally on all segments except A-IX and A-X. In contrast, the larvae of *Urodontus* (*U. mesemoides* Louw, *U. scholtzi*) have a hypognathous, well sclerotised head; small antennae; broad, sharply bidentate mandibles; a large clypeus and labrum; a crescentic, evenly thickly cylindrical body slightly thinning in the posterior third, with A-IX small and short and A-X very small with indistinct, unsclerotised anal lobes; all segments without ambulatory ampullae and very sparsely setose (a few long dorsal setae on A-VI to A-IX). The larva of *Urodontidius* (*U. enigmaticus*) differs significantly from both these forms, having the head hypognathous, unsclerotised except for the small, narrow, finely bidentate mandibles; the body only slightly curved, flattened (broader than high), in lateral view gradually thickening from T-1 to A-V, then thinning more rapidly posteriad, with A-VIII and A-IX small, short, and A-X with a weak transverse cleft, the anus possibly closed; segments A-I to A-VII laterally with thick, elongate ambulatory ampullae, the largest on A-V, dorsally without ampullae but the membranes between the abdominal segments medially eversible; the entire body without any macrosetae. This represents a distinct third body type in Urodontinae.

The life history of these larval types also differs. The larvae of *Bruchela* and *Urodontellus* are very mobile, moving in and between seed capsules [9] on their backs using the dorsal ampullae [10], and they are able to spin threads from the tip of their abdomen. The larvae of *Bruchela* use these threads to close the open *Reseda* seed capsule in which they feed, as a suspension when they drop to the ground for pupating and for constructing a cocoon in the soil [10]. The larvae of *Urodontellus* pupate in the closed seed capsules in which they feed, but they also spin a silken cocoon, inside the capsule. They can also drop from the capsule suspended by a silken thread secreted from the abdomen, and they can, in fact, roll up the thread again with the tip of their abdomen in a gyrating motion to ascend back into the capsule (Stefan Neser, pers. obs. 2007, 2018). This action is evidently performed by the peculiarly sclerotised anal lobes of the larva. The silk of *Urodontellus* is proteinacous in nature, with high percentages of amino acids and its infrared absorption spectrum showing strong amide peaks (Andrew Walker, pers. com. 2012), suggesting that the silk is derived from the Malpighian tubules, which are pink in colour [11]. The larvae of *Urodontus* instead are not or are only very weakly mobile, apparently spending their entire development inside their feeding cells in the soft vegetative tissues of the stems and flower heads of their hosts or, in the case of *Urodontus scholtzi* and *U. tesserus* Louw, in woody stem galls. The larvae of *Urodontidius* appear to be a more specialised version of the latter, feeding in hard cells in large succulent galls, in which they probably move around using their lateral segmental ampullae and the eversible dorsal intersegmental membranes. A detailed morphological comparison of all known urodontine larvae remains to be carried out.

Gall formation in succulents is unusual, possibly because of the high levels of fluids in tissues, although gall midges (Cecidomyiidae) have been recorded to induce them in various Aizoaceae in southern Africa [12]. As far as we could ascertain, however, this is the first record of a beetle galling a succulent plant. Besides this being an unusual habit for insects in general, it is atypical for Urodontinae in that gall formation in the subfamily is very rare (only known for three out of 91 species), and galling a succulent plant adds to the uniqueness of the phenomenon.

5. A Tribute to Schalk Louw

Schalk van der Merwe Louw was born on the 28th March 1952 in Windhoek, Namibia (then South West Africa). He attended school in Windhoek from 1959 to 1970 and subsequently enrolled in undergraduate studies at the University of Pretoria in South Africa, from 1972 to 1975, obtaining a B. Sc. degree in Zoology and Entomology in 1974 and a first-class B. Sc. Honours degree in Entomology in 1975. He then proceeded with studies at the same university for an M. Sc. degree, which he obtained in 1979 in arid-zone insect ecology, and for a Ph. D. degree in weevil systematics, which was bestowed on him in 1985. He started his research career in 1976 as Curator of Invertebrates at the State Museum in Windhoek, where he remained until 1981, when he moved to South Africa to become Head of the Entomology Department at the National Museum in Bloemfontein. In 1992 he took up a position as Senior Lecturer in the Department of Zoology and Entomology at the University of the Orange Free State in Bloemfontein, where he remained for the rest of his career, rising to Associate Professor in 1996 and to full Professor in 2002. Over his career he was a member of many learned societies, served on the executive committees and editorial boards of several scientific societies and associations and played an active role in the entomology scene in southern Africa.

He married Elizabeth Susanna van Niekerk on the 7th January 1978, and the couple had two children: a daughter, Sarita, born on the 27th September 1979, and a son, Schalk Merwe, born on the 16th June 1983.

Schalk's love of beetles was kindled and nurtured by the eminent southern African coleopterists of the 1970s and 1980s, in particular Mary-Louise Penrith, with whom he worked at the State Museum in Windhoek, and the late Sebastian Endrödy-Younga of the former Transvaal Museum in Pretoria and Erik Holm, who played a leading role in invertebrate research at the then Namib Desert Research Station at Gobabeb in Namibia and later became Professor of Entomology at the University of Pretoria. Like Mary-Lou and Sebastian, Schalk initially also studied Tenebrionidae, but on the collecting expeditions he undertook with them into remote parts of the Namib and Kalahari deserts he soon discovered his love for weevils, which remained with him for the rest of his life. The cryptic, terricolous desert weevils held a particular fascination for him, which led to several research papers, among them his classic revisions of the genus *Hyomora* (Cyclominae) and the subfamily Microcerinae, the latter flowing from his Ph. D. research and becoming his *magnum opus* in weevil systematics.

When Schalk moved to Bloemfontein, his interest in weevils widened to include their associations with plants, and he set up an experimental site at Glen, just outside the city, where several brachycerine and cyclomine species appeared annually to munch on the leaves of the numerous geophytic lilies that sprung to life after rains. His paper of the life history and immature stages of the huge *Brachycerus ornatus* remains a benchmark study on the biology of this iconic African weevil genus. From Bloemfontein he explored the weevil fauna of the Orange Free State, in particular at the museum's research station at Florisbad and at Krugersdrift Dam but also on expeditions into the drier sandy areas of the north-western Orange Free State and the Northern Cape province.

In 1991 Schalk went small in weevil terms, embarking on a study of the enigmatic and taxonomically badly neglected southern African fauna of the anthribid subfamily Urodontinae, which is particularly species-rich in the Namaqualand and Richtersveld regions of South Africa and Namibia. His 1993 revision drastically increased the known diversity and host range of this group, with the description of four new genera and twenty new species and an analysis of their host associations. These weevils also ignited a new passion in Schalk, that of studying galls and the evolution of galling behavior in weevils, on which he published a number of papers and book chapters and delivered talks at international symposia in Russia and Hungary. While his employ at the university allowed him few opportunities to continue his studies of weevil systematics and forced him to engage in various other research fields, he managed to keep a connection to weevils in his research on new crops, in particular on pigweed (*Amaranthus*), which is loved by a lixine weevil (*Hypolixus haerens*) in South Africa, and this research also enabled Schalk to pursue an interest in tritrophic associations between plants, insects and parasites/pathogens.

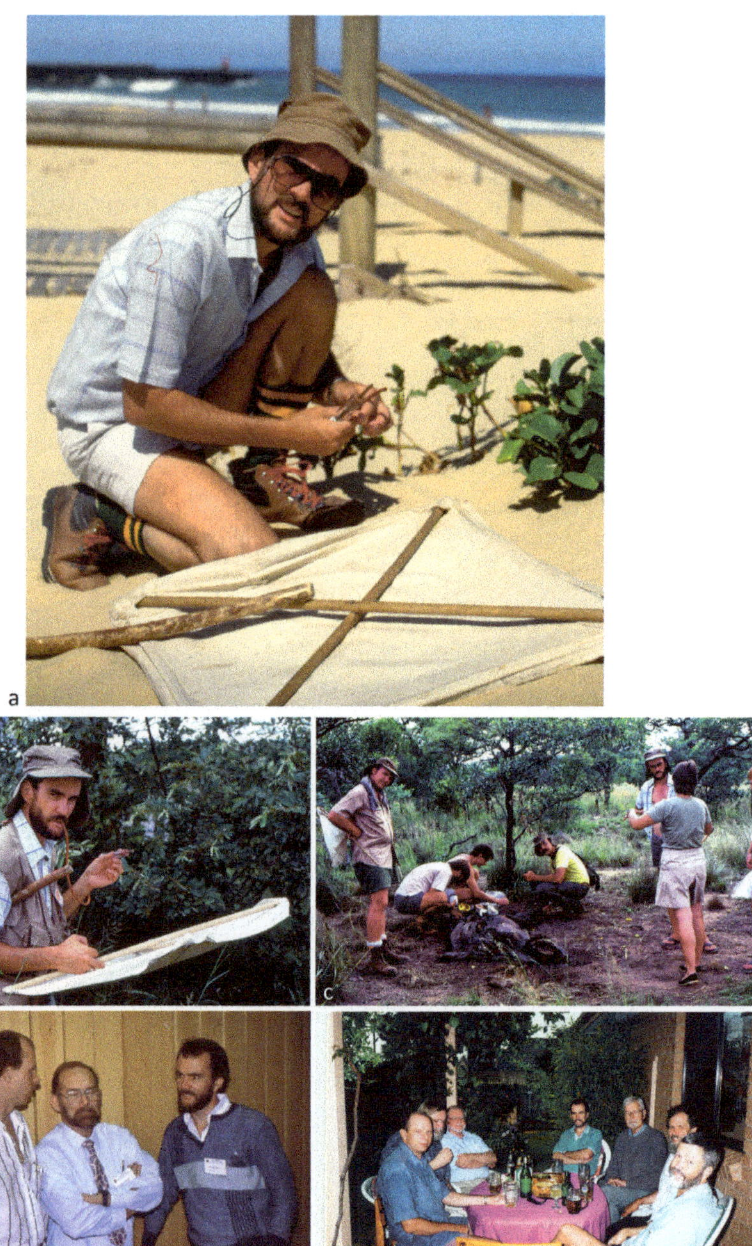

Figure 5. Schalk Louw: (**a**) collecting weevils at Port Alfred, November 1992; (**b**) collecting weevils at Lapalala, January 1987; (**c**) with colleagues at a rhino carcass, Lapalala, January 1987; (**d**) with Rolf Oberprieler (left) and Elbert Sleeper at the weevil symposium during the XVIIIth International Congress of Entomology, Vancouver, July 1988; (**e**) at dinner with colleagues during the John Lawrence Beetle Symposium, Canberra, December 1999 (f.l.t.r. Rolf Oberprieler, Jyrki Muona, John Lawrence, Schalk Louw, Willy Kuschel, Vladimir Zherikhin, Clarke Scholtz). (Photos RGO except d).

Schalk retired from his university at the end of 2017 but maintained an association with it as a Research Fellow. He had plans to re-engage in weevil systematics and biology, first of which was to complete a global perspective on the diversity and pattern of gall induction in weevils, for which he had unpublished data from Horace Burke available and which he intended to contribute to this Special Issue on weevils. Alas, ill health prevented him from completing this project and finally even the smaller, present paper as he had conceived in its place.

Schalk was one of the "Young Turks" at the first international weevil symposium, held in 1988 during the XVIIIth International Congress of Entomology in Vancouver, Canada, where we gathered to learn of weevils from the previous generation of gurus such as Willy Kuschel, Katsura Morimoto, Horace Burke, Elbert Sleeper, Stephen Wood, Charlie O'Brien and Anne Howden (Figure 5d). In 1996 Schalk co-organised the weevil symposium at the XXth International Congress of Entomology in Florence, Italy, together with Enzo Colonnelli and Giuseppe Osella, and co-edited the proceedings from it, and at the XXIst International Congress of Entomology, held in 2000 in Foz de Iguassu, Brasil, he co-organised a symposium on biodiversity and biogeographical research in Africa with Wolfram Mey (Berlin). In 1999 he visited Australia to participate in the John Lawrence Beetle Symposium in Canberra, Australia, keen to meet up with old and new colleagues, discuss various weevil matters, and enjoy the company (Figure 5e).

Apart from his numerous and wide-ranging contributions to the knowledge of weevils and other beetles (Appendix A), we also remember Schalk as a pleasant and valuable companion on numerous collecting trips in southern Africa (Figure 5a–c). He had a knack for finding cryptic beetles on the ground, a skill he had honed early during his many expeditions in Namibia, and he was always ready to show and share his daily haul with others. During Willy Kuschel's visit to South Africa in 1992, Schalk took great pride in showing Willy his monster *Brachycerus* weevils at Glen and the multitude of terricolous genera at Krugersdrift Dam that were wholly unknown to Willy [13]. Another memorable episode was a joint 1993 expedition to the Richtersveld [14], where Schalk not only showed everyone how to really hunt for ground weevils in the desert but also enthusiastically joined in the smoking of thick cigars to keep the annoying blackflies at bay that descended onto our camp in the afternoon, just when we could sit down to sort and admire our daily catches and wash down the dust in our throats with a beer.

Schalk's untimely passing leaves a large gap both in the entomological community in South Africa and in international weevil systematics and ecology.

Author Contributions: Both authors contributed equally to the design, analysis and writing of the paper.

Funding: This research received no external funding.

Acknowledgments: We thank Charles Haddad of the University of the Free State for tracing the specimens of *Urodontidius enigmaticus* in Schalk Louw's office and sending them to RGO in Australia for study, and for forwarding the draft manuscript of this paper as Schalk was working on at the time of his death. Burgert Muller of the National Museum in Bloemfontein kindly provided photographs of the holotype and paratype of *U. enigmaticus* housed in his museum. Debbie Jennings (ANIC), Carmen Jacobs (University of Pretoria) and Hennie de Klerk are thanked for the photographs of the adults and larvae of *U. enigmaticus*, and we acknowledge Arthur V. Evans (Virginia, U.S.A.) for permission to publish the photograph in Figure 5d. We also like to acknowledge Stefan Neser (Pretoria) for sharing photographs and video clips of silk-climbing and cocoon-spinning *Urodontellus* larvae.

Conflicts of Interest: The authors declare no conflict of interest.

Appendix A

Publications on weevils by Schalk van der Merwe Louw

1. Louw, S. v. d. M. (1976) The genus *Brachycerus* (Coleoptera: Curculionidae). *State Museum Newsletter*, **5**, 4–6.
2. Louw, S. v. d. M. (1981) Revision of the genus *Hyomora* Pascoe, 1865 (Coleoptera: Curculionidae: Rhytirrhininae). *Cimbebasia (A)*, **5**, 225–250.
3. Louw, S. v. d. M. (1982) The occurrence of Microcerinae (Coleoptera: Curculionidae) in Botswana. *Botswana Notes and Records*, **14**, 11–22.
4. Louw, S. v. d. M. (1983) A new species of *Hyomora* Pascoe (Coleoptera: Curculionidae: Rhytirrhininae) with notes on the distribution of the genus. *Navorsinge van die Nasionale Museum, Bloemfontein*, **4** (6), 169–175.
5. Louw, S. v. d. M. (1985a) African Curculionidae in collections in Europe and England. *Curculio*, **18**, 7–8.
6. Louw, S. v. d. M. (1985b) The status of *Hyomora adversaria occidentalis* Louw (Coleoptera: Curculionidae: Rhytirrhininae). *Journal of the Entomological Society of Southern Africa*, **48** (2), 342–343.
7. Oberprieler, R. G., & Louw, S. v. d. M. (1985) Curculionoidea. Pp. 270–280. In: Scholtz, C. H., & Holm, E. (Eds.), *Insects of Southern Africa*. Butterworths, Durban, 502 pp.
8. Louw, S. v. d. M. (1986a) Revision of the Microcerinae (Coleoptera: Curculionidae) with an analysis of their phylogeny and zoogeography. *Memoirs van die Nasionale Museum, Bloemfontein*, **21**, 1–331.
9. Louw, S. v. d. M. (1986b) Curculionidae collections in southern Africa. *Curculio*, **21**, 3–4.
10. Louw, S. v. d. M. (1987) *In situ* predation by ants (Hymenoptera: Formicidae) on the eggs of *Brachycerus ornatus* Drury (Coleoptera: Curculionidae: Brachycerinae). *The Coleopterists Bulletin*, **41** (2), 180.
11. Louw, S. v. d. M. (1988a) Weevils in bird diets. *Rostrum*, **19**, 4.
12. Louw, S. v. d. M. (1988b) Snuitkewers in die maaginhoude van voëls. *Nasionale Museum Nuus*, **34**, 24–25.
13. Louw, S. v. d. M. (1988c) Notes on adult overwintering of Entiminae and Microcerinae (Coleoptera: Curculionidae) in southern Africa. *The Coleopterists' Bulletin*, **42** (2), 155–156.
14. Louw, S. v. d. M. (1988d) Arboreal Coleoptera associated with *Leucosidea sericea* (Rosaceae) at the Golden Gate Highlands National Park. *Koedoe*, **31**, 53–70.
15. Louw, S. v. d. M. (1988e) Taxonomic and nomenclatorial notes on Rhytirrhininae (Coleoptera: Curculionidae). *The Coleopterists' Bulletin*, **42** (3), 217–218.
16. Louw, S. v. d. M. (1990a) The life-history and immature stages of *Brachycerus ornatus* Drury (Coleoptera: Curculionidae). *Journal of the Entomological Society of Southern Africa*, **53** (1), 27–40.
17. Louw, S. v. d. M. (1990b) General processing and storage of a Curculionidae (Coleoptera) larval collection. Pp. 49–52. **In**: Herholdt, E. M. (Ed.). *Natural History Collections: their Management and Value.* Transvaal Museum Special Publication No. 1, Transvaal Museum, Pretoria.
18. Louw, S. v. d. M. (1991) A new species of *Breviurodon* Strejcek (Coleoptera: Urodontidae) from Zaire and its bearing on urodontid phylogeny. *African Journal of Zoology*, **105**, 323–329.
19. Louw, S. v. d. M. (1993a) Systematics of the Urodontidae (Coleoptera: Curculionoidea) of southern Africa. *Entomology Memoirs of the Department of Agriculture, Republic of South Africa*, **87**, 1–92.
20. Louw, S. v. d. M. (1993b) Breeding populations of *Lixus carinerostris* Boheman and *Calodemas prolixus* Faust (Coleoptera: Curculionidae) co-existing on *Mesembryanthemum*. *The Coleopterists Bulletin*, **47** (4), 335–339.

21. LOUW, S. v. d. M. (1993c) Seed-feeding Urodontidae weevils and the evolution of the galling habit. Pp. 186–193. **In:** PRICE, P. W., MATTSON, W. J., & BARANCHIKOV, Y. N. (Eds.). *The Ecology and Evolution of Gall-forming Insects.* United States Department of Agriculture, Forest Service, North Central Forest Experiment Station. General Technical Report NC-174.
22. KOK, O. B., & LOUW, S. v. d. M. (1994) Bird and mammal predators of curculionid and tenebrionid beetles in semi-arid regions of South Africa. *Journal of African Zoology*, **108**, 555–563.
23. LOUW, S. v. d. M., VAN EEDEN, C. F., & WEEKS, W. J. (1995) Weevil infestation on cultivated *Amaranthus* in South Africa. *African Crops Science Journal*, **3** (1), 93–98.
24. LOUW, S. v. d. M. (1995) Systematics and biogeography of the subfamily Microcerinae (Coleoptera: Curculionidae): A re-evaluation based on larval morphology. *Memoirs of the Entomological Society of Washington*, **14**, 169–174.
25. VAN EEDEN, C. F., WEEKS, W. J., & LOUW, S. v. d. M. (1996) Insects associated with wild and cultivated *Amaranthus* spp. (Amaranthaceae) in South Africa. *Proceedings of 2nd Crop Science Conference for Eastern and Southern Africa, University of Malawi, Blantyre.*
26. PRICE, P. W., & LOUW, S. v. d. M. (1996) Resource manipulation through resource modification of the host plant by a galling weevil, *Urodontus scholtzi* Louw (Coleoptera: Urodontidae). *African Entomology*, **4** (2), 103–110.
27. LOUW, S v. d. M. (1998a) Weevils systematics in the 21st Century. *Atti Museo Regionale di Scienze Naturale Torino*, pp. 7–17.
28. LOUW, S. v. d. M. (1998b) Solving the riddle: Combining life-history analysis and morphological comparison in weevil phylogenetics. *Atti Museo Regionale di Scienze Naturale Torino*, pp. 19–26.
29. LOUW, S. v. d. M. (1998c) The gall-inhabiting weevil (Coleoptera) community on *Galenia africana* (Aizoaceae): co-existence or competition? Pp. 122–126. **In:** CSÓKA, G., MATTSON, W. J., STONE, G. N., & PRICE, P. W. (Eds.). *The Biology of Gall-inducing Arthropods.* United States Department of Agriculture, Forest Service, North Central Forest Experiment Station, General Technical Report NC-199.
30. BLODGETT, J. T., SWART, W. J., KLOPPERS, F. J., & LOUW, S. v. d. M. (1998) Identification of fungi associated with *Hypolixus haerens* in *Amaranthus hybridus* stems. *South African Journal of Science*, **94** (11), xxv–xxvi.
31. LOUW, S. v. d. M., SWART, W. J., HONIBALL, S. J., & CHEN, W. (2002) Weevil-fungus interaction on *Amaranthus hybridus* (Amaranthaceae) in South Africa. *African Entomology*, **10** (2), 361–364.
32. KIGGUNDU, A., GOLD, C. S., LABUSCHAGNE, M. T., VUYLSTEKE, D., & LOUW, S. v. d. M. (2003) Levels of resistance to Banana Weevil (*Cosmopolites sordidus*) in *Musa* germplasm in Uganda. *Euphytica*, **133**, 267–277.
33. LOUW, S. v. d. M. (2004) Microcerini Lacordaire, 1863 (Coleoptera, Curculionoidea). Pp. 905–935. In: SFORZI, A., & BARTOLOZZI, L. (Eds.). 2004. Brentidae of the World. *Monographia di Museo Regionale di Scienzi Naturali, Torino*, **38**, 971 pp.
34. BLODGETT, J. T., SWART, W. J., & LOUW, S. v. d. M. (2004) Identification of fungi and fungal pathogens associated with *Hypolixus haerens* and decayed and cankered stems of *Amaranthus hybridus*. *Plant Disease*, **88**, 333–337.
35. KOK, O. B., LOUW, S. v. d. M., & KOK, A. C. (2005) Snuitkewers in die diëet van die Swart Korhaan: meer as net voedsel? *Suid-Afrikaanse Tydskrif vir Natuurwetenskap en Tegnologie*, **24** (4), 118–123.

References

1. Kuschel, G. A phylogenetic classification of Curculionoidea to families and subfamilies. *Mem. Entomol. Soc. Wash.* **1995**, *14*, 5–33.
2. Louw, S. Systematics of the Urodontidae (Coleoptera: Curculionoidea) of southern Africa. *Entomol. Mem. Dept. Agric.* **1993**, *87*, 1–92.
3. Osella, G.; Colonnelli, E.; Zuppa, A.M. Mediterranean Curculionoidea with southern African affinities. In *Taxonomy, Ecology and Distribution of Curculionoidea (Coleoptera: Polyphaga), Proceedings of the XX. International Congress of Entomology, Florenze, Italy, 28 August 1996*; Colonnelli, E., Louw, S., Osella, G., Eds.; Museo Regionale de Scienze Naturali: Torino, Italy, 1998; pp. 221–265.
4. Mucina, L.; Rutherford, M.C. The vegetation of South Africa, Lesotho and Swaziland. *Strelitzia* **2006**, *19*, 1–807.
5. Germishuizen, G.; Meyer, N.L. Plants of southern Africa: An annotated checklist. *Strelitzia* **2003**, *14*, 1–1231.
6. Price, P.W.; Louw, S.v.d.M. Resource manipulation through architectural modifications of the host plant by a gall-forming weevil, *Urodontus scholtzi* (Coleoptera: Anthribidae). *Afr. Entomol.* **1996**, *4*, 103–110.
7. Strejcek, J. *Breviurodon africanus* get. et sp. n. aus Congo-Brazzaville (Coleoptera, Urodonidae). *Annls. Hist.-Nat. Mus. Nat. Hung.* **1981**, *73*, 203–205.
8. Louw, S. A new species of *Breviurodon* Stejcek (Coleoptera: Urodontidae) from Zaire and its bearing on urodontid phylogeny. *J. Afr. Zool.* **1991**, *105*, 323–329.
9. Bailey, P.; Sagliocco, J.-L.; Vitou, J.; Cooke, D. Prospects for biological control of cutleaf mignonette, *Reseda lutea* (Resedaceae), by *Baris picicornius* and *Bruchela* spp. in Australia. *Aust. J. Exp. Agric.* **2002**, *42*, 185–194. [CrossRef]
10. Buddeberg, K.D. Beobachtungen über Lebensweise und Entwicklungsgeschichte einiger bei Nassau vorkommender Käfer: *Mecinus janthinus* Germ., *Baris morio* Schh., *Phlocosinus thujae* Perris, *Urodon conformis* Suffr. *Jahrb. Nassau. Ver. Naturkd.* **1883**, *36*, 124–144.
11. May, B.M. Larvae of Curculionoidea (Insecta: Coleoptera): A systematic overview. *Fauna N. Z.* **1993**, *28*, 1–223.
12. Dorchin, N.; Harris, K.M.; Jaschhof, M. 22. Cecidomyiidae (gall midges). In *Manual of Afrotropical Diptera. Volume 2. Nematocerous Diptera and Lower Brachycera. Suricata 5*; Kirk-Spriggs, A.H., Sinclair, B.J., Eds.; South African National Biodiversity Institute: Pretoria, South Africa, 2017; pp. 581–599.
13. Louw, S.; Oberprieler, R.G. Willy Kuschel visits South Africa. *Curculio* **1993**, *34*, 5–6.
14. Oberprieler, R.G.; Mansell, M.W.; Louw, S. Richtersveld—An insectile paradise / Richtersveld se insekte. *Custos* **1993**, *21*, 30–35.

Article

Unveiling the History of a Peculiar Weevil-Plant Interaction in South America: A Phylogeographic Approach to *Hydnorobius hydnorae* (Belidae) Associated with *Prosopanche americana* (Aristolochiaceae)

Andrea S. Sequeira [1,*,†], Nicolás Rocamundi [2,†], M. Silvia Ferrer [1,3,4], Matias C. Baranzelli [2] and Adriana E. Marvaldi [5]

1 Department of Biological Sciences, Wellesley College, Wellesley, MA 02481, USA; sisiferrer@hotmail.com
2 Laboratorio de Ecología Evolutiva y Biología Floral, Instituto Multidisciplinario de Biología Vegetal, Universidad Nacional de Córdoba, CONICET, FCEFyN, Córdoba X5016GCA, Argentina; nicolasrocamundi@gmail.com (N.R.); matiasbaranzellibc@gmail.com (M.C.B.)
3 Instituto Argentino de Investigaciones de Zonas Áridas, Consejo Nacional de Investigaciones Científicas y Técnicas, C.C. 507, Mendoza 5500, Argentina
4 Present address: Laboratorio de Salud Pública, Ministerio de Salud, Desarrollo Social y Deportes de la Provincia de Mendoza, Talcahuano 2194, Godoy Cruz, Mendoza 5501, Argentina
5 División Entomología, Facultad de Ciencias Naturales y Museo, Universidad Nacional de La Plata, CONICET, Paseo del Bosque s/n, B1900FWA La Plata, Buenos Aires, Argentina; marvaldi@fcnym.unlp.edu.ar
* Correspondence: asequeir@wellesley.edu; Tel.: +1-781-283-3376
† Equal contributions.

Received: 28 March 2018; Accepted: 4 May 2018; Published: 6 May 2018

Abstract: Interspecific interactions take place over both long and short time-frames. However, it is not completely understood if the interacting-partners persisted, migrated, or expanded in concert with Quaternary climate and landscape changes. We aim to understand whether there is concordance between the specialist weevil *Hydnorobius hydnorae* and its parasitic host plant, *Prosopanche americana* in space and time. We aim to determine whether *Prosopanche* had already established its range, and *Hydnorobius* later actively colonized this rare resource; or, if both host plant and herbivore expanded their range concomitantly. We performed population genetic, phylogeographic and Bayesian diffusion analysis of Cytochrome B sequences from 18 weevil localities and used paleodistribution models to infer host plant dispersal patterns. We found strong but uneven population structure across the range for *H. hydnorae* with weak signals of population growth, and haplotype network structure and SAMOVA groupings closely following biogeographic region boundaries. The ancestral areas for both *Hydnorobius* and *Prosopanche* are reconstructed in San Luis province within the Chaco Biogeographic province. Our results indicate a long trajectory of host-tracking through space and time, where the weevil has expanded its geographic range following its host plant, without significant demographic growth. We explore the past environmental changes that could underlie the boundaries between locality groups. We suggest that geographic dispersal without population growth in *Hydnorobius* could be enabled by the scarcity of the host plant itself, allowing for slow expansion rates and stable populations, with no need for significant demographic growth pulses to support range expansion.

Keywords: spatio-temporal diffusion; specialist weevils; parasitic plants; co-dispersal through space and time; stable populations

---

1. Introduction

The specialized interactions evolved between phytophagous insects and their host plants are deemed to have profoundly affected their mutual diversification and the overall biodiversity on Earth [1]. Interspecific interactions take place over both relatively long evolutionary and short ecological time-frames (e.g., [2,3]) and current research on biological interactions has been strengthened, thanks to the development of phylogeography, by considering evolutionary processes in a dynamic spatial context [4]. For example, from this perspective, numerous studies are suggesting that climatic oscillations during the last million years (i.e., the Late Quaternary) had a major impact not only on species distribution, but also on their demography and diversification [5]. However, it is less clear how biological interactions may modulate organisms' responses to these climatic changes (but see [6]). Although the importance of interspecific interactions in the evolution of the species at a geographical scale is recognized [7,8], it is not completely understood whether or not the interacting-partners responded in the same way to Quaternary climate and landscape changes (that is if they survived or expanded together in the landscape [6]).

Among phytophagous insects, the weevils (Coleoptera: Curculionoidea) are, as highlighted by Kuschel [9], not only impressive for their taxonomic diversity but also for their diverse associations with plants. Weevils offer a myriad of examples of persistent interactions with plants, many of which are highly specialized. A most intriguing association with plants is presented by weevils of the genus *Hydnorobius* Kuschel (Curculionoidea: Belidae: Oxycoryninae). The South American genus *Hydnorobius* was created by Kuschel [10] to harbor the species *H. hydnorae* (Pascoe), *H. helleri* (Bruch) and *H. parvulus* (Bruch), all of which are associated with parasitic angiosperms within the genus *Prosopanche* (R. Br.) Baillon [11] in the Aristlochiaceae family [12–15]. Species of *Prosopanche* are very peculiar plants in arid and semiarid regions, being holoparasites that lack chlorophyll and attach to their hosts' roots by special "haustorial" structures on their rhizomes [12,16]. Flowers are visited by nitidulid beetles and belid weevils that become dusted with pollen. The nitidulids are the main pollinators as only they can enter into the stigmatic chamber beneath the anther body [17]. Two species of *Prosopanche* are known to host belid weevils in South America: *P. americana*, which parasitizes *Prosopis* spp. (Fabaceae), is host-plant of *H. hydnorae* and *H. helleri*, while *P. bonacinai* Spegazzini, which parasitizes a broader range of dicots [17], is the host-plant of *H. parvulus* and can also harbor *H. hydnorae* [11].

Prosopanche americana is distributed along an "arid diagonal" in the Monte, Chaco and Pampean biogeographic provinces in Argentina and Paraguay [18–20]. The weevil genus *Hydnorobius* is also mostly distributed in Argentina, with some populations having dispersed into Paraguay. The life histories of *Hydnorobius* weevils seem to be synchronized with that of their parasitic hostplants. According to information provided by Bruch [21] and Marvaldi [22], during the summer time and usually after rainfall has supplied moisture to the arid soil, the flowers of *Prosopanche* emerge by breaking through the soil and open, attracting adults of *H. hydnorae* (Figure 1A–C). The weevils become dusted with pollen as feeding, mating and oviposition occur in the flower. The female weevil lays eggs into holes that she prepares with the rostrum in the fleshy tepals and anther body. The larvae develop feeding progressively on parenchymatous tissue inside the flower and subterranean fruit. Plant tissues are often decayed by the time the larva finishes growing (Figure 1D) and pupation occurs (Figure 1E) leading to emerged adult weevils (Figure 1F). Such symbiotic interaction between weevil and plant seems to span from commensalism (which is common, since larval feeding may not damage the seeds) to parasitism (particularly when larval infestation is high).

A phylogeographic study would help in the understanding of the geographic and historical aspects of the relationship between *H. hydnorae*, the most abundant species of the genus, and its host plant *Prosopanche*. By using genetic data of individuals of the species, analyzed in conjunction with its geographic distribution [23], we may provide insights into one of the enigmatic evolutionary paths that occurred in the Oxycoryninae. Despite the fact that in the last years phylogeographic studies on South American organisms have increased (revised in [24,25]), some places on the continent, such as arid regions, are still unexplored from a phylogeographic perspective (revised in [26]). In this sense, the phylogeographic study of *H. hydnorae* represents the first one undertaken for an insect species endemic to the Monte desert/Chacoan regions.

Figure 1. The plant *Prosopanche americana* (**A–C**) and its associated weevil *Hydnorobius hydnorae* (**D–E**). (**A**) Flower showing the open perianth and perigonial tube on top of the inferior ovary attached to a rhizome bearing haustorial roots; (**B**) Open flower showing the dome-like androecium releasing pollen and bearing a pair of mating *H. hydnorae*; (**C**) Development of the *Prosopanche* flower from bud (left) to open flower in stigmatic (middle) and staminal (right) phases (a, anther body; f, fenestrae connecting a and c; c, stigmatic chamber; s, receptive stigma; p, pollen); (**D**) *H. hydnorae* full-grown larva in decaying plant tissue; (**E**) *H. hydnorae* pupa in pupal cell and (**F**) Dorsal view of adult *H. hydnorae*.

We aim to pinpoint the geographical origin of the association of the weevil *H. hydnorae* with *P. americana*, and to provide a better understanding of the evolution of host choices in an ancient weevil group like oxycorynine belids. The high specificity of the interaction between *H. hydnorae* and *P. americana*, with the weevil's life cycle tightly connected to that of its host-plant, is suggestive of a hypothesis of the weevil coevolving, or at least co-dispersing with the plant, and it could then be expected that the niche of the weevil would be coincident with that of its host-plant. Our general approach is to perform a phylogeographic study of *H. hydnorae* to elucidate the timing and geographic origin of the association with its host plant, and to study the weevil's ancestral range and possible range expansions. Our interest was, specifically, in the relative timing of the geographic and demographic expansions of the weevil populations in conjunction with its elusive and rare host plant. In a sense, we aim to determine if *Prosopanche* had already established its range, and *Hydnorobius* later actively colonized this available although rare resource; alternatively, it is plausible that both host plant and herbivore have expanded their range concomitantly, and that weevils track their own host plant, *P. americana*, while *Prosopanche* is itself expanding its range. In the first scenario, we would expect to find that the structuring, demographic and range expansion models in the genetic data of *H. hydnorae* are independent in time and space from the paleodistribution models (PDMs) of the host plant. In the second scenario a geographic and temporal association between the genetic patterns of *H. hydnorae* and distribution models of *P. americana* is expected.

2. Materials and Methods

2.1. Study Area and Sample Collection

With the objective of detecting the ancestral area of *Hydnorobius hydnorae* and exploring the origin of the association of this weevil with *Prosopanche americana*, specimens were collected from 18 localities in Argentina and Paraguay along the Western Chacoan district, Northern Monte district and Espinal district of Chaco, Monte and Pampean biogeographic provinces (as defined in [27]) (Table 1, Figure 2A). The localities sampled extend over the distribution range of the species, covering most of its northern, central and southern areas of occurrence. The collecting strategy was to search for the host-plant, *P. americana*, which can usually be found emerging from the soil, under different species of *Prosopis* trees (*P. flexuosa, P. alpataco, P. caldenia, P. chilensis, P. ruscifolia,* and others). Once the plant was located, it was inspected for the presence of adult specimens. Adults were deposited in vials with 96% ethanol, separated by locality. When there were no adults, plants were dug up and preserved in bags separated by locality, until processed in the laboratory. Once in the laboratory, plants were inspected for the presence of larvae, most of which were preserved in 80% ethanol. A few larvae were left alive in the plants (kept in rearing jars) until adults emerged, in order to re-confirm the identification of the species and to have adult voucher specimens. All ethanol-preserved specimens were kept at −20 °C until processed. Specimens of the remaining species of *Hydnorobius* (*H. parvulus* and *H. helleri*) were also collected and used as outgroups. Voucher specimens of the three *Hydnorobius* species used in this study are deposited in the Entomological collection of the Museo de La Plata (MLP, La Plata, Argentina).

Table 1. Locality information for all sampled specimens of *Hydnorobius hydnorae* organized by geographic area and by locality groupings from the SAMOVA $K = 3$ analysis. Population codes reflect province and locality name as in Figure 2. N indicates sample size per locality.

	Province and Locality Name	Population Code	Coordinates	N
North	La Rioja: San Ramón	LSR	S 30° 22.002′; W 66° 52.002′	4
	San Luis: Quines	SLQUI	S 32° 17.472′; W 65° 50.982′	5
	Córdoba: Chancaní	CCH	S 31° 22.584′; W 65° 28.812′	4
	Córdoba: Árbol blanco	CAB	S 30° 9.06′; W 64° 4.404′	5
	Paraguay: Hayes	PHAY	S 23° 4.818′; W 59° 13.272′	3
	Santiago del Estero: Aurora-Huayapampa	SAUHU	S 27° 25.416′; W 64° 16.686′	5
	Salta: La Unión	SLU	S 23° 45.918′; W 63° 4.914′	1
	Chaco: Taco Pozo	CHTP	S 25° 40.74′; W 63° 7.8′	2
Central	Catamarca: Pio Brizuela	CBR	S 27° 49.878′; W 66° 12,16′	3
	Catamarca: Tinogasta	CTI	S 28° 4.99′; W 67° 34.002′	4
	San Juan: Bermejo	SBE	S 31° 31.998′; W 67° 24′	3
	San Juan: Huaco	SHU	S 30° 9′; W 68° 34.998′	3
South	Mendoza: Divisadero	MDI	S 33° 11.4′; W 67° 51′	5
	Mendoza: Reserva de la Biósfera Ñacuñán	MNA	S 34° 3′; W 67° 57′	4
	Mendoza: Paso del Loro	MPD	S 35° 39.672′; W 67° 33,492′	5
	La Pampa: Chacharramendi	LPCHA	S 37° 24.372′; W 65° 18.39′	1
	La Pampa: El Durazno	LPDZO	S 36° 40.448′; W 65° 17.346′	3
	La Pampa: La Maruja	LPMAR	S 35° 37.62′; W 64° 50.418′	3

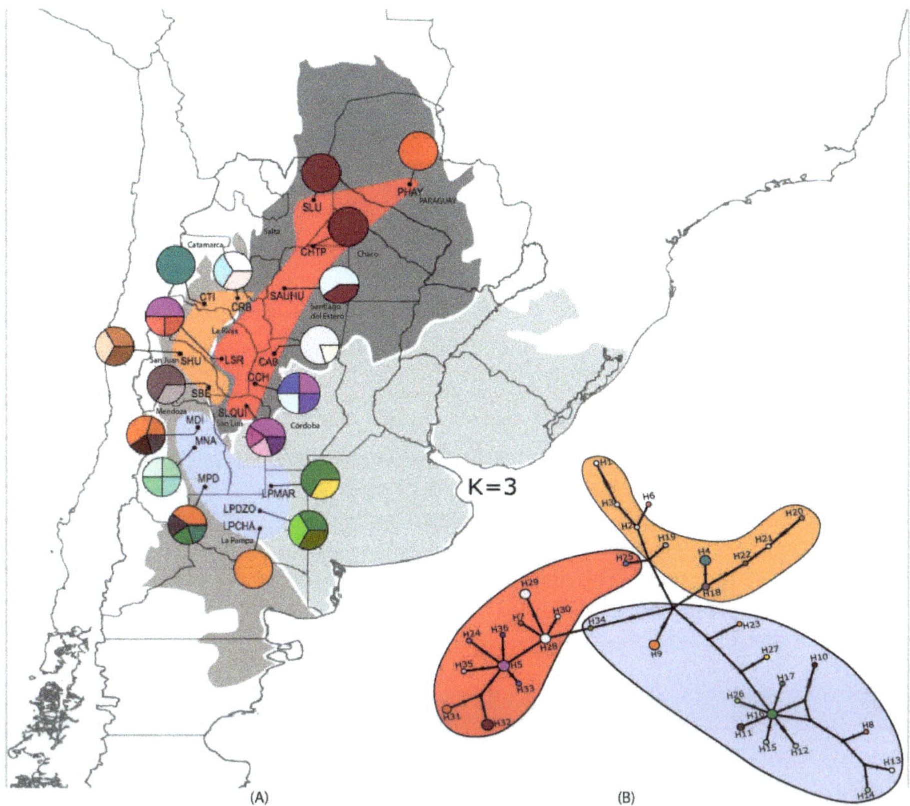

Figure 2. (**A**) Northwestern region of Argentina with all *Hydnorobius hydnorae* collecting localities. Details associated with locality codes can be found in Table 1. The larger shaded areas correspond to biogeographic provinces as in Morrone [27]. Pie charts indicate proportional presence of distinct Cytochrome B (Cyt-B) haplotypes in each locality. (**B**) Minimum spanning network showing all Cyt-B haplotypes for the 18 localities. Haplotype IDs are indicated by the circles and the size of the circles is proportional to the frequency of each haplotype. Dashes on the haplotype connections indicate the number of mutational steps between them. Sections of the network are colored according to the locality groups suggested by the SAMOVA analysis ($K = 3$). Those same colors are then depicted on the map grouping localities in three areas: North (red), Central (orange) and South (blue).

2.2. DNA Isolation, PCR Amplification and Sequencing

Total genomic DNA was extracted from thoracic tissues of adult and larval ethanol-preserved voucher specimens using the DNeasy Blood and Tissue Kit (QIAGEN, MD, USA). Tissue was processed from the legs and thorax in adult individuals and from part of thorax in larval individuals. The extracted DNA was stored at −20 °C. Mitochondrial DNA is well suited for phylogeographic studies given its preponderant cytoplasmatic non-Mendelian inheritance and its general lack of recombination [28,29]. Gene sequences from the mitochondrial genome present a high degree of polymorphism, a desirable property for intraspecific levels of study, specifically to resolve aspects about its history and population structure [23,30]. Amplification and sequencing of regions of the mitochondrial gene Cytochrome B (Cyt-B) were performed using an array of primer combinations; Cb1: 5′-TAT GTA CTA CCA TGA GGA CAA ATA TC-3′ and Cb2: 5′-ATT ACA CCT CCT AAT TTA TTA GGA AT-3′, CytB.B1 5′-TTA ATT ATT CAA ATT GCA ACA GGA TTA TTT-3′ and CytB.A1 5′-AAG

TTT AAA ATT CTA YCC AAT TAA TCA A-3′ [31,32] and in some instances in combination with a primer designed for this study CytB 110 5′-GAG GAG GAT TTT CAG TTG AC-3′. PCR conditions for amplification with Cb1-Cb2 were an initial denaturation step of 3 min at 95 °C followed by 5 cycles of 60 s at 95 °C, 30 s at 42 °C, 90 s at 72 °C, then 34 cycles of 60 s at 95 °C, 30 s at 45 °C, 90 s at 72 °C and a final elongation step of 5 min at 72 °C. PCR conditions for amplification with CytB.B1-CytB.A1 were an initial denaturation step of 3 min at 95 °C followed by 2 cycles of 30 s at 95 °C, 30 s at 58 °C, 60 s at 72 °C with seven repeats of the above steps decreasing the annealing temperature by 2 °C every two cycles, followed by 20 cycles of 30 s at 95 °C, 40 s at 42 °C, 40 s at 72 °C and a final elongation step of 10 min at 72 °C. The PCR products were purified and bi-directionally sequenced. Electropherograms were edited using ChromasPro v.1.5 (Technelysium Pty Ltd., (Brisbane, Australia), Sequencher v.5 (GeneCodes Corp., Ann Arbor, USA) and BioEdit v.7.0.9.0 [33]. Sequences were edited for disagreements between fragments and checked for the presence of an open reading frame (ORF). All sequences are available in Genbank under accession numbers: MH119874-MH119936.

2.3. Haplotype Network, Population Genetic Structure and Genetic Diversity

We obtained haplotypes using DnaSP v.5 [34]. The haplotype network was constructed using the median-joining algorithm [35] implemented in PopARTv.1.7 [36].

Genetic structure at a geographical scale was explored using spatial analysis of molecular variance (SAMOVA) in Samova v.1.0 [37]. Analyses were run with *K* values (# of genetic groups) ranging from 2 to 9, using 10,000 independent annealing processes. To select the optimal number of genetic groups we used the among-group component (F_{CT}) of the overall genetic variance. The selected *K* value and proposed groupings were used to conduct independent analyses of molecular variance (AMOVA) in ARLEQUIN v.3.5 [38].

Additional hierarchical analyses of molecular variance were performed in order to explore other levels of population structure. (1) Among all localities without groupings; (2) among regional groups as single large populations and considering the cladistic biogeographic regionalization proposed by Flores and Roig-Juñent [39] based on six genera of insects and one genus of plant; and (3) with the same groupings as 2 but also incorporating locality information. AMOVAs were performed with groupings 2 and 3 in order to investigate if the genetic information of this ancient weevil species recovers the vicariant pattern exhibited by its habitat.

In order to further explore the spatial structure of genetic diversity, pairwise F_{ST} (an indicator of population differentiation due to population structure) among all localities and among localities within each of the SAMOVA determined groups were also calculated in ARLEQUIN v.3.5. The number of migrants that localities exchange (*Nm*) were also estimated using the inverse of pairwise F_{ST} values.

Finally, to measure the degree of polymorphism at different geographical scales for each locality and for the main genetic groups, diversity indices were calculated using DNAsp v.4.10 and ARLEQUIN v.3.5. The three indexes estimated were: haplotype diversity (*h*), that provides information on the number and frequencies of different alleles at a locus, regardless of the differences in nucleotide sequences; nucleotide diversity (π) that measures sequence divergence between individuals in a population, regardless of the number of different haplotypes; and mean number of pairwise differences (*K*).

2.4. Phylogenetic Relationships among Haplotypes

We inferred the phylogenetic relationships among haplotypes and outgroups *H. helleri* and *H. parvulus* using two different types of analysis with different criteria: Bayesian inference (BI) and Maximum Likelihood (ML). For the first analyses we used BEAST v.1.7.5 [40]. BI analyses were run for 5×10^7 generations, with the HKY+G model (selected in jModelTest v.2.1.4 [41] following the Akaike Information Criterion; AIC), selecting a Yule tree prior and two Monte Carlo Markov chains (MCMC), starting with a random tree and sampling parameters every 5000 steps to obtain 10,000 trees for each run. For each one, the first 25% of trees prior to stationarity were excluded, and high values of effective sample sizes (ESS > 200) and convergence of estimated parameters were verified using Tracer v.1.6 [42]. The resulting files (i.e., the *.log* file with estimated parameters and *.trees* with phylogenetics relationships) were combined using LogCombiner v.1.7.5 [39] and topologies were assessed using TreeAnnotator v.1.7.5 [40]. FigTree v.1.6.1 [43] was used to estimate Bayesian posterior probabilities (PP). The ML analyses used the online platform PhyML 3.0 [44] with the same substitution model used in the BI analyses. The robustness of the phylogenetic relationships was evaluated through 1000 bootstrap replications (BP).

2.5. Demographic History Analysis

Tajima's D test and Fu's F test were calculated using ARLEQUIN v.3.5, under the assumption that the markers used are selectively neutral. These neutrality tests also assume that a population has been in mutation-drift balance for a long period of evolutionary time [45]. These test statistics are expected to be significantly negative when genetic structure has been influenced by rapid range expansion [46]. Mismatch analyses were also used as a way of measuring the frequency of the number of differences between pairs of haplotypes. To compare observed frequencies of pairwise differences with those expected under a model of demographic expansion, mismatch distributions were generated using ARLEQUIN v.3.5 for each locality group as determined by SAMOVA and for all the samples together. In the absence of population size changes (i.e., the population is subdivided or in demographic equilibrium), a multimodal distribution is expected; however, if sudden demographic expansions have occurred, unimodal distributions are expected. In addition, 1000 coalescent simulations under the sudden expansion model were used to test the significance of the raggedness statistic (r) for each mismatch distribution. Populations that have undergone expansions will exhibit smooth, unimodal mismatch distributions with low raggedness values, whereas more ragged mismatch distributions tend to result from large, stable populations [47].

To complement the results derived from the previous analysis and to obtain an estimation of the timing of demographic events, we performed a Bayesian Skyline Plot (BSP) analysis for each genetic group recovered with SAMOVA using BEAST v.1.7.5. Unlike previous demographic analyses, BSP does not use a specific, particular model to estimate the population size over time. To run the BSP we set the number of group intervals to 10, with a piece-wise constant model and selecting the maximum time in the root height as Median. Moreover, the HKY+I model was selected for the Northern group, HKY for the Central group and HKY+G for the Southern Group (see Results) following the AIC criteria implemented in JmodelTest. Two MCMC starting with a random tree were run for 5×10^7 generations, with parameters sampled every 5000 steps. To calibrate these BSPs, we employed an uncorrelated lognormal relaxed clock that allows for rate heterogeneity among lineages with a normal prior distribution (mean = 0.0645 substitutions/My; SD = 0.01) on the substitution rate of mDNA, following recent estimates for Belidae [48,49]. The chain convergence check and *.log* and *.trees* combinations were used as previously described for the BI haplotype tree reconstruction. The demographic profiles were constructed with Tracer v.1.6.

2.6. Bayesian Spatio-Temporal Diffusion Analyses

We used BEAST v.1.7.5 to analyse the Cyt-B data using a continuous spatial diffusion model ("Relaxed Random Walk", RRW; [50]) in order to infer the geographical origin and the spatial expansion of *H. hydnorae* lineages during diversification. These continuous-diffusion Bayesian analyses allow reconstruction of ancestral distributions and the diffusion of lineages continuously through space and time, using the latitude and longitude coordinates of each genealogical terminal, while taking into account genealogical uncertainty [50]. This Bayesian phylogeographic approach has the power of both estimating and distinguishing between demographic expansion and spatial expansion (i.e., between population growth and geographic range expansion) [51]. Continuous-diffusion models are analogous to those for relaxed-molecular clocks, allowing the rate of spatial expansion to vary along the branches of the phylogeny [52]. This is considered convenient particularly for species with large geographical ranges, like *H. hydnorae*, where it is expected that favorable conditions for spatial expansion were not even over time [50]. This analysis was done using a subsampled data set, including one representative individual of each haplotype per locality (n = 45; e.g., [26,53]). JmodelTest v.2 selected HKY+I for this data matrix and the same parameters described for the previous Bayesian analyses were set for clock rate, clock model, chain convergence check and tree annotation, but for this particular analysis a population coalescent Bayesian Skyride model for the prior tree, and a normal distribution for the diffusion rate were set. We used the *jitter* option on statistical Trait Likelihood with a parameter of 0.01 to add variation to sequences with the same geographic location. We examined lineage diversification through the landscape using SPREAD v.1.0.7 [54], having as input the MCC tree obtained under the continuous diffusion model.

2.7. Paleodistribution Models of the Host Plant *Prosopanche americana*

Distribution models are useful to obtain the potential distribution of a species using different algorithms that relate the climatic conditions of the current collection sites with the potential geographic distribution of the species, assuming that this set of environmental variables will reflect the ecological niche of the species [55]. An advantage of these spatial predictions is that they can be projected under different past (and future) environmental scenarios, producing habitat suitability maps for the species over the time and inferring its historical distribution limits [56]. In this study the past distribution of the host plant *P. americana* was estimated by georeferencing and mapping the presently known localities of this plant, which were then used to model their past distribution and dispersal, between 120,000 years ago (kya) and the present, via predictive methods based on paleodistribution models (PDM). This approach is particularly important to meet the objectives of the paper, since a phylogeographic study of *P. americana*, comparing its genetic information at the geographical scale with that of the weevil cannot be attempted at this time. Being holoparasitic, nonphotosynthetic plants with extremely reduced plastid genome [12,57], the markers conventionally used in plant phylogeography are missing. We used 51 trustable georeferenced *P. americana* occurrence points obtained mainly from the field between 2015–2017, but also completed from herbarium records (CORD, MERL) and the literature [16]. From these georeferenced locations, current climatic data with grid cell resolution of 0.25 degrees (~5 km^2 cell) were downloaded from the WorldClim database v.1.4 [58,59] represented by 19 bioclimatic variables constructed with the variation in precipitation and temperatures throughout the year. All bioclimatic layers were cropped to span from 15.15° S to 44.57° S and from 57.20° W to 77.44° W, a spatial range that contains the current range of *P. americana*. To estimate the species distribution model to the current condition (average 1950–2000), we used the Maximum Entropy algorithm implemented in MaxEnt v.3.3.3 [60] and visualized it in DIVA-GIS v.7.5 [59]. To run MaxEnt we set the random test percentage in 25, the convergence threshold in 0.00001, a maximum number of iterations in 1000, the regularization multiplier was selected at 0.75 (to avoid over dispersion of the projected models outside the current distribution range known for the species) [61] and selected the autofeatures option. Finally, we reported the averaged across 10 bootstrap runs. We used the lowest value of probability of occurrence among the 51 trustable points as the threshold value for each prediction. Area under the receiver operating characteristic curve (AUC) was used as a performance characterization of the model, namely as the probability that a random positive instance and a random negative instance are correctly ordered by the classifier [60].

To estimate how *P. americana* distributions may have changed through time, and to evaluate if this may have impacted the demographic history and spatial distribution of genetic diversity of the *H. hydnorae* weevils feeding on them, this current model was projected for each of the palaeoclimatic scenarios, from the Last Interglacial period (LIG; 130–114 kya; [62]), the Last Glacial Maximum (LGM; 21 kya) and the mid Holocene (6 kya), based on the Community Climate System Model (CCSM4). Additionally, for the last two periods, the distribution was also reconstructed based on the Model for Interdisciplinary Research on Climate (MIROC-ESM).

3. Results

3.1. Strong but Unevenly Distributed Population Structure Across the Range for Hydnorobius hydnorae

Analyzing a 460 base-pair fragment of Cyt-B in 64 *H. hydnorae* weevil specimens, 36 mitochondrial DNA (mDNA) haplotypes were detected forming a single network with three main groups following a latitudinal pattern: The 'southern group' (SG), 'central group' (CG) and 'northern group' (NG; Figure 2B). The SG is composed of 14 haplotypes distributed exclusively south of 33° S. Within SG, one of the most frequent and widespread haplotypes (H16) appears in an internal position at the core of a star-like network topology. Haplotype 11 is connected by one step to the central haplotype H16 and is shared by two sampling sites. Haplotype 9 is shared by two localities too, but it is connected by many steps to the central haplotype, H16. The rest of the haplotypes of the SG are exclusive to single localities (i.e., H12, H13, H26, H27). The CG consists of nine haplotypes distributed in the central-west area in the septentrional Monte. Each locality presented exclusive haplotypes; even the most frequent haplotype H4 is found in a single locality near to the northwest Monte boundary limit. All of the haplotypes of NG are present in the Chaco province (23–32° S). The most frequent NG haplotype, H5, is present in three southern localities of Chaco, appears at an internal position and forms the core of another star-like network topology. Haplotypes located at the center of star-like structures and with a wide geographic distribution are considered ancestral and good indicators of potential ancestral areas [23]. Haplotypes 31 and 32 are exclusively found in the north of the distribution.

The SAMOVA structure analysis showed an optimal partition of genetic diversity of $K = 3$ ($F_{CT} = 0.45$, $p < 0.0001$), revealing three genetic groups mostly in concordance with the haplotype network (Figure 2A,B). The exception is the position of H25, a unique haplotype from Chancaní (CCH) that is grouped with other 'northern group' haplotypes in the SAMOVA analysis; however, it appears to be only a few mutational steps away from 'central group' haplotypes in the network.

Results for all AMOVA combinations of analyses are presented in Table 2. The significantly high variation found among localities ($\Phi_{ST} = 0.5345$) without hierarchical levels assigned can be interpreted as a signal of population structure. This means that populations are highly differentiated and with infrequent migration (average migration rate, $Nm = 0.109$). More specifically, the structure appears as significant between the regional groupings as determined by SAMOVA, either considering or not considering the localities as units. It is not unexpected that when analyzing the groupings as large populations, a significant amount of variation is found between the groups ($\Phi_{ST} = 0.4709$), however, a substantial amount of variation (52.91%) is also found within each of these groups of localities. The analysis among the locality groupings maintaining the locality delimitations indicates that a significant 44.8% of variation ($\Phi_{CT} = 0.4488$) is found among area groupings. The rest of the variation is divided between an average 15.32% among localities within each area, and 39.8% within populations. In summary, AMOVA results suggest that this weevil species presents population structure, which contains a geographic signal and is most likely represented by the phylogenetic relationships of the haplotypes.

Table 2. Hierarchical analysis of molecular variance (AMOVA) for *H. hydnorae* across 18 localities. Tests were performed for regional groups as determined by the SAMOVA analysis (northern; central and southern), and for all localities without hierarchical levels. Asterisks (*) denote significance level ($p \leq 0.05$).

Source of Variation	% of Variation	Fixation Indices (Φ-Statistics)
Among all localities without hierarchical levels	53.45	$\Phi_{ST} = 0.5345$ *
Within Localities	46.55	
Among regional groups as single large populations	47.09	$\Phi_{ST} = 0.4709$ *
Within regional groups	52.91	
Among SAMOVA proposed groupings	44.88	$\Phi_{CT} = 0.4488$ *
Among localities within groups	15.32	$\Phi_{SC} = 0.2780$ *
Within localities	39.80	$\Phi_{ST} = 0.6020$ *

The pairwise F_{ST} values among localities detailed in Table 3 and Figure 3, as well as the pairwise F_{ST} values among the three areas without the locality distinctions, illustrate the patterns of structure at smaller scales and reveal that the degree of isolation and structuring between localities is not homogeneous across the range. Pairwise F_{ST} values between the three areas each as a single unit are all significant, however the Northern area is the most distinct, with higher differentiation with the Central and Southern areas (F_{ST} N-C 0.5493; F_{ST} N-S 0.4999), while the haplotypes in Southern and Central areas are not as differentiated (F_{ST} S-C 0.3023). Values of pairwise F_{ST} between localities in different areas (inter-area values in Figure 3) are all larger than those among localities within any of the three areas, and a high proportion of them are significant (61.40%). In addition to being the most distinct, the Northern locality area displays the wider range of pairwise F_{ST} values between localities, including some of the larger pairwise differentiation indexes, and the largest proportion of within-area significant values (30%). The Central locality area displays lower differentiation values, with only 10% of them being significant. Finally, the Southern locality area appears the most homogeneous, with low and not significant pairwise F_{ST} values.

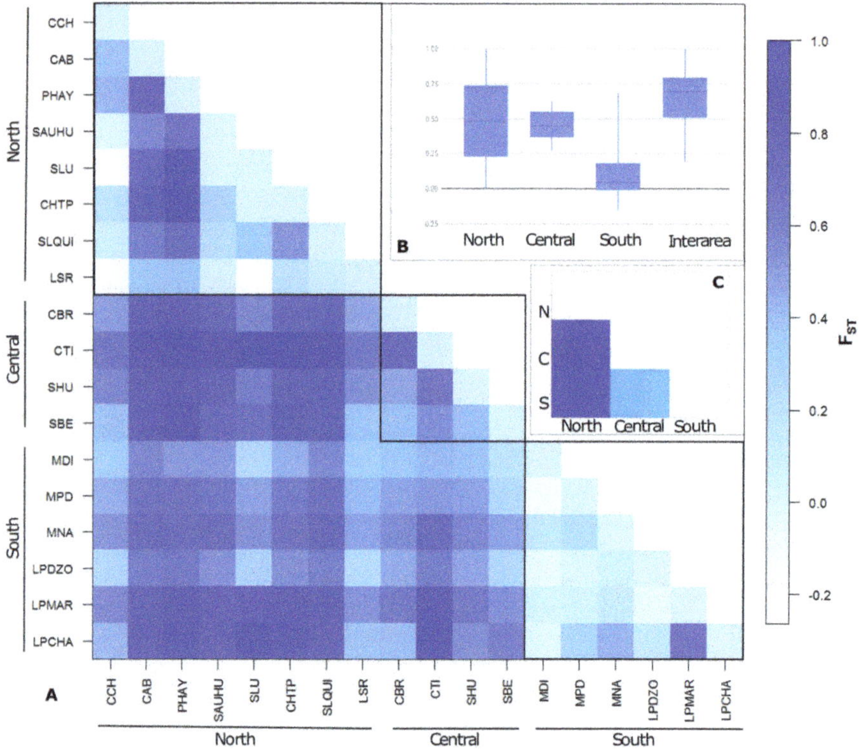

Figure 3. Genetic differentiation between localities and locality areas. (**A**) Graphical depiction of pairwise Fst values between individual localities; (**B**) Box plots contrasting the distribution of pairwise Fst values between localities within each locality area, and those between areas (interarea), horizontal bars represent mean values for each group; (**C**) Graphical depiction of pairwise Fst values between locality areas.

Table 3. Pairwise FST estimates among 18 Localities of *H. hydnorae* sampled. Inter-area contrasts are not shaded while those within areas are shaded by locality area as defined through SAMOVA analysis (Northern: light gray; Central: dark gray; Southern: medium gray). Asterisks (*) denote significance level values ($p \leq 0.05$).

CAB	0.398 *	0															
PHAY	0.436 *	0.793 *	0														
SAUHU	−0.013	0.553 *	0.681 *	0													
SLU	−0.220	0.715	1	0	0												
CHTP	0.178	0.779 *	1	0.286	0	0											
SLQUI	0.06	0.581	0.68	0.2	0.335	0.508	0										
LSR	−0.239	0.33	0.353 *	0	−0.2	0.152	0.06	0									
CBR	0.13	0.033	0.032	0.045	0.098	0.05	0.04	0.152	0								
CTI	0.07	0.013	0	0.02	0	0	0.018	0.076	0.047	0							
SHU	0.1	0.031	0.028	0.043	0.09	0.047	0.039	0.119	0.145	0.065	0						
SBE	0.2	0.033	0.023	0.053	0.066	0.036	0.044	0.223	0.212	0.118	0.194	0					
MDI	0.27	0.1	0.129	0.132	0.511	0.177	0.105	0.304	0.223	0.184	0.199	0.556	0				
MPD	0.17	0.06	0.068	0.078	0.15	0.087	0.065	0.186	0.144	0.124	0.14	0.386	−0.154	0			
MNA	0.12	0.05	0.055	0.056	0.113	0.065	0.05	0.146	0.106	0.056	0.101	0.132	0.136	0.215	0		
LPDZO	0.4	0.09	0.08	0.115	0.369	0.114	0.091	0.533	0.161	0.091	0.151	0.407	−0.051	0.005	0.065	0	
LPMAR	0.09	0.028	0.017	0.034	0.033	0.02	0.03	0.121	0.08	0.022	0.072	0.109	0.036	0.054	0.116	−0.059	0
LPCHA	0.18	0.022	0	0.034	0	0	0.031	0.213	0.197	0	0.107	0.083	−0.014	0.3	0.446	0.141	0.686

Results of the pairwise calculations of the number of migrants that localities exchange (*Nm*, Supplementary Table S1) are the inverse of those found for the pairwise F_{ST} values and support the same pattern of isolation between areas, with the Northern area more distinct and less homogeneous than the others. Similarly, even though the *Nm* values are highly variable among all localities, inter-area values are much lower than within-area values supporting less connectivity and therefore genetic structure among the three areas. Some pairwise *Nm* values (*inf*) may indicate that those locality pairs are behaving as a single population.

Phylogenetic relationships among haplotypes inferred by ML and BI are shown in Figure 4. Both reconstructions retrieved all *H. hydnorae* haplotypes nested in a strongly supported clade (BP = 89, PP = 0.99) and are congruent with the topology resulting from the haplotype network, and largely in agreement with SAMOVA groupings (Figure 2A,B). These analyses provide high support for a split between most of the haplotypes of the NG (except for H24, H25, H29 and H30). The rest of haplotypes appeared in a clade with moderate support that corresponds with the groupings for CG and SG.

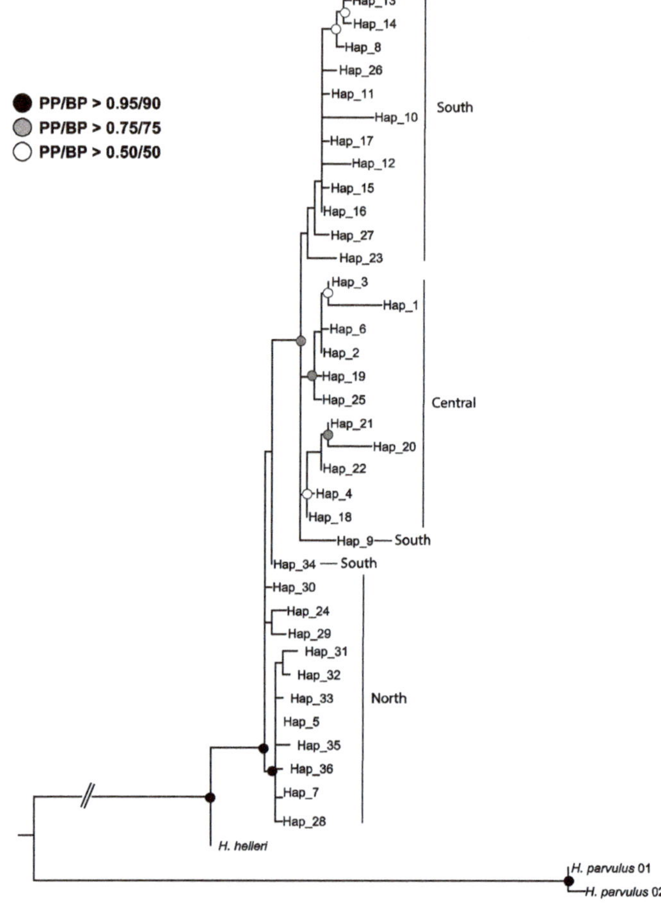

Figure 4. Bayesian inference (BI) and maximum likelihood (ML) topology illustrating relationships between individual haplotypes for *H. hydnorae*. Labeling of haplotypes by geographic group follows the regions depicted in Figure 2 and Table 1. *Hydnorobius helleri* and *H. parvulus* are used as outgroups. PP: posterior probabilities, PB: bootstrap resampling.

3.2. Weak Signals of Population Expansion for the Hydnorobius Hydnorae Population as a Whole

Considering all localities as a single population, the haplotype diversity (h) was 0.5714 and the nucleotide diversity (π) was 0.0185. This pattern of substantial haplotype diversity with moderate nucleotide diversity could be a signature of population growth from a smaller ancestral population size. The suggestion might be that since the origin of *H. hydnorae*, enough time has elapsed to produce some haplotype variation via mutation (h) but not enough for the accumulation of large differences in sequence (π). Table 4 shows the haplotype and nucleotide diversity for each locality. All localities present high values of haplotype diversity and low values of nucleotide diversity.

Table 4. Genetic diversity and Neutrality tests (Tajima's D_s and Fu's F_s) per *H. hydnorae* locality and per SAMOVA defined locality group. (*n*: number of individuals per locality; *h*: haplotype diversity; π: nucleotide diversity; *K*: average number of pairwise differences).

					Tajima's D_s		Fu's F_s	
Area/Population	*n*	*h*	*K*	π	D_s	*p* Value	F_s	*p* Value
LSR	4	0.8333	2.6667	0.0351	−0.3145	0.533	0.8114	0.568
SLQUI	5	0.8000	2.0000	0.0048	−1.1240	0.071	−1.0116	0.114
CCH	4	1.0000	5.1667	0.0123	−0.5281	0.453	−0.4805	0.205
CAB	5	0.4000	1.6000	0.0038	−1.0938	0.080	2.2024	0.830
PHAY	3	0.3333	0.0000	0.0000	0.0000	1	N/A	N/A
SAUHU	5	0.4000	1.8000	0.0043	1.5727	0.965	2.4285	0.859
SLU	1	N/A	N/A	N/A	N/A	N/A	N/A	N/A
CHTP	2	0.5000	0	0	0	1	N/A	N/A
North	29	0.4482	3.8276	0.0091	−1.0018	0.161	−3.7314	0.086
CBR	3	1.0000	4.0000	0.0526	0.0000	0.551	2.3031	0.543
CTI	4	0.5000	2.5000	0.0329	−0.7968	0.166	2.5980	0.859
SBE	3	0.6667	0.6667	0.0088	0.0000	1	0.0000	N. A.
SHU	3	0.8333	12.0833	0.1590	−1.4104	0.078	3.0688	0.092
Central	13	0.6923	5.4359	0.0129	−0.8445	0.208	−1.5152	0.201
MDI	5	0.9000	3.2000	0.0421	−0.8173	0.149	−1.0124	0.106
MNA	4	0.7000	1.4000	0.0184	0.0000	0.948	1.0609	0.561
MPD	5	0.8000	1.6000	0.0210	0.6990	0.785	0.2764	0.523
LPCHA	1	N/A	N/A	N/A	N/A	N/A	N/A	N/A
LPDZO	3	1	6.0000	0.0142	0.0000	1	0.5878	0.400
LPMAR	3	0.6667	2.0000	0.0047	0.0000	1	1.6094	0.701
South	22	0.6667	6.2905	0.0149	−0.9651	0.172	−3.4286	0.007
ALL	63	0.5714	7.8218	0.0185	−1.5263	0.111	−16.064	<0.05

Results of Tajima's D and Fu's F tests are shown in Table 4. For the entire sample, both Tajima's and Fu's neutrality tests were negative, with only the F_S index being significant ($D_S = -1.5263$, $p = 0.111$; $F_S = -16.064$, $p < 0.05$). Some of the individual localities presented negative values for these statistics, but none of them were significant. Given that many of the individual locality sample sizes are small, estimates of population size changes will be more accurately derived, in this case, from the analysis of locality groupings. Analyses of localities grouped according to the three SAMOVA groupings also produce negative neutrality indexes, with only one significant F_s value for SG.

The mismatch distribution analysis performed for the locality groupings presents clear evidence of stable demographic history (Figure 5A) with multimodal mismatch patterns and non-significant raggedness values (NG: $r = 0.087$, $p = 0.35$; CG: $r = 0.091$, $p = 0.59$; SG: $r = 0.009$, $p = 0.91$). However, the mismatch distribution analysis for the entire sample of this weevil species does not immediately suggest a stable demographic history, since it showed a tendency to a unimodal distribution (Figure 5A). Even though the raggedness value was low and non-significant ($r = 0.0063$; $p = 0.87$), the shape of the distribution could suggest that, as a whole, populations of this weevil species are not in demographic stability, probably reflecting some demographic expansion, as also suggested by the results of Fu's F tests for the whole group.

For the three genetic groups detected (NG, CG, SG), the BSPs suggest an initial period of stability between 600 and 200 kya, followed by weak growth in population size since ~200 kya until ~30 kya where a decrease in the effective size is detected (Figure 5B). This pattern is common for the three groups but more pronounced in SG and CG.

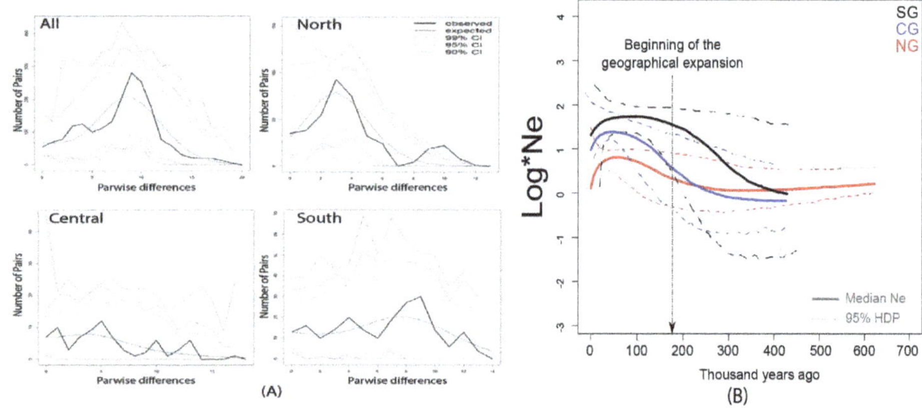

Figure 5. Estimates of demographic expansion in *Hydnorobius hydnorae*. (**A**) Mismatch distributions for all samples as a single population (ALL) and for each locality group analyzed as a single unit (North, Central, South). To construct the curves of expected values, 10,000 datasets were simulated under a coalescent algorithm using estimated parameters based on a sudden demographic expansion [63]; (**B**) Bayesian skyline plots for all locality groups including confidence intervals. Arrow indicates the time estimate for geographic expansions derived from the Bayesian spatial diffusion analysis.

3.3. Area of Origin and North-South Axis of Spatio-Temporal Diffusion for H. hydnorae

The spatial diffusion rate for the Cyt-B matrix for *H. hydnorae* was estimated as 2175 km/My (95% HPD = 1040 km/My, 3301 km/My). The RRW diffusion model inferred a first step of the expansion consisting of two simultaneous colonization paths towards the North and South from the area of origin in northwestern San Luis Province (32°44′ S 66°55′ W) beginning at around 206–143 kya (Figure 6A). The Northern colonization route split into two independent areas: towards the West reaching the Northern Monte, and towards the East reaching the southern edge of the Chaco biogeographic province. The Southern colonization route established the ancestors in the austral part of Northern Monte (Figure 6B).

After this initial expansion around 128–79 kya, the Northern groups expanded into multiple areas and at a faster rate than those from the South; this is especially true for those from the Northeast (Figure 6C).

Around 79 kya, the ancient Northern group would have covered most of Western Chaco and the Northern Monte regions, while the Southern group would have reached the southern end of the current distribution of *H. hydnorae* (Figure 6D). Since 63 kya to the present, the colonizations were directed to the Northwest of Argentina and from the center of the Argentinian Chaco to the center of Paraguay. During this last period there are no expansions towards the South, but instead re-colonizations of the austral areas of Northern Monte from the areas near the center of dispersion of the species (Figure 6E,F).

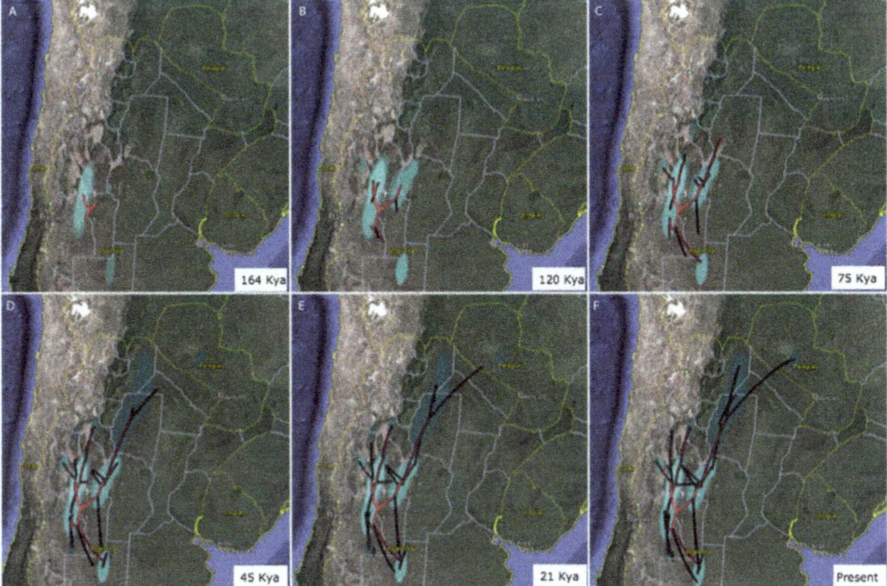

Figure 6. Reconstructed spatio-temporal diffusion of *H. hydnorae* in South America, shown at six time slices: (**A**) 164 kya. (**B**) 120 kya. (**C**) 75 kya. (**D**) 45 kya. (**E**) 21 kya. (**F**) Present time. Lines represent branches of the maximum clade credibility tree (MCC), estimated with a Bayesian phylogeographic analysis in BEAST using a "Relaxed Random Walk" (RRW) model of continuous diffusion through time and space. Map data ©2018 Google Imagery ©2018 NASA, TerraMetrics.

3.4. North-South Range Expansion in Prosopanche Americana, the Host of Hydnorobius hydnorae during 120 Kya

The average AUC obtained for the current climatic model (1959–2000) for *P. americana* was 0.9192 ($\pm$ 0.031), indicating an optimal performance of the MaxEnt algorithm. The paleodistribution obtained for *P. americana* at 120 Kya during the LIG suggests a very restricted distribution in the Northern Monte (Figure 7A). Predictions at the LGM, under the CCSM4 simulation, suggest a southeastern range expansion in the southern portion of Northern Monte and west of Pampean biogeographic province with high-suitability values, while the MIROC-ESM climatic model suggests a northeastern range expansion in a fragmented scenario in the Chaco biogeographic province (Figure 7B,C). During the Mid-Holocene, both models suggest continuous expansions to the Northeast and Southeast (Figure 7D,E) approaching the current distribution of *P. americana* (Figure 7F). The PDM to current climatic conditions occupies a larger high-favorability area than at the LGM but smaller than in Mid-Holocene, suggesting range fragmentation mainly in the northern portion of the distribution in the Chaco region (Figure 7F).

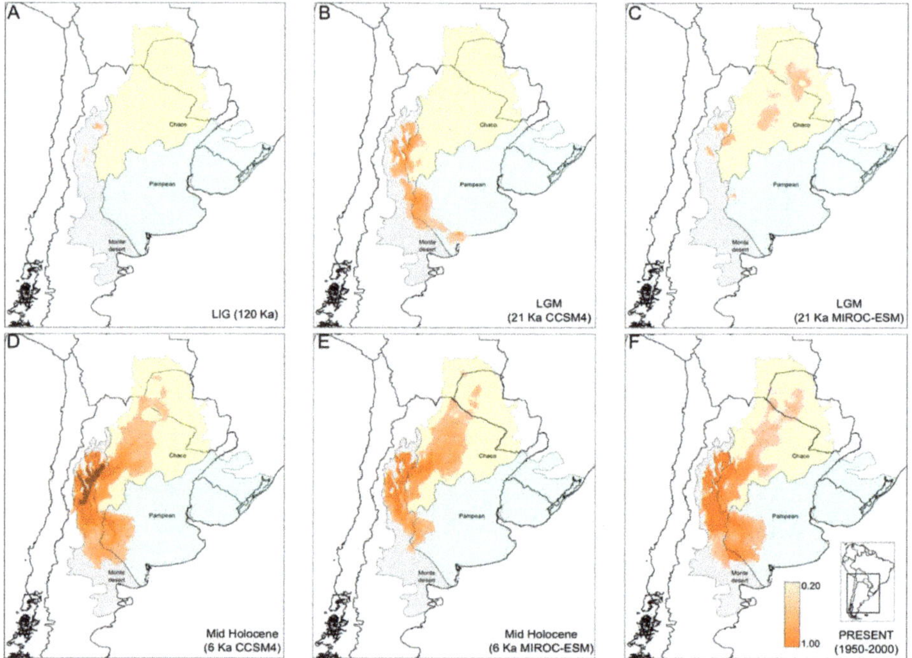

Figure 7. Spatial projections of *Prosopanche americana* climatic niche across several Quaternary climatic scenarios. (**A**) Last Interglacial maximum (LIG; 120 kya; CCMS); (**B**) Last Glacial Maximum (LGM; 21 kya; CCMS4); (**C**) Last Glacial Maximum (LGM; 21 kya; MIROC-ESM); (**D**) Mid-Holocene (6 kya, CCMS4); (**E**) Mid-Holocene (6 kya, MIROC-ESM); (**F**) Current conditions (average 1950–2000). Dotted lines indicate biogeographical regionalization and orange hues signals the climatic suitability for *P. americana*.

4. Discussion

4.1. Genetic Structure and Geographic Expansions without Major Demographic Change Across the Range of Hydnorobius hydnorae

The weevil *H. hydnorae* is a univoltine beetle specifically associated to its host plant *P. americana*, [11,64] which in turn is parasite on the roots of *Prosopis* spp. ("Algarrobo" trees, Fabaceae) in arid and semiarid regions of the Monte, Chaco and Pampean biogeographic provinces. Although weevils of this species can fly, they do not present high vagility. Likewise, the dispersal capacity of *P. americana* may also be rather low, by means of endozoochory, carried by nocturnal mammals that eat the fruits [17]. Emergence of adults of *H. hydnorae* starting a new generation coincides with the emergence of new plants. The weevil's low vagility and restrained biological habits may provide an explanation for the high degree of genetic structuring found across the range of *H. hydnorae*. Interestingly, the Northern haplogroup is the most structured and most distinct from the Central and Southern haplogroups, while it is also the one showing some early signals of rather weak population growth. However, most of the geographic expansion occurred after the growth pulses (220 kya), when there was effectively no growth occurring or even in the face of size reductions (30 kya). We find it intriguing that geographic expansion from ancestral areas to the edges of the distributions (see below) is uncoupled in time from any substantial demographic expansion. Studies on other species in the Monte desert describe the opposite pattern: demographic expansions occurring after sustained periods of range expansion [53].

The vicariant event described by Flores and Roig-Juñent [18] proposes a split of the Northern Monte from the remaining Southern areas, isolating Patagonia from Northern areas in Argentina. They attribute this event to marine transgressions that occurred 9.55 to 9.11 Mya, which are suggested by several sources of evidence [65]. Molecular phylogenetics for the Belidae estimated the age of origin of this weevil at about ~10 Mya [66,67]; thus, marine transgressions that occurred at ~9 Mya could have affected the distribution of *H. hydnorae*, generating a northern and southern pattern of distribution. This pattern of distribution could have also been later affected by volcanic and glacial periods, as has been proposed for other animals [68].

The sampled range of *H. hydnorae* spans the defined districts of Northern Monte, Western or Dry Chaco and Espinal of the Monte, Chaco and Pampean biogeographical provinces. Further subdivisions of the Monte area have been proposed based on vegetation [19] and in concordance with entomological evidence [18]. The phylogenetic relationships between *H. hydnorae* haplotypes, as well as the current structuring of the variation of *H. hydnorae* haplotypes into the three locality groups of NG, CG and SG, are quite concordant with current biogeographic regionalization. The Northern haplogroup occurs exclusively in the Chaco province, namely in the Dry Chaco, while the Central haplogroup resides mostly in the Northern Monte [19] and the Southern haplogroup occurs southward and eastward in Northern Monte and Espinal of the Pampean biogeographic province. These areas show a climatic transition from subtropical to temperate [69]. Separations such as the one we observe for *H. hydnorae* between the Central and Southern locality groups within the Monte region, north and south of 35° S, are common to a varied array of Monte inhabitants such as lizards, parrots and plants [26,70,71]. This is, to our knowledge, the first such example from an insect. The co-occurrence of such a genetic break in a disparate set of taxa has prompted suggestions of a shared regional history underlying these microevolutionary patterns, as well as those at a macroevolutionary level [26,72]. The Quarternary climatic history of the region includes severe glaciation patterns and shifts in the boundaries of ecotones [73]. On the other hand, the separation between the Northern and Central locality groups, each housed in the Dry Chaco and Northern Monte, respectively, could be mediated, in part, by the presence of the Famatina–Sañogasta Mountains, as observed for other Monte dwellers [68,74] and therefore be more independent of climatic patterns. However, a study on turtles that also finds the Northern Monte/Dry Chaco split has linked the separation to a vicariant event rooted in Plio-Pleistoce climatic changes [75].

Alternatively, the Northern area could have acted as a northern refuge from glacial periods (Pleistocene: 1.81–0.01 Mya) or Miocene periods of volcanic activity [76–78]. The idea of the Northern localities acting as refugia may be supported by the location of the ancestral haplotypes in the area, as well as by the location of the ancestral area selected by the Bayesian diffusion analysis.

4.2. Ancestral Weevil Haplotypes and Ancestral Areas for Hydnorobius hydnorae and Its Host Plant

In populations that present scarce or limited gene flow, the oldest and ancestral haplotypes are expected to be those with the most widespread geographic range [23]. Conversely, the haplotypes restricted to a single area are considered to have a recent origin [79]. Haplotype networks for *H. hydnorae* indicate that the most probable ancestral haplotypes are either H16 or H5, given that they are widespread across multiple localities and at the center of star-like topologies in the haplotype network in the SG and NG respectively. H16 is the most widespread SG southern haplotype distributed from the localities of *El Durazno* and *La Maruja* in La Pampa to *Paso del Loro* at the southern tip of Mendoza province (Figure 2A). H5 also is at the center of a star like structure (six other haplotypes derive from it) and is the most widely distributed NG haplotype present in Quines in San Luis, Chancaní in Córdoba and San Ramón in La Rioja, all localities within the Chaco Biogeographic province.

Bayesian diffusion analysis provides additional clues regarding the area of first expansion of *H. hydnorae*. The RRW model suggests that diffusion started at around 206–143 kya from northwestern San Luis province located within the Northern area, in agreement with the location of the more Northern putative ancestral haplotypes (H5). Despite the potential ancestral condition of H16, the RRW analyses that explicitly integrate geographic locations and coalescent history of haplotypes indicate that the ancestral area is further north with a higher posterior probability, compared to that of the area of prevalence of H16.

The predicted past distribution of the host plant *P. americana* during the LIG 130 kya shows that the most probable area of occurrence for this species in the north of the biogeographic region of Monte desert and close to the western edge of the Chaco region. The past distribution of *P. americana* and the ancestral range of *H. hydnorae* show a high degree of overlap suggesting an ancient association within concordant ancient ranges, as was expected under a coevolutionary or co-dispersal scenario between the weevil and its host plant.

In addition, information from paleoclimatic conditions and the fossil record indicates that plants of genus *Prosopis* (Fabaceae), the host of *P. americana*, were abundant during the Miocene (5–23 Mya) and Lower Pliocene (1–5 Mya) [80,81] in Northern Argentina. In essence, evidence suggests that the evolution and diversification of *Prosopis* took place jointly with the expansion of arid areas in the American continent, after the Andean uplift in the late Miocene [82]. This is because the uplift of the Andes caused the blocking of the more humid winds [83,84] and the expansion of xeric habitats. This provides evidence for the long-term persistence of a "three-way" interspecies interaction (*Prosopis-Prosopanche-Hydnorobius*) in the Monte desert. Some other enduring two-way associations between insects and South American plants have been reported for weevils and beetles feeding on relictual ancient conifers [85–89], and for two oil-collecting bee species and the perennial endemic herbs they pollinate [6]. More generally, an old history of specialized insect-plant interactions has been suggested as a main contributing factor to current biodiversity in the Patagonian region [90].

4.3. Concordant Weevil and Host Plant Diffusion-Expansion Patterns

Despite the caveats of the particular models used to generate either dispersal patterns or niche projections [91], we see the dispersal of weevil and host plant across space and time as interestingly synchronized. Both members of the interaction have concomitantly broadened their range from a common ancestral area following a Northern and a Southern track. If we were to think of *Prosopanche* flowers as *Hydnorobius* weevils' habitat, then a specialist species adapted to this moving habitat must track its habitat spatially if it is to persist [92]. Nevertheless, tracking the host plant in its dispersal does not preclude the original locations from continuing being occupied, so in essence what we are detecting are not range shifts but matching ranges, not an unusual result in specialist interactions [93].

However, long-term co-dispersal seems to be less common. Passive co-dispersal of mutualists has been observed in other insect-plant interactions, such as ants and mealybugs [94]. There seems to be no evidence that *Hydnorobius* adults or larvae are passively dispersed with *Prosopanche*. Given the weevil life cycle, so tightly linked to the host-plant as an obligate relationship at least for the weevil (which is entirely dependent on the *Prosopanche* for feeding, mating and larval development), we are inclined to suggest that active host tracking by *Hydnorobius* adults has led to the observed matching ranges. Similar generation times and modest dispersal capacity for both interacting species can further contribute to concordant dispersal patterns through space and time [95].

Even though episodes of environmental change have been suggested to create opportunities for host-switching during geographical expansion in host-parasite systems [96], it appears that host switching during the recorded history of geographic expansion and environmental changes in *Hydnorobius* did not occur, or occurred just to an extremely similar and phylogenetically close species such as *P. bonacinai* [11]. Rather, what we observe is a long trajectory of host-tracking through space and time, where the weevil has expanded its geographic range following its host plant, but without significant demographic growth. Other monophagous insects show local population size closely following the cover of its food plant, so that host plant density could be a reliable prognosticator of the population size of the specialist insect [97,98]. Additionally, insect expansion rates have been suggested to be likely to increase with habitat availability [99]. One potential explanation for geographic dispersal without any substantial population growth in *Hydnorobius* could be that the scarcity of the host plant itself allows for maintenance of slow expansion rates and stable populations, with no need for significant demographic growth pulses to support geographic range expansion.

5. Conclusions

Genetic structuring of *H. hydnorae* populations across Northern Argentina appears to have arisen through a combination of the weevil's low vagility and specialist larval feeding habits, with cycles of historical climatic changes. Such an obligate association has persisted across glacial cycles, generating a close match in the dispersal histories of *H. hydnorae* and its host plant *P. americana* in space and time, illustrating a long standing dependent association. Similarly, aspects of the population biology, ecology and life history of the host plant itself appear to influence the historical demography of the weevil allowing for range expansion without any substantial population growth.

Supplementary Materials: The following are available online at http://www.mdpi.com/1424-2818/10/2/33/s1, Table S1: Pairwise *Nm* estimates among 18 localities of *H. hydnorae*. Inter-area contrasts are not shaded while those within areas are shaded by locality area as defined through SAMOVA analysis (North: light gray; Central: dark gray; South: medium gray).

Author Contributions: A.S.S., A.E.M. and M.S.F. conceived and designed the experiments; M.S.F. and N.R. performed the experiments; A.S.S., M.S.F., N.R. and M.C.B. analyzed the data; A.E.M. and A.S.S. contributed reagents and materials; A.S.S., N.R., A.E.M., M.C.B. and M.S.F. wrote the paper.

Acknowledgments: We gratefully acknowledge the laboratory assistance of M. Sijapati, M. Maraorgarti, P. Mandal (Wellesley College) and L. Caeiro (Laboratorio de Biología Molecular, Universidad Nacional de Córdoba). This work was supported with Brachman Hoffman funds through Wellesley College (M.S.F. and A.S.S.) and through a travel award to M.S.F. from the Society of Systematic Biologists. This research received continued support from the National Scientific and Technical Research Council (CONICET, Argentina) through doctoral (to N.R. and M.S.F.) and postdoctoral (M.C.B.) fellowships, and research grants (PIPs 6766 and 00162 to A.E.M. and PIP 00765 to A. Cocucci) and by the National Agency of Promotion of Science (ANPCyT, Argentina) through grant PICTs 2011-2573 and 2016-2798 (to A.E.M.) and PICT-2015-3325 (to A. Cocucci). Additional aid to field and lab work of N.R. and M.C.B. was received from ANPCyT grant PICT 2015-3089 (to A. Sércic). We are very thankful to A. Cocucci for providing Figure 1C and assisting with questions regarding flower morphology and pollination. Our gratitude also goes to many people who assisted in field trips and/or provided specimens for molecular work, especially M. F. Fernández-Campón, D. R. Maddison, M. B. Maldonado, F. C. Ocampo, S. A. Roig-Juñent, G. Salazar. All authors acknowledge research facilities and support from their respective institutions.

Conflicts of Interest: The authors declare no conflict of interest. The founding sponsors had no role in the design of the study; in the collection, analyses, or interpretation of data; in the writing of the manuscript, and in the decision to publish the results.

References

1. Futuyma, D.J.; Agrawal, A.A. Macroevolution and the biological diversity of plants and herbivores. *Proc. Natl. Acad. Sci. USA* **2009**, *106*, 18054–18061. [CrossRef] [PubMed]
2. Toju, H.; Sota, T. Phylogeography and the geographic cline in the armament of a seed-predatory weevil: Effects of historical events vs. natural selection from the host plant. *Mol. Ecol.* **2006**, *15*, 4161–4173. [CrossRef] [PubMed]
3. De-la-Mora, M.; Piñero, D.; Oyama, K.; Farrell, B.; Magallón, S.; Núñez-Farfán, J. Evolution of Trichobaris (Curculionidae) in relation to host plants: Geometric morphometrics, phylogeny and phylogeography. *Mol. Phylogenet. Evol.* **2018**, *124*, 37–49. [CrossRef] [PubMed]
4. Alvarez, N.; Kjellberg, F.; Mckey, D.; Hossaert-McKey, M. Phylogeography and historical biogeography of obligate specific mutualisms. In *The Biogeography of Host-Parasite Interactions*; Oxford University Press: Oxford, UK, 2010; pp. 31–39. ISBN 978-0-19-956134-6.
5. Gavin, D.G.; Fitzpatrick, M.C.; Gugger, P.F.; Heath, K.D.; Rodríguez-Sánchez, F.; Dobrowski, S.Z.; Hampe, A.; Hu, F.S.; Ashcroft, M.B.; Bartlein, P.J. Climate refugia: Joint inference from fossil records, species distribution models and phylogeography. *New Phytol.* **2014**, *204*, 37–54. [CrossRef] [PubMed]
6. Sosa-Pivatto, M.; Cosacov, A.; Baranzelli, M.C.; Iglesias, M.R.; Espíndola, A.; Sérsic, A.N. Do 120,000 years of plant–pollinator interactions predict floral phenotype divergence in *Calceolaria polyrhiza*? A reconstruction using species distribution models. *Arthropod-Plant Interact.* **2017**, *11*, 351–361. [CrossRef]

7. Thompson, A.R.; Thacker, C.E.; Shaw, E.Y. Phylogeography of marine mutualists: Parallel patterns of genetic structure between obligate goby and shrimp partners. *Mol. Ecol.* **2005**, *14*, 3557–3572. [CrossRef] [PubMed]
8. Valiente-Banuet, A.; Rumebe, A.V.; Verdú, M.; Callaway, R.M. Modern Quaternary plant lineages promote diversity through facilitation of ancient Tertiary lineages. *Proc. Natl. Acad. Sci. USA* **2006**, *103*, 16812–16817. [CrossRef] [PubMed]
9. Kuschel, G. *Oxycorynus missionis* spec. nov. from NE Argentina, with key to the South American species of Oxycoryninae (Coleoptera: Belidae). *Acta Zool. Lilloana* **1995**, *43*, 45–48.
10. Kuschel, G. Nemonychidae, Belidae y Oxycorynidae de la fauna chilena, con algunas consideraciones biogeográficas. *Investig. Zool. Chile* **1959**, *5*, 229–271.
11. Ferrer, M.S.; Marvaldi, A.E. New host plant and distribution records for weevils of the genus Hydnorobius (Coleoptera: Belidae). *Rev. Soc. Entomol. Argent.* **2010**, *69*, 271–274.
12. Nickrent, D.L.; Blarer, A.; Qiu, Y.-L.; Soltis, D.E.; Soltis, P.S.; Zanis, M. Molecular data place Hydnoraceae with Aristolochiaceae. *Am. J. Bot.* **2002**, *89*, 1809–1817. [CrossRef] [PubMed]
13. Naumann, J.; Salomo, K.; Der, J.P.; Wafula, E.K.; Bolin, J.F.; Maass, E.; Frenzke, L.; Samain, M.-S.; Neinhuis, C.; Wanke, S. Single-copy nuclear genes place haustorial Hydnoraceae within Piperales and reveal a Cretaceous origin of multiple parasitic angiosperm lineages. *PLoS ONE* **2013**, *8*, e79204. [CrossRef] [PubMed]
14. Massoni, J.; Forest, F.; Sauquet, H. Increased sampling of both genes and taxa improves resolution of phylogenetic relationships within Magnoliidae, a large and early-diverging clade of angiosperms. *Mol. Phylogenet. Evol.* **2014**, *70*, 84–93. [CrossRef] [PubMed]
15. Byng, J.W.; Chase, M.W.; Christenhusz, M.J.; Fay, M.F.; Judd, W.S.; Mabberley, D.J.; Sennikov, A.N.; Soltis, D.E.; Soltis, P.S.; Stevens, P.F. An update of the Angiosperm Phylogeny Group classification for the orders and families of flowering plants: APG IV. *Bot. J. Linn. Soc.* **2016**, *181*, 1–20.
16. Cocucci, A.E. Estudios en el género Prosopanche (Hydnoraceae). I. Revisión taxonómica. *Kurtziana* **1965**, *2*, 53–74.
17. Cocucci, A.E.; Cocucci, A.A. Prosopanche (Hydnoraceae): Somatic and reproductive structures, biology, systematics, phylogeny and potentialities as a parasitic weed. In *Congresos y Jornadas-Junta de Andalucía (España)*; JA, DGIA: New York, NY, USA, 1996.
18. Roig-Juñent, S.; Flores, G.; Claver, S.; Debandi, G.; Marvaldi, A. Monte Desert (Argentina): Insect biodiversity and natural areas. *J. Arid Environ.* **2001**, *47*, 77–94. [CrossRef]
19. Roig, F.A.; Roig-Juñent, S.; Corbalán, V. Biogeography of the Monte desert. *J. Arid Environ.* **2009**, *73*, 164–172. [CrossRef]
20. Vogt, C. Composición de la flora vascular del Chaco Boreal, Paraguay III. Dicotyledoneae: Gesneriaceae–Zygophyllaceae. *Steviana* **2013**, *5*, 5–40.
21. Bruch, C. Coleópteros fertilizadores de "Prosopanche Burmeisteri" De Bary. *Physis* **1923**, *7*, 82–88.
22. Marvaldi, A.E. Larval morphology and biology of oxycorynine weevils and the higher phylogeny of Belidae (Coleoptera, Curculionoidea). *Zool. Scr.* **2005**, *34*, 37–48. [CrossRef]
23. Avise, J.C. *Phylogeography: The History and Formation of Species*; Harvard University Press: Cambridge, MA, USA, 2000; ISBN 0-674-66638-0.
24. Sérsic, A.N.; Cosacov, A.; Cocucci, A.A.; Johnson, L.A.; Pozner, R.; Avila, L.J.; Sites, J.W.; Morando, M. Emerging phylogeographical patterns of plants and terrestrial vertebrates from Patagonia. *Biol. J. Linn. Soc.* **2011**, *103*, 475–494. [CrossRef]
25. Turchetto-Zolet, A.C.; Pinheiro, F.; Salgueiro, F.; Palma-Silva, C. Phylogeographical patterns shed light on evolutionary process in South America. *Mol. Ecol.* **2013**, *22*, 1193–1213. [CrossRef] [PubMed]
26. Baranzelli, M.C.; Cosacov, A.; Ferreiro, G.; Johnson, L.A.; Sérsic, A.N. Travelling to the south: Phylogeographic spatial diffusion model in *Monttea aphylla* (Plantaginaceae), an endemic plant of the Monte Desert. *PLoS ONE* **2017**, *12*, e0178827. [CrossRef] [PubMed]
27. Morrone, J.J. Biogeographical regionalisation of the Neotropical region. *Zootaxa* **2014**, *3782*, 1–110. [CrossRef] [PubMed]
28. Rokas, A.; Ladoukakis, E.; Zouros, E. Animal mitochondrial DNA recombination revisited. *Trends Ecol. Evol.* **2003**, *18*, 411–417. [CrossRef]
29. Kraytsberg, Y.; Schwartz, M.; Brown, T.A.; Ebralidse, K.; Kunz, W.S.; Clayton, D.A.; Vissing, J.; Khrapko, K. Recombination of human mitochondrial DNA. *Science* **2004**, *304*, 981. [CrossRef] [PubMed]

30. Avise, J.C.; Arnold, J.; Ball, R.M.; Bermingham, E.; Lamb, T.; Neigel, J.E.; Reeb, C.A.; Saunders, N.C. Intraspecific phylogeography: The mitochondrial DNA bridge between population genetics and systematics. *Annu. Rev. Ecol. Syst.* **1987**, *18*, 489–522. [CrossRef]
31. Crozier, R.H.; Crozier, Y.C. The cytochrome b and ATPase genes of honeybee mitochondrial DNA. *Mol. Biol. Evol.* **1992**, *9*, 474–482. [PubMed]
32. Simon, C.; Frati, F.; Beckenbach, A.; Crespi, B.; Liu, H.; Flook, P. Evolution, weighting, and phylogenetic utility of mitochondrial gene sequences and a compilation of conserved polymerase chain reaction primers. *Ann. Entomol. Soc. Am.* **1994**, *87*, 651–701. [CrossRef]
33. Hall, T.A. BioEdit: A user-friendly biological sequence alignment editor and analysis program for Windows 95/98/NT. In *Nucleic Acids Symposium Series*; Information Retrieval Ltd.: London, UK, 1999; Volume 41, pp. 95–98.
34. Rozas, J.; Sánchez-DelBarrio, J.C.; Messeguer, X.; Rozas, R. DnaSP, DNA polymorphism analyses by the coalescent and other methods. *Bioinformatics* **2003**, *19*, 2496–2497. [CrossRef] [PubMed]
35. Bandelt, H.-J.; Forster, P.; Röhl, A. Median-joining networks for inferring intraspecific phylogenies. *Mol. Biol. Evol.* **1999**, *16*, 37–48. [CrossRef] [PubMed]
36. Leigh, J.W.; Bryant, D. Popart: Full-feature software for haplotype network construction. *Methods Ecol. Evol.* **2015**, *6*, 1110–1116. [CrossRef]
37. Dupanloup, I.; Schneider, S.; Excoffier, L. A simulated annealing approach to define the genetic structure of populations. *Mol. Ecol.* **2002**, *11*, 2571–2581. [CrossRef] [PubMed]
38. Excoffier, L.; Lischer, H.E.L. Arlequin suite ver 3.5: A new series of programs to perform population genetics analyses under Linux and Windows. *Mol. Ecol. Resour.* **2010**, *10*, 564–567. [CrossRef] [PubMed]
39. Flores, G.E.; Roig-Juñent, S. Cladistic and biogeographic analyses of the Neotropical genus Epipedonota Solier (Coleoptera: Tenebrionidae), with conservation considerations. *J. N. Y. Entomol. Soc.* **2001**, *109*, 309–336. [CrossRef]
40. Drummond, A.J.; Rambaut, A. BEAST: Bayesian evolutionary analysis by sampling trees. *BMC Evol. Biol.* **2007**, *7*, 214. [CrossRef] [PubMed]
41. Darriba, D.; Taboada, G.L.; Doallo, R.; Posada, D. jModelTest 2: More models, new heuristics and parallel computing. *Nat. Methods* **2012**, *9*, 772. [CrossRef] [PubMed]
42. Tracer. Available online: http://tree.bio.ed.ac.uk/software/tracer/ (accessed on 23 March 2018).
43. FigTree. Available online: http://tree.bio.ed.ac.uk/software/figtree/ (accessed on 23 March 2018).
44. Guindon, S.; Dufayard, J.-F.; Lefort, V.; Anisimova, M.; Hordijk, W.; Gascuel, O. New Algorithms and Methods to Estimate Maximum-Likelihood Phylogenies: Assessing the Performance of PhyML 3.0. *Syst. Biol.* **2010**, *59*, 307–321. [CrossRef] [PubMed]
45. Nei, M.; Kumar, S. *Molecular Evolution and Phylogenetics*; Oxford University Press: Oxford, UK, 2000; ISBN 0-19-535051-0.
46. Tajima, F. Statistical method for testing the neutral mutation hypothesis by DNA polymorphism. *Genetics* **1989**, *123*, 585–595. [PubMed]
47. Harpending, H.C. Signature of ancient population growth in a low-resolution mitochondrial DNA mismatch distribution. *Hum. Biol.* **1994**, *66*, 591–600. [PubMed]
48. Mckenna, D.D.; Wild, A.L.; Kanda, K.; Bellamy, C.L.; Beutel, R.G.; Caterino, M.S.; Farnum, C.W.; Hawks, D.C.; Ivie, M.A.; Jameson, M.L.; et al. The beetle tree of life reveals that Coleoptera survived end-Permian mass extinction to diversify during the Cretaceous terrestrial revolution. *Syst. Entomol.* **2015**, *40*, 835–880. [CrossRef]
49. Zhang, S.-Q.; Che, L.-H.; Li, Y.; Pang, H.; Ślipiński, A.; Zhang, P. Evolutionary history of Coleoptera revealed by extensive sampling of genes and species. *Nat. Commun.* **2018**, *9*, 205. [CrossRef] [PubMed]
50. Lemey, P.; Rambaut, A.; Welch, J.J.; Suchard, M.A. Phylogeography Takes a Relaxed Random Walk in Continuous Space and Time. *Mol. Biol. Evol.* **2010**, *27*, 1877–1885. [CrossRef] [PubMed]
51. Pybus, O.G.; Tatem, A.J.; Lemey, P. Virus evolution and transmission in an ever more connected world. *Proc. R. Soc. B* **2015**, *282*, 20142878. [CrossRef] [PubMed]
52. Drummond, A.J.; Ho, S.Y.W.; Phillips, M.J.; Rambaut, A. Relaxed Phylogenetics and Dating with Confidence. *PLoS Biol.* **2006**, *4*, e88. [CrossRef] [PubMed]

53. Camargo, A.; Werneck, F.P.; Morando, M.; Sites, J.W.; Avila, L.J. Quaternary range and demographic expansion of *Liolaemus darwinii* (Squamata: Liolaemidae) in the Monte Desert of Central Argentina using Bayesian phylogeography and ecological niche modelling. *Mol. Ecol.* **2013**, *22*, 4038–4054. [CrossRef] [PubMed]
54. Bielejec, F.; Rambaut, A.; Suchard, M.A.; Lemey, P. SPREAD: Spatial phylogenetic reconstruction of evolutionary dynamics. *Bioinformatics* **2011**, *27*, 2910–2912. [CrossRef] [PubMed]
55. Guisan, A.; Thuiller, W. Predicting species distribution: Offering more than simple habitat models. *Ecol. Lett.* **2005**, *8*, 993–1009. [CrossRef]
56. Graham, C.H.; Ron, S.R.; Santos, J.C.; Schneider, C.J.; Moritz, C. Integrating phylogenetics and environmental niche models to explore speciation mechanisms in dendrobatid frogs. *Evolution* **2004**, *58*, 1781–1793. [CrossRef] [PubMed]
57. Naumann, J.; Der, J.P.; Wafula, E.K.; Jones, S.S.; Wagner, S.T.; Honaas, L.A.; Ralph, P.E.; Bolin, J.F.; Maass, E.; Neinhuis, C. Detecting and characterizing the highly divergent plastid genome of the nonphotosynthetic parasitic plant *Hydnora visseri* (Hydnoraceae). *Genome Biol. Evol.* **2016**, *8*, 345–363. [CrossRef] [PubMed]
58. WorldClim—Global Climate Data | Free climate data for ecological modeling and GIS. Available online: http://www.worldclim.org/ (accessed on 22 March 2018).
59. Hijmans, R.J.; Cameron, S.E.; Parra, J.L.; Jones, P.G.; Jarvis, A. Very high resolution interpolated climate surfaces for global land areas. *Int. J. Climatol.* **2005**, *25*, 1965–1978. [CrossRef]
60. Phillips, S.J.; Anderson, R.P.; Schapire, R.E. Maximum entropy modeling of species geographic distributions. *Ecol. Model.* **2006**, *190*, 231–259. [CrossRef]
61. Anderson, R.P.; Gonzalez, I., Jr. Species-specific tuning increases robustness to sampling bias in models of species distributions: An implementation with Maxent. *Ecol. Model.* **2011**, *222*, 2796–2811. [CrossRef]
62. Otto-Bliesner, B.L.; Marshall, S.J.; Overpeck, J.T.; Miller, G.H.; Hu, A. Simulating Arctic climate warmth and Icefield retreat in the last interglaciation. *Science* **2006**, *311*, 1751–1753. [CrossRef] [PubMed]
63. Schneider, S.; Excoffier, L. Estimation of past demographic parameters from the distribution of pairwise differences when the mutation rates vary among sites: Application to human mitochondrial DNA. *Genetics* **1999**, *152*, 1079–1089. [PubMed]
64. Marvaldi, A.E.; Oberprieler, R.G.; Lyal, C.H.C.; Bradbury, T.; Anderson, R.S. Phylogeny of the Oxycoryninae sensu lato (Coleoptera: Belidae) and evolution of host-plant associations. *Invertebr. Syst.* **2006**, *20*, 447–476. [CrossRef]
65. Werneck, F.P. The diversification of eastern South American open vegetation biomes: Historical biogeography and perspectives. *Quat. Sci. Rev.* **2011**, *30*, 1630–1648. [CrossRef]
66. Ferrer, M.S. Molecular Systematics and Evolution of Belidae, with Special Reference to Oxycoryninae (Coleoptera: Curculionoidea). Ph.D. Thesis, Universidad Nacional de Cuyo, Mendoza, Argentina, 2011.
67. Ferrer, M.S.; Sequeira, A.S.; Marvaldi, A.E. Host associations in ancient weevils: A phylogenetic perspective on Belidae and Nemonychidae. *Diversity*. Under preparation.
68. Yoke, M.M.; Morando, M.; Avila, L.J.; Sites, J.W., Jr. Phylogeography and genetic structure in the *Cnemidophorus longicauda* complex (Squamata, Teiidae). *Herpetologica* **2006**, *62*, 420–434. [CrossRef]
69. Morrone, J.J. Biogeographic Areas and Transition Zones of Latin America and the Caribbean Islands Based on Panbiogeographic and Cladistic Analyses of the Entomofauna. *Annu. Rev. Entomol.* **2006**, *51*, 467–494. [CrossRef] [PubMed]
70. Morando, M.; Avila, L.J.; Baker, J.; Sites, J.W., Jr. Phylogeny and phylogeography of the *Liolaemus darwinii* complex (Squamata: Liolaemidae): Evidence for introgression and incomplete lineage sorting. *Evolution* **2004**, *58*, 842–861. [CrossRef] [PubMed]
71. Masello, J.F.; Quillfeldt, P.; Munimanda, G.K.; Klauke, N.; Segelbacher, G.; Schaefer, H.M.; Failla, M.; Cortés, M.; Moodley, Y. The high Andes, gene flow and a stable hybrid zone shape the genetic structure of a wide-ranging South American parrot. *Front. Zool.* **2011**, *8*, 16. [CrossRef] [PubMed]
72. Vuilleumier, B.S. Pleistocene changes in the fauna and flora of South America. *Science* **1971**, *173*, 771–780. [CrossRef] [PubMed]
73. Markgraf, V. Late and postglacial vegetational and paleoclimatic changes in subantarctic, temperate, and arid environments in Argentina. *Palynology* **1983**, *7*, 43–70. [CrossRef]

74. Rivera, P.C.; González-Ittig, R.E.; Robainas Barcia, A.; Trimarchi, L.I.; Levis, S.; Calderón, G.E.; Gardenal, C.N. Molecular phylogenetics and environmental niche modeling reveal a cryptic species in the *Oligoryzomys flavescens* complex (Rodentia, Cricetidae). *J. Mammal.* **2018**, *99*, 363–376. [CrossRef]
75. Sánchez, J. Variabilidad Genética, Distribución y Estado de Conservación de Las Poblaciones de Tortugas Terrestres *Chelonoidis chilensis* (Testudines: Testudinidae) Que Habitan en la República Argentina. Ph.D. Thesis, Universidad Nacional de La Plata, Buenos Aires, Argentina, 2013.
76. Nullo, F.E.; Stephens, G.C.; Otamendi, J.; Baldauf, P.E. El volcanismo del Terciario superior del sur de Mendoza. *Rev. Asoc. Geol. Argent.* **2002**, *57*, 119–132.
77. Sruoga, P.; Guerstein, P.; Bermudez, A. Riesgo volcánico. In *XII Congreso Geológico Argentino*; Ramos, V., Ed.; Asociación Geológica Argentina: Mendoza, Argentina, 1993; pp. 659–667.
78. Ortiz-Jaureguizar, E.; Cladera, G.A. Paleoenvironmental evolution of southern South America during the Cenozoic. *J. Arid Environ.* **2006**, *66*, 498–532. [CrossRef]
79. Neigel, J.E.; Ball, R.M.; Avise, J.C. Estimation of single generation migration distances from geographic variation in animal mitochondrial DNA. *Evolution* **1991**, *45*, 423–432. [CrossRef] [PubMed]
80. Anzótegui, L.M.; Garralla, S.S.; Herbst, R. Fabaceae de la Formación El Morterito (Mioceno superior) del valle del Cajón, provincia de Catamarca, Argentina. *Ameghiniana* **2007**, *44*, 183–196.
81. Anzótegui, L.M.; Horn, Y.; Herbst, R. Paleoflora (Fabaceae y Anacardiaceae) de la Formación Andalhuala (Plioceno Inferior), provincia de Catamarca, Argentina. *Ameghiniana* **2007**, *44*, 525–535.
82. Catalano, S.A.; Vilardi, J.C.; Tosto, D.; Saidman, B.O. Molecular phylogeny and diversification history of Prosopis (Fabaceae: Mimosoideae). *Biol. J. Linn. Soc.* **2008**, *93*, 621–640. [CrossRef]
83. Pascual, R.; Ortiz Jaureguizar, E.; Prado, J.L. Land mammals: Paradigm for Cenozoic South American geobiotic evolution. *Münch. Geowiss. Abh.* **1996**, *30*, 265–319.
84. Alberdi, M.T.; Bonnadona, F.P.; Ortiz Jaureguizar, E. Chronological correlation, paleoecology, and paleobiogeography of the late Cenozoic South American Rionegran land-mammal fauna: A review. *Rev. Esp. Palent.* **1997**, *12*, 249–255.
85. Kuschel, G.; Poinar, G.O. *Libanorhinus succinus* gen. & sp. n. (Coleoptera: Nemonychidae) from Lebanese amber. *Insect Syst. Evol.* **1993**, *24*, 143–146.
86. Kuschel, G.; May, B.M. Discovery of Palophaginae (Coleoptera: Megalopodidae) on Araucaria araucana in Chile and Argentina. *N. Z. Entomol.* **1996**, *19*, 1–13. [CrossRef]
87. Farrell, B.D. "Inordinate fondness" explained: Why are there so many beetles? *Science* **1998**, *281*, 555–559. [CrossRef] [PubMed]
88. Sequeira, A.S.; Normark, B.B.; Farrell, B.D. Evolutionary assembly of the conifer fauna: Distinguishing ancient from recent associations in bark beetles. *Proc. R. Soc. Lond. B Biol. Sci.* **2000**, *267*, 2359–2366. [CrossRef] [PubMed]
89. Sequeira, A.S.; Farrell, B.D. Evolutionary origins of Gondwanan interactions: How old are *Araucaria* beetle herbivores? *Biol. J. Linn. Soc.* **2001**, *74*, 459–474. [CrossRef]
90. Wilf, P.; Labandeira, C.C.; Johnson, K.R.; Cúneo, N.R. Richness of plant–insect associations in Eocene Patagonia: A legacy for South American biodiversity. *Proc. Natl. Acad. Sci. USA* **2005**, *102*, 8944–8948. [CrossRef] [PubMed]
91. Peterson, A.T.; Soberón, J.; Pearson, R.G.; Anderson, R.P.; Martínez-Meyer, E.; Nakamura, M.; Araújo, M.B. *Ecological Niches and Geographic Distributions (MPB-49)*; Princeton University Press: Princeton, NJ, USA, 2011; ISBN 1-4008-4067-8.
92. Pease, C.M.; Lande, R.; Bull, J.J. A model of population growth, dispersal and evolution in a changing environment. *Ecology* **1989**, *70*, 1657–1664. [CrossRef]
93. Keane, R.M.; Crawley, M.J. Exotic plant invasions and the enemy release hypothesis. *Trends Ecol. Evol.* **2002**, *17*, 164–170. [CrossRef]
94. Gaume, L.; Matile-Ferrero, D.; Mckey, D. Colony foundation and acquisition of coccoid trophobionts by *Aphomomyrmex afer* (Formicinae): Co-dispersal of queens and phoretic mealybugs in an ant-plant-homopteran mutualism? *Insectes Soc.* **2000**, *47*, 84–91. [CrossRef]
95. Waltari, E.; Perkins, S.L. In the hosts footsteps? Ecological niche modeling and its utility in predicting parasite distributions. In *The Biolgeography of Host-Parasite Interactions*; Oxford University Press: Oxford, UK, 2010; pp. 145–155, ISBN 978-0-19-956134-6.

96. Hoberg, E.P.; Brooks, D.R. A macroevolutionary mosaic: Episodic host-switching, geographical colonization and diversification in complex host–parasite systems. *J. Biogeogr.* **2008**, *35*, 1533–1550. [CrossRef]
97. Krauss, J.; Steffan-Dewenter, I.; Tscharntke, T. Landscape occupancy and local population size depends on host plant distribution in the butterfly *Cupido minimus*. *Biol. Conserv.* **2004**, *120*, 355–361. [CrossRef]
98. León-Cortés Jorge, L.; Lennon Jack, J.; Thomas Chris, D. Ecological dynamics of extinct species in empty habitat networks. 2. The role of host plant dynamics. *Oikos* **2003**, *102*, 465–477. [CrossRef]
99. Thomas, C.D.; Bodsworth, E.J.; Wilson, R.J.; Simmons, A.D.; Davies, Z.G.; Musche, M.; Conradt, L. Ecological and evolutionary processes at expanding range margins. *Nature* **2001**, *411*, 577–581. [CrossRef] [PubMed]

Article

Molecular and Morphological Phylogenetic Analyses of New World Cycad Beetles: What They Reveal about Cycad Evolution in the New World

William Tang [1,*], Guang Xu [2], Charles W. O'Brien [3], Michael Calonje [4], Nico M. Franz [5], M. Andrew Johnston [5], Alberto Taylor [6], Andrew P. Vovides [7], Miguel Angel Pérez-Farrera [8], Silvia H. Salas-Morales [9], Julio C. Lazcano-Lara [10], Paul Skelley [11], Cristina Lopez-Gallego [12], Anders Lindström [13] and Stephen Rich [2]

1. USDA APHIS PPQ South Florida, P.O. Box 660520, Miami, FL 33266, USA
2. Department of Microbiology, University of Massachusetts, Amherst, MA 01003, USA; xuguang54@hotmail.com (G.X.); smrich@umass.edu (S.R.)
3. 2313 W. Calle Balaustre, Green Valley, AZ 85622, USA; cobrien6@cox.net
4. Montgomery Botanical Center, 11901 Old Cutler Road, Miami, FL 33156, USA; michaelc@montgomerybotanical.org
5. School of Life Sciences, P.O. Box 874501, Arizona State University, Tempe, AZ 85287-4501, USA; nico.franz@asu.edu (N.M.F.); ajohnston@asu.edu (M.A.J.)
6. Departamento de Botánica, Universidad de Panamá, Estafeta Universitaria, Panamá City, Panamá; sidney@cwpanama.net
7. Instituto de Ecología, A.C., Red de Ecología Evolutiva, Apartado Postal 63, Antigua a Coatepec 351, El Haya, Xalapa 91070, Veracruz, Mexico; andrew.vovides@inecol.mx
8. Herbario Eizi Matuda (HEM), Facultad de Ciencias Biológicas, Universidad de Ciencias y Artes de Chiapas, Libramiento Norte Poniente 1150, Col. Lajas Maciel, Tuxtla Gutierrez CP29039, Chiapas, Mexico; perezfarreram@yahoo.com.mx
9. Sociedad Para el Estudio de los Recursos Bióticos de Oaxaca, A.C. Camino Nacional No. 80-b, San Sebastián Tutla CP71246, Oaxaca, Mexico; sschibli@hotmail.com
10. Departamento de Biología, Universidad de Puerto Rico, Río Piedras, P.O. Box 23360, San Juan 00931-3360, Puerto Rico; jlazcano1@yahoo.com
11. Florida State Collection of Arthropods, Florida Department of Agriculture—DPI, 1911 SW 34th St., P.O. Box 147100, Gainesville, FL 32614-7100, USA; Paul.Skelley@FreshFromFlorida.com
12. Instituto de Biologia, Universidad de Antioquia, AA 1226 Medellin, Colombia; mariac.lopez@udea.edu.co
13. Nong Nooch Tropical Botanical Garden, 34/1 Moo 7 Na Jomtien, Sattahip 20250, Chonburi, Thailand; ajlindstrom71@gmail.com

* Correspondence: william.tang@aphis.usda.gov

Received: 8 March 2018; Accepted: 17 May 2018; Published: 23 May 2018

Abstract: Two major lineages of beetles inhabit cycad cones in the New World: weevils (Curculionoidea) in the subtribe Allocorynina, including the genera *Notorhopalotria* Tang and O'Brien, *Parallocorynus* Voss, *Protocorynus* O'Brien and Tang and *Rhopalotria* Chevrolat, and beetles in the family Erotylidae, including the genus *Pharaxonotha* Reitter. Analysis of the 16S ribosomal RNA (rRNA) mitochondrial gene as well as cladistic analysis of morphological characters of the weevils indicate four major radiations, with a probable origin on the cycad genus *Dioon* Lindl. and comparatively recent host shifts onto *Zamia* L. Analysis of the 16S rRNA gene for erotylid beetles indicates that an undescribed genus restricted to New World *Ceratozamia* Brongn. is the most early-diverging clade, and this lineage is sister to a large radiation of the genus *Pharaxonotha* onto *Zamia*, with apparent host shifts onto *Dioon* and *Ceratozamia*. Analysis of beetles are in accord with current models of continental drift in the Caribbean basin, support some proposed species groupings of cycads, but not others, and suggest that pollinator type may impact population genetic structure in their host cycads.

Keywords: Belidae; Oxycoryninae; Erotylidae; Pharaxonothinae; cycad pollination

1. Introduction

Over the last three decades, evidence has accumulated that insect pollination is widespread in New World cycads. This evidence includes wind and insect exclusion experiments on ovulate cones of three species in the cycad genus *Zamia*, detailed observations of the life cycle and behavior of the beetles that inhabit them [1–5], as well as observations on other New World cycad genera *Ceratozamia*, *Dioon*, and *Microcycas* (Miq.) A. DC. [6–8]. Similar experiments and observations on other continents indicate the same for other genera of cycads [9–19]. In a recent seminal work on guidelines for cycad classification, insect symbionts of cycads were identified as having a potentially important impact on cycad classification: "Insects appear to be the primary vectors for pollination [...] evidence is accumulating to suggest coevolutionary processes between cycads and their pollinators. Once these processes are uncovered, resulting data will probably have a significant impact on how cycad taxa are classified" [20]. During the 6th International Conference on Cycad Biology, a coordinated global effort was organized to collect and study the insect pollinators of cycads [21]. One of the explicit goals was to use this information to understand cycad evolution. In this paper, we report on results of this insect survey effort in the New World, present phylogenetic analyses of the insects found, and discuss implications for cycad taxonomy.

The majority of cycads in the New World host more than one species of beetle in their cones and some host as many as three species. These beetles fall into two distinct and not closely related groups: (1) Weevils of the subtribe Allocorynina (Coleoptera: Curculionoidea: Belidae: Oxycoryninae: Oxycorinini; higher-level classification follows Bouchard et al. [22]) associated with *Dioon* and *Zamia*; these include the genera *Notorhopalotria*, *Parallocorynus*, *Protocorynus*, and *Rhopalotria* [8,23] (see Figure 1A–D) and (2) Erotylidae (Coleoptera: Cucujoidea) in the subfamily Pharaxonothinae associated with *Ceratozamia*, *Dioon*, *Microcycas*, and *Zamia* (Cycadales: Zamiaceae; classification follows Calonje et al. [24]); these include the genus *Pharaxonotha* and an undescribed genus [7,25–27] (see Figure 1E,F).

O'Brien and Tang [8] recently described or reviewed all known species of Allocorynina, but they did not present a detailed phylogenetic analysis of the species. All known species inhabit and develop in cones of New World cycads. Six species of New World Pharaxonothinae have been described, but many remain undescribed [7,27]. New World forms are closely related to the recently described genus *Cycadophila* found on the Asian cycad genus *Cycas* [26,28]. The lack of phylogenetic frameworks for the New World groups of cycad beetles hinders the proper allocation of biological information and host/beetle associations and limits what can be interpreted from them. For instance, Maldonado-Ruiz and Flores-Vazquez [29] catalogued beetles found with a species of *Dioon* in Mexico, however, due to lack of keys for identification or prior phylogenetic work, they were not able to assess how many species of Allocorynina and Pharaxonothinae they were dealing with. In this paper, phylogenetic analyses were conducted on both morphological characters and DNA data of Allocorynina and DNA data of Pharaxonothinae beetles collected from cycad cones. The resulting trees from these analyses are used to generate hypotheses about cycad biogeography and evolutionary patterns at genus, species, and population levels [30,31].

Figure 1. Representatives of the six major lineages of Coleoptera inhabiting New World cycads, dorsal views: (**A**) *Protocorynus bontai* O'Brien and Tang, male; (**B**) *Notorhopalotria montgomeryensis* O'Brien and Tang, male; (**C**) *Rhopalotria* (*R.*) *dimidiata* Chevrolat, male; (**D**) *Parallocorynus* (*Neocorynus*) *schiblii* Tang and O'Brien, male; (**E**,**F**) *Pharaxonotha* sp. and Erotylidae, undescribed genus inhabiting a male cone of *Ceratozamia vovidesii* Pérez-Farr. and Iglesias; scale bars = 1 mm.

2. Materials and Methods

Beetles were collected from cycad cones in habitat by the authors and other cooperators and include previously described cycad-associated beetle species, as well as many undescribed beetle taxa [8,27]. Total number of cycad taxa sampled include 3 of the 31 recognized species of *Ceratozamia*, 13 of the 15 species of *Dioon*, and 29 of the 77 currently recognized species of *Zamia* [24]. In total, 89 cycad populations or localities yielded beetles with useable DNA for this study. The monotypic Cuban cycad genus *Microcycas* was not sampled, however, a species of *Pharaxonotha* has been described from this host [7]. For Allocorynina, institutions for deposition of specimens are listed in O'Brien and Tang [8]. Pharaxonothinae used are deposited in the Florida State Collection of Arthropods, Division of Plant Industry, Gainesville, FL, USA.

2.1. Morphological Analysis

For the Allocorynina weevils, external morphology and morphology of genitalia were studied and photographed using Nikon® SMZ1500 stereoscopic and Eclipse 80i compound microscopes mounted with Nikon® DS-Fi1 digital cameras. For morphological characters, taxonomic description of species and specimens used, see Appendix A and O'Brien and Tang [8]. For the outgroup, *Oxycraspedus cornutus* Kuschel (Belidae: Oxycorininae: Oxycornini: Oxycraspedina), currently placed in the same subfamily and tribe as the Allocorynina [32], was chosen. A matrix of 89 characters based mainly on morphology, but also host-associations and behavior such as diet and pupation sites, was built (see Figure A1 in Appendix B). A phylogenetic tree was generated using maximum parsimony (MP) implemented by TNT [33] using default settings, and a strict consensus tree was generated from the ten most parsimonious trees found. Bootstrap support values were generated based on 1000 replicates. Multi-state characters were treated as non-additive; and all characters were weighed equally in the analysis. Morphological analyses of cycad-associated erotylid taxa are under way [27], but are not presented here.

2.2. DNA Analysis

The quality of DNA preservation of the beetles available for this study varied and was often poor, therefore, only short sections of DNA could be consistently sequenced in the samples available. We selected a fragment of the mitochondrial 16S ribosomal RNA (rRNA) gene with a combined sequence length ranging from 311–316, varying with additions and deletions of sections. The aligned data set contained 318 sites, with 222 constant, 93 variable, and 68 parsimony informative sites. As seen in other arthropods, the 16S rRNA gene is highly AT-rich with average nucleotide frequencies of thymine (T) 42.9%, cytosine (C) 7.9%, adenine (A) 35.0%, and guanine (G) 14.3%. The 16S rRNA gene has been used widely in insect molecular systematics and its utility in discerning species groups and deeper divisions in beetles and other holometabolous insects is well-founded [34–38]. It has been proposed for use as a standard for insect phylogenetics [39]. Total DNA was extracted from individual beetles, either adults or larvae, using Epicenter Master Complete DNA and RNA Purification Kits (Epicenter Technologies, Madison, WI, USA) following the manufacturer protocols and dissolved in 30 μL H_2O. The mitochondrial 16S rRNA was amplified using the following primers: 73Forward-AGATAGAAACCARCCTGGCT, 98Forward-CGGTYTRAACTCAGATCATGTA, and 430Reverse-AAGACGAGAAGACCCTATAG [26,28]. Reactions were carried out in 25 μL volumes containing 1 μL DNA, 5 μL 5X buffer, 4 μL of 25 μM $MgCl_2$, 1 μL of 10 mM dNTPs, 1 μL of 10 μM of each primer, and 0.2 μL 5 U/μL of Taq polymerase (Promega, Madison, WI, USA). PCR was performed using an Eppendorf ep mastercycler (Eppendorf, Westbury, NY, USA) using the following DNA denaturation, annealing and replication protocol: 94 °C for 1 min, then 40 cycles of 94 °C for 15 s, 50 °C for 15 s, and 72 °C for 40 s. Amplified products were cleaned up with the ExoSAP-IT kit (USB, Cleveland, OH, USA) and sequenced bidirectionally on an ABI 3130XL Genetic Analyzer (Applied Biosystems, Foster City, CA, USA).

DNA sequences were deposited into GenBank with accession numbers MF990634-MF990709 for weevils and KR005722, KR005724, KR005725, KY365240, KY365243, and MG256677-MG256758 for Erotylidae. Specimens of *Oxycraspedus cornutus*, used as the outgroup for the morphological analyses, did not yield usable DNA, therefore, for the phylogenetic analysis for weevils using DNA, we chose as outgroups *Hypera postica* (Curculionidae), *Ischnopterapion virens* (Brentidae), *Rhinotia haemoptera* (Belidae), *Anthribus albinus*, and *Anthribus nebulosus* (Anthribidae), with 16S sequences obtained from GenBank, accession numbers: U16967.1, KY084146.1, AJ495455.1, AJ495448.1, and AJ495449.1, respectively.

For weevils, nucleotide sequences were aligned using MAFFT version 7 [40]. The aligned sequences were then phylogenetically analyzed via: (1) maximum parsimony (MP) implemented by TNT [33] using default settings: a strict consensus tree was generated from the ten most parsimonious trees found; (2) maximum likelihood (ML) implemented in RAxML version 8 [41]: the best tree from 20 independent searches was selected and bipartition bootstrap values were written from 500 bootstrap replicates; (3) bayesian inference (BI) implemented in Mr Bayes version 3.2 [42], which was run with two simultaneous searches each using four chains for 1 million generations: trees were sampled every 100 generations, and the first 25% were discarded as burnin.

For erotylid beetles, two Asian species of Pharaxonothinae that inhabit the genus *Cycas* were used as outgroups: *Cycadophila (C.) debaonica* and *C. (Strobilophila) tansachai* [26,28], GenBank accession numbers KR005715, KY365223, respectively. Multiple-sequence alignments were conducted with CLUSTAL W [43]. Phylogenetic trees were generated using maximum parsimony (MP), neighbor joining (NJ), and maximum likelihood (ML) methods as implemented in PAUP 4.0b10 [44] and MEGA5 [45]. Bootstrap support values were generated based on 1000 replicates. The best fit model of sequence evolution employed in the ML analysis was the Tamura 3-parameter+G model with log likelihood −1448.98 and Gamma distribution 0.2653.

3. Results and Discussion

3.1. *Allocorynina Trees*

The phylogenetic tree based on MP analysis of the morphological, behavioral, and host-association data set is displayed in Figure 2. This tree generally supports the genera and subgenera recognized by O'Brien and Tang [8]. Genera *Protocorynus*, *Neocorynus*, and *Parallocorynus* and subgenera *Rhopalotria*, *Dysicorynus*, *Neocorynus*, *Eocorynus*, and *Parallocorynus* are monophyletic. Only the monophyly of genus *Rhopalotria* and its subgenus *Allocorynus* is not supported. The three molecular trees for the Allocorynina generated from MP, ML, and BI analyses of the 16S rRNA data set were almost identical, and any slightly conflicting clades are linked to very low support values. Therefore, in Figure 3, only the molecular tree from the BI analysis is presented, with ML support values annotated underneath the branches. Also, in this tree, each analyzed sample displays corresponding host cycad species and the geographic region where they were collected. All genera (and most subgenera) are well supported, but the relationships between them are not. The main difference between the two trees is that in the molecular tree subgenus *Rhopalotria* is paraphyletic with respect to *Allocorynus*, whereas in the morphology tree, the situation is reversed and subgenus *Allocorynus* is paraphyletic, but neither of them refute the monophyly of the other analysis. Another major difference in the molecular tree is the strong support for *Protocorynus* to be sister to the remaining three genera. These differences between the two trees may, in part, be attributed to different choices in outgroups.

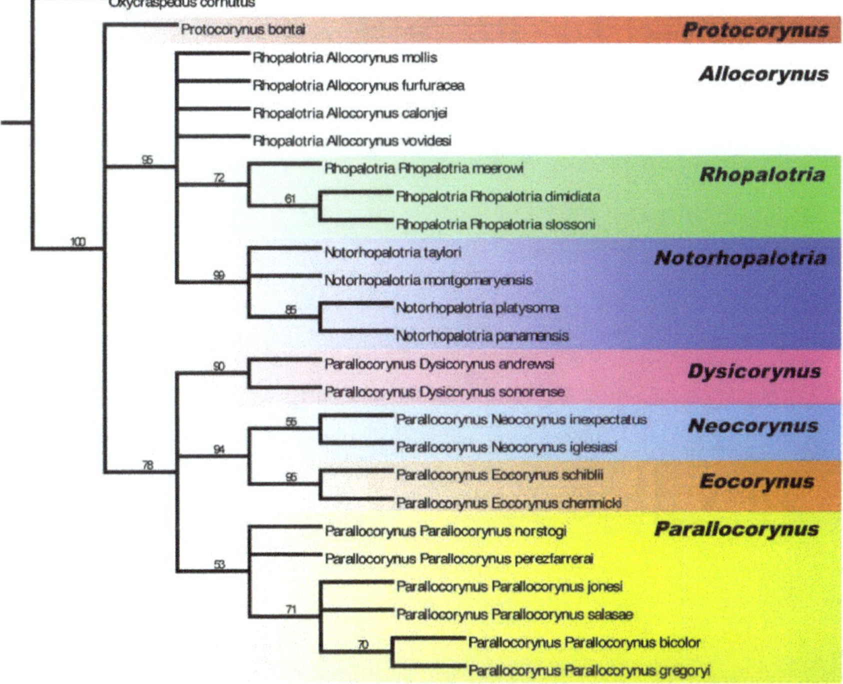

Figure 2. Phylogenetic tree for the Allocorynina based on maximum parsimony analysis of the matrix of morphological, behavioral, and host characters in Appendix B; *Oxycraspedus cornutus* is used as the outgroup; numbers are bootstrap values.

The morphological analysis and the resultant tree are valuable in answering how each of these weevil lineages differ and what may be driving evolution in this group. Each recognized genus and subgenus is distinguished by differences in their genitalia and also in the spination of their profemora. Behavioral observations [3,4] indicate that the profemora are used during courtship battles between males. The spination on their profemora appear to function in grasping an opposing male during these mating struggles. Sexual selection, larval feeding sites, pupation sites, as well as host genus (*Dioon* vs. *Zamia*) appear to account for many of the morphological differences between the genera and subgenera. For details on synapomorphic characters for major clades see Appendix C.

This phylogenetic analysis of the Allocorynina is the most extensive to date and greatly expands upon previous efforts [31,32], while generally reaffirming the sub-/generic concepts erected by O'Brien and Tang [8].

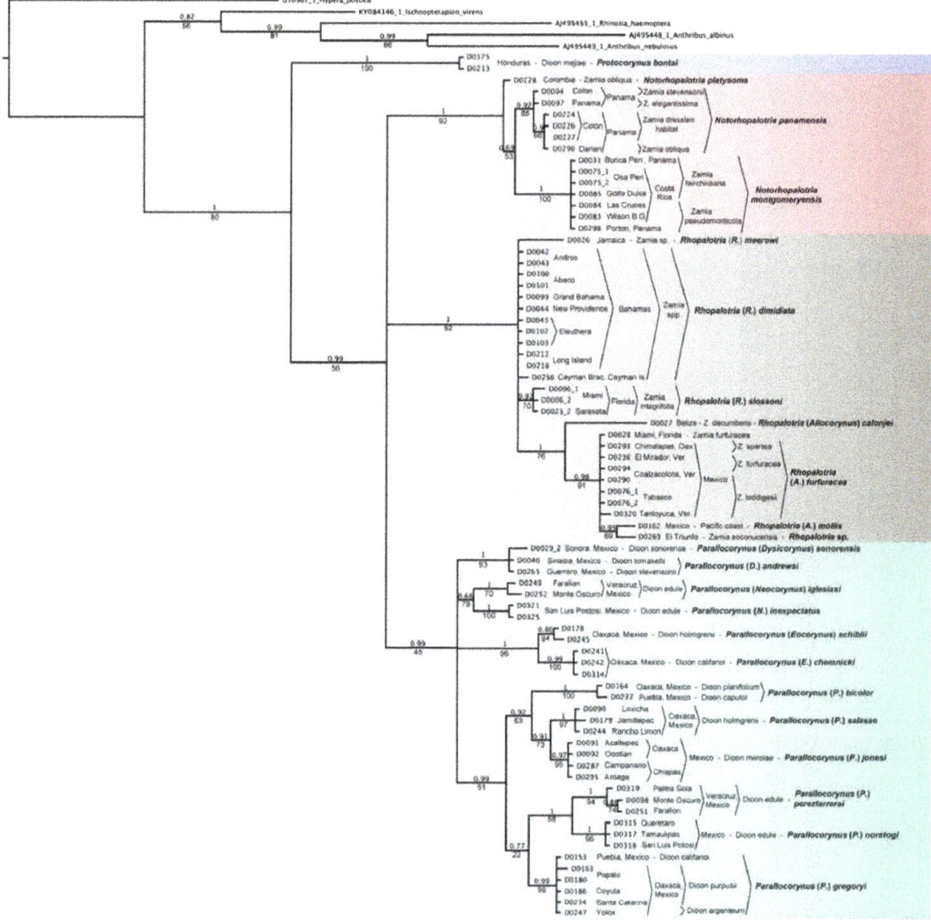

Figure 3. Phylogenetic tree for the Allocorynina synthesized from maximum likelihood (ML) and bayesian inference (BI) analyses of 16S ribosomal RNA (rRNA) gene sequences; numbers above the branches are posterior probabilities from Mr Bayes and the numbers below the branches are bootstrap support from RAxML. Localities and host cycad species are indicated for each sample.

3.2. Erotylidae Tree

The trees produced from MP, NJ, and ML analyses of 16S rRNA sequences for Erotylidae were similar, therefore, only the ML tree is displayed in Figure 4. The MP and NJ trees are provided in Appendix D as Figures A2 and A3. The ML tree, as well as the MP and NJ trees, can be divided into four sections: (1) The most early-diverging clade is an undescribed genus residing in cones of *Ceratozamia* (purple color in Figure 4); the remainder of the New World taxa are sister to this clade and are tentatively assigned to the genus *Pharaxonotha*. (2) This *Pharaxonotha* clade can be broadly assigned to three sections; the first of these are labeled "Early-diverging lineages Mexico to South America" (blue color in Figure 4). Within this division, the most early-diverging branches reside in cones of *Zamia onan-reyesii* C. Nelson and Sandoval, an aerial-stemmed species from Honduras and subterranean-stemmed *Z. cunaria* Dressler and D. W. Stev. and *Z. pyrophylla* Calonje, D. W. Stev. and A. Lindstr., in the eastern Panama-northern Colombia region. Also among these early diverging lineages are species of *Pharaxonotha* that reside in other *Ceratozamia*, *Dioon*, and *Zamia* hosts. (3) A third broad division consist of *Pharaxonotha* inhabiting cones of the *Zamia pumila* L. species complex in the Caribbean, specifically on islands of the Greater Antilles, the Bahamas archipelago, and the Florida peninsula (green color in Figure 4). This division consists of two clades; one in the easternmost section of the Greater Antilles on Puerto Rico and Hispaniola and the other clade in the western Greater Antilles, Florida, and the Bahamas. (4) Lastly, there is a fourth major division, consisting of a more recent radiation of *Pharaxonotha* that also extends from Mexico to South America and inhabits *Zamia* and *Dioon* (red color in Figure 4). Although bootstrap support for many of these clades are weak, they fall consistently into these four broad categories in the ML, MP, and NJ trees, with only one exception: the *P. kirschii* Reitter lineage, discussed below.

Although no morphological analyses are presented in this paper for the Erotylidae, Xu et al. [28], Skelley et al. [26], and Tang et al. [27] have identified a number of external morphological and genital characters which support many of the clades revealed here (Figure 4) through phylogenetic analysis of the 16S rRNA gene. All species discussed herein are assigned to a single subfamily, the Pharaxonothinae, of the Erotylidae. Of special interest is the taxon *Pharaxonotha kirschii*, the type species for the genus and the only known Pharaxonothinae in the New World that is a generalist feeder. Although it has been found with *Zamia*, it is the only known member of the New World Pharaxonothinae that is not an obligate inhabitant of a cycad during its life cycle and is widely distributed in Central America [26]. Our DNA analysis indicates that the *Pharaxonotha* associated with *Z. inermis* (specimen D0057 in Figure 4), while a distinct species, is related to *P. kirschii*. The specimen used for DNA analysis was a larva and an adult female associated with that larva matches the morphology of *P. kirschii*. This specimen and *Pharaxonotha kirschii* may be part of a species complex, with different branches of this complex inhabiting either cycads or non-cycad hosts. In the ML tree, this *P. kirschii* clade is associated with the Caribbean group, however, in the MP and NJ trees (Figures A2 and A3), these two taxa are associated with the early diverging lineages from Mexico to South America.

While all *Zamia* that have been closely sampled have yielded *Pharaxonotha* beetles from their cones, this is not so for all *Dioon* and *Ceratozamia* populations examined. The cycad genus *Dioon* was especially well-sampled in this study and in one species, *Dioon mejiae* Standl. and L.O.Williams, no erotylids were detected in three separate samples consisting of a total of many hundreds of beetles (only Allocorynina weevils found). Furthermore, the erotylid beetles sampled here from seven other *Dioon* species cluster in three disparate branches in the erotylid trees, suggesting separate colonization events of *Dioon* by this group of beetles.

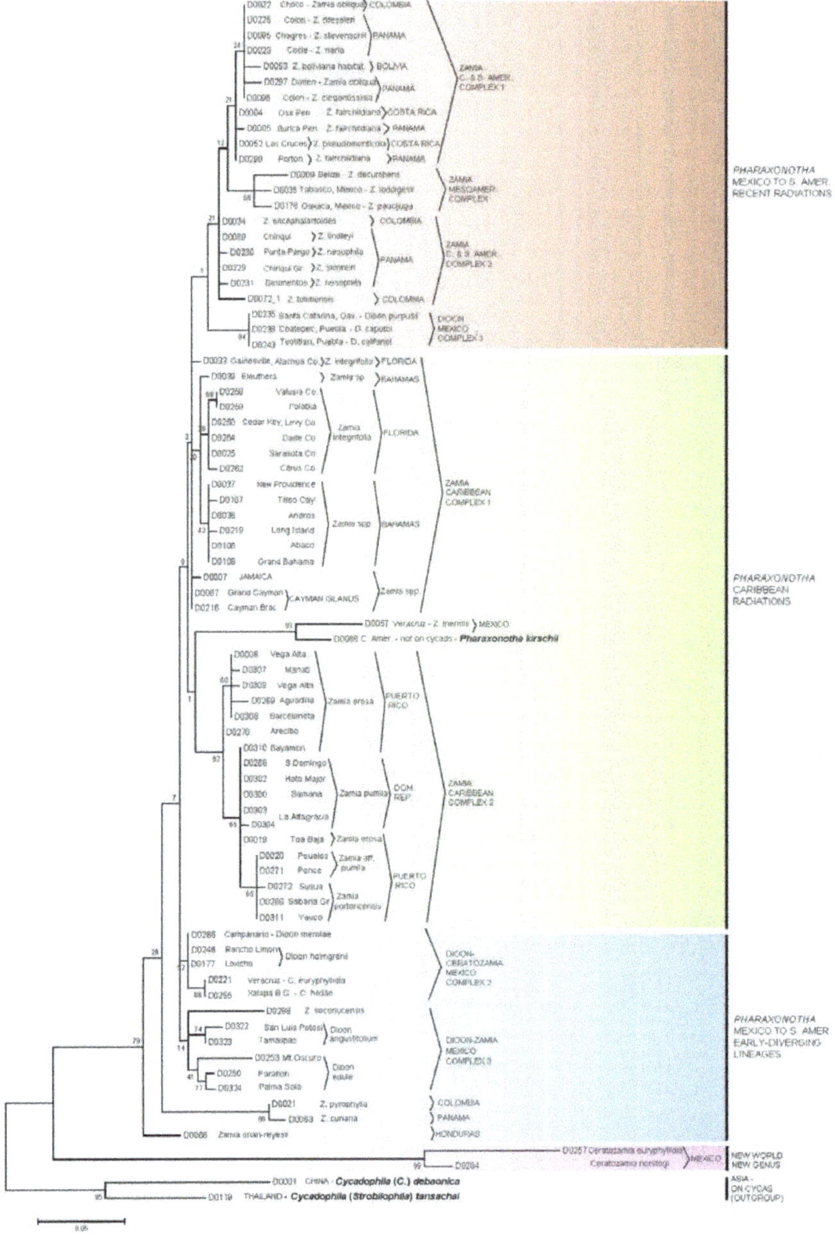

Figure 4. Phylogenetic tree for the Erotylidae: Pharaxonothinae on the cycads of the New World based on maximum likelihood analysis of 16S rRNA gene sequences; scale bar indicates base pair substitutions per nucleotide position; numbers on branches are bootstrap values.

3.3. Implications for the Evolution of Cycad Hosts

3.3.1. Using Beetle Trees to Generate Hypotheses of Cycad Evolution

While the 16S rRNA gene is considered to provide fairly accurate phylogenies for beetles and other insects, the search for plastid and nuclear genes of similar utility in the construction of phylogenies for New World cycads has had equivocal results [46–49]. Genetic analysis of beetles that have coevolved with cycads may reveal evolutionary patterns in cycads that are much more difficult to discern from direct examination of cycad genomes. The following discussions will be confined mainly to the generation of hypotheses about cycad evolution in the New World based on the interpretation of the beetle trees produced in this study.

Before proceeding, we offer some caveats and limitations as to what can or cannot be deduced about cycad evolution from coevolving insect pollinators. Cycads typically form colonies or groves widely dispersed from other populations of conspecific or closely related species [50], and in such situations, flight distances of seed dispersers or pollinators between populations may be typically measured in kilometers or tens of kilometers. Cycad seeds are relatively large and heavy compared to available animal dispersers, and field studies indicate that the great majority of cycad seeds fall within two meters of their mother [51–53], suggesting that gene flow between distant populations via seed dispersal is rare. These population structures and seed dispersal constraints may configure insect pollinators as the primary gatekeepers of gene flow in most cycad populations. Thus, when two cycad populations host different species of beetles believed to be involved in pollination, we can use this to infer that the cycads in question may be genetically isolated, as little or no gene flow is likely to occur via exchange of pollinators. When the scenario is reversed and cycad populations share the same pollinator, we cannot automatically assume gene flow is occurring between the cycad populations via pollinators. Studies by Donaldson et al. [9] and Terry [13] show that pollen loads on insects that leave cycad cones drop rapidly with distance and time from the source cones. Crowson [54,55] illustrated deep antennal pockets in the antennae of Allocoynina weevils and hypothesized that these might be structures that function to hold pollen. Closer examination of these pockets with SEM [8] indicate that their entrances are occluded with hairs with a probable sensory function and that these pockets are unlikely to facilitate pollen transport. While a cycad pollinator may be able to fly between populations of different host cycad species and interbreed with other populations of its own species, during this process, it may not be transferring any cycad pollen between distant host cycad populations. A phylogenetic study of thrips pollinators of the Australia cycad *Macrozamia* suggests that both cycad and pollinator populations fragment together as a result of climate change and aridification [56] and that co-diversification of the host and pollinator in allopatry appears to be the process affecting diversity. Furthermore, that study concluded that distinct cycad species separated by short distances may continue to share the same pollinator species. Similarly, in New World cycads, O'Brien and Tang [8] recognized that some cycad beetles inhabit more than one host cycad species: (1) *Parallocorynus* (*P.*) *gregoryi* O'Brien and Tang inhabits the cones of three closely related *Dioon* species, *D. argenteum* T.J.Greg., Chemnick, Salas-Mor. and Vovides, *D. califanoi* De Luca and Sabato and *D. purpusii* Rose; (2) *Rhopalotria* (*R.*) *dimidiata* inhabits cones of many species of the *Zamia pumila* complex living in the Bahamas, Cuba, and Cayman Islands, and (3) *Notorhopalotria panamensis* O'Brien and Tang inhabits at least three species of *Zamia* in central Panama with adjoining geographic distributions, *Z. dressleri* D. W. Stev., *Z. elegantissima* Schutzman, Vovides and R. S. Adams and *Z. stevensonii* A. S. Taylor and Holzman.

Examination of these DNA-based trees indicates that host-shifts of pollinating beetles have occurred during the evolution of cycads. We cannot assume strict co-speciation of beetle lineages with their corresponding host cycad lineage. Host-shifts of cone beetles between cycad genera and between cycad species within a genus can occur. Although this limits our ability to make inferences about coevolution of cycads and their beetles, the host-shifts in themselves are interesting and informative about evolutionary processes in cycads [57].

3.3.2. Cycad Hypotheses Based on the Allocorynina Trees

From the DNA-based tree for Allocorynina (Figure 3) in combination with morphological data (Appendix A, [8]) we generate these hypotheses about cycads:

(1) The presence of one or two species of Allocorynina in all species of *Dioon* sampled compared with its absence from many *Zamia* species and its complete absence in the other New World cycad genera *Ceratozamia* and *Microcycas* suggests that *Dioon* is the earliest host lineage colonized by Allocorynina weevils, with one or possibly two host-shifts onto *Zamia*. In this hypothesis, Allocorynina are the original pollinators in the genus *Dioon*, while erotylid beetles are later colonists in *Dioon*.

(2) Based on the morphological and genetic analyses of its pollinator *Dioon mejiae*, located on the Chortis block, a tectonic terrane roughly corresponding to the country of Honduras [58], is hypothesized to be one of the earliest bifurcating lineages within *Dioon*.

(3) Species of narrow-leaflet *Dioon* in Mexico form four lineages with distinct biogeographic distributions: (A) western Mexico lineage along the Pacific drainage of the Sierra Madre Occidental from Sonora to Guerrero consisting of *D. sonorense* (De Luca, Sabato and Vázq.Torres) Chemnick, T.J.Greg. and Salas-Mor., *D. stevensonii* Nic.-Mor. and Vovides and *D. tomasellii* De Luca, Sabato and Vázq.Torres; (B) eastern Mexico lineage along the Sierra Madre Oriental from Nuevo Leon to Veracruz consisting of *D. angustifolium* Miq. and *D. edule* Lindl.; (C) south central Mexico lineage consisting of *D. argentium*, *D. califanoi*, and *D. purpusii*; (D) southern Mexico lineage along the Pacific drainage of Oaxaca to Chiapas, consisting of *D. caputoi* De Luca, Sabato and Vázq.Torres, *D. holmgrenii* De Luca, Sabato and Vázq.Torres, *D. merolae* De Luca, Sabato and Vázq.Torres and *D. planifolium* Salas-Mor., Chemnick and T. J. Greg.

(4) For the eastern Mexico lineage (group 3B above) *Dioon edule* (including *D. angustifolium*) north of the Trans-Mexican Volcanic Belt in the states of Nuevo Leon, Queretaro, San Luis Potosi, and Tamaulipas is likely a distinct species from *D. edule* south of this mountain range in Veracruz; the two lineages of Allocorynina and the lineage of erotylids inhabiting this cycad species group support this division.

(5) The absence of Allocorynina in the periphery of the geographic range of *Zamia* (e.g., eastern part of the Greater Antilles and much of Panama and South America) suggests that colonization of *Zamia* by Allocorynina is relatively recent and perhaps an ongoing ecological and evolutionary process. The alternate hypothesis is that there may have been a more widespread distribution on *Zamia*, but these weevils have suffered selective extinction in parts of their range.

(6) The shift of Allocorynina from *Dioon* onto *Zamia* may have occurred during major tectonic events in the formation of Central America when landmasses were moving through the region, emerging from the sea, and/or colliding with Mesoamerica [59,60]; during this time, lineages of cycads (including *Zamia* and/or possibly other extinct cycad lineages) and the Allocorynina associated with them migrated in three directions: (A) south into Central America, (B) east into the Caribbean islands, and (C) within Mesoamerica.

(7) *Zamia obliqua* A. Braun in the Choco of Colombia is likely a different species from *Z. obliqua* in the Darien of Panama based on genetic differences in the Allocorynina inhabiting their respective cones.

3.3.3. Cycad Hypotheses Based on the Erotylidae Tree

From the DNA-based tree of the Erotyidae (Figure 4) we generate these hypotheses:

(I) The most early-diverging lineage of cycad-associated erotylid pharaxonothine beetles in the New World is confined to the genus *Ceratozamia*. This branching pattern is consistent with fossil evidence indicating that the *Ceratozamia* lineage may have first evolved in Europe in the mid Cenozoic and then migrated to North America prior to the complete separation of these continents [61]. In addition, the apparent close relation between cycad-associated erotylids of the New World with those found on Cycas in Asia, suggest that these beetles may have an ancient Laurasian association with cycads that predates the breakup of Laurasia.

(II) Two early-diverging erotylid beetle lineages associated with *Zamia* are located in: (a) Honduras on *Z. onan-reyesii* and (b) The northern South America-Darien region on *Zamia cunaria* and *Z. pyrophylla*. We may hypothesize that these host lineages of *Zamia* are among the earliest to diverge for the genus and are likely relicts from an earlier radiation of *Zamia* throughout these regions.

(III) The presence of erotylid beetles in the cones of all species of *Zamia* that have been carefully sampled suggests that these were the original pollinators of *Zamia*. In this hypothesis, the Allocorynina weevils are later colonists of *Zamia*.

(IV) In addition to old relictual clades in hypothesis II, the existence of three separate derived clades of *Pharaxonotha* beetles on *Zamia* suggests that at least three recent and separate radiations of *Zamia* have occurred in the following regions: (a) A radiation into the eastern islands of the Greater Antilles, which includes Hispaniola and Puerto Rico, probably beginning when these landmasses were more closely associated with Central America [60]; (b) A more recent radiation into the western islands of the Greater Antilles, including Cuba, Cayman Islands and Jamaica and neighboring landmasses of the Bahamas and Florida; and (c) Sister to these two Caribbean lineages, a recent radiation onto *Zamia* in Mesoamerica, Central America, and northern South America.

(V) The separation of *Pharaxonotha* beetles, that inhabit *Dioon* cones, into three distinct and not closely related clades and the absence of erotylids on one species, *D. mejiae* in Honduras, at the periphery of the range of *Dioon*, suggests that erotylids colonized *Dioon* from the *Zamia* lineage in three separate host shift events and that these host shifts have occurred relatively recently compared to the radiation of Allocorynina in *Dioon*.

(VI) At least one recent host-shift of *Pharaxonotha*, originating from *Zamia*, have occurred onto *Ceratozamia*. A larger and wider sampling of *Ceratozamia* beetles may reveal more than one host shift. These coexist within *Ceratozamia* cones with the more ancient erotylids beetles discussed in hypothesis I, so that now two disparate lineages of Pharaxonothinae coinhabit *Ceratozamia* cones. This host-shift radiation is allied with those in *Dioon*, suggesting important watershed periods in cycad evolution when exchange of pollinators occurred among cycad genera in the New World. Deeper study of these periods may be crucial in understanding the relatively recent resurgence of cycad evolution that have been proposed [62–64].

(VII) The pattern of population genetic variation of *Pharaxonotha* beetles that presumably pollinate *Zamia* in Puerto Rico and Hispaniola mirrors to a great extent the population genetic variation exhibited by the *Zamia* on those islands [65]. These mirroring patterns suggest that the mobility and/or abundance of a cycad's pollinator may influence gene flow in its host cycad and consequently the speciation pattern of its host. For example, observations [66] suggests that unlike other cycad beetles, this lineage of *Pharaxonotha* is highly sensitive to human disturbance of its vegetative habitat and easily becomes rare or locally extinct as a result. This susceptibility to disturbance and the low ability to recolonize its host from nearby populations suggests low mobility and low ability to mediate gene flow in its host *Zamia*. The resulting effect is to produce local reproductive isolation of cycad populations that appear on casual observation to have continuous distributions.

(VIII) Based on population genetic variation of beetles discussed in hypothesis VII, the *Zamia* populations near Bayamon and Toa Baja, Puerto Rico may be conspecific with *Zamia pumila* in Hispaniola; furthermore, *Z. pumila* populations in Hispaniola may be recent colonists from an ocean dispersal event originating from Bayamon and Toa Baja.

3.3.4. Independent Tests of Beetle Generated Hypotheses

Many of these beetle-generated hypotheses about cycad evolutionary patterns in the New World may be tested with data sets of genetic and/or morphological characters from their host *Ceratozamia*, *Dioon*, or *Zamia*. Our tree for erotylid beetles (Figure 4) contains samples from 61 populations of *Zamia* along with collection localities. A phylogenetic tree for *Zamia* based on sequences of one gene combined with morphological data published by Caputo et al. [67] is available for comparison. Their tree contains 22 species of *Zamia*, of which only ten correspond with populations in our tree, nevertheless, some broad tests can be made of our erotylid hypotheses II and IV. Their tree shows some congruence with the erotylid trees presented here. For instance, in their tree as in ours, the Caribbean *Z. pumila* group is sister to many Mesoamerican and South American *Zamia*. Their tree differs significantly from our erotylid tree, however, in that a large branch of the Central American *Zamia* forms a clade that is sister to the previously described clades. Our erotylid tree suggests that their Central American clade should form part of their Mesoamerican and South American *Zamia* clade. In addition, our erotylid tree suggests that *Z. soconuscensis* Schutzman, Vovides and Dehgan belongs in a clade that is an early diverging relative to the Caribbean clade, however, *Z. soconuscensis* appears as a member of their Mesoamerican and South American clade that is sister to the Caribbean clade. If both beetle and *Zamia* host phylogenies are accurate, we would conclude that host and symbiont beetles are not co-speciating in a perfectly parallel fashion and that host shifts have occurred at important junctures in both *Zamia* and beetle radiations. The converse may be true, that the beetle phylogenies do reflect the evolutionary patterns of their host *Zamia* accurately, and that a more extensive genetic data set for *Zamia* may be required for a more accurate comparison.

A maximum likelihood tree produced by Nagalingum et al. [68] using three genes is also available for testing of hypothesis IV. Their tree contains 35 species of *Zamia*, of which 17 correspond with host populations in our tree and is broadly congruent with ours. In their tree, the Caribbean clade is sister to other Mesoamerican and Central and South America clades. None of our identified early diverging host *Zamia* lineages, however, were sampled in their study, so hypothesis I cannot be tested with their dataset.

Hypotheses may be tested by comparing trees from different beetle groups that are cohabiting in the same hosts. For example, we can test hypotheses 3A–D generated using Allocorynina weevils that there are four species groups of narrow-leafed *Dioon* that inhabit four distinct geographic regions. Our erotylid tree, based on less extensive samples from *Dioon* than that for Allocorynina, exhibits three distinct clades for *Dioon* erotylids that largely correspond to the three Allocorynina clades in hypotheses 3B–D. The only exception is the erotylid beetle on *D. caputoi*, which suggest a single host shift has occurred between our proposed regions 3C and 3D. Also, no usable DNA was extracted for erotylids from region 3A, so no comparison can be made for hypothesized region 3A. A recently published phylogeny for *Dioon* based on DNA by Gutiérrez-Ortega et al. [49] also provides a test of some of the Allocorynina generated hypotheses. Their tree [49] supports Allocorynina hypothesis 1 that *D. mejiae* is one of the early diverging lineages of the genus and the geographic regions proposed in hypotheses 3A–B. Their tree, however, does not support the distinction between hypothesized *Dioon* regions 3C and 3D, and in their tree, the *Dioon* species of 3C and 3D form an integrated clade. Their tree also does not support our hypothesis 4, that *D. edule* and *D. angustifolium* north of the Trans-Mexican Volcanic Belt form a distinct clade from *D. edule* south of the belt. Another published phylogeny for *Dioon* based on combined molecular and morphological data by Moynihan et al. [48] supports our hypotheses 3A–D, except their tree shows *D. caputoi* as a distinct branch separate from the other four.

A test of erotylid hypothesis VII, that mobility of a pollinator may influence the population genetic structure of its host, can be partly tested with the beetles inhabiting *Zamia* in the six island groups of the Bahamas. Although the Allocorynina beetles sampled from these islands all display the same 16S rRNA haplotype, suggesting panmixia or alternatively recent introduction to the islands from a single source, the *Pharaxonotha* beetles on three of these islands, Eleuthera, Long Island, and Tiloo Cay (near Abaco), exhibit haplotypes distinct from the rest. The hosts of these three *Pharaxonotha* haplotypes,

Z. angustifolia Jacq., *Z. lucayana* Britton, and Z. sp. "Tiloo Cay", possess distinctly narrower, broader, or more coriaceous leaflet phenotypes than *Zamia* on other Bahamian islands. Genetic analysis of Bahamian *Zamia* populations [69] supports the possibility of extended genetic isolation of the Long Island *Z. lucayana* population, but not of the Eleuthera and Tiloo Cay *Zamia* populations. In this case, genetic patterns in *Pharaxonotha* mirror the phenoptypic traits displayed by their host *Zamia* better than the genetic analysis of the *Zamia* populations themselves. Possibly, restricted gene flow mediated by the *Pharaxonotha* may have been masked by gene flow mediated by Allocorynina pollinators at a later stage, since both pollinators are present in these *Zamia* populations.

Hypothesis VII also predicts that low mobility of cycad pollinators may result in high genetic variation among cycad populations that are in relatively close proximity. Lazcano-Lara (66) demonstrated that successful pollination of *Z. portoricensis* Urb. in Puerto Rico decreases dramatically when a female plant is beyond 1.9 m of a male plant, suggesting that its sole beetle pollinator, *Pharaxonotha portophylla* Franz and Skelley, provides relatively ineffective long range pollination, and this appears to contribute to the high degree of genetic differentiation exhibited between neighboring *Zamia* populations on this island [65,66].

4. Conclusions

It is widely understood that insect pollination is a critical facet of cycad biology and conservation [30,31,70,71]. In addition to their importance in cycad reproductive biology, the morphological and genetic analysis of pollinators can provide evidence for supporting or refuting hypotheses about cycad taxonomy and biogeography. As more extensive phylogenetic studies of New World cycads become available, the hypotheses presented here can be tested in more depth. Our hypotheses and analyses can also be refined in future work with a broader sampling of mitochondrial as well as nuclear genes from cycad beetles and a wider geographic sample of beetles that corresponds more closely with the host cycads on which phylogenetic studies have been conducted. Also, analyses of beetles can be improved with divergence time estimates, which require fossils. Recently, putative fossil relatives in both Allocorynini and Pharaxonothinae have been described from amber deposits [72,73], however, with only crude age estimates ranging from Eocene to Miocene. Hypotheses about cycad evolution generated through study of their associated insects can be novel and unexpected, with examples above, and provide new insights into the evolutionary history of New World cycads. This first attempt at reciprocal illumination of New World cycads through study of their pollinators indicates that this is a fruitful avenue of endeavor.

Author Contributions: W.T. and C.W.O.B. conceived and designed the study; W.T., C.W.O.B., M.C., A.T., A.P.V., M.A.P.-F., S.H.S.-M., J.C.L.-L., P.S., C.L.-G., A.L., and S.R. contributed materials; G.X., W.T., C.W.O.B., M.C., N.M.F., and M.A.J. analyzed the data; W.T., N.M.F., and C.W.O.B. wrote the paper.

Acknowledgments: We thank Jeff Chemnick, Wes Field, Tim Gregory, Chip Jones, Oscar Moreno, Limei Tang, and Maria Ocasio Torres for assistance in the field and Jeff Chemnick, Si-Lin Yang, and SeqGen, Inc. for financial support of field work and molecular analysis. We thank the Florida Department of Agriculture and Consumer Services—Division of Plant Industry for their support on this contribution.

Conflicts of Interest: The authors declare no conflicts of interest.

Appendix A

Morphological, behavioral, and host characters of Allocorynina used in the matrix in Appendix B.

(1) Labial palp: (0) 3 segments; (1) 2 segments. The presence of 2-segmented labial palps is a synapomorphy for the *Notorhopalotria-Rhopalotria* clade.

(2) Mean male rostral length/pronotal length (RL/PL): (0) <1.0; (1) >1.0, and <1.30; (2) >1.30. High male RL/PL are characteristic of *Parallocorynus* subgenus *Eocorynus* and the subgenus *Parallocorynus bicolor-jonesi-salasae-gregoryi* clade, while low male RL/PL (rostral length < pronotal length) is found in the genus *Notorhopalotria* and two species of *Rhopalotria* subgenus *Allocorynus*.

(3) Mean female RL/PL: (0) <1.25; (1) >1.25, and <1.50; (2) >1.50. High female RL/PL are characteristic of *Parallocorynus* subgenera *Eocorynus* and *Neocorynus*, but has arisen independently in females of other taxa including *Protocorynus bontai*, *Notorhopalotria montgomeryensis*, *Rhopalotria vovidesi*, *Parallocorynus* (*Parallocorynus*) *bicolor*, and *P.* (*P.*) *gregoryi*.

(4) Interocular width/head width at eye: (0) = or >0.5; (1) = or <0.4; (2) >0.4, and <0.5. Short interocular widths are indicative of large eyes and are found in the genera *Protocorynus* and *Notorhopalotria* and in *Rhopalotria* in the *mollis-furfuracea* species group; long interocular distances are indicative of smaller eyes and are characteristic of *Parallocorynus* in the subgenus *Eocorynus* and the subgenus *Parallocorynus bicolor-jonesi-salasae-gregoryi* clade.

(5) Male: mean post-ocular head width/head width at eye (POW/HW): (0) = or <0.95; (1) >0.95 and = or <1.0; (2) >1.0. Wide male post-ocular head width is characteristic of *Notorhopalotria*, *Parallocorynus* subgenus *Eocorynus*, the subgenus *Parallocorynus P. bicolor-jonesi-salasae-gregoryi* clade, and *Rhopalotria calonjei*.

(6) Female: mean POW/HW: (0) = or < 0.95; (1) >0.95 and = or < 1.0; (2) >1.0. Medium to short female postocular head widths are characteristic of the Allocorynina compared with the narrow width in the outgroup genus *Oxycraspedus*.

(7) POW/HW: sexually dimorphic (no overlap): (0) No (dimorphism absent); (1) Yes (dimorphism present). Strong sexual dimorphism in post-ocular head width is characteristic in *Parallocorynus* subgenus *Eocorynus*, but has arisen six separate times in other genera and subgenera.

(8) Transverse postocular groove: (0) Absent; (1) Present. Character 1 is a synapomorphy for the genus *Parallocorynus*.

(9) Antennal insertion shape: (0) Foveiform to slightly oval; (1) Sulciform.

(10) Antennal club, connection of antennomeres: (0) Distinct, 9–10 and 10–11 loosely connected; (1) Distinct, 9–10 loosely connected, 10–11 tightly joined.

(11) Number of pockets on each side club antennomeres: (0) 1; (1) 2; (2) 3. One pocket is characteristic for *Parallocorynus* and *Protocorynus*; two pockets is characteristic of *Rhopalotria* subgenus *Rhopalotria*.

(12) Antennal club pocket shape: (0) Half circle (autapomorphy for *Protocorynus*); (1) Oval to round; (2) Elongate oval with irregular outline. Character state 2 is a synapomorphy for *Rhopalotria* subgenus *Rhopalotria*.

(13) Funiclular antennomere 1 in females: (0) Approximately symmetrical; (1) Strongly asymmetrical. Synapomorphy for the *Parallocorynus bicolor-jonesi-salasae-gregoryi* clade.

(14) Mean scape length in males: (0) >1.1X and <1.8X eye length; (1) <1.1X eye length; (2) >1.8X eye length. Scape length shorter than 1.1X eye length in males is characteristic for *Notorhopalotria* and *Rhopalotria*. Scape length relative to eye length is also a synapomorphy for the *Parallocorynus bicolor-jonesi-salasae-gregoryi* clade; this is the only clade where the scape length routinely exceeds 2X eye length.

(15) Mean scape length in males: (0) <1.27X length of funicular antennomeres 1 and 2; (1) >1.27X length of funicular antennomeres 1 and 2. Scape length shorter than 1.27X length of funicular antennomeres 1 and 2 in males is characteristic for *Rhopalotria*.

(16) Gular suture: (0) Entirely separated; (1) Fused.

(17) Sulcus at posterior margin of eye: (0) Absent; (1) Present and extending around dorsal margin of eye.

(18) Collar on anterior pronotal margin: (0) Present; (1) Absent. Character shared between *Protocorynus* and *Parallocorynus*.

(19) Pronotal apex: (0) Without constriction; (1) With constriction.

(20) Lateral margin of pronotum: (0) Not carinate; (1) With carinae.

(21) Lateral pronotal margin: (0) Not crenulate; (1) Crenulate.

(22) Shape of prothorax: (0) Anterior lateral angles not produced; (1) Anterior lateral angles produced forward.

(23) Male: mean pronotal width/pronotal length (PW/PL): (0) < or = 1.35; (1) >1.35 and = or < 1.5; (2) >1.5. Within the Allocorynina a relatively narrow pronotum in males is characteristic for *Parallocorynus* except for an inferred reversal in the subgenus *Dysicorynus*.
(24) Female: mean PW/PL: (0) <1.25; (1) >1.25 and <1.45; (2) >1.45. Within the Allocorynina, a relatively wide pronotum is characteristic in *Protocorynus* and *Rhopalotria* (except *R. vovidesi*).
(25) PW/PL: (0) Overlap between sexes; (1) No overlap between sexes. Strong sexual dimophism in this character is characteristic of *Parallocorynus* subgenus *Eocorynus*, but has arisen independently twice in *Notorhopalotria* and *Rhopalotria*.
(26) Anterior pronotal setal fringe: (0) Present; (1) Obsolete between eyes (not protruding beyond margin). Shared character between *Notorhopalotria* and *Rhopalotria* (except for *R. vovidesi*).
(27) Fovea on pronotum: (0) Absent; (1) Present.
(28) Notopleural suture reaching anterior margin of pronotum: (0) Yes; (1) No. Characteristic for *Rhopalotria* and with a reversal for *Parallocorynus chemnicki*.
(29) Mean distance from procoxa to anterior margin of prosternum/distance from procoxa to posterior margin of prosternum: (0) >2.2 and <3.8; (1) <2.2; (2) >3.8. High ratios indicate that the procoxae are inserted on the posterior side of the prosternum and is a synapomorphy for *Parallocorynus* subgenera *Eocorynus* and *Neocorynus*.
(30) Procoxae separated by: (0) Broad sclerite; (1) Sclerotized septum; (2) Not separated. The lack of septum is a synapomophy for *Parallocorynus*.
(31) Forecoxae: (0) Partially open laterally; (1) Completely closed laterally.
(32) Male profemur: (0) Not enlarged; (1) Enlarged. Enlarged profemora in males appears to have arisen independently in *Notorhopalotria*, *Rhopalotria*, and the *Eocorynus-Neocorynus* clades.
(33) Male profemur granular field: (0) Absent (1) Present. Presence is a synapomorphy for the *Eocorynus-Neocorynus* clade.
(34) Male profemoral ventrodistal spine number: (0) Absent; (1) One; (2) Two; (3) More than two. Profemoral spines in males appear to have arisen three times independently in the *Notorhopalotria*, *Rhopalotria*, and the *Eocorynus* clades.
(35) Male profemoral spine position at ventrodistal pit: (0) Absent; (1) At proximal apex; (2) Lateral.
(36) Male profemoral spine location from margin of ventrodistal pit: (0) Absent; (1) At margin; (2) Away from margin. Spine location away from pit is a synapomorphy for *Notorhopalotria*.
(37) Male profemur with a longitudinal ventroproximal ridge: (0) Absent; (1) Present. Synapomorphy for *Notorhopalotria*.
(38) Male: profemora with ventrodistal angulation: (0) Absent; (1) Present. Synapomorphy for subgenus *Neocorynus*.
(39) Meso- and metafemora: (0) Not conspicuously compressed; (1) Strongly compressed.
(40) Meso- and metafemora: (0) Without dorsal crenulation; (1) With dorsal crenulation.
(41) Tibial spurs: (0) Present and articulated; (1) Present but fused.
(42) Basal tarsal segment: (0) Subequal to second segment; (1) Much shorter than second and almost concealed.
(43) Pronotum, frons, and dorsal surfaces of profemora with fine reticulation: (0) No; (1) Yes. Synapomorphy for *Notorhopalotria-Rhopalotria* clade (except for a reversal in *R. vovidesi*).
(44) Pronotum compressed: height < 0.4X width: (0) No; (1) Yes.
(45) Pronotum consistently bicolored: (0) No; (1) Yes. Characteristic of *Protocorynus*, with one independent reversal in *N. montgomeryensis*.
(46) Punctures on elytra: (0) Irregularly distributed; (1) Ordered longitudinally but not in perfect striae.
(47) Elytra: (0) With wing locking mechanism, closing to apices, concealing pygidium; (1) Without wing locking mechanism, rounded at apex, pygidium visible.

(48) Elytra bicolored: (0) No; (1) Yes. This character has arisen independently three times within Allocorynina.
(49) Color of frons black (vs. brown): (0) No; (1) Yes. This character has arisen twice in Allocorynina in *Protocorynus* and *Parallocorynus chemnicki*.
(50) Metasternum color black (vs. brown): (0) No; (1) Yes. Within the Allocorynina, this character has arisen once in the *Parallocorynus* subgenus *Parallocorynus bicolor-jonesi-salasae-gregoryi* clade.
(51) Meso- and metafemur always black: (0) No; (1) Yes.
(52) Tibia and femur colors often differ: (0) No; (1) Yes. This character is found in *Parallocorynus* in the subgenus *Eocorynus*, and in the subgenus *Parallocorynus bicolor-jonesi-salasae-gregoryi* clade.
(53) Rostrum color: (0) Brown; (1) Black. A black rostrum has arisen independently twice in *Parallocorynus*.
(54) Mesoventrite: (0) Flat with intercoxal process strongly projected ~45° angle; (1) Slightly proclinate with intercoxal process on same level.
(55) Metaventrite: (0) Convex; (1) Disk flattened.
(56) Metaventrite, latero-posteriorly: (0) Gently rounded; (1) Sharply declined.
(57) Wing vein rm: (0) Not sclerotized (obsolete); (1) Sclerotized.
(58) Wing vein Mr spur: (0) Present; (1) Absent.
(59) Wing vein $1A_2$ length: (0) $>1A_1$; (1) $<1A_1$; (2) Missing. Missing vein is a synapomorphy for the *Notorhopalotria* and *Rhopalotria* clade.
(60) Wing vein $1A_1$: (0) Present; (1) Missing. Missing $1A_1$ vein is a synapomorphy for the *Notorhopalotria* and *Rhopalotria* clade.
(61) Wing vein 3A: (0) Extends beyond confluence with 2A; (1) Obsolete beyond confluence with 2A. Character state 1 is a synapomorphy for the *Notorhopalotria*.
(62) Aedeagus apex subtruncate: (0) No; (1) Yes. Synapomorphy for *Rhopalotria*.
(63) Aedeagus apex length: (0) Approximately equal to own width; (1) Twice own width. Character state 1 is a synapomorphy for *Parallocorynus* subgenus *Dysicorynus*.
(64) Gonopore with sclerotized knob: (0) No; (1) Yes. Synapomorphy for the *Parallocorynus norstogi-perezfarrerai* clade.
(65) Gonopore position: (0) Dorsal; (1) Ventrolateral.
(66) Aedeagus internal sac with ventral strut: (0) No; (1) Yes. Synapomorphy for the *Parallocorynus* subgenus *Parallocorynus*.
(67) Aedeagus internal sac with transfer apparatus: (0) No; (1) Yes. Synapomorphy for *Rhopalotria*.
(68) Aedeagus internal sac w/dart: (0) No; (1) Yes. Synapomorphy for *Parallocorynus*.
(69) Aedeagus internal sac with dorsal pleats: (0) Absent; (1) Present. Synapomorphy for *Parallocorynus* subgenus *Parallocorynus*.
(70) Aedeagus with prominent sclerotized transverse bridge: (0) No; (1) Yes. Synapomorphy for *Parallocorynus*.
(71) Aedeagus shape: (0) Trough-shaped; (1) Flattened.
(72) Tegmen dorsal bridge length from base to its junction with the apical plate extends <1/2 length of apical plate (vs. greater than): (0) No; (1) Yes.
(73) Tegmen apical setae: (0) Absent or length < width of apical plate; (1) Length > width of apical plate. Character state 1 is a synapomorphy for *Notorhopalotria*.
(74) Tegmen apical visor: (0) Absent; (1) Present. Synapomorphy for *Rhopalotria*, with the character arising independently in *Protocorynus* where the visor extends across lateral and part of ventral margin.
(75) Tegmen apical visor curled laterally: (0) No; (1) Yes.
(76) Tegmen apical plate curls transversely: (0) No; (1) Yes. Synapomorphy for *Notorhopalotria*.
(77) Tegmen apodeme height: (0) <width of apical plate; (1) >width of apical plate.

(78) Female: sternum VIII distal half of arms: (0) Strongly converge; (1) Mostly parallel. Character state 1 is a synapomorphy for the *Notorhopalotria* and *Rhopalotria* clade.

(79) Female: sternum VIII arm length: (0) About equal to length of apodeme; (1) >1.5 length of apodeme. Character state (0) is found only in the *R. furfuracea-R. mollis* clade and has arisen independently in *Protocorynus*.

(80) Female: sternum VIII arms: (0) Curved evenly; (1) With sharp angulate bend. Angulate bends is characteristic of *Parallocorynus* with a reversal in the subgenera *Eocorynus* and *Neocorynus*.

(81) Female sternum VIII: junction of arms: (0) Diverging at angle $<90^\circ$; (1) Forming transverse bar.

(82) Female: spermathecal tube length: (0) <sternum VIII length; (1) >sternum VIII length. Long tube is characteristic of *Notorhopalotria* and *Parallocorynus* subgenus *Dysicorynus*.

(83) Spermatheca: (0) Present and falciform; (1) Absent.

(84) Spermathecal gland: (0) Tapering to spermathecal duct; (1) Forming common tube with duct.

(85) Larval feeding site: (0) Female cone; (1) Male sporophyll; (2) Male cone axis.

(86) Pupation site: (0) Female cone; (1) Male cone; (2) Outside of cone. Pupation site outside of cone is a synapomorphy for the *Eocorynus-Neocorynus* clade.

(87) Host plant family: (0) Araucariaceae; (1) Zamiaceae. Synapomorphy for Allocorynina.

(88) Host genus: (0) *Araucaria*; (1) *Dioon*; (2) *Zamia*.

(89) Adult gut contents: (0) Mainly cone tissues other than pollen; (1) Predominately pollen. Character state 1 is a synapomorphy for *Parallocorynus*.

Appendix B

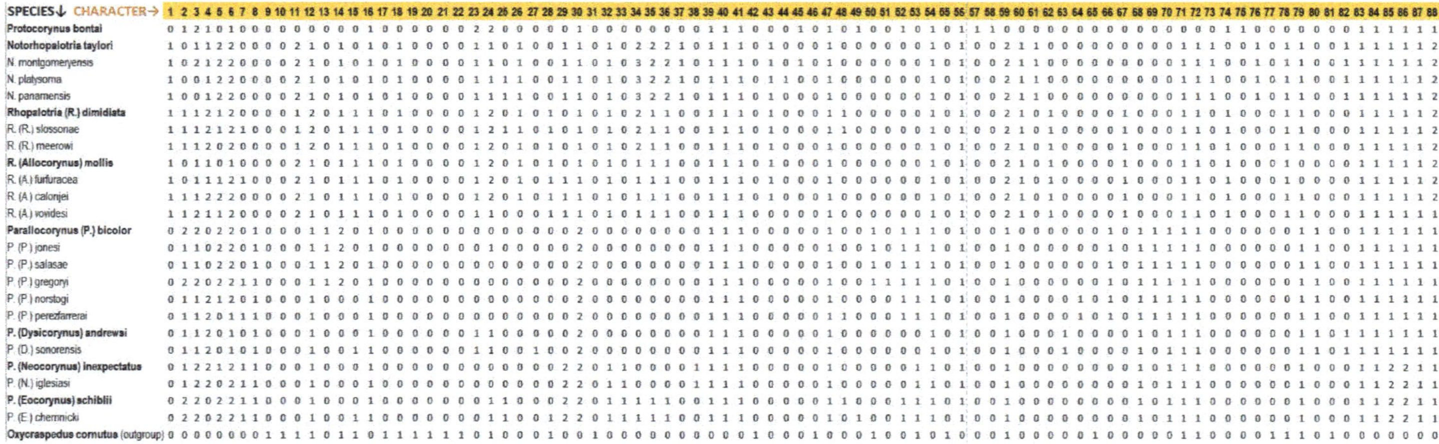

SPECIES↓ CHARACTER→	1	2	3	4	5	6	7	8	9	10	11	12	13	14	15	16	17	18	19	20	21	22	23	24	25	26	27	28	29	30	31	32	33	34	35	36	37	38	39	40	41	42	43	44	45	46	47	48	49	50	51	52	53	54	55	56	57	58	59	60	61	62	63	64	65	66	67	68	69	70	71	72	73	74	75	76	77	78	79	80	81	82	83	84	85	86	87	88	89
Protocorynus bontai	0	1	2	1	0	1	0	0	0	0	0	0	0	0	0	1	0	0	0	0	0	0	2	2	0	0	0	0	0	1	0	0	0	0	0	0	0	0	1	1	1	0	0	0	1	0	1	0	1	0	0	1	0	1	0	1	1	1	0	0	0	0	0	0	0	0	0	0	0	0	0	0	0	1	1	0	0	0	0	0	0	0	1	1	1	1	1	1	0
Notorhopalotria taylori	1	0	1	1	2	2	0	0	0	0	2	1	0	1	0	1	0	1	0	0	0	0	1	1	0	1	0	0	1	1	0	1	0	2	2	2	1	0	1	1	1	0	1	0	0	0	1	0	0	0	0	0	0	1	0	1	0	0	2	1	1	0	0	0	0	0	0	0	0	0	1	1	1	0	0	1	0	1	1	0	0	1	1	1	1	1	1	2	0
N. montgomeryensis	1	0	2	1	2	2	0	0	0	0	2	1	0	1	0	1	0	1	0	0	0	0	1	1	0	1	0	0	1	1	0	1	0	3	2	2	1	0	1	1	1	0	1	0	1	0	1	0	0	0	0	0	0	1	0	1	0	0	2	1	1	0	0	0	0	0	0	0	0	0	1	1	1	0	0	1	0	1	1	0	0	1	1	1	1	1	1	2	0
N. platysoma	1	0	0	1	2	2	0	0	0	0	2	1	0	1	0	1	0	1	0	0	0	0	1	1	1	1	0	0	1	1	0	1	0	3	2	2	1	0	1	1	1	0	1	1	0	0	1	0	0	0	0	0	0	1	0	1	0	0	2	1	1	0	0	0	0	0	0	0	0	0	1	1	1	0	0	1	0	1	1	0	0	1	1	1	1	1	1	2	0
N. panamensis	1	0	0	1	2	2	0	0	0	0	2	1	0	1	0	1	0	1	0	0	0	0	1	1	1	1	0	0	1	1	0	1	0	3	2	2	1	0	1	1	1	0	1	0	0	0	1	0	0	0	0	0	0	1	0	1	0	0	2	1	1	0	0	0	0	0	0	0	0	0	1	1	1	0	0	1	0	1	1	0	0	1	1	1	1	1	1	2	0
Rhopalotria (R.) dimidiata	1	1	1	2	1	2	0	0	0	0	1	2	0	1	1	1	0	1	0	0	0	0	1	2	0	1	0	1	0	1	0	1	0	2	1	1	0	0	1	1	1	0	1	0	0	0	1	1	0	0	0	0	0	1	0	1	0	0	2	1	0	1	0	0	0	0	1	0	0	0	1	1	0	1	0	0	0	1	1	0	0	0	1	1	1	1	1	2	0
R. (R.) slossonae	1	1	1	2	1	2	1	0	0	0	1	2	0	1	1	1	0	1	0	0	0	0	1	2	1	1	0	1	0	1	0	1	0	2	1	1	0	0	1	1	1	0	1	0	0	0	1	1	0	0	0	0	0	1	0	1	0	0	2	1	0	1	0	0	0	0	1	0	0	0	1	1	0	1	0	0	0	1	1	0	0	0	1	1	1	1	1	2	0
R. (R.) meerowi	1	1	1	2	0	2	0	0	0	0	1	2	0	1	1	1	0	1	0	0	0	0	1	2	0	1	0	1	0	1	0	1	0	2	1	1	0	0	1	1	1	0	1	0	0	0	1	0	0	0	0	0	0	1	0	1	0	0	2	1	0	1	0	0	0	0	1	0	0	0	1	1	0	1	0	0	0	1	1	0	0	0	1	1	1	1	1	2	0
R. (Allocorynus) mollis	1	0	1	1	0	1	0	0	0	0	2	1	0	1	1	1	0	1	0	0	0	0	1	2	0	1	0	1	0	1	0	1	0	1	1	1	0	0	1	1	1	0	1	0	0	0	1	0	0	0	0	0	0	1	0	1	0	0	2	1	0	1	0	0	0	0	1	0	0	0	1	1	0	1	0	0	0	1	0	0	0	0	1	1	1	1	1	2	0
R. (A.) furfuracea	1	0	1	1	1	2	1	0	0	0	2	1	0	1	1	1	0	1	0	0	0	0	1	2	0	1	0	1	1	1	0	1	0	1	1	1	0	0	1	1	1	0	1	0	0	0	1	0	0	0	0	0	0	1	0	1	0	0	2	1	0	1	0	0	0	0	1	0	0	0	1	1	0	1	0	0	0	1	0	0	0	0	1	1	1	1	1	2	0
R. (A.) calonjei	1	1	1	2	2	2	0	0	0	0	2	1	0	1	1	1	0	1	0	0	0	0	1	2	0	1	0	1	1	1	0	1	0	1	1	1	0	0	1	1	1	0	1	0	0	0	1	0	0	0	0	0	0	1	0	1	0	0	2	1	0	1	0	0	0	0	1	0	0	0	1	1	0	1	0	0	0	1	1	0	0	0	1	1	1	1	1	2	0
R. (A.) vovidesi	1	1	2	1	1	2	0	0	0	0	2	1	0	1	1	1	0	1	0	0	0	0	1	1	0	0	0	1	1	1	0	1	0	1	1	1	0	0	1	1	1	0	0	0	0	0	1	0	0	0	0	0	0	1	0	1	0	0	2	1	0	1	0	0	0	0	1	0	0	0	1	1	0	1	0	0	0	1	1	0	0	0	1	1	1	1	1	1	0
Parallocorynus (P.) bicolor	0	2	2	0	2	2	0	1	0	0	0	1	1	2	0	1	0	0	0	0	0	0	0	0	0	0	0	0	0	2	0	0	0	0	0	0	0	0	1	1	1	0	0	0	0	0	1	0	0	1	0	1	1	1	0	1	0	0	1	0	0	0	0	0	0	1	0	1	1	1	1	1	0	0	0	0	0	0	1	1	0	0	1	1	1	1	1	1	1
P. (P.) jonesi	0	1	1	0	2	2	0	1	0	0	0	1	1	2	0	1	0	0	0	0	0	0	0	1	0	0	0	0	0	2	0	0	0	0	0	0	0	0	1	1	1	0	0	0	0	0	1	0	0	1	0	1	1	1	0	1	0	0	1	0	0	0	0	0	0	1	0	1	1	1	1	1	0	0	0	0	0	0	1	1	0	0	1	1	1	1	1	1	1
P. (P.) salasae	0	1	1	0	2	2	0	1	0	0	0	1	1	2	0	1	0	0	0	0	0	0	0	0	0	0	0	0	0	2	0	0	0	0	0	0	0	0	1	1	1	0	0	0	0	0	1	0	0	1	0	1	1	1	0	1	0	0	1	0	0	0	0	0	0	1	0	1	1	1	1	1	0	0	0	0	0	0	1	1	0	0	1	1	1	1	1	1	1
P. (P.) gregoryi	0	2	2	0	2	2	1	1	0	0	0	1	1	2	0	1	0	0	0	0	0	0	0	0	0	0	0	0	0	2	0	0	0	0	0	0	0	0	1	1	1	0	0	0	0	0	1	0	0	1	1	1	1	1	0	1	0	0	1	0	0	0	0	0	0	1	0	1	1	1	1	1	0	0	0	0	0	0	1	1	0	0	1	1	1	1	1	1	1
P. (P.) norstogi	0	1	1	2	1	2	0	1	0	0	0	1	0	0	0	1	0	0	0	0	0	0	0	0	0	0	0	0	0	2	0	0	0	0	0	0	0	0	1	1	1	0	0	0	0	0	1	0	0	0	0	1	1	1	0	1	0	0	1	0	0	0	0	1	0	1	0	1	1	1	1	1	0	0	0	0	0	0	1	1	0	0	1	1	1	1	1	1	1
P. (P.) perezfarrerai	0	1	1	2	0	1	1	1	0	0	0	1	0	0	0	1	0	0	0	0	0	0	0	0	0	0	0	0	0	2	0	0	0	0	0	0	0	0	1	1	1	0	0	0	0	0	1	1	0	0	0	1	1	1	0	1	0	0	1	0	0	0	0	1	0	1	0	1	1	1	1	1	0	0	0	0	0	0	1	1	0	0	1	1	1	1	1	1	1
P. (Dysicorynus) andrewsi	0	1	1	2	0	1	0	1	0	0	0	1	0	0	0	1	0	0	0	0	0	0	1	1	0	0	0	0	0	2	0	0	0	0	0	0	0	0	1	1	1	0	0	0	0	0	1	0	0	0	0	0	0	1	0	1	0	0	1	0	0	0	1	0	0	0	0	1	0	1	1	1	0	0	0	0	0	0	1	1	0	1	1	1	1	1	1	1	1
P. (D.) sonorensis	0	1	1	2	0	1	0	1	0	0	0	1	0	0	1	1	0	0	0	0	0	0	1	1	0	0	1	0	0	2	0	0	0	0	0	0	0	0	1	1	1	0	0	0	0	0	1	0	0	0	0	0	0	1	0	1	0	0	1	0	0	0	1	0	0	0	0	1	0	1	1	1	0	0	0	0	0	0	1	1	0	1	1	1	1	1	1	1	1
P. (Neocorynus) inexpectatus	0	1	2	2	1	2	1	1	0	0	0	1	0	0	0	1	0	0	0	0	0	0	0	0	0	0	0	0	2	2	0	1	1	0	0	0	0	1	1	1	1	0	0	0	0	0	1	0	0	0	0	0	1	1	0	1	0	0	1	0	0	0	0	0	0	0	0	1	0	1	1	1	0	0	0	0	0	0	1	0	0	0	1	1	2	2	1	1	1
P. (N.) iglesiasi	0	1	2	2	0	2	1	1	0	0	0	1	0	0	0	1	0	0	0	0	0	0	0	0	0	0	0	0	2	2	0	1	1	0	0	0	0	1	1	1	1	0	0	0	0	0	1	0	0	0	0	0	1	1	0	1	0	0	1	0	0	0	0	0	0	0	0	1	0	1	1	1	0	0	0	0	0	0	1	0	0	0	1	1	2	2	1	1	1
P. (Eocorynus) schiblii	0	2	2	0	2	2	1	1	0	0	0	1	0	0	0	1	0	0	0	0	0	0	0	1	1	0	0	0	2	2	0	1	1	1	1	1	0	0	1	1	1	0	0	0	0	0	1	1	0	0	0	1	1	1	0	1	0	0	1	0	0	0	0	0	0	0	0	1	0	1	1	1	0	0	0	0	0	0	1	0	0	0	1	1	2	2	1	1	1
P. (E.) chemnicki	0	2	2	0	2	2	1	1	0	0	0	1	0	0	1	1	0	0	0	0	0	0	0	1	1	0	0	1	2	2	0	1	1	1	1	1	0	0	1	1	1	0	0	0	0	0	1	0	1	0	0	1	1	1	0	1	0	0	1	0	0	0	0	0	0	0	0	1	0	1	1	1	0	0	0	0	0	0	1	0	0	0	1	1	2	2	1	1	1
Oxycraspedus cornutus (outgroup)	0	0	0	0	0	0	0	0	1	1	1	1	0	1	1	0	1	1	1	1	1	1	0	1	0	0	0	1	0	0	1	0	0	0	0	0	0	0	0	0	0	1	0	0	0	1	0	0	0	1	0	0	1	0	1	0	0	0	1	0	0	0	0	0	1	0	0	0	0	0	1	1	0	0	0	0	1	1	1	0	1	0	0	0	0	0	0	0	0

Figure A1. Character matrix for Allocorynina based on morphological, behavioral, and host-association characters (see Appendix A for description of characters states).

Appendix C

Synapomorphic characters separating genera and subgenera of Allocorynina.

The monophyly of *Notorhopalotria* is supported by the following synapomorphies:

(1) Male profemur with a ventroproximal ridge.
(2) Male profemoral spine(s) located distantly from margin of profemoral apical pit.
(3) Wing veins $1A_1$, $1A_2$, and 3A obsolete and not reaching margin of wing.
(4) Tegmen apical setae longer than width of tegmen.
(5) Tegmen apical plate curls transversely.

The monophyly of *Parallocorynus* is supported by the following synapomorphies:

(1) Transverse postocular groove.
(2) Antennal insertion pointing ventrad.
(3) One oval sensory pocket on each side of club antennomeres 1 and 2.
(4) Procoxae not separated by septum.
(5) Wing vein $1A_1$ present and shorter than $1A_2$.
(6) Aedeagal internal sac with a dart.
(7) Adults feeding primarily on pollen.

The monophyly of monotypic *Protocorynus* is supported by the following autapomorphies:

(1) Single semicircular-shaped pit on each side of club antennomeres 1 and 2.
(2) Pronotal maculation that extends to base of pronotum.
(3) Tegmen with an apical visor that extends from the dorsal region to part of ventral margin.
(4) Aedeagus dorsoventrally flattened.
(5) Spermathecal tube covered with filaments (versus smooth in other Allocorynina [8]).

The monophyly of *Rhopalotria* is supported by these synapomorphies:

(1) Wing vein $1A_1$ missing, but $1A_2$ and 3A retained.
(2) Aedeagal apex subtruncate.
(3) Aedeagal internal sac with transfer apparatus.
(4) Tegmen with a dorso-lateral apical visor.

Within the genus *Rhopalotria* two subgenera are distinguished by the following combinations of characters:

Subgenus Rhopalotria:

(1) Male profemora with a single spine at base of profemoral apical pit.
(2) Two elongate oval sensory pits on each side of club antennomeres 1 and 2.
(3) Average scape length < length of funicular antennomeres 1 and 2.

Subgenus Allocorynus:

(1) Male profemora with pair of spines at either side of base of the profemoral apical pit.
(2) Three round sensory pits on either side of club antennomeres 1 and 2.

Within the genus *Parallocorynus*, four subgenera are supported by the following combination of characters:

Subgenus Dysicorynus:

(1) Profemora without granules or spines.
(2) Female RL/PL >1.27 and <1.44.

(3) Length of aedeagal apex twice own width.
(4) Larvae feed and pupate inside of male cone sporophylls.

Subgenus *Eocorynus*:

(1) Male profemora with granular field and spine.
(2) Female RL/PL >1.77 and <1.95.
(3) Larvae feed along cone axis.
(4) Pupation outside of cone.

Subgenus *Neocorynus*:

(1) Male profemora with granular field and no spine.
(2) Female RL/PL >1.55 and <1.76.
(3) Larvae feed along cone axis.
(4) Pupation outside of cone.

Subgenus Parallocorynus:

(1) Profemora without granules or spines.
(2) Female RL/PL >1.27 and <1.65.
(3) Aedeagus with internal sac with ventral strut and dorsal pleats.
(4) Larvae feed and pupate inside of male cone sporophylls.

Appendix D

Additional phylogenetic trees for Erotylidae.

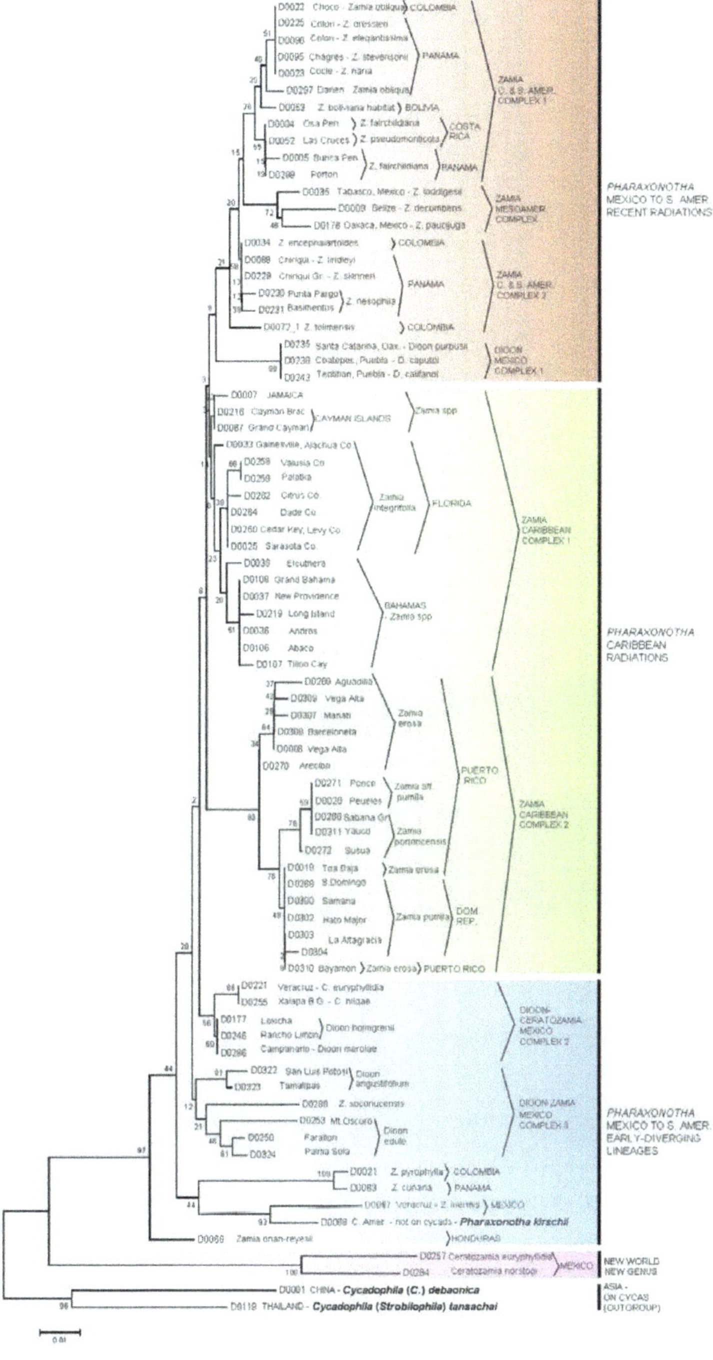

Figure A2. Phylogenetic tree for the Erotylidae: Pharaxonothinae on the cycads of the New World based on neighbor joining (NJ) analysis of 16S rRNA gene sequences; scale bar indicates base pair substitutions per nucleotide position; numbers on branches are bootstrap values.

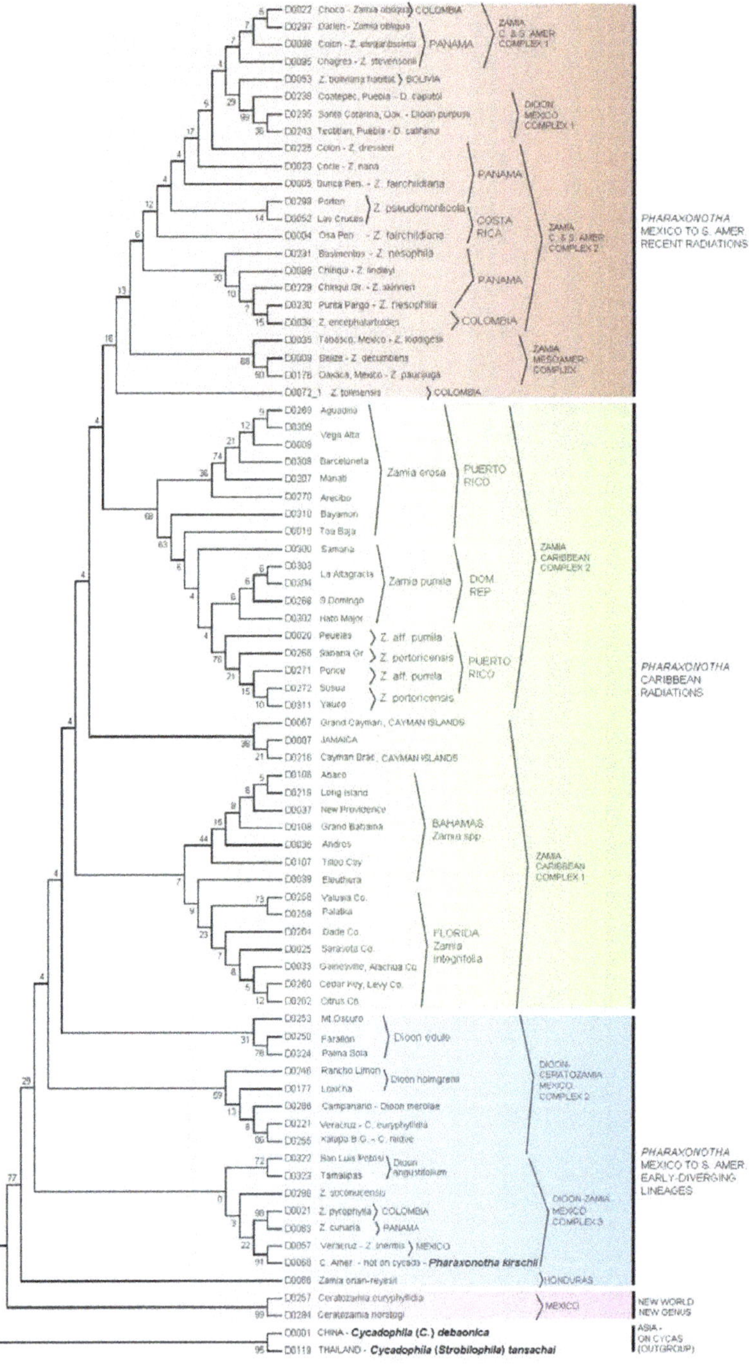

Figure A3. Phylogenetic tree for the Erotylidae: Pharaxonothinae on the cycads of the New World based on maximum parsimony (MP) analysis of 16S rRNA gene sequences; numbers on branches are bootstrap values.

References

1. Norstog, K.; Stevenson, D.W.; Niklas, K.J. The role of beetles in the pollination of *Zamia furfuracea* L.fil. (Zamiaceae). *Biotropica* **1986**, *18*, 300–306. [CrossRef]
2. Norstog, K.; Fawcett, P.K.S.; Vovides, A.P. Beetle pollination of two species of *Zamia*: Evolutionary and ecological considerations. *Palaeobotanist* **1992**, *41*, 149–158.
3. Norstog, K.; Fawcett, P.K.S. Insect-cycad symbiosis and its relation to the pollination of *Zamia furfuracea* (Zamiaceae) by *Rhopalotria mollis* (Curculionidae). *Am. J. Bot.* **1989**, *76*, 1380–1394. [CrossRef]
4. Tang, W. Insect pollination in the cycad *Zamia pumila* (Zamiaceae). *Am. J. Bot.* **1987**, *74*, 90–99. [CrossRef]
5. Valencia-Montoya, W.A.; Tuberquia, D.; Guzmán, P.A.; Cardona-Duque, J. Pollination of the cycad *Zamia incognita* A. Lindstr. and Idárraga by *Pharaxonotha* beetles in the Magdalena Medio Valley, Colombia: A mutualism dependent on a specific pollinator and its significance for conservation. *Arthropod-Plant Interact.* **2017**, *11*, 717–729. [CrossRef]
6. Vovides, A.P. Insect symbionts of some Mexican cycads in their natural habitat. *Biotropica* **1991**, *23*, 102–104. [CrossRef]
7. Chaves, R.; Genaro, J. A new species of *Pharaxonotha* (Coleoptera: Erotylidae) probable pollinator of the endangered Cuban cycad, *Microcycas calocoma* (Zamiaceae). *Insecta Mundi* **2005**, *19*, 143–150.
8. O'Brien, C.; Tang, W. Revision of the New World cycad weevils of the subtribe Allocorynina, with description of two new genera and three new subgenera (Coleoptera: Belidae: Oxycoryninae). *Zootaxa* **2015**, *3970*, 1–87. [CrossRef] [PubMed]
9. Donaldson, J.S.; Nanni, I.; De Wet Bosenberg, J. The role of insects in the pollination of *Encephalartos cycadifolius*. In *Proceedings of the Third International Conference on Cycad Biology*; Vorster, P., Ed.; Cycad Society of South Africa: Stellenbosch, South Africa, 1995; pp. 423–434. ISBN 0620192283.
10. Donaldson, J.S. Is there a floral parasite mutualism in cycad pollination? The pollination biology of *Encephalartos villosus* (Zamiaceae). *Am. J. Bot.* **1997**, *84*, 1398–1406. [CrossRef]
11. Tang, W.; Oberprieler, R.; Yang, S. Beetles (Coleoptera) in cones of Asian *Cycas*: Diversity, evolutionary patterns, and implications for *Cycas* taxonomy. In *Proceedings of the Fourth International Conference on Cycad Biology*; Chen, C., Ed.; International Academic Publishers: Beijing, China, 1999; pp. 280–297.
12. Tang, W. Cycad insects and pollination. In *Vistas in Palaeobotany and Plant Morphology: Evolutionary and Environmental Perspectives Professor D.D. Pant Memorial Volume*; Srivastava, P.C., Ed.; U.P. Offset: Lucknow, India, 2004; pp. 383–394.
13. Terry, I. Thrips and weevils as dual, specialist pollinators of the Australian cycad *Macrozamia communis* (Zamaiceae). *Int. J. Plant Sci.* **2001**, *162*, 1293–1305. [CrossRef]
14. Terry, I.; Tang, W.; Taylor, A.; Donaldson, J.; Singh, R.; Vovides, A.; Cibrián Jaramillo, A. An overview of cycad pollination studies. *Mem. N. Y. Bot. Gard.* **2012**, *106*, 352–394.
15. Wilson, G.W. Pollination in the cycad genus *Bowenia* Hook. Ex Hook. f. (Stangeriaceae). *Biotropica* **2002**, *34*, 438–441. [CrossRef]
16. Hall, J.A.; Walter, G.H.; Bergstrom, D.M.; Machin, P. Pollination ecology of the Australian cycad *Lepidozamia peroffskyana* (Zamiaceae). *Aust. J. Bot.* **2004**, *52*, 333–343. [CrossRef]
17. Kono, M.; Tobe, H. Is *Cycas revoluta* (Cycadaceae) wind- or insect pollinated? *Am. J. Bot.* **2007**, *94*, 847–855. [CrossRef] [PubMed]
18. Procheş, Ş.; Johnson, S.D. Beetle pollination of the fruit-scented cones of the South African cycad, *Stangeria eriopus*. *Am. J. Bot.* **2009**, *96*, 1722–1730. [CrossRef] [PubMed]
19. Suinyuy, T.N.; Donaldson, J.S.; Johnson, S.D. Insect pollination in the African cycad, *Encephalartos friderici-guilielmi* Lehm. *S. Afr. J. Bot.* **2009**, *75*, 682–688. [CrossRef]
20. Walters, T.; Osborne, R.; Decker, D. 'We hold these truths'. In *Cycad Classification Concepts and Recommendations*; Walters, T., Osborne, R., Eds.; CABI Publishing: Oxfordshire, UK, 2004; pp. 1–11. ISBN 0851997414.
21. Chemnick, J.; Oberprieler, R.; Donaldson, J.; Terry, I.; Osborne, R.; Tang, W.; Forster, P. Insect pollinators of cycads A report from a cycad pollination workshop held in Thailand, 2002 with a protocol for collecting and studying cycad pollinators. *Cycad Newsl.* **2004**, *27*, 3–7.
22. Bouchard, P.; Bousquet, Y.; Davies, A.E.; Alonso-Zarazaga, M.A.; Lawrence, J.F.; Lyal, C.H.C.; Newton, A.F.; Reid, C.A.M.; Schmitt, M.; Ślipińsk, S.A.; et al. Family-group names in Coleoptera (Insecta). *ZooKeys* **2011**, *88*, 1. [CrossRef] [PubMed]

23. Tang, W.; O'Brien, C.W. Distribution and evolutionary patterns of the cycad weevil genus *Rhopalotria* (Coleoptera: Curculionoidea: Belidae) with emphasis on the fauna of Panama. *Mem. N. Y. Bot. Gard.* **2012**, *106*, 335–351.
24. Calonje, M.; Stevenson, D.W.; Stanberg, L. The World List of Cycads. Available online: http://www.cycadlist.org (accessed on 2 May 2018).
25. Franz, N.M.; Skelley, P. *Pharaxonotha portophylla* (Coleoptera: Erotylidae), new species and pollinator of *Zamia* (Zamiaceae) in Puerto Rico. *Caribb. J. Sci.* **2008**, *44*, 321–333. [CrossRef]
26. Skelley, P.; Xu, G.; Tang, W.; Lindström, A.; Marler, T.; Khuraijam, J.S.; Singh, R.; Radha, R.; Rich, S. Review of *Cycadophila* Xu, Tang and Skelley (Coleoptera: Erotylidae: Pharaxonothinae) inhabiting *Cycas* (Cycadaceae) in Asia, with descriptions of a new subgenus and thirteen new species. *Zootaxa* **2017**, *4267*, 1–63. [CrossRef] [PubMed]
27. Tang, W.; Skelley, P.; Thomas, M.C.; Perez-Farrera, M.A. Beetles inhabiting the cycad genus Ceratozamia (Cycadales: Zamaiceae). in preparation.
28. Xu, G.; Tang, W.; Skelley, P.; Liu, N.; Rich, S. *Cycadophila*, a new genus (Coleoptera: Erotylidae: Pharaxonothinae) inhabiting *Cycas debaoensis* (Cycadaceae) in Asia. *Zootaxa* **2015**, *3986*, 251–278. [CrossRef] [PubMed]
29. Maldonado-Ruiz, M.F.; Flores-Vázquez, J.C. Entomofauna Asociada a *Dioon* sp. nov. y actividad de los insectos polinizadores en San Jerónimo Taviche, Oaxaca, Mexico. *Mem. N. Y. Bot. Gard.* **2012**, *106*, 295–301.
30. Oberprieler, R.G. The weevils (Coleoptera: Curculionoidea) associated with cycads. 2. Host specificity and implications for cycad taxonomy. In *Proceedings of the 3rd International Conference of Cycad Biology, Conservation through Cultivation; Vorster*; Vorster, P., Ed.; Cycad Society of South Africa: Stellenbosch, South Africa, 1995; pp. 335–365. ISBN 0620192283.
31. Oberprieler, R.G. Evil weevils—The key to cycad survival and diversification? In *Proceedings of the Sixth International Conference on Cycad Biology*; Lindstrom, A., Ed.; Nong Nooch Tropical Garden: Bangkok, Thailand, 2004; pp. 170–194. ISBN 9749235916.
32. Marvaldi, A.; Oberprieler, R.; Lyal, C.; Bradbury, T.; Anderson, R. Phylogeny of the Oxycoryninae *sensu lato* (Coleoptera: Belidae) and evolution of host-plant associations. *Invertebr. Syst.* **2006**, *20*, 447–476. [CrossRef]
33. Goloboff, P.A.; Farris, J.A.; Nixon, K.C. TNT, a free program for phylogenetic analysis. *Caldistics* **2008**, *24*, 774–786. [CrossRef]
34. Wink, M.; Mikes, Z.; Rheinheimer, J. Phylogenetic relationships in weevils (Coleoptera: Curculionoidea) inferred from nucleotide sequences of mitochondrial 16S rDNA. *Naturwissenschaften* **1997**, *84*, 318–321. [CrossRef] [PubMed]
35. Whitfield, J.B.; Cameron, S.A. Hierarchical analysis of variation in the mitochondrial 16S rRNA gene among Hymenoptera. *Mol. Biol. Evol.* **1998**, *15*, 1728–1743. [CrossRef] [PubMed]
36. Hosoya, T.; Araya, K. Phylogeny of Japanese stag beetles (Coleoptera: Lucanidae) inferred from 16S mtrRNA gene sequences, with reference to the evolution of sexual dimorphism of mandibles. *Zool. Sci.* **2005**, *22*, 1305–1318. [CrossRef] [PubMed]
37. Sobti, R.S.; Sharma, V.L.; Kumari, M.; Gill, T.K. Genetic relatedness of six North-Indian butterfly species (Lepidoptera: Pieridae) based on 16S rRNA sequence analysis. *Mol. Cell. Biochem.* **2007**, *295*, 145–151. [CrossRef] [PubMed]
38. Aruggoda, A.G.B.; Shunxiang, R.; Baoli, Q. Molecular phylogeny of ladybird beetles (Coccinellidae: Coleoptera) inferred from mitochondrial 16S rDNA sequences. *Trop. Agric. Res.* **2010**, *21*, 209–217. [CrossRef]
39. Caterino, M.S.; Cho, S.; Sperling, F.A.H. The current state of insect molecular systematics: A thriving tower of babel. *Ann. Rev. Entomol.* **2000**, *45*, 1–54. [CrossRef] [PubMed]
40. Katoh, K.; Standley, D.M. MAFFT multiple sequence alignment software version 7: Improvements in performance and usability *Mol. Biol. Evol.* **2013**, *30*, 772–780. [CrossRef] [PubMed]
41. Stamatakis, A. RAxML version 8: A tool for phylogenetic analysis and post-analysis of large phylogenies. *Bioinformatics* **2014**, *9*, 1312–1313. [CrossRef] [PubMed]
42. Ronquist, F.; Teslenko, M.; van der Mark, P.; Ayres, D.L.; Darling, A.; Höhna, S.; Larget, B.; Liu, L.; Suchard, M.A.; Huelsenbeck, J.P. MrBayes 3.2: Efficient bayesian phylogenetic inference and model choice across a large model space. *Syst. Biol.* **2012**, *61*, 539–542. [CrossRef] [PubMed]
43. Larkin, M.A.; Blackshields, G.; Brown, N.P.; Chenna, R.; McGettigan, P.A.; McWilliam, H.; Valentin, F.; Wallace, I.M.; Wilm, A.; Lopez, R.; et al. Clustal W and Clustal X version 2.0. *Bioinformatics* **2007**, *23*, 2947–2948. [CrossRef] [PubMed]

44. Swofford, D.L. *PAUP*: Phylogenetic Analysis Using Parsimony (*and Other Methods), version 4.0*, Sinauer Associates, Inc.: Sunderland, MA, USA, 2002.
45. Tamura, K.; Peterson, D.; Peterson, N.; Stecher, G.; Nei, M.; Kumar, S. MEGA5: Molecular evolutionary genetics analysis using maximum likelihood, evolutionary distance, and maximum parsimony methods. *Mol. Biol. Evol.* **2011**, *28*, 2731–2739. [CrossRef] [PubMed]
46. Nicolalde-Morejon, F.; Vergara-Silva, F.; Gonzalez-Astoraga, J.; Stevenson, D.W. Character based, population-level DNA barcoding in Mexican species of *Zamia* L. (Zamiaceae: Cycadales). *Mitochondrial DNA* **2010**, *21*, 51–59. [CrossRef] [PubMed]
47. Moynihan, J.; Vovides, A.P.; González-Astorga, J.; Francisco-Ortega, J. Population genetic diversity in the *Dioon edule* Lindl. species complex (Zamiaceae, Cycadales): Evidence from microsatellite data. *Mem. N. Y. Bot. Gard.* **2012**, *106*, 224–250.
48. Moynihan, J.; Stevenson, D.W.; Lewis, C.E.; Vovides, A.P.; Caputo, P.; Francisco-Ortega, J. A phylogenetic study of *Dioon* Lindl. (Zamiaceae, Cycadales), based on morphology, nuclear ribosomal DNA, a low-copy nuclear gene, and plastid RFLPs. *Mem. N. Y. Bot. Gard.* **2012**, *106*, 448–479.
49. Gutiérrez-Ortega, J.S.; Yamamoto, T.; Vovides, A.P.; Pérez-Farrera, M.A.; Martínez, J.F.; Molina-Freaner, F.; Watano, Y.; Kajita, T. Aridification as a driver of biodiversity: A case study for the cycad genus *Dioon* (Zamiaceae). *Ann. Bot.* **2018**, *121*, 47–60. [CrossRef] [PubMed]
50. Hall, J.A.; Walter, G.H. Seed dispersal of the Australian cycad *Macrozamia miquelii* (Zamiaceae): Are cycads megafauna-dispersed "grove forming" plants? *Am. J. Bot.* **2013**, *100*, 1127–1136. [CrossRef] [PubMed]
51. Burbridge, A.H.; Whelan, R.J. Seed dispersal in acycad, *Macrozamia riedlei*. *Aust. J. Ecol.* **1982**, *7*, 63–67. [CrossRef]
52. Ballardie, R.T.; Whelan, R.J. Masting, seed dispersal and seed predation in the cycad *Macrozamia communis*. *Oecologia* **1986**, *70*, 100–105. [CrossRef] [PubMed]
53. Tang, W. Seed dispersal in the cycad *Zamia pumila* in Florida. *Can. J. Bot.* **1989**, *67*, 2066–2070. [CrossRef]
54. Crowson, R.A. On the systematic position of *Allocoryninae* (Coleoptera: Allocorynidae). *Coleopt. Bull.* **1986**, *40*, 243–244.
55. Crowson, R.A. The relations of Coleoptera to Cycadales. In *Advances in Coleopterology*; Zunino, M., Belle's, X., Blas, M., Eds.; Asociacion Europea de Coleopterologia: Torino, Italy, 1991; pp. 13–28.
56. Brookes, D.R.; Hereward, J.P.; Terry, L.I.; Walter, G.H. Evolutionary dynamics of a cycad obligate pollination mutualism—Pattern and process in extant Macrozamia cycads and their specialist thrips pollinators. *Mol. Phylogenet. Evol.* **2015**, *93*, 83–93. [CrossRef] [PubMed]
57. Franz, N.M.; Engel, M.S. Can higher-level phylogenies of weevils explain their evolutionary success? A critical review. *Syst. Entomol.* **2010**, *35*, 596–606. [CrossRef]
58. Rogers, R.D.; Mann, P.; Emmet, P.A. Tectonic terranes of the Chortis block based on integration of regional aeromagnetic and geologic data. In *Geologic and Tectonic Development of the Caribbean Plate in Northern Central America*; Mann, P., Ed.; Geological Society of America: Boulder, CO, USA, 2007; Volume 428, pp. 65–88. [CrossRef]
59. Coates, A. The forging of Central America. In *Central America: A Natural and Cultural History*; Coates, A., Ed.; Yale Univ. Press: New Haven, CT, USA, 1997; pp. 1–37. ISBN 0300068298.
60. Tang, W. Evolutionary history of cycads in North America. *Cycad Newsl.* **2012**, *35*, 7–13.
61. Kvaček, Z. A noteworthy cycad, *Ceratozamia hofmannii* Ettingshausen 1887, from the Lower Miocene of Austria re-examined. *Neues Jahrb. Geol. Paläontol. Monatshefte* **2004**, 111–118.
62. Nagalingum, N.; Marshall, C.; Quental, T.; Rai, H.; Little, D.; Matthews, S. Recent synchronous radiation of a living fossil. *Science* **2011**, *334*, 796–799. [CrossRef] [PubMed]
63. Salas-Leiva, D.; Meerow, A.W.; Calonje, M.; Griffith, M.P.; Francisco-Ortega, J.; Stevenson, D.W.; Nakamura, K.; Lewis, C.E.; Namoff, S. Phylogeny of the cycads based on multiple single copy nuclear genes: Congruence of concatenation and species tree inference methods. *Ann. Bot.* **2013**, *112*, 1263–1278. [CrossRef] [PubMed]
64. Condamine, F.; Nagalingum, N.; Marshall, C.R.; Morlon, H. Origin and diversification of living cycads: A cautionary tale on the impact of the branching process prior in Bayesian molecular dating. *BMC Evol. Biol.* **2015**, *15*, 65. [CrossRef] [PubMed]

65. Meerow, A.W.; Francisco-Ortega, J.; Calonje, M.; Griffith, M.P.; Ayala-Silva, T.; Stevenson, D.W.; Nakamura, K. *Zamia* (Cycadales: Zamiaceae) on Puerto Rico: Asymmetric genetic differentiation and the hypothesis of multiple introductions. *Am. J. Bot.* **2012**, *99*, 1828–1839. [CrossRef] [PubMed]
66. Lazcano-Lara, J.C. The Reproductive Biology of Zamia (Cycadales: Zamaiaceae) in Puerto Rico: Implications for Patterns of Genetic Structure and Species Conservation. Ph.D. Thesis, University of Puerto Rico, Rio Píedras, Puerto Rico, May 2015.
67. Caputo, P.; Cozzolino, S.; De Luca, P.; Moretti, A.; Stevenson, D.W. Molecular Phylogeny of *Zamia* (Zamiaceae). In *Cycad Classification Concepts and Recommendations*; Walters, T., Osborne, R., Eds.; CABI Publishing: Oxfordshire, UK, 2004; pp. 149–157. ISBN 085199741.
68. Nagalingum, N.; Marshall, C.; Quental, T.; Rai, H.; Little, D.; Matthews, S. Supporting Online Material for Recent Synchronous Radiation of a Living Fossil. *Science* **2011**, 1–38. [CrossRef]
69. Salas-Leiva, D.E.; Meerow, A.W.; Calonje, M.; Francisco-Ortega, J.; Griffith, M.P.; Nakamura, K.; Sánchez, V.; Knowles, L.; Knowles, D. Shifting Quaternary migration patterns in the Bahamian archipelago: Evidence from the *Zamia pumila* complex at the northern limits of the Caribbean island biodiversity hotspot. *Am. J. Bot.* **2017**, *104*, 757–771. [CrossRef] [PubMed]
70. Norstog, K.; Nicholls, T. *The Biology of the Cycads*; Cornell Univ. Press: Ithaca, NY, USA; London, UK, 1997; 383p, ISBN 080143033X.
71. Donaldson, J. *Cycads Status Survey and Conservation Action Plan*; IUCN: Gland, Switzerland; Cambridge, UK, 2003; 86p, ISBN 2831706998.
72. Poinar, G.; Legalov, A.A. *Pleurambus strongylus* n. gen., n. sp. (Coleoptera: Belidae) in Dominican amber. *Hist. Biol.* **2014**, *26*, 670–674. [CrossRef]
73. Alekseev, V.I.; Bukejs, A. First fossil representatives of Pharaxonothinae Crowson (Coleoptera: Erotylidae): Indirect evidence for cycads existence in Baltic amber forest. *Zootaxa* **2017**, *4337*, 413–422. [CrossRef] [PubMed]

Article

The Weevil Fauna Preserved in Burmese Amber—Snapshot of a Unique, Extinct Lineage (Coleoptera: Curculionoidea)

Dave J. Clarke [1,*], Ajay Limaye [2], Duane D. McKenna [1] and Rolf G. Oberprieler [3,*]

1 Department of Biological Sciences, University of Memphis, 3700 Walker Ave, Memphis, TN 38152, USA; dmckenna@memphis.edu

2 National Computational Infrastructure, Australian National University, 143 Ward Road, Acton, Canberra, A.C.T. 2601, Australia; Ajay.Limaye@anu.edu.au

3 CSIRO, Australian National Insect Collection, G.P.O. Box 1700, Canberra, A.C.T. 2601, Australia

* Correspondence: djclarke@memphis.edu (D.J.C.); rolf.oberprieler@csiro.au (R.G.O.); Tel.: +1-773-573-2000 (D.J.C.)

http://zoobank.org/urn:lsid:zoobank.org:pub:9DAC77A0-D2F8-432D-B0DB-782B9F122C8D

Received: 15 November 2018; Accepted: 10 December 2018; Published: 20 December 2018

Abstract: Only a few weevils have been described from Burmese amber, and although most have been misclassified, they show unusual and specialised characters unknown in extant weevils. In this paper, we present the results of a study of a much larger and more diverse selection of Burmese amber weevils. We prepared all amber blocks to maximise visibility of structures and examined these with high-magnification light microscopy as well as CT scanning (selected specimens). We redescribe most previously described taxa and describe 52 new species in 26 new genera, accompanied by photographs. We compare critical characters of these weevils with those of extant taxa and outline the effects of distortion on their preservation and interpretation. We conclude that only two weevil families are thus far represented in Burmese amber, Nemonychidae and a newly recognised family, Mesophyletidae, which appears closely related to Attelabidae but cannot be accommodated in this family. The geniculate antennae and long rostrum with exodont mandibles of most Mesophyletidae indicate that they were highly specialised phytophages of early angiosperms preserved in the amber, likely ovipositing in flowers or seeds. This weevil fauna appears to represent an extinct mid-Cretaceous ecosystem and fills a critical gap in the fossil record of weevils.

Keywords: Curculionoidea; Mesophyletidae; Cretaceous; taxonomy; morphology; CT scanning; amber preparation; angiosperm associations

1. Introduction

The beetle superfamily Curculionoidea (weevils) comprises one of the largest diversifications of phytophagous insects and the largest in beetles, and with more than 62,000 species in about 5800 genera [1,2] it is one of the most diverse groups of metazoans. Weevils occur worldwide in all terrestrial habitats with vegetation and feed on all plant parts, some also on fungi associated with decaying plant material [3–5]. An elongated snout, or rostrum, with terminally positioned mouthparts is the quintessential feature of weevils and likely one of several 'key innovations' to which the astonishing taxonomic diversity of the group can be attributed. Weevils have been the subject of numerous recent and ongoing [6] phylogenetic and evolutionary studies, their evolutionary success generally ascribed to their intimate and complex ecological associations with plants (reviewed by [2,7]), and these inferences and insights are increasingly reliant on discoveries and robust interpretations of extinct forms (fossils).

The first phylogenetic classification system for weevils was devised by G. Kuschel [1], who identified six major lineages (families). This classification differs from non-phylogenetic systems, which generally recognise a larger number of families and subfamilies, e.g., [8–11], but is now consistently reflected in the results of most modern phylogenetic analyses based on morphological or molecular data (e.g., [5,12–15]). The most recent higher-level classification of weevils [6] recognises eight extant families (Cimberididae, Nemonychidae, Anthribidae, Belidae, Attelabidae, Caridae, Brentidae and Curculionidae), raising the previous nemonychid subfamily Cimberidinae to family level. Differences among phylogenetic studies now seem to involve the placement of individual taxa rather than the basal branching patterns [2,13,16] (but see [6]). While few of these eight lineages have identifiable synapomorphies, their morphological definitions are robust, allowing for fossils to be reliably assigned to family group taxa provided sufficient characters are visible and interpretable.

Achieving a robust and densely taxon-sampled phylogeny is crucial for deciphering the complex evolutionary history of weevils [15,17], but such attempts are equally reliant on a robust evaluation of morphological characters—the process of 'painstaking character analysis' [2]—as they are on phylogenomics and the promise these new data bring to fleshing out the weevil phylogeny [6,18]. This will provide the greatest chance of successfully incorporating the relatively rich fossil record of weevils into the bigger picture of weevil evolution as well as accurately testing evolutionary hypotheses involving morphological features and life history traits at scales appropriate to the questions being addressed.

The oldest definitive weevil fossils are known from the Upper Jurassic Daohugou and Karatau *Lagerstätten* in the Northern Hemisphere [19] and the Talbragar Fish Bed of Australia in the Southern Hemisphere (spanning 164–151 Ma in age) [16]. More *Lagerstätten* yielding weevils are known from the Lower Cretaceous period, the most productive ones being the Yixian Formation in China, the Pedrera de Rubies Formation in Spain and the Crato Formation in Brazil, of Barremian to Aptian age (129–113 Ma), but the number and diversity of weevils from these deposits is much smaller than that of Karatau. Weevil fossils are considerably rarer still in the Upper Cretaceous, few specimens only having been described from the Cenomanian and Turonian stages (100–90 Ma), in French and New Jersey amber and from the kimberlite diatreme at Orapa in Botswana. From the later Upper Cretaceous stages (Coniacian to Maastrichtian) weevil fossils are almost completely unknown, except for a very few poorly preserved specimens. Although weevil fossils are thus known from just before and just after the evolutionarily critical boundary period between the Lower and the Upper Cretaceous, these are far too few and not well enough preserved (most are compression fossils) to allow an assessment of the nature and characteristics of the weevil fauna that existed when the angiosperms began their great radiation [13]. This is changing now that more and more weevils are being recovered from Burmese amber [19], which, dated as 99 Ma [20], was formed just at this time and has also yielded a number of angiosperm fossils that paint a portrait of the flora among which these mid-Cretaceous weevils lived. Moreover, the often exquisite preservation of the Burmese amber weevils allows a much better assessment of their adaptations to this flora than would be possible from compression fossils.

The weevil fossils from the Upper Jurassic and early Lower Cretaceous periods seemingly all represent Cimberididae and/or Nemonychidae, although the identity of groups such as Baissorhynchini remains somewhat uncertain [16]. Anthribidae, Belidae and Caridae are not yet authentically documented from the Mesozoic era and Attelabidae are known only by a single specimen preserved in New Jersey amber [14,16], whereas fossils assigned to Brentidae and Curculionidae are known from the Lower-Cretaceous (Aptian) Crato Formation in Brazil and the early Upper-Cretaceous (Turonian) kimberlite diatreme at Orapa, Botswana [16]. The rich weevil fauna preserved in Burmese amber is also of critical importance in this respect, as it can potentially contribute more authentic early records of several weevil families from the middle of the Cretaceous period.

Burmese amber, known as burmite, is found mainly in the Hukawng Valley in Kachin State, northern Myanmar, although it is also known from several other sites in northern and central Myanmar [21,22]. Most commercially available burmite with inclusions originates from the Noije

Bum mine near the Tanai township in the Hukawng Valley, and zircons from matrix sediments associated with amber from this mine have been U-Pb-dated as 98.79 $\pm$ 0.62 Ma in age [20]. This date therefore only applies to amber from this mine. Burmite has been mined for many centuries, but the significance of its rich animal and plant inclusions, specifically arthropods, only became recognised two decades ago [23,24]. Since then, Burmese amber has yielded an extraordinary number and diversity of insects and other arthropods (e.g., [25], including several extinct ordinal-level taxa, the earliest known representatives of many extant taxa [26] and the greatest diversity of Coleoptera recorded for any Cretaceous amber deposit [27]. Surprisingly, however, few weevils have been described from Burmese amber until now.

The first specimen in burmite to be described as a weevil was named *Cryphalites rugosissimus* and interpreted as a species of Scolytinae [28]. Its description and illustration are far too poor to ascertain its true identity, but a colour photograph published later supports the assessment that it is not a weevil but likely a colydiine zopherid [23]. The first definite weevil described from burmite is *Mesophyletis calhouni* Poinar, 2006, an unusual species in possessing geniculate antennae with loosely articulated clubs, exodont mandibles, serrulate tibiae with modified spurs, elongate-slender tarsi with very deeply bilobed third tarsites, and claws with bizarre dentition. Its generic name expresses the difficulty of placing it in any extant family, and a monotypic subfamily, Mesophyletinae, was erected for it, in a family called Eccoptarthridae [29]. Eleven further genera and species were described subsequently in isolated publications, in five different families: *Burmonyx zigrasi* Davis & Engel, 2014, *Aepyceratus hyperochus* Poinar, Brown & Legalov, 2017 and *Burmomacer kirejtshuki* Legalov, 2018 in Nemonychidae; *Burmocorynus jarzembowskii* Legalov, 2018 in Belidae; *Mekorhamphus gyralommus* Poinar, Brown & Legalov, 2016 and *Habropezus plaisiommus* Poinar, Brown & Legalov, 2016 in Caridae (as 'Ithyceridae'); *Anchineus dolichobothris* Poinar & Brown, 2009 in Curculionidae; *Burmorhinus georgei* Legalov, 2018 in Curculionidae: Erirhininae; *Palaeocryptorhynchus burmanus* Poinar, 2009 in Curculionidae: Cryptorhynchinae; *Microborus inertus* Cognato & Grimaldi, 2009 in Curculionidae: Scolytinae and *Palaeotylus femoralis* Poinar, Vega & Legalov, 2018 in Platypodidae. However, very few of these family assignments are credible, because relevant family apomorphies were either not suitably demonstrated or actually misinterpreted. Many of these new taxa are similar to *Mesophyletis* in having geniculate antennae with loosely articulated clubs and are equally difficult to classify. Most intriguing are the species placed in extant subfamilies of Curculionidae, because dated phylogenetic analyses of weevils indicate these subfamilies to be younger than the middle of the Cretaceous [6,13,14,30]. None of these burmite weevil fossils have been re-examined and had their identities and/or provenance verified, which is of key importance for ongoing evolutionary studies of weevils, particularly for divergence time estimation, and some have unfortunately been used uncritically in divergence dating analyses (e.g., [31,32]), with likely significant impacts on resulting evolutionary scenarios [6].

The aim of the present contribution is to document the significant morphological and taxonomic diversity of weevils preserved in Burmese amber. We examine, reinterpret and in many cases redescribe 10 of the 13 Burmese amber fossils previously described as weevils, and we describe and illustrate 52 new species in 26 new genera, the majority belonging to a newly recognised family, Mesophyletidae, characterised by geniculate antennae with loosely articulated 4-segmented clubs and usually exodont and uniquely functioning mandibles, among other characters. A summary of all weevil taxa now known from Burmese amber is presented in an Appendix A. Integral to our goals is synthesising the main morphological characters useful to properly describe and classify weevil fossils and those that are also typically visible in amber fossils, with the intent of providing a crucial resource for future workers seeking to describe weevil fossils. We also touch on other character systems that are critical in the higher classification of weevils but only rarely visible in fossils, and we provide a general classification of morphological artefacts observable in Burmese amber weevils (and applicable to all amber inclusions) resulting from various kinds of degradative processes during and after entrapment and eventual preservation. This is coupled with an overview of the kinds of obscurities present in Burmese amber weevil fossils and their impact on observational accuracy. In a second paper we will conduct a phylogenetic analysis of the genera we identify herein, as well as of the family Mesophyletidae.

2. Materials and Methods

2.1. Materials

The study is based on 95 inclusions (93 examined by us) in as many amber pieces, deposited in the following collections:

AMNH American Museum of Natural History, New York, U.S.A.
ANIC Australian National Insect Collection, Canberra, Australia
CNUB Capital Normal University, Beijing, China
FMNH Field Museum of Natural History, Chicago, Illinois, U.S.A.
GPIH Geological-Palaeontological Institute and Museum, University of Hamburg, Hamburg, Germany
ISEA Institute of Systematics and Ecology of Animals, Novosibirsk, Russia
NIGP Nanjing Institute of Geology and Palaeontology, Chinese Academy of Sciences, Nanjing, China
PACO Poinar Amber Collection, maintained at Oregon State University, Corvalis, U.S.A.

We obtained on loan for this study 56 specimens from Professor Bo Wang (NIGP), one from Professor Ren Dong (CNUB), one from Carsten Gröhn (GPIH) and four from private collectors in China, communicated by Yu-Lingzi Zhou (ANIC). One other specimen was purchased online by one of us (R.G.O.; in ANIC), and of another specimen housed in NIGP we received several high-quality images from Chenyang Cai (NIGP) that enabled us to describe the species (*Nugatorhinus chenyangi*) without having studied the specimen in person. The types of all new species described from these specimens will be deposited in these institutions after study, and the type locality for all of them is the Noije Bum Summit mine, located in the Hukawng Valley, Kachin State, Myanmar.

We also studied the type specimens of all hitherto described species of weevils preserved in Burmese amber, except those of *Mesophyletis calhouni* and *Palaeotylus femoralis*. We received the type specimens of *Burmonyx zigrasi* and *Microborus inertus* on loan from David Grimaldi (AMNH) and those of *Mekorhamphus gyralommus*, *Burmorhinus georgei*, *Burmomacer kirejtshuki* and *Burmocorynus jarzembowskii* from Andrei Legalov (ISEA). The types of *Aepyceratus hyperochus*, *Anchineus dolichobothris*, *Habropezus plaisiommus* (holotype and paratype) and *Palaeocryptorhynchus burmanus* were made available to us for study at Oregon State University. We were unfortunately unsuccessful in obtaining the holotype of *Mesophyletis calhouni* for study. This specimen was deposited in the amber collection of Ron Buckley, Sumter Ridge, Florence, KY, U.S.A., but in reply to our request to examine or borrow the specimen from Mr. Buckley, we were advised that it had been sold to Deniz Eren in Turkey. Our repeated requests to her to borrow the specimen remained unanswered, and thus it is effectively unavailable for scientific study and verification of the characters of the species. We also did not study the holotype of *Paleotylus femoralis*, as the description of this species was only published very recently, when our manuscript was already complete; besides, it is amply clear from the photographs and characters of the specimen that it is not a platypodine weevil (see below).

We received another nine specimens from the AMNH and nine more from the FMNH (Shuhei Yamamoto) when this study was already well underway. We examined these to gain a more comprehensive understanding of the diversity of taxa and characters of the Burmese amber weevil fauna, but their amber blocks need to be cut down for a proper study of the specimens, which is planned for a follow-up paper. One of the FMNH specimens evidently belongs in *Mesophyletis*, thus allowing us to redescribe at least this genus.

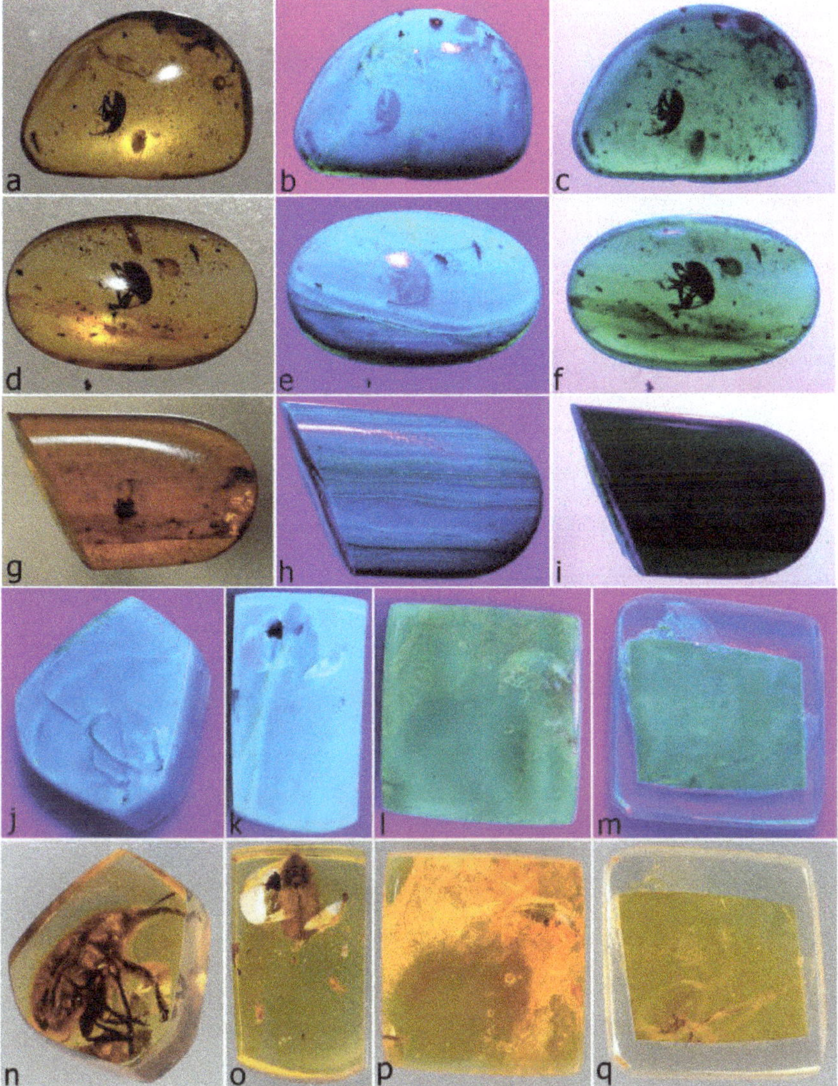

Figure 1. Amber in white light (**a**,**d**,**g**,**n**–**q**), incident near-ultraviolet (UVA) (**b**,**e**,**h**,**j**–**m**) and reflected UVA light (**c**,**f**,**i**). Amber piece containing specimen of *Bowangius* sp. 2 (**a**–**c**); amber piece containing specimen of *Mekorhamphus beatae* (**d**–**f**); Burmese amber offcut (**g**–**i**); amber piece containing specimen of *Opeatorhynchus comans* (**j**,**n**); amber piece containing holotype of *Mekorhamphus gyralommus* (**k**,**o**); Baltic amber piece containing a specimen of *Baltocar* sp. (**l**,**p**); amber piece containing holotype of *Microborus inertus* (**m**,**q**).

We checked the authenticity of all amber blocks available for study under ultraviolet light. Burmese amber is known to fluoresce in UV light, producing a pale milky blue colour when photographed [33,34]. The peak light absorbance of Burmese amber occurs at a wavelength of 380 nm [35], which is near the upper end of the range of UVA (near-ultraviolet) light (315–400 nm). As most commercially available UV lamps ('black lights') emit light in this range, they can be used conveniently to test Burmese amber. We used a 3 W 365 nm LED UV torch (TECH LIGHT) with a zoom lens and a 395 nm LED torch (SCORPION MASTER) for this purpose. The amber pieces illuminated with incident light from these torches appeared milky yellowish to bluish to the eye, but the images produced by a camera are pale bluish in colour. We photographed the specimens with an Olympus E-M1 digital camera, without using a flash. The amber pieces were placed on a non-reflective glass plate about 100 mm above the table surface, photographed first under white room light (Figure 1a,d,g,n,o), then under incident UV light (Figure 1b,e,h,j–m) and finally with UV light reflected from the table surface beneath the glass plate (Figure 1c,f,i). The amber produced the characteristic pale milky (opaque) blue colour only under incident light, whereas under reflected light it also fluoresced in blue but remained translucent. All amber pieces we tested with incident UV light in this way fluoresced in the characteristic milky blue colour, except those of Dominican amber, Baltic amber (Figure 1l,p) and *Microborus inertus* (Figure 1m,q), which all remained a dull yellowish-green under UV light, depending on the strength of the light, never fluorescing milky blue. As the *Microborus* amber piece is embedded in a thin block of artificial resin, we also tested a known burmite specimen (enclosing a staphylinid beetle) embedded in resin (and more deeply so, ~2.0 mm), and this glowed in the same characteristic blue as our Burmese amber weevil specimens did, indicating that the resin does not interfere with the fluorescence of the amber.

2.2. Specimen Preparation

After initial inspection and photography, almost all amber pieces were further cut or ground down in places to remove bubbles and impurities in the amber, correct planes or curvatures that obscured or distorted critical features of the specimens or just reduce the thickness of the surrounding amber and improve the visibility of important characters. The best shape to study the specimens from all angles generally proved to be a rectangular block (cuboid) aligned with the dorsal, ventral, lateral, cephalic, and caudal sides of the specimen. The results of such trimming were often astonishing, rendering a specimen almost impossible to study (Figure 2a–c,e,j) into one revealing most of its critical structures (Figure 2d,f–i,k). Coarser cutting was done using a Dremel 200 or 4000 rotary tool, at low speed, fitted with a Dremel 545 Diamond Wheel (22.2 × 0.62 mm) for most cuts, or with a EZ545 blade (38.1 × 0.62 mm) when cutting larger pieces (we also used a ~1 mm blade), and mounted to a Dremel Workstation 220 in horizontal position so that the cutting could be viewed under a microscope. The edge and surfaces of these cutting wheels are impregnated with diamond and can be used both for cutting and for initial coarse grinding needs (other abrasives will also work for coarse grinding). Finer cutting was done by hand using an adjustable jeweller's saw (3″ throat depth) fitted with a #7/0 gauge blade (smallest blade size available for this saw). Surfaces were ground down (with or without prior cutting) using emery paper of 240 or 400 grit and subsequently smoothened using wet/dry emery paper of 800, 1200, 2000, 3000, 5000 and 7000 grit, usually under water and under the microscope at low magnifications. For final polishing we used either acrylic polishing paste (as used for polishing artificial resin; Vosschemie) or Dremel 421 polishing compound. A few inclusions had been cut or ground into during the original preparation; the resulting cavities were cleaned and then filled with clear polyester casting resin (Diggers Casting and Embedding Resin, Recochem Inc., Australia, http://www.recochem.com.au/files/downloads/Cons_Casting__Embedding_Resin_PDS_Apr11.pdf, and Castin' Craft® liquid plastic) so as to seal and properly preserve the specimens. All this work was undertaken under a microscope, generally a Leica EZ4 and a Wild M3C instrument with Leica/Wild Planapo 1.0× objectives.

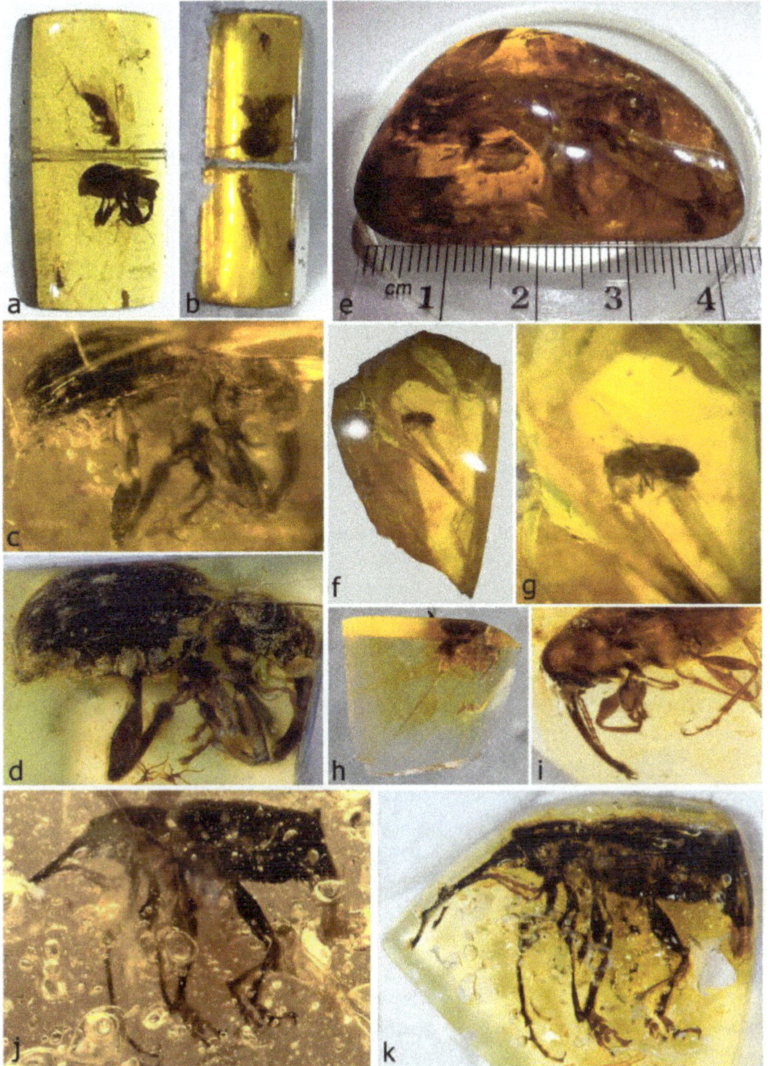

Figure 2. Block trimming to reveal inclusions in amber blocks and remove obstructions obscuring inclusions. Amber block containing holotype of *Compsopsaros reneae* showing sawcut between specimen and a second inclusion obscuring dorsal side (**a**); same, caudal view (**b**); same, lateral view of specimen prior to trimming (**c**); same, lateral view of specimen after removing most amber from side (**d**); original large amber piece containing holotype of *Burmonyx zigrasi* (**e**); resulting wedge containing specimen cut out of block, prior to final trimming (**f**); detail of specimen (**g**); final trimmed block (**h**); close-up of specimen in final block (**i**); amber piece containing holotype of *Cyrtocis gibbus* prior to trimming (**j**); block after trimming with specimen now largely free of bubbles and cracks (**k**).

2.3. Specimen Study

As typical for amber inclusions, critical structures and morphological characters were often not or only imperfectly visible in the amber pieces as received. This could be due to poor preservation of the specimen (compression, distortion or decomposition) or to impurities in the amber, such as bubbles of varying sizes and opacity, pieces of dark foreign matter (sometimes of a metallic nature), flow lines or shearing planes or cracks in the amber (sometimes visible by different colours of the amber) (Figure 3a) or a general cloudiness of the amber, which sometimes appeared as a greenish haze around the specimen (Figure 3h). It could also be due simply to the thickness of even relatively clear amber (without debris or obscurities) surrounding the specimen, which can cause it to appear blurred (especially at higher magnifications) due to the varying internal structure of the amber, the resulting varying optical clarity and available viewing angles. This alone, combined with potentially inadequate magnification and viewing techniques (dry vs. immersed in a liquid, the latter far superior), is likely an important factor in the omission or misinterpretation of several features in previously described amber weevil fossils and one for which different viewing techniques cannot compensate. Larger air bubbles can impede the view of some specimens due to the reflection of light from the inner bubble surfaces. Similarly, turbid (milky) impurities in the amber can restrict the visibility of specimens, and larger impurities and shearing planes close to the specimen sometimes can make it difficult to isolate structures (e.g., in *Myanmarus caviventris*; Figure 3g). A thin, continuous or fragmented film of air partially or substantially covers several specimens directly over the integument (Figure 3c,f), similar to the familiar *Verlumung* in Baltic amber but never of the same opaque, white, emulsion-like nature (Figure 3d). These layers may stem from gases released by the insect body or from slight movements (particularly of appendages) after embedding in the amber, and they were particularly troublesome in the study of the specimens as they completely obscure the underlying features and are not removable by cutting down the amber, as larger obstructions further away from the specimen often are. Additional difficulties in assessing the characters of the specimens arose from the lenticular or cabochon shape of most amber pieces as received. While this usually allowed a good view of the specimen from two sides (the flat sides of the lens), characters on the other four sides were often impossible to study due to the distortions created by the narrowly rounded edges of the lens and to the thickness of the amber along the sides. Many of these difficulties and obstructions could be overcome by trimming the amber block down into a cuboid aligned with the sides of the specimen, and sometimes difficulties associated with block curvature could be compensated for, in part, by viewing the specimen immersed in a liquid. Large bubbles approaching or reaching the specimen could sometimes be opened from the outside, cleaned out under ethanol using a very fine brush and then filled with clear polyester resin. Essentially, it proved impossible to properly study any specimen without trimming down the amber block as far as possible. The difficulty of handling very small amber pieces can be overcome by embedding them in a larger block of clear artificial resin, a method employed routinely at the AMNH [36] (this can be done with or without the aid of a vacuum pump). Such embedding also provides added protection for the specimen. Mounting small and fragile amber pieces between cover slips [37] will also stabilise them but has the significant disadvantage of precluding adequate views of the entire specimen.

All specimens were first photographed as received, and further photographs were taken after cutting down and repolishing the amber block, especially of critical structures and characters. Photographs were mainly taken using a Leica DFC500 camera attached to a Leica M205 microscope, with the specimens immersed in mineral oil (of similar refractive index as the amber) to eliminate light reflections. The required viewing angle was obtained either by supporting the amber block in a bed of small glass beads in the oil or, when these reflected too much light, by tilting the container (a small glass beaker) with the amber block resting on its bottom. Multiple photographs taken at different focus levels were combined into single images using the software Leica Application Suite V4.3, and the images were enhanced (for brightness, contrast and sharpness) and cropped as necessary using the Adobe Photoshop CS6 software.

For study and illumination of the specimens under the microscope, a cold-light (LED) illuminator with two goose-neck arms proved to be best, generally with the light from one arm reflected up from the surface beneath the amber from one side and that from the other arm directed onto the specimen from the other side. Manipulating the tips of the arms allowed for fine adjustment of the direction and intensity of both lights so as to correctly illuminate the structures of interest. Such fine light adjustment was not possible during photography, however.

After initial study and photography, several specimens were scanned at the National Laboratory for X-ray Micro-Computed Tomography (CT Lab) of the Australian National University in Canberra (https://ctlab.anu.edu.au/capabilities/micro-ct.php) to reveal surface structures obscured by deposits or amber impurities too close to the specimens to be removable by grinding down the amber. The instrument used was a high-resolution micro-CT system (ANU4) with a double helical trajectory spanning 100 mm in height, a camera length of 315.4 mm and a detector resolution of 3k × 3k. The X-rays were generated by a Hama-L source, set at 80 kV and 55–60 μA beam current. The sample distance was 12 mm for small to medium-sized specimens and 13.2 mm for larger ones, and the field-of-view (FOV) dimensions were 13.5 mm width by 26.1–28.0 mm length for the smaller specimens and 14.8 × 21.2 mm for the larger ones, giving a voxel size of 5.3 μm for small to medium-sized specimens and 5.9 μm for the larger ones. Scanning time varied from 23 h for the larger to 32 h for the smaller specimens. Similar-sized specimens (amber blocks) were scanned together in a single tube of suitable diameter. The resulting 3D datasets were explored and rendered using the Volume Exploration and Presentation Tool of the open-source scientific visualisation software Drishti v2.6.4 [38], designed at the National Computational Infrastructure's VizLab (https://github.com/nci/drishti). Drishti is a graphics, hardware-based, direct-volume rendering application for real-time exploration and presentation of volumetric data. It comprises a series of modules (Import, Render and Paint) and allows researchers to colour, render, cut, slice, explore and animate a dataset and then prepare images and videos for presentation and publication. The NetCDF files generated by the CT scanner were first imported into Drishti using the Import module, in which the amber pieces scanned in the same tube were digitally separated from each other. The individual data sets were read at low (8-bit) resolution, and contrast was incremented with the help of histograms and slides were filtered. The information thus generated is written as *.pvl.nc (xml) files, which were read in the module Render to generate volumetric images, using the transfer function to adjust a gradient of density until the specimen was visible and the image suitably cropped around it. The transfer function of Render was then used in high-resolution mode to tune visualisation as well as lighting and shading effects. For most of the specimens, it was necessary to combine different transfer functions. Images of different views and video clips (Videos S1–S7) were saved for each specimen where possible.

The quality of images obtained from CT scanning varied significantly between specimens. Ironically, specimens well visible under a light microscope apart from small obscuring structures could often not be visualised satisfactorily from CT scans, whereas specimens very poorly visible under a light microscope, such as *Calyptocis brevirostris* and *Petalotarsus oxycorynoides*, yielded far better pictures and details from CT scans compared with photographs taken under a light microscope. The main reason for poor results from CT scanning was the lack of density contrast between the amber matrix and the specimen. This was especially true for finer structures such as appendages and setae. Sometimes different body parts also had different densities, so that the CT scans would reveal a different structure or layer of the specimen than is visible under the light microscope. For example, in *Petalotarsus orycorynoides* the CT scans revealed a severely cracked deeper integument layer than is visible externally and also showed the long single gular suture of the specimen much better than is visible via light microscopy. The lack of visibility and/or resolution of fine details (such as setae) in the CT scans was also due to the voxel size chosen for the scanning (5.3–5.9 μm) being too large, especially for small specimens. It is likely that a smaller voxel size (1–2 μm) would yield better and finer pictures of small structures, such as setae and teeth on the tarsal claws (e.g., in *Calyptocis brevirostris*), but it increases the scanning time and thus the costs. These structures are nevertheless clearly visible in photographs.

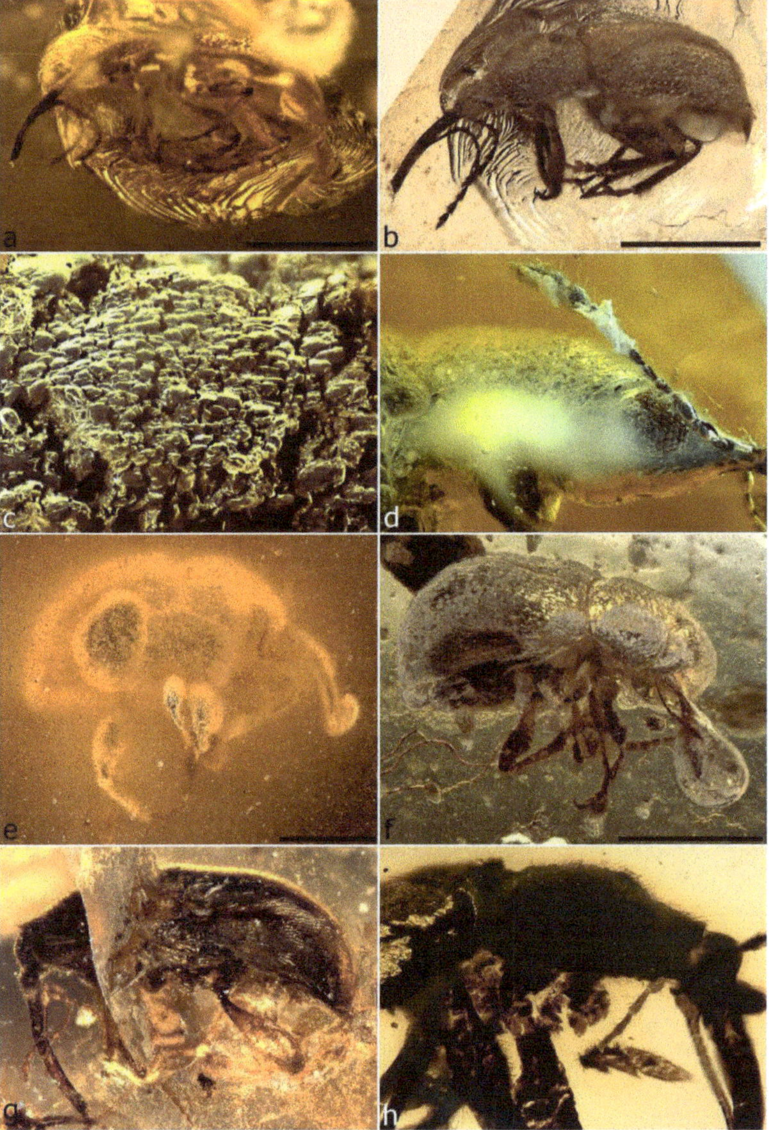

Figure 3. Obstructions in amber pieces obscuring critical taxonomic features of inclusions. *Habropezus tenuicornis* obscured by bubbles, flow lines and shearing plane (**a**); same, left side exposed after removal of excess amber and some bubbles and flow lines (reflection of shearing plane on right side of specimen reduced by different lighting) (**b**); *Calyptocis brevirostris*, irregular film of air obscuring elytra (**c**); *Baltocar* sp., white *Verlumung* obscuring right side of prothorax and structure of notosternal suture (**d**); *Gnomus brevis*, hazy cloud of organic particles shrouding specimen (**e**); *Gnomus spinipes*, imprint of specimen in amber after specimen pulled away from surrounding amber (**f**); *Myanmarus caviventris*, large fracture plane obscuring middle of left side and fine film of debris obscuring ventroposterior aspect (**g**); *Mekorhamphus* sp., badly distorted and compressed specimen with milky greenish film around thorax (**h**). Scale bars: 1.0 mm (a,b,e,f).

2.4. Measurements

All dimensions were measured using a graticule inserted into an ocular of the microscope. Length given is the standard length (SL) for weevils, measured in lateral view from the apex of the pronotum to the apex of the elytra. Due to distortions or obscurities in some specimens, it was not always possible to measure the length or other dimensions accurately.

3. Results

3.1. Effects of Decomposition and Distortion on the Interpretation of Morphological Characters

As in all insect fossils, the distortion and deformation of body parts can also be a significant source of misinterpretation of morphological characters of weevils preserved in Burmese amber. These processes can create seemingly authentic-looking structures and pose the risk of misinterpreting artefacts resulting from decomposition and deformation as legitimate taxonomic characters. One advantage of having examined so many specimens in our study is that of encountering a wide range of preservation quality, from near-perfect to very poor and resulting in unclassifiable specimens, and thus being able to compare the same structures in well and poorly preserved specimens. This has allowed the reliable identification of aberrations and of characters likely affected by these processes and, in most cases, their associated causes.

These processes, particularly decomposition, can have a significant effect on the structures of inclusions, in some cases completely destroying them, modifying them beyond recognition or generally preventing proper evaluation of some characters by affecting their natural positions in relation to other structures.

3.1.1. Decomposition and Erosion

Decomposition of inclusions may have occurred before a specimen was embedded and can continue in the amber after entrapment. Several fossils in our study are significantly modified by decomposition (usually in association with deformation), in some cases making it difficult or impossible to classify the specimen, depending on the characters affected. Of chief concern is when decomposition is minimal and seemingly affects only certain structures, in particular when the appendages have partially decomposed and some structures are potentially modified and/or missing. An example is the structure of the outer edge of the tibiae having seemingly lost the serrulation or crenulation due to decomposition. In the poorly preserved holotype of *Bowangius glabratus*, the protibiae are clearly carinate but the meso- and metatibiae serrulate, a combination otherwise rare and known mainly from other poorly preserved specimens. The legs of the holotype are clearly partially decomposed and the teeth of the protibiae may have been lost as a result. In the holotype of *Anchineus dolichobothris*, by contrast, remnant denticles are discernible on the protibiae but not on the others (which are all costate/carinate). In the *Habropezus plaisiommus* holotype, the basal half of the protibiae is carinate but the distal half distinctly serrulate (but the teeth irregularly separated), suggesting that the teeth on the basal half and some on the distal half have been lost. A complication in the interpretation of such inconsistencies is that some structures may also be affected by wear in life. In *Mekorhamphus gracilipes* and *Periosocerus crenulatus*, the protibiae and the meso- and metatibiae, respectively, are weakly crenulate, this being particularly difficult to see in the holotype of *M. gracilipes* and also on the other tibiae of this specimen.

3.1.2. Depression, Compression and Crumpling Deformations

As tree resin is deformable by changes in pressure and possibly heat during fossilisation, well preserved mummified inclusions may become modified subsequent to entrapment, even if decomposition was minimal. A common type of deformation is depression (dorsoventral flattening). In extreme cases the entire specimen is significantly affected, ranging from mild (e.g., the holotypes of *Mekorhamphus gyralommus* and *Periosocerus deplanatus*) to severe (e.g., *Cetionyx ursinus*, the whole body substantially compressed and the dorsal surface of the elytra concave) to uneven (e.g., *Anchineus dolichobothris* and *Ocriocis binodosus*, depression and crumpling affecting certain parts but not others). Localised depressions can considerably alter structures and make it difficult to interpret them. For example, in the holotype of *Habropezus plaisiommus* the rostrum is depressed and, as a result, the antennal insertions, mandible articulations and lateral structures are displaced ventrad and not properly visible. Depression can create the appearance of flattened structures or accentuate naturally flattened structures, such as tarsi and antennae (e.g., *Cyrtocis gibbus*). It can also cause the collapse of body cavities, e.g., of the ventrites into the abdominal cavity in *Habropezus plasiommus*, *Anchineus dolichobothris*, *Palaeocryptorhynchus burmanus* and *Cetionyx batiatus*. Localised depressions can cause artificial concavities in structures, such as the concave ventrites of *Bowangius cyclops*, and can also potentially alter the junctions of structures, such as by causing the separation of two sclerites at a suture, e.g., partly open coxal cavities. Depression of leg segments is evidently much rarer than compression but is evident in the holotype of *Cyrtocis gibbus*. Depression can also produce artefactual novel characters (see below).

Compression is lateral deformation and seemingly more common than depression. Examples are the compression of the elytra and abdomen of *Cyrtocis gibbus* and *Bowangius tanaops*, in which the elytra appear too narrow in dorsal view. The head and rostrum are frequently compressed, potentially creating artificial grooves or ridges along natural lines of weakness. The rostrum of *Rhadinomycter perplexus*, and potentially the head and prothorax, are distinctly compressed, and at least the rostrum in the basal part is deformed by compression. When leg segments become compressed, it can be difficult to interpret dorsal structures such as ridges, as the compression creates a sharp edge that can hide the presence of a true ridge or make the femur appear falsely carinate or costate.

In extreme cases, a specimen has become crumpled as a result of both depression and compression, the entire body collapsed and folded in on itself, as in *Bowangius glabratus*, *Bowangius* sp. 4, *Ocriocis binodosus* and *Compsopsarus* sp. In most cases these specimens are unclassifiable, but sometimes sufficient characters remain clearly visible so that it is still possible to make a genus and, in some cases, species assignments.

3.1.3. Distortion

Distortion is a kind of deformation in which the natural form and symmetry of the specimen is altered due to forces such as twisting or stretching in the amber. It can cause a general asymmetry of the body or localised asymmetry or asymmetrical placement of structures about the body axis. For example, in *Rhadinomycter perplexus* the left eye and antennal insertion are more anteriorly positioned than the right ones, and a similar altered form and position of the eyes occurs in the specimen of *Rhynchitomimus chalybeus* and several others. Distortion can also asymmetrically affect the junctions of structures (e.g., sutures) by pulling them apart. In *Myanmarus caviventris* and a few others, for example, the notosternal suture is slightly open on one side but fully closed on the other (as it is in congeners), further demonstrating the need to have a complete view of a specimen in order to check for these kinds of asymmetrical deformations and not misinterpret critical characters.

3.1.4. Eruption of Internal Materials

A common artefact in Burmese amber weevils is the occurrence of cuticular structures formed by the release of substances, apparently usually gases, from inside the body cavity. Such eruptions can create tubercle-like structures or protuberances, including structures resembling spines. In *Cyrtocis gibbus* the right elytron has a number of spine-like protuberances that are not present on the left, and the rostrum has two large asymmetrical humps. After block repreparation we could clearly determine that these processes are all associated with gas bubbles that erupted through the body wall and created these artificial structures. The holotypes of *Habropezus plaisiommus* and *Leptopezus rastellipes* also have notable examples of expulsion, but these are more obviously identifiable as mere eruptions because they have locally damaged the specimen (right profemur in *H. plaisiommus*; abdomen and legs in *L. rastellipes*) but not produced natural-looking false structures.

3.1.5. Inauthentic Characters

Any of the above processes can also have the effect of producing structures that appear to be characters of the specimen but, on closer inspection, are revealed to be artefacts because they are not present symmetrically. Usually this becomes apparent only after the amber block is trimmed and the entire specimen and details of its preservation become visible. For example, in the holotype of *Burmonyx zigrasi* a curved carina is evident on the upper side of the right profemur, but no trace of such a character is visible on the left one. Similarly, in *Acalyptopygus elongatus* the right side of ventrite 5 has a distinct notch, but this is absent from the left side. In the description of *Anchineus dolichobothris*, the claw bases are shown to have two apical processes between them, but these are artefacts resulting from the apex having split due to depression. A similar condition occurs in *Burmocorynus longus*. Trapped air layers over the surface of a specimen can also create surficial artefacts that can make the surface appear differently coloured or patterned. The specimen of *Compsopsarus reneae* has several patches of such surface layers, in places seemingly divided into scale-like chambers, and this artefact may also occur in the holotype of *Mesphyletis calhouni* (see there). These layers are nearly always not bilaterally symmetrical.

3.1.6. Exaggeration of Structures

Some stuctures or characters may become exaggerated (e.g., enlarged) by the processes of deformation and distortion. This is most frequently evident on the elytra, in which compression can amplify the distinctness of interstriae and the depth of striae. Usually (but not always) this artefact can also be identified by asymmetry between the two elytra.

3.1.7. Identification of Artefacts

In paired structures and body parts all these possible artefacts can usually be identied by determining whether they occur on both sides of the body, but this is not always possible, e.g., when the other side is obscured or differently deformed. It is also useful to check multiple conspecific specimens, or closely similar ones, as we have been able to do in several cases, but usually there will not be material available to do so (always an issue with isolated descriptions).

Trimming down the amber block is usually also essential for the accurate recognition of taxonomically valuable characters, as this often reveals whether a putative character is a legitimate one or merely an artefact of preservation. As a clear example, cutting the holotype of *Burmonyx zigrasi* out of its large original block revealed that what was drawn in the original publication as a wing is merely a crack in the amber.

3.2. *Morphological Characters*

In the description and classification of the Burmese amber weevils, as of all fossils, accurate identification and correct interpretation of morphological characters in comparison with those of relevant extant taxa is critical. Unfortunately this is often not included in published descriptions, leading to incorrect or doubtful assignment of fossils to higher taxa and subsequent untenable scenarios of evolutionary ages, distributions and biotic associations of taxa. Proper assessment of morphological characters is also hamstrung in studies of isolated specimens, as critical structures are often obscured or distorted. Only longer series of specimens can compensate for this, potentially revealing structures that are obscured in some specimens. From our studies of 93 specimens, the following assessment of the salient characters of the Burmese amber weevils was possible.

3.2.1. Mouthparts

Although all mouthparts are important in the classification of weevils, usually only the labrum and the mandibles are sufficiently preserved and discernible in amber inclusions to be of general use in the assignment of specimens to higher taxa.

A distinct ('free') labrum occurs in weevils only in the basal ('lower') families Cimberididae, Nemonychidae (Figure 4a) and Anthribidae (Figure 4b). In Belidae, Attelabidae (Figure 4g), Caridae (Figure 4c), Brentidae and Curculionidae the apex of the rostrum is dorsally formed by a short epistome, which is firmly fused to the rostrum, without a transverse suture delimiting it posteriorly, and often hardly distinguishable. In Attelabidae the epistome is often apically spinose (Figure 4g), and in some Curculionidae (e.g., *Meriphus* Erichson, Eugnomini) it is also large and elongated beyond the mandibles (Figure 4j). In the Burmese amber weevils of this study, a distinct labrum is only discernible in *Burmonyx zigrasi*, *Burmomacer kirejtshuki* and *Guillermorhinus longitarsis* (Figure 4e,f), whereas all other specimens have an epistome (Figure 4d,m).

The shape of the mandibles differs widely in weevils, although the more basal families have fairly characteristic types. In Cimberididae and Nemonychidae the mandibles are typically long, narrow and falcate, with no or only few inner teeth (Figure 4a), and similar but broader and flatter mandibles occur in most Anthribidae (Figure 4b). In these families the mandibles also typically have a large setiferous groove or depression on the outside. In Belidae and Caridae the mandibles are generally small and strongly dentate apically or on the inside, without external setae (Figure 4c). This type also occurs in some Burmite specimens (Figure 4d). In Attelabidae the mandibles are typically exodont with two large outer teeth, the basal one often double (two cusps above each other) and the apical one aligned with a similar internal one to form a stout, anvil-shaped apex (Figure 4g). Exodont mandibles also occur in other weevils, e.g., in curculionine and eugnomine Curculionidae, and they generally have a different action (cutting rather than crushing) and hence a different shape (flattened), as well as a different socket and articulation plane. In Attelabidae their socket is elongate, narrow and in an apicolateral position (Figure 4h), allowing the mandibles to open widely and cut during the opening action, in a horizontal plane (Figure 4g). Similarly flat and sharply exodont, horizontal mandibles occur in several Eugnomini, e.g., *Meriphus*, but with an outwardly sickle-shaped apex (Figure 4j,n). In the curculionine genus *Ergania* Pascoe, the mandibles are also exodont and flattened but not anvil-shaped (Figure 4j), and their narrow socket and articulation plane are oblique (Figure 4p), the cutting or slicing action evidently occurring during a slanting, upward stroke. In other genera of the tribe Curculionini, the articulation plane is completely vertical and the mandibles are simplified into a single triangular tooth. Similar vertical mandibles also occur in a few other weevil taxa, e.g., the oxycorynine genus *Rhopalotria* Chevrolat and some species of the brentid genus *Antliarhinus* Schoenherr.

In the Burmese amber weevils, the flattened, exodont anvil shape of mandibles is particularly common, but it differs significantly from the attelabid type both in shape and in orientation and action. These mandibles typically have three small, triangular teeth on the outside and three on the inside, the two proximal ones being much larger and often recurved (especially the larger basal tooth) and the small apical ones together forming a narrow 'anvil' or 'T' (Figure 4k,l). The mandibles also close in a horizontal plane but open into an obligue to vertical one (Figure 4o), their narrow sockets apparently curved in apical view and allowing the mandibles to rotate from a horizontal into a vertical position as they open, so that the large inner teeth face upwards and the smaller outer ones downwards in the open position (Figure 4o). This extraordinary orientation and movement of the mandibles appears unique to Burmese amber weevils, not having been recorded for any extant ones. It seems to enable the mandibles to cut a semicircular or, if the weevil rotated the rostrum by 180°, a circular groove into plant tissues, similar to that cut by a hole saw. However, narrower, horizontally articulating exodont mandibles occur in some taxa (Figure 4m), and the full range of mandible shape and articulation in these weevils is in need of closer study under higher magnifications or with other imaging techniques, such as higher-resolution CT scanning.

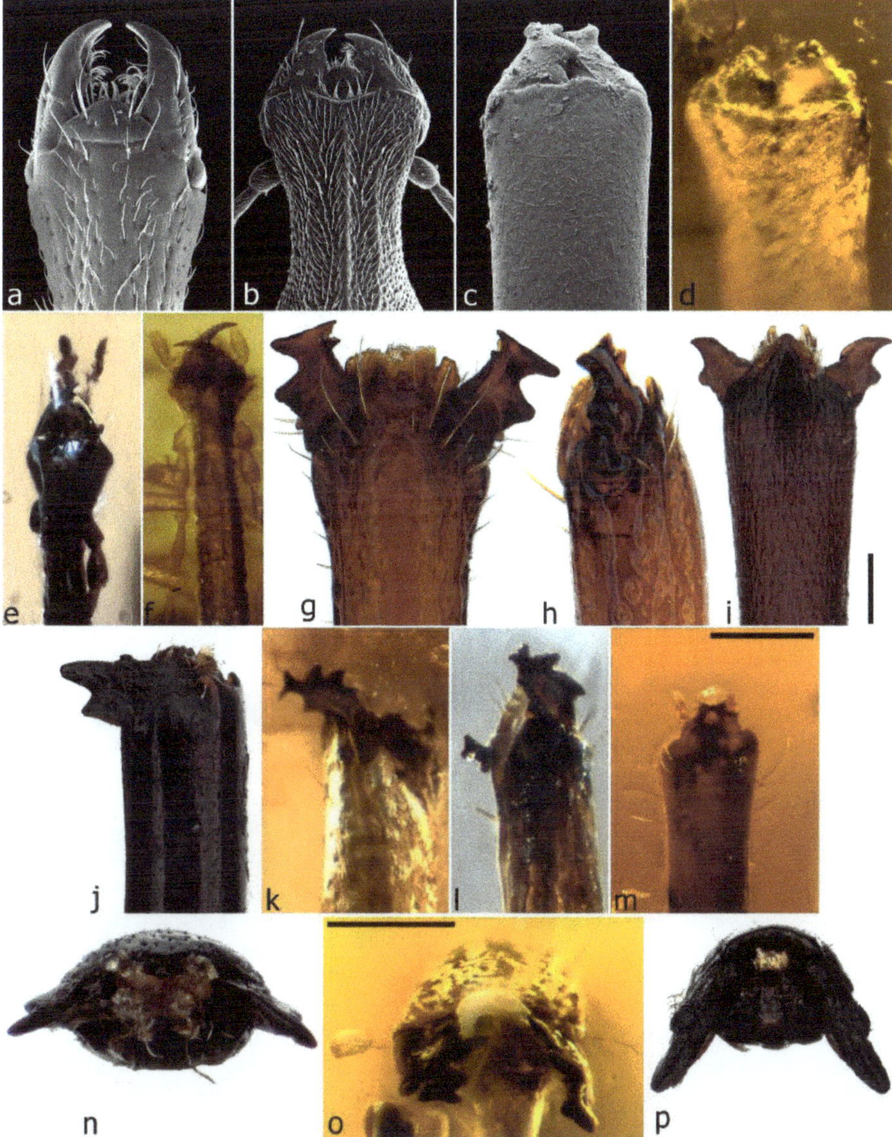

Figure 4. Apical part of rostrum showing mouthpart structures relevant to the classification of Burmese amber weevils. *Basiliorhinus araucariae* (Nemonychidae), dorsal view (**a**); *Telala* sp. (Anthribidae), dorsal view (**b**); *Car cf. condensatus* (Caridae), dorsal view (**c**); *Petalotarsus* sp. (Mesophyletidae), dorsal view (**d**); *Guillermorhinus longitarsis* (Nemonychidae), dorsal view (**e**); *Burmonyx zigrasi* (Nemonychidae), dorsal view (**f**); *Rhodocyrtus cribripennis* (Attelabidae), dorsal view (**g**); *Rhodocyrtus cribripennis* (Attelabidae), lateral view (**h**); *Meriphus fullo* (Curculionidae: Eugnomini), dorsal view (**i**); *Ergania gibba* (Curculionidae: Curculionini), dorsolateral view (**j**); *Acalyptopygus brevicornis* (Mesophyletidae), dorsolateral view (**k**); *Louwiocis megalops* (Mesophyletidae), lateral view (**l**); *Elwoodius conicops* (Mesophyletidae), dorsal view (**m**); *Meriphus fullo* (Curculionidae: Eugnomini), apical view (**n**); *Habropezus kimpulleni* (Mesophyletidae), apical view (**o**); *Ergania gibba* (Curculionidae: Curculionini), apical view (**p**). Scale bars: 0.1 mm (o); 0.2 mm (i,m).

3.2.2. Antennae

Several characters of the antennae are important in the classification of weevil fossils, namely the shape of the antenna overall (straight or geniculate), the length of the basal segments (scape and first funicle segments) relative to each other, the shape of the club (the segments loosely articulated or compact) and the insertion on the rostrum (in a lateral or ventral position).

In the more primitive weevil families the antenna is straight, in that the scape is short and the pedicel (the first funicle segment) inserts into it in an apical position, its long axis aligning with that of the scape (Figure 5a). The term orthocerous ('straight-horned') is used to described weevils with this type of antenna (it applies to the weevil, not to the antenna). In some genera of these families, in particular in Belidae and Caridae, the scape is slightly elongate (usually as long as the first two or three funicle segments) and its articulation with the scape more flexible (Figure 5j). This type of antenna is here termed subgeniculate, and it also occurs in several Burmese amber weevils (Figure 5b,c). In the large family Curculionidae, the scape is typically about as long as the entire funicle and the latter is held at about a right angle to the scape (Figure 5d). This type of antennae is referred to as geniculate ('elbowed'), and weevils possessing it are termed gonatocerous. In the geniculate antenna of Curculionidae, the insertion of the pedicel in the scape is shifted into a ventral position, which allows the funicle to rotate forwards beneath the scape, and the ventral position means that the articulation socket is not visible ('closed') in apical view (Figure 5e). Even though curculionid antennae vary widely in shape, the ventral insertion of pedicel into scape generally remains, even in secondarily straight antennae (e.g., Brachycerini [39]). Geniculate antennae also occur in one other group of extant weevils, the subfamily Nanophyinae of Brentidae, but in it the geniculation is different in that the insertion of the pedicel in the scape remains in a more or less apical position, rendering the articulation socket visible ('open') in apical view (Figure 5f). The weevil fauna preserved in Burmese amber is remarkable for being the only other group of weevils with conspicuously geniculate antennae (it is by far the dominant type of antenna in the group), and these geniculate antennae are also of the 'open' type as in Nanophyinae (Figure 5g,h), not the 'closed' one of Curculionidae. The difference in the articulation type between the geniculate antenna in Curculionidae and in Nanophyinae and Mesophyletidae indicates that such antennae have evolved independently at least three times in weevils.

The shape of the antennal club also differs between weevil families, in that the club segments are distinct from each other (loosely articulated) (Figure 5a,j) in the more basal families but fused together (non-articulating) in Curculionidae (Figure 5d) and in some subfamilies of Brentidae (Ithycerinae, Microcerinae and Apioninae). In the Burmese amber weevils, the clubs are nearly always loose (Figure 5b,c,g,h), the only exception being the genus *Petalotarsus*, in which they are subcompact (Figure 5i), the segments not articulating though not as tightly fused together as they are in Curculionidae. As in weevils in general, the clubs of Burmese amber weevils are also four-segmented (in addition to a seven-segmented funicle) but generally much more distinctly so, with the terminal segment clearly inserted in the third (Figure 5b) and often conspicuously thinner (e.g., *Habropezus tenuicornis*, Figure 78), its base not being merely a constriction of the third. Even in species with subcompact clubs (e.g., *Petalotarsus oxycorynoides*), the four segments are clearly discernible (Figure 5i). In extant weevils, similar distinctly four-segmented clubs occur in some Anthribidae (e.g., *Euciodes* Pascoe), Caridae (e.g., *Car pini* Lea, Figure 5j) and even Curculionidae (e.g., *Trichodocerus* Chevrolat), whereas in Nemonychidae, Belidae and Attelabidae (Figure 5a) the distal two segments are seemingly always firmly fused and the presence of the fourth, apical segment is usually detectable only by the ring of sparse, long, erect sensory setae that occurs in about the distal quarter of all club segments, including the fourth. Even in the compact clubs of Brentidae and Curculionidae, the four segments are usually well discernible (Figure 5d).

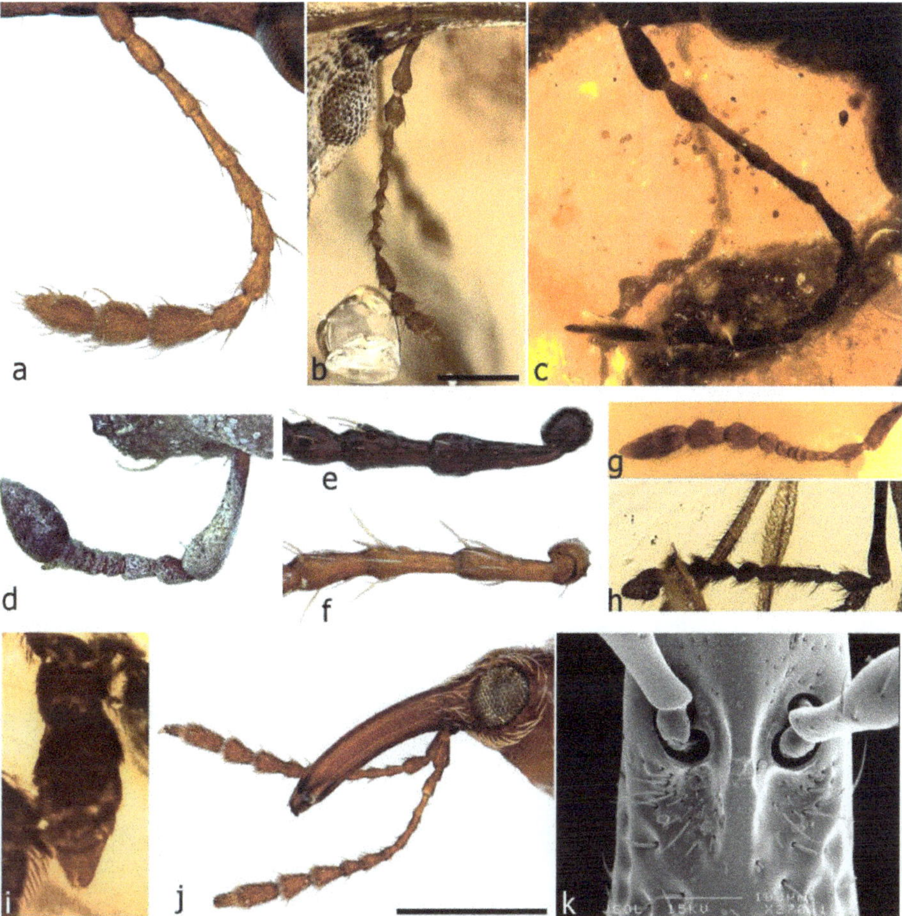

Figure 5. Structure of antennae in extant and Burmese amber weevils. Straight antenna, *Rhodocyrtus cribripennis* (Attelabidae), dorsolateral view (**a**); subgeniculate antenna, *Acalyptopygus brevicornis* (Mesophyletidae), dorsolateral view (**b**); subgeniculate antenna, *Platychirus beloides* (Mesophyletidae), dorsolateral view (**c**); geniculate antenna with compact club, *Byrsops deformis* (Curculionidae), dorsal view (**d**); ventral scape-pedicel articulation, *Tournotaris granulipennis* (Curculionidae), apical view (**e**); apical scape-pedicel articulation, *Ctenomerus* sp. (Nanophyinae), apical view (**f**); geniculate antenna, *Bowangius cyclops* (Mesophyletidae) (**g**); geniculate antenna, *Habropezus crenulatus* (Mesophyletidae), (**h**); subcompact antennal club, *Petalotarsus oxycorynoides* (Mesophyletidae) (**i**); ventrally inserted antennae, *Car pini* (Caridae), lateral view (**j**); reniform ventral antennal insertion sockets without scrobes, *Car cf. condensatus* (Caridae) (**k**). Scale bars: 0.2 mm (b); 0.5 mm (j).

The insertion of the antenna on the rostrum is typically in a lateral position (Figure 5a–c), and with geniculate antennae the socket opens posteriorly into a long groove (the scrobe) that runs along the side of the rostrum to about the eye and into which the scape recedes when the antenna is folded back against the head. With non-geniculate antennae there is no scrobe, and with subgeniculate ones there may at most be a shallow elongate depression on the side of the rostrum. In Caridae the antennal insertions are unusual and characteristic in that they are in a ventral position, the sockets being foveiform to reniform and no scrobes being present (Figure 5j,k). Such ventral antennal insertions do not occur in Burmese amber weevils, as known, the antennae always inserted laterally and the scapes usually folding back into a long scrobe.

The position of the antennal insertions in relation to the length of the rostrum is of less importance, as it depends primarily on the length of the scape. When the scape is short, the antennae are often inserted near the base of the rostrum, but when it is long (in geniculate antennae), the insertion has to be at least near the middle of the rostral length or is closer to the apex. In many weevils this position also differs between the sexes, the antennae generally inserted closer to the apex of the rostrum in males and nearer to the middle in females. This sexual dimorphism needs to be taken in consideration when delimiting species among weevil fossils (not only in amber inclusions but also in sedimentary compressions).

3.2.3. Gular Sutures

On the ventral side of the head, the development of the median sclerite, the gula, and its bordering sutures is important in the classification of families [16]. A distinct gula is only present in the families Cimberididae, Nemonychidae and Belidae, laterally bordered by a pair of sutures stretching from the base of the head forwards to the posterior tentorial pits. In Cimberididae the gula and its sutures are long and extend beneath the eyes to the base of the rostrum, the sutures converging anteriad (Figure 6a). In Nemonychidae they are vestigial and hidden under the anterior part of the prosternum, and in Anthribidaethey are absent. In Belidae the gula is generally smaller than in Cimberididae and its sutures are shorter, not extending anteriad to the rostrum but ending much further back (Figure 6b), and sometimes they meet anteriorly in a single tentorial pit beneath the posterior margin of the eyes. In Attelabidae the gula is lost towards the back of the head, and the gular sutures are united into a single suture that stretches from the back of the head forwards to beneath the eyes (Figure 6c), where it usually joins the subgenal sutures on the underside of the rostrum (Figure 6d). In the three remaining families of extant weevils, Caridae, Brentidae and Curculionidae, the gula is also absent and a single gular suture remains, but it is short and ends in a more or less conspicuous tentorial pit beneath the posterior margin of the eyes (Figure 6d) or even further back. In Nanophyinae, however, in which the eyes and the posterior tentorial pit are placed further forward, the gular suture is also relatively long, sometimes extending shortly beyond the pit towards the base of the rostrum but not joining the subgenal sutures.

In the Burmese amber weevils, which nearly always have a porrect head, the underside is often well visible and shows a single, long gular suture (Figure 6f) as it occurs among extant weevils only in Attelabidae. Even when such a suture is not visible on the surface, it appears to be present underneath, as shown by CT scan images of the specimen of *Petalotarsus oxycorynoides*, which reveal a long gular suture in the denser inner layer of the cuticula (Figure 6d, in grey colour) but not on the less dense outer one (orange in Figure 6d). Failure to discern such a gular surface in a Burmese amber weevil therefore does not mean that it is not present.

3.2.4. Coxal Cavities

Important features of the pro- and mesocoxal cavities lie in their lateral closure. In most weevils the procoxal cavities are laterally (above the procoxae) closed by the juncture of the anterolateral part of the prosternum and the posteroventral part of the hypomeron (the hypomeral lobe), marked by the notosternal suture that typically extends more or less vertically for a short distance and then bends or curves anteriad towards the anterior prothoracic margin (Figure 7a,b,h). When the anterior prosternal and the posterior hypomeral lobes meet along their entire length along the vertical part of the notosternal suture, the suture (and the procoxal cavity) is termed closed. In most Belidae and Rhynchitinae of Attelabidae, however, the horizontal branch of the notosternal suture is obsolete and the vertical branch opened up, forming a narrowly triangular cleft that exposes the procoxal trochantin (or pleurotrochantin) (Figure 7c,d). In the vast majority of Burmese amber weevils, the notosternal suture is closed (Figure 7e), but in a few specimens it is distinctly open (Figure 7f,g).

Figure 6. Gular sutures in extant and Burmese amber weevils. Long paired sutures, *Cimberis elongata* (Cimberididae) (**a**); short paired sutures, *Hadrobelus undulatus* (Belidae) (**b**); long single suture, *Rhodocyrtus cribripennis* (Attelabidae) (**c**); long single suture confluent with subgenal sutures, *Merhynchites bicolor* (Attelabidae) (**d**); short single suture, *Car pini* (Caridae) (**e**); long single suture, *Elwoodius conicops* (Mesophyletidae) (**f**); long single suture beneath surface, *Petalotarsus oxycorynoides* (Mesophyletidae) (**g**).

The position of the procoxal cavities (medially confluent or separate; in the middle of the prothorax or closer to the anterior or posterior margin) is also an important character but only on lower taxonomic levels (the generic mainly). In most weevils the procoxal cavities are medially confluent (the procoxae contiguous), separated cavities generally occurring in dorsoventrally flattened weevils or in those with a prosternal channel into which the rostrum recedes. The separation is caused by a median anterior prosternal and a median posterior hypomeral process (the latter originally paired) intruding between the cavities and meeting between them, but sometimes the processes do not quite meet and leave the cavities narrowly confluent in their middle. Procoxal cavities can only be termed separated when these processes are joined and completely separate the cavities, and separated procoxae do not always signify separated cavities as well. In the vast majority of Burmese amber weevils, the procoxal cavities are confluent (Figure 8a), distinctly separated ones occurring only in the genera *Aepyceratus, Cetionyx, Burmocorynus* and *Petalotarsus* (Figure 8b). In the descriptions of some Burmese amber weevils, the procoxae are said to be separated by a narrow septum, but this is in need of confirmation as it is usually difficult to properly discern the area between the procoxae, due to their position or to distortion or compression of the specimen, and bases of prosternal and hypomeral processes being visible does not necessarily mean that they meet between the procoxae and truly separate the coxal cavities. The typical shape of the body (high) and of the procoxae (deep and prominent) of most Burmese amber weevils suggests that the procoxal cavities are then always confluent; only in dorsoventrally flattened specimens can they be expected to be separate (if not discernible).

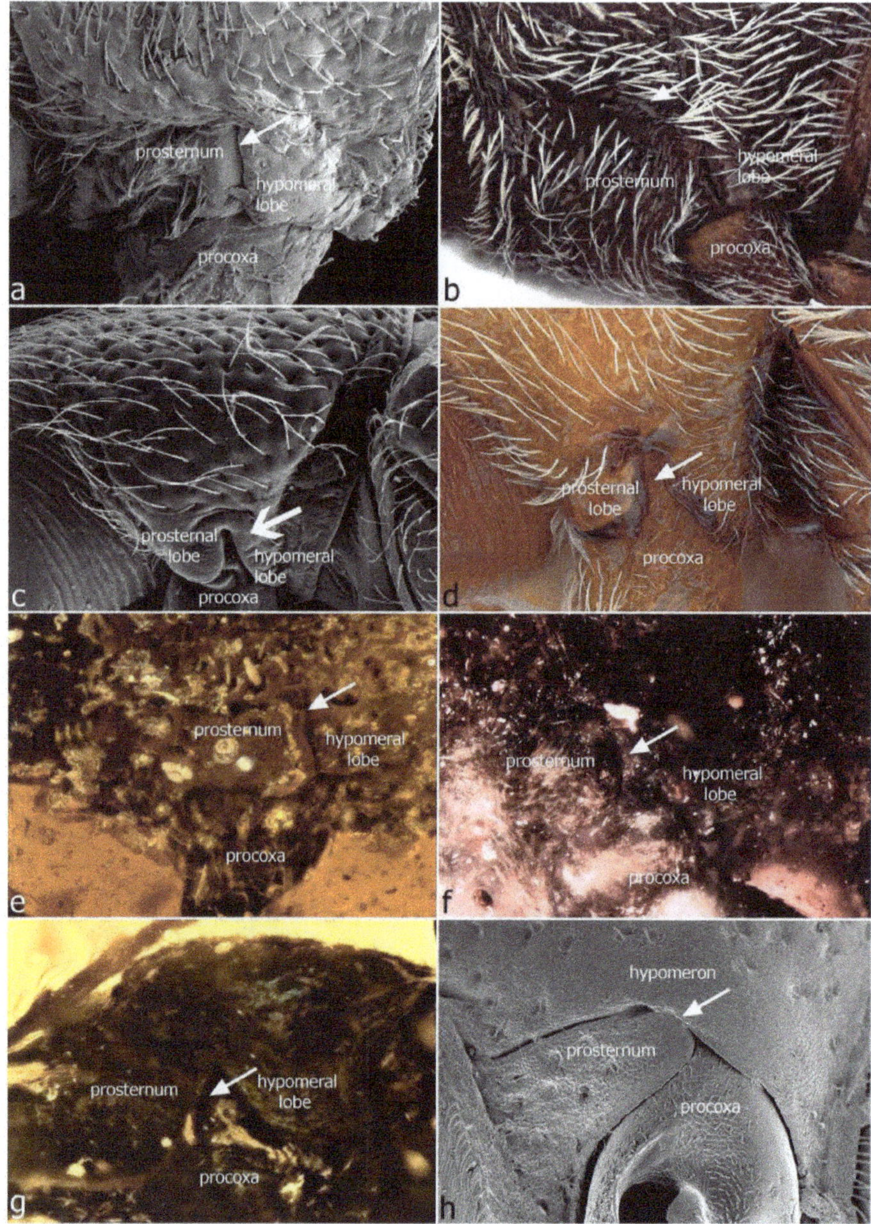

Figure 7. Procoxal cavities and notosternal sutures (arrows) in extant and Burmese amber weevils. Closed cavity, *Car cf. condensatus* (Caridae) (**a**); closed cavity, *Notomacer araucariae* (Nemonychidae) (**b**); open cavity, *Metopum* sp. (Attelabidae) (**c**); open cavity, *Rhodocyrtus cribripennis* (Attelabidae) (**d**); closed cavity, *Electrocis dentitibialis* (Mesophyletidae) (**e**); open cavity, *Platychirus beloides* (Mesophyletidae) (**f**); open cavity, *Rhynchitomimus chalybeus* (Mesophyletidae) (**g**); closed cavity, *Rhopalotria slossonae* (Belidae) (**h**).

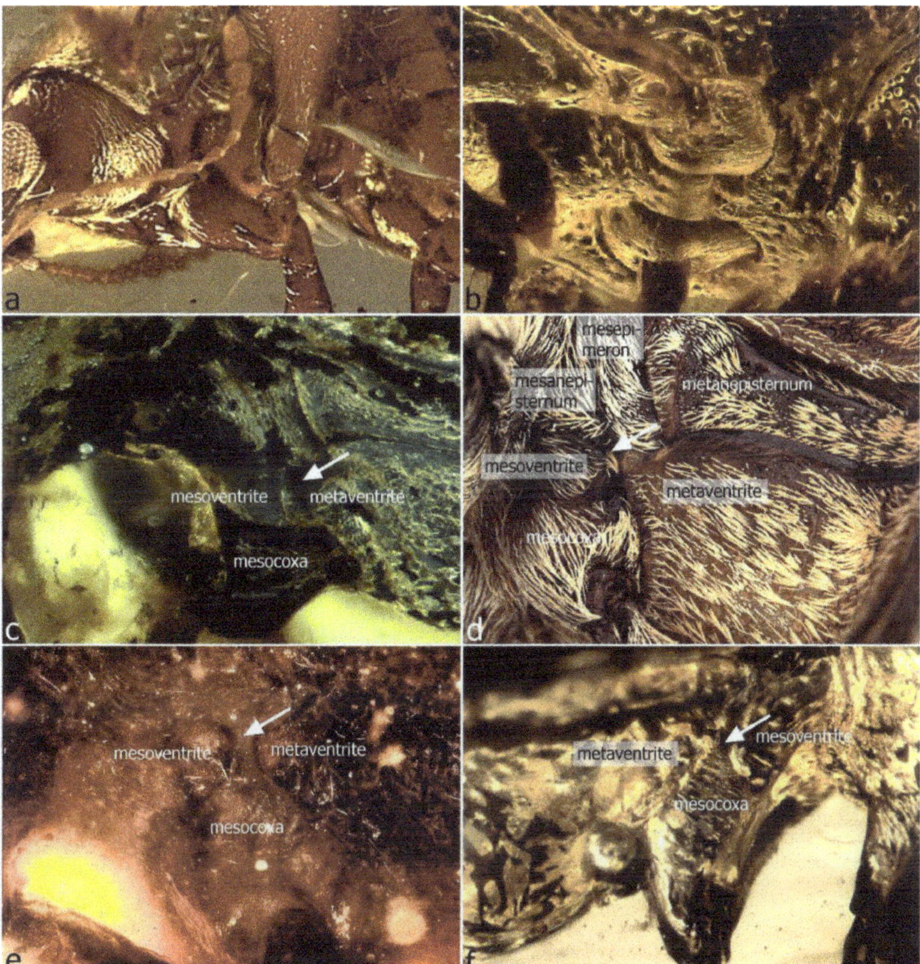

Figure 8. Procoxal and mesocoxal cavities in Burmese amber weevils. Confluent proxocal cavities (contiguous procoxae), *Elwoodius conicops* (Mesophyletidae) (**a**); separated proxocal cavities (procoxae), *Petalotarsus oxycorynoides* (Mesophyletidae) (**b**); closed mesocoxal cavity (suture arrowed), *Acalyptopygus brevicornis* (Mesophyletidae) (**c**); open mesocoxal cavity (suture arrowed), *Hadrobelus undulatus* (Belidae) (**d**); open mesocoxal cavity (suture arrowed), *Platychirus beloides* (Mesophyletidae) (**e**); open mesocoxal cavity (suture arrowed), *Rhynchitomimus chalybeus* (Mesophyletidae) (**f**).

The position of the procoxae along the longitudinal axis of the prothorax can also be of taxonomic importance. They are inserted between the anterior prosternum and the posterior hypomeron (the lateral portions of the prothorax that extend ventrad in weevils to close the procoxal cavities posteriorly), and they may be placed in the middle of the prothorax (prosternum and hypomeron being equally long, Figure 8b), closer to the anterior ventral prothoracic margin (the prosternum reduced in length) or closer to the posterior ventral prothoracic margin (the hypomeron reduced in length). In most Burmese amber weevils the prothorax is proclinate, the dorsum (pronotum) being much longer than the venter and the anterior lateral margins slanting backwards ventrad, so that the prosternum and/or hypomeron are very short (Figure 8a).

The mesocoxal cavities in weevils are laterally typically broadly closed by the juncture of the meso- and metaventrites, marked by a more or less vertical suture above the mesocoxa. This condition occurs in most Burmese amber weevils too (Figure 8c). In most Belidae and rhynchitine Attelabidae, however, the lobes of the meso- and metaventrites do not quite meet above the mesocoxae, leaving a narrow gap and their cavities thus laterally open (Figure 8d) (without the mesanepisternum and/or mesepimeron closing the cavity). This condition occurs in a few Burmese amber specimens too (Figure 8e,f), and in the same taxa in which the procoxal cavities are likewise open. It also occurs in the extant genus *Nemonyx* Redtenbacher (Nemonychinae).

The transverse metacoxae laterally usually fit against the posterior part of the metanepisternum, thus completely separating the thoracic metaventrite from abdominal ventrite 1. Only when the metacoxae are shortened (along their horizontal axis), as occurs in compact forms (e.g., Cryptoplini, some Cryptorhynchini and Entiminae), can metaventrite and ventrite 1 meet laterally below the metanepisternum (which is then often reduced or fused to the metaventrite). The metepimera are usually small and hidden under the elytra, but when they are exposed below the elytra, they also touch the lateral edge of the metacoxa. Exposed metepimera occur in a number of Belidae and Attelabidae (e.g., *Rhodocyrtus*) and also in some Curculionidae (e.g., Eugnomini). In Mesophyletidae as examined, the metacoxae always meet the metanepisternum, and metepimera are never exposed.

3.2.5. Elytra and Scutellary Strioles

The elytra in weevils are typically punctostriate, with the punctures aligned into ten longitudinal striae (Figure 9a,b) stretching from near the base of the elytra to the apex (Figure 9c). This condition also occurs in the vast majority of Burmese amber weevils (Figure 9d). In the more basal families Nemonychidae, Anthribidae, Belidae and Attelabidae, each elytron usually has an additional short striole behind the scutellar shield, stretching for only ca. 20 % of the elytral length, next to the median suture (Figure 9a,b). The first two complete striae then often bend around this scutellary striole, so are not straight as the more lateral ones are. When the punctures are small and the elytra fairly densely setose, as in Nemonychidae, the scutellary strioles can be difficult to discern. The presence of scutellary strioles therefore is of critical importance in assigning a fossil to one of the above four families, although their absence does not necessarily exclude the specimen from them, as these strioles are lost in some extant members of all of them.

In the Burmese amber weevils studied, distinct scutellary strioles are only present in *Burmonyx* (though not as long as illustrated in the original description) and *Guillermorhinus*, and probably also in *Burmomacer* (not discernible). In all other specimens there are ten full striae where these are distinct and discernible, but in some specimens the striae are very faint and almost obsolete or covered by a vestiture of dense setae. As discernible, and as in Attelabidae and Caridae, the setae arise from the interstriae between the punctures and from the interstriae, not from the punctures, and they can be aligned into rows or irregularly scattered over the elytra.

The lateral edge of the elytra, beyond the 10th stria, can be inflexed to form an epipleural flange, which is often broader anteriorly and narrows more or less rapidly behind the abutment of the metacoxa onto the elytron. The upper margin of this flange, just beneath the 10th stria, can be produced into a costa or sharp carina, as occurs in Cimberididae, Nemonychidae, many Anthribidae, very weakly in some Belidae, Attelabidae and Caridae and also in some Curculionidae (e.g., Eugnomini). This condition is also present in many Burmese amber weevils, with the narrow posterior part of the flange often depressed to form a narrow groove. Most Burmese amber weevils also possess a distinct notch in the lateral margin just behind the base. In all specimens having this notch, an anterodorsal lobe or process of the metanepisternum fits into it. In association with the laterally folded ventrites fitting tightly against the inner elytral margins, this structural interaction between elytron and metanepisternum has been related to the locking of the elytra to the body in a resting position [40]. We are not certain that this structure has a locking function (it may actually serve an opposite function) and for reference prefer to simply call it an anterior marginal notch.

Figure 9. Elytral striae in extant and Burmese amber weevils. Scutellary strioles present (arrow), *Aragomacer leai* (Nemonychidae) (**a**); scutellary strioles present (arrow), *Rhodocyrtus cribripennis* (Attelabidae) (**b**); all striae complete (scutellary strioles absent), *Car cf. condensatus* (Caridae) (**c**); all striae complete (scutellary strioles absent), *Louwiocis megalops* (Mesophyletidae) (**d**).

3.2.6. Tibiae

In the families Belidae, Attelabidae and Caridae, the tibiae and femora often carry a conspicuous row of short black pegs along their outer edges. In Belidae such crenulations occur in the tribes Agnesiotidini and Belini of Belinae (but not in Pachyurini) and in most Oxycoryninae, in Attelabidae they occur in many Rhynchitinae (Figure 10a,b) and some Attelabinae, and in Caridae they are present in the genus *Carodes* Zimmerman (Figure 10c,d). In Belidae and Attelabidae these pegs are modified setae (Figure 10e), in several cases (especially in Belini) still ending in a stout seta, but in Agnesiotidini and Rhynchitinae setal remnants are only sometimes present in some pegs on the femora. In *Carodes*, by contrast, the pegs are formed from the integument above the setae, these remaining in their normal shape and conspicuously visible between the pegs (Figure 10d,f).

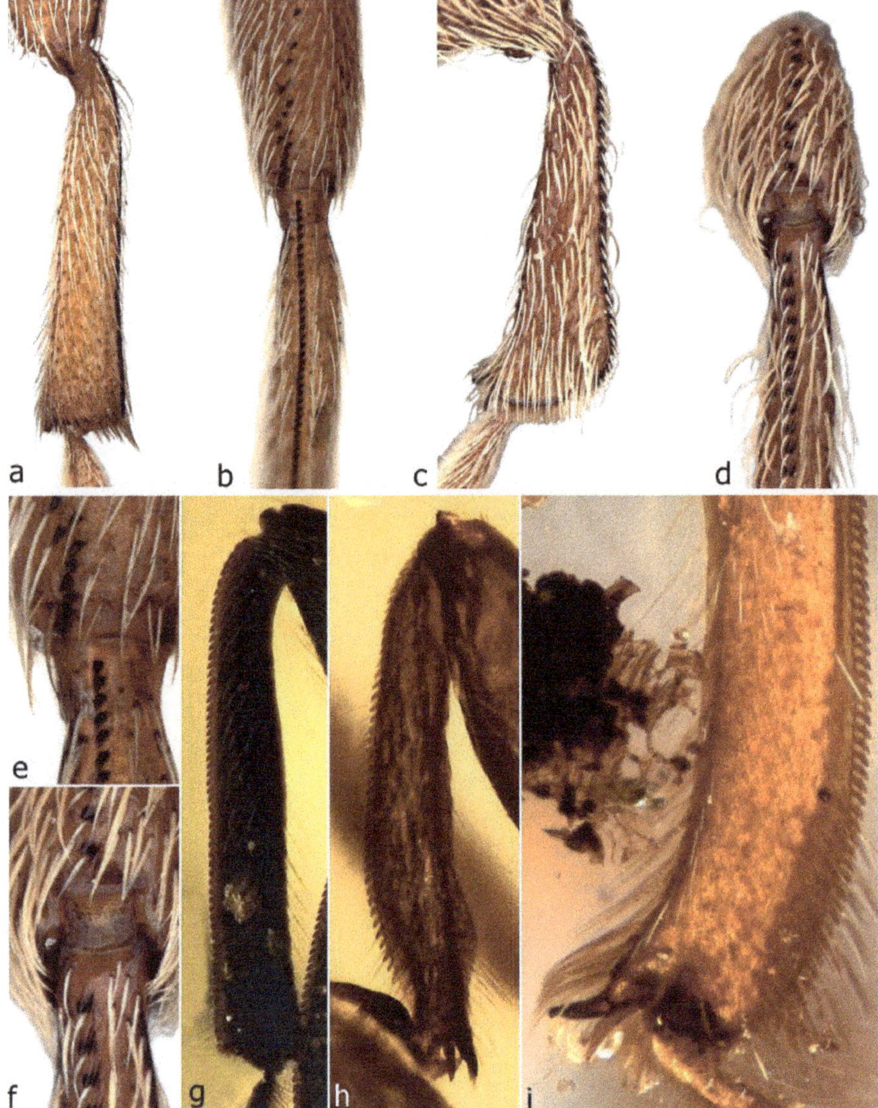

Figure 10. Tibial and femoral edges in extant and Burmese amber weevils. Serrulate metatibia and femur, *Rhodocyrtus cribripennis* (Attelabidae), lateral and outer view (**a**,**b**,**e**); serrulate metatibia and femur, *Carodes revelatus* (Caridae), lateral and outer view (**c**,**d**,**f**); crenulate metatibia, *Mekorhamphus beatae* (Mesophyletidae), lateral view (**g**); serrulate metatibia, *Bowangius cyclops* (Mesophyletidae), lateral view (**h**); pectinate metatibia, *Elwoodius conicops* (Mesophyletidae), lateral view (**i**).

Similar crenulations also occur in many Burmese amber weevils, but the pegs appear to be always embedded in a low, flat carina rather than directly in the integument and not derived from setae. When this carina is regularly finely notched and the pegs are rounded-curved and basally contiguous, the tibiae and femora are termed crenulate (Figure 10g), when it is toothed and the pegs are subtriangular and basally separated or subcontiguous, the tibiae are termed serrulate (Figure 10h), and when the pegs are longer, slender and subparallel, the tibiae are termed pectinate (Figure 10i). The difference between these types is not always clear, however, especially in small, poorly preserved or distorted specimens. In some specimens there appears to be only a smooth or finely scalloped carina present, not divided into distinct pegs. When such crenulations occur, they are present on the meso- and metatibiae and often also on the protibiae, and they may then occur on the femora as well, at least in the distal part. Although the tibial and femoral crenulations in Belidae, Attelabidae, Caridae and the Burmese amber weevils are very similar in appearance, they are evidently not homologous and have evolved (and probably been lost) several times.

The tibiae of most Burmese amber weevils carry the typical pair of apical spurs as it occurs in many weevils. These spurs are usually large and clearly discernible among the smaller setae fringing the apical surface of the tibiae, but occasionally they are smaller and difficult to discern. The distribution of the spurs between the three tibiae (the spur formula) is usually difficult to impossible to assess, as the spurs are rarely clearly visible on all three pairs of tibiae. In those specimens in which they are, the spur formula is generally 2-2-2, but it appears as if it may be 1-2-2 in some species. In *Nugatorhinus* and *Petalotarsus*, spurs are apparently consistently absent. In *Elwoodius conicops* and apparently *Mesophyletis calhouni* the inner spur on the meso- and metatibiae is fused to the tibia and broadened and flattened (Figure 11l,m). A similar condition occurs in the extant *Carodes revelatus* (Caridae).

3.2.7. Tarsi

The tarsi of the Burmese amber weevils are characteristically long and the tarsites (tarsal segments) apically deeply excised to bilobed, not only tarsites 3 but also 2 and even 1 (Figure 11a–i). Tarsites 3 are often so deeply and narrowly bilobed that the two lobes become narrowly stalked (pedunculate) and basally connected only by a very short bridge (Figure 11c,d). They appear to become dislodged easily, i.e., during decomposition (Figure 11a); in the holotype of *Echogomphus viridescens* only one of the twelve lobes is still attached to the tarsus. This structure must have given the tarsi extreme flexibility and, together with the large, usually dentate claws and ventral pulvilli of dense, fine or stiff, sharp setae (Figure 11e), superb adhesion on smooth surfaces. In the genus *Petalotarsus* the tarsi are shorter and broader, apparently adapted to allow the robust and flattened weevils of this genus to tightly cling to the substrate on which they walked or sheltered. In *Platychirus* and *Burmorhinus*, tarsites 1 are much larger and broader than the others, similar to the condition in some extant members of Belidae (Figure 11j) and Caridae (Figure 11k). This probably also increased the adhesion ability of the tarsi, but as this enlargement of tarsites 1 is sexually dimorphic in *Stenobelus* Zimmerman [41], it may have played a role in mating or oviposition rather than in walking or feeding.

3.2.8. Tarsal Claws

The terminal tarsite (the onychium) in weevils generally carries an apical pair of claws, which are typically simple, i.e., smooth and evenly curved downwards and slightly divergent (Figure 12d). Modifications involve both the direction and the shape of the claws. They may become spread apart to align into a plane and point in opposite directions (termed divaricate; Figure 12c,i,g), or they may become pushed together to align parallel and point in the same direction, which can lead to various degrees of fusion (connate claws) until only a single claw remains. The shape may be altered by the development of an additional tooth on each claw, but the origin of this tooth differs and must be properly assessed; a simple description of claws as being 'toothed' or 'appendiculate' is misleading and not very meaningful. In addition, many claws have a long stout seta near the base, inserted in a ventral position slightly on the outside of the claw. This claw seta, here termed the ventrobasal seta, persists in many weevils and is sometimes characteristically modified, such as in the tribe Sitonini of Entiminae, in which it is long, flattened and slightly twisted and parallel to the claw on its outside. The presence of this seta is critical in understanding the dentition of tarsal claws.

Figure 11. Tarsi in Mesophyletidae and some extant weevils. *Calyptocis brevirostris*, mesotarsus (CT scan) (**a**); *Opeatorhynchus comans*, protarsus (ventral view) (**b**); *Cetionyx batiatus*, protarsus (CT scan) (**c**); *Echogomphus viridescens*, protarsus (**d**); *Habropezus incoxatirostris*, protarsus, lateral view (**e**); *Myanmarus caviventris*, protarsus (**f**); *Petalotarsus oxycorynoides*, protarsus (**g**); *Platychirus brevirostris*, mesotarsus (**h**); *Burmorhinus setosus*, mesotarsus (**i**); *Stenobelus testaceus* (Belidae), protarsus (**j**); *Car cf. condensatus* (Caridae), metatarsus (**k**); *Elwoodius conicops*, mesotibial apex with fixed inner spur (**l**); *Elwoodius conicops*, metatibial apex with fixed inner spur (**m**); *Burmonyx zigrasi* (Nemonychidae), left metatarsus (dorsal view) (**n**).

A fundamental difference exists between claws that have a basal tooth on the underside (here termed dentate) and those that have a secondary tooth on the inside of the claw (termed bifid). The tooth of the dentate claw arises from an angled swelling of the ventral edge of the claw at the insertion of the ventrobasal seta (Figure 12a), whereas that of the bifid claw arises from a split on its inside (Figure 12b,e), not at the insertion of the ventrobasal seta, which is nearly always absent in bifid claws. When bifid claws become divaricate, the inner tooth is usually broadened and resembles that of the dentate claw (Figure 12c,f), but it is distinguishable by its position and the narrow angle of the separation from the claw and also by the absence of the ventrobasal seta. This claw is termed laminate, and it may differ between males and females of the same species [42]. The conclusion that these two types of toothed claws are not homologous is aptly confirmed by the Burmese amber species *Myanmarus diversiunguis*, whose front claws are dentate as well as bifid (Figure 12l). We know of no such example among extant weevils.

The distribution of these claw types differs characteristically between the weevil families. Simple or basally slightly angled claws with a ventrobasal seta occur in Cimberididae, Caridae (Figure 12a), the brentid subfamilies Microcerinae (except ventrobasal seta absent in *Gyllenhalia* Aurivillius and *Microcerus* Schoenherr), Eurhynchinae, Nanophyinae (in Nanophyini claws connate with small ventrobasal seta; in Corimaliini free, divergent, without seta) and Brentinae and in most subfamilies of Curculionidae, in particular Brachycerinae, Erirhininae, Dryophthorinae, Cyclominae, Entiminae and Molytinae (though ventrobasal seta also absent in many genera and even within some genera, e.g., *Brachycerus* Olivier). In Belidae the claws are also simple or basally angled but always without a ventrobasal seta. Dentate claws with a ventrobasal seta occur in most Apioninae of Brentidae and in many taxa of the curculionid subfamily Curculioninae. Bifid or laminate claws (without ventrobasal seta) are characteristic of Nemonychidae (Nemonychinae and Rhinorhynchinae), Anthribidae (Anthribinae and Urodontinae) and Attelabidae (Rhynchitinae) as well as the monotypic brentid subfamily Ithycerinae (Figure 12e) (this exceptionally with a ventrobasal seta as well as 1–2 shorter ones above it) and in several tribes of Curculionidae (e.g., Anthonomini, Ceutorhynchini, Cleogonini). It thus appears that the ventrobasal seta is a plesiomorphic character that is dragged along throughout the weevils but has been lost numerous times, consistently in some taxa (Belidae) but irregularly in others (Brentidae and Curculionidae). In bifid claws (in Nemonychidae, Anthribidae and Rhynchitinae) it always seems to be absent, except for *Ithycerus* Schoenherr and some curculionid genera, e.g., *Conotrachelus* Dejean of Cleogonini.

The ventrobasal seta is also present in the vast majority of Burmese amber weevils (the Mesophyletidae), whose claws are mostly strongly dentate (Figure 12i,j) but sometimes only slightly angled (Figure 12h) to simple (Figure 12g). Among our sample, bifid claws only occur in *Burmonyx, Guillermorhinus* and *Burmomacer* (all Nemonychidae) as well as on the protarsi in *Myanmarus diversiunguis*, in which they are bifid and dentate (only dentate on the meso- and metatarsi). In their dentate, divaricate claws, the mesophyletids therefore differ significantly from all weevil families other than Brentidae and Curculionidae. Dentate claws (and bifid/laminate ones) appear associated with an arboreal life style as they are consistently absent in terricolous forms (unlike the ventrobasal seta, which also occurs in terricolous species), so they probably evolved convergently a number of times in different weevil groups. Those Mesophyletidae with dentate claws are therefore likely to have led a more specialised arboreal life than those with only simple or slightly swollen claws.

3.2.9. Abdomen

Structures of the abdomen relevant for classification concern the sclerotisation and exposure of tergite VII and/or VIII beyond the elytra to form a visible pygidium and the level of fusion of the ventrites. In extant weevils such a pygidium (formed by tergite VII) occurs characteristically in Anthribidae (Figure 13a,b), in most Attelabidae (Figure 13c) and in some Apioninae (fully exposed, partly exposed or covered) and also in several tribes of Curculionidae (Acalyptini, Ceutorhynchini, Curculionini, Ectemnorhinini, Mecinini, Metatygini, Microstylini, Trigonocolini, in males of some Derelomini and Tychiini). In males of Urodontinae (Anthribidae) tergite VIII is also exposed beyond VII [43], and in males of *Ithycerus* (Brentidae) tergite VIII is strongly sclerotised and pouch-like bent over to be visible from ventral view; in females it is hidden under tergite VII. An exposed pygidium formed by tergite VII also occurs in a number of Burmese amber weevils, in the genera *Acalyptopygus* (Figure 13d,e,g), *Echogomphus* (Figure 13h) and *Calyptocis* (Figure 13i,j). It has also been described for *Mesophyletis*, but this appears to be an error, as the apex of the abdomen is flexed down in the specimen and seems to just reveal a normal, sclerotised tergite VII that is covered by the elytra when the abdomen is in its normal position.

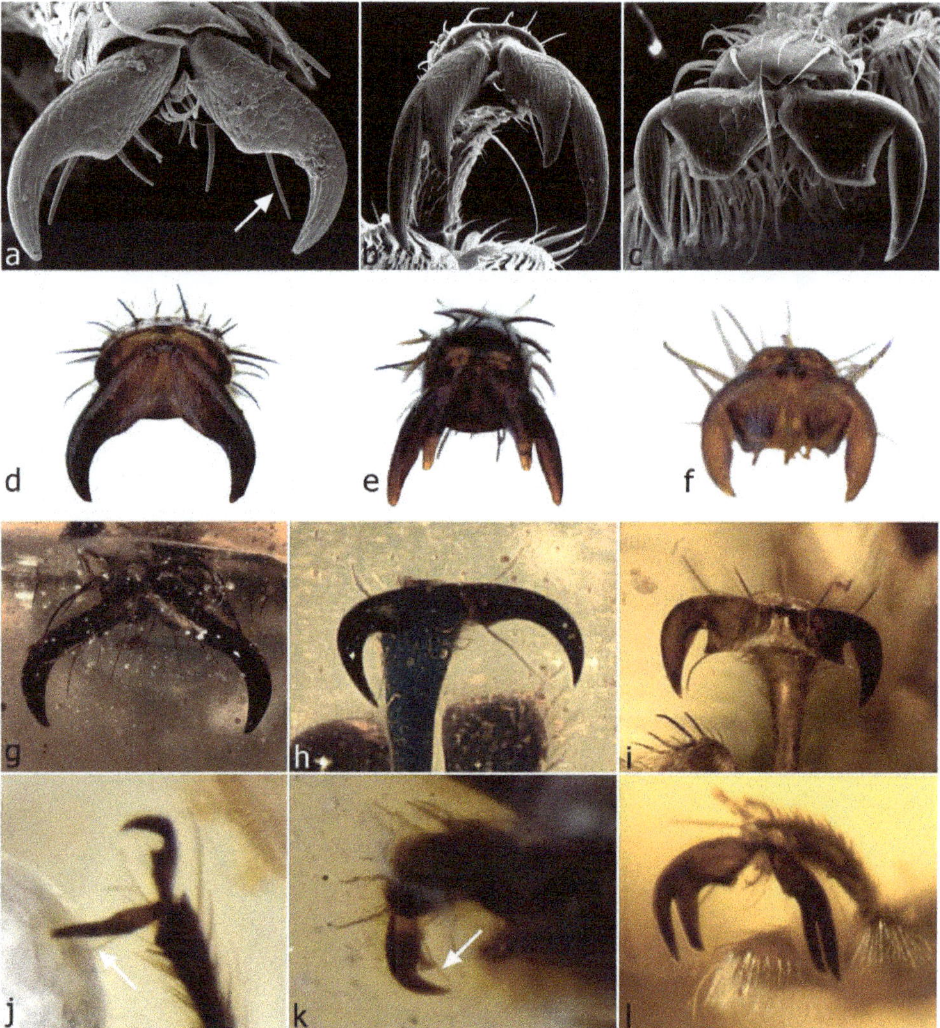

Figure 12. Tarsal claws in extant and Burmese amber weevils. Angulate, with ventrobasal seta (arrow), *Car cf. condensatus* (Caridae) (**a**); bifid, without ventrobasal seta, *Telala* sp. (Anthribidae) (**b**); laminate, without ventrobasal seta, *Basiliorhinus araucariae* (Nemonychidae) (**c**); simple, without ventrobasal seta, *Hadrobelus undulatus* (Belidae) (**d**); bifid, with ventrobasal seta, *Ithycerus noveboracensis* (Brentidae: Ithycerinae) (**e**); laminate, without ventrobasal seta, *Rhodocyrtus cribripennis* (Attelabidae) (**f**); simple, with ventrobasal seta, *Cetionyx ursinus* (Mesophyletidae) (**g**); angulate, with ventrobasal seta, *Opeatorhynchus comans* (Mesophyletidae) (**h**); dentate, with ventrobasal seta, *Mekorhamphus gracilipes* (Mesophyletidae) (**i**); dentate, with ventrobasal seta (arrow), *Rhynchitomimus chalybeus* (Mesophyletidae) (**j**); dentate, with ventrobasal seta (arrow), *Calyptocis brevirostris* (Mesophyletidae) (**k**); bifid and dentate, with ventrobasal seta, *Myanmarus diversiunguis* (Mesophyletidae) (**l**).

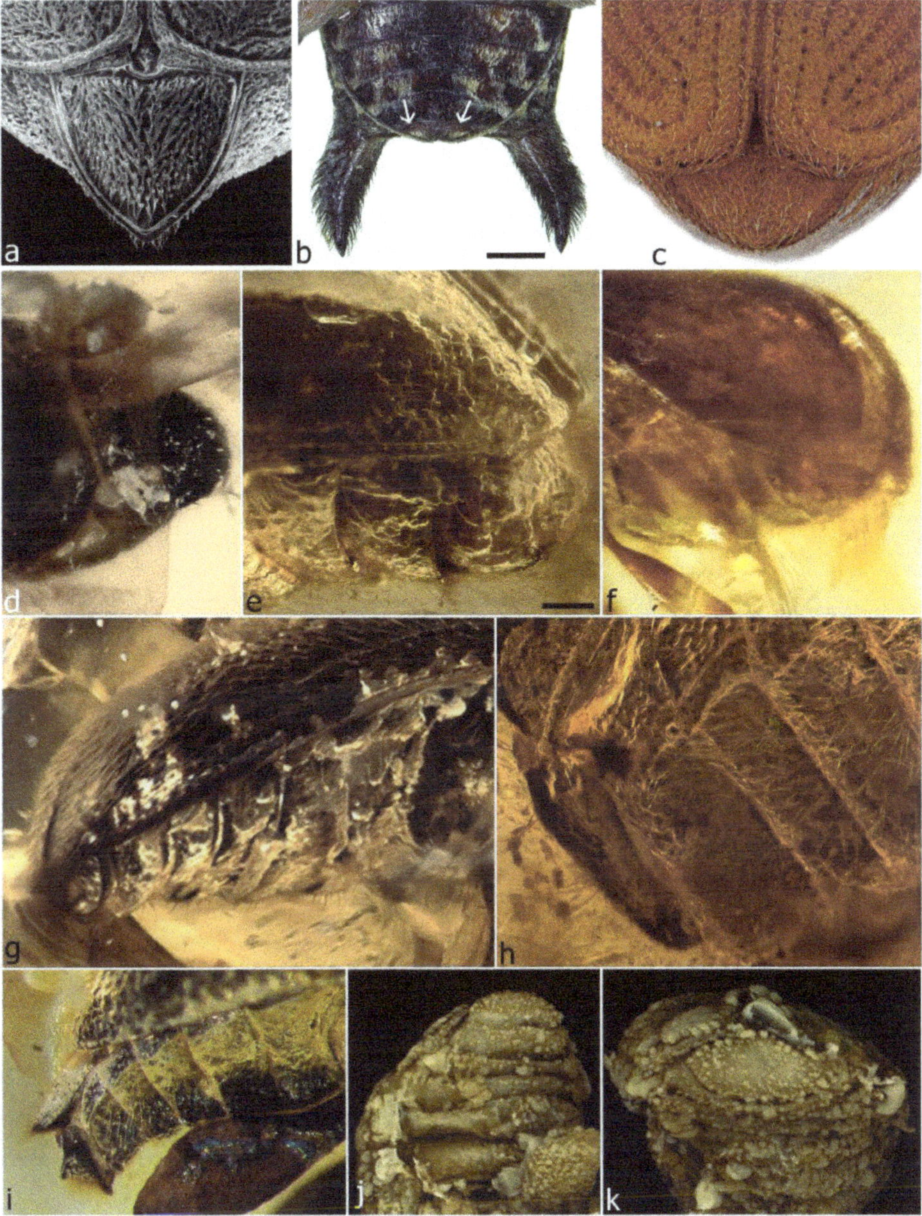

Figure 13. Pygidia and abdomens in extant and Burmese amber weevils. *Araecerus fasciculatus*, female (Anthribidae), caudal view (**a**); *Telala* sp. (Anthribidae), ventral view (arrows indicating long border with truncate ventrite 5) (**b**); *Rhodocyrtus cribripennis* (Attelabidae), caudal view (**c**); *Acalyptopygus brevicornis* (Mesophyletidae), dorsal view (**d**); *A. lingziae*, lateral view (**e**); *Burmonyx zigrasi* (Nemonychidae), caudolateral view (**f**); *Acalyptopygus astriatus* (Mesophyletidae), ventrolateral view (**g**); *Echogomphus viridescens* (Mesophyletidae), ventral view (**h**); *Bowangius tanaops* (Mesophyletidae), right lateral view (**i**); *Calyptocis brevirostris* (Mesophyletidae), ventral view (**j**); same, caudal view (**k**). Scale bars: 1.0 mm (b); 0.1 mm (e).

A sclerotised apical tergite is also visible in *Burmonyx zigrasi* (Figure 13f) but does not form an exposed pygidium. In amber fossils in which the abdomen is so flexed down or the apex of the elytra is obscured, the shape of ventrite 5 can serve as an indicator of whether the pygidium is exposed, as an exposed pygidium abuts broadly onto this ventrite, so that its posterior margin is truncate, not rounded or angled. The triangular shape of ventrite 5 as depicted in the crude drawing of *Mesophyletis calhouni* indicates that a pygidium is not exposed in this species. In a few specimens of *Bowangius*, and possibly also in *Anchineus dolichobothris*, tergite VIII appears to be permanently exposed as well, in a vertical position between tergite VII and ventrite 5 (Figure 13i), suggesting that these specimens are males and that tergite VIII is exposed in some species of Mesophyletidae.

In extant weevils the ventrites are all free (separated from each other by an extendable membrane) in Cimberididae, Nemonychidae, Belinae of Belidae and Caridae, whereas in Oxycoryninae of Belidae, Attelabidae, Brentidae and Curculionidae ventrites 1 and 2 are braced or fused (not moveable against each other) and in Anthribidae ventrites 1 to 4 are so braced or fused [1]. In the Burmese amber weevils, five free ventrites occur only in the genera *Burmonyx*, *Burmomacer* and *Guillermorhinus*, whereas in all others as studied the basal two ventrites are fused, the suture between them thin though distinct, and only the three distal ventrites are free, separated by a more or less distinct membrane (Figure 13e,g,h) and usually slightly stepped (the anterior margin higher than the posterior margin of the preceding ventrite, Figure 13e). The difference between the suture separating ventrites 1 and 2 and those separating the other ventrites is usually readily visible (Figure 13g,i), except when the abdomen is obscured or distorted.

3.3. Keys to Weevil Families and to Genera and Species of Burmese Weevils

3.3.1. Key to Families of Weevils

Following recent changes to the family classification of extant weevils [6] and the addition of another family of extinct weevils here, a revised key to the families of Curculionoidea is required. The key below is intended to be applicable to fossils as well and therefore does not include characters of internal structures, genitalia and larvae. It does, however, rely on some key features that are not always readily observable in fossils (e.g., labrum, gular sutures, scutellary strioles, tarsal claws), but these are critical in distinguishing the extant families and without them such a key would become much more complex and fragmented. Additionally, in a group as large as weevils there are always exceptions, which cannot all be accommodated in a key such as this. If critical features as used in this key are not preserved or observable in fossils, their classification to families must be undertaken with great caution and possibly not attempted at all. Misclassifications can have grave consequences for evolutionary assessments of extinct taxa or faunas.

1 Labrum free, separated from frontoclypeus by a suture; antennae usually inserted apically, rostrum expanded in front of them; mandibles usually long, falcate to blade-like, with external setiferous groove (Figure 4a,b) 2

– Labrum absent (fused to frontoclypeus, without clypeolabral suture); antennae usually inserted medially to basally, rostrum not expanded in front of them; mandibles usually small, scoop-shaped with strong internal teeth or flat, exodont, without external setiferous groove (Figure 4c,d) 4

2(1) Prothorax with bracteate basal and lateral carinae (except Urodontinae); tibiae mostly without spurs; abdomen mostly with tergite VII vertical, exposed as pygidium (Figure 13a,b); ventrites 1–4 braced or fused .... **Anthribidae**

– Prothorax without bracteate carinae; all tibiae with a pair of spurs; abdomen with tergite VII horizontal, hidden under elytra; all ventrites free 3

3(2) Tarsal claws simple, with ventrobasal seta (Figure 12g); gular sutures paired, long, reaching between eyes (Figure 6a); elytra with punctation irregularly scattered, not aligned in striae . . . . **Cimberididae**

– Tarsal claws bifid/laminate, without ventrobasal seta (Figure 12b,c); gular sutures paired, short, hidden under prosternal margin or vestigial; elytra with punctation aligned in striae, with scutellary striole (Figure 9a) (except irregular in *Nemonyx*) . . . . **Nemonychidae**

4(1) Antennae non-geniculate (straight), scape shorter than funicle 5

– Antennae geniculate, scape about as long as funicle 9

5(4) Gular sutures paired, short (Figure 6b); protibiae with apical grooming device in a broad shallow groove; tarsal claws simple, without ventrobasal seta . . . . **Belidae**

– Gular suture single (Figure 6c–g); protibiae lacking grooming device; tarsal claws simple, dentate with ventrobasal seta or bifid/laminate without ventrobasal seta 6

6(5) Gular suture long, extending from base of head between eyes onto rostrum 7

– Gular suture short, extending from base of head to tentorial pit beneath posterior margin of eyes 8

7(6) Tarsal claws bifid/laminate, without ventrobasal seta (simple in some *Auletes* and Attelabinae, in latter connate); elytra punctostriate, mostly with scutellary striole, or irregularly punctate (Figure 9b) . . . . **Attelabidae**

– Tarsal claws dentate, with ventrobasal seta; elytra punctostriate, always without scutellary striole (Figure 9c) . . . . **Mesophyletidae: Aepyceratinae**

8(6) Abdomen flat, all ventrites free, in lateral view at same level; first two ventrites not or slightly longer than 3; antennal insertions ventral (Figure 5j,k) . . . . **Caridae**

– Abdominal segments in lateral view uneven, first two ventrites bulging downwards, fused; ventrites 1 and 2 distinctly longer than 3; antennal insertions lateral . . . . **Brentidae** except Nanophyinae

9(4) Geniculation of antennae (articulation of pedicel in scape) ventral, closed in apical view (Figure 5e); antennal clubs compact, segments pressed together (Figure 5d) . . . . **Curculionidae**

– Geniculation of antennae (articulation of pedicel in scape) apical, open in apical view (Figure 5f); antennal clubs loose, segments free (Figure 5a–c,g,h,j) 10

10(9) Basal margin of elytra crenulate; funicles 4- to 6-segmented; mandibles not exodont; interstriae 8 usually with crenulate carina; scutellar shield absent (not exposed); trochanters large, usually elongate, separating femur from coxa; tibiae without spurs; tarsal claws simple, usually connate (free, divergent in Corimaliini); ventrites 1 and 2 at lower level than 3–4 . . . . **Brentidae: Nanophyinae**

– Basal margin of elytral simple; funicles 7-segmented; mandibles usually strongly exodont; interstriae 8 never with crenulate carina; scutellar shield exposed; trochanters short, oblique, not separating femur from coxa; tibiae with spurs; tarsal claws always free, divaricate, simple, angulate or dentate; ventrites 1 and 2 at same level as 3–4 . . . . **Mesophyletidae: Mesophyletinae**

3.3.2. Key to the Genera and Species of Burmese Amber Weevils

The genus *Anchineus* keys out in two different places because of the likelihood that it has serrulate tibiae (see treatment of *Anchineus*, below). The unnamed species of *Mekorhamphus* and *Bowangius* species 3 and 4 are excluded from the key, because they are too poorly preserved to accurately classify.

1 Labrum present; mandibles long (apically exposed in repose), simple, falcate; antennae non-geniculate (scape shorter than first funicular segment), inserted at apical quarter of rostral length; tarsal claws bifid; all ventrites free . . . . **Nemonychidae: Rhinorhynchinae** . . . 2

–	Labrum absent; mandibles short (may be exposed or concealed in repose), dentate internally and/or externally; antennae geniculate (scape about as long as funicle) or subgeniculate (scape longer than first funicular segment), inserted near middle or in posterior half of rostral length, or if antemedian then scapes much longer than first funicular segment; tarsal claws simple, basally angulate or dentate; ventrites 1 and 2 fused . . . . **Mesophyletidae** . . . 4
2 (1)	Rostrum shorter than pronotum, 2.0 × longer than wide in middle, dorsally with 2 grooves; anterolateral corners of postmentum not extended into long processes . . . . *Burmomacer kirejtshuki*
–	Rostrum longer than pronotum, >> 2.0 × longer than wide in middle, dorsally with 4 grooves; anterolateral corners of postmentum extended into long processes 3
3 (2)	Elytra coarsely punctostriate; striae distinct, coarse; tarsi with lobes of tarsite 3 broad, tarsites 5 narrow; claws strongly bifid, not basally angulate . . . . *Burmonyx zigrasi*
–	Elytra weakly punctostriate; striae indistinct, fine; tarsi with lobes of tarsite 3 digitate, tarsites 5 broad and flat; claws shortly bifid, preapical tooth small, basally angulate . . . . *Guillermorhinus longitarsis*
4 (1)	Antennae subgeniculate (scape longer than first funicular segment) . . . . **Aepyceratinae** . . . 5
–	Antennae geniculate (scape about as long as entire funicle); if antennae not visible, rostrum folding down into prosternal channel . . . . **Mesophyletinae** . . . 14
5 (4)	Antennal insertions in middle of rostrum; antennae with scape only slightly (1.2 ×) longer than first funicular segment, mandibles non-exodont 6
–	Antennal insertions behind middle of rostrum, in basal third; antennae with scape about twice as long as first funicular segment; mandibles exodont 8
6 (5)	Pygidium exposed; vestiture even, without distinct setal patches . . . . *Calyptocis brevirostris*
–	Pygidium hidden under elytra; vestiture with distinct patches of coloured setae . . . . *Nugatorhinus* . . . 7
7 (6)	Setal patches on body and legs orange-brown; rostrum shorter than pronotum; body length 4 mm . . . . *Nugatorhinus chenyangi*
–	Setal patches on body and legs brilliant white; rostrum about as long as pronotum; body length 5 mm . . . . *Nugatorhinus albomaculatus*
8 (5)	Body moderately flattened; elytral margins explanate; pronotum margined and weakly toothed at least basally, strongly inflexed ventrally; protibiae with elongate patch of dense subequally long setae (antennal cleaning brush); meso- and metatibiae with coarse teeth on distal half of inner margins . . . . *Aepyceratus hyperochus*
–	Body more evenly convex; elytra and pronotum not margined as above; protibiae lacking brushes; meso- and metatibiae smooth along inner margins 9
9 (8)	Eyes finely facetted, nearly smooth; funicle segment 2 ca. twice longer than 1; pro- and mesocoxal cavities widely open laterally (with V-shaped suture) 10
–	Eyes coarsely facetted; funicle segment 2 shorter than 1; pro- and mesocoxal cavities closed (with linear suture) . . . . *Acalyptopygus* . . . 11
10 (9)	Rostrum at least twice longer than pronotum, slender; tarsites 1 and 2 subequal in length and width; claws dentate (with large subtriangular basal tooth), with ventrobasal seta . . . . *Rhynchitomimus chalybeus*

– Rostrum only slightly longer than pronotum, stout; tarsites 1 enlarged, much longer and wider than 2; claws simple, slightly basally angulate, without ventrobasal seta ... ... *Platychirus beloides*

11 (9) Elytra without striae ... ... *Acalyptopygus astriatus*

– Elytra with more or less distinct striae 12

12 (11) Dorsum of head with dark median impunctate costa; scapes shorter than eye; funicle segments 2–7 subequal ... ... *Acalyptopygus brevicornis*

– Dorsum of head without costa; scapes as long as eyes; funicle segments 2–7 not all subequal 13

13 (12) Elytra distinctly punctostriate; funicle segments 2–4 subequal; body setae long, also with distinct erect setae throughout; metatibiae with single very long apical spur ... ... *Acalyptopygus elongatus*

– Elytra weakly punctostriate; funicle segments 2–4 not subequal, 3 distinctly shorter than 2 or 4; body setae short, erect setae absent; metatibiae with two short spurs ... ... *Acalyptopygus lingziae*

14 (4) Tarsal claws simple (at most with angle or swelling at position of ventrobasal seta) 15

– Tarsal claws dentate (with distinct ventral tooth at position of ventrobasal seta) 36

15 (14) Prothorax with prosternal channel for reception of rostrum, anterior lateral margins drawn out into ocular lobe; tarsi narrow, tarsites 3 deeply lobed but not pedunculate 16

– Prothorax without prosternal channel, ocular lobes absent; tarsi very broad, flattened, tarsites 3 distinctly pedunculate 19

16 (15) Body squamose (with subcircular appressed scales); mesothorax with receptacle for apex of rostrum in repose ... ... *Palaeocryptorhynchus burmanus*

– Body setose (setae erect); mesothorax without receptacle 17

17 (16) Mesoventrite in ventral view with forward-directed acute process (appearing as fin-like carina in lateral view); tibiae with 1 spur ... ... *Rhadinomycter perplexus*

– Mesoventrite without process; tibiae with 2 spurs ... ... *Burmorhinus* ... 18

18 (17) Body elongate-slender (elytra ca. 2.0 × longer than wide, pronotum slightly longer than wide); vestiture dense, on pronotum confusedly multidirectional; tibiae with inner apical tooth and 2 prominent tibial spurs; tarsites 5 shorter than 1– 3 ... ... *Burmorhinus georgei*

– Body broader (elytra ca. 1.75 × longer than wide, pronotum about as long as wide); vestiture sparser, on pronotum largely projecting anteromesad; tibiae without inner apical tooth, with 2 indistinct tibial spurs; tarsites 5 as long as 1–3 ... ... *Burmorhinus setosus*

19 (15) Procoxae narrowly or widely separated by connected or nearly connected intercoxal processes of prosternum and hypomera (forming bridge between coxae; cavities separate or nearly so) 20

– Procoxae contiguous, intercoxal processes of prosternum and hypomera short, pointed, not connected between coxae (cavities broadly confluent) 28

20 (19) Antennal clubs distinctly subcompact (individual segments still visible); body generally elongate, usually somewhat flattened; prothorax much longer than wide; first funicular segment not distinctly shorter than second 21

– Antennal clubs loose; body more robust; prothorax wider than long or subequal; first funicular segment distinctly shorter than second 26

21 (20) Funicle and club segments flattened, the former distinctly widening towards club; prothorax not proclinate, anterior lateral margins vertical in lateral view; setae on protarsi extremely long (longer than width of tarsus) and wavy (possibly a male trait) ... ... *Burmocorynus* ... 22

– Funicle and club segments rounded, the former not or weakly widening towards club; prothorax weakly to strongly proclinate, anterior lateral margins oblique in lateral view; setae on protarsi shorter than width of protarsi ... ... *Petalotarsus* ... 23

22 (21) Antennal clubs short, apical segment subequal to segment 3; apices of tibiae with long but weak fringing setae (coarser on metatibiae); meso- and metatibiae setose on only outer sides of distal half; mesotibiae with modified setae (robust, translucent spines) on inner edge of mesotibiae ... *Burmocorynus jarzembowskii*

– Antennal clubs elongate, apical segment longer than segment 3; apices of tibiae with short and very coarse fringing setae; meso- and metatibiae setose on all sides of distal half; mesotibiae without modified setae on inner edge ... ... *Burmocorynus longus*

23 (21) Elytra with inner and outer stripes of whitish (or paler) setae ... 24

– Elytra without setal stripes ... 25

24 (23) Prothorax laterally deplanate; elytra with setal stripes distinct ... *Petalotarsus oxycorynoides*

– Prothorax laterally rounded; elytra with setal stripes indistinct ... *Petalotarsus curculionoides*

25 (23) Eyes round, weakly protruding; body subcylindrical, especially prothorax; tarsites 2 triangular, apically subtruncate; tarsal claws basally angulate ... *Petalotarsus cylindricus*

– Eyes elongate, dorsoventrally compressed, distinctly protruding; body flattened, especially prothorax; tarsites 2 very deeply cleft, V-shaped, almost bilobed; tarsal claws simple ... ... *Petalotarsus* sp.

26 (20) Rostrum strongly curved; antennal insertions median; procoxae widely separated; apices of elytra angulate ... *Cetionyx ursinus*

– Rostrum substraight, antennal insertions in basal third of rostral length; procoxae only narrowly separated; apices of elytra rounded ... 27

27 (26) Antennal clubs indistinct, not much wider than funicles; pronotum and elytra not close-fitting, both rounded basally ... ... *Cetionyx terebrans*

– Antennal clubs distinctly wider than funicles; pronotum and elytra close-fitting ... ... *Cetionyx batiatus*

28 (19) Tarsi very broad, flattened; tarsites 2 deeply bilobed, 3 strongly pedunculate, with bases very slender; claws large, robust ... 29

– Tarsi narrower; tarsites 2 apically truncate, slightly emarginate or apicolaterally angled; claws small, slender ... 30

29 (28) Pygidium broadly exposed; vestiture with scattered green iridescent setae; meso- and metatibae each with 1 spur and long, inwardly projecting spike at apex ... ... *Echogomphus viridescens*

– Pygidium not exposed; vestiture without green iridescent setae; all tibiae with 2 short, distadly projecting spurs or none ... *Opeatorhynchus comans*

30 (28) Body black, sparsely setose; ventrites 1 and 2 elongate, 3–4 each half as long as 1 or 2; ventrites 3–5 at higher level 31

– Body testaceous, or paler; ventrites 1 and 2 not distinctly longer than others ... 33

31 (30) Inner tibial edge with 3–4 widely spaced denticles in distal half ... *Electrocis dentitibialis*

– Inner tibial edge smooth along distal half ... 32

32 (31) Elytral apices acutely rounded; outer edge of tibiae rounded, without tubercles ... ... *Cyrtocis gibbus*

– Elytral apices broadly rounded; outer edge of meso- and metatibiae with widely spaced flat tubercles ... *Ocriocis binodosus*

33 (30) Eyes conically protruding, facing forwards; outer edge of protibiae with small, sharp, widely spaced tubercles . . . . *Debbia gracilirostris*

– Eyes subspherical, barely protruding, lateral; protibiae without outer tubercles 34

34 (33) Lobes of tarsites 3 short and broad . . . . *Gnomus brevis*

– Lobes of tarsites 3 finger-like 35

35 (34) Rostrum stout, nearly straight; tibial apices with 1 spur and a stout mucro directed perpendicular to tibial axis; elytra asetose . . . . *Gnomus* sp.

– Rostrum long, curved; tibial apices with 2 spurs, no mucro; elytra setose . . . *Gnomus spinipes*

36 (14) Head with pair of tubercles between eyes 37

– Head without tubercles between eyes 43

37 (36) Rostrum about as long as pronotum; tibiae carinate or costate on outside; ventrites straight, distinctly stepped, sutures not deeply incised . . . . *Compsopsarus* . . . 38

– Rostrum much longer than pronotum; meso- and metatibiae crenulate on outside; ventrites curved, subflatly aligned, sutures deeply incised between 2 and 5 . . . . *Mekorhamphus* . . . 39

38 (37) Pronotum and elytra with loose whitish setal patches; sides of elytra in dorsal view rounded; pygidium exposed beyond elytral apices; ventrite 5 apically truncate . . . . *Compsopsarus reneae*

– Pronotum and elytra with uniform vestiture; sides of elytra in dorsal view distinctly emarginate; pygidum concealed by elytra; ventrite 5 elongate broadly triangular . . . . *Compsopsarus* sp.

39 (37) Prothorax laterally with small dentiform process at anterior third of length; protibiae on outside distinctly curved in apical half, inner margin straight, width greater in distal half 40

– Prothorax laterally without dentiform process at anterior third of length; protibiae slender, outer and inner margins substraight, subequal in width 41

40 (39) Rostrum very long, >2. × longer than pronotum, slender; eyes slightly elongate; mesocoxae globular, moderately projecting, about as long as metaventrite; ventrite 5 as long as 3 and 4 together . . . . *Mekorhamphus gyralommus*

– Rostrum shorter, about 1.5× longer than pronotum, thicker; eyes round; mesocoxae subflat, elongate, longer than metaventrite; ventrite 5 very short, hidden . . . . *Mekorhamphus poinari*

41 (39) Elytra sparsely covered in long, thin suberect setae; tarsites 2 bilobed (very deeply excised) . . . . *Mekorhamphus tenuicornis*

– Elytra without long, thin suberect setae; tarsites 2 shallowly excised, not bilobed 42

42 (41) Ventrites weakly curved; sutures between ventrites straight at sides; tibiae distinctly crenulate along entire length; metatibiae apically truncate or only weakly notched dorso-apically; tarsi stout, short . . . . *Mekorhamphus beatae*

– Ventrites distinctly curved; sutures between ventrites kinked at sides; tibiae much less distinctly crenulate along entire length, most distinct in basal fifth; metatibiae distinctly notched dorso-apically; tarsi thin, long . . . . *Mekorhamphus gracilipes*

43 (36) Body moderately depressed; prothorax elongate, not proclinate, anterior lateral margins vertical in lateral view; prosternum elongate, about as long as procoxal cavity diameter; antennae thick, inserted in apical third (possibly a male trait); clubs short, with segments apically strongly oblique . . . . *Periosocerus* . . . 44

– Body robust, not depressed; prothorax transverse or elongate but variously proclinate, anterior lateral margins oblique in lateral view; prosternum short to very slender, shorter than procoxal cavity diameter; antennae slender, inserted closer to middle or in basal third of rostrum; clubs usually long to very long, with segments apically truncate to weakly oblique 45

44 (43) Tibiae crenulate-serrulate, with thickend spur-like seta at outer apical margin; protarsi short and broad, ca. half as long as tibiae; antennal clubs elongate, almost as long as funicle .... *Periosocerus crenulatus*

– Tibiae not crenulate or serrulate, outside rounded, without distinctly thicker seta at outer apical margin; protarsi elongate, more than half as long as tibiae; antennal clubs slightly longer than half length of funicle .... *Periosocerus deplanatus*

45 (43) Tibiae carinate or rounded along outer edge, without any teeth forming a row 46

– At least pro- or meso- and metatibiae crenulate or serrulate along outer edge 48

46 (45) Tibiae cylindrical, without crest or ridge on outer side; tarsites 3 broadly bilobed; mandibles horizontal; length > 3.0 mm .... *Hukawngius crassipes*

– Tibiae crested or carinate on outer side; tarsites 3 digitate; mandibles vertical; length < 3.0 mm 47

47 (46) Antennae inserted at or slightly in front of middle of rostrum; mandibles with indistinct teeth; scapes, first 2 funicle segments and last 2 club segments distinctly pale, depigmented; length < 2.0 mm .... *Anchineus dolichobothris*

– Antennae inserted just behind middle of rostrum; mandibles with sharp inner and outer teeth; antennae uniformly coloured; length > 2.0 mm .... *Euryepomus lophomerus*

48 (45) Meso- and metatibae distinctly crenulate or pectinate-carinate on outer side, teeth indistinctly separated 49

– Meso- and metatibae distinctly serrulate on outer side, teeth distinctly separated 55

49 (48) Body darkly pigmented, appendages distinctly paler; at least one ventrite impressed or with distinct concavity .... *Myanmarus* ... 50

– Body and legs not distinctly differently pigmented, ventrites not impressed or concave 53

50 (49) Pronotum with sharp lateral tooth slightly in front of middle 51

– Pronotum smooth laterally, without tooth 52

51 (50) Tibiae with two large flattened spurs; protarsal claws dentate, with large triangular tooth; metafemora subequal to others; ventrites 1–3 grooved with posterior angles elevated into blunt setose tubercles .... *Myanmarus dentifer*

– Tibiae with two normal spurs; protarsal claws deeply bifid and dentate; metafemora more swollen than others; ventrites 1–3 broadly shallowly impressed with posterior angles elevated into low, asetose tubercles .... *Myanmarus diversiunguis*

52 (50) Elytral setae black but with whitish setae in loose patches; ventrites smooth except 5 with distinct concavity, setose on either side; outer side of meso- and metatibiae emarginate before apex; outer side of metafemora crenulate in distal third .... *Myanmarus caviventris*

– Elytral setae whitish; ventrites shallowly concavely depressed on disc, 5 without concavity; outer side of meso- and metatibiae straight, not emarginate before apex; outer side of metafemora carinate in distal third .... *Myanmarus robustus*

53 (49) Eyes distinctly conically protruding; meso- and metatibiae with strong carinate-pectinate ridge extending full length of tibia, and with large uneven apical spurs, the inner one fixed (fused) . . . . *Elwoodius conicops*

– Eyes elongate, protruding, slanting backwards; meso- and metatibiae crenulate but not to apex, spurs normal . . . . *Leptopezus* . . . 54

54 (53) Elytra distinctly punctostriate, interstriae broad and flat; rostrum much longer than pronotum, strongly curved (possibly a female trait), laterally with slender short setae in front of antennal insertions; outer edge of meso- and metafemora crenulate in distal third . . . . *Leptopezus rastellipes*

– Elytral striae and interstriae indistinct; rostrum slightly longer than pronotum, moderately curved, thick basally, laterally with thick long setae in front of antennal insertions; outer edge of femora carinate in distal third . . . . *Leptopezus barbatus*

55 (48) Tarsites 2 apically deeply excised, appearing almost bilobed; eyes flatly hemispherical . . . . *Aphelonyssus latus*

– Tarsites 2 apically truncate to weakly incised, entire; eyes protruding or hardly protruding and not hemispherical 56

56 (48) Eyes vertically elongate, hardly prominent; outer edge of protibia carinate; mesotibiae distinctly bent inwards at apex . . . . *Louwiocis megalops*

– Eyes round or compressed, variously prominent; outer edge of protibiae serrulate; mesotibiae straight to at most slightly curved inwards 57

57 (56) Tarsite 3 lobes digitate; mandibles with indistinct teeth; scapes, first 2 funicle segments and last 2 club segments distinctly pale, depigmented . . . . *Anchineus dolichobothris*

– Tarsite 3 lobes broadly bilobed or pedunculate; mandibles with distinct inner and outer teeth; antennae uniformly coloured 58

58 (57) Pro- and mesotibiae with long inner spur (probably fixed) and diminutive outer one, metatibiae with 2 subequal small spurs; tarsite 3 lobes strongly pedunculate; ventrites flatly at same level . . . . *Mesophyletis calhouni*

– All tibiae with 2 subequal spurs; tarsite 3 lobes long but broader basally; ventrites not entirely at same level (slightly stepped posteriorly) 59

59 (58) Eyes moderately to distinctly depressed, flatly elongate; forehead flat to shallowly concave, broad, distance between eyes expanding posteriad . . . . *Habropezus* . . . 60

– Eyes roundly protuberant or slightly elongate; forehead forming a narrow groove, distance between eyes subequal to rostral width, not increasing posteriad . . . . *Bowangius* . . . 63

60 (59) Rostrum in front of antennal insertions with row of distinct thick elongate setae; tibiae on outside at most sparsely serrulate, teeth widely spaced 61

– Rostrum in front of antennal insertions without row of setae, at most a few near apex or setae indistinct; tibiae on outside closely serrulate or crenulate, teeth closely spaced 62

61 (60) Rostrum bent downwards in middle (perhaps a distortion); protibiae carinate; meso- and metatibiae weakly serrulate, teeth minute and widely separated (probably setal sockets); tarsi slender, lobes of tarsites 3 long, slender at base . . . . *Habropezus incoxatirostris*

– Rostrum evenly curved to nearly straight; all tibiae distinctly serrulate, teeth large, closely spaced; tarsi robust, lobes of tarsites 3 short, broad at base . . . . *Habropezus tenuicornis*

62 (60) Outer sides of meso- and metafemora (and possibly profemora) with crenulate or serrulate ridge in distal third; tibiae with sparse fine setae in distal half, serrulate on outside, teeth distinctly separate, sharp; meso- and metatibae straight, not apically excised on outside, apically narrowly rounded . . . . *Habropezus plaisiommus*

– Outer sides of femora rounded or carinate; tibiae with dense coarse setae in distal third, crenulate on outside, teeth contiguous, rounded; meso- and metatibae apically excised on outside, distinctly obliquely truncate apically . . . . *Habropezus kimpulleni*

63 (59) Body length > 5.0 mm; elytra shining, without sculpture, setae very short, inconspicuous; protibiae carinate, others serrulate ... ... *Bowangius glabratus*
– Body length < 3.0 mm; elytra dull, rugosely sculptured, setae long distinct; all tibiae serrulate 64
64 (63) Eyes elongate in lateral view, slightly compressed 65
– Eyes round in lateral view, hemispherical 66
65 (64) Rostrum substraight, on ventral side in front of eyes with row of several rounded teeth with intervening short setae on either side of gular suture; pronotal setae distinctly directed anteriad; procoxae closer to anterior prosternal margin ... ... *Bowangius zhenuai*
– Rostrum curved, flat on ventral side in front of eyes; pronotal setae directed anteromesad; procoxae closer to posterior hypomeral margin ... ... *Bowangius tanaops*
66 (64) Pronotal setae distinctly directed anteriad; outer side of femora rounded, without carina or teeth; inner edge of metatibiae in apical third with rounded fin-like carina; tarsites 1 excised, 2 more deeply excised ... ... *Bowangius cyclops*
– Pronotal setae directed anteromesad; outer side of femora with carina in distal third to half; inner edge of metatibiae straight; tarsites 1 and 2 subtruncate to truncate 67
67 (66) Humeri convex, not flatly extended ... ... *Bowangius* sp. 1
– Humeri flatly extended ... ... *Bowangius* sp. 2

3.4. Descriptions

Family NEMONYCHIDAE

Following Kuschel [1], the usual concept of this family has included three subfamilies, Nemonychinae, Cimberidinae and Rhinorhynchinae (e.g., [9,44]). Although these three taxa share several characters, most of them are plesiomorphic and occur in other families too, and the three synapomorphies proposed for them [1,44,45] are not particularly strong (one applying to the male terminalia, two to the larva). The only molecular phylogenetic analysis to date that included genera of all three subfamilies [13] did not recover them as a monophyletic family, finding Nemonychinae and Rhinorhynchinae more closely related to Anthribidae and Cimberidinae forming the sister taxon of this clade. A recent molecular analysis [6] based on a larger sample of DNA data (but fewer taxa) yielded a similar result, except that Cimberidinae formed the sister group of all other weevils, and as the nodal support for these relationships was maximal, the authors raised Cimberidinae to family level and restricted Nemonychidae to include only Rhinorhynchinae and Nemonychinae (though the latter was not represented in the analysis). In this narrower concept, Nemonychidae can be more easily characterised by possessing bifid or laminate tarsal claws, without a ventrobasal seta, and very short gular sutures, hidden under the prosternal margin to being obsolete. There are, however, several differences between Nemonychinae and Rhinorhynchinae [1], and their precise relationships require further study. In fossils (if well preserved), Cimberididae can be distinguished from Rhinorhynchinae by the lack of elytral striae, simple tarsal claws with a ventrobasal seta, apically lobed tarsites 2, long gular sutures and, if visible, the lack of mesonotal stridulatory files.

Among the Burmese amber weevils known to us, none represent Cimberididae, but three species (*Burmonyx zigrasi*, *Burmomacer kirejtshuki* and *Guillermorhinus longitarsis*) are classifiable in Rhinorhynchinae due to their free labrum, distinct elytral striation and bifid claws.

Subfamily Rhinorhynchinae

With 19 genera and 57 species, this subfamily comprises the bulk of the extant fauna of Nemonychidae. Apart from the small genus *Atopomacer* Kuschel in Mexcio and the U.S.A., it is restricted to the Southern Hemisphere and particularly diverse in the Australo-Pacific region. It is divided into three tribes, Mecomacerini, Rhinorhynchini and Rhynchitomacrini, which differ in small features [45] mostly unobservable in fossils and may be easier characterised by their host plants, respectively Araucariaceae, Podocarpaceae and Nothofagaceae. All three nemonychid genera in

Burmese amber, *Burmonyx*, *Burmomacer* and *Guillermorhinus*, possess the characters of Mecomacerini, i.e., apical maxillary palp segments as long as the scapes, a long prosternum in front of the procoxae (as long as the coxae) and a head deeply constricted behind the eyes. The apparent araucariaceous origin of the Burmese amber would support this assessment. A separate tribe was proposed for *Burmomacer*, based on its seemingly separated procoxal cavities [46]. Separated procoxal cavities do not occur in any extant nemonychid, not even in *Brarus* Kuschel, in which the prosternal and hypomeral intercoxal processes intrude more strongly between the procoxae but do not meet, thus separating the procoxae but not their cavities. Although at least the prosternal process is very broad and long in *Burmomacer* and clearly separates the procoxae, it is not possible to assess whether it also meets a hypomeral process and actually separates the procoxal cavities. Even if it does, this difference hardly warrants recognition of a different tribe, and we therefore treat Burmomacrini as a synonym of Mecomacerini (**syn. n.**). In its original description [46], the name Burmomacrini was malformed as Burmomaceratini; the stem of the name of the type genus is *Burmomacr-*, not *Burmomacerat-* (for details of the stem of names ending in *-macer* and the retention of the name Mecomacerini due to prevailing usage, see [11]).

A further alleged nemonychid tribe, Oropseini (originally misspelled as Oropsini), was proposed for the genus *Oropsis* Legalov & Kirejtshuk, described from Lebanese amber [47]. However, its combination of characters as described (short but visible gular sutures, simple tarsal claws, striate elytra with short scutellary strioles) does not accord with either Nemonychidae or Cimberididae, and several other features visible in the photographs of the specimen (compact antennal clubs, elongate depressed eyes, low head, rough (rugose) elytral sculpture without distinctly visible striae, strongly elongated ventrite 1) also militate against its inclusion in either of these families. If it truly has a labrum as described, it cannot fit into any other currently recognised weevil family either, and it needs reexamination to ascertain whether it is a weevil at all. In particular, the segmentation of the tarsi, described as being pseudotetramerous (as "pseudoquadri segmented") but not visible in the published photographs, has to be verified. Another extinct genus described as a nemonychid and subsequently placed in Oropseini [47], *Arra* Peris, Davis & Delclòs [48], has pentamerous rather than pseudotetramerous tarsi and evidently belongs in the tenebrionoid family Salpingidae.

Genus ***Burmonyx*** Davis & Engel, 2014

Burmonyx Davis & Engel, 2014: 129 [49] (type species, by original designation: *Burmonyx zigrasi* Davis & Engel, 2014)

Redescription. Size. Length 2.33 mm, width 0.96 mm. **Head** short, transverse, globular, constricted behind eyes. **Eyes** large, lateral, strongly but flatly protruding, directed somewhat anteriad, coarsely facetted, dorsally separated by width of rostrum anteriorly but much further posteriorly; forehead flat, without tubercles between anterior margin of eyes; ventrally with anterolateral processes of postmentum. **Rostrum** about 1.5 × longer than pronotum, slender, flattened, substraight, apically strongly expanded; antennal insertions in apical quarter, dorsolateral, without scrobes behind them; anterolateral corners of postmentum extended into long processes. **Antennae** non-geniculate, long; scapes short, globular, about half as long as funicle segment 1; funicles 7-segmented, segment 1 subequal in width to scape, segments 2–5 thinner, 6–7 broader, subequal; clubs large, much shorter than funicle, loosely articulated, 4-segmented, segment 4 acute, shorter than 3. **Mouthparts**. Labrum present. Mandibles long, falcate, non-exodont, articulation horizontal. Maxillary palps prominent, elongate, projecting well beyond rostral apex, apparently 3-segmented. Labial palps elongate, slender, projecting well beyond rostral apex apparently 3-segmented. **Thorax**. Prothorax robust, not proclinate, anterior lateral margins vertical in lateral view. Pronotum convex, broad, only slightly narrower than elytra, laterally rounded, without tooth, disc flatter, posterior corners indistinct, fitting closely onto elytra; surface rugose, sparsely setose, setae reclinate, directed anteriad; notosternal sutures closed ventrally, not evidently continuing anteriad. Prosternum long, prosternal process apically slightly broader, shallowly excised, projecting obliquely ventrad over procoxae, not contacting hypomeral process (this not visible); procoxal cavities medially confluent, appearing separated anteriorly and between coxae, closer to posterior margin of hypomeron. Scutellar shield sparsely setose. Mesocoxal cavities closed laterally by meso- and metaventrite. Metanepisterna distinct, sparsely setose, without anterodorsal lobe. Metaventrite long, slightly convex. **Elytra** elongate, not basally lobed over base of pronotum, with broadly rounded humeri, posteriorly declivous, lateral margin sinuate to roundly emarginate in middle, anterior marginal notch absent, apically conjointly rounded, not exposing pygidium; sutural flanges narrow, equal; surface punctostriate, with scutellary striole (6-punctate on right elytron, 7–8-punctate on left), extending obliquely posteriad to about basal third of sutural length; interstriae flat, finely but distinctly punctosetose, setae short, thin, black, reclinate, directed caudad. Hindwings not visible. **Legs**. Procoxae small, globular, medially seemingly separated by intrusion of prosternal process but probably contiguous beneath this; mesocoxae subglobular, broadly separated; metacoxae flat, transversely elongate. Trochanters short, oblique. Femora long, weakly inflated, outer side rounded. Tibiae straight, compressed, outer side smooth, somewhat edged, with dense long stiff setae in distal half, apex obliquely truncate, with 2 spurs. Tarsi elongate, slender, almost as long as tibiae; tarsite 1 apically truncate, 2 shorter, subtriangular, 3 deeply bilobed, lobes broad, somewhat flattened, 5 long, apically expanded; claws divaricate, strongly bifid (inner tooth spiniform, almost as long as outer claw), without ventrobasal seta. **Abdomen** with all ventrites free, slightly stepped; 1–4 progressively shorter, 5 about as long as 3 + 4, broadly rounded.

Derivation of name. The gender of the name *Burmonyx* was not specified by its authors and is not inferable from the name of the single species placed in the genus (a noun in genitive case). The suffix *-nyx* in scientific names is normally derived from the masculine Greek noun *onyx* (G: *onychos*), a claw (e.g., in the comparable name *Nemonyx*, meaning 'divided claw'), but in *Burmonyx* it is said to be derived from the feminine Greek noun *nyx* (G: *nyktos*) [49], the night (actually the goddess of the night), and thus the gender of *Burmonyx* must be taken to be feminine (Art. 30.1. of the International Code of Zoological Nomenclature) and the stem of the name to be *Burmonyct-*, not *Burmonych-*.

Remarks. Although the long, apically expanded rostrum, falcate mandibles, bifid claws and coarsely punctostriate elytra depicted in the original description of *Burmonyx* are consistent with a placement of the genus in Nemonychidae: Rhinorhynchinae as given [49], this assignment was unconvincing as the rostrum and antennae are not properly visible in the photographs provided and the alleged scutellary striole as indicated is too long for such a striole as it occurs in all extant weevils. Considering their possible misinterpretation of the scutellary striole (the optimal viewing angle being unavailable in the original block) and the similarity of the "appendiculate" claws of *Burmonyx* to the "divaricate bifid" ones described for *Mesophyletis calhouni* [29] (see there for further discussion), the authors conceded a "superficial resemblance" of *Burmonyx* to Mesophyletinae [49]. After cutting the specimen out of its large original amber block, we could ascertain that it indeed has a free labrum, falcate mandibles (see below) and non-geniculate antennae as well as scutellary strioles (but of normal length, not the one originally indicated, which is the inner edge of the left sutural stria) (Figure 14l). Together with the striate elytra and verified bifid tarsal claws, these features indicate that the placement of the genus in Nemonychidae is correct. The large right falcate mandible as drawn by the authors is in fact the right maxillary palp, which in the original block resembled a large mandible but was blurred by the thickness of the amber; the right mandible is tightly folded under the left one (Figure 14d,e). Whereas the insertion of the scape in the rostrum is correctly indicated in Figure 1 of the orginal description [49], most of the left antenna and the apical part of the rostrum were obscured in the original block and the actual length of the scape and funicle was therefore unclear. *Burmonyx* agrees with *Guillermorhinus* in having a long, thin rostrum and the anterolateral corners of the postmentum extended into long processes, but it differs from it in its coarsely punctostriate elytra, narrower tarsi (especially tarsites 5), broader lobes of tarsites 3 and more strongly bifid, not basally angulate tarsal claws. From *Burmomacer* it differs readily in its long rostrum, postmental extensions and the coarsely punctostriate elytra. It is represented by a single species.

Burmonyx zigrasi Davis & Engel, 2014 (Figure 14)

Burmonyx zigrasi Davis & Engel, 2014: 129 [49]

Redescription. Size. Length 2.33 mm (excl. rostrum, not SL), width 0.96 mm. Small, robust; orange-brown. **Head** broad, convex posteriorly; sparsely, shortly punctosetose, punctures umbilicate, setae directed anteriad. **Eyes** directed forwards, anterior margin much longer than posterior margin; coarsely facetted; without interfacettal setae. **Rostrum** dorsally carinate, with median, paramedian and dorsolateral carinae, these very sinuate, variously obsolete before apex; setose, with curved setae along most of length; ventrally with carinate edge extending from before antennal insertions to apex; with setae in space in front of insertions between dorso- and ventrolateral carina. Mandibular articulations lateral; dorso-apical rostral apex subsinuate, slightly emarginate in middle, slightly apicolaterally lobed. **Antennae** with club segments 1–3 distinctly obconical, subequal, 2 slightly wider than 1, 4 short, acute. **Mouthparts.** Labrum transverse, broadly convex; clypeolabral suture present but indistinct. Mandibles without internal teeth; articulation plane horizontal. Maxillary palps apparently 3-segmented (visible in dorsal view); terminal segment subfusiform, nearly 4× longer than penultimate one, this seemingly longer externally, segment 2 (?) seemingly subequal in length to and narrower than 3 (?). Labial palps with apical segment longer than penultimate one. **Thorax.** Prothorax with anterior margin extending seemingly evenly down sides to prosternum. Pronotum with white, subrecurved setae, basal margin not distinctly beaded. Prosternum transverse, much longer than hypomeron, about 3.0× longer than procoxae, minutely punctosetose like pronotum, setae whitish, anterior margin straight, prosternal process projecting subventrally, apically slightly bilobed. Mesothoracic sutures largely indistinct (mesoventrite and mesanepisterna possibly party fused); mesanepisterna and mesepimera sparsely setose. Metaventrite sparsely shallowly punctate; posterolaterally at same level as metacoxae; not protruding between meso- and metacoxae. Metanepisterna without anterodorsal lobe (dorsal edge straight). **Elytra** with 10 thin punctate striae; sutural interstriae more raised than others, setose distally near apex, forming slender, low carina. Elytral bases broadly arcuate, extending evenly across middle; subcarinate, continuous with anterior scutellar margin; not submarginally concave to receive basal pronotal margin; margin seemingly doubled, with carinate ridge delimiting basal margin, submarginally with secondary ridge appearing as dark arcuate line. Elytral sides with marginal groove gradually attenuating posteriad, becoming approximate with ventral edge of elytron near apex; without anterior marginal notch, dorsal edge forming distinct lateral carina upturned near humeri. **Legs.** Coxae small, indistinct. Tibiae sparsely setose, setae longer in distal half; apically with short flanges, with long, coarse, loose fringing setae. Tarsites 1 much longer than wide, subcylindrical, gradually expanding apicad, 2 slightly shorter, 3 lobate, lobes subflattened but concave along inner edges, cryptotarsite distinct, articulating with 3 basomedially, 5 about as long as 1 + 2, slender, only very weakly curved. Protarsi slightly shorter, narrower than others; with lobes of tarsites 3 seemingly shorter than on other legs. **Abdomen** with terminal tergite sclerotised, visible above ventrite 5; setose. Ventrites moderately densely setose, sutures broadly arcuate.

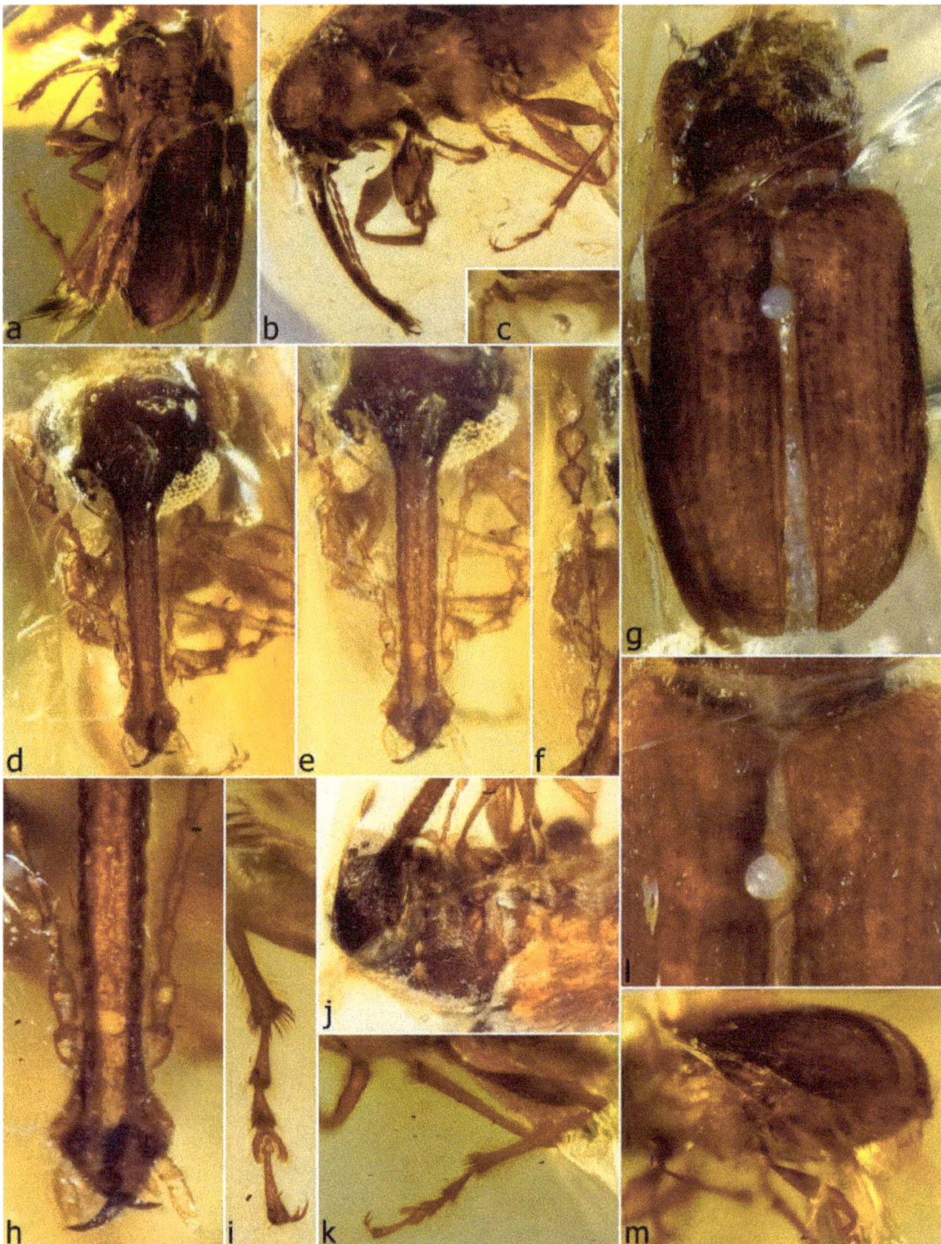

Figure 14. *Burmonyx zigrasi*, holotype. Habitus, left lateral (**a**); head and prothorax, left lateral (**b**); left antennal club, left lateral (**c**); head, dorsal (**d**); same, showing details of head and eyes (**e**); right antenna, dorsal (**f**); habitus, dorsal (**g**); detail of rostrum and mandibles, dorsal (**h**); left metatarsus, dorsal (**i**); head, prothorax and mesothorax showing closed coxal cavities, right lateroventral (**j**); left hindleg, anterior (outer) (**k**); detail of elytral striation showing scutellary striole, dorsal (**l**); elytral and abdominal apex, laterocaudal (**m**).

Material examined. Holotype (AMNH JZC-Bu228): well preserved intact specimen, with mouthparts and appendages visible (right antennal club severed from funicle), prothorax and metaventrite distorted (or teratological; somewhat bulging, cuticle seemingly modified); re-prepared from much larger pebble-shaped amber piece from which the original description was prepared, now at end of flattened wedge 6.3 × 6.9 × 0.3–2.1 mm, with two large flat faces and two oblique edge-faces exposing dorsal side and dorsum of head and rostrum; amber clear yellow, without any debris, with fracture running transversely through middle of inclusion near elytral bases, other less significant fractures in surrounding matrix.

Remarks. Even following extraction of this specimen from the original large amber piece, numerous details of it are partly or wholly obscured. These include the ventral side of the head and rostrum, the ventral prothorax (particularly the intercoxal and hypomeral areas) and much of the meso- and metathorax and ventrites. These details may become visible if the block were cut along the ventral side, but the existing fractures would necessitate stabilising the block with embedding resin prior to further cutting, ideally by the method described by Nascimbene and Silverstein [36]. Some distortion artefacts have created differences between the left and right sides of structures, particularly of the antennae, prothorax anterolaterally (the right side seemingly is the undistorted one) and metaventrite (unclear which side is less distorted). A bubble emanating from the elytral suture partly obscures the posterior endpoint of the scutellary strioles and locally distorts the sutural margins, probably facilitating the appearance of a long scutellary striole in the specimen. A distinct, curved, ridge-like structure on the upper side of the right profemur (Figure 14b) superficially seems unlikely to be a result of distortion, but it is seemingly absent on the left femur, thus casting doubt on the taxonomic value of this very distinctive feature. Contrary to the original drawing of the specimen [49], the left hindwing is not exposed behind the elytron; the structure drawn is merely a crack in the amber. Surface obscurities, mainly of the ventral side, may pose an additional difficulty when attempting to correctly identify further species or specimens of *Burmonyx*, as proportions of structures and other useful surface characteristics are obscured in places.

Perhaps the most noticeable differences between *B. zigrasi* and *Guillermorhinus longitarsis* are the much longer and coarser tibial fringing setae and very long spurs. Additional differences from *G. longitarsis* include the distinctly scalloped dorsal carinae of the rostrum, the longer scutellary strioles, the longer, slenderer funicle segments and the shorter apical club segments.

Genus ***Guillermorhinus*** Clarke & Oberprieler, **gen. n.**

Type species: *Guillermorhinus longitarsis* Clarke & Oberprieler, **sp. n.**

Description. Size. Length 2.1 mm, width 0.93 mm. **Head** porrect, short, largely not visible (retracted into prothorax). **Eyes** not visible, evidently small, directed somewhat forward, coarsely facetted, dorsally separated by width of rostrum anteriorly. **Rostrum** slightly longer than pronotum, slender, subflattened, weakly curved, apically expanded; antennal insertions in apical quarter, dorsolateral, with shallow scrobes indicated in front and behind them; anterolateral corners of postmentum extended into long processes (Figure 15f). **Antennae** non-geniculate, slightly longer than rostrum; scapes short, globular, subequal in length to funicle segment 1; funicles 7-segmented, segment 1 broader than scape, 2–7 thinner, shorter, subequal; clubs large, much shorter than funicle, loosely articulated, 4-segmented, apical segment short, acute, about as long as 3. **Mouthparts**. Labrum present. Mandibles long, falcate, non-exodont, articulation horizontal. Maxillary palps prominent, elongate, projecting well beyond rostral apex, apparently 3-segmented. **Thorax**. Prothorax narrower than elytra, proclinate, with anterior lateral margins oblique in lateral view. Pronotum laterally without tooth, posterior corners indistinct, fitting closely onto elytra; surface finely punctate, sparsely setose, setae reclinate, directed anteromesad; notosternal sutures closed ventrally, vertical, not evidently continuing anteriad. Prosternum long; procoxal cavities probably separated at least anteriorly but poorly visible (crumpled), closer to posterior hypomeral margin. Scutellar shield densely setose. Mesocoxal cavities closed laterally. Metanepisterna not visible. Metaventrite long, lightly convex. **Elytra** elongate, not basally lobed over base of pronotum, with broadly rounded humeri, posteriorly hardly declivous, lateral margin nearly straight, anterior marginal notch absent, apically closing evenly, conjointly rounded, not exposing pygidium; sutural flanges narrow, equal; surface weakly, shallowly punctostriate, with short scutellary striole, interstriae flat, finely punctosetose, setae short, thin, whitish, reclinate, directed caudad. **Legs**. Procoxae globular, not projecting, medially apparently separated by intrusion of prosternal process; mesocoxae globular, broadly separated; metacoxae flat, transversely elongate. Trochanters short, oblique. Femora long, weakly inflated at middle, outer side rounded. Tibiae straight, compressed, slightly widening distad, outer edge smooth, apices obliquely truncate, with 2 spurs. Tarsi elongate, slender, flattened, almost as long as tibiae; tarsites 1 and 2 subequal, widening distad, apically truncate, 3 deeply bilobed, lobes digitate, 5 long, apically expanded; claws divaricate, bifid, also with basal angulation, without ventrobasal seta. **Abdomen** with all ventrites free, slightly stepped; 1–5 subequal, 5 about as long as 3 + 4, broadly rounded.

Derivation of name. The name of the genus is composed of the first name of the eminent weevil specialist Guillermo ('Willy') Kuschel, to whom the Special Issue of *Diversity* in which this paper appears is dedicated, and the Greek noun *rhis* (G: *rhinos*), meaning nose or snout; its gender is masculine.

Remarks. *Guillermorhinus* agrees with *Burmonyx* in having the anterolateral corners of the postmentum extended into long processes (Figure 15f) but differs from it in its weakly, finely punctostriate elytra, broader, flat tarsites 5, narrow, digitate lobes of tarsites 3 and shortly bifid, basally angulate tarsal claws. It is also represented by a single species. Like *Burmonyx*, it differs from all extant genera of Rhinorhynchinae in its long postmental processes and narrow, finger-like lobes of tarsites 3. Its long terminal maxillary palp segments, long prosternum and deeply constricted head behind the eyes indicate that it belongs in the tribe Mecomacerini. From *Burmomacer* it also differs in its long rostrum and the postmental extensions.

Guillermorhinus longitarsis Clarke & Oberprieler, **sp. n.** (Figure 15)

Description. Size. Length 2.1 mm, width 0.93 mm. **Head** largely invisible. **Eyes** evidently small. **Rostrum** subequal in width for most of length, expanded apically. **Mouthparts.** Mandibles prominent, long, slender, acutely pointed (overlapping and exposed in repose). **Antennae** with club segments 1–2 shortly obconical, subequal in length and width, 2 slightly wider than long, 3 subequal in width to 2. **Thorax.** Pronotum finely punctosetose; with short, recurved, mesadly or anteromesadly directed setae. Prosternum moderately elongate, seemingly about as long as procoxae, anterior margin slightly emarginate. Mesoventrite short. Metaventrite long, flat. **Elytra** subparallel-sided, in dorsal view shallowly concave in front of middle; surface densely covered with short, acutely pointed setae directed caudad. **Legs** Metafemora more strongly inflated in middle than others; tibiae apically with short fringing setae; inner tooth of bifid claws curved. **Abdomen** with 5 subequal, free ventrites. Ovipositor elongate, weakly sclerotised (structural details not clearly visible).

Material examined. Holotype (NIGP154201), female: heavily distorted (depressed and crumpled) but otherwise largely intact and well visible specimen, with head mostly recessed into prothorax (this and eyes barely visible inside prothoracic cavity), prothorax and elytra collapsed and many thoracic details not visible or interpretable, rostrum, mouthparts, antennae and surface details well visible, right hindleg severed at trochantero-femoral joint, left metatibia fractured near base; near centre of short pyramidal block with rounded base, 5.0 × 4.5 × 2.8 mm; amber clear but with yellow flow bands, free of fractures and organic debris, with few silvery diffuse bubbles.

Derivation of name. The species is named for its elongate legs, especially the tarsi.

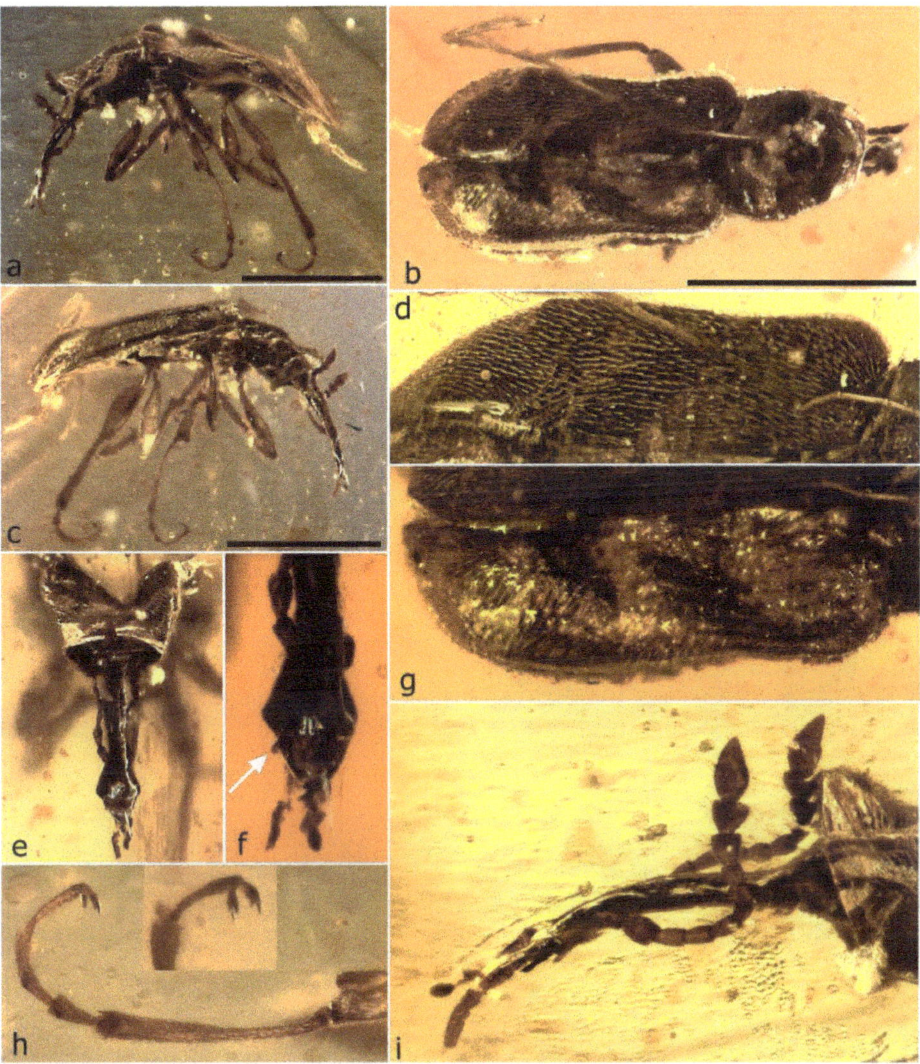

Figure 15. *Guillermorhinus longitarsis* sp. n., holotype. Habitus, left lateral (**a**); habitus, dorsal (**b**); habitus, right lateral (**c**); left elytron, dorsal (**d**); rostrum, dorsal (**e**); apex of rostrum and mandibles showing exposed postmental process (arrow), dorsal (**f**); right elytron showing faint striae, dorsal (**g**); left mestotibia and tarsus, insert showing detail of tarsal claw (**h**); head, rostrum and antenna, left lateral (**i**). Scale bars: 1.0 mm (a–c).

Remarks. In addition to its remarkable and diagnostic claws, this species is characterised most readily by its very faintly and relatively finely punctostriate elytra. The elytra of *Burmomacer kirejtshuki* are seemingly coarsely punctostriate at the sides and fine on the disc, but the air layer covering most of the surface prevents a clear view of the surface and the more distinct lateral interstriae may be a result of compression at the sides. The postmental processes (Figure 15f,i) are more slender than those of *Burmonyx zigrasi* and very elongate, visible even in dorsal view as slight protrusions. The poor preservation of the holotype of *G. longitarsis* may make comparison with future specimens difficult, as neither body proportions and outlines nor other details useful for species distinctions may be readily comparable. Better views of the head, rostrum (especially the ventral side), thoracic venter and abdomen in additional specimens or species will hopefully aid in a more complete characterisation of the species and its affinities.

Genus ***Burmomacer*** Legalov, 2018

Burmomacer Legalov, 2018: 2 [46] (type species, by original designation: *Burmomacer kirejtshuki* Legalov, 2018)

Redescription. Size. Length 3.52 mm, width 1.56 mm. **Head** porrect, short, broad, constricted behind eyes. **Eyes** large, lateral, directed forwards, coarsely facetted, dorsally separated by slightly less than width of rostrum anteriorly but much further separated posteriorly; forehead flat, without tubercles between anterior margin of eyes. **Rostrum** short, subequal in length to pronotum, broad, robust, weakly curved, apically expanded; antennal insertions at apex, behind mandibular articulations, dorsolateral, without scrobes in front or behind them. **Antennae** non-geniculate, slightly longer than rostrum; scapes short, globular, slightly longer than funicle segment 1; funicles 7-segmented, segment 1 narrower than scape, 2–7 thinner, shorter, subequal; clubs large, almost as long as funicle, loosely articulated, 4-segmented, apical segment about as long as 3, acute. **Mouthparts**. Labrum present. Mandibles long, falcate, non-exodont, articulation horizontal. Maxillary palps prominent, elongate, projecting well beyond rostral apex, apparently 3-segmented. **Thorax**. Prothorax about as broad as elytra, not or only weakly proclinate, with anterior lateral margin slightly oblique in lateral view. Pronotum laterally without tooth, posterior corners distinct, angulate, fitting closely onto elytra; surface finely punctate, sparsely setose; notosternal sutures closed, curved anteriad; procoxal cavities broadly separated anteriorly and medially (possibly confluent posteriorly nearer hind margin of procoxae), situated nearer to posterior prothoracic margin. Scutellar shield densely setose. Mesocoxal cavities closed laterally by meso- and metaventrite. Metanepisterna distinct, sparsely setose, without anterodorsal lobe. Metaventrite long, lightly convex. **Elytra** elongate, not basally lobed over base of pronotum, with broadly rounded humeri, posteriorly hardly declivous, lateral margin nearly straight, anterior marginal notch absent, apically conjointly rounded, not exposing pygidium; sutural flanges not visible (only left apically); surface weakly, shallowly punctostriate, scutellary striole not visible, interstriae flat, finely punctosetose, setae short, thin, whitish, reclinate, directed caudad. **Legs**. Procoxae globular, not projecting, widely separated by broad prosternal process; mesocoxae globular, broadly separated; metacoxae flat, transversely elongate. Trochanters short, oblique. Femora long, subcylindrical, weakly inflated at middle, dorsally rounded. Tibiae straight, subcylindrical, slightly widening distad, outer edge smooth, apices obliquely truncate, with probably 2 spurs (2 on metatibiae, 1 discernible on mesotibiae). Tarsi elongate, slender, subcylindrical, almost as long as tibiae; tarsite 1 longer than 2, widening distad, apically subtruncate, 3 deeply bilobed, lobes short, narrow, 5 long, apically expanded; claws divaricate, strongly bifid (inner tooth spiniform, almost as long as outer claw), without ventrobasal seta. **Abdomen** with free ventrites, slightly stepped, 1 slightly longer than 2, 2–4 subequal, 5 subequal to 3 + 4, broadly rounded, shortly notched at middle of posterior margin.

Remarks. The genus was not specifically described, only a reference given to the diagnosis of the tribe Burmomacrini. Its most unusual feature is the broad separation of the procoxae, but it cannot be ascertained whether the procoxal cavities are also completely separated, because the posterior part of the prosternum is not visible (hidden beneath the folded tarsi). It is thus not determinable whether the broad prosternal process meets a similar hypomeral one to fully separate the cavities. In the extant genus *Brarus* Kuschel the procoxae are also separated by a broad prosternal process, but this narrows rapidly posteriad and does not meet the opposing hypomeral process, leaving the coxal cavities shortly confluent in their middle. As the prosternal process in *Burmomacer* is longer and not narrowing, extending nearly to the posterior hypomeral margin, it is somewhat similar to that of *Nemonyx*, which extends to the posterior end of the procoxae but does not abut against any hypomeral process and thus leaves the coxal cavities broadly confluent posteriorly, almost open. *Burmomacer* differs from *Burmonyx* and *Guillermorhinus* in its much shorter, broader rostrum with only two grooves dorsally and in that the anterolateral corners of the postmentum are not extended into long processes. It agrees with *Burmonyx* in its similarly strongly bifid claws but has narrower lobes of tarsites 3. It also appears to fit into the tribe Mecomacerini, but more details of its prosternal structure need to be known to ascertain this. It contains a single species as known.

Burmomacer kirejtshuki Legalov, 2018

Burmomacer kirejtshuki Legalov, 2018: 2 [46]

Material examined. Holotype (ISEA no. MA 2017/2): well preserved specimen missing only claws of right hindleg, otherwise intact, minimally depressed with elytra partly forced open, length 3.52 mm, width 1.56 mm, largely separated from surrounding amber, leaving most visible detail as an impression in amber matrix, well visible on all sides without surface obscurities; towards one end of rectangular cuboid 7.1 × 1.8–2.5 × 1.2–1.6 mm, rounded off over head, with sides parallel to block faces; amber clear yellow-brown with diffuse tiny organic particles, free of fractures near specimen, with one large flat bubble obscuring basal pronotal and elytral areas.

Remarks. In view of the redescription of the genus above, the original description of the species is adequate for the moment. This species is similar to *Burmonyx zigrasi* in having bifid claws with long slender inner teeth but is much larger and has a shorter broader rostrum with only two dorsal grooves. The unusual preservation of the holotype makes it difficult to determine the true form of the specimen. It has clearly separated from the amber and partially decomposed in the resulting cavity, but some areas and appendages are evidently still intact and visible. The inside surface of the cavity has preserved a perfect impression of the original weevil surface, but the robustness and flatness of the specimen seems to be exaggerated.

Family MESOPHYLETIDAE Poinar, 2008 **stat. n.**

Mesophyletinae Poinar, 2006: 879 [29] (not available; no type genus designated)

Mesophyletinae Poinar, 2008: 262 [50] (type genus: *Mesophyletis* Poinar, 2006)

Description. Head porrect, short, usually subglobular. **Eyes** moderately sized to very large in relation to head, hemispherical to subglobular to elongate conical to dorsoventrally flattened, usually strongly protruding and coarsely facetted, sometimes flatter and finely facetted; rarely with distinct interfacettal setae (*Compsopsarus reneae*). **Rostrum** slightly shorter to much longer than pronotum, subcylindrical, usually long and evenly thin, slightly downcurved, sometimes shorter and stouter, rarely flexible into a prosternal channel (*Burmorhinus*, *Rhadinomycter*); antennal insertions usually median, sometimes subbasal, rarely in apical third or quarter, behind them usually with long narrow scrobe extending to eye, in front of them usually with a lateral row of sparse, long erect setae. Gular suture (where discernible) single, long, extending from base of head onto underside of rostrum. **Antennae** subgeniculate to geniculate, of 'open' type (funicle inserted apically into scape); scapes usually about as long as entire funicle, at least slightly longer than funicle segment 1; funicles 7-segmented; clubs loosely articulated to subcompact, 4-segmented, apical segment usually distinctly inserted in or set off from penultimate one. **Mouthparts**. Labrum absent. Mandibles mostly strongly exodont and then often flattened, sometimes non-exodont, articulation horizontal to oblique. Maxillary palps (where discernible) 4-segmented, robust, short, not projecting beyond mandibles. Labial palps 3-segmented, inserted apically in prementum. **Thorax**. Prothorax usually proclinate, with anterior lateral margins oblique in lateral view, straight, rarely drawn out into ocular lobe. Pronotum subquadrate to transversely trapezoidal, rarely elongate, often narrowing apicad but apically not constricted or collared, laterally slightly to strongly rounded, posterior corners usually slightly produced, fitting closely onto elytra; notosternal sutures mostly closed, rarely open. Prosternum variable; procoxal cavities usually medially confluent, rarely separated. Mesocoxal cavities medially separated, laterally usually closed by meso- and metaventrite, rarely open (mesanepisternum and mesepimeron also not reaching coxa); metacoxal cavities elongate transverse, separated. Metanepisterna usually distinct, suture without sclerolepidia. Mesoventrite short, usually steeply declivous. Metaventrite longer, flat to slightly convex, often with rounded transverse weals before metacoxae. **Elytra** usually elongate, subparallel-sided, anterior margin straight to slightly arcuate or bisinuate, humeri broadly rounded; lateral edge (beyond stria 10) often inflexed into epipleural flange, without upper costa but with narrow posterior part (behind metacoxa) sometimes forming a narrow groove, lateral margin anteriorly sometimes notched to receive anterodorsal process of metanepisternum; apices usually individually rounded, sometimes exposing short pygidium; sutural flanges narrow, equal; surface distinctly to indistinctly punctostriate, with 10 striae but no scutellary striole; usually moderately setose, setae on striae between punctures and on interstriae, not in strial punctures, short, thin, sharp, suberect to reclinate, directed caudad. **Legs**. Procoxae mostly elongate and prominent, rarely globular, usually medially contiguous, rarely separated by prosternal and hypomeral processes meeting between them; mesocoxae globular, narrowly separated; metacoxae flat, transversely elongate, medially narrowly separated. Femora subcylindrical to flattened, inflated in distal half, dorsally rounded or with distal carina or crenulation, almost always unarmed. Tibiae subcylindrical to strongly flattened, outer edge often sharply crenulate to serrulate, sometimes rounded, distally usually expanded, sometimes slightly excised, and with dense, long, stiff setae, apex usually with 2 spurs, inner one sometimes fixed and broadened, rarely without spurs. Tarsi elongate, loose; tarsite 1 elongate to broadly triangular, apically often excised to even sublobate, 2 more narrowly triangular, apically usually deeply excised, sometimes bilobed, 3 deeply lobed, lobes long and stalked, often pedunculate, 5 elongate, basally narrow but distally broadened; claws divaricate, simple to basally angulate or dentate, nearly always with long ventrobasal seta. **Abdomen** with ventrites 1 and 2 fused, each longer than 3 and 4, 5 usually subtriangular but rectangular and apically truncate when last tergite (apparently VII) exposed as pygidium. Ovipositor with gonocoxites long, slender, with thin dorsal and ventral baculi (sclerotised rods), apically finely setose, with small cylindrical apical stylus with apical tuft of setae.

Remarks. This taxon was described as a subfamily of 'Eccoptarthridae', the family name then sometimes used for Caridae [9], although Poinar excluded the carids from his concept of Eccoptarthridae and restricted this family to the extinct genus *Eccoptarthrus* Arnoldi (an eobeline nemonychid) and Baissorhynchinae [29], which were probably Nemonychidae or Cimberididae as well [16]. A comparison between *Mesophyletis* and *Eccoptarthrus* was not made, but Poinar [29] placed the baissorhynchine genus *Cretonanophyes* Zherikhin in Mesophyletinae as it seemingly shared the diagnostic characters of *Mesophyletis*, namely geniculate antennae with loosely segmented clubs, elongate trochanters, tarsi with dentate claws and pedunculate tarsite 3 lobes and an exposed pygidium. He compared *Mesophyletis* with Caridae, Apioninae and Nanophyinae and could, correctly, not assign it to any of these taxa, but he did not make a comparison with Attelabidae, presumably because of their non-geniculate antennae.

With the much larger diversity of mesophyletid taxa now known and a much better assessment of their characters, it is evident that this group cannot be accommodated in any of the extant families without a significant widening of their concepts. The absence of a labrum readily distinguishes the mesophyletids from the 'lower' weevil families Cimberididae, Nemonychidae and Anthribidae. From Belidae they differ foremost in their geniculate antennae, long single gular suture and angulate to dentate tarsal claws with a ventrobasal seta, and although a few genera with subgeniculate antennae share features such as open coxal cavities and the antennal configuration with some Belidae, their protibiae also do not possess the apical antenna cleaner that is an autapomorphy for Belidae [1] (see Aepyceratinae below). With Attelabidae the mesophyletids share the long single gular suture, but they differ in many other important characters (geniculate or subgeniculate antennae, distinctly four-segmented clubs, consistent absence of scutellary strioles, divaricate dentate claws with ventrobasal seta, usually closed pro- and mesocoxal cavities), even in characters that are superficially similar (exodont mandibles, crenulate femora and tibiae), so that they cannot be included in this family. From the family Caridae the mesophyletids differ mainly in their long gular suture, their geniculate, laterally inserted antennae, fused ventrites 1 and 2 and (again) their differently exodont mandibles and crenulate femora and tibiae, so that they cannot be regarded as carids in the current concept of this family. The long single gular suture also distinguishes the Mesophyletidae from the 'higher' families Brentidae and Curculionidae, and from Brentidae other than Nanophyinae they further differ in their mostly geniculate antennae, from Nanophyinae also mainly in their short oblique trochanters, seven-segmented funicles and divaricate, mostly dentate tarsal claws (see key to families for other differences) and from Curculionidae additionally in their loose antennal clubs and differently (apically) geniculate antennae. Their unique combination of characters thus dictates the recognition of Mesophyletidae as a separate, ninth family of weevils that apparently became extinct without leaving any extant relatives. They are evidently most closely related to the 'middle' weevil families, specifically to Attelabidae and Caridae, although the similarly geniculate antennae and longer gular sutures of Nanophyinae may indicate a relationship to this subfamily too, despite the many differences between the two taxa. The precise relationships of Mesophyletidae to all these family taxa require further study.

The concept of Mesophyletidae is diffused to some degree by the few aberrant genera that have subgeniculate antennae and sometimes open pro- and mesocoxal cavities, but these have the same dentate tarsal claws, similarly exodont mandibles and single gular sutures (where discernible) and are therefore here included in Mesophyletidae as well, albeit in a different subfamily, Aepyceratinae.

Subfamily Aepyceratinae Poinar, Brown & Legalov, 2017

Aepyceratinae Poinar, Brown & Legalov, 2017: 75 [51] (type genus, by original designation: *Aepyceratus* Poinar, Brown & Legalov, 2017)

Diagnosis. Head porrect, short. **Eyes** relatively small, hemispherical to subconical, slightly protruding, finely to coarsely facetted. **Rostrum** slightly longer or shorter than pronotum; antennal insertions subbasal to median, sometimes with weak short scrobe behind them. Gular suture single, long (where discernible). **Antennae** subgeniculate; scapes as long to twice as long as funicle segment 1; funicles 7-segmented; clubs loosely articulated, 4-segmented. **Mouthparts**. Labrum absent. Mandibles mostly exodont but sometimes non-exodont, articulation horizontal. **Thorax**. Pronotum subquadrate to transverse, laterally rounded, posterior corners usually extended, fitting closely onto elytra; notosternal sutures open or closed; procoxal cavities medially confluent; mesocoxal cavities open or closed. **Elytra** with broadly rounded humeri, apically individually rounded, in *Acalyptopygus* and *Calyptocis* exposing short pygidium; surface distinctly to indistinctly punctostriate, without scutellary striole. **Legs**. Procoxae subglobular to elongate, mostly prominent, medially contiguous; mesocoxae globular, narrowly separated; metacoxae flat, transversely elongate. Femora inflated in distal half, outside rounded; unarmed. Tibiae subcylindrical, outer edge rounded, apex with or without spurs. Tarsi elongate; tarsite 1 elongate to broadened, 3 deeply lobed; claws divaricate, dentate, with ventrobasal setae except in *Platychirus*. **Abdomen** with ventrites 1 and 2 fused, longer than 3 + 4; last tergite exposed as pygidium in *Acalyptopygus* and *Calyptocis*.

Remarks. This taxon was described as a new subfamily of Nemonychidae in the mistaken observation that the type genus and species, *Aepyceratus hyperochus*, possesses a free labrum (along with non-geniculate antennae, free abdominal ventrites and tibial spurs) [51], but our examination of the type specimen confirmed that it has no labrum (as indeed evident in Figures 2C and 2D of Poinar et al. [51]). As non-geniculate antennae also occur in all other weevil families except Curculionidae, free ventrites also in Belinae and Caridae [1] and tibial spurs in all other families except Anthribidae, none of the other three characters specifically relates *Aepyceratus* to Nemonychidae (and Cimberididae). Moreover, *Aepyceratus* has ventrites 1 and 2 fused, not free (movable) as the remaining three, and subgeniculate antennae (the scape twice as long as funicle segment 1), unlike the condition in Nemonychidae. These features as well as its exodont mandibles and dentate tarsal claws strongly indicate that *Aepyceratus* and five other genera with subgeniculate antennae (*Acalyptopygus, Calyptocis, Nugatorhinus, Platychirus, Rhynchitomimus*) also belong in Mesophyletidae, in which they form a group that may for now be treated as a subfamily Aepyceratinae. Apart from the subgeniculate antennae there is no character in evidence to suggest that this subfamily constitutes a monophylum, and differences in the conditions of some characters (subbasal and median antennal insertions, exodont and non-exodont mandibles, open and closed pro- and mesocoxal cavities, presence and absence of tibial spurs, pygidium exposed or not) suggest that it may be a paraphyletic group with respect to the genera with geniculate antennae combined in the subfamily Mesophyletinae. A phylogenetic analysis is needed to test the concept of Aepyceratinae as here proposed.

Genus ***Aepyceratus*** Poinar, Brown & Legalov, 2017

Aepyceratus Poinar, Brown & Legalov, 2017: 75 [51] (type species, by original designation: *Aepyceratus hyperochus* Poinar, Brown & Legalov, 2017)

Redescription. Size. Length 6.9 mm. **Head** short, transversely globular. **Eyes** large, flatly protuberant-subconical, lateral, finely facetted, dorsally separated by 1.5 × basal width of rostrum anteriorly but much further separated posteriorly; forehead flat, without tubercles or ridges between eyes. **Rostrum** much longer than pronotum, weakly curved, subcylindrical (slightly flattened); antennal insertions subbasal, lateral, with faint scrobe indicated in front of them. **Antennae** subgeniculate, long; scapes oblong-fusiform, apically narrowed, about twice as long as funicle segment 1; funicles 7-segmented, segment 1 narrower than scape, 2 about 3.0 × longer than 1, 2–7 subequal, slender, apically slightly wider, 7 about 0.67 × as long as 6; clubs large, long, loosely articulated, broader than funicle, 4-segmented, apical segment acute, about as long as 3. **Mouthparts**. Labrum absent. Mandibles large, exodont and endodont, articulation horizontal. **Thorax**. Pronotum transversely convex, narrower anteriorly, without lateral tooth, posterior corners distinctly angulate, fitting closely onto elytra; surface densely tomentose; notosternal sutures open ventrally. Prosternum moderately long; procoxal cavities medially confluent, approximately in middle of prothorax. Scutellar shield densely tomentose. Mesocoxal cavities closed laterally. Metanepisterna distinct. Metaventrite elongate, nearly flat, slightly convex in front of metacoxae. **Elytra** elongate, basally excised to receive base of pronotum, with weakly rounded, subflat humeri, posteriorly declivous, lateral margins broadly explanate, apically weakly individually rounded, not exposing pygidium; surface not or faintly punctostriate, without scutellary striole, densely tomentose, setae confusedly multidirectional. **Legs**. Procoxae large, subglobular but expanded laterally, medially subcontiguous; mesocoxae large, subglobular, narrowly separated; metacoxae transversely elongate. Trochanters large, sublobate, not recessed into coxae. Femora long, distally inflated, outside rounded, inside excavate in distal half, receiving tibiae in repose. Tibiae distally expanded, outer edge rounded, densely setose, apices obliquely truncate, with 2 spurs, meso- and metatibiae distinctly upturned apicad. Tarsi slightly longer than tibiae; strongly flattened; tarsite 1 broad, apically exicsed, 2 shorter, strongly lobate, 3 very deeply, narrowly bilobed (subpedunculate), 5 long, apically expanded; claws divergent, strongly dentate, without ventrobasal seta (or this not visible). **Abdomen** with ventrites 1–5 progressively shorter, 1 and 2 at same level, fused, 3–5 articulated, each at slightly higher level; sutures between ventrites substraight.

Remarks. This presently monotypic genus known from the single specimen of *Aepyceratus hyperochus* is among the most distinctive of Burmese amber weevils, both in habitus and in characters. Although sharing a generally similar tarsal structure with *Cetionyx*, *Burmocorynus*, *Opeatorhynchus* and *Petalotarsus* (but differing in important details) and rostral and antennal characters with other genera of Aepyceratinae, its overall flattened shape and combination of characters sets it apart from all other Burmese amber weevils. Among these characters (all of which are unique among known Burmese amber weevils) are the distinctly tomentose dorsal vestiture (ventrally with long, more usual setae), the explanate, strongly inflexed elytral margins and ridged (including basally toothed) pronotal margins, the sublobate trochanters (edges not continuous with femoral edges) that are not recessed into the coxae (possibly what Poinar et al. [51], p. 76, meant by "trochanters not separating femora and coxae"), the distinct brush of dense short subequal setae along the distal inner sides of the protibiae (possibly a grooming device), the row of cuticular teeth along the inner sides of the meso- and metatibiae, the strong black erect tibial setae, the outwardly projecting cuticular spine on the outer apical angles of the tibiae and the structure of the tarsi (particularly the enlarged first tarsites, subpedunculate second tarsites and large flat basal tooth on the claws). Other notable differences include the much greater and posteriadly increasing interocular distance (larger than rostral width at base), similar to that of

Debbia (Mesophyletinae), the laterally open procoxal cavities (shared with *Rhynchitomimus*, *Platychirus* and *Nugatorhinus*) and the laterally closed or possibly slightly open mesocoxal cavities (otherwise only in *Nugatorhinus*), the unusual configuration of the procoxae and surrounding prothoracic structures (as described) and the exodont but distinctly robust mandibles (not flattened, blade-like), in which the inner and outer sides are concave surfaces delimited by dorsal and ventral edges and the inner teeth are situated on dorsal and ventral edges of the inner side.

Aepyceratus hyperochus Poinar, Brown & Legalov, 2017

Aepyceratus hyperochus Poinar, Brown & Legalov, 2017: 76 [51]

Redescription. Size. Length 6.9 mm (excl. rostrum, not SL); moderately dorsoventrally flattened; dorsally subtomentose, covered in dense, multidirectional, subappressed, brownish and whitish setae. **Head** somewhat bulbous dorsally and ventrally; subtomentose, setae short; surface behind eyes rugose, sparsely setose; between eyes densely setose, setae whitish and brown, coarsely minutely punctate. **Rostrum** elongate, 2.9 mm long (Poinar et al., 2017); junction with head dorsally slightly concave; densely setose, with basal setae longer than distal ones; dorso-apically sparsely setose, shining; broadening apicad from about middle, apically ca. 2 × broader than in middle. Apex dorsally broadly convex, laterally somewhat lobate; mandibular articulations shallow. **Antennae.** Scapes short, narrow basally at insertion, densely setose; clubs distinct, densely setose (tomentose), with longer setae apically, segments 1 and 2 obconical, 1 basally much thinner than 2, apically subequal, 2 slightly shorter than 1, 3 elongate, expanding apicad. **Mouthparts.** Mandibles with 2 large pointed external teeth (one basally, one apically); externally with dorso- and ventrolateral edges, the intervening surface slightly concave, without setae; internally with dorso- and ventromedial edges, the intervening surface slightly concave, with 3 smaller teeth, mesal and apical ones arising from dorsomedial edge and apical one arising from ventral edge, apically truncate, anvil-shaped with outer and inner angles formed by external and dorso-internal apical teeth, respectively. **Thorax**. Pronotum broadest at base, gradually narrowing anteriad, sides sinuate, outwardly curved until about apical third, then slightly concavely curved to anterior margin (appearing somewhat collared anterolaterally); sides forming lateral ridge (subcarinate) clearly demarcating pronotum from inflexed hypomeron, finely toothed in basal third; basal margin nearly straight, submarginally forming distinct lip separated from higher basal surface by narrow groove, lip thickest in middle third, fitting under inflexed basal elytral margins, medially with slight emargination to receive anterior margin of scutellar shield; setae mixed whitish/brownish, more whitish basally and laterally. Prosternum elongate, densely setose, setae longer than on pronotum, anterior margin slightly broadly emarginate, prosternal process abruptly convexly curved posteroventrad between inner apical margins of procoxae (not projecting between them), broadened apicad, not connected to hypomeral process, this apically subtruncate, vertically disjunct from hypomeral region by about procoxal length. Scutellar shield broadly transverse, anteriorly broadly rounded (fitting into emargination of pronotal margin), posterior margin medially pointed. Meso- and metaventrites densely setose, at most finely punctate; each with narrow intermesocoxal process. Mesocoxal cavities large. **Elytra** somewhat flattened; subrectangular; bases slightly sinuate, nearly straight; humeri weakly rounded, somewhat concavely extended; striae or interstriae absent or indistinct, sutural striae absent; surface finely punctate, punctation indistinct due to extremely dense short subappressed (tomentose) vestiture, setae confusedly multidirectional; explanate margin of sides dorsally densely setose as disc, with narrow band of whitish setae seemingly situated in a broad, stria-like channel along inner edge of margin and with seemingly evenly spaced sparse tufts of fine, curved setae; laterally with distinct and setose marginal groove (on inflexed portion, adjacent to edge contacting body). Elytral apices with small but distinct medial emargination (carinate explanate edge not continuing flushly from side to side). **Legs** generally short, robust; densely setose, covered in subtomentose vestiture, with different types of setae; setae generally shorter on forelegs. Procoxae apparently at most only narrowly separated behind prosternal process; very densely setose, especially posteriorly; setae golden, longer than dorsal setae. Trochanters robust, prominent, not recessed into coxae. Tibiae with long setae on inner faces, scattered elsewhere, apically denser near outer apical edge; protibiae subequal in width and depth for most of length, outer side near apex with several coarser blackish spine-like setae, also with longer whitish setae, inner side in distal half broadly concave, with area of dense subequal setae increasing in density apicad, apically with inner edge broadly convex, faintly produced to a subdued point, then concavely continuing along dorso-apical edge, with moderately dense setae partly obscuring outer spine, apical edge lined with coarser black setae; meso- and metatibiae with row of cuticular teeth along entire inner edge, teeth larger than on mesotibia, increasing in size apicad, each indistinctly separated by basal width of a tooth, with numerous distinctly coarse black semi-erect setae projecting almost perpendicularly to long axis of tibia (absent internally), with long whitish setae concentrated on inner faces, especially towards apex, metatibiae with 2 or 3 small cuticular teeth on outer apical edge. Tarsi strongly flattened, densely setose, with longer setae dorsally; ventrally with thick tenant setae forming dense setal pads; tarsites 1–3 progressively broader, 1 cordiform, progressively shorter from fore- to hindlegs, 2 strongly lobate, lobes slender and very narrow basally on meso- and metatarsi (subpedunculate), medial length less than half that of 1, tarsite 3 with lobes elongate, about half as long as 5, strongly broadening apicad, continuous basally (not individually flexible), 4 (cryptotarsite) distinct, 5 somewhat flattened, ventrally setose, setae long and irregularly positioned and sized, ventroapically lobate, with pair of fine, curved setae projecting distad; claws with basal tooth large and explanate, inner edge almost flat, outer edge curved forming deep, curved notch between outer claw and tooth. **Abdomen.** Ventrites very densely setose, setae longer than elytral setae, distinctly patterned; ventrite 1 slightly longer than 2, 3 about 0.67 × as long as 2, 3, and 4 with differently coloured setae, 4 slightly shorter than 3, 5 subequal to or shorter than 4.

Material examined. Holotype (PACO, with curatorial #B-C-50): exceptionally well preserved, intact specimen, not depressed, largely undistorted, well visible with exception of the ventral side partly obscured by thickness of amber and presumed plant debris, surface partly obscured by fragmented whitish coating; at corner of long irregular pyramidal block, with three large, mostly flat faces, one smaller flat face; with several pieces of possibly woody plant material obscuring clear view of ventral side; amber clear-brown, with minimal other debris (see also [51]).

Remarks. The holotype shows little sign of distortion associated with the preservation process, and aspects of the structure of the specimen seemingly related to dorsoventral flattening are symmetrical and do not appear to be aberrations. Nevertheless, we were unfortunately unable to borrow and further prepare the amber block, which would result in much clearer views of the ventral side of the inclusion and of several structures that are currently either not or insufficiently visible because of removable obscurities (debris, bubbles and amber). This includes the ventral side of the head (and thus the condition of the gular sutures) and ventral mouthparts and the ventral surface of (especially) the prothorax and the tarsal claws. Through our examination of the holotype, we could not confirm whether the basal teeth of the claws also possess a ventrobasal seta as we define herein. Only one of the visible claws has a seta visible in the correct general location, but as only the distal part of this seta could be viewed, we could not assess its insertion point. In all Mesophyletidae with dentate claws, the ventrobasal seta arises from the back face of and usually at or near the apex of the tooth and generally projects ventrad (and thus often appearing as an apical seta), and we therefore surmise that these setae are also present in *A. hyperochus* but simply obscured in the holotype in the current amber block in both possible viewing angles and by the enlarged basal teeth of the claws in this species (much larger and broader than in other genera).

Genus ***Platychirus*** Clarke & Oberprieler, **gen. n.**

Type species: *Platychirus beloides* Clarke & Oberprieler, sp. n.

Description. Size. Length ca. 5 mm, width ca. 2 mm. **Head** short, broad, transverse, subglobular. **Eyes** large, strongly protruding, lateral, finely facetted, dorsally without tubercles between them. **Rostrum** short, stout, cylindrical, antennal insertions lateral, subbasal, behind them without scrobes, in front of them without lateral row of setae. No gular suture visible. **Antennae** subgeniculate, long; scapes short but elongate, cylindrical, slightly thickening distad; funicles 7-segmented, longer than scape, segment 1 shorter than scape, medially inflated, 2 longer than 1, segments 3 and 4 also elongate, 5–7 shorter; clubs 4-segmented, very loosely articulated (hardly differentiated from funicle), segment 4 narrow, elongate. **Mouthparts**. Labrum absent; mandibles narrow, elongate, exodont, articulation horizontal. **Thorax**. Prothorax proclinate, with anterior lateral margins oblique in lateral view. Pronotum short, broad, convex, rounded laterally, posterior corners rounded, not fitting closely onto elytra; surface finely punctate, sparsely setose, setae pale, directed anteriad; notosternal sutures widely open. Prosternum moderately long; procoxal cavities apparently medially confluent, in middle of prothorax. Scutellar shield not discernible. Mesocoxal cavities laterally narrowly open (meso- and metaventrite not meeting above coxae). Metanepisternal sutures distinct. Mesoventrite short, anteriorly sloping. Metaventrite long, convex. **Elytra** shortly elongate, with weakly, broadly rounded humeri, apically jointly truncate, only slightly individually rounded, not exposing pygidium; surface not punctostriate, finely setose. **Legs**. Procoxae short, globular, apparently medially contiguous; mesocoxae subglobular, narrowly separated; metacoxae flat, transversely elongate. Trochanters short, oblique. Femora short, thick, robust, subcylindrical, inflated through most of length, unarmed, outside rounded. Tibiae short, straight, robust, subcylindrical, outer edge rounded, apex truncate, without spurs; protibiae on inside without apical brush but a few black setae. Tarsi almost as long as tibiae; tarsite 1 long, very broad, flat, 2 triangular, insertion of 3 dorso-apical, 3 shortly bilobed, 5 as long as 1, narrow between lobes but broadening apicad; claws divaricate, ventrally bluntly dentate, apparently without ventrobasal seta. **Abdomen** with ventrites 1 and 2 fused, each slightly longer than 3.

Derivation of name. *Platychirus* is named for its broad flat first tarsites, the name formed from the Greek adjective *platys*, broad, and noun *cheir* (G: *cheiros*), a hand, and being masculine in gender.

Remarks. This genus has several characters of extant Belidae, in particular the broad, short head with large, round eyes, the lengths of the antennal segments (scape shortly elongate, funicle segment 1 short, the others elongate) and the open pro- and mesocoxal cavities, but it has no protibial brushes (antenna cleaners) and its mandibles are narrow, elongate, with a blunt, apical external tooth. Its remarkably broadened and flattened basal tarsites are also similar to those of some Belidae (e.g., *Stenobelus* [41]), but such tarsites occur in *Car* Blackburn (Caridae) as well. Despite these character agreements and the overall similarity of *Platychirus* to Belidae, the lack of protibial brushes, which are considered a synapomorphy for extant Belidae [1], and the elongate, exodont mandibles militate against assigning the genus to Belidae, and we consider it more plausible to group it together with the other Burmese amber genera with subgeniculate antennae and open coxal cavities in the subfamily Aepyceratinae of Mesophyletidae. In this group it is most similar to *Rhynchitomimus*, which has very similar antennae and also similar claws, although these carry a ventrobasal seta. If additional specimens of *Platychirus* are discovered, its taxonomic affinities may be able to become better understood. *Platychirus* is currently monotypic.

Platychirus beloides Clarke & Oberprieler, **sp. n.** (Figure 16)

Description. Size. Length 5.12 mm, width ca. 2.25 mm. **Head** constricted behind eyes. **Eyes** subglobular, dorsally separated by about their width, forehead as broad as rostrum at base, flat. **Rostrum** as long as pronotum, slightly curved, abruptly thinner in front of antennal insertions, these subbasal; behind antennal insertions with sparse, short, erect setae dorsally and laterally, surface coarsely granulose. **Antennae.** Scapes reaching anterior margin of eye in repose; funicles much longer than scape, segment 1 half as long as scape, spindle-shaped, in middle as thick as scape, segment 2 ca. 1.5 × longer than 1, thinner, slightly thicker apically, 3 similar but shorter, 4 similar but slightly shorter again, 5–7 shorter, bulbous at apex; clubs flat, basal segments broadening distad, segment 4 narrow, elongate. **Mouthparts**. Mandibles with 1 blunt apical external tooth. Maxillae and labium not discernible. **Thorax**. Pronotum short, roundly trapezoid, strongly convex, tumescent in basal half; surface minutely punctate, sparsely setose. **Elytra** shortly elongate, posteriorly sloping down and abruptly declivous apically; surface relatively densely very finely setose, setae black, recumbent, directed caudad. **Legs**. Metafemora not quite reaching posterior margin of ventrite 2. Tibiae slightly shorter than femora, sparsely setose in apical half. Tarsi with tarsite 1 as long as 2 + 3, apically slightly emarginate, 2 with apex slightly excised, 3 with lobes short, broad, broadly connected basally, 5 as long as 1; claws as for genus. **Abdomen** as for genus.

Material examined. Holotype (NIGP154202): very well preserved, intact specimen, not compressed or distorted, well visible under strong light; in rectangular block 7.2 × 6.4 mm, drop-shaped in cross-section, 5.8 mm thick across the weevil body; amber slightly cloudy with many small impurities, especially over back of elytra and oblique crack along left side of prothorax.

Derivation of name. The species name is an adjective formed for the genus name *Belus* and the suffix *-oides*, like, in reference to the similarity between the species and extant Belidae.

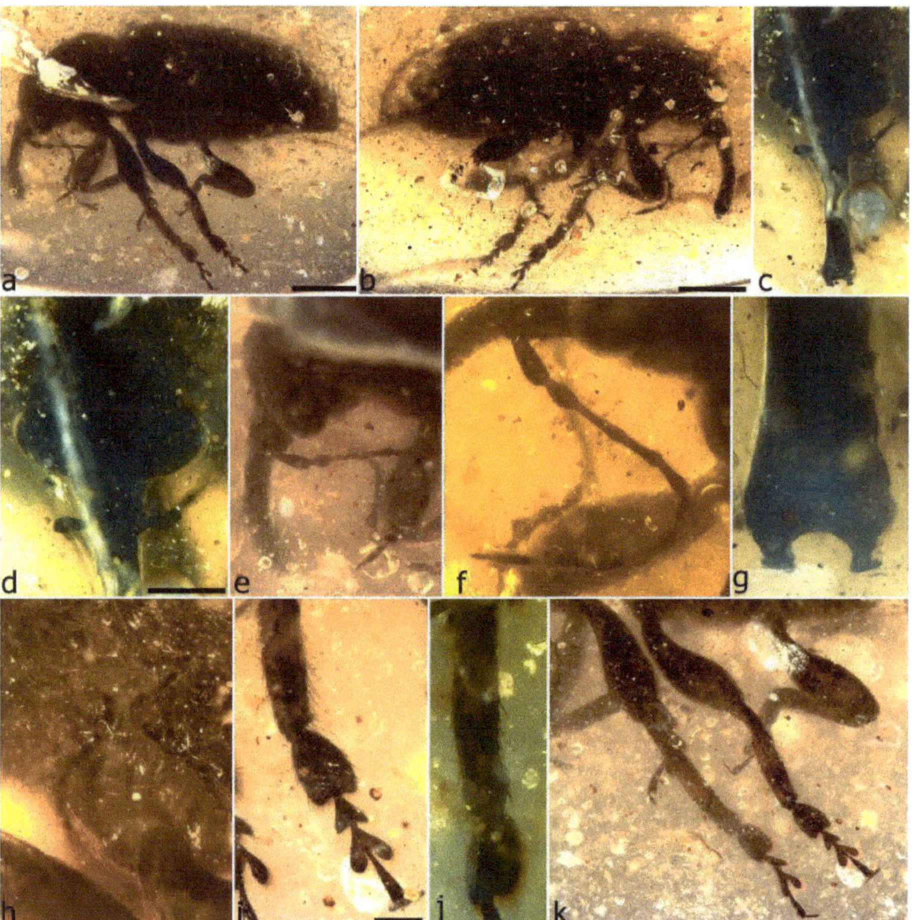

Figure 16. *Platychirus beloides* sp. n., holotype. Habitus, left lateral (**a**); habitus, right lateral (**b**); head, dorsal (**c**); eyes, dorsal (**d**); head and antenna, left lateral (**e**); detail of antenna and eyes, left lateral (**f**); apex of rostrum and mandibles, dorsal (**g**); open mesocoxal cavities (**h**); left mesotibiae, dorsal (**i**); left protibia, dorsal (**j**); left legs, dorsal (**k**). Scale bars: 1.0 mm (a,b); 0.5 mm (d); 0.2 mm (i).

Remarks. The species is very distinctive due to its rostrum being strongly constricted at the subbasally inserted antennae, its broadened first tarsites and its weakly dentate claws apparently without a ventrobasal seta. The constricted rostrum suggests that the single specimen known of it to date may be a female.

Genus ***Rhynchitomimus*** Clarke & Oberprieler, **gen. n.**

Type species: *Rhynchitomimus chalybeus* Clarke & Oberprieler, sp. n.

Description. Size. Length 2.94 mm, width 1.2 mm. **Head** long, broad, subglobular. **Eyes** large, lateral, strongly protruding, finely facetted, dorsally separated by slightly more than width of rostrum anteriorly but much further posteriorly; forehead weakly convex, without any tubercles. **Rostrum** very long (more than half body length), slender, subcylindrical, substraight; antennal insertions subbasal, lateral, with scrobes behind them, in front of them laterally without setae. Gular suture single, long. **Antennae** subgeniculate, long; scapes slender, cylindrical, apically slightly inflated, articulation with funicle segment 1 apical but open, free, about 3 times longer than funicle segment 1; funicles 7-segmented, segment 1 slightly roundedly obconical, basally thin and bent, apically subequal in width to scape, segments 2–6 subequal, nearly twice longer than 1, subcylindrical but 6 widened apically, 7 half as long, obconical shorter towards club; clubs large, loosely articulated, 4-segmented, without long sensory setae, apical segment acute, slightly shorter than 3. **Mouthparts**. Labrum absent. Mandibles small, exodont and endodont, articulation horizontal. **Thorax**. Prothorax strongly proclinate, with anterior lateral margins oblique in lateral view. Pronotum convex, laterally rounded, without lateral tooth, posterior corners slightly angulate, fitting closely onto elytra; surface irregularly transversely rugose, sparsely setose, setae short, recurved anteriad and anteromesad; notosternal sutures widely open (forming open triangle), curved anteriad. Prosternum short, about as long as hypomeron; procoxal cavities medially confluent, in middle of prothorax. Scutellar shield transverse, glabrous. Mesocoxal cavities laterally widely open (forming open triangle). Metanepisterna distinct, glabrous, dorsal margin straight anteriorly, without lobe. Mesoventrite short, anteriorly sloping. Metaventrite longer, convex. **Elytra** elongate, basal margins extended into short broad lobe overlapping posterolateral angles of pronotum, with broad, obtusely rounded humeri, lateral margin weakly sinuate in middle, without anterior marginal notch; surface punctostriate, without scutellary striole but sutural stria slightly curved outwards at base, interstriae indistinct, rugose, very sparsely setose, setae long, thin, reclinate, directed caudad, uniformly coloured. **Legs** long slender. Coxae large, pro- and mesocoxae prominent, subconical; procoxae medially contiguous; mesocoxae nearly separated (process of meso- and metaventrites projecting between coxae but not contacting); metacoxae flat, transversely elongate. Trochanters short, oblique. Femora long, subcylindrical, strongly inflated in apical half, unarmed, outside rounded. Tibiae terete, nearly straight, not expanded distally, outer edge narrow but not carinate or crenulate, inner edge with sparse long stiff setae, apex subtruncate, with 2 spurs. Tarsi long, thin; tarsites 1 and 2 subequal in length, gradually widening apicad, apex subtruncate/lobed but extended beyond socket of following tarsite, 3 deeply bilobed, 5 longer than 3, narrowly cylindrical at base but broadening apicad; claws large, slender divaricate, with broad sharp flat basal tooth and long stiff ventrobasal seta arising from underside of tooth. **Abdomen** not preserved (cut away).

Derivation of name. The name of the genus is composed of the generic name *Rhynchites* and the Greek noun *mimos* (G: *mimou*), an imitator or actor, in reference to the similarity of the genus with those of Rhynchitinae; the gender of the name is masculine.

Remarks. The subbasally inserted, subgeniculate antennae, widely open notosternal sutures, single long gular suture and large, protruding eyes of this genus are features of rhynchitine Attelabidae, but its elongate scapes (more than twice as long as funicle segment 1) and four-segmented clubs (without any long sensory setae) as well as the slender, divaricate and dentate tarsal claws bearing a long ventrobasal seta rule out a placement in Attelabidae. In all Rhynchitinae examined, the scapes are subequal in length to funicle segment 1 and the articulation between them is apical and narrow, and the tarsal claws are divergent but not divaricate, typically strongly bifid (rarely laminate) and always without a ventrobasal seta. Simple claws do occur in some species of *Auletes* [52] as well as in *Baltocar*, placed in Attelabidae: Sayrevilleinae [40], but at least in the latter again without a ventrobasal seta (no Auletini with simple claws examined). *Baltocar* also possesses scutellary strioles, which are absent in *Rhynchitomimus* but also in several extant Rhynchitinae. Given that elongate scapes, a single long gular suture, protruding eyes and tarsal claws with a ventrobasal seta occur widely in Mesophyletidae, it is evident that *Rhynchitomimus* belongs in this family too, in which its subgeniculate antennae place it in Aepyceratinae as here delineated. Open pro- and mesocoxal cavities also occur in *Platychirus* and partly in *Aepyceratus* and *Nugatorhinus* and must be interpreted as a convergence with those of Attelabidae. *Rhynchitomimus* currently includes only one species.

Rhynchitomimus chalybeus Clarke & Oberprieler, **sp. n.** (Figures 17 and 18)

Description. Size. Length 2.94 mm (apex of abdomen cut off), width 1.2 mm. **Head** porrect but flexible downwards, elongate from base of rostrum to near occipital foramen; vertex more finely punctate than pronotum; venter transversely strigate. **Eyes** bulbous, slightly dorsoventrally compressed, maximally separated dorsally by distance of one eye diameter. **Rostrum** subequal in width in basal half but widening apicad from middle, gradually curved dorsoventrad; scrobes receiving only basal half of scape in repose. **Antennae** subgeniculate; scapes slender, slightly longer than funicle segments 1 + 2, reaching anterior margin of eye in repose, slender in basal 2/3 but inflated in apical 1/3; funicles with segment 1 with narrow, bent stalk; clubs with segments elongate, subequal in length (2 slightly shorter), widening apicad, segment 4 distinct, slightly shorter than 3. **Mouthparts.** Mandibles short, exodont, outside with 2 small acute teeth, inside with at least one apical internal tooth, apex broadly T-shaped. Palpi obscured by debris around apex of rostrum. **Thorax.** Pronotum irregularly transversely rugose, sparsely covered with very fine, short setae recurved anteriad; with slight anterior collar; sides broadly rounded, posterolateral corners slightly angulate. Scutellar shield subrectangular, with rounded corners, transverse. Mesoventrite small, depressed, strongly sloping to between mesocoxae. Mesanepisterna large, raised above mesoventrite, distal end rounded, overlapping mesocoxa. Mesepimera small, narrowly triangular. Metaventrite large, bulging, with precoxal groove tracking margin of metacoxal cavity. Metanepisterna long, broad, with anteroventral hook. **Elytra** with 10 complete indistinct striae of large open punctures; side with marginal groove subequal in width for entire length, dorsally delimited by thin, distinct keel, margin without anterior marginal notch. **Legs.** Metacoxae flat, slanting forwards laterally. **Abdomen.** Not preserved (cut away with amber).

Material examined. Holotype (CNUB, #CNU-COL-MA-0444): very well preserved and well visible specimen, abdomen cut off at apex, much of legs cut away with amber (right protarsus from apex of tarsite 1, left middle leg at femorotibial joint, right mesotarsus and apical part of tibia, both hindlegs near trochanters), only moderately distorted; in wedge, 5.1 × 2.0 × 2.5 mm, rounded at one end, with all sides visible; amber around specimen clear, colourless, rest yellow, with large mass of organic material obscuring the prosternum and venter of head, with few bubbles but without other major impurities; with section of amber broken out, exposing but not damaging part of abdomen, this and a cavity in right metafemur left from cutting away amber filled in with casting resin (see Section 2.2).

Derivation of name. The species is named for its bluish-black metallic coloration, from the Greek noun *chalyps* (G. *chalybos*) (hardened iron, steel); this metallic coloration is evident under different lightings.

Remarks. In addition to the unusual generic characters summarised above, the combination of the metallic and generally shining lustre with sparse, pale and generally inconspicuous setae, the very long, slender, substraight rostrum and legs (especially the protibiae) and the elongate elytra with weakly concave sides and coarsely rugose sculpture makes this one of the more distinctive species among our sample. The metallic blue cuticle is so far a unique feature in Burmese amber weevils. Several features of the mesothorax of the holotype are indicative of distortion (as seen more clearly in the head) and must be interpreted with caution. They may have resulted from the seemingly strongly depressed mesoventrite and include the form of the mesanepisterna (raised above the mesoventrite, overlapping the mesocoxae) and the medially confluent mesocoxal cavities, between which the pointed apices of the meso- and metaventral intercoxal processes do not touch (a unique feature among the studied specimens). These three features are likely not real or somewhat exaggerated character states, and better-preserved specimens are needed to assess them properly; such may also enable description of the ventrites and hindlegs, which are cut away in the holotype.

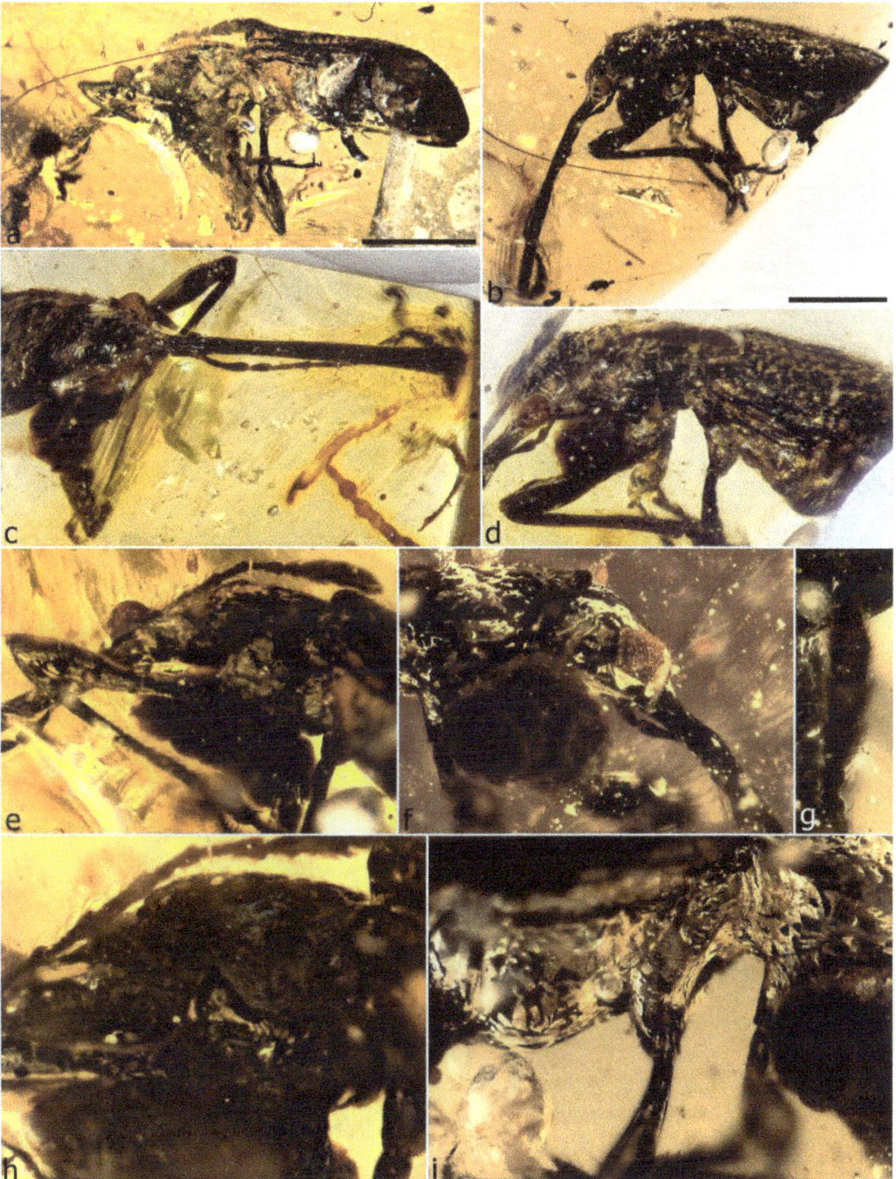

Figure 17. *Rhynchitomimus chalybeus* sp. n., holotype. Habitus, right lateral oblique (**a**); habitus, left lateral (**b**); head and antenna, dorsal (**c**); head, prothorax and pterothorax, left lateral (**d**); prothorax and antenna, left lateral oblique (**e**); head and prothorax (**f**); antennal club (**g**); prothorax showing widely open notosternal suture (**h**); meso- and metathorax and mesocoxa, right lateral (**i**). Scale bars: 1.0 mm (a,b).

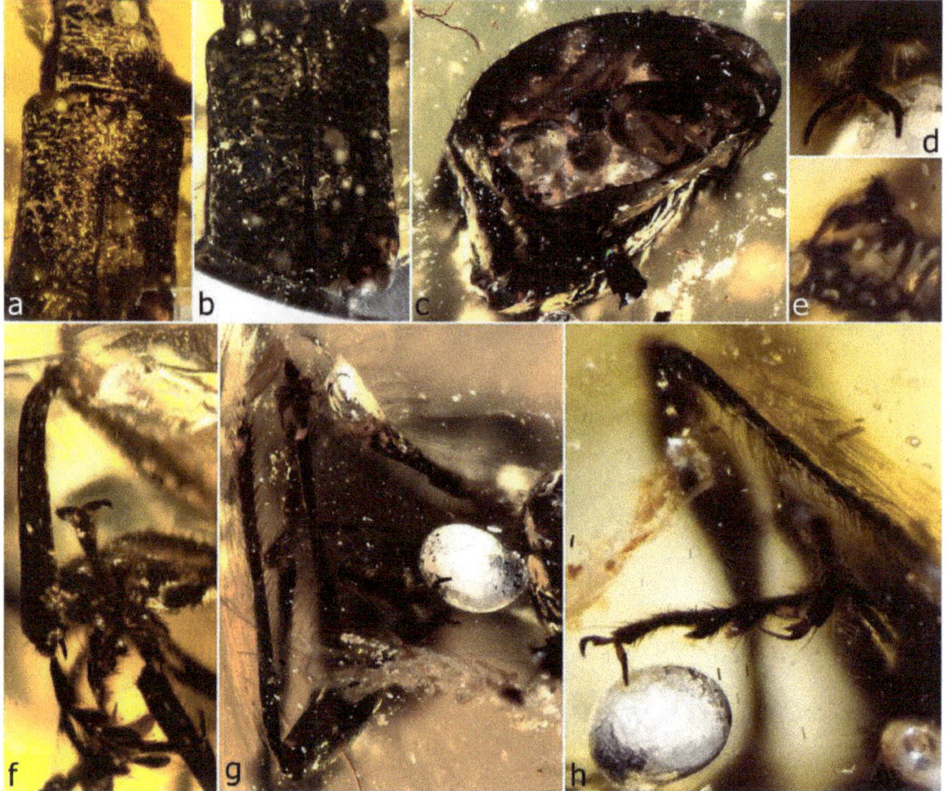

Figure 18. *Rhynchitomimus chalybeus* sp. n., holotype. Pronotum and elytra, dorsal (**a**); elytra (**b**); cut-away end of abdomen (**c**); tarsal claw, doso-apical (**d**); tarsal claw, apical (**e**); tibiae and details of legs (**f**–**h**).

Genus ***Nugatorhinus*** Clarke & Oberprieler, **gen. n.**

Type species: *Nugatorhinus chenyangi* Clarke & Oberprieler, sp. n.

Description. Size. Length 3.58–4.82 mm, width 1.32–2.4 mm. **Head** short, subquadratic transverse, slightly flattened. **Eyes** large, strongly protruding, coarsely facetted, dorsally separated by nearly width of rostrum anteriorly but further posteriorly; forehead flat, with a pair of low, transverse curved ridges between anterior margin of eyes and an elongate tuft or patch of coloured setae above eyes. **Rostrum** as long as or shorter than pronotum, stout, subcylindrical, substraight; antennal insertions lateral, without scrobes behind them, in front of them laterally with a few long erect setae. Apparently a single long gular suture present (not clearly discernible). **Antennae** subgeniculate, long; scapes stout, cylindrical, apically only slightly inflated, slightly longer than funicle segment 1; funicles 7-segmented, segment 1 subequal to scape, others thinner and progressively shorter towards club; clubs large, loosely articulated, flattened, 4-segmented, apical segment acute, about as long as 3. **Mouthparts**. Labrum absent. Mandibles small, non-exodont, articulation horizontal. **Thorax**. Prothorax proclinate or not. Pronotum slightly convex, laterally rounded, without tooth, posterior corners slightly extended, fitting closely onto elytra; surface coarsely tuberculate, sparsely setose, setae reclinate, directed anteromesad, disc with 2 pairs of patches of dense coloured setae, a small anterior one just before middle and a larger, elongate one just behind middle, laterally with patch of long coloured setae anteriorly and another posteriorly; notosternal sutures open ventrally but then closed, curved anteriad. Prosternum moderately long; procoxal cavities medially confluent, in middle of prothorax. Scutellar shield densely setose. Mesocoxal cavities closed or possible slightly open. Metanepisterna distinct, at least posteriorly densely setose. Mesoventrite short, anteriorly sloping. Metaventrite longer, flat or slightly convex. **Elytra** elongate, basally lobed over base of pronotum, with broadly rounded humeri, posteriorly declivous, lateral margin strongly sinuate to roundly emarginate in middle, apically individually rounded, not exposing pygidium; sutural flanges narrow, equal; surface punctostriate, without scutellary striole, interstriae convex, finely tuberculate, setose, setae long, thin, reclinate, directed caudad, interstriae 3 with row of 4 large, spaced, subcircular patches of dense setae, interstriae 7 with similar row of 3 setal patches placed slightly further back, interstriae 5 with 1 or 2 much smaller anterior setal patches, interstriae 9 with 4 or 5 similar patches, 1 or 2 anteriorly and 2–3 smaller ones at declivity. **Legs**. Procoxae large, prominent, medially contiguous; mesocoxae subglobular, narrowly separated; metacoxae flat, transversely elongate. Trochanters short, oblique. Femora long, subcylindrical, inflated in distal half, outside rounded, apically with patch of long coloured setae, inside excavate in distal half, receiving tibiae in repose, walls of groove at apex flatly tooth-like extended (meso- and metafemora). Tibiae straight, compressed, distally expanded, outer edge rounded, with dense long stiff setae in distal half, apex obliquely truncate, without spurs. Tarsi almost as long as tibiae; tarsite 1 apically excised, 2 shorter, triangular, apically excised, 3 deeply bilobed, 5 long, apically expanded; claws divaricate, dentate with ventrobasal seta at apex of tooth. **Abdomen** with ventrites 1 and 2 fused, each slightly longer than 3, 3 slightly longer than 4, 5 as long as 3, apically broadly rounded.

Derivation of name. *Nugatorhinus* is named for the funky patches of coloured setae that adorn its head and body, the name formed from the Latin noun *nugator*, a joker or jester (clown), and the Greek noun *rhis* (G: *rhinos*), a nose or snout, and being masculine in gender.

Remarks. *Nugatorhinus* is distinguishable from all other Burmese amber weevils by the large, dense, coloured setal patches on its head, body and legs, especially the large round to oval ones on the elytra. It is one of only six genera of Mesophyletidae with subgeniculate antennae, the scapes being elongate but only slightly longer than funicle segment 1. From *Aepyceratus*, *Platychirus*, *Rhynchitomimus* and *Acalyptopygus* it also differs in having small, non-exodont mandibles, a thick, straight rostrum and a pair of crescentic ridges between the eyes, and from *Calyptocis* it is further distinguishable by not having an exposed pygidium. It currently includes two species.

Nugatorhinus chenyangi Clarke & Oberprieler, **sp. n.** (Figures 19 and 20)

Description. Size. Length 3.58 mm, width 1.32 mm. **Head** slightly constricted behind eyes. **Eyes** hemispherical, forehead with a pair of elongate tufts of long, orange-brown setae above eyes. **Rostrum** shorter than pronotum, straight; basal 2/3 of length dorsally sparsely covered with long, orange-brown setae directed caudad; antennal insertions in middle of rostral length, in front of them with lateral row of a few long, erect setae, epistome flanked by 3 pairs of long, erect setae. **Antennae.** Funicles with segment 1 not inflated, 2 subequal in length but thinner, 3–7 progressively shorter towards club; clubs slightly flattened, with apical segment distinct, narrow, as long as 3. **Mouthparts**. Mandibles with 3 inner teeth. Maxillae and labium not discernible. **Thorax**. Prothorax proclinate, with anterior lateral margins oblique in lateral view. Pronotum elongate, 1.5 × longer than broad in middle, laterally rounded; surface sparsely setose, setae long, thin, disc with dense orange-brown setal patches; notosternal sutures shortly, broadly open ventrally, then closed. Scutellar shield small, rounded, convex, covered with dense orange-brown setae. Mesocoxal cavities laterally possibly open (not clearly discernible). Metaventrite flat. **Elytra** narrowly elongate, posteriorly very gently declivous, lateral margin strongly sinuate; interstriae densely setose, setal patches orange-brown, interstriae 5 with 2 smaller anterior setal patches, interstriae 9 with a larger patch and 4 smaller ones spaced to near apex. **Legs**. Femora apically with patch of long orange-brown setae. Tibiae slightly flattened, basally strongly curved, with dense long stiff setae in distal half. Tarsi with tarsite 1 subcylindrical, 2 apically excised, 3 with lobes narrow, not pedunculate, 5 about as long as 1; claws as for genus. **Abdomen** as for genus.

Material examined. Holotype (NIGP168266): extremely well preserved, intact specimen, not compressed or distorted, only a large clear bubble obscuring a small part of right side; in centre of rectangular block ca. 5.15 × 3.6 × 3.4 mm; amber very clear, without major impurities.

Figure 19. *Nugatorhinus chenyangi* sp. n., holotype. Habitus, dorsal (**a**); habitus, ventral (**b**); habitus, right lateral (**c**); habitus, left lateral (**d**); head and antenna, left lateral (**e**); rostrum, dorsal (**f**); head showing detail of setal tufts, dorsal (**g**); elytra, dorsal (**h**); elytra, lateral (**i**); pronotum, dorsal (**j**); apex of rostrum and mandibles (**k**); detail of elytra showing setal tufts and deep striae (**l**). Scale bars: 1.0 mm (a–d,h); 0.2 mm (e,i,j); 0.1 mm (f,g,k,l).

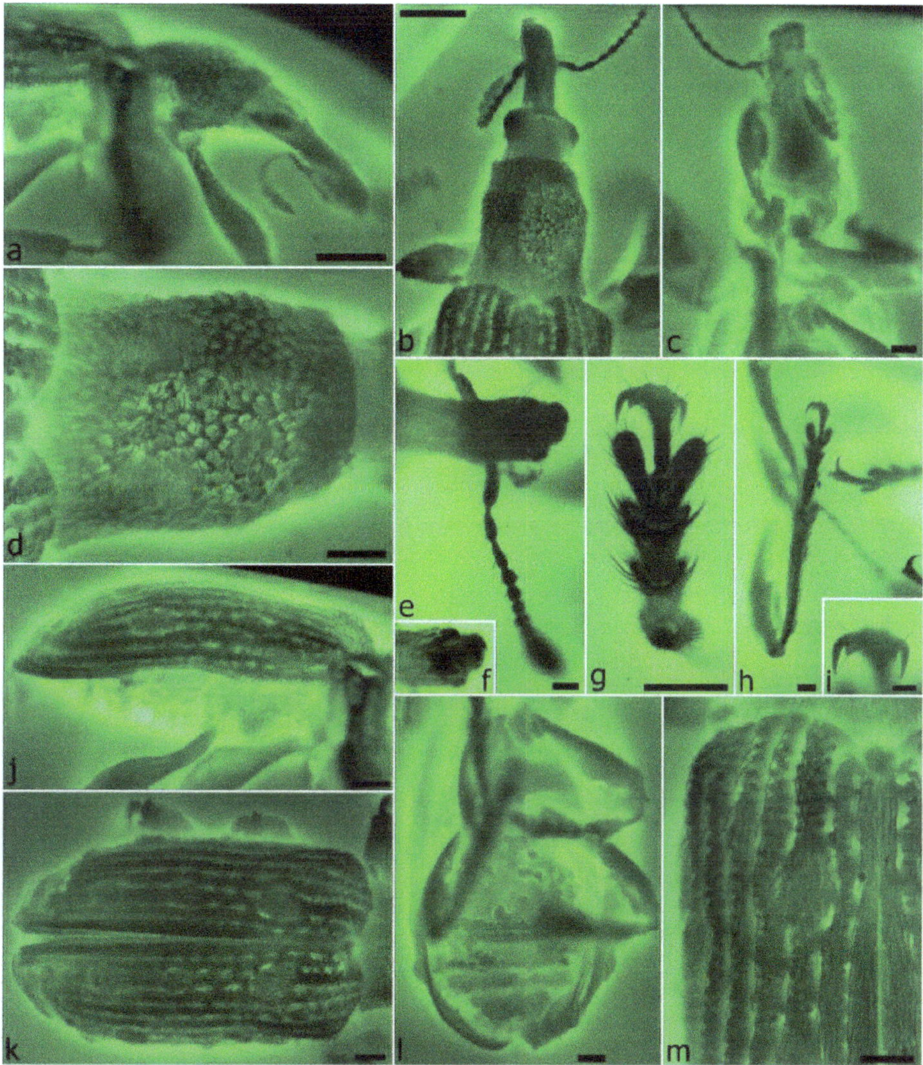

Figure 20. *Nugatorhinus chenyangi* sp. n., holotype. Images taken under fluorescent green light. Head and prothorax, right lateral (**a**); same, dorsal (**b**); same, ventral (**c**); pronotum, dorsal (**d**); rostrum and antenna, dorsal (**e**); apex of rostrum and mandibles, dorsal (**f**); tarsus, dorsal (**g**); right metatibia and tarsus, dorsal (**h**); tarsal claw, apical (**i**); elytra, right lateral (**j**); elytra, dorsal (**k**); ventrites (**l**); detail of left elytron (**m**). Scale bars: 0.5 mm (a,b); 0.2 mm (c,d,j–m); 0.1 mm (e,g,h); 0.05 mm (i).

Derivation of name. The species is cordially named for Chenyang Cai (NIGP), who made a series of good photographs of the specimen available to us, from which we compiled the description. The dedication also recognises Chenyang's contributions to the study of beetle fossils, from Burmese amber and other *Lagerstätten*.

Remarks. Due to its conspicuous patches of orange-brown setae adorning its body, *Nugatorhinus chenyangi* is one of the most readily recognisable Burmese amber weevils. The single specimen so far known is remarkably well preserved and allows a complete characterisation of the species. The other known species, *N. albomaculatus*, has a similar arrangement of setal patches, but silvery white in colour, and a longer rostrum and larger body size.

Nugatorhinus albomaculatus Clarke & Oberprieler, **sp. n.** (Figure 21)

Description. Size. Length 4.82 mm, width 2.4 mm. **Head** not constricted behind eyes. **Eyes** elongate, dorsoventrally flattened, slanting backwards, forehead with a pair of elongate patches of dense white setae above eyes. **Rostrum** about as long as pronotum, hardly curved; basal part dorsally not conspicuously setose, only few fine setae; antennal insertions slightly before middle of rostral length, in front of them with a few long, erect, lateral setae. **Antennae.** Scapes longer than funicle segment 1; funicles with segment 1 slightly inflated at apex, others not clearly visible; clubs with apical segment acute, longer than 3. **Mouthparts**. Mandibles with 2 visible inner teeth, apical one sharp, other blunter. Maxillae and labium not discernible. **Thorax**. Prothorax not evidently proclinate. Pronotum subquadratic, laterally straight; surface coarsely tuberculate, tubercles anteriorly elongate, confluent, forming short longitudinal ridges, very sparsely setose, setae small, thin, disc with dense white setal patches. Scutellar shield large, trapezoidal, convex, covered with dense white setae. Metanepisterna posteriorly with dense white setae. Metaventrite slightly convex. **Elytra** moderately elongate, posteriorly strongly declivous, lateral margin strongly roundly emarginate above metacoxae; interstriae sparsely setose, setal patches white, interstriae 5 with a small anterior patch, interstriae 9 with a larger and a small one anteriorly and 2 smaller ones at declivity. **Legs**. Femora apically with patch of dense white setae. Tibiae with median third of length girdled with silvery-white setae, distal third with dense, long, black, suberect setae; protibiae longer than others. Tarsi large, flat, protarsi only half as long as protibiae; tarsite 1 triangular, 2 apically deeply excised, 3 with lobes flat, 5 about as long as 1 + 2; claws divaricate, apparently simple (not clearly discernible). **Abdomen** with ventrites 1 and 2 as for genus, rest not properly discernible.

Material examined. Holotype (NIGP154203): reasonably well preserved, intact specimen with compressed legs, left protarsus lying over apex of rostrum, poorly visible due to numerous small white bubbles on dorsal surface and a thick layer of small to minute bubbles below ventral surface, rostrum and legs protruding through it, also several small cracks along both sides; situated on right side of roundedly rectangular cuboid ca. 6.6 × 4.9 × 4.3 mm with rounded edges and corners, dorsal surface slightly convex; amber clear with few small impurities apart from the numerous bubbles.

Derivation of name. The species is named for its conspicuous silvery-white, dense, setal patches on especially the elytra, the name being an adjective.

Remarks. *Nugatorhinus albomaculatus* is also an easily recognisable Burmese amber weevil due to the conspicuous white setal patches on its body. Although the single specimen so far known is not very well preserved and many of its features are obscured by bubbles, it clearly represents the same genus as *N. chenyangi*, from which it differs not only in the white colour of its setal patches but also in having a longer rostrum, a larger body size and various other subtle characters.

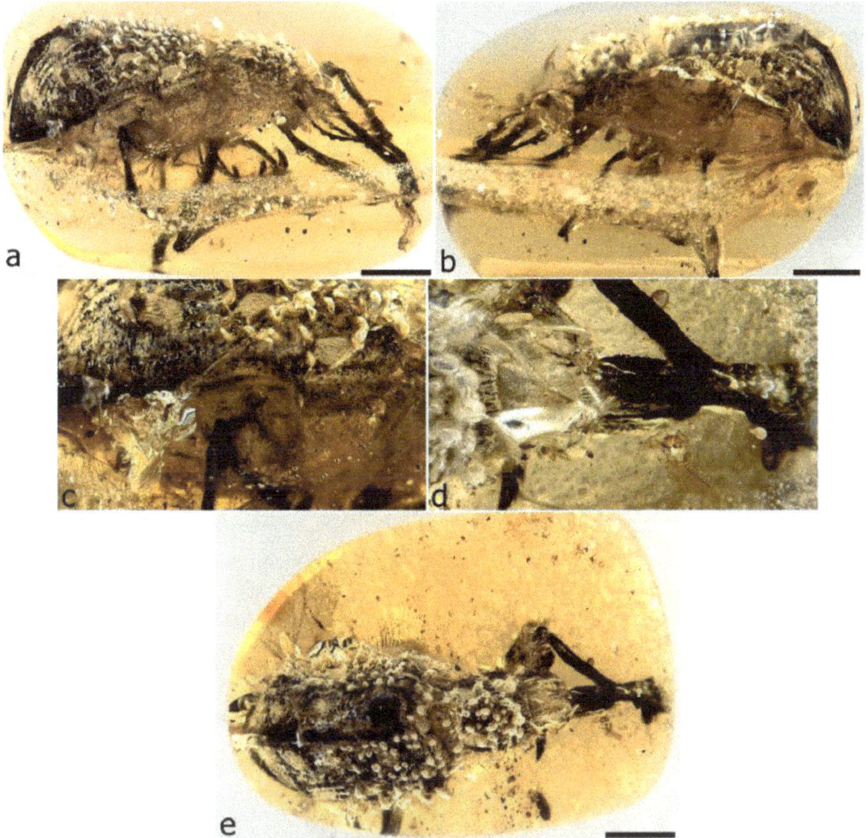

Figure 21. *Nugatorhinus albomaculatus* sp. n., holotype. Habitus, right lateral (**a**); habitus, left lateral (**b**); detail of elytra (**c**); head, dorsal (**d**); habitus, dorsal (**e**). Scale bars: 1.0 mm (a,b,e).

Genus ***Calyptocis*** Clarke& Oberprieler, **gen. n.**

Type species: *Calyptocis brevirostris* Clarke & Oberprieler, sp. n.

Description. Size. Length 5.6 mm, width 3.19 mm. **Head** porrect, short, not constricted behind eyes. **Eyes** relatively small, hemispherical but only slightly protruding, coarsely facetted, dorsally separated by about half basal rostral width anteriorly but further posteriorly; forehead flat, without tubercles above eyes. Gular suture single, long, from base of head to about base of rostrum. **Rostrum** subequal in length to pronotum, stout, subcylindrical, substraight; antennal insertions lateral, slightly antemedian, with scrobes behind them directed obliquely ventrad beneath eye. **Antennae** subgeniculate, long; scapes short, subcylindrical, slightly longer than funicle segment 1; funicles 7-segmented, segments 1–2 elongate, subequal in length, 3 slightly shorter, apically rounded, 4–7 progressively shorter, obconical; clubs large, loosely articulated, 4-segmented, apical segment long, acute, slightly longer than segment 3. **Mouthparts**. Labrum absent. Mandibles small, non-exodont, articulation horizontal. Maxillary palps robust but short, not projecting beyond mandibles, 4-segmented; segment 2 shorter and narrower than 1, 3 shorter than 2, 4 subequal in length to 3, obconical and narrower than 3. Labial palps attached apically to prementum, 3-segmented; segment 1 slightly longer and broader than 2, 3 subequal in length to and narrower than 2, obconical. **Thorax**. Prothorax strongly proclinate, with anterior lateral margins oblique in lateral view. Pronotum evenly convex, laterally rounded, without tooth, posterior corners not extended, fitting closely onto elytra; surface seemingly coarsely rugose; notosternal sutures closed. Prosternum moderately short; procoxal cavities medially confluent, slightly closer to anterior margin of prosternum. Mesocoxal cavities seemingly closed. Mesoventrite short, anteriorly strongly sloping down. Metaventrite longer, convex. **Elytra** short and broad, seemingly basally lobed over base of pronotum, with broadly rounded humeri, strongly sloping posteriorly, lateral margin strongly sinuate to roundly emarginate in middle, apically individually rounded, exposing short pygidium; surface indistinctly punctostriate, interstriae seemingly subflat, rugose or shallowly tuberculate. **Legs**. Procoxae globular, only slightly projecting, medially contiguous; mesocoxae globular, narrowly separated; metacoxae flat, transversely elongate. Trochanters short, oblique, apparently not recessed into coxae. Femora long, inflated in distal half but constricted before apex, outside rounded, on inside of constriction with large acute tooth shearing against basal curved part of tibia. Tibiae subcylindrical, widening apicad, pro- and mesotibiae straight, metatibiae curved backwards, outer edge rounded (not carinate or crenulate), all with dense erect setae in distal half, especially on posterior surface and possibly forming antennal cleaner on protibiae, apex subobliquely truncate, without spurs (or too small to discern). Tarsi elongate, about 0.67 × as long as tibae, robust; tarsite 1 elongate-triangular, apically truncate to weakly lobed, 2 shorter and broader, apically slightly emarginated, 3 broader still, deeply emarginate with lobes broad, 4 forming distinct globular cryptotarsite recessed into base of 3, 5 slightly longer than 3, strongly widening distad; claws divaricate, dentate, inner tooth small, acutely and inwardly curved, with long ventrobasal seta, almost reaching outer tip of claw. **Abdomen** with ventrites 1 and 2 subequal in length, 3–4 progressively shorter, 5 subequal to 3 and 4; last tergite exposed as distinctly cupular pygidium.

Derivation of name. The name of the genus is composed from the Greek adjective *kalyptos* (covered) and noun *kis* (G: *kios*) (weevil or beetle); its gender is masculine.

Remarks. The absence of a labrum, the subgeniculate lateral antennae (the scapes only slightly longer than funicle segment 1), the single long gular suture and the exposed pygidium conform with the characters of Attelabidae, but *Calyptocis* differs from this family in its distinctly dentate tarsal claws with a ventrobasal seta and must therefore also be placed in the subfamily Aepyceratinae of Mesophyletidae. In this subfamily it agrees with *Acalyptopygus* in having an exposed pygidium, but this genus differs in having strongly exodont mandibles and the scapes apically clavate and about twice as long as the first funicle segment. *Calyptocis* includes only one species.

Calyptocis brevirostris Clarke & Oberprieler, **sp. n.** (Figures 22–24, Video S1)

Description. Size. Length 5.6 mm, width 3.19 mm. **Head** with dorsal outline continuing evenly from base of rostrum, without sinus; ventrally bulging. **Rostrum** short (but longer than exposed part of head), depth slightly increasing apicad, coarsely rugose. **Thorax.** Pronotum slightly longer than head, slightly narrower than elytra basally. Lateral pronotal margins broadly rounded; setation not visible. Prosternum thin, shorter than hypomeron. Scutellar shield prominent. **Elytra** punctostriate; striae thin, linear, interstriae broad, subflat; concave behind humeri.

Material examined. Holotype (NIGP154204): well preserved specimen, minimally distorted or compressed but most surface details including vestiture, surface sculpture and mouthparts obscured by nearly unbroken layer of bubbles and debris, missing part of left protarsus and left mesotarsal claw (cut away with amber) and left antenna; in block with two flat faces and one large curved face, 11 × 7.9 × 6.4 mm; amber clear yellow-brown, with diffuse gritty impurities forming cloud surrounding much of specimen, obscuring especially ventral structures, with large cavity on flat side, two smaller cavities on curved side (above elytra) infilled with resin (see Section 2.2).

Derivation of name. The species is named for its short, stout rostrum, which resembles that of the extant Australian belid genera *Pachybelus* Zimmerman and *Pachyura* Hope.

Remarks. This species is unique among Burmese amber weevils in its large size and robust body, short rostrum, subgeniculate antennae and pygidium. Although the specimen is largely covered with a film of tiny bubbles that obscures many of its features, several critical ones are discernible, especially in the CT scans we had done (Figures 23 and 24, Video S1).

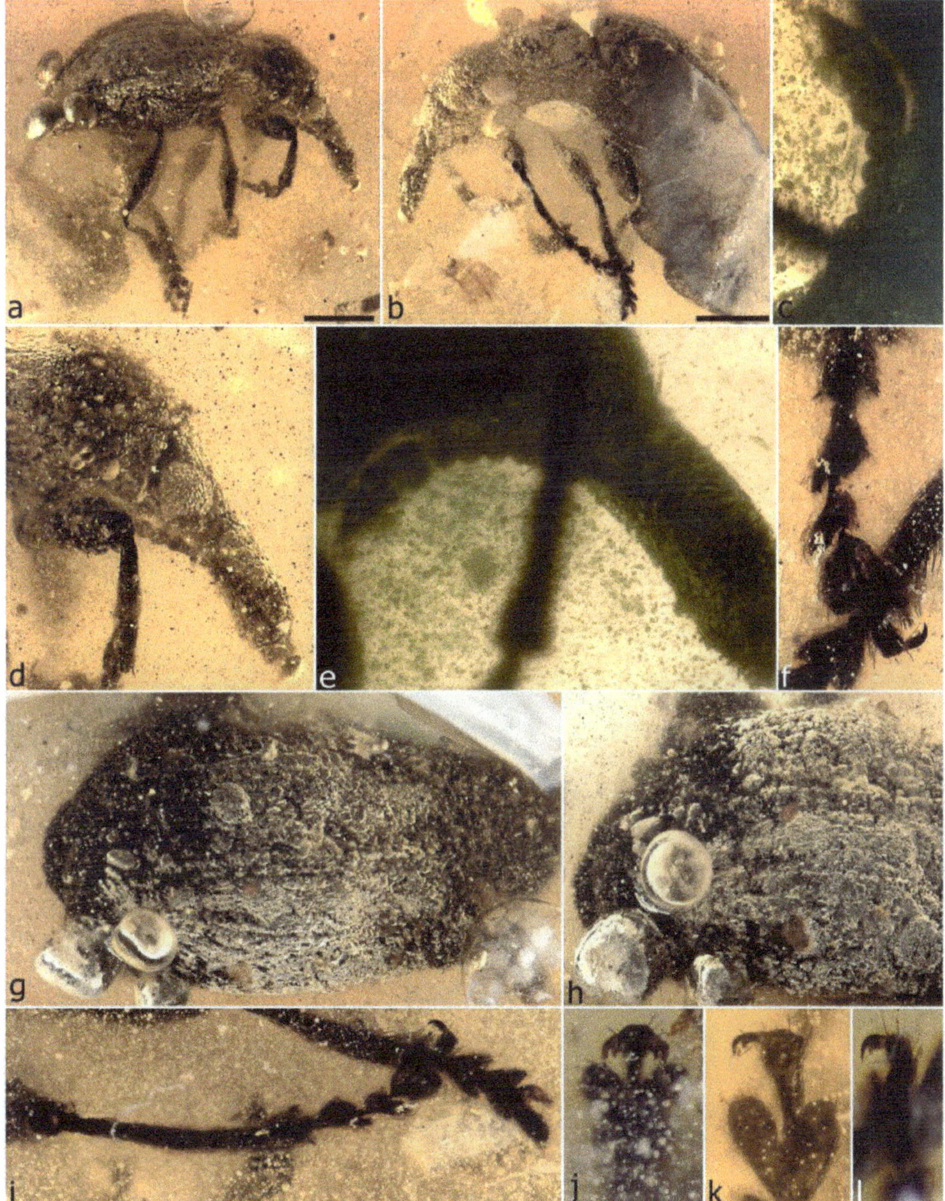

Figure 22. *Calyptocis brevirostris* sp. n., holotype. Habitus, right lateral (**a**); habitus, left lateral (**b**); detail of antenna, right lateral (**c**); head and prothorax, right lateral (**d**); rostrum and antenna, right lateral (**e**); protarsus (**f**); elytra, dorsal (**g**); elytra and pygidium, dorsoposterior (**h**); forelegs (**i**); claw (**j–l**). Scale bars: 1.0 mm (a,b).

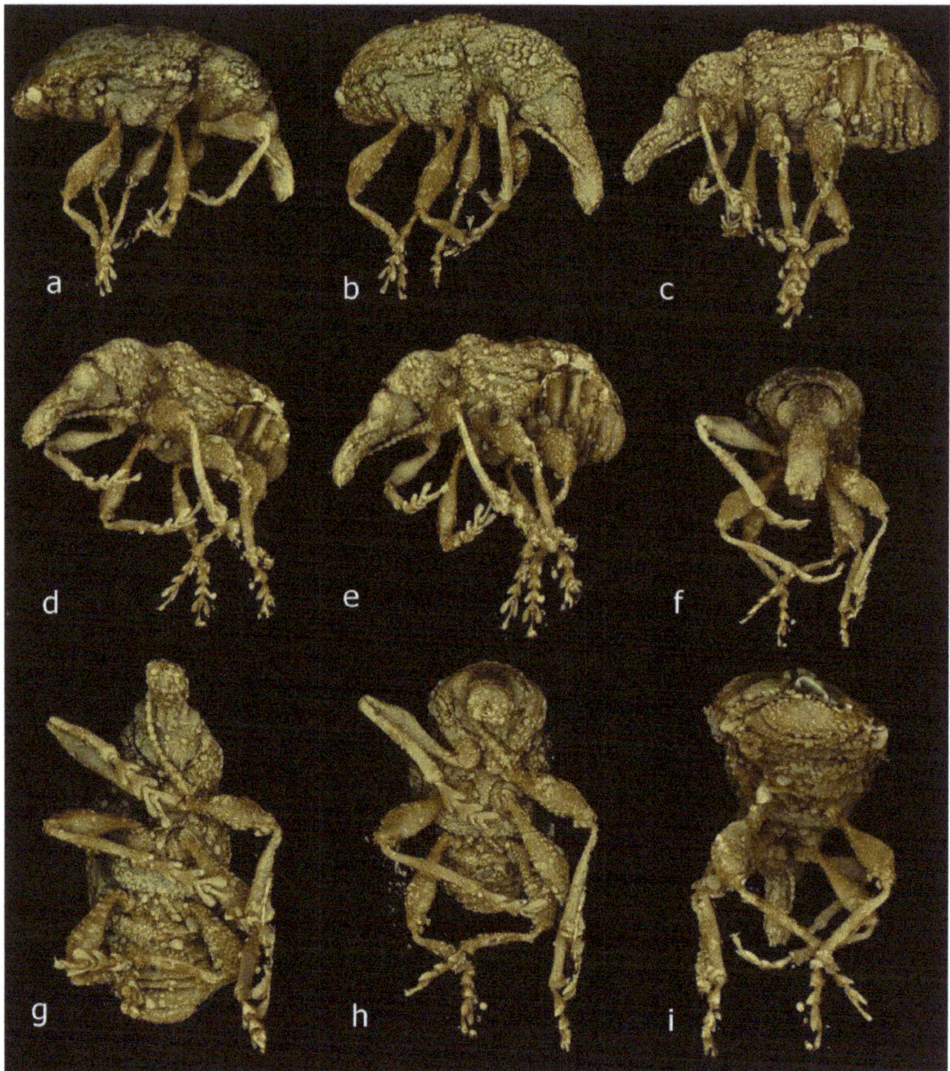

Figure 23. *Calyptocis brevirostris* sp. n., holotype. Habitus images extracted from micro-CT scanning reconstruction (see also Video S1). Right lateral (**a**); right lateral oblique (**b**); left ventral oblique (**c**); left lateral oblique (**d**); left lateral (**e**); frontal (**f**); ventral (**g**); frontal oblique (**h**); posterior oblique (**i**).

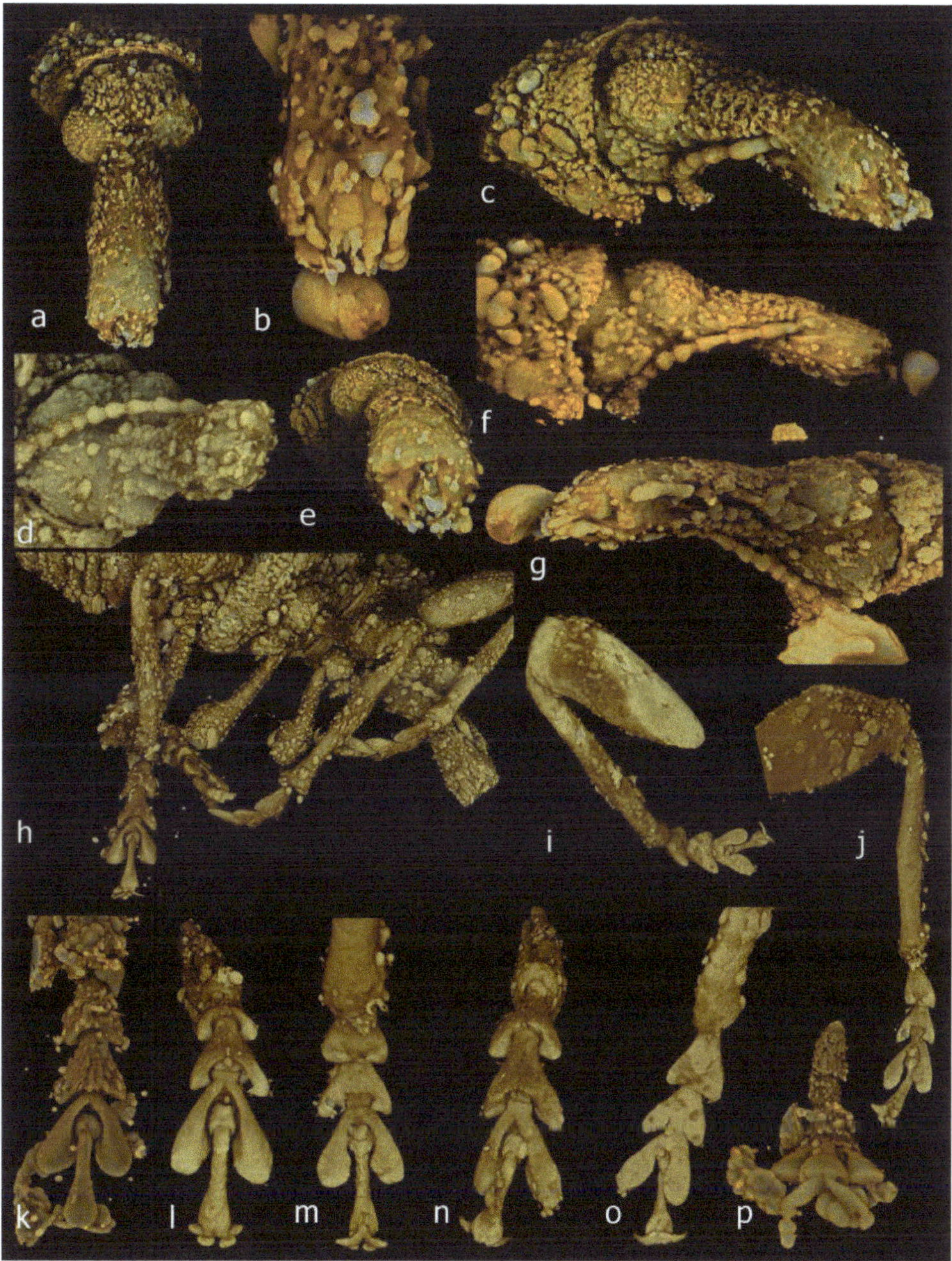

Figure 24. *Calyptocis brevirostris* sp. n., holotype. Morphological details extracted from micro-CT scanning reconstruction (see also Video S1). Head, dorsal (**a**); apical part of rostrum and mouthparts, ventral (**b**); head, right lateral oblique (**c**); head, ventral (**d**); head, frontal oblique (**e**); head, right lateral (**f**); head left lateral (**g**); head, thorax and legs, ventral oblique (**h**); legs (**i**,**j**); tarsi (**k**–**p**).

Genus ***Acalyptopygus*** Clarke & Oberprieler, **gen. n.**

Type species: *Acalyptopygus brevicornis* Clarke & Oberprieler, sp. n.

Description. Size. Length 2.0–2.75 mm, width 0.8–1.1 mm. **Head** short, porrect, subconical, not constricted behind eyes. **Eyes** large, elongate, somewhat compressed, strongly protruding, coarsely facetted, dorsally separated by width of rostrum at base, forehead subtrapezoid, without tubercles. **Rostrum** as long as pronotum, stout, slightly downcurved, subcylindrical; antennal insertions lateral, in basal quarter to third of rostral length or less, behind them with short scrobes extending to eye, in front of them without lateral row of setae. Gular suture single, long, from base of head to base of rostrum. **Antennae** subgeniculate; scapes straight, subcylindrical, distally strongly clavate, twice longer than funicle segment 1; funicles nearly 3 × longer than scape, 7-segmented, segment 1 as thick as apex of scape, other segments half as thick as 1, subequal; clubs long, very loosely articulated, especially segment 1, 4-segmented, segment 4 distinct, broad, long. **Mouthparts**. Labrum absent. Mandibles small, flat, horizontal, exodont, with T-shaped apex, articulation oblique. Maxillary palps 3-segmented, elongate, slender. **Thorax**. Prothorax slightly proclinate, with anterior lateral margins oblique in lateral view. Pronotum roundly subrectangular to trapezoidal, laterally broadly rounded, without tooth, weakly convex, punctate, sparsely setose, setae short, fine, suberect, pale, directed mesad to anteromesad; notosternal sutures closed. Prosternum moderately long; procoxal cavities medially confluent, in middle of prothorax or closer to anterior margin of prosternum. Scutellar shield small, sometimes indistinct. Mesocoxal cavities laterally closed (by meso- and metaventrite). Metanepisternal sutures distinct. Mesoventrite short, anteriorly strongly sloping. Metaventrite raised into transverse weals before metacoxae. **Elytra** elongate, bases tightly abutting pronotum but not extending over its base, with weak, broadly rounded humeri closely fitting with pronotal corners, posteriorly declivous, apically individually rounded, exposing pygidium in repose; sutural flanges apically visible, thin, equal; surface punctostriate, rarely astriate, without scutellary striole, sparsely setose, setae short, stout, acute, reclinate, directed caudad. **Legs**. Procoxae large, elongate, prominent, medially contiguous; mesocoxae subglobular, narrowly separated; metacoxae transversely elongate, reaching elytra. Trochanters short, oblique. Femora long, subcylindrical, strongly inflated in distal half, unarmed, outside rounded. Tibiae long, straight, flattened, outer edge rounded, apically with long dense setae, apex obliquely truncate, with 2 fine spurs (or 1 on metatibiae). Tarsi narrow, about 0.67× or more as long as tibiae; tarsite 1 elongate, apically weakly excised, 2 shorter, deeply excised, 3 deeply bilobed, 5 as long as 1 + 2; claws divaricate, dentate with very narrow space between outer edge of basal tooth and inner edge of claw, with or without ventrobasal seta at apex of tooth. **Abdomen** with ventrites 1 to 2 fused, each slightly longer than 3 and 4, 4 shorter than 3, 5 longer than 4, with long, slightly curved apical margin fitting onto ventral margin of pygidium.

Derivation of name. The genus is named for its exposed pygidium, the name derived from the Greek adjective *akalyptos*, meaning uncovered, and noun *pyge* (G: *pygos*), rump or buttocks, and being masculine in gender.

Remarks. *Acalyptopygus* differs from the other genera placed in Aepyceratinae except *Calyptocis* foremost by its exposed pygidium, from *Aepyceratus*, *Platychirus* and *Rhynchitomimus* also by having closed pro- and mesocoxal cavities and from *Nugatorhinus* by its exodont mandibles and uniform vestiture, not featuring distinct coloured setal patches on the head, pronotum, elytra and legs. From *Calyptocis* it differs in its elongate, slender rostrum with exodont mandibles, long (at least twice longer than pedicel), apically strongly clavate scapes and very long clubs with a distinct and usually narrow apical segment. Uniquely among Aepyceratinae, *Acalyptopygus* is the only genus having the type of exodont mandibles typical of the majority of Mesophyletinae, being flattened with several large inner and outer teeth and horizontal in repose but opening into a vertical position via oblique articulation sockets. It currently includes four species. In *A. brevicornis* and *A. lingziae* the tarsal claws lack the ventrobasal seta that is so characteristic of and almost ubiquitous in Mesophyletidae.

Acalyptopygus brevicornis Clarke & Oberprieler, **sp. n.** (Figure 25)

Description. Size. Length 2.25 mm, width 0.81 mm. Dark blackish-brown; antennae and legs paler. **Head** subporrect; dorsally slightly convex, with median, slightly raised, flat, impunctate and glabous costa extending from base of rostrum to hind part of head; finely punctosetose, denser behind and between eyes. **Eyes** lateral, subelongate, dorsally separated by width of rostrum anteriorly, further posteriorly. **Rostrum** slightly downcurved, inserted in dorsal half of head, basally with dorsal outline continuing onto head, with weak sinus before eyes but ventral outline forming large sinus with strongly bulging head; dorsally and laterally without carinae; antennal insertions in basal quarter of rostral length, in front of them with deep scrobes extending to slightly below front margin of eye; mandibular articulations oblique. **Antennae.** Scapes ca. 2.0 × longer than funicle segment 1, extending to below front margin of eye, apex truncate; funicles with segment 1 oval-shaped, 2 nearly 0.67 × as long as 1 but much narrower, segments 2–7 subcylindrical, subequal; clubs with segments obconical, apically oblique, 1–3 subequal, obconical, 4 narrow, acute, slightly shorter than 3. **Mouthparts.** Mandibles with 2 teeth on outside, a larger rounded basal and a smaller rounded apical one, and 3 teeth on inside, 2 large basal and a smaller rounded apical one with short slender apical part, apical teeth forming T. Maxillary palps with segments 1 and 2 subequal, 3 ca. 1.5 × longer than 2. **Thorax.** Prothorax with indistinct lateral edge. Pronotum widest just behind middle, narrowing anteriorly, only slightly narrower than elytra; not constricted anterolaterally; densely punctosetose, punctures small, distinct; base broadly sinuate, shortly lobate at middle, closely abutting bases of elytra and scutellar shield; corners slightly extended, closely fitting with humeri. Prosternum short, about as long as hypomeron. Scutellar shield at same level as elytral bases, subquadrate, with straight anterior margin, densely setose. Metaventrite punctate, setose. **Elytra** densely setose, setae short, obliquely subrecurved; surface rugose, weakly punctostriate, striae indistinct; interstriae basally and laterally slightly raised above striae, indistinct from striae on disc, without any prominent lateral striae; bases broadly rounded; humeri flat, weakly produced; marginal groove subequal in width along entire length. **Legs.** Procoxae conical; mesocoxae subglobular, very prominent. Tibiae sparsely setose, with longer denser setae in distal half; apically with long slender fringing setae and 2 spurs; protibiae slender, apically slightly expanded and with outer edge obliquely truncate, somewhat produced distad to form angulate lobe; mesotibiae with outer edge continuing as rounded apical flange, with small mucro; metatibiae with setation as on mesotibiae and outer edge produced to lobe as in protibiae, ventrally produced to acute point. Tarsi with tarsite 1 slightly widening apicad, 2 deeply excised, 3 broadly lobate, lobes subpedunculate, about half as long as 5, 5 slender; claws without ventrobasal seta. **Abdomen**. Ventrites 1 and 2 subflatly aligned, with indistinct suture between them, subequal in length, 3–5 slightly stepped, 3 slightly shorter than 2, 4 slightly shorter than 3, 5 longer than 4.

Material examined. Holotype (NIGP154205): exceptionally well preserved and well visible specimen, with left fore and middle legs cut off at femorotibial joint, mandibles and most aspects of antennae and legs visible (some obscured due to being folded beneath body), surface details also well visible through fragmented coating of whitish debris, left wing partly extended; in centre of cuboid 6.1 × 2.2 × 1.2 mm, rounded off at one corner, with sides subparallel to flat faces of block and dorsal side parallel to curved edge; amber clear yellow, with few impurities but large fracture on right side of specimen and few other smaller fractures and minimal debris in vicinity of specimen.

Derivation of name. The species is named for its short antennae, especially the scapes, which are only 2 × longer than funicle segments 1 but still reach the eyes in repose (antennae inserted subbasally). The name is a Latin adjective.

Remarks. The species differs from *A. astriatus* in the rugose elytra with indistinct striae and the weak sinus between rostrum and head in lateral view, and from *A. elongatus* and *A. lingziae* it is readily distinguishable by having the scapes shorter than the eyes and subequal funicle segments. A distinctive feature of the holotype is the broad, slightly raised, smooth and impunctate median costa on the head, a feature so far restricted to *Acalyptopygus* and shared at least with *A. astriatus* and possibly with *A. lingziae*. Another remarkable feature of the species is the heterogenous structure of the tibial apex, in the pro- and metatibiae produced into a distinct truncate outer flange (in the latter so developed that, in concert with the spurs, the apex appears claw-like in lateral view) but in the mesotibiae forming simple rounded flanges equipped on the inside with a short sharp mucro. This mesotibial mucro also occurs in *A. elongatus* and one undescribed species known to us. As in *A. lingziae*, the tarsal claws of *A. brevicornis* lack the ventrobasal seta.

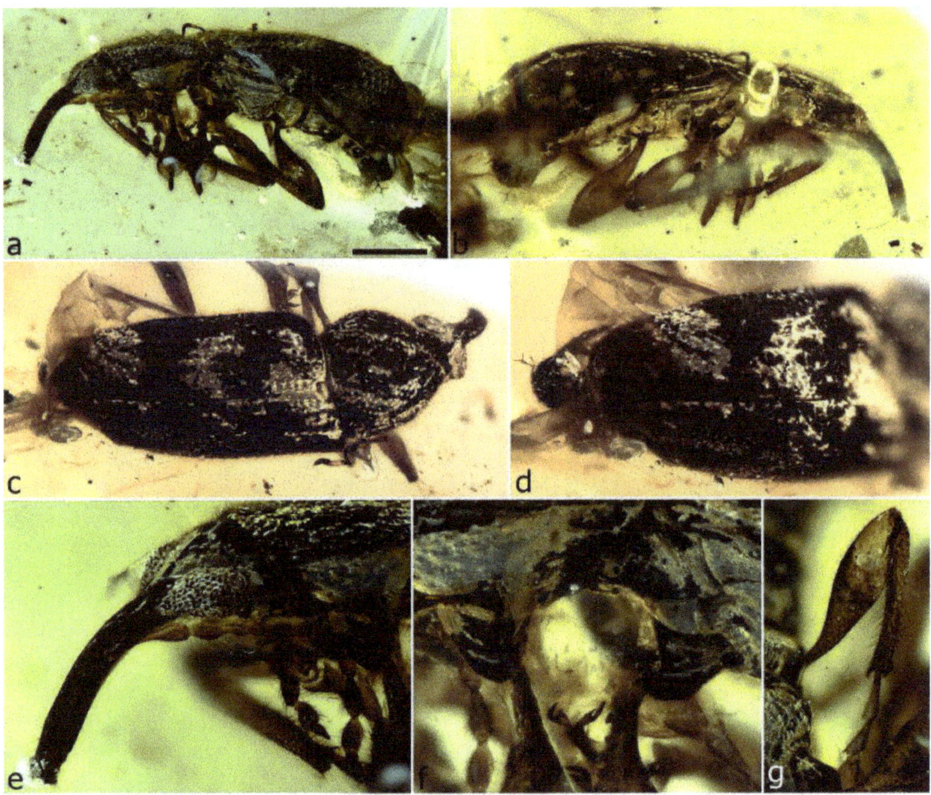

Figure 25. *Acalyptopygus brevicornis* sp. n., holotype. Habitus, left lateral (**a**); right lateral (**b**); dorsal (**c**); elytral apices and pygidium, dorsal (**d**); head and prothorax, showing median costa of forehead (**e**); pro- and mesocoxae, showing closed coxal cavities (**f**); left metatibia (**g**). Scale bars: 0.5 mm (a).

Acalyptopygus lingziae Clarke & Oberprieler, **sp. n.** (Figure 26)

Description. Size. Length 2.75 mm, width 1.1 mm. **Eyes** only slightly slanting backwards, protruding, forehead between them narrowly triangular, apparently with dark median costa (head somewhat compressed). **Rostrum** distinctly downcurved, inserted in dorsal half of head but basally with dorsal outline not continuing onto head, forming a shallow sinus before eyes, ventral outline curved evenly onto that of head, not forming a conspicuous sinus. **Antennae.** Scapes as long as eye; funicles with segments 2 to 4 subequal, slightly shorter than 1, subcylindrical, 5 slightly shorter and apically swollen, 6 shorter, swollen in middle and thicker, 7 longer, subapically swollen; clubs with segments 1 to 3 subequal in length. **Thorax**. Pronotum broadly roundly trapezoidal, laterally slightly expanded but not explanate, posterior corners distinctly narrowly extended to fit closely onto elytral humeri. Metaventrite longer than mesoventrite, distinctly raised into transverse weals. **Elytra** posteriorly strongly declivous, well exposing pygidium in repose; surface faintly punctostriate, without scutellary striole, vestiture not discernible. **Legs**. Tibiae with 2 relatively stout spurs. Tarsal claws without ventrobasal seta. **Abdomen** with ventrite 5 longer than each of 3 and 4.

Material examined. Holotype (NIGP154206): very well preserved, intact specimen, not compressed or distorted, well visible but with thin layer of air over most of surface; in centre of elongate rectangular cuboid ca. 5.3 × 2.7 × 2.7 mm; amber clear, with many small impurities and four large but clear bubbles on right side of specimen.

Derivation of name. The species is named after Yu-Lingzi Zhou, presently at ANIC, for obtaining specimens for this study and for her various help and discussions about this interesting extinct fauna.

Remarks. This species also differs from *A. astriatus* by possessing faint but broad elytral striae and a slight dorsal sinus between the head and the rostrum. From *A. brevicornis* it is distinguishable by its longer scapes (as long as the eye) and from *A. elongatus* by having two tibial spurs. With *A. brevicornis* it agrees in lacking the ventrobasal seta on the tarsal claws.

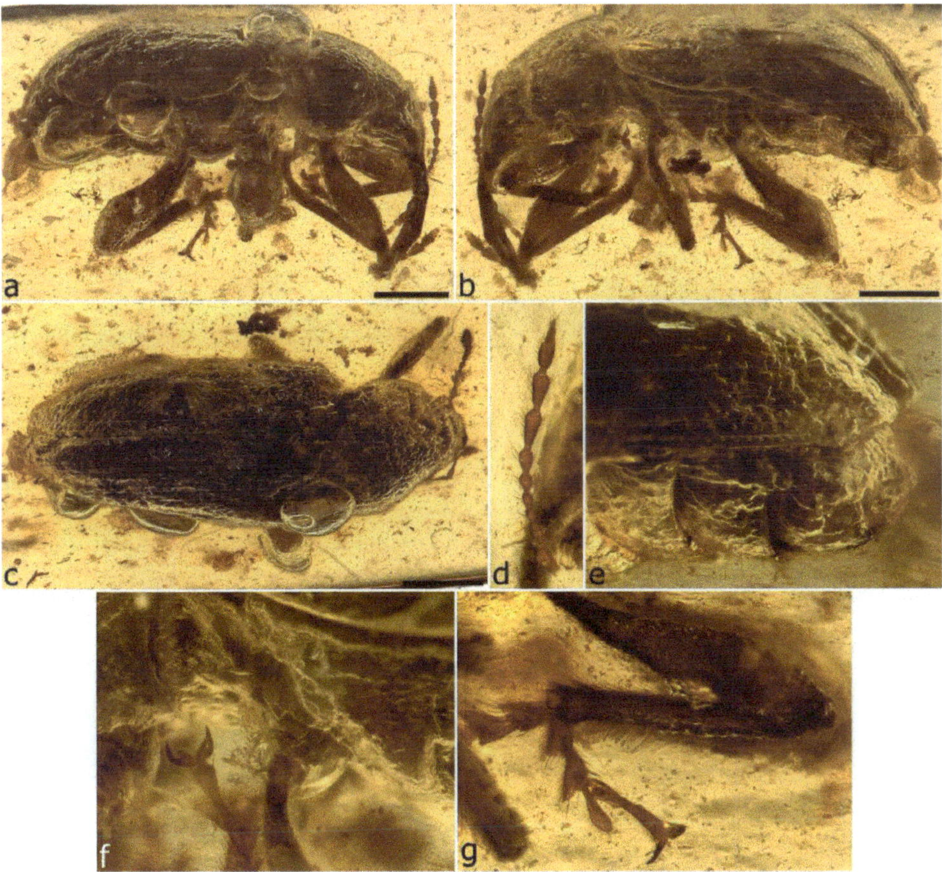

Figure 26. *Acalyptopygus lingziae*, holotype. Habitus, right (**a**); left (**b**); dorsal (**c**); antenna (**d**); elytral apices and ventrites 3–5 (**e**); thorax, left lateral (**f**); hindleg (**g**). Scale bars: 0.5 mm (a–c).

Acalyptopygus elongatus Clarke & Oberprieler, **sp. n.** (Figure 27)

Description. Size. Length 1.99 mm, width 0.82 mm. **Head** setose, setae short, recurved, some erect; dorsally convex, dorsal outline nearly continuously curved from base of rostrum; ventrally less convex; between eyes separated by about a basal width of rostrum. **Eyes** lateral; moderately protuberant, subspherical, slightly elongate. **Rostrum** slightly longer than pronotum; moderately downcurved, inserted in dorsal half of head, basally with dorsal outline continuing onto head, without any sinus before eyes but ventral outline forming large sinus with strongly bulging head; dorsally and laterally without carinae; antennal insertions in basal quarter of rostral length; with scrobes behind them weakly delimited; mandibular articulations deep, oblique. **Antennae.** Scapes elongate, about as long as eye, reaching front margin of eye, apically truncate; funicles with segment 1 slightly longer than 2, broader, 2 ca. 2.0 × longer than 3, 4–7 progressively slightly broader and longer. **Mouthparts**. Mandibles with 2 teeth on outer side, one forming long outer apical tooth; inner edge with 3 large subequal teeth; apically forming T, slightly emarginate at middle; articulation plane oblique. Right maxilla apically setose; maxillary palps 3-segmented, subtelescoped (2 within 1), segments progressively narrower; projecting obliquely ventrally. Labial palps short, apically and closely inserted, projecting obliquely. **Thorax.** Pronotum widest at about middle, slightly narrower than elytra; coarsely punctate; densely setose, setae long, reddish; sides rounded; somewhat constricted anteriorly; basal margin broadly lobate in middle, closely fitting to elytra, corners rounded. Scutellar shield small, at same level as elytra; densely setose. Metaventrite distinctly concave medially between narrow, very prominent transverse weals. **Elytra** distinctly punctostriate; setose, setae reddish, suberect-recumbent except for scattered long erect setae; striae ca. 2.0 × wider than interstriae; interstriae prominent, interstria 8 forming rounded ridge, prominent in dorsal view, 7 and 8 confluent; anteriorly forming prominent slightly produced humeri; lateral margin sinuate; with marginal groove subequal in width along entire length, punctosetose, with anterior marginal notch. **Legs.** Mesocoxae globular, very prominent, longer than wide. Tibiae apically without distinct flanges, apex lined with long slender fringing setae, dorso-apical edge with several elongate thin setae; protibiae with 2 short spurs; mesotibiae with small mucro and 2 short spurs; metatibiae ventrally with single long, probably fixed spur and long, slender seta, ca. 2.0 × longer than spur. Tarsi elongate, about as long as tibiae; tarsite 1 elongate, ca. 2.0 × longer than 2, apically slightly excised, 2 similar, narrower than 1, 3 deeply lobate, lobes subpedunculate, concave along inner side, ventrally with dense setal pads, cryptotarsite distinct, 5 about as long as 1 + 2, setose dorsally and ventrally; claws with ventrobasal seta. **Abdomen.** Tergites VII and VIII strongly sclerotised, forming cupular pygidium with apparent inflexed median lip. Ventrites sparsely setose, denser laterally; sutures substraight; 1 and 2 subequal in length, 3–4 subequal (4 very slightly shorter), 5 longer than 4, with posterior edge broadly rounded to substraight.

Material examined. Holotype (NIGP154207): well preserved, undistorted and well visible specimen, with left protarsus missing tarsites 3–5, appendages and rostrum otherwise intact but legs folded under specimen and right antenna obscured by bubble, right maxilla projecting obliquely from rostral apex (displaced), pygidium dislodged and partly severed, one wing partly extended, with sparse coating of whitish debris partly obscuring surface details; in irregular rectangular block with large curved and large flat face and two smaller flat faces, 6.7 × 1.9 × 1.7 mm; amber clear yellow, with flat bubbles and small fractures partly obscuring ventral side (mainly head and thorax).

Derivation of name. The species is named for its elongate shape, the name being an adjective.

Remarks. This is the smallest of the four known species of *Acalyptopygus* and unique in the genus in possessing a single elongate spur on the metatibiae. It is also distinguishable from *A. astriatus* by having distinct elytral striae, but it agrees with this species and differs from the other two in having a ventrobasal seta on the teeth of the tarsal claws. It agrees with *A. brevicornis* (and one undescribed species) in having a mesotibial mucro but has differently shaped pro- and metatibiae and scapes as long as the eyes. From *A. lingziae* it differs additionally in its conspicuous erect body setae and progressively longer, more uniform funicle segments. The holotype appears to be a female, as a pair of long, flat, basally contiguous processes with dense and quite long setae on the inner and apical edges is faintly discernible inside the cupular pygidium.

Figure 27. *Acalyptopygus elongatus* sp. n., holotype. Habitus, left lateral (**a**); head and anterior leg, left (**b**); habitus, right ventrolateral (**c**); habitus, left dorsolateral (**d**); habitus, right lateral (**e**); hindleg, ventral (**f**); pronotum and elytra (**g**). Scale bars: 0.5 mm (a,c–e).

Acalyptopygus astriatus Clarke & Oberprieler, **sp. n.** (Figure 28)

Description. Size. Length 2.20 mm, width 0.95 mm. **Eyes** slanting backwards, forehead between them trapezoidal, with narrow bare (black) median costa widening basad. **Rostrum** slightly downcurved, inserted in dorsal half of head, basally with dorsal outline continuing onto head, without any sinus before eyes, but ventral outline forming large sinus with strongly bulging head. **Antennae.** Scapes shorter than eye; funicles with segments 2 to 7 subequal, 7 slightly thicker; clubs with segment 1 elongate, spindle-shaped, 2 and 3 shorter and thicker. **Mouthparts.** Mandibles with 2 teeth on outer side, one forming long outer apical tooth, another sharper broader one at middle; inner edge with small basal tooth, 2 large subequal teeth at middle and a smaller apical one; apically forming T, slightly emarginate at middle. **Thorax**. Pronotum roundly subrectangular, widest in middle of length, laterally roundly explanate (sides of prothorax steeply sloping down), posterior corners angulate but not extended to fit closely onto elytral humeri; surface finely granulose. Metaventrite short, indistinctly convex. **Elytra** posteriorly evenly declivous, only shortly exposing pygidium in repose; surface not punctostriate, sparsely irregularly punctosetose, intervals between punctures flat, dull. **Legs**. Tibiae with 2 long fine spurs. Tarsal claws with distinct ventrobasal seta. **Abdomen** with ventrite 5 as long as 3.

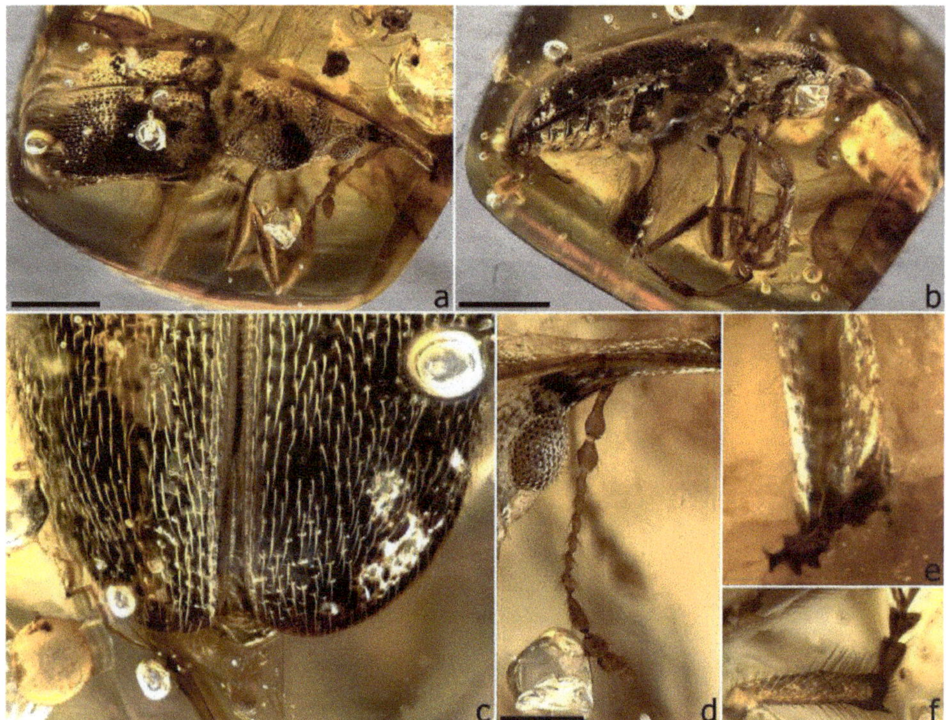

Figure 28. *Acalyptopygus astriatus* sp. n., holotype. Habitus, dorsal (**a**); habitus, right lateral (**b**); elytral declivity, showing no striae (**c**); head and right antenna, lateral oblique (**d**); apex of rostrum and right mandible, right lateral (**e**); left metatibia, tarsites 1–2 (**f**). Scale bars: 0.5 mm (a,b); 0.2 mm (d).

Material examined. Holotype (NIGP154208): very well preserved, intact specimen, not compressed or distorted, well visible; on right side of rectangular cuboid 2.95 × 2.6 × 2.2 mm with rounded edges and corners; amber clear, with few large but clear bubbles at front on left side of specimen, and large vertical shearing plane across left side of specimen.

Derivation of name. The species is named for the absence of striae on its elytra, the name being a Latin adjective.

Remarks. This species differs from all other *Acalyptopygus* species in its lack of elytral striae, the setiferous punctures being aligned into irregular rows. From *A. brevicornis* and *A. lingziae*, it is also distinguishable by the absence of a dorsal sinus between the rostrum and the head, the granulose pronotum and the unequal club segments, and from *A. elongatus* it further differs in having two metatibial spurs (a single long one in *A. elongatus*) but no sparse long erect body setae. As in *A. elongatus*, however, its tarsal claws possess a distinct ventrobasal seta.

Subfamily Mesophyletinae Poinar, 2008

Mesophyletinae Poinar, 2008: 262 [50] (type genus: *Mesophyletis* Poinar, 2006)

Anchineini Poinar & Legalov, 2015: 558 [53] (type genus: *Anchineus* Poinar & Brown, 2009) **syn. n.**

Mekorhamphini Poinar, Brown & Legalov, 2016: 158 [54] (type genus: *Mekorhamphus* Poinar, Brown & Legalov, 2016) **syn. n.**

Burmocorynini Legalov, 2018: 5 [46] (type genus: *Burmocorynus* Legalov, 2018) **syn. n.**

Diagnosis. Head porrect, short. **Eyes** mostly large to very large, hemispherical to conical, mostly strongly protruding, coarsely facetted. **Rostrum** slightly to much longer than pronotum, usually long and thin, sometimes shorter and stouter, rarely flexible into a prosternal channel (*Burmorhinus*, *Rhadinomycter*); antennal insertions mostly subbasal to median, rarely in apical quarter, behind them with long narrow scrobe extending to eye. Gular suture single, long. **Antennae** geniculate of 'open' type; scapes about as long as funicle; funicles 7-segmented; clubs loosely articulated to subcompact, 3–4-segmented. **Mouthparts**. Labrum absent. Mandibles mostly exodont but sometimes non-exodont, articulation horizontal to oblique or quasivertical. **Thorax**. Pronotum subquadrate to trapezoidal, laterally rounded, posterior corners usually extended, fitting closely onto elytra; notosternal sutures closed; procoxal cavities usually medially confluent, rarely separated; mesocoxal cavities closed. **Elytra** with broadly rounded humeri, apically mostly individually rounded, rarely subtruncate, sometimes exposing short pygidium; surface distinctly to indistinctly punctostriate, without scutellary striole, usually densely setose. **Legs**. Procoxae mostly elongate and prominent, medially contiguous, rarely subglobular and medially separated; mesocoxae globular, narrowly separated; metacoxae flat, transversely elongate. Femora inflated in distal half, outside often carinate to crenulate in distal half, almost always unarmed. Tibiae subcylindrical to flattened, outer edge usually serrulate or crenulate, apex usually with 2 spurs. Tarsi long; tarsite 1 usually elongate triangular, 3 deeply lobed, lobes often pedunculate, claws divaricate, simple to angulate to dentate, with ventrobasal setae. **Abdomen** with ventrites 1 and 2 fused, longer than 3–4; last tergite sometimes exposed as pygidium.

Remarks. The subfamily Mesophyletinae comprises the bulk of the known mesophyletid genera and species and is characterised foremost by its geniculate antennae, by which it differs from Aepyceratinae as here delineated. Other conspicuous characters present in most taxa are the serrulate or crenulate tibiae and the strongly dentate tarsal claws. On the basis of the tarsal claws, the genera of Mesophyletinae may be divided into two groups, those with simple to basally slightly swollen or angulate tarsal claws and those with strongly dentate ones. The first group currently includes eleven genera (*Burmocorynus*, *Burmorhinus*, *Cetionyx*, *Cyrtocis*, *Echogomphus*, *Electrocis*, *Debbia*, *Gnomus*, *Opeatorhynchus*, *Petalotarsus* and *Rhadinomycter*) and the second fourteen (*Anchineus*, *Aphelonyssus*, *Bowangius*, *Compsopsarus*, *Elwoodius*, *Euryepomus*, *Habropezus*, *Hukawngius*, *Leptopezus*, *Louwiocis*, *Mekorhamphus*, *Mesophyletis*, *Myanmarus* and *Periosocerus*). Whether these two groups may constitute natural entities in some concept (not necessarily in the generic aggregations as above) and can be recognised as tribes will have to be determined from a comprehensive phylogenetic analysis of their relevant characters; they are here used merely as a convenient way of grouping the genera for identification purposes. In extant weevils toothed claws are often used as a generic character, e.g., to distinguish *Storeus* Schoenherr (dentate) from *Emplesis* Pascoe ('simple'), but such a distinction is frequently blurred by conditions of bluntly toothed and basally swollen claws. In Mesophyletidae the dentition of the claws is clearly significant at a higher level, and intermediary states are unknown (occur to some degree only in *Gnomus*), but whether it is suitable to characterise generic groups remains to be seen.

This subfamily subsumes the tribes Mesophyletini, Anchineini and Mekorhamphini as erected by Legalov and Poinar [53] and Poinar et al. [54]. These are not maintained here, as the differences between their type genera are trivial in comparison with the greater character disparities exhibited by the large, robust forms with simple or angulate tarsal claws, in particular the *Burmocorynus-Cetionyx-Petalotarsus* group but also a number of isolated smaller genera. If the provisional generic group with simple to angulate tarsal claws as here circumscribed can be found to be monophyletic, the tribal name Burmocorynini may be applied to it, but this is not evident at present and we therefore also subsume the tribe Burmocorynini into Mesophyletinae. A tribal classification of Mesophyletinae is premature at this stage and should only be considered when the characters of the group have been studied in more detail and in a phylogenetic context.

Genus ***Cetionyx*** Clarke & Oberprieler, **gen. n.**

Type species: *Cetionyx batiatus* Clarke & Oberprieler, sp. n.

Description. Size. Length 8.2–10.0 mm, width 3.8–5.7 mm. **Head** short, subconical, not or slightly constricted behind eyes. **Eyes** small, elongate, somewhat compressed, strongly protruding, coarsely facetted, dorsally separated by width of rostrum, further expanded posteriorly; forehead flat, without tubercles. **Rostrum** longer than pronotum, slender, cylindrical, slightly to strongly downcurved; antennal insertions lateral, in or behind middle of rostral length, behind them with scrobes extending to eye, in front of them without lateral row of setae. Gular suture not discernible. **Antennae** geniculate, long, very slender; scapes long, thin, cylindrical, apically slightly thickened; funicles shorter than scape, 7-segmented, segment 1 less than half as long as 2, 2–4 long, progressively shorter, 5–7 distinctly shorter; clubs long, weakly differentiated, loosely articulated, 4-segmented, segments subcyclindrical, apical segment bluntly triangular. **Mouthparts**. Labrum absent. Mandibles small, exodont, with at least 2 external teeth, articulation horizontal. **Thorax**. Prothorax slightly to strongly proclinate, with anterior lateral margins oblique to nearly vertical in lateral view. Pronotum roundly subrectangular to trapezoidal, laterally weakly to broadly rounded, without tooth; weakly convex, finely to coarsely punctate, sparsely to densely setose, setae short, fine, suberect, pale, directed mesad; notosternal sutures closed. Prosternum moderately long, prosternal process meeting hypomeral process or almost so; procoxal cavities narrowly to broadly separate or nearly so, in about middle of prothorax. Scutellar shield indistinct, small. Mesocoxal cavities laterally closed (by meso- and metaventrite). Metanepisterna distinct. Mesoventrite short, anteriorly strongly sloping. Metaventrite raised into transverse weals before metacoxae. **Elytra** shortly to moderately elongate, with weak, broadly rounded or sharply angulate humeri, posteriorly declivous, apically individually rounded, not exposing pygidium in repose; sutural flanges narrow, equal; surface weakly coarsely punctostriate, without scutellary striole, sparsely to densely setose, setae short, stout, reclinate, directed caudad. **Legs**. Front legs often longer than middle and hindlegs. Procoxae subglobular, prominent, medially narrowly to broadly separated; mesocoxae subglobular, broadly separated; metacoxae transversely elongate, reaching elytra. Trochanters short, oblique. Femora long, subcylindrical, inflated in distal half, unarmed, outer side rounded, inner side excavate in distal quarter, receiving tibiae in repose, walls of groove at apex flatly roundly extended. Tibiae long, straight, subcylindrical, outer edge rounded, apically with long dense setae, apex obliquely truncate, with 2 small spurs (sometimes absent on protibiae or possibly on metatibiae). Tarsi 0.67 $\times$ or more as long as tibiae, broad, flat, densely setose, fringed with long setae; tarsite 1 elongate, apically deeply excised to lobate, 2 shorter, apically very deeply excised, 3 deeply bilobed, lobes strongly pedunculate, attachment with basal plate extremely slender in both dimensions, apically broad and flat, 5 slightly longer than 1 + 2; claws divaricate, basally swollen to angulate (not dentate) with ventrobasal seta at apex of swelling or angle. **Abdomen** flat, ventrites 1 and 2 subequal, slightly longer than 3 and 4, 5 broadly rounded, subequal or longer than 4.

Derivation of name. The genus is named for its large tarsi and claws, latinised from the Greek adjective *keteios* (large, monstrous) and noun *onyx* (G: *onychos*), a claw or talon; the gender of the name is masculine.

Remarks. This genus includes the largest weevils known from Burmese amber, reaching more than 11 mm in body length and with a rostrum that can attain three-quarters of the body length, so another 8 mm. *Cetionyx* is mainly characterised by its large, broad, shaggy tarsi with large claws, which are similar only to those of *Opeatorhynchus* and (less so) of *Burmocorynus* (and partly to those of *Petalotarsus* but longer and more loosely articulated), but also by its distinctive antennae, being very slender with a short funicle segment 1 (much shorter than segment 2) and indistinct clubs that are scarcely broader than the funicles. *Cetionyx* is readily distinguishable from *Burmocorynus* and *Petalotarsus* by its higher and broader body, long slender antennae and loosely articulated clubs. From *Opeatorhynchus* it is also distinguishable by its antennal structure (the short funicle segments 1 and indistinct clubs) as well as by its separated procoxae. It agrees with *Opeatorhynchus* in having the most strongly pedunculate lobes of tarsite 3 among our sample, the basal part extremely thin in both dorsal and lateral view, especially at the point of the very flexible attachment to a distinct basal plate. This basal structure of tarsites 3 differs from that of *Burmocorynus*, in which the basal connection of the lobes is broader and thicker in dorsal and lateral view. Among the material available for this study, *Cetionyx* is represented by three species. Unfortunately none of the specimens currently available are sufficiently well preserved to allow a detailed study and depiction of all characters. However, our CT reconstruction of *C. batiatus* (Video S2) gives a good impression of what these large Cretaceous weevils looked like and at least in this specimen demonstrably compensates for the lack of detail observable under light microscopy.

Cetionyx batiatus Clarke & Oberprieler, **sp. n.** (Figures 29–31, Video S2)

Description. Size. Length 8.22 mm, width 4.62 mm. **Head** bulbous dorsally and ventrally; not constricted behind eyes (posteriorly); punctate, punctures small. **Rostrum** elongate, perhaps as long as entire body, inserted at front of head (with dorsal sinus); subcylindrical; moderately curved. Scrobes lateral, narrow, sharply delimited, reaching eyes. **Antennae** sparsely setose; scapes reaching to just below eye, apically slightly bent, articulation with pedicel oblique; funicles with segment 1 ca. half as long as 2, 2 elongate, subparallel, only slightly widening apicad, 3 ca. 0.67 × as long as 2, 3 and 4 subequal, 5 subequal in width, about 0.67 × as long as 4, 6 slightly shorter than 5, 6 and 7 subequal; clubs with segments 1 and 2 subequal, 4 distinct, broadly inserted in 3. **Thorax.** Prothorax not or slightly proclinate, with anterior lateral margins seemingly nearly vertical in lateral view. Pronotum narrower than elytra; strongly transverse; densely and coarsely punctate; rugosely sculptured, with minute channels between punctures; sparsely setose, setae short, recurved, directed mesad to anteromesad; sides rounded, with slight anterior collar; basal pronotal margin broadly curved, closely fitting with elytra. Prosternum short, about 0.33 × as long as procoxae, anterior margin broadly emarginate, prosternal process broadly pointed, contacting apex of hypomeral process (Figures 30e and 31e). **Elytra** broad, moderately strongly convex; striae and interstriae indistinct, striae thin, interstriae wide; sculptured as pronotum; sparsely setose (appearing almost glabrous), bases straight, subcarinate, slightly raised, medially continuous with anterior scutellar edge, tightly fitting with pronotum, forming broadly obtuse angle at elytral suture; humeri indistinct, cupulate, strongly concave to receive posterolateral angles of pronotum; sides broadly rounded, basally curved evenly to humeri, with marginal groove distinct, finely punctate, broadened anteriorly forming lateral ridge; anterior marginal notch present; apices individually rounded. Scutellar shield rectangularly rounded, glabrous, flush with elytral surface. Mesocoxal cavities widely separated by less than half of mesocoxal width; meso- and metaventral intercoxal processes broad, flat, abutting, separated by straight suture. Meso- and metaventrite coarsely punctate; with apparent discrimen. **Legs.** Tibiae long, slender; clothed with distally increasingly dense setae; tibiotarsal articulation surfaces oblique to concave; spur formula 0-2-2; protibiae elongate-slender, distinctly curved outward, apically expanded, inside longer than outside, in apical two-thirds with dense increasingly longer setae, anterior and posterior apical flanges densely setose, apical fringing setae not lining edges; meso- and metatibiae slightly curved outward, on inner side in apical half with distinct elongate patch of dense, long setae, basal- and distal-most setae longer than intervening ones, apically with narrowly rounded anterior flange, spurs closely situated, robust. Tarsi with tarsite 2 wider and shorter than 1, lobes of 3 ca. 0.67 × as long as 5, very narrow basally, protarsi elongate, ca. half as long as protibia, longer than meso- and metatarsi, metatarsi with tarsite 2 more distinctly Y-shaped than on other legs; claws of meso- and metatibae less basally angulate than those of protarsi. **Abdomen** Ventrites slightly stepped, finely setose, more finely punctate than meso- and metaventrites; ventrite 3 shorter than 2, 4 longer than 3, 5 elongate, ca. 1.5 × longer than 4, apically broadly rounded, 4 and 5 with very short, fine, subappressed setae, impunctate; sutures straight or nearly so, not forming wide/deep gaps between ventrites.

Material examined. Holotype (NIGP154209): poorly preserved, partly decomposed but mostly visible specimen, with ventral parts of head and prothorax and ventrites partly collapsed and variously broken, rostrum distorted, apices of right femora, left metatarsal claws and apical part of rostrum cut away with amber, fore- and middle legs severed at trochanters; in high-domed irregular cabochon, 14.5 × 11.5 × 6.5 mm, orientated with dorsal side subparallel to curved face; amber brownish-yellow, largely transparent but with gritty impurities and larger debris particles, with only small oblique fracture over elytra and pronotum.

Derivation of name. The species is named for its large, robust shape, reminiscent of an ancient Roman gladiator, after Batiatus, the owner of a gladiatorial school from which the Spartacus rebellion arose in 73 B.C.E.; the name is a noun in apposition.

Remarks. The poor preservation and partial distortion of the holotype made it difficult to assess many of the characters of this species under the light microscope, but fortunately the specimen became well visible from CT scanning, which allowed a more or less complete assessment of its structural details (Figures 30 and 31, Video S2), in particular those of the ventrites and the separation of the procoxal cavities (by narrow, adjoining intercoxal processes), which correlates with the much more broadly separated mesocoxae. All these areas are difficult to impossible to properly discern on the holotype under a light microscope. *Cetionyx batiatus* differs from *C. ursinus* mainly in its sparse, indistinct vestiture (appearing almost glabrous), straighter rostrum and rounded elytral apices and from *C. terebrans* in its more distinct antennal clubs (wider than the funicles), more closely fitting pronotum and elytra and the absence of spurs on the protibiae.

Cetionyx terebrans Clarke & Oberprieler, **sp. n.** (Figure 32)

Description. Size. Length 10.00 mm, width 3.8 mm. **Eyes** small, distorted, seemingly flattened and pushed up. **Rostrum** ca 2 × longer than pronotum, slender, only gently downcurved; antennal insertions just before basal third of length. **Antennae.** Funicles with segment 1 elongate, thin, distinctly shorter than 2, 2 and 3 subequal, 4 slightly shorter, 5–7 shorter, subequal; clubs long, segment 1 slightly longer than funicle segment 7 but similar, 2 thicker, obconical, 3 as long but narrower, 4 slightly shorter than 3. **Mouthparts**. Mandibles obliquely cut away, apparently scoop-shaped, not exodont; articulation horizontal. Maxillae and labium not discernible. **Thorax**. Prothorax slightly proclinate, with anterior lateral margins oblique in lateral view. Pronotum (poorly visible, obscured by while bubbly layer and cloudy amber) elongate, slightly convex, laterally hardly rounded, posterior corners rounded, not fitting closely onto elytra; surface not visible. Scutellar shield seemingly short and broad. **Elytra** (largely obscure, only basal half of left elytron properly visible), shortly elongate, laterally substraight, posteriorly gently declivous, apically rounded; surface weakly punctostriate, interstriae sparely setose, setae short, robust, reclinate (visible laterally). **Legs.** Front legs longer than middle and hindlegs. Procoxae slightly protruding and diverging, possibly separated, others and trochanters not properly visible; metacoxae appearing subglobular. Femora long, subcylindrical, slightly sinuate, profemora only slightly inflated in middle, meso- and metafemora more so. Tibiae densely setose in apical quarter, meso- and metatibiae apically broadened, bent inwards (possibly a preservation artefact); apex with 2 short, stout spurs (only one visible on protibiae). Tarsi about as long as tibiae; tarsite 1 elongate, broadly triangular, apically deeply excised, 2 slightly shorter, apically very deeply excised, 3 very deeply bilobed, lobes pedunculate, 5 very narrow between lobes of 3, rapidly broadened beyond them; claws divaricate, basally swollen, with ventrobasal seta. **Abdomen** not visible.

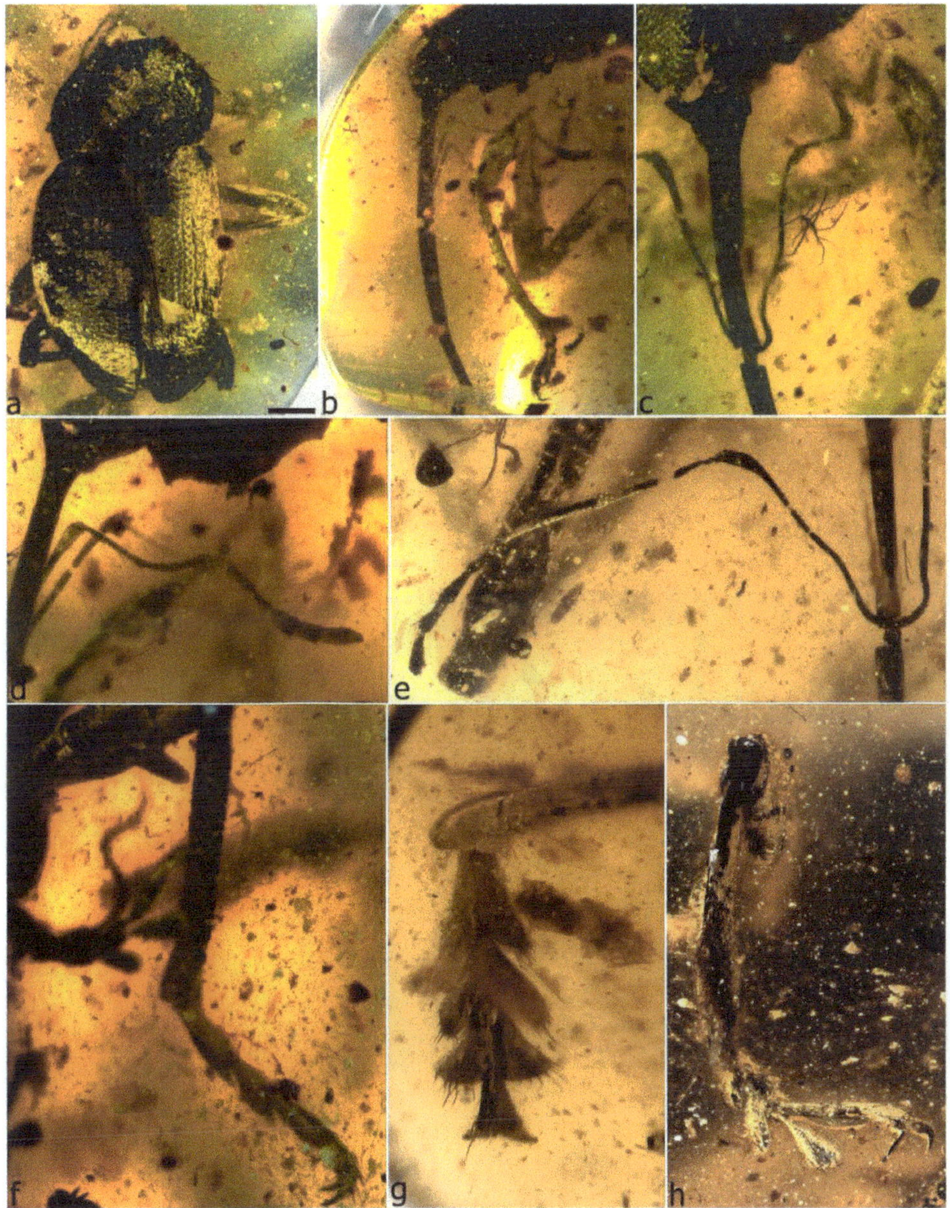

Figure 29. *Cetionyx batiatus* sp. n., holotype. Habitus, dorsal (**a**); head, left lateral (**b**); head and antenna, dorsal oblique (**c**); detail of left antenna, dorsal oblique (**d**); detail of right antenna, dorsal (**e**); mesotibia and -tarsus (**f**); mesotarsus, dorsal (**g**); mesotibia and -tarsus (**h**). Scale bar: 1.0 mm (a).

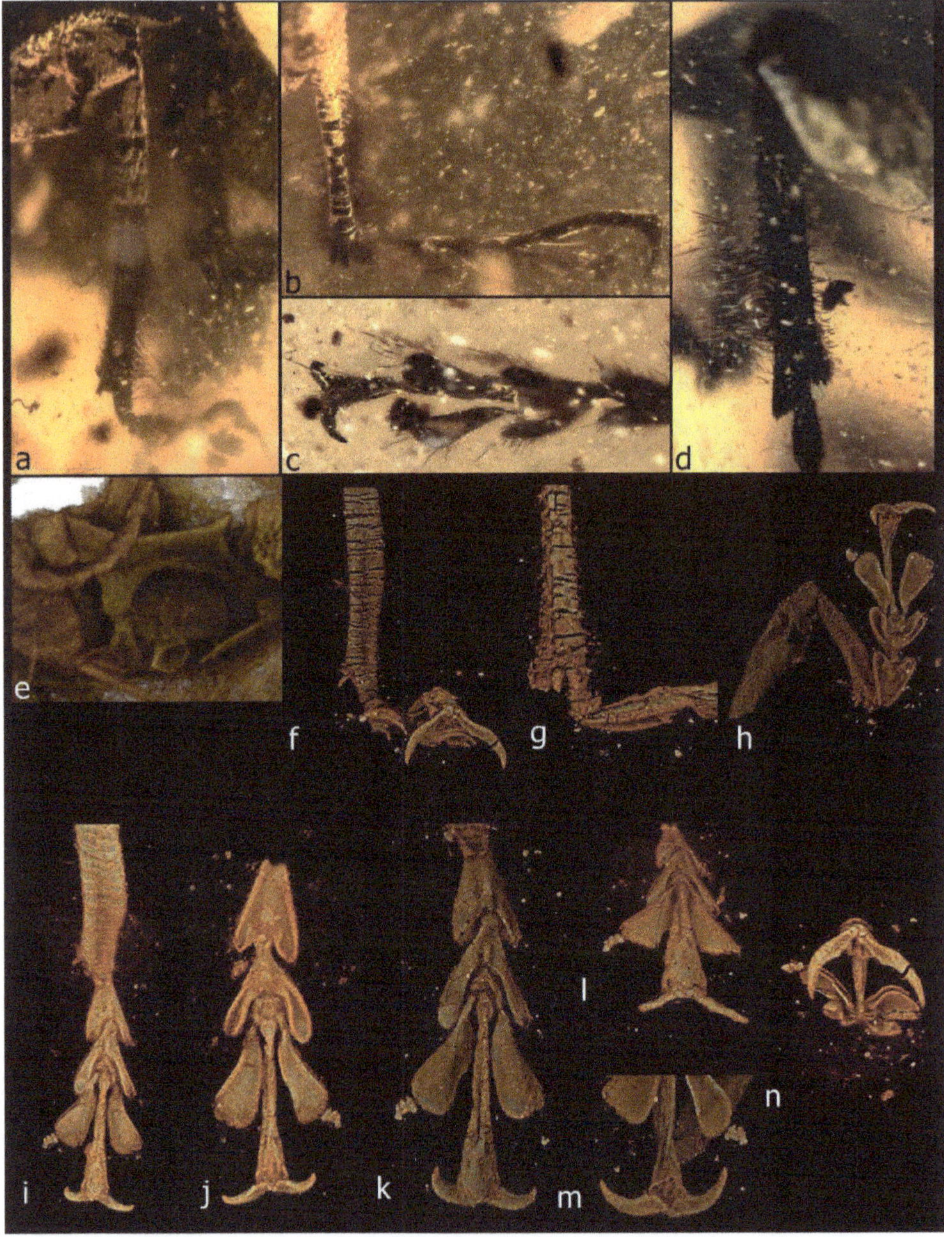

Figure 30. *Cetionyx batiatus* sp. n., holotype. Mesotibia showing spurs (**a**); mesotibia and -tarsus (**b**); protarsus (**c**); metatibia showing spurs (**d**); details of prothorax, ventral (**e**); micro-CT scan reconstructions showing details of tibiae and tarsi (**f–n**). See also Video S2.

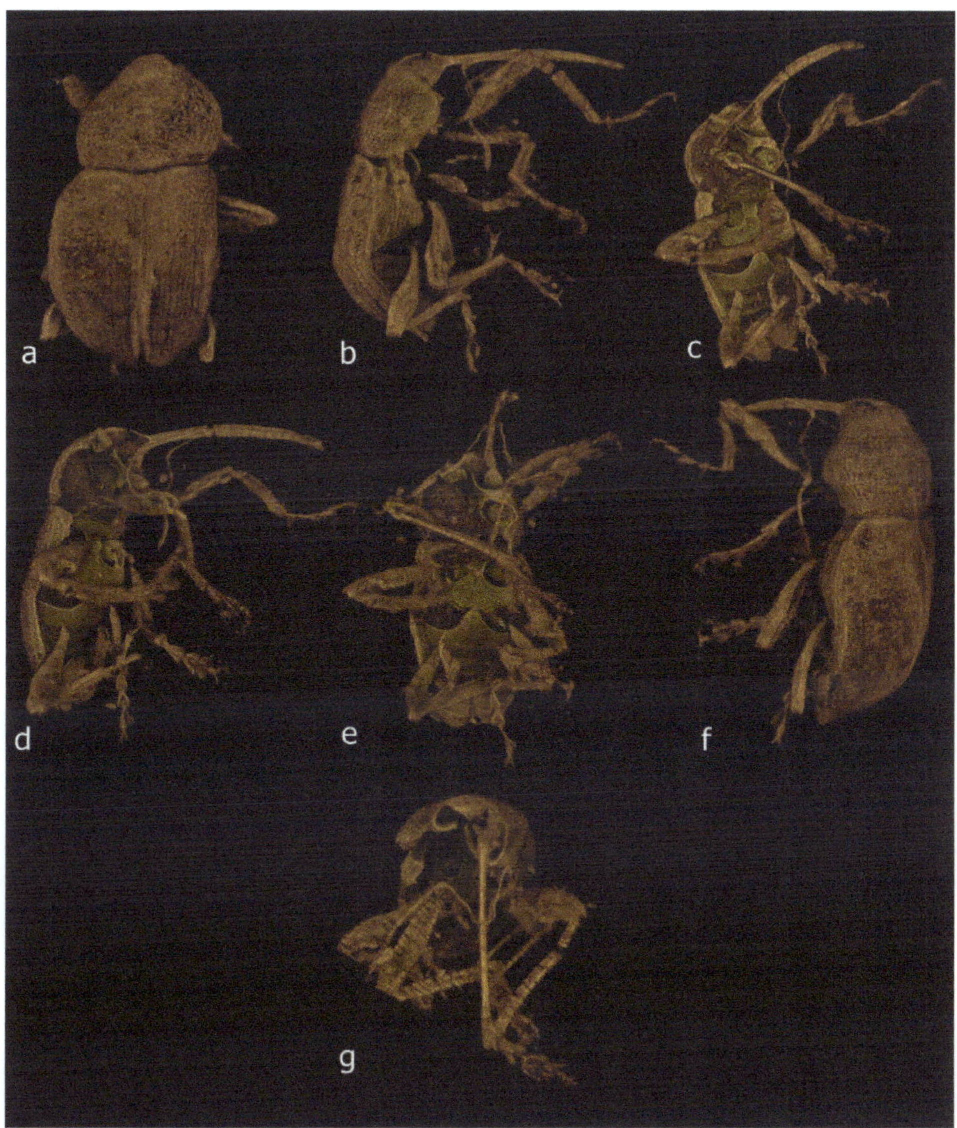

Figure 31. *Cetionyx batiatus* sp. n., holotype. Micro-CT scan images showing different views of the whole body, particularly useful for revealing obscured ventral details and the general effects of deformation and distortion on the shape and visibility of structures (**a–g**). See also Video S2.

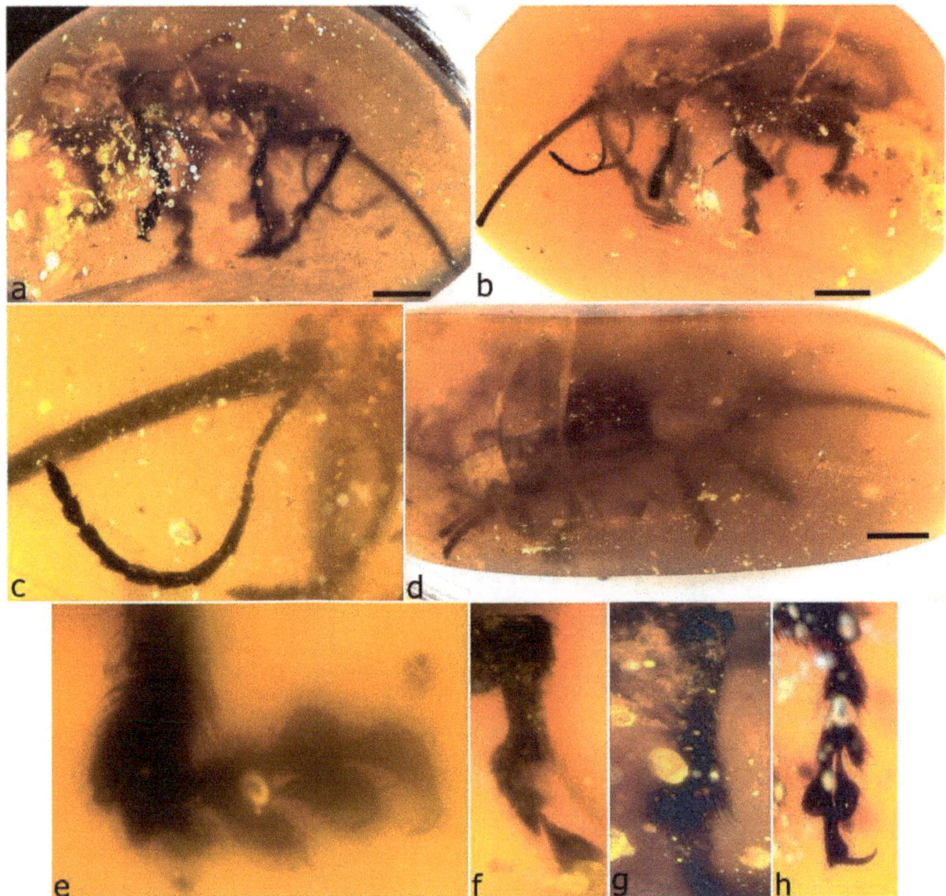

Figure 32. *Cetionyx terebrans* sp. n., holotype. Habitus, right lateral (**a**); left lateral (**b**); left antenna (**c**); habitus, dorsal (**d**); left metatarsus, dorsal (**e**); right mesotibia showing curvature (**f**); right mesotibia showing dorso-apical setal patch (**g**); right metatarsus, dorsal (**h**). Scale bars: 2.0 mm (a,b,d).

Material examined. Holotype (NIGP154210): well preserved, intact specimen, with slightly compressed legs and eyes, dorsal surface obscured by layer of white encrustation except over apical half of elytra, apex of rostrum slightly cut off on left side; at slight angle in middle of irregular block 8.7 mm thick with flat oval surfaces ca. 19.7 × 14.2 mm; amber dark orange-brown, slightly opaque, with 2 irregular transverse vertical cracks around specimen, a thin warped horizontal layer of bubbles and other debris at back, and various other small bubbles and impurities throughout.

Derivation of name. The species is named for its long, stout, straight rostrum, which is evidently suited to piercing plant material. The name is a Latin participle meaning 'the piercer' or 'the borer', treated as a noun in apposition.

Remarks. The single specimen of this species is the largest of all weevil fossils in Burmese amber studied by us. It differs from *C. ursinus* in having a stouter, almost straight rostrum, a narrower prothorax with less closely fitting pronotum and elytra, a sparser vestiture and differently elongate antennal segments. *Cetionyx batiatus* has a similarly straight rostrum but longer, thinner antennae, especially the clubs, and also a shorter, broader prothorax and no spurs on the protibiae. A more detailed comparison of these two species is prevented by the poor preservation of and visibility of both holotypes, but that of *C. terebrans* may also become better visible with CT scanning.

Cetionyx ursinus Clarke & Oberprieler, **sp. n.** (Figure 33)

Description. Size. Length 9.71 mm, width 5.69 mm. **Head** punctate, ventrally bulbous. **Eyes** slightly protruding, seemingly longer than wide. **Rostrum** extremely long, subequal to body length, strongly curved, dorsally with longitudinal wavy carinae or ridges; setose, setae short, directed anteromesad. **Antennae** setose; scapes reaching front margin of eye, apical articulation with pedicel oblique; funicles with segments 1–4 progressively gradually widening, 5–7 progressively shorter towards club, 2 elongate, 3 and 4 subequal, ca. 0.67 × as long as 2, 5 slightly more than half as long as 4, expanded apically, 6 ca. 0.67 × as long as 5, 7 slightly shorter than 6; clubs with segments 1 and 2 subequal in length, obconical, 2 slightly wider than 1, 3 slightly shorter than 2, 4 broadly inserted into 3, about 0.67 × as long as 3. **Mouthparts.** Mandibles large, robust, flattened but thick, probably bilaterally asymmetrical, on outer side with 2 small triangular teeth, on inner side with one very large, somewhat dorsadly directed tooth with complex structure (possibly bicuspid), apex anvil-shaped with inner and outer teeth small, inner apex of left mandible appearing to form concave receptacle; articulation sockets seemingly oblique. **Thorax.** Prothorax strongly proclinate, with anterior lateral margins oblique in lateral view. Pronotum much broader than long, narrower than elytra, widest in front of base, more strongly narrowing anteriad; sides rounded. Prosternum about as long as procoxae, anterior margin broadly emarginate, prosternal process broadly rounded, separating procoxal cavities anteriorly but not posteriorly (not contacting hypomeral process). Pronotum densely and coarsely punctorugose; densely setose; anterior margin straight, posterior margin nearly straight, closely fitting to elytra. Meso- and metaventral intercoxal processes broad, separated by transverse suture. Metaventrite posterolaterally densely setose. **Eytra.** Bases straight, obtusely angulate, tightly fitting with pronotum, carinate from scutellar shield to junction with mesepimeron; humeri not strongly projecting, distinctly concave to receive basal pronotal corners; sides broadly curved, widest just before base, narrowing anteriad; lateral marginal groove distinct, broad anteriorly forming lateral ridge, from middle very narrow to elytral apex, region in groove densely setose, anterior marginal notch present; apices not meeting evenly at suture, individually bluntly angulate (apices together shallowly emarginate). **Legs** long, robust; densely setose, setae mostly subappressed; in distal third with longer setae on outer sides; forelegs distinctly longer than middle and hindlegs. Pro- and mesocoxae broadly separated, globular, not strongly projecting; metacoxae subglobular, reaching elytra. Tibiae on outer side of distal third with longer denser setae, at apex posteriorly well emarginate to allow basal tarsites to bend up; with pronounced inner (protibiae) and outer (meso- and metatibae) apical flanges lined with fringing setae, tarsal articulation surfaces oblique-concave, spur formula 0-2-2; protibiae with apical edge lined with short, spine-like fringing setae, longer setae near apex, seemingly without spurs; mesotibiae more robust and distinctly shorter than protibiae, apical edge with short dense fringing setae, spurs short, outer one distinctly longer and thicker than inner one; metatibiae along outer side in distal third with area of very dense and much longer setae, setae longer basally and apically, shorter in middle, spurs short. Meso- and metatarsi about 0.67 × as long as protarsi; tarsite 1 strongly lobate apicolaterally, 2 very deeply excised, 3 distinctly pedunculate, lobes about 0.67 × as long as 5, each strongly broadened apicad, 5 about as long as 1 + 2, curved ventrad. **Abdomen.** Ventrites slightly stepped; densely setose; 1 and 2 subequal in length, each longer than 3 and 4, 1 with broad intercoxal process, 5 about as long as 4, broadly rounded; sutures straight, not forming wide/deep gaps between ventrites.

Material examined. Holotype (NIGP154211): poorly preserved, partly decomposed and poorly visible specimen (especially ventrally), heavily but symmetrically distorted, with elytra and pronotum depressed (caved in) forming broadly concave surface, right middle leg severed and missing, rostrum and other appendages intact, well visible though mandibles partly obscured by bubble; in irregular cuboid 16.5 × 11.5 × 9.0 mm, with three curved faces; amber clear yellow-brown, cloudy due to granular or gritty impurities, with many larger impurities partly obscuring specimen, with small fracture over right protarsus and several conspicuous flow bands, a more conspicuous one separating most legs from rest of body.

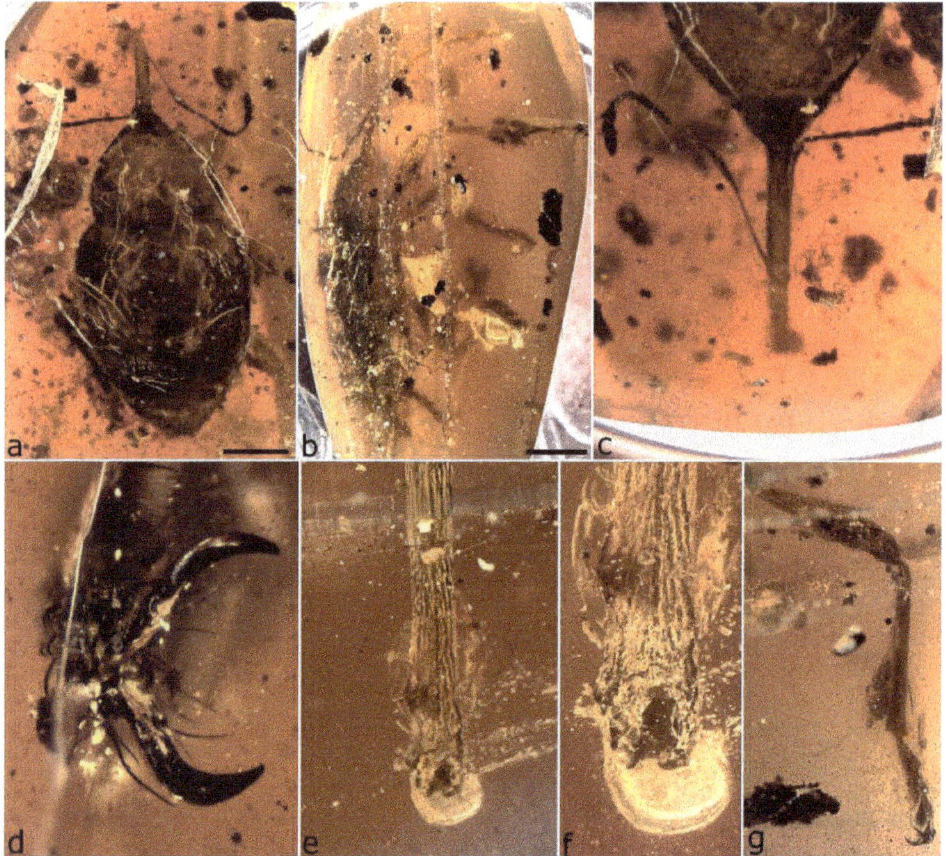

Figure 33. *Cetionyx ursinus* sp. n., holotype. Habitus, dorsal (**a**); habitus, right lateral (**b**); head and antennae, dorsal (**c**); protarsal claws (**d**); apical part of rostrum, dorsal (**e**); mandibles (**f**); right anterior leg (**g**). Scale bars: 2.0 mm (a,b).

Derivation of name. The species is named for its bear-like appearance (from the Latin adjective *ursinus*, derived from the noun *ursus*, a bear), being very large with a dense brown vestiture, large shaggy tarsi and sharp claws.

Remarks. This species is almost as large as *C. terebrans* but not as robust. It differs from this species as well as from the slightly smaller *C. batiatus* mainly due to its longer, much more strongly curved rostrum, denser vestiture and distinctly angulate elytral apices. From the latter species it also differs in not having the procoxal cavities fully separated (the prosternal and hypomeral process not meeting). The long, thin, strongly curved rostrum of the single known specimen suggests that it is a female and able to drill holes into thick plant organs or tissues.

Genus ***Burmocorynus*** Legalov, 2018

Burmocorynus Legalov, 2018: 5 [46] (type species, by original designation: *Burmocorynus jarzembowskii* Legalov, 2018)

Redescription. Size. Length 6.15–6.7 mm, width 1.2–1.8 mm. **Head** elongate, slightly flattened dorsally, strongly bulging ventrally, slightly constricted behind eyes. **Eyes** elongate, flatly protruding, evidently not facetted (surface smooth), dorsally separated by slightly more than width of rostrum anteriorly but slightly further separated posteriorly; forehead flat, without tubercles between eyes. **Rostrum** evidently at least as long as pronotum, cylindrical, substraight; antennal insertions lateral, with scrobes behind them, in front of them laterally with line of numerous long erect setae. Single long gular suture present. **Antennae** geniculate, long, flattened; scapes cylindrical, apically not or only slightly inflated, not reaching eyes; funicles 7-segmented, segments 1 and 2 subequal, elongate, obconical, 3–7 shorter, progressively wider towards club; clubs large, subcompact, flattened, 4-segmented, segment 4 narrowly rounded, longer than 3. **Mouthparts** not visible. **Thorax.** Prothorax elongate, not proclinate, anterior lateral margins vertical in lateral view. Pronotum slightly convex, laterally rounded, without tooth, posterior corners truncate, fitting closely onto elytra; surface coarsely rugose, sparsely setose, setae reclinate, golden and black, directed anteromesad on disc; notosternal sutures closed, curved anteriad. Prosternum moderately long; procoxal cavities widely separated, in middle of prothorax. Scutellar shield elongate-rectangular, densely setose. Mesocoxal cavities closed laterally. Metanepisterna distinct, densely setose. Mesoventrite long, flat. Metaventrite longer, flat or slightly convex, raised into low transverse weals. **Elytra** elongate, basally lobed over base of pronotum, with narrowly truncate humeri, posteriorly weakly declivous, lateral margin nearly straight, apically individually rounded, not exposing pygidium, extending beyond apex of ventrite 5; sutural flanges narrow, equal; surface coarsely punctostriate, without scutellary striole, interstriae indistinct, subflat, coarsely rugose, setose, setae short, thin, reclinate, directed caudad. **Legs.** Pro- and mesocoxae subequal, longer than wide; procoxae large, globular, prominent, widely separated; mesocoxae globular, very widely separated; metacoxae subglobular, shortly transverse. Trochanters short, oblique. Femora long, subcylindrical, inflated in distal half, outside rounded. Tibiae substraight, compressed, distally expanded, outer edge rounded, with dense long stiff setae in distal half, apex strongly oblique, with 2 spurs. Tarsi almost as long as tibiae (metatarsi slightly more than half length of tibia); tarsite 1 apically deeply excised, 2 shorter, very deeply excised (almost bilobed), 3 deeply bilobed, lobes pedunculate but broadly connected basally, 5 as long as 1 + 2, apically expanded, not flattened; claws divergent-divaricate, basally angulate with ventrobasal seta at apex of tooth. **Abdomen** with ventrites 1 and 2 each about as long as 3 + 4, 3 and 4 subequal, 5 about as long as 3 + 4, apically narrowly rounded.

Remarks. The genus was not specifically described, only a reference given to the diagnosis of the tribe Burmocorynini. The tribe was placed in the family Belidae because of the alleged non-geniculate, basally inserted antennae, protibial grooves (the antennal cleaners of Belidae) and paired gular sutures of the single specimen, but none of these character interpretations is correct, as we could ascertain from our examination of the specimen (after trimming the amber block) as well as of another similar, evidently congeneric specimen we received too late for inclusion in this paper. The antennal insertions of the holotype of *Burmocorynus jarzembowskii* are not visible (cut off with the rostrum), and the seemingly short scapes are in fact the clavate apices of longer thinner ones, as is the case in our other specimen, which has clearly geniculate antennae. A very similar, elongate but larger specimen illustrated by Xia et al. [55], p. 115, with very similar eyes and tarsi, also has geniculate antennae and probably represents the same genus, and also the less well-preserved specimen here described as *Burmocorynus longus* has geniculate antennae. The protibiae of *Burmocorynus jarzembowskii* are apically dilated and densely setose as they are in *Cetionyx* and many other mesophyletids, but they do not have antennal cleaning brushes as they occur in Belidae, and there are no short paired gular sutures as alleged (as drawn in Figure 14, [46]). *Burmocorynus* therefore cannot be classified in Belidae but clearly belongs in Mesophyletidae and specifically in the vicinity of *Cetionyx* and *Petalotarsus*, with which it shares the separated procoxae and the large, loose tarsi with deeply lobed tarsites 2 and 3. We will assess the relationships of these genera in more detail with the description of our additional specimen of *Burmocorynus*. The second part of the genus name (-*corynus*) is poorly chosen, as the taxon has nothing to do with the belid genus *Oxycorynus* and tribe Oxycorynini, to which it was related by its author. *Burmocorynus* presently comprises two species, but it is likely that our additional specimen and the one depicted by Xia et al. [55] represent different further species.

Burmocorynus jarzembowskii Legalov, 2018

Burmocorynus jarzembowskii Legalov, 2018: 5 [46]

Material examined. Holotype (ISEA no. MA 2018/1); length 6.7 mm, width 1.78 mm: very well preserved and well visible specimen, with a fragmented film of organic material over body but surface still well visible, most of rostrum, antennal scapes and part of right funicle, apex of left profemur and right leg at femorotibial joint cut away with amber, missing outer claw of left protarsus; re-prepared from much larger discoidal amber piece from which the original description was prepared, now in elongate irregular cuboid 7.5 × 3.0 × 4.1 mm with end closest to head of specimen rounded; amber hazy brown with greenish hue (under microscope light), with diffuse microscopic particles, with 2 fractures over dorsal side of specimen, one larger one ventrally obscuring right side of prothorax.

Remarks. In view of the redescription of the genus above, the original description of the species is adequate for the moment. Additional characters warranting mention here (and not present in *B. longus*) include the unusual modified thick setae on the inner side of the mesotibiae, the dense long setae on the ventrites, the extension of the elytral apices beyond ventrite 5 and the pale setae on the sides of the pronotum and elytral interstriae.

Burmocorynus longus Clarke & Oberprieler, **sp. n.** (Figure 34)

Description. Size. Length 6.15 mm, width 1.91 mm. **Head.** Antennal insertions median. **Antennae.** Scapes elongate, not extending to eye; funicles about as long as scape, segments 1–2 subcylindrical, subequal, 3–7 broadly subobconical, progressively longer and broader towards club, 6–7 slightly narrower than club; clubs with segment 1–2 broadly flatly obconical, 2 ca. half as long as 1, 3 ca. half as long as 2, much narrower, 4 ca. 2.5 × longer than 3, elongately narrowly flatly conical, broadly inserted into 3. **Mouthparts.** Not preserved. **Thorax.** Pronotum elongate, longer than wide; basal pronotal margin sinuate, with weak median process. Prosternum elongate, longer than hypomeron. **Elytra.** Elytra punctostriate; sparsely setose, punctures not well defined; bases and humeri strongly concave to receive base of pronotum; bases widely obtuse, with notch in middle to receive median process of base of pronotum; margins with broad shallow anterior notch, with marginal groove gradually thickening anteriad but lacking lateral carina; apices individually rounded. **Legs.** Tibiae apically lined with coarse fringing setae, tarsal articulation surfaces suboblique, with 2 short spurs, shorter on protibiae; meso- and metatibiae with in distal half with denser, longer setae on outer and inner side. Tarsi broad, setae distinctly elongate, especially on protarsi; tarsite 1 elongate-triangular, acutely lobate, 2 broader, strongly lobate; 3 distinctly pedunculate, lobes 0.67 × as long as 5, 5 elongate, slender, slightly curved. **Abdomen.** Ventrites sparsely setose, 1 and 2 subequal in length, 3 and 4 successively slightly shorter, 5 about as long as 3 + 4, narrowly rounded.

Material examined. Holotype (NIGP154212): poorly preserved, partly decomposed specimen, with most of head and thorax heavily distorted, appendages somewhat compressed, in places fragmented, head partly retracted into prothorax well visible, vestiture possibly partly abraded, apical part of rostrum and of right elytron cut away, appendages intact, with large crystalline mass erupting from elytral suture in about middle, causing local distortion of cuticle, hindwings partly extended; in centre of irregular block with three flat sides and one long curved side over dorsal side of specimen, 8.7 × 2.5–4.0 × 3.3 mm; amber clear yellow with flowbands of greenish hue (under microscope light) and minimal impurities, with few discoidal fractures below head and prothorax.

Derivation of name. The species is named for its long, narrow shape, from the Latin adjective *longus*.

Remarks. Despite the substantial distortion of this specimen, particularly of the head and ventral side, sufficient detail can be seen in it to conclude that it belongs in *Burmocorynus*. Of key importance is the visibility of the procoxae, which are broadly separated by adjoining prosternal and hypomeral processes. The species agrees with *B. jarzembowskii* in its size and very elongate body, differing mainly by the characters given in the key. The surface of the body seems to have had much of the setae abraded, but in places pale setae are visible, and this species may have had a vestiture similar to that of *B. jarzembowskii*.

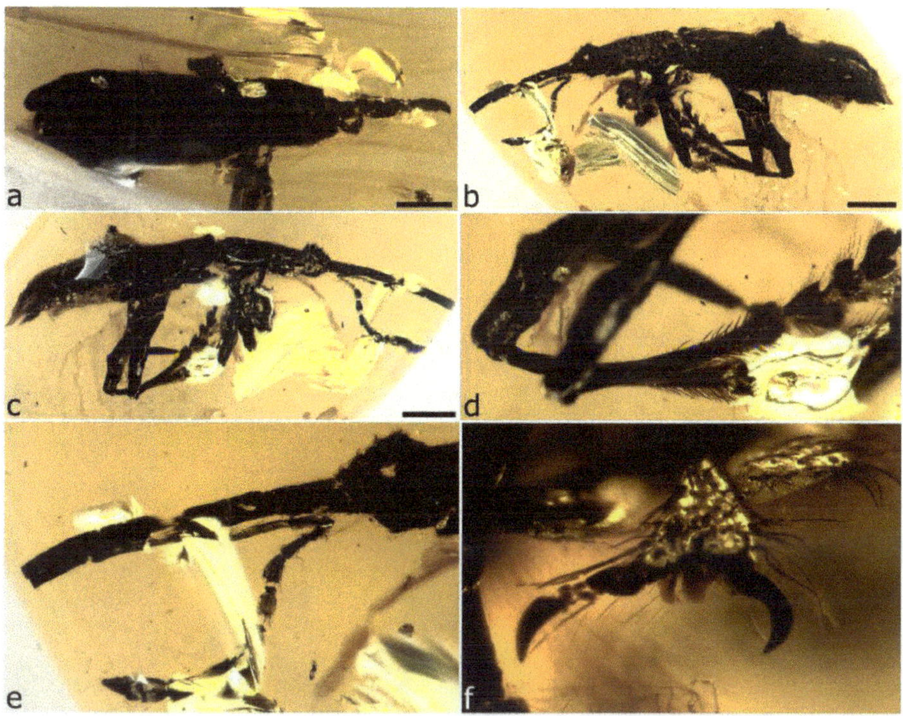

Figure 34. *Burmocorynus longus* sp. n., holotype. Habitus, dorsal (**a**); habitus, left lateral (**b**); habitus, right lateral (**c**); right metatibia and tarsites 1–3, showing pair of short spurs between coarse short rows of inner and outer apical fringing setae (**d**); head and left antenna, left lateral (**e**); left protibial claws, showing lobate ventral apex of claw segment (artefact) (**f**). Scale bars: 1 mm (**a**–**c**).

Genus ***Petalotarsus*** Clarke & Oberprieler, **gen. n.**

Type species: *Petalotarsus oxycorynoides* Clarke & Oberprieler, sp. n.

Description. Size. Length 4.1–7.25 mm, width 1.7–2.2 mm. **Head** short, not or weakly constricted behind eyes. **Eyes** protruding, coarsely facetted, dorsally separated by about width of rostrum at base, without tubercles between them, forehead flat to slightly impressed. **Rostrum** as long as or slightly longer than pronotum, stout, cylindrical; antennal insertions lateral, in or near middle of rostral length, behind them with scrobes extending to eye, in front of them with or without lateral row of erect setae. Long single gular suture indicated. **Antennae** geniculate, robust; scapes elongate, cylindrical, distally inflated; funicles about as long as scape, 7-segmented, segment 1 elongate, thick, 2 about as long as 1, rest progressively shorter and thicker; clubs 4-segmented, subcompact but segments still distinct, progressively wider towards club, apical segment acute. **Mouthparts**. Labrum absent. Mandibles small, seemingly exodont with 1 blunt outer tooth, articulation horizontal. **Thorax**. Prothorax slightly to distinctly proclinate, with anterior lateral margins oblique in lateral view. Pronotum elongate, narrowly trapezoidal to rectangular, laterally rounded in dorsal view, usually deplanate but not carinate, without lateral tooth, posterior corners extended, fitting closely onto elytra; surface setose, setae short, acute, directed mesad to anteromesad; notosternal sutures closed. Prosternum moderately long; procoxal cavities separated, in middle of prothorax. Scutellar shield small. Mesocoxal cavities laterally closed (by meso- and metaventrite). Metanepisternal sutures distinct. Mesoventrite short, anteriorly sloping. Metaventrite longer, flat, not raised into weals before metacoxae. **Elytra** elongate, bases extended anteriad to fit over base of pronotum, with broadly rounded humeri, posteriorly gently declivous, apically individually rounded, not exposing pygidium; sutural flanges narrow, equal; surface punctostriate, without scutellary striole, sparsely setose, setae sometimes concentrated into longitudinal bands, directed caudad. **Legs**. Procoxae subglobular, not prominent, medially separated; mesocoxae subglobular, separated; metacoxae flat, transversely elongate. Trochanters short, oblique. Femora straight, subcylindrical, inflated in distal half, unarmed, outside rounded. Tibiae straight, subcylindrical, outer edge rounded to subcostate, distally strongly setose, apex truncate, with 1 or 2 spurs. Tarsi large and broad, almost as long as tibiae, strongly setose; tarsite 1 shortly triangular, apically excised, 2 shorter, apically deeply excised to bilobed, 3 deeply bilobed, lobes broad, subpedunculate, 5 about as long as 1 + 2, apically broad; claws divaricate, basally swollen or angulate, with ventrobasal seta. **Abdomen** with ventrites 1 and 2 longer than each of 3 and 4.

Derivation of name. The genus is named for its broad flat tarsi, the name derived from the Greek adjective *petalos* (broad, flat) and noun *tarsos* (G: *tarsou*), the flat of the foot, and being masculine in gender.

Remarks. This genus is remarkable among the Burmese amber weevil fauna in that its species have broadly separated procoxae, broad and flat tarsi and subcompact antennal clubs, and most also having a flattened body. *Petalotarsus* differs from the similar genus *Burmocorynus* mainly by the flatter body and the antennal structure (rounded segments, less robust funicles, clubs more distinctly differentiated from funicles) but also in the proclinate prothorax. Three of the five known specimens are distinguishable from *Burmocorynus* also by the distinctive pale setal stripes on the elytra, although this feature is evidently not always as distinct as in *P. oxycorynoides* and may be an unrecognised feature of the other species as well (their elytra not properly visible). Most of the included species, especially *P. oxycorynoides*, possess features related to the flattened form, including the elongate prothorax, broadly separated procoxae and mesocoxae, horizontal and flatly aligned mesoventrite, metaventrite and hypomera, flattened apically expanded tibiae, broad flat and short tarsi and long subflatly aligned ventrites. Like the other genera with large species, most of the available material of *Petalotarsus* is poorly preserved or visible, and better material of most included species is needed before a detailed comparison with *Burmocorynus* is possible. Currently none of the available specimens allow a clear view of the mandibles (or other mouthparts), and those of *Burmocorynus* have been cut away with the amber in all known specimens. Only the mandibles of the largest known *Petalotarsus* species are partly visible, and their short, seemingly triangular and non-exodont form (Figure 4d) indicates a possible difference between these genera and the more typical exodont form of the mandibles as known from other genera of this group (*Opeatorhynchus*, *Cetionyx*). Whereas the subcompact clubs may suggest an affinity of *Petalotarsus* with the family Curculionidae, the lack of any scales and the broadly divaricate, basally swollen or angulate tarsal claws with a long ventrobasal seta (as typical for Mesophyletidae) indicate that the genus belongs in this family as well, just representing seemingly specialised forms adapted to living in narrow spaces (as extant Curculionidae of a similar body shape do). *Petalotarsus* is currently represented by four species, three of which are described here.

Petalotarsus oxycorynoides Clarke & Oberprieler, **sp. n.** (Figures 35–37, Video S3)

Description. Size. Length 7.01 mm, width 1.72 mm. **Head** subporrect, dorsally coarsely punctate in front of eyes; setose, setae directed anteriad; ventrally bulging behind eyes. **Eyes** large, prominent, slightly elongate, dorsolateral. **Rostrum** slightly shorter than pronotum; coarsely punctate basally; punctures well defined; antennal insertions median. **Antennae.** Scapes extending to front margin of eye, apically truncate; funicles with segment 2 nearly 2.0 × longer than 3, 3 globular, ca. half as long as 2, 4–6 longer, subequal in length, 7 slightly longer and wider than 6; clubs with segments 1–4 gradually shortening apically, apical one obconical, rounded at apex. **Mouthparts.** Mandibles visible but blurred. **Thorax.** Prothorax slightly proclinate, tightly fitting onto both mesoventrite and elytral humeri, posterolateral margin narrowly emarginate to receive lobe of mesothorax (see below). Pronotum laterally rounded, narrowing anteriad and basad, anterior margin broadly curved, basal margin subangulate at middle, tightly fitting to elytral bases; disc raised into central platform; surface punctosetose, punctures round, distinct, setae yellowish-white; notosternal sutures obliquely curved anteriad. Prosternum elongate, flattened, slightly longer than hypomeron, behind anterior margin with distinct slender groove tracking margin, medially forming broad V, prosternal process very broad, ca. half as wide as procoxa, with semi-longitudinal rugose sculpturing, similar to ventral head sculpturing but finer; hypomeral process with median suture, this and prosternal process separated by sternellum-like structure comprising pair of indistinct sclerites, posterior margin broadly arcuate. Mesoventrite flat, at same level as metaventrite, elongate in front of mesocoxae; lateral part and inner anterior part of mesanepisterna forming anterolateral lobe fitting into prothorax. Mesanepisterna with row of irregular punctures tracking suture between mesoventrite and mesanepisterna. Metaventrite disc distinctly punctate, punctures deep; sparsely setose, setae short. **Elytra.** Disc with broad stripe of dense, subappressed, yellowish-white setae adjacent to suture, gradually fading posteriad, with another stripe just laterally of middle of disc; intervening setae brownish, sparser; intervening surface finely sculptured, impunctate; surface along suture subsmooth, laterally of outermost setal stripe with at least 4 visible striae, with tiny punctures; bases obtusely angulate, extended to fit over pronotum; humeri submarginally concave to receive pronotal corners; sides moderately densely setose, setae directed mesad, with marginal groove narrowing posteriad, dorsal edge forming distinct lateral carina, obsolete before humerus, with distinct anterior marginal notch. **Legs.** Coxae slightly longer than wide, globular, widely separated. Femora narrow basally, gradually inflated in distal half, distally narrowed ventrally, inside concave to receive tibiae in repose; sparsely setose. Tibiae flattened, dorso-apically with several elongate setae; protibiae apparently flattened (distorted?), thicker apically, sparsely setose, apically with large inner flange, fringing setae similar to others (not coarse); with at least 1 spur; meso- and metatibiae densely setose in distal half, apically with coarse fringing setae, shorter on outer apical edge; without (meso-) or with 1 spur (metatibiae). Tarsi with long setae apicolaterally, ventrally with dense pulvilli of whitish setae on tarsites 1–3; tarsite 1 narrower than 2, 5 shorter than median lengths of 1–3. **Abdomen.** Tergites weakly sclerotised. Ventrites densely setose, 1 and 2 at approximately same level, suture between them weakly angulate medially, fainter than others, 3–5 more distinctly stepped, sutures between them broadly arcuate, 1 slightly shorter than 2, densely setose medially, sparser laterally; 2–4 progressively shorter; 5 about twice length of 4, rounded posteriorly.

Material examined. Holotype (NIGP154213): poorly preserved, strongly distorted specimen, especially around head and prothorax, with elytra spread, apparently both hindwings extended (left seemingly separated), appendages and rostrum intact, distal half of rostrum and mandibles obscured by large bubble, other areas and some appendages pulled away from amber leaving an impression in surface; in drop-shaped slab, 17.1 × 9.2 × 4.2 mm; amber clear yellow, unfractured, with few large debris masses near specimen.

Derivation of name. The species is named for its similarity to the belid genus *Oxycorynus*, which is also flattened with a laterally deplanate prothorax, the name being an adjective composed from *Oxycorynus* and the suffix *-oides* (similar).

Remarks. This species is similar to *P. curculionoides* in having interstriae 2 and 8 covered with dense yellowish setae, but they are more distinctive in this species and it is also larger and has the disc of the pronotum elevated and the sides more strongly deplanate. Other seemingly unique features as observed include the distinct row of short pale setae on the basal underside of mesotarsites 5 and the medial angulate emargination of the suture between ventrites 1 and 2. The CT scans (Figure 37c; Video S3) confirm the presence of a single long gular suture, but this is not so distinctly evident under a light microscope.

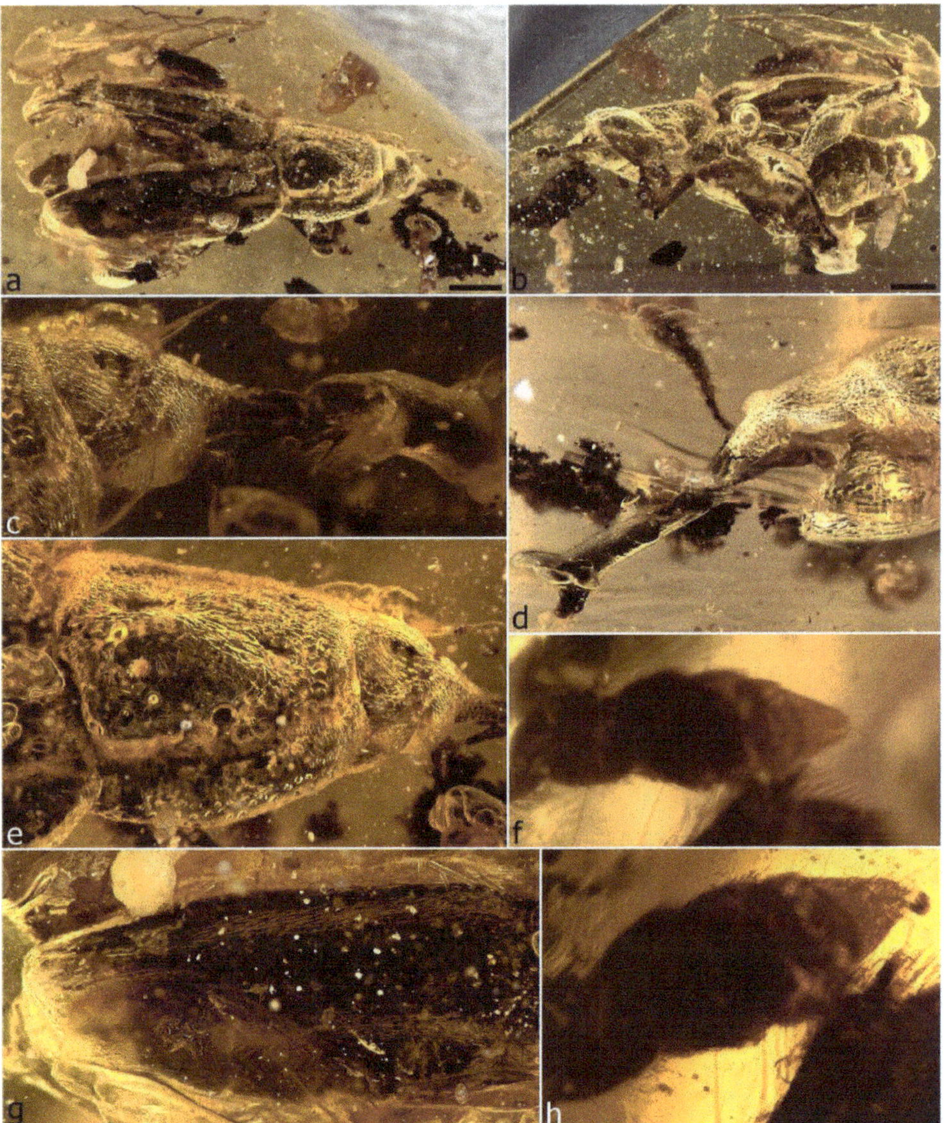

Figure 35. *Petalotarsus oxycorynoides* sp. n., holotype. Habitus, dorsal (**a**); habitus, left ventrolateral (**b**); head, right dorsolateral (**c**); head, left dorsolateral (**d**); prothorax, dorsal (**e**); antennal club (**f**); right elytron showing setal stripes, dorsal (**g**); antennal club enlarged (**h**). Scale bars: 1.0 mm (a,b).

Figure 36. *Petalotarsus oxycorynoides* sp. n., holotype. Prothorax, ventrolateral (**a**); thorax, ventrolateral (**b**); edge of elytra, left hindleg and probable wing (**c**); left foreleg (**d**); right protarsus, dorsal (**e**); protibia (**f**); metatarsus, dorsal (**g**); right meso- and metatarsal claws (**h**,**i**).

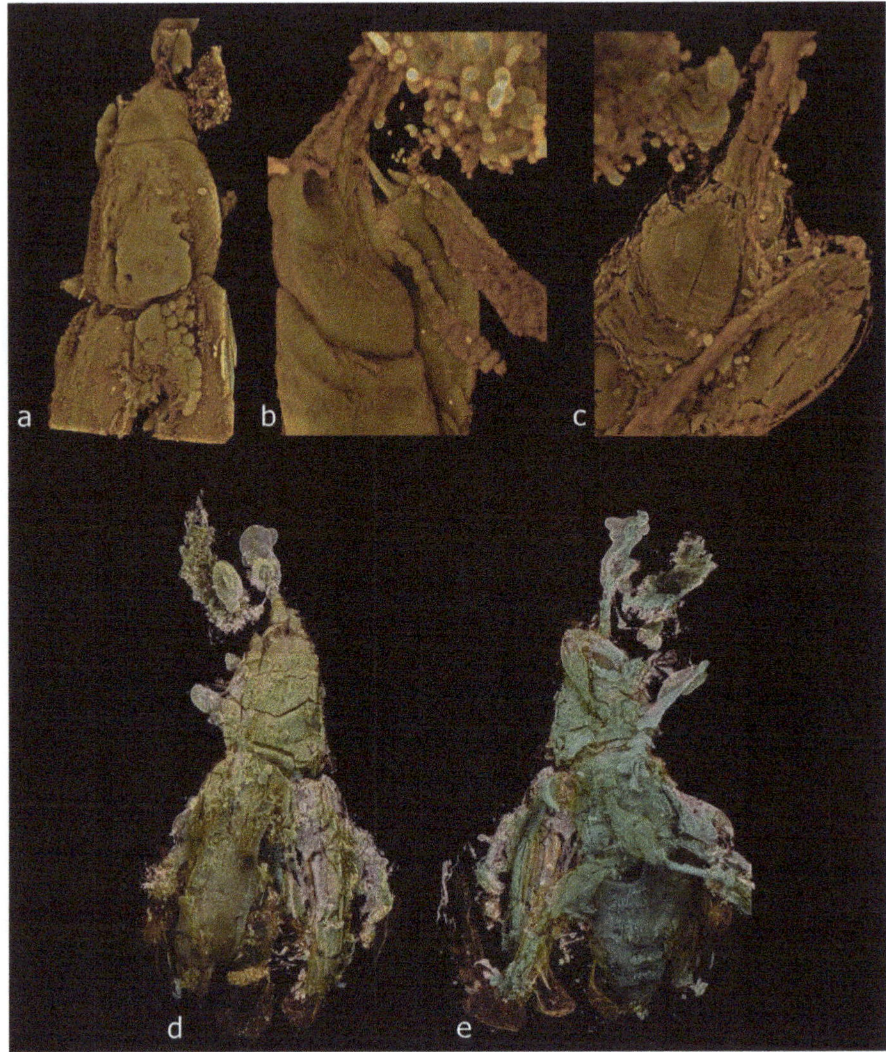

Figure 37. *Petalotarsus oxycorynoides* sp. n., holotype. Habitus and detail images extracted from micro-CT scanning reconstruction (see also Video S3). Head and pronotum, dorsal (**a**); same, lateral, showing antenna (**b**); same, ventral (**c**); habitus, dorsal (**d**); habitus, ventral (**e**).

Petalotarsus curculionoides Clarke & Oberprieler, **sp. n.** (Figures 38 and 39)

Description. Size. Length 4.1–4.4 mm, width 1.6–1.7 mm. **Head** transverse, weakly constricted behind eyes. **Eyes** large, elongate, subglobular, somewhat compressed, strongly protruding, forehead between them flat, triangular. **Rostrum** as long as pronotum, slightly curved; antennal insertions in middle of rostral length, in front of them with lateral row of 7–10 long, erect setae. **Antennae** moderately long; scapes thin, gradually inflated in distal third; funicles with segment 1 elongate, thick, 2 slightly shorter and thinner, 3–5 progressively shorter and thicker, 6 and 7 as thick as club; clubs with segment 4 long, thin, narrowly obconical, longer than 3. **Mouthparts**. Mandibles obscured by fine debris, dentition evidently weak; tips of labial palps visible. **Thorax**. Prothorax slightly proclinate.

Pronotum narrowly roundly trapezoidal, longer than broad at base, laterally explanate, slightly convex; surface punctorugose, sparsely setose, setae pale, directed mesad. Scutellar shield small, subquadatic. **Elytra** with 2 stripes of moderately dense setae along length of elytra, one on interstria 2, the other on interstria 8, other interstriae with few small setae directed caudad, fully developed. **Legs**. Procoxae subglobular, not prominent, medially separated by less than their width; mesocoxae separated by about their width. Femora long, strongly inflated in distal half, thickest in distal quarter. Tibiae shorter than femora, outer edge subcostate, distally strongly setose, apically flatly expanded, at apex with 2 small spurs (at least on metatibiae). Tarsi almost as long as tibiae, protarsi broader than meso- and metatarsi, strongly setose; tarsite 1 very shortly, broadly triangular, apically deeply excised, 2 shorter, deeply bilobed, 3 with lobes short and broad, subpedunculate, with dense pulvilli of silvery white setae, 5 short and broad; claws basally swollen with ventrobasal seta. **Abdomen** with ventrites 1 and 2 ca. 1.5 × longer than each of 3 and 4, 5 narrowly semicircular.

Material examined. Holotype (NIGP154214; Figure 38), length 4.10 mm, width 1.7 mm: well preserved, intact specimen, not compressed or distorted, mostly well visible except for left side of pronotum, with apex of left hindwing exposed; in centre of elongate rectangular cuboid 7.4 × ca. 4.8 × 3.7 mm, dorsal surface convex; amber dark but clear, on left side of pronotum with large and small clear bubble, on right side with small white bubble over metathorax, with large transparent brown ring beneath rostrum, also a transverse vertical crack on left side of rostrum, few larger impurities. Paratype (NIGP154215; Figure 39), length 4.38 mm, width 1.66 mm: poorly preserved, strongly distorted specimen covered by layer of bubbles and debris, with elytra separated from abdomen and from each other, exposing hindwings and membranous abdominal tergites, claw segment of right hindleg missing, left antenna and rostral apex not visible (obscured by bubbles); in rod-shaped block with three flat sides and one long curved side, tapered at one end, flat at other, 9.1 × 1.5–4.0 × 3.2 mm; amber clear yellow, with diffuse microscopic particles and minimal debris.

Derivation of name. The species is named for the similarity of its compact antennal clubs to those of the family Curculionidae, the name being an adjective composed of the stem of the name Curculionidae and the suffix *-oides* (similar).

Remarks. This species is similar to *P. oxycorynoides* in having interstriae 2 and 8 conspicuously densely setose (though less distinct in the paratype), but it is smaller and less flattened and has an evenly, slightly convex pronotum (the disc not elevated) as well as somewhat narrower tarsi and spurs on the meso- and metatibiae. The two specimens available for study differ slightly from each other but are considered to be conspecific. The main differences are (states in paratype): shorter and basally broader elytra with straight sides, only curved posteriorly (longer and narrower elytra with sides more evenly curved from humerus to apex), distinct setal stripes (faint, indistinct, outer one much broader), distinct, coarse, subevenly aligned fringing setae, at least on the outer flanges of the tibial apices (only weak apical fringing setae) and broad tarsi (narrow, longer).

Petalotarsus cylindricus Clarke & Oberprieler, **sp. n.** (Figure 40)

Description. Size. Length 5.3 mm, width 1.9 mm. **Head** subconical, not constricted behind eyes. **Eyes** medium-sized, round, slightly protruding, forehead between them slightly impressed. **Rostrum** 1.5 × longer than pronotum, only very slightly curved; antennal insertions behind middle of rostral length, almost at basal third, in front of them seemingly without lateral row of setae. **Antennae** relatively short; scapes straight, apically clavate; funicles about as long as scape, segment 1 slightly inflated, 2 as long as 1 but thinner, 3–5 similar, 6–7 shorter; clubs short, with apical segment small, acute, about as long as 3. **Mouthparts**. Mandibles with 1 blunt tooth on outside. **Thorax**. Prothorax proclinate. Pronotum roundly rectangular, longer than broad at base, slightly convex, laterally not explanate; surface densely rugose, sparsely setose, setae very fine, thin, recumbent, directed anteromesad. Prosternum moderately long; procoxal cavities medially narrowly separated or confluent (not clearly discernible). Scutellar shield subquadratic, slightly convex. Metaventrite weakly tumescent before metacoxae. **Elytra** basally rounded (shallowly lobed); surface very sparsely setose, setae not clearly visible. **Legs**. Procoxae slightly elongate, prominent, medially narrowly separated;

mesocoxae narrowly separated. Femora long, strongly inflated in distal half, thickest in distal quarter. Tibiae shorter than femora, outer edge rounded, at apex with 1 spur visible on mesotibiae (but possible 2 present), 2 spurs on metatibiae. Tarsi about 0.67 × as long as tibiae; tarsite 1 apically slightly excised, 2 as long as 1, apically strongly excised, 3 with lobes short, 5 slightly shorter than 1 + 2, broadening apicad; claws basally angulate, with ventrobasal seta at apex of angulation. **Abdomen** with ventrites 1 and 2 fused, longer than each of 3 and 4, 5 not visible.

Figure 38. *Petalotarsus curculionoides* sp. n., holotype. Habitus, right lateral (**a**); habitus, left lateral (**b**); habitus, dorsal (**c**); elytra, dorsal (**d**); head, dorsal (**e**); antennae, ventral (**f**); right antenna (**g**); right antennal club detail (**h**); protarsus (**i**). Scale bars: 1.0 mm (a–d); 0.5 mm (i); 0.1 mm (h).

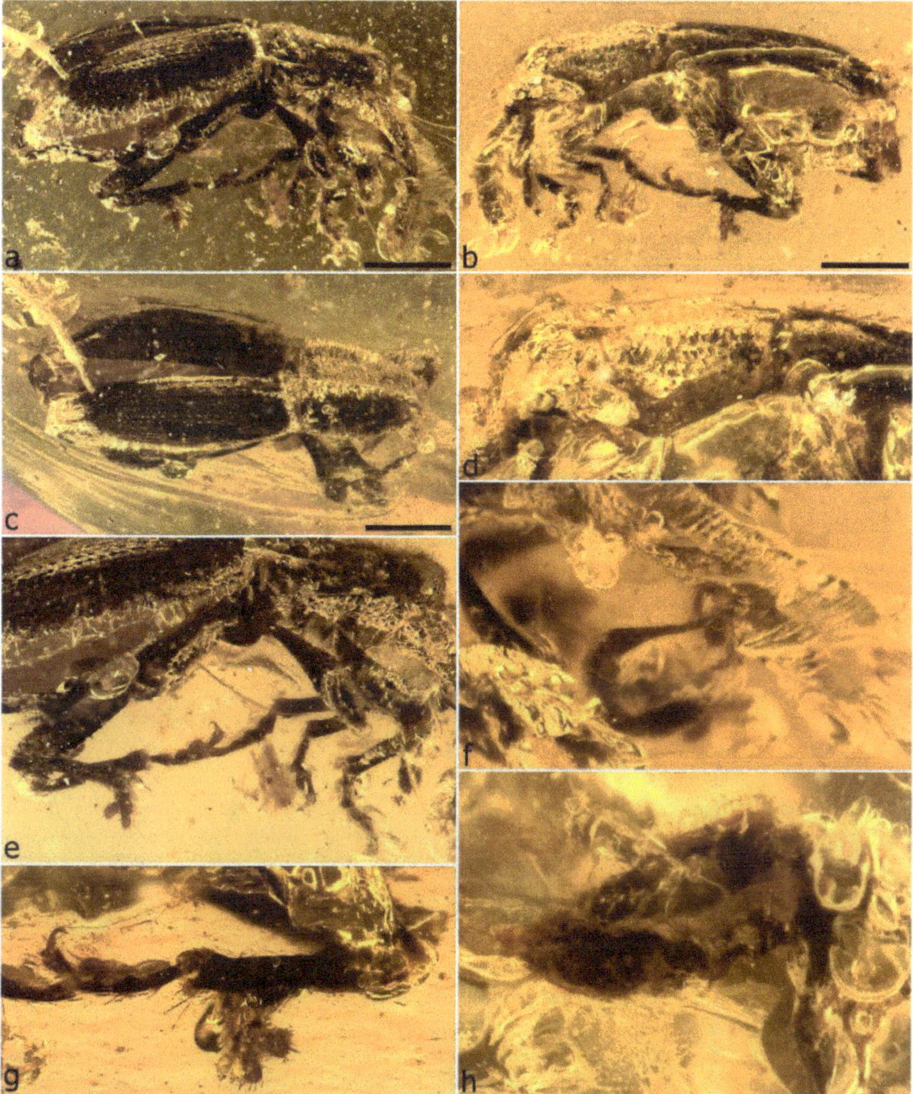

Figure 39. *Petalotarsus curculionoides* sp. n., paratype. Habitus, right lateral (**a**); habitus, left lateral (**b**); habitus, dorsal (**c**); head, prothorax and elytral humerus, left lateral (**d**); elytra and legs, right lateral (**e**); right antenna (**f**); left metatibia and -tarsus (**g**); antennal club (**h**). Scale bars: 1.0 mm (a–c).

Material examined. Holotype (NIGP154216): reasonably well preserved, intact specimen, not compressed or distorted, not well visible; diagonally in centre of thin rectangular cuboid 7.5 × 7.2 × 3.5 mm; amber dark, clear on left side of specimen but cloudy on right side, with large warped film posteriorly on ventral side, wrapping over middle and hindlegs, and with large crack along middle of elytra and right around specimen, with many small impurities especially on right side of specimen.

Derivation of name. The species is named for its cylindrical body shape, the name being a Latin adjective.

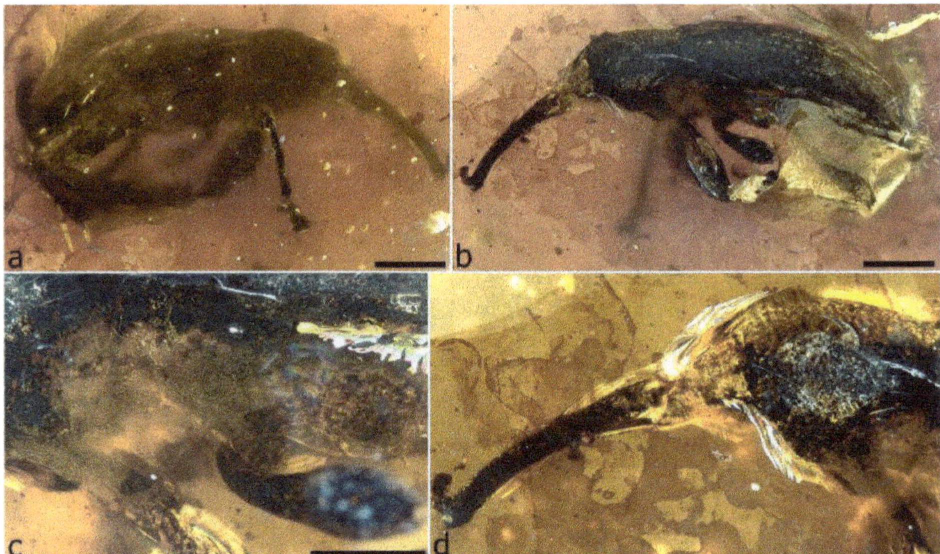

Figure 40. *Petalotarsus cylindricus* sp. n., holotype. Habitus, right lateral (**a**); left lateral (**b**); pro- and mesothorax, left lateral (**c**); head and prothorax, left lateral (**d**). Scale bars: 1.0 mm (a,b); 0.5 mm (c).

Remarks. *Petalotarsus cylindricus* is atypical of the genus in that its body is not flattened and the tarsi are not as conspicuously flat and broad, and the procoxae are also not distinctly separated (possibly even contiguous), but the antennal clubs are evidently subcompact and no other significant differences from the other species of *Petalotarsus* are apparent. However, the single specimen is not well preserved and visible, and if CT scanning of it is attempted and successful and further differences become evident, or if further specimens are found, its status in *Petalotarsus* may need to be reassessed.

Petalotarsus sp. (Figure 41)

Material examined. One specimen (in private collection of Mr. Wei Ma, China), 7.25 mm long, 2.2 mm wide: very well preserved, intact, not compressed or distorted, reasonably well visible from left and dorsal sides, underside completely obscured by layer of white and black impurities and right side obscured by large dense cylinder cut along length; placed at angle in left side of lenticular block ca. 14.2 × 11.0 × 5.8 mm; amber clear but much debris on right and underside of specimen, much of which can be removed by cutting away the long halfpipe (already half trimmed away) above specimen.

Remarks. This large specimen is quite well visible in dorsal view, but its other sides are obscured by dense debris and the curvature of the amber piece (a cabochon). For proper study and description of the specimen, the cabochon needs to be trimmed into a cuboid, which would also remove a lot (though not all) of the dense matter obscuring the right side of the specimen. We do not describe this species as the specimen is housed in a private collection, in which accessibility of a holotype to science cannot be guaranteed. We provisionally treat it as belonging in *Petalotarsus*, but in some characters it agrees better with *Burmocorynus*, having flattened funicles and clubs, the funicle segments widening towards the club and similar protarsi but with shorter setae. Its procoxal cavities are distinctly separated, as in *Burmocorynus* and *Petalotarsus*.

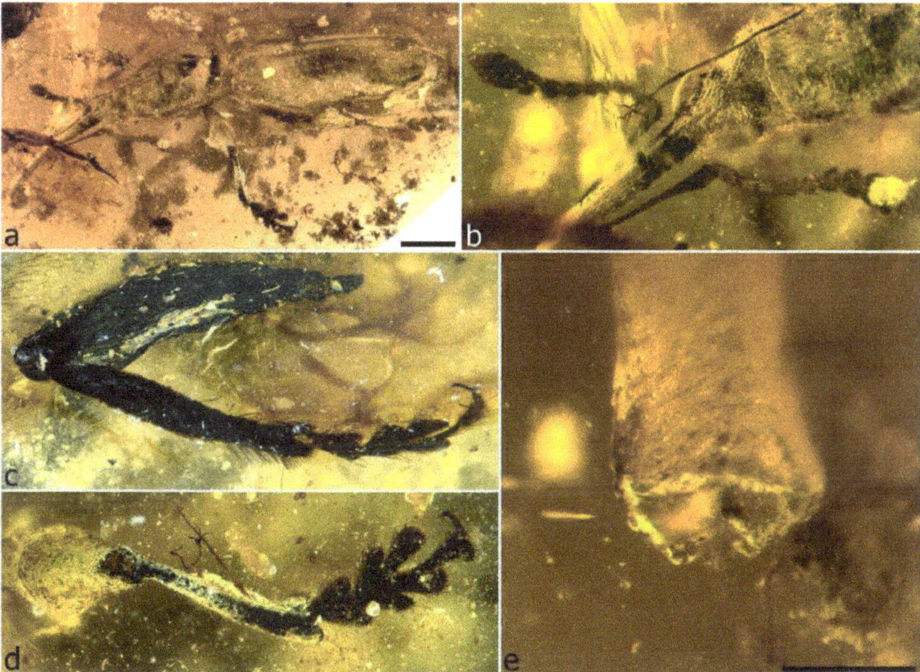

Figure 41. *Petalotarsus sp.* Habitus, left lateral (**a**); head and antennae, dorsal (**b**); right hindleg (**c**); left foreleg, dorsal (**d**); apex of rostrum and mandibles, dorsal (**e**). Scale bars: 1.0 mm (a); 0.2 mm (e).

Genus ***Opeatorhynchus*** Clarke & Oberprieler, **gen. n.**

Type species: *Opeatorhynchus comans* Clarke & Oberprieler, sp. n.

Description. Size. Length 5.68 mm, width 2.06 mm. **Head** short, porrect, not constricted behind eyes, dorsally continuous with rostrum, ventrally bulging. **Eyes** elongate, flatly protruding, coarsely facetted, dorsally separated by about width of rostrum at base, further separated posteriorly, without tubercles between them, forehead flat to slightly convex. **Rostrum** relatively short, slightly longer than pronotum, subcylindrical; antennal insertions lateral, in middle of rostral length, behind them with scrobes reaching eye, in front of them without lateral row of erect setae. Long single gular suture present. **Antennae** geniculate, long, slender; scapes elongate, reaching anterior margin of eye in repose, cylindrical, distally inflated; funicles about as long as scape, 7-segmented, segments progressively thicker towards club, segment 1 elongate, thick, 1–4 subequal in length, 5 shorter than 4, 6 and 7 subequal, slightly longer than 5; clubs 4-segmented, loose, thicker than funicle, segment 4 large, broadly inserted into 3. **Mouthparts**. Labrum absent. Mandibles small, exodont with 2 large teeth on outer side, articulation oblique. Maxillary palps robust, slightly projecting from apex of rostrum, 3-segmented. **Thorax**. Prothorax slightly proclinate, with anterior lateral margins oblique in lateral view. Pronotum elongate, subrectangular, convex, laterally weakly rounded in dorsal view, without lateral tooth, posterior corners rounded, fitting closely into humeri; surface densely setose, partly obscuring integument, setae short, acute, directed mesad to anteromesad; notosternal sutures closed. Prosternum moderately long; procoxal cavities confluent, in middle of prothorax. Scutellar shield not visible. Mesocoxal cavities laterally closed (by meso- and metaventrite). Metanepisterna distinct, setose at least anteriorly. Mesoventrite short, anteriorly sloping. Metaventrite behind mesocoxae short, about half as long as mesocoxa, raised into slight weals before matacoxae. **Elytra** elongate, base extended to fit over pronotum, with narrow acute humeri, closely fitting to and not much broader than pronotal corners, posteriorly gently declivous, apically slightly individually rounded, not exposing pygidium; sutural flanges narrow, equal, broader apically; surface punctostriate, without scutellary striole, densely setose, partly obscuring integument, setae short, thin, sharp, reclined, directed caudad. **Legs**. Procoxae large, subglobular, prominent, contiguous; mesocoxae subglobular, separated; metacoxae flat, shortly transversely elongate. Trochanters short, oblique. Femora subcylindrical, gradually inflated in distal half, unarmed, outside rounded. Tibiae straight, subcylindrical, outer edge rounded, densely setose in distal half, apex obliquely truncate, with 2 spurs. Tarsi large and broad, almost as long as tibiae, strongly setose; tarsite 1 shortly triangular, apically excised, 2 shorter, narrowly bilobed, 3 deeply bilobed, lobes strongly pedunculate, attachment with basal plate extremely slender in both dimensions, 5 much longer than 1 + 2, apically broad; claws divaricate, basally angulate, with ventrobasal seta. **Abdomen** with ventrites 1 and 2 longer than each of 3 and 4.

Derivation of name. The genus is named for its awl-shaped body and rostrum in lateral view, the name formed from the Greek nouns *opeas* (G: *opeatos*), an awl, and *rhynchos* (G: *rhyncheos*), a snout, and being masculine in gender.

Remarks. This genus is distinguishable from other genera with broad tarsi and pedunculate lobes of tarsites 3 by the combination of the antennal structure, including the broad, very loosely articulated clubs, and the medially confluent procoxal cavities. It is most similar to *Cetionyx*, differing from it mainly in these characters. It is one of only three specimens in the group of genera including *Burmocorynus*, *Petalotarsus*, *Cetionyx* and *Echogomphus* in which the mandibles are relatively distinct, being exodont with two teeth on the outer edge.

Opeatorhynchus comans Clarke & Oberprieler, **sp. n.** (Figures 42 and 43, Video S4)

Description. Size. Length 5.68 mm, width 2.06 mm. Large, robust, densely setose; especially on ventrites. **Head** coarsely punctate, punctures small; setose. **Eyes** with deep narrow groove tracking posterior margin (possibly an artefact). **Rostrum** slightly downcurved, setose from base to antennal insertions, apicodorsally with 2 elongate setae. Mandibular articulations deep, about as long as mandible length. **Antennae.** Scapes reaching slightly below front margin of eye, apically truncate; funicles with segments 1–7 subequal, 2–4 slightly expanded apically, 6–7 slightly thicker; clubs with segments subequal in width, 1 slightly longer than 2, 2–3 subequal in length, 4 slightly shorter than and broadly inserted onto 3, paler than rest of antenna, apically narrowly rounded. **Mouthparts.** Mandibles on inner and outer edges with 2 teeth, basal ones much larger than apical ones, these forming apical T. Maxillary palps with basal segment short, broad, apical segment elongate, tapering. **Thorax.** Pronotum widest just before middle, densely, coarsely punctate, punctures larger than on head; setose, setae mixed dark and paler brownish, on sides distinctly patterned with browner setae mixed with seemingly golden setae. Pronotum strongly convex, sides rounded; posterior angles rounded; base tightly fitting to elytra. Prosternum elongate, about as long as procoxae, anterior margin substraight in middle, prosternal process short, acute; hypomeron well-developed, about as long as prosternum. Mesothorax distinctly, coarsely punctate. Metaventrite coarsely punctate, densely setose, especially posterolaterally; punctures large, dense, distinct. **Elytra** with striae much narrower than interstriae, punctures distinct, shallow; densely setose, denser just before humeri, setae laterally forming paler thick stripe; interstriae broad, flat; bases subsinuate, forming obtuse angle, carinate to junction with mesepimeron, humeri distinctly cupulate, receiving posterior pronotal corners; lateral marginal groove very narrow in apical half, abruptly broadened anteriorly to humerus, dorsally forming epipleural carina, area inside groove densely setose; lateral margin with anterior notch. **Legs** long, slender, densely setose. Procoxae large, projecting, subconical, elongate, contiguous; mesocoxae narrowly separated, large, globular, prominent; metacoxae very large, prominent, subglobular (not distinctly transverse), longer than ventrite 1. Tibiae straight, apically expanded; tarsal articulation surfaces concave, with pronounced apical flanges, inner and outer edges lined with coarse fringing setae, dorso-apical edge well emarginate to allow basal tarsites to bend up, with 5–6 elongate slender setae; protibiae with outer side slightly elongately concave, lined with dense setae, with 2 short thin indistinct apical spurs; meso- and metatibiae very densely setose in apical half, on outer side situated in broad emargination, this more distinct on metatibiae, with 2 larger spurs. Tarsi elongate, flattened, (protarsus) at least 0.67 × as long as protibia; tarsite 1 triangular, apically excised, of protarsi longer than of mesotarsi, 2 slightly narrower than 1, shortly bilobed, lobes narrow, lobes of 3 ca. 0.67 × as long as 5, 4 distinct, recessed into basal plate of 3, 5 long, exceeding 3 by third its length, very slender and strongly widening apicad, curved dorsoventrally. **Abdomen.** Tergites (III–VI) weakly sclerotised, seemingly glabrous; VII large, VIII small, both strongly sclerotised and densely setose, setae apically on each longer than rest; VII basally on either side of middle with orange, distinctly transversely irregularly strigose wing-binding patch, also with parasclerite. Ventrites slightly stepped (not at same level), 3–5 free, at progressively slightly higher levels, densely setose, setae fine, long and erect; punctate/rugosely sculptured; sutures straight; ventrite 1 slightly shorter than 2, 3 ca. 0.67 × as long as 2, 4 very slightly shorter than 3, 5 shorter than 4, broadly shallowly impressed, with transverse line of long dark setae at apex, appearing as posterolateral tufts on either side of impressed margin.

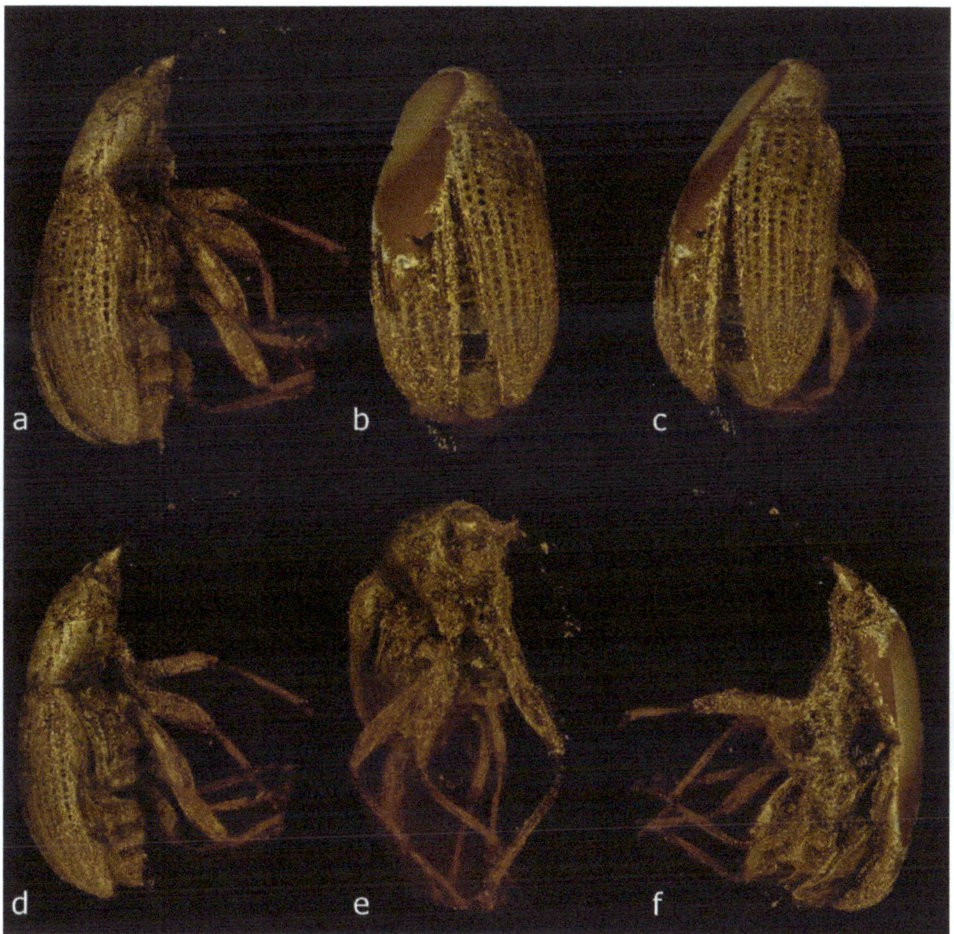

Figure 42. *Opeatorhynchus comans* sp. n., holotype. Habitus images extracted from a micro-CT scanning reconstruction (see also Video S4). Right lateral oblique (**a**); dorsal (**b**); right dorsolateral (**c**); right lateral (**d**); frontoventral (**e**); left lateral (**f**).

Figure 43. *Opeatorhynchus comans* sp. n., holotype. Habitus, lateral (**a**); head and prothorax, left lateral (**b**); mandibles, dorsal (**c**); habitus, dorsal (**d**); head and antenna, left lateral (**e**); protarsus, dorsal (**f**); elytra, dorsal (**g**); right antenna (**h**); left protibibia and -tarsus (**i**); legs, lateral (**j**); metatibia, showing dense setae in emargination of outer side (**k**); middle leg, dorsal (**l**). Scale bars: 1.0 mm (a,d).

Material examined. Holotype (NIGP154217) (probably a male): extremely well preserved and well visible specimen, largely intact but with left side of pronotum, anterior half of left elytron, scutellar shield and part of left protarsal claw partly cut away during preparation, subsequently infilled with resin (see Section 2.2), left side partly obscured by fracture planes and large flattish murky bubble; in irregular amber block 8.2 × 7.3 × 4.0 mm, with large flat and large curved face and five smaller flat faces; amber clear yellow with diffuse tiny particles; with several large fractures connected to specimen.

Derivation of name. The species is named for its hairy appearance, the name being the Latin participle *comans* (covered with hair).

Remarks. This species is distinctive among those of *Cetionyx* and *Burmocorynus*, also with broad tarsi and shaggy legs, in having the distal half of the outer sides of the meso- and metatibiae emarginate and lined with dense long thick setae and in also having a shorter rostrum. The specimen was submitted for CT scanning, but the resulting images and video are not too clear (Figure 42, Video S4).

Genus ***Echogomphus*** Clarke & Oberprieler, **gen. n.**

Type species: *Echogomphus viridescens* Clarke & Oberprieler, sp. n.

Description. Size. Length 5.5 mm, width 1.22 mm. Large, densely setose, with scattered iridescent green setae everywhere. **Head** elongate, subporrect, only slightly bulging ventrally, not constricted behind eyes. **Eyes** flattened, elongate, protruding, coarsely facetted, dorsally separated by approximately width of rostrum anteriorly but further separated posteriorly; forehead flat, without tubercles between eyes. **Rostrum** long, terete (apex cut off), weakly downcurved; antennal insertions lateral, inserted in about posterior third of rostrum; without scrobes behind them, in front of them laterally with a few long erect setae. Apparently a single long gular suture present (not clearly discernible). **Antennae** geniculate, very long and thin; scapes reaching anterior margin of eye in repose, cylindrical, apically only slightly inflated, shorter than funicles; funicles 7-segmented, segment 1 short, obconical, ca. 0.33 × as long as 2, 2 long, others progressively shorter towards club; clubs ill-defined, long, loosely 4-segmented, apical segment about as long as 3, acute, broadly inserted, others slightly expanding apicad. **Mouthparts** not visible. **Thorax**. Prothorax slightly proclinate, with anterior lateral margins oblique in lateral view. Pronotum slightly convex, laterally rounded, without tooth, posterior corners distinctly angulate, fitting closely onto elytra; surface finely punctate, densely setose, setae reclinate, directed caudad; notosternal sutures closed, vertical, abrupty deflected anteriad. Prosternum moderately long; procoxal cavities medially confluent, in middle of prothorax. Scutellar shield not visible. Mesocoxal cavities laterally not discernible. Metanepisterna distinct, densely setose. Mesoventrite short, anteriorly sloping. Metaventrite longer, shape not discernible. **Elytra** elongate, basally extended over pronotum; humeri concavely produced, posteriorly declivous, lateral margin sinuate to roundly emarginate in middle; apically individually rounded, exposing broad pygidium; surface punctostriate, without scutellary striole; interstriae broad, flat, rugose, setose, setae short, thin, reclinate, directed caudad. **Legs**. Procoxae rather short (not protruding), possibly slightly separated; meso- and metacoxae small, globular, separated. Trochanters short, oblique. Femora long, inflated at about middle, outside rounded, inside excavate in distal half, receiving tibiae in repose, walls of groove at apex roundly flatly extended. Tibiae substraight, compressed, distally slightly expanded, outer edge rounded, with dense long stiff setae in distal half, apex obliquely truncate; protibiae with 2 short spurs, meso- and metatibiae at outer edge emarginate in apical third with row of long setae, at inner apical angle with sharp inner spike and 1 outer spur at base of spike, mesotibial spike straight, smooth, metatibial spike flattened, curved at apex and with brush of long setae on underside. Tarsi with tarsite 1 elongate (2.0 × longer than 2), 1 and 2 excised, 3 deeply bilobed, lobes strongly pedunculate (all but 1 of 12 lobes broken off!), 5 long, apically expanded; claws divaricate, basally angulate with ventrobasal seta. **Abdomen** with ventrites broad, subequal in length except 4 shorter, 5 ca. 2.0 × longer than 4, broadly subtruncate at apex; broad pygidium exposed, without median groove.

Derivation of name. The name of the genus is composed of the Greek verb *echo* (to have) and noun *gomphos* (G: *gomphou*), a peg or nail, in reference to the stout spike at the apex of the tibiae; its gender is masculine.

Remarks. This is a seemingly isolated genus, unique in the group of Mesophyletidae with non-dentate tarsal claws in having a large exposed pygidium and further distinguishable from all other Burmese amber weevils by the combination of its size, iridescent green setae and long apical spike on the inner angle of the meso- and metatibiae. Whereas the smooth, straight mesotibial spike appears as if it may represent the inner, fixed and enlarged spur, the setiferous nature of the larger, curved metatibial spike suggests that it may be a secondary outgrowth of the tibial apex, equivalent to the mucro in some Curculionidae. A specimen of another, seemingly related species that we received too late for inclusion in this paper has a similar but much longer metatibial spike (ca. 0.25 × as long as metatibia). Similarly modified meso- and metatibial spurs occur in *Elwoodius*, but the metatibial one without setae. *Echogomphus* also has a unique tarsal structure, with elongate, flattened, apically emarginate tarsites 1 and 2 together with the strongly pedunculate but relatively short lobes of tarsites 3 and very long tarsites 5. It seems most closely related to *Opeatorhynchus*, with which it shares confluent procoxal cavities and similarly emarginate outer sides of the meso- and metatibiae lined with dense setae (a similar setation occurs in *Cetionyx* but the tibiae are not or only indistinctly emarginate). However, better-preserved additional specimens are required to assess its affinities in Mesophyletidae. *Echogomphus* contains a single known species, but a specimen of an undescribed species known to us may also belong in it as well; it lacks iridescent green setae but possesses an extraordinarily long apical spike perpendicular to the tibial axis and also with a setal brush on the ventral side, as occurs in *Echogomphus viridescens*.

Echogomphus viridescens Clarke & Oberprieler, **sp. n.** (Figures 44 and 45)

Description. Size. Length 5.5 mm, width 1.22 mm. Body blackish, concolorous, tarsi slightly paler, translucent. **Head** subporrect, elongate, not expanded dorsally, slightly bulging ventrally; not constricted behind eyes dorsally or laterally; distance between eyes at base of rostrum approximately one rostral width, expanding in width posteriorly. **Eyes** bulbous, flatly elongately protuberant; coarsely facetted. **Rostrum** at least as long as pronotum; antennal insertions in basal third. **Antennae.** Scapes elongate, apically slightly oblique; funicles with segments very sparsely setose, segment 1 flask-shaped, expanding distad, 2–7 elongate, progressively shorter; clubs densely, shortly setose, segments obconical, 1 and 2 subequal, apically more than 2.0 × wider than at base, 3 shorter than 2, 4 elongate, acutely pointed, much narrower than and distinctly articulating with 3. **Thorax**. Pronotum densely setose, setae directed caudad; seemingly with slight change in contour anterolaterally. Prosternum and hypomeron very short, seemingly much less than half procoxal length. **Elytra** densely setose, whitish and brown setae interspersed with iridescent green setae; striae linear, thin; interstriae broad, flat; laterally with clear marginal groove, gradually widening anteriad to form distinct lateral carina extending to humeri; anterior marginal notch present just before humeri. **Legs** densely setose, increasingly so distally. Tibiae apically with coarse fringing setae; protibiae with 2 short spurs and tiny burr on outer side of apex; meso- and metatibiae apically with shortly rounded outer flanges, metatibiae with setae in outer apical emargination denser and longer than on mesotibiae, tarsal articulation surfaces concavely truncate, inside with at least 1 (meta-) or 2 (mesotibiae) thicker setae, spike unarticulated, directed diagonally to tibial axis, on mesotibiae straight but bent at apex, on metatibiae subflattened and with dense long setal brush on underside, outer spur much smaller and directed distad. Tarsi elongate, flattened but not laterally expanded; tarsite 1 ventrally with dense, long setae, on protarsi longer and broader than on meso- and metatarsi, tarsite 2 ca. 0.33 × as long as 1, ventrally with dense, long setae, on protarsi slightly longer and broader than on meso- and metatarsi, more flattened, ventrally with dense pad-like setae, tarsite 3 with lobes shortly and abruptly pedunculate, slightly more than 2.0 × longer than 5, very narrow basally (threadlike peduncle) then abruptly elongately lobate, tarsite 5 subequal in length to 1 + 2 (slightly less in metatarsi), curved, not flattened, with longer setae dorso-apically, setose ventrally. **Abdomen**. Ventrites densely setose; 1 slightly shorter than 2, 2 slightly longer than 3, 4 shorter than 3, 5 nearly 2.0 × longer than 4, apically subtruncate to broadly rounded.

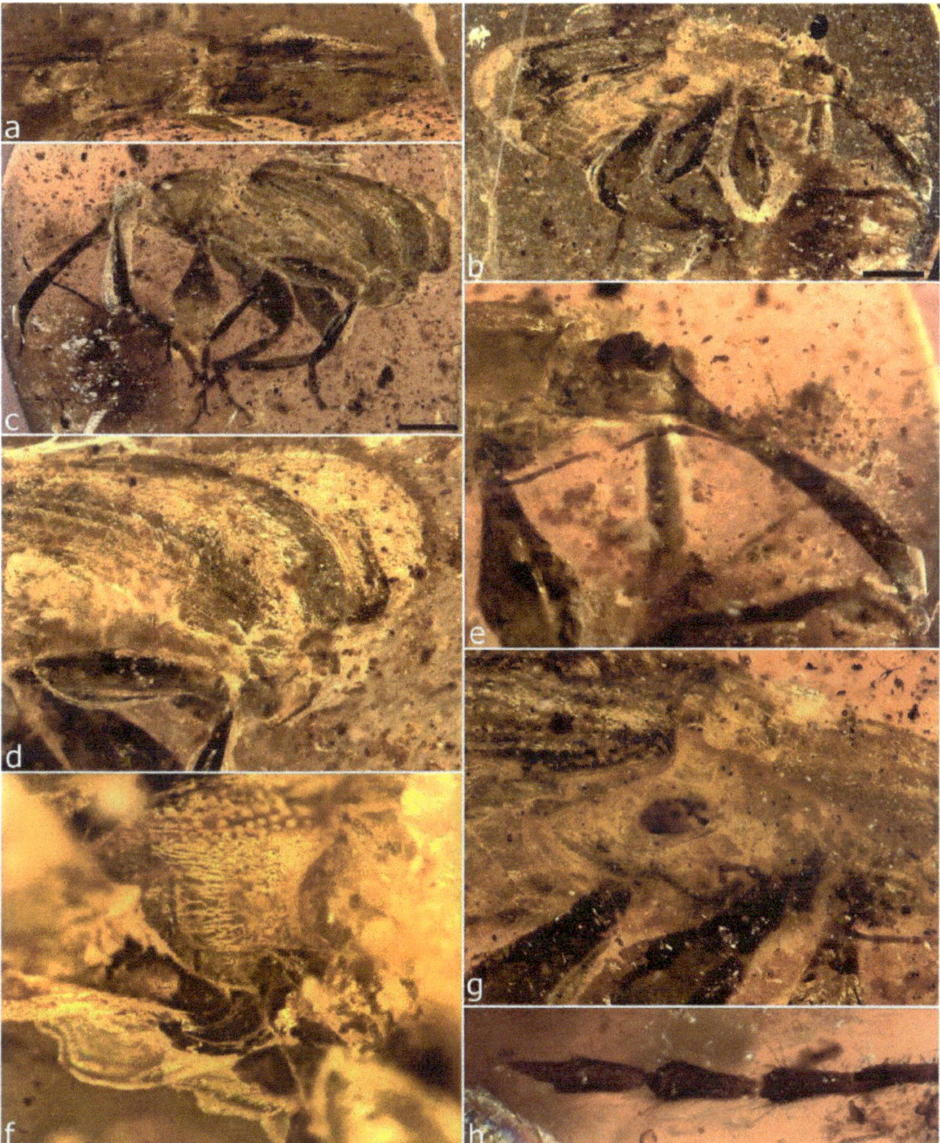

Figure 44. *Echogomphus viridescens* sp. n., holotype. Habitus, dorsal, showing effects of lateral distortion (**a**); habitus, right lateral (**b**); habitus, left lateral (**c**); apices of elytra and detail of pygidium, laterocaudad (**d**); head and antenna, right lateral (**e**); prothorax, showing notosternal suture (**f**); prothorax and elytral humerus, right lateral (**g**); antennal club (**h**). Scale bars: 1.0 mm (b,c).

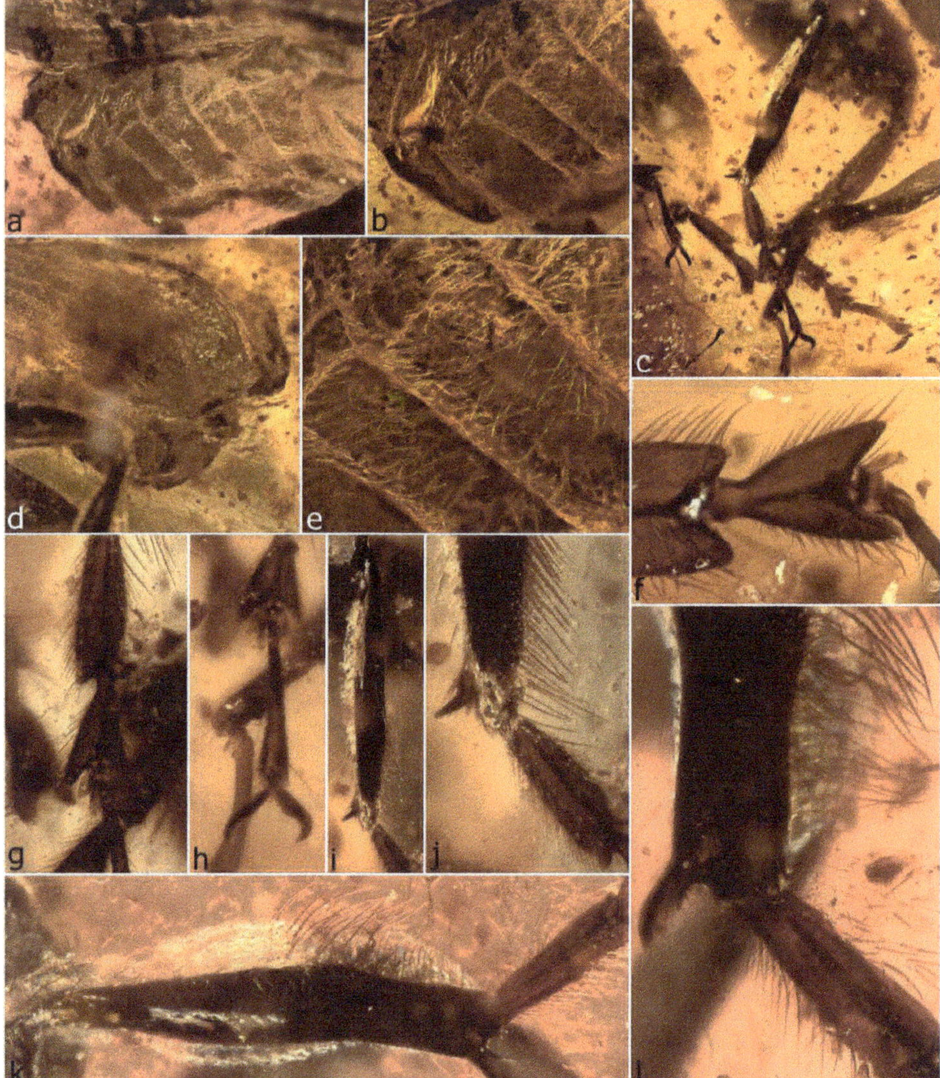

Figure 45. *Echogomphus viridescens* sp. n., holotype. Ventrites, ventrolateral (**a**); detail of ventrites, ventrolateral (**b**); legs (**c**); apex of elytra and pygidium, ventrolateral (**d**); detail of ventrites and iridescent setae (**e**); apex of tarsite 1, tarsites 2 and 3 and base of 5, showing broken peduncle of 3, dorsal (**f**); left mesotarsites 1 and 2, dorsal (**g**); left mesotarsal claws, dorsal (**h**); left mesotibia showing apical spur (**i**); same, apical part (**j**); left metatibia showing larger apical spike and smaller spur (**k**); detail of metatibial spike and spur (**l**).

Material examined. Holotype (NIGP154218): moderately well preserved but somewhat damaged and heavily distorted specimen, with body, rostrum and legs compressed and right elytron displaced, not well visible (width measurement probably not accurate), most of body covered with layer of air partly obscuring surface details and most ventral details of head and thorax invisible or not properly interpretable, right antenna severed between funicular segments 2 and 3 and 5 and 6, left antenna apparently missing except for possibly funicle segment 7 and club near block surface at left of rostrum, onychium of right protarsus missing and that of left metatarsus separated from tarsus, all tarsite 3 lobes missing except inner one of right mesotarsus, apical part of rostrum cut away with amber; in irregular cabochon 17.0 × 11.9 × 3.0 mm, with flat top and one end obliquely cut to expose dorsum of specimen; amber clear yellow-brown with dense diffuse particles and few larger loose masses of organic material, with one large fracture behind posterior end of specimen.

Derivation of name. The species is named for its scattered iridescent green setae, which appear to be unique among the more primitive weevil families.

Remarks. This species has several unusual features of the legs, the most notable being the pair of indistinct short spurs of the protibiae contrasted with only one such spur and a conspicuous large spike on the meso- and metatibiae (on the latter broad, flat and ventrally lined with setae), and a small outer apical bur on the protibiae. The antennal clubs are also distinctively long and loosely articulated, with the long narrow fourth segment distinctly articulating with the third. The holotype was submitted for CT scanning, but the contrast between the specimen and the amber was too low to permit any meaningful visualisation of the specimen.

Genus *Cyrtocis* Clarke & Oberprieler, **gen. n.**

Type species: *Cyrtocis gibbus* Clarke & Oberprieler, sp. n.

Description. Size. Length 2.92 mm, width 0.93 mm. **Head** elongate, subconical, dorsally slightly convex, ventrally bulging. **Eyes** large, elongate, strongly (possibly flatly) protruding, coarsely facetted, dorsally separated by nearly width of rostrum anteriorly but further separated posteriorly; forehead flat, without tubercles between anterior margin of eyes. **Rostrum** ca. 2.0 × longer than pronotum, slender, subcylindrical, downcurved; antennal insertions lateral (possibly lateroventral), with scrobes behind them, in front of them laterally without long erect setae. Apparently a single long gular suture present (not clearly discernible). **Antennae** geniculate, long; scapes elongate, cylindrical, apically slightly inflated, about as long as funicle segments 1–5; funicles 7-segmented, segment 1 broader than others, 2–5 subequal, elongate, apically widened, 6–7 subequal, shorter, thinner, apically oblique; clubs large, loosely articulated, subcompressed, 4-segmented, segments apically strongly oblique, 4 flattened, acute, about as long as 3. **Mouthparts**. Labrum absent. Mandibles small, exodont, articulation probably horizontal. Maxillary and labial palps 3-segmented. **Thorax**. Prothorax proclinate, with anterior lateral margins oblique in lateral view. Pronotum slightly convex, laterally nearly straight, without tooth, posterior corners truncate, fitting closely to elytra; surface densely, coarsely punctate, tuberculate, sparsely setose, setae reclinate, directed anteromesad; notosternal sutures closed, abruptly curved anteriad. Prosternum moderately long; procoxal cavities seemingly medially confluent, in middle of prothorax. Scutellar shield densely setose. Mesocoxal cavities laterally closed by meso- and metaventrite. Metanepisterna distinct, sparsely setose. Mesoventrite short, anteriorly sloping. Metaventrite longer, posteriorly raised into transverse weals. **Elytra** elongate, basally concave to receive pronotum, with broadly rounded humeri, posteriorly very strongly, abruptly declivous (forming nearly right angle), with distinct hump or prominence separating disc from declivity; lateral margin weakly sinuate to roundly emarginate in middle, apically individually narrowly rounded, not exposing pygidium; sutural flanges narrow, seemingly equal; surface punctostriate, without scutellary striole, interstriae ill-defined, narrow, finely tuberculate, setose, setae long, thin, reclinate, directed caudad. **Legs**. Procoxae large, conical, medially contiguous; mesocoxae subglobular, prominent, narrowly separated; metacoxae flat, transversely elongate. Trochanters long, oblique. Femora long, subcylindrical, inflated in distal half, outside rounded, inside excavate in distal fifth to half, receiving tibiae in repose, walls of groove at apex flatly rounded, not extended. Tibiae straight, probably terete (heavily distorted), distally expanded, outer edge rounded, with dense long stiff setae in distal half, apex obliquely truncate, with 2 spurs. Tarsi about 0.75 × as long as tibiae; moderately broad, flat; tarsite 1 apically subtruncate, 2 shorter, triangular, apically excised, 3 deeply bilobed, 5 long, apically expanded; claws divaricate, basally angulate with ventrobasal seta. **Abdomen** with ventrites 1 and 2 much longer than 3, 3 slightly longer than 4, 5 as long as 3, apically narrowly rounded.

Derivation of name. The name of the genus is derived from the Greek adjective *kyrtos* (humped) and noun *kis* (G: *kios*) (weevil or beetle); its gender is masculine.

Remarks. *Cyrtocis* is most similar to *Electrocis* but larger and with protruding eyes, broader tarsi (with broader lobes of tarsite 3) and stouter antennae with a shorter and inflated funicle segment 1 and apically oblique club segments. It also differs from the latter genus in having exodont mandibles with horizontal or oblique articulations (non-exodont and vertical in *Electrocis*) and different ventrites. A proper interpretation of further seemingly distinct characters will require better-preserved specimens and structures, including possibly partly elongate trochanters.

Cyrtocis gibbus Clarke & Oberprieler, **sp. n.** (Figure 46)

Description. Size. Length 2.92 mm, width 0.93 mm. **Head**. Head dorsally densely, coarsely punctate, sparsely and shortly setose. **Eyes** elongate-oval, weakly protruding. **Rostrum** long, dorsally and ventrally with dense long curved setae in basal half; antennal insertions median. **Antennae.** Scapes slender, not apically clavate, gradually widening. **Mouthparts.** Mandibles small, with 2 small sharp teeth on outer side. Maxillary palps with segments progressively narrower toward apex, with terminal segment elongate, slender, much narrower than penultimate segment. Labial palps short, stout, segments subequal in width. **Thorax.** Pronotum elongate, laterally weakly rounded, nearly straight; pronotum rugose. Scutellar shield elongate, prominent. **Elytra** rugose, sparsely covered with short, blunt, recumbent reddish setae, weakly punctostriate, surface coarsely punctostriate, striae wider than interstriae, punctures large, deep, well defined; with single large hump at top of declivity; sides with marginal groove distinct, gradually widening anteriad, with anterior marginal notch; apices very narrowly rounded, almost spine-like. Hindwings seemingly fully developed. **Legs.** Procoxae elongate, very prominent, contiguous. Femora long, slender, subcylindrical, inside notched before apex (in lateral view), apical excavation extending to distal half on profemora, distal fifth on meso- and metafemora. Tibiae straight, with long stiff setae in distal half, apex with long, coarse fringing setae; protibiae with small, indistinct spurs; meso- and metatibiae with larger and prominent spurs, robust and slightly curved on metatibae. Tarsi long, robust, protarsi seemingly with much thicker, denser setal pads; tarsite 3 deeply, broadly bilobed. **Abdomen.** Tergites VII and VIII densely setose, setae long. Ventrites densely, coarsely punctate, setose, setae laterally very long.

Material examined. Holotype (NIGP154219): moderately well preserved, largely intact and well visible specimen, partly decomposed and distorted, with several artefactual cuticular protrusions usually associated with bubbles erupted from specimen (e.g., two large humps on rostrum, these possibly real in part, numerous small tubercle-like protrusions on right elytron and tarsi), legs largely compressed or depressed, missing left protarsus, left wing partly extended; in irregular 7-sided block $4.0 \times 3.5 \times 0.9$ mm; amber clear yellow with large oblique fracture above weevil pronotum, posteriorly and ventrally with numerous bubbles largely obscuring those sides, otherwise with minimal impurities and an insect wing below legs, several bubbles exposed during block preparation infilled with resin (see Section 2.2).

Derivation of name. The species is named for its conspicuously humped elytra, the Latin adjective *gibbus* meaning humped.

Remarks. Distinctive characteristics of this species include the elytral hump at the top of the declivity (though this appears partly exaggerated by a cuticular eruption), the strongly acute elytral apices, straight-sided pronotum and distinctive rugose sculpture. The first three features and several others may have been affected by an overall compression and distortion that is most obvious in dorsal view (the holotype seems disproportionately narrow), and additional specimens are needed to confirm them. The specimen may be a male (the apical part of what appears to be an aedeagus is visible in the abdominal apex) and is overall rather obscured by numerous close and large bubbles. Because of this it was submitted for CT scanning, but the contrast between the specimen and the amber was too low to permit any meaningful visualisation of the specimen.

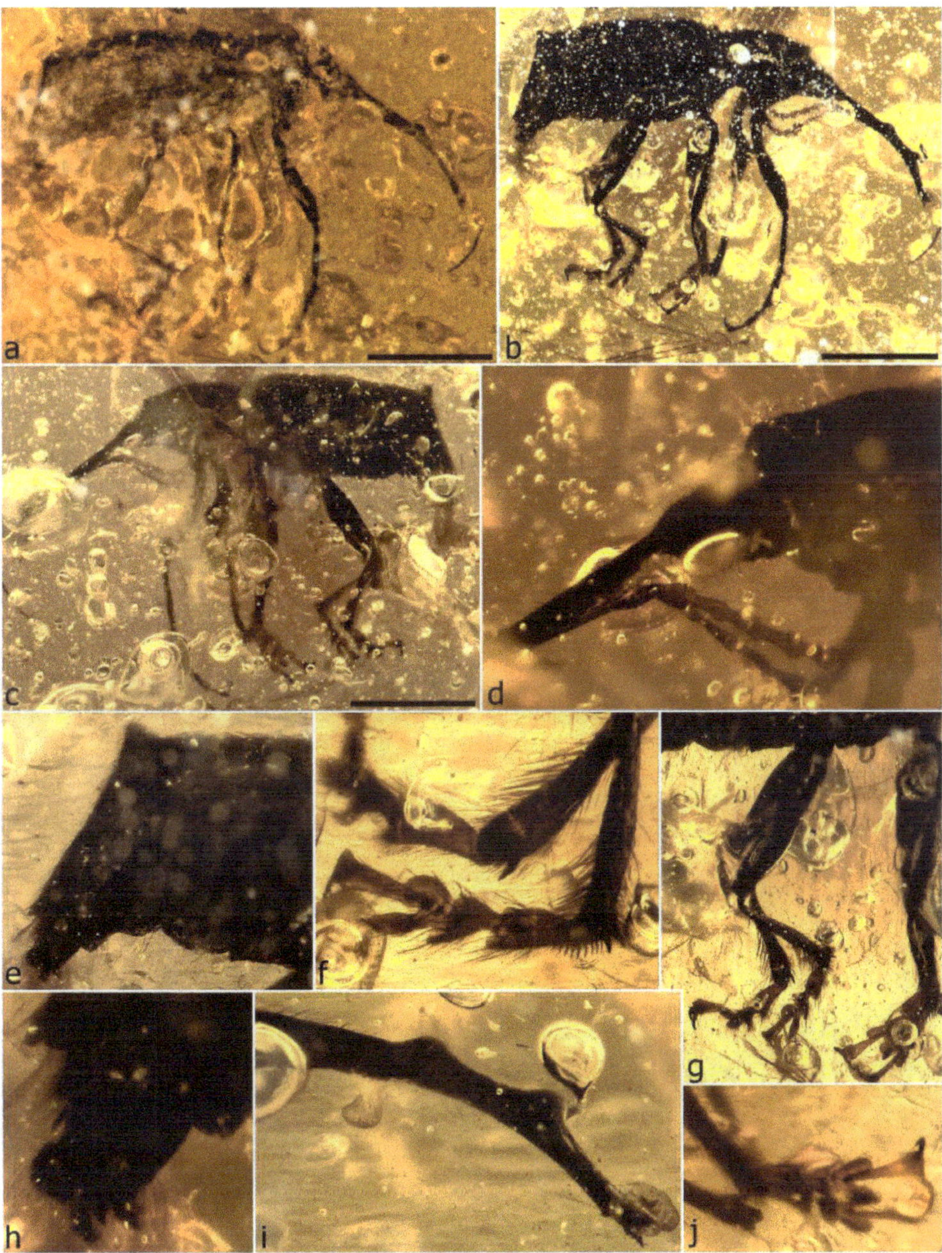

Figure 46. *Cyrtocis gibbus* sp. n., holotype. Habitus, right lateral (**a**); habitus, right lateral (different lighting) (**b**); habitus, left lateral (**c**); head, prothorax, and antenna, left lateral (**d**); posterior part of abdomen, right lateral (**e**); metatibiae and -tarsi (**f**); hindlegs, right (**g**); apex of abdomen, right lateral (**h**); rostrum, right lateral (**i**); mesotarsus, left (**j**). Scale bars: 1.0 mm (a–c).

Genus ***Ocriocis*** Clarke & Oberprieler, **gen. n.**

Type species: *Ocriocis binodosus* Clarke & Oberprieler, sp. n.

Description. Size. Length 1.97 mm, width 0.78 mm. **Head** elongate, subconical, dorsally slightly convex, ventrally bulging. **Eyes** large, round, strongly (possibly flatly) protruding, coarsely facetted, dorsally separated by nearly width of rostrum anteriorly but further separated posteriorly; forehead flat, without tubercles between anterior margin of eyes. **Rostrum** ca. 1.25 × longer than pronotum, slender, subcylindrical, downcurved; antennal insertions lateral, with scrobes behind them, in front of them laterally with long erect setae. Apparently a single long gular suture present. **Antennae** geniculate, long; scapes elongate, cylindrical, apically slightly inflated, longer than funicle; funicles 7-segmented, very slender, segment 1 broader than others, 2–4 subequal, elongate, apically widened, 5–7 subequal, shorter, broader, apically truncate; clubs large, loosely articulated, subcompressed, 4-segmented, segments apically straight, segment 4 flattened, acute, slightly shorter than 3. **Mouthparts**. Labrum absent. Mandibles small, exodont, articulation probably oblique. **Thorax**. Prothorax seemingly proclinate, with anterior lateral margins oblique in lateral view. Pronotum slightly convex, laterally broadly rounded, without tooth, posterior corners truncate, fitting closely onto elytra; surface coarsely tuberculate, sparsely setose, setae reclinate, directed anteromesad; notosternal sutures closed, abruptly curved anteriad. Prosternum moderately long; procoxal cavities medially confluent, in middle of prothorax. Scutellar shield densely setose. Mesocoxal cavities closed laterally by meso- and metaventrite. Metanepisterna distinct, sparsely setose. Mesoventrite short, anteriorly sloping. Metaventrite longer, flat or slightly convex. **Elytra** elongate, basally extended over pronotum, with broadly rounded humeri, posteriorly strongly, abruptly declivous, probably with hump or prominence on side of disc of each elytron separating disc from declivity; lateral margin weakly sinuate to roundly emarginate in middle; apically conjointly rounded, apices broad, not exposing pygidium; sutural flanges narrow, equal; surface punctostriate, without scutellary striole, interstriae distinct, broadly convex, tuberculate, setose, setae long, thin, suberect, directed caudad. **Legs**. Procoxae large, conical, medially contiguous; mesocoxae subglobular, prominent, narrowly separated; metacoxae flat, transversely elongate. Trochanters short, oblique. Femora long, subcylindrical, inflated in distal half, outside rounded, profemora inside excavate in distal quarter, receiving tibiae in repose, walls of groove at apex flatly rounded, not extended, other femora not ventrally excavate apically. Tibiae substraight, flattened, distally expanded, outer edge costate (protibiae) or tuberculate (meso- and metatibiae), with dense long stiff setae in distal half, apex obliquely truncate, with 2 spurs. Tarsi almost as long as tibiae, moderately broad, flat; tarsite 1 apically subtruncate, 2 shorter, triangular, apically excised, 3 deeply bilobed, 5 long, apically expanded; claws divaricate, basally angulate with ventrobasal seta. **Abdomen** with ventrites 1 and 2 longer than 3, 3 slightly shorter than 4, 5 nearly as long as 3 + 4, apically very broadly rounded.

Derivation of name. The name of the genus is derived from the Greek nouns *okris* (G: *okrios*), (roughness) and *kis* (G: *kios*) (weevil or beetle), in reference to the granulose and generally rough surface of the weevil; its gender is masculine.

Remarks. This genus is unique in the group of genera with non-dentate tarsal claws in having a row of distinct, widely spaced oval slightly pointed carinulae on the outer edges of the meso- and metatibiae (Figure 47g) but a costate edge on the protibiae, as well as a low hump at the top of each elytral declivity and distinctive antennae, with very slender funicles but short, broad, subcompressed clubs with shortly obconical segments. In the non-serrate/crenulate cuticular projections on the tibiae it is similar to *Debbia* and *Hukawngius*, but in these genera the projections are differently shaped and in the former seemingly occurring only on the protibiae and in the latter not forming a row in dorsal view. *Hukawngius* also has dentate tarsal claws. *Ocriocis* also shares a similar tarsal structure and long rostrum with *Debbia* and *Cyrtocis* (tarsites 1 apically subtruncate, 2 distinctly excised, 3 deeply broadly bilobed). From *Cyrtocis* it can also be distinguished by the lateral row of long erect setae in front of the antennal insertions.

Ocriocis binodosus Clarke & Oberprieler, **sp. n.** (Figure 47)

Description. Size. Length 1.97 mm, width 0.78 mm. **Antennae.** Club segments shortly obconical, broad basally, segment 4 broadly inserted into 3. **Mouthparts.** Mandibles on outer side with 2 small broadly rounded teeth, on inner side with 2 large acute teeth, distal part slender, apex forming T, with small rounded inner and outer apical tooth. **Elytra.** Surface tuberculate on disc, more granulose on declivity. **Legs.** Protibiae carinate on outer edge, meso- and metatibae with evenly widely spaced, subflat tubercles, roundly pointed at apex. Tibial spurs short, thin, indistinct on protibiae. Tarsi with tarsites 1 longer than 2, subequal in width, 3 broadly lobate. **Abdomen.** Tergite VII (or VIII?) broadly convex, posteriorly densely setose, setae long, subequal in length to apical setae of ventrite 5.

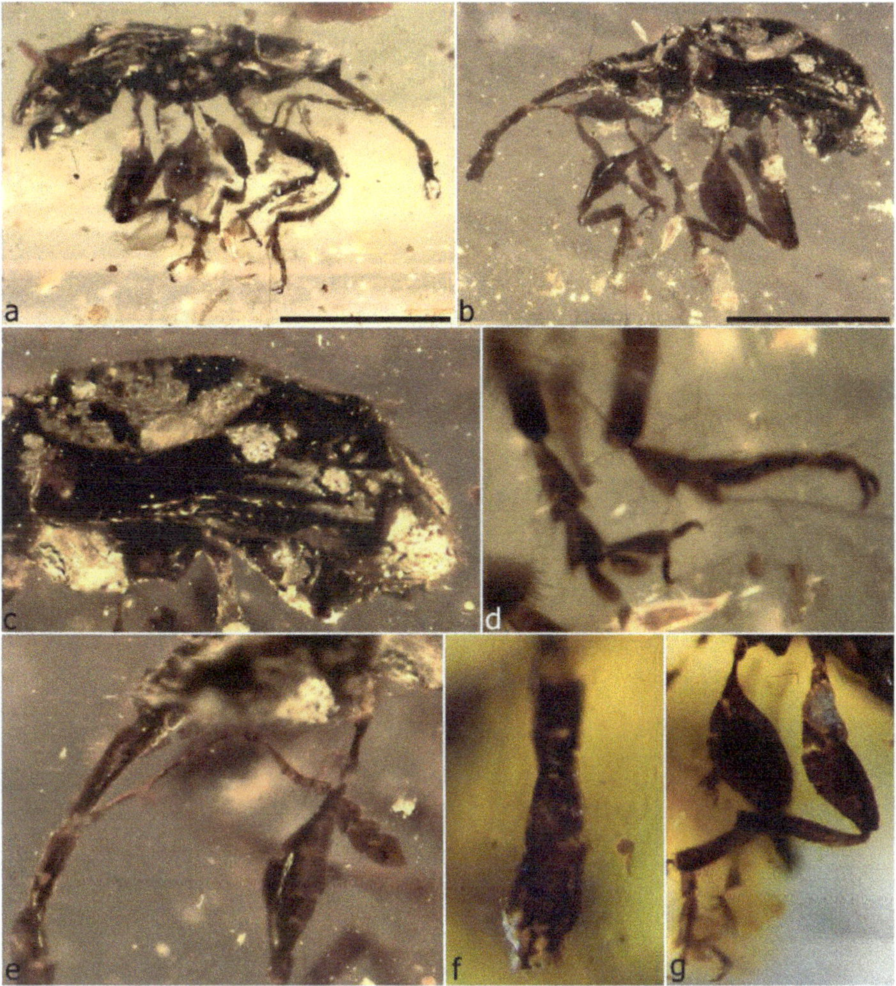

Figure 47. *Ocriocis binodosus.* Habitus, right lateral (**a**); habitus, left lateral (**b**); elytra, left lateral (**c**); protarsi (**d**); rostrum and left antenna, left lateral (**e**); apex of rostrum and mandibles, dorsal (**f**); middle and hindlegs, showing spurs and carinulae (**g**). Scale bars: 1.0 mm (a,b).

Material examined. Holotype (NIGP154220): poorly preserved, heavily distorted (crumpled) and somewhat decomposed specimen, with head retracted into prothorax, rostrum and antennae fragmented, right middle leg and hindlegs with tarsites 3–5 separated (some missing), right wing exposed, fragmented, surface details largely unobstructed except for fragmented coating of whitish debris; near centre of cuboid, 3.2 × 2.1 × 1.8 mm; amber clear yellow, with gritty particles not obscuring specimen.

Derivation of name. The species is named after the pair of humps on the top of its elytral declivity, the name being a Latin adjective.

Remarks. The poorly preserved holotype nevertheless preserves sufficient structural details to reveal the specimen as being generically distinct from all others in our sample. Additional specimens assignable to this genus should also be identifiable to species, as details of the elytra, ventrites, legs, antennae and mandibles of *O. binodosus* are variably well preserved.

Genus ***Electrocis*** Clarke & Oberprieler, **gen. n.**

Type species: *Electrocis dentitibialis* Clarke & Oberprieler, sp. n.

Description. Size. Length 2.16 mm, width 0.91 mm. **Head** long, porrect, broadening posteriad, not constricted behind eyes. **Eyes** elongate, subflattened (not strongly protruding), coarsely facetted, dorsally separated by twice basal width of rostrum anteriorly but further separated posteriorly; forehead flat, without pair of tubercles between anterior margin of eyes. **Rostrum** ca. 1.5 × longer than pronotum, slender, subcylindrical, weakly curved; antennal insertions lateral or possibly lateroventral, with scrobes behind them, in front of them laterally without any erect setae. Gular suture present, apparently long. **Antennae** geniculate, long; scapes elongate, cylindrical, apically clavate, longer than funicles; funicles 7-segmented, segments 1–3 elongate, others shorter, obconical; clubs short, loosely articulated, 4-segmented, segment 4 acute, longer than 3. **Mouthparts**. Labrum absent. Mandibles small, vertical, slightly scoop-like, non-exodont but with 2 upturned apical teeth, articulation plane vertical. **Thorax**. Prothorax not proclinate, with anterior lateral margins vertical in lateral view, narrow, elongate, coarsely punctate. Pronotum strongly convex, laterally rounded, without tooth, posterior corners obsolete, fitting closely onto elytra; sparsely setose, setae erect, directed anteromesad; notosternal sutures closed, vertical above coxal cavities. Prosternum moderately long; procoxal cavities medially confluent, in about middle of prothorax. Scutellar shield densely setose. Mesocoxal cavities closed. Metanepisterna distinct, sparsely setose. Mesoventrite short, anteriorly sloping. Metaventrite longer, raised into subglobular weals. **Elytra** elongate, weakly extended over pronotum, with broadly rounded humeri, posteriorly steeply declivous, lateral margin slightly sinuate to roundly emarginate in middle, apically individually rounded, not exposing pygidium; sutural flanges not visible; surface coarsely punctostriate, punctures large, open, without scutellary striole, interstriae convex, finely tuberculate, very sparsely setose, setae short, thin, reclinate, directed caudad, interstriae without dense patches of coloured setae. **Legs**. Procoxae, prominent, medially contiguous at least at base; mesocoxae subglobular, narrowly separated; metacoxae flat, shortly transversely elongate. Trochanters short, oblique. Femora long, slender, subcylindrical, inflated in distal half, dorsally rounded, ventrally excavate in distal quarter, receiving tibiae in repose, walls of groove at apex flatly rounded. Tibiae substraight, compressed, distally expanded, outer edge costate, very sparsely setose, inner edge dentate; apex obliquely truncate, with dentate mucro at inner apex, with 2 spurs. Tarsi nearly as long as tibiae, slender; tarsites 1–2 subcylindrical, apically subtruncate, 2 nearly half length of 1, 3 deeply bilobed, lobes short, digitate, 5 elongate, length subequal to 1–3; claws robust, divergent, very slightly basally angulate, with ventrobasal seta. **Abdomen** with ventrites 1 and 2 very elongate, fused, subflatly aligned, with suture less distinct than others, subequal in length, each longer than 3 + 4, 3 and 4 subequal, each less than half of 1 or 2.

Derivation of name. The name of the genus is derived from the Greek nouns *elektron* (G: *elektrou*) (amber) and *kis* (G: *kios*) (weevil or beetle); its gender is masculine.

Remarks. Seemingly an isolated genus, *Electrocis* is distinguishable from all other genera with non-dentate tarsal claws by its combination of non-exodont, vertically articulating mandibles, ventrally dentate and mucronate tibiae (with modified translucent setae in ventral preapical emargination), the elongate, strongly porrect head with vertical (not proclinate) anterior prothoracic margin and elongate prosternum and the slender tarsi with narrow digitate lobes of tarsites 3 (only tarsus of left hindleg preserved). These characters also distinguish it from the other genera with very long and subflatly aligned ventrites 1 and 2 (*Burmorhinus* and *Rhadinomycter*).

Electrocis dentitibialis Clarke & Oberprieler, **sp. n.** (Figure 48)

Description. Size. Length 2.16 mm, width 0.91 mm. Body and legs black. **Head.** Dorsal outline of head in lateral view nearly continuous with rostrum (without distinct sinus); distance between eyes increasing to ca. 3.5 times rostral width at posterior margin of eyes; ventrally moderately bulging. **Eyes** lateral, longer than wide. **Rostrum** evenly weakly curved; antennal insertions median or slightly antemedian; scrobes reaching eye. **Antennae**. Scapes reaching just below front margin of eye, apically oblique; funicles with segment 1 ca. 1.5 × longer than 2, slightly broader, 2–5 slightly expanding distally, 4 and 5 more abruptly so, 2 and 3 subequal in length, 4 ca. half as long as 3, 5–7 subequal, slightly shorter than 4; clubs with segments 1 and 2 subequal, obconical, 1 subconvex apically, 2 flat apically, 3 shorter than 2, slightly narrower, 4 acute, broadly inserted into 3, about 1.5 × longer than 3. **Mouthparts.** Mandibles with 2 apicovertically orientated cusps. **Thorax.** Pronotum narrower than elytra, widest just before middle, gradually narrowing anteriad and posteriad; coarsely punctate. Prosternum elongate, about half as long as procoxae, anterior margin straight, prosternal process narrowly pointed; hypomeron about 1.25 × longer than prosternum. Setae on scutellar shield pale or whitish. Metaventrite narrowly concave between weals. **Elytra** seemingly fused along suture and to thorax and abdomen; bases obtusely angulate, weakly sinuate; humeri narrowly rounded, with subserrate edges (3 teeth visible on left, 2 on right), margins indistinct from sides of thorax and ventrites 1 and 2, without marginal groove, seemingly without anterior marginal notch. **Legs.** Procoxae subconical; mesocoxae narrowly separated, moderately projecting, subglobular; metacoxae broadly separated. Femora slender, widening gradually distad, preapically swollen, abruptly constricted before apex, sparsely covered in whitish setae. Tibiae subequal in length, subequal in width, widest at ca. middle, gradually narrowed towards base and apex, on inner edges with ca. 8 denticles, in weak emargination distal third with ca. 1–4 coarse modified setae; apical edges with coarse fringing setae lining edges of weakly developed inner and outer flanges, on meso- and metatibiae fringing setae extending partly along outer apical edge of tibia. Tarsi (metatarsus) elongate, nearly as long as tibia; tarsite 1 ca. 2.0 × longer than 2, lobes of 3 distinctly concave along inner margins, broadly connected basally. **Abdomen.** Ventrites 3–5 at higher level than 1–2, somewhat recessed into elytra (perhaps in part depression artefact); 3 and 4 subequal in length; 5 longer than 3 + 4, apically rounded; sutures between 2 and 5 distinct, deeply grooved.

Material examined. Holotype (NIGP154221): very well preserved, largely undistorted specimen but with surface details partly obscured by surface debris and tiny bubbles, antennae intact but anterior, middle and right metatarsi missing tarsites 3–5 (only left metatarsus intact), left side of rostral apex (including left mandible) and parts of tarsi of anterior and middle legs cut away with amber during block preparation, other tarsites apparently severed prior to fossilisation; in irregular wedge with two large curved faces and three smaller edge faces, 3.4 × 2.4 × 1.0 mm; amber imperfectly clear yellow-brown with greenish hue (under scope light), with dense granular impurities mainly below specimen, with small oblique fracture partly obscuring left ventral side.

Derivation of name. The name of the species is an adjective derived from the Latin nouns *dens* (G: *dentis*) (tooth) and *tibia* (G: *tibiae*) (shin).

Remarks. This species is distinctive in its black, heavily punctate integument. It is one of only a few probably flightless species of Burmese amber weevils, characterised by minimally tightly closed elytra but seemingly fused along the suture and to the thorax and ventrites 1 and 2. In several respects this species resembles those of the next group of genera, especially *Burmorhinus georgei*, which also has two spurs and a mucro, and all these species share subcylindrical, apically expanded then constricted femora with retractable tibiae and slender tarsi, especially the lobes of tarsites 3, and very long, flatly aligned ventrites 1 and 2 followed by very short ventrites 3–5.

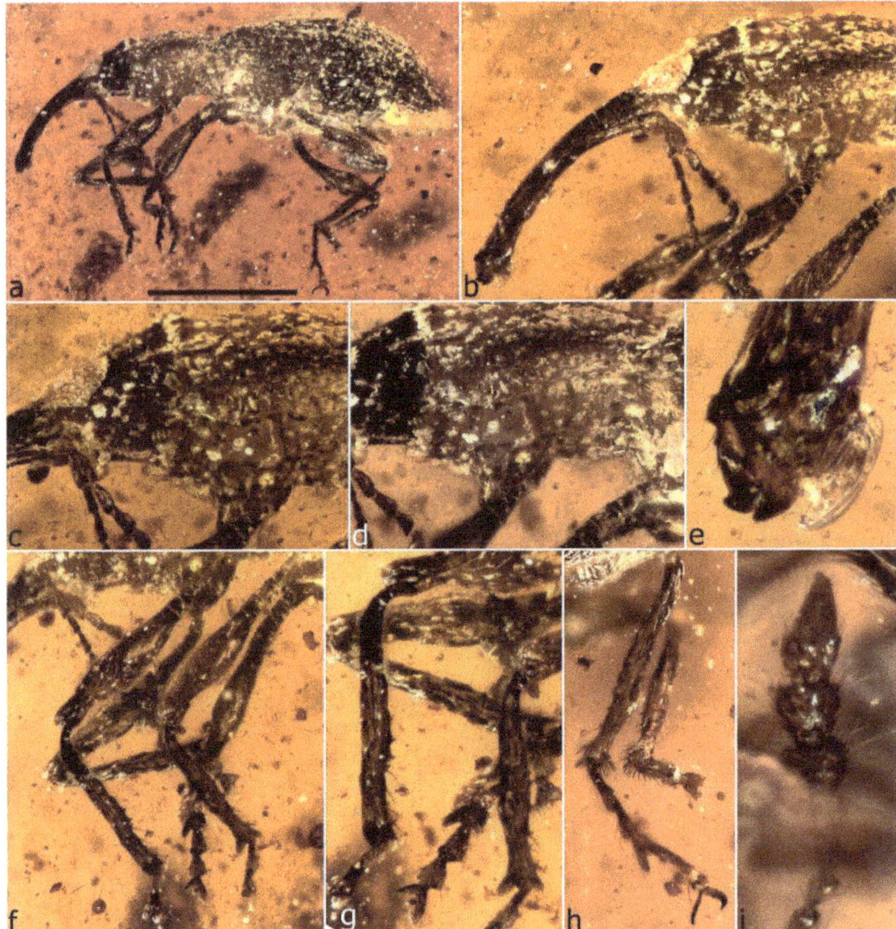

Figure 48. *Electrocis dentitibialis* sp. n., holotype. Habitus, left lateral (**a**); head and prothorax, left lateral (**b**); same, detail (**c–d**); apex of rostrum, left lateral, showing inner face of right mandible with two blunt apical cusps (left mandible cut away) (**e**); fore and middle legs (**f**); pro- and mesotibiae (**g**); hindlegs, showing complete left metatarsus (**h**); antennal club (**i**). Scale bars: 0.5 mm (a).

Genus ***Debbia*** Clarke & Oberprieler, **gen. n.**

Type species: *Debbia gracilirostris* Clarke & Oberprieler, sp. n.

Description. Size. Length 2.58 mm, width 1.13 mm. **Head** short, subspherical, slightly flattened. **Eyes** small, conically protuberant, forward-facing, coarsely facetted, dorsally separated by slightly more than width of rostrum anteriorly but further separated posteriorly; forehead flatly convex, without tubercles between eyes. **Rostrum** about 1.75 × longer than pronotum, very thin, subcylindrical, distinctly curved; antennal insertions lateral, median, with scrobes behind them, in front of them laterally with a few long erect setae. Single gular suture present. **Antennae** geniculate, long; scapes elongate, slender, cylindrical, apically slightly inflated, about as long as funicle, not reaching eye; funicles thin, 7-segmented, segment 1 about as long as 2, wider, others progressively shorter towards club; clubs thin, large, loosely articulated, 4-segmented, segment 4 about as long as 3, acute. **Mouthparts**. Labrum absent. Mandibles small, flat, exodont, articulation oblique. **Thorax**. Prothorax slightly proclinate, with anterior lateral margins oblique in lateral view. Pronotum convex, laterally rounded, without tooth, posterior corners rounded, fitting closely onto elytra; surface punctate, densely setose, setae reclinate, directed anteromesad; notosternal sutures closed, abruptly curved anteriad. Prosternum moderately long; procoxal cavities medially confluent, in middle of prothorax. Scutellar shield prominent. Mesocoxal cavities laterally closed. Metanepisterna distinct, densely setose. Mesoventrite short, anteriorly sloping. Metaventrite longer, raised into strong transverse weals. **Elytra** elongate, basally obtusely straight, with weakly rounded humeri, posteriorly declivous, lateral margin weakly sinuate in middle; apically individually rounded, not exposing pygidium; surface weakly punctostriate, without scutellary striole, interstriae narrow, setose, setae long, thin, reclinate, directed caudad. **Legs**. Procoxae large and conical, prominent, medially contiguous; mesocoxae subglobular, narrowly separated; metacoxae flat, transversely elongate. Trochanters short, oblique. Femora long, strongly inflated in distal half, outside with thin black carina along most of length. Tibiae long, slender, subterete, outer edge rounded, on protibia with sparse subevenly spaced tubercles, apex obliquely truncate, with 2 spurs (visible on left pro- and mesotibia). Tarsi about 0.75 × as long as tibiae; tarsite 1 subtriangulate, 2 shorter, apically excised, 3 deeply bilobed, lobes pedunculate, 5 very long and slender, apically expanded; claws divaricate, basally angulate with ventrobasal seta. **Abdomen** with ventrites 1 and 2 subequal, 3 and 4 slightly shorter, 5 subtriangular; ovipositor with long slender gonocoxites, each with a small elongate apical stylus.

Derivation of name. The genus is cordially named for our colleague Debbie Jennings, in recognition of all her help with and dedication to the study of these weevil fossils, in particular her superb photographs of them; the name of the genus is feminine.

Remarks. Among the genera with basally angulate claws, *Debbia* is distinctive in having conically protruding, widely separated eyes, long slender tibiae and tarsi, the latter with excised second tarsites, and small rounded tubercles along the outer edge of the protibiae. It is also one of the few genera in this group characterised by the type of exodont mandibles that is typical of the majority of Mesophyletinae, being flattened with large inner and outer teeth and horizontal in repose but opening into a vertical position via oblique articulation sockets. This mandible type seems to be shared with *Cyrtocis* and *Ocriocis*, which also have similar tarsi, but better-preserved specimens of these genera are needed to further understand their possible affinities.

Debbia gracilirostris Clarke & Oberprieler, **sp. n.** (Figures 49 and 50, Video S5)

Description. Size. Length 2.58 mm, width 1.13 mm. **Head** sparsely, shallowly punctosetose; punctures minute, sparse, denser between eyes. **Rostrum** slightly widened before ape; dorsally with pair of fine grooves extending between base and apex; with elongately strigulose sculpture; scrobes extending to front of eyes. **Antennae**. Scapes apically subtruncate (slightly oblique); funicles with segment 1 subfusiform, wider than 2, 2–4 elongate, slightly expanded apically, with sparse long setae, 5–7 thinner, progressively shorter, 7 subglobular; clubs with segments 1 and 2 distinctly obconical, densely setose, with numerous longer setae, 2 ca. 0.67 × as long as 1, 3 slightly shorter, 4 slightly longer than 3, broadly inserted into 3. **Mouthparts**. Mandibles individually subsymmetrical; with at least 3 teeth on inner and outer edges, 2 larger blunt teeth basally (also seemingly third large tooth at base of inner edge), apical part of mandible slender, with smaller blunt inner and outer apical teeth forming weak T. Maxillary palps projecting. Labial palps apically projected (2 segments visible); apical segment narrower and subequal in length to penultimate one, with minute tufted apical setae. **Thorax.** Pronotum densely setose, setae directed anteromesad, punctate, punctures coarser than on head; narrowed but not constricted anteriorly or posteriorly. **Elytra** densely setose; striae ca. 2.0 × wider than interstriae, coarsely punctate, punctures large; laterally with distinct marginal groove, slightly increasing in width anteriad, with distinct anterior marginal notch. **Legs.** Femora sparsely setose, setae long, whitish; profemora on inside slightly notched before apex (in lateral view). Tibiae slender, slightly bent inwards near apex, densely setose with longer setae in distal half, setae whitish; apically with tarsal articulation surfaces concavely oblique, with short, indistinct flanges lined with long coarse loose fringing setae, outer edge apically with several elongate slender setae, spurs short, narrow, indistinct; protibiae with outer edge sparsely tuberculate for most of length, tubercles minute but distinct, subconical, apically rounded, widely and subevenly spaced; meso- and metatibiae shorter, probably without tubercles along outer edge, mesotibiae apically on outside with short row of slightly less coarse setae continuous with outer apical fringing setae. Tarsi densely, shortly setose dorsally and ventrally, with relatively short apicolateral setae; tarsite 3 lobes about half as long as 5, 5 very long, slightly shorter than 1–3, very slender basally. **Abdomen.** Ventrites 1 and 2 subequal in length, 2–4 progressively slightly shorter, 5 longer than 4.

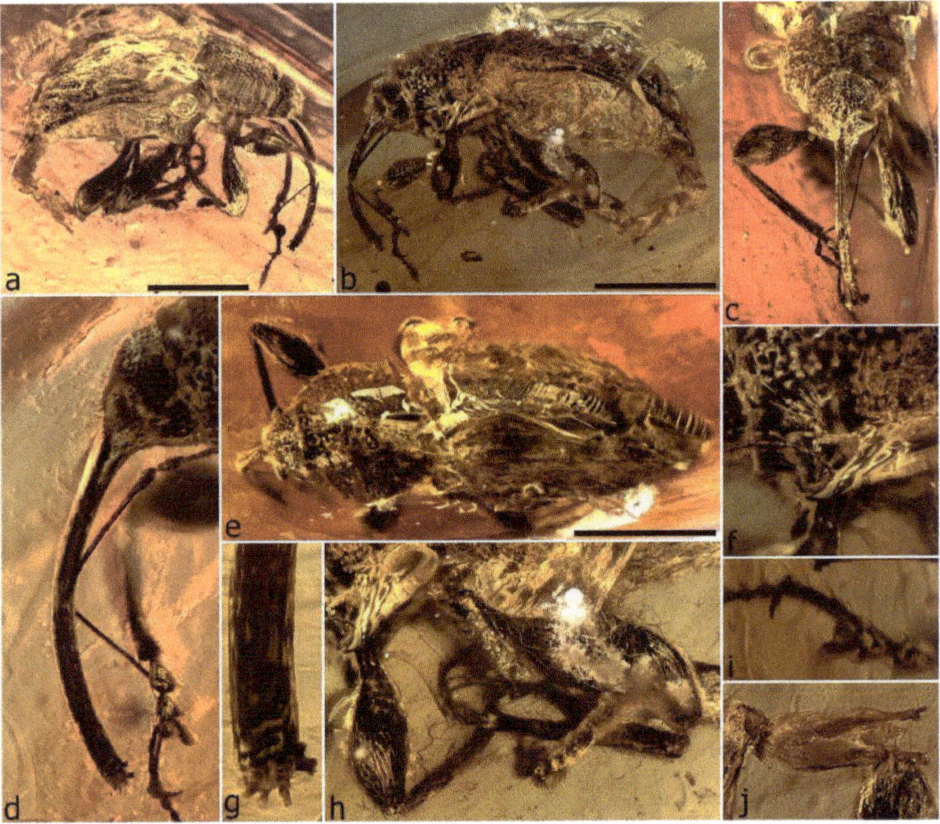

Figure 49. *Debbia gracilirostris* sp. n., holotype. Habitus, right lateral (**a**); habitus, left lateral (**b**); head, frontal (**c**); head and left protarsus, left lateral (**d**); habitus, dorsal (**e**); prothorax, showing notosternal suture, left lateral (**f**); apex of rostrum showing mandibles and maxillary palps, left lateral (**g**); legs, left lateral (**h**); right protarsus (**i**); ovipositor (**j**). Scale bars: 1.0 mm (a,b,e).

Material examined. Holotype (NIGP154222), female: well preserved, intact but poorly visible specimen, much of body surface obscured by whitish coating (seemingly mixed fungal hyphae, debris and air), rostrum and appendages well visible, with ovipositor fully extruded; in cuboid, 5.5 × 5.1 × 1.1 mm, with one large face rounded to edge; amber clear yellow with several curved flow bands parallel to curved block face, with several small fractures in vicinity of and partly obscuring surface of specimen.

Derivation of name. The species is named for its long and slender rostrum, the name being a Latin adjective.

Figure 50. *Debbia gracilirostris* sp. n., holotype. Habitus images extracted from a micro-CT scanning reconstruction (see also Video S5). Right lateral (**a**); left lateral (**b**); dorsal (**c**); ventral (**d**); ventrolateral right (**e**); ventrolateral left (**f**); frontoventral (**g**); frontal (**h**).

Remarks. This species is characterised by the long, slender, curved, cylindrical rostrum with distinctive, symmetrically guitar-shaped mandibles. The general symmetry of the mandibles, with similarly sized teeth on both edges, is unusual in Mesophyletidae; normally either the inner or the outer teeth are significantly larger than those on the opposite edge. This difference may reflect a differentiation in specific details of mandible function among such taxa. The mandibles of *Debbia* are also extraordinary in having a very slender apical part in front of the teeth; this again may reflect a functional aspect of the mandible form and is comparable with the apically similarly slender mandibles of several other unrelated species (e.g., *Bowangius cyclops*) but contrasted with those mandibles with a robust anvil- or T-shaped apex, in which the apical part is often thick and the teeth can sometimes be as large as or larger than the basal ones. We are aware of one other undescribed species likely related to *D. gracilirostris*, of general similarity and with similarly guitar-shaped mandibles but a shorter rostrum and without the conical eyes that are so distinctive of *D. gracilirostris*.

The holotype was submitted for CT scanning, with astonishing results (Figure 50, Video S5) that demonstrate the inherent variability in the success of this imaging technique among samples. Much of the surface of the specimen (especially dorsally) is obscured by seemingly dense hyphal growths and other debris, and many ventral structures are also nearly entirely obscured from view under a light microscope. These surfaces were rendered very clear with CT scanning, the confluent procoxal cavities, narrowly separated mesocoxal cavities and distinctly punctostriate elytra being well visible.

Genus ***Burmorhinus*** Legalov, 2018

Burmorhinus Legalov, 2018: 13 [56] (type species, by original designation: *Burmorhinus georgei* Legalov, 2018)

Redescription. Size. Length 2.45–2.94 mm, width 1.0–1.05 mm. **Head** short, subglobular-transverse, strongly convex dorsally, not constricted behind eyes. **Eyes** large, weakly protruding, coarsely facetted, dorsally separated by basal width of rostrum anteriorly, similarly or further separated posteriorly; forehead concave or flat, without paired tubercles between anterior margin of eyes. **Rostrum** about as long as pronotum, stout, compressed behind antennal insertions, dorsoventrally flattened in front of them, strongly downcurved; antennal insertions lateral, in apical quarter (possibly a male trait), with scrobes behind them extending obliquely ventrad to below eye, in front of them laterally with a few long erect setae. Single long gular suture indicated. **Antennae** geniculate, long; scapes elongate, cylindrical, apically only slightly inflated, about as long as funicle segments 1–4, reaching front margin of eye; funicles 7-segmented, segments 1–4 progressively shorter towards club, 5–7 subglobular; clubs long, loosely 3-segmented but with weakly set-off segment 4, segments 1 and 2 obconical. **Mouthparts**. Labrum absent. Mandibles small to large, flat, horizontal, non-exodont, with single large inner tooth, articulation plane horizontal. **Thorax**. Prothorax slightly proclinate, with anterior lateral margins sinuate in lateral view, drawn out into distinct or weak ocular lobe. Pronotum slightly convex, laterally rounded, without tooth, posterior corners rounded and obsolete or angulate and slightly extended, fitting closely onto elytra; surface coarsely punctorugose, sparsely setose, setae recurved, multidirectional or directed anteromesad, not forming coloured patches; notosternal sutures closed, obliquely vertical. Prosternum with precoxal channel; procoxal cavities medially confluent, closer to posterior margin of hypomeron. Scutellar shield densely setose. Mesocoxal cavities closed laterally. Mesoventrite short, anteriorly sloping. Metaventrite about 3 times longer, flatly concave. Metanepisterna distinct, possibly fused to metaventrite, setose. **Elytra** elongate, narrow, basally extended over pronotum, with narrowly rounded humeri, posteriorly declivous, lateral margin nearly straight or weakly sinuate, apically conjointly rounded, not exposing pygidium; sutural flanges narrow, subequal; surface punctostriate, without scutellary striole, interstriae convex, subcostate, setose, setae long, thin, recurved, directed caudad, interstriae without dense patches of coloured setae. **Legs**. Procoxae large and subconical to subglobular, prominent, medially contiguous; mesocoxae globular, narrowly separated; metacoxae flat, transversely elongate, widely separated by broadly rounded process of ventrite 1. Trochanters short, oblique. Femora long, subcylindrical, slightly inflated in middle, outside rounded, inside excavate in distal quarter to half, receiving tibiae in repose, walls of groove at apex flatly, roundly extended, shearing against basal part of tibia. Tibiae straight, subcylindirical to subcompressed, distally expanded, outer edge rounded to subcostate, densely setose, with longer stiff setae in distal half, apex obliquely truncate to subconcave, with 2 spurs and with or without a small mucro. Tarsi almost as long as or slightly longer than tibiae, narrow; tarsite 1 apically subtruncate to slightly rounded or apicolaterally roundly lobate, 2 shorter, apically truncate to slightly rounded or excised, 3 deeply but shortly bilobed, lobes narrow, 5 very long, slender, apically expanded; claws divergent or divergent-divaricate, basally angulate with ventrobasal seta. **Abdomen** with ventrites 1 and 2 elongate, fused (with suture less distinct than others), subflatly aligned, each about as long as 3 + 4, 3–4 subequal, 5 longer, apically broadly rounded.

Remarks. This genus was described in the family Curculionidae based on its geniculate antennae, uncinate tibiae, elongate ventrite 1 fused with 2 and the ventrites lying in one plane [56], but none of these characters are exclusive to this family. Three imaginal characters are generally considered to be synapomorphies of Curculionidae (e.g., [57]), and the states of two of them can be readily assessed in the holotypes of *B. georgei* and *B. setosus*. Although their antennae are clearly geniculate, the geniculation is of the 'open' type as in Mesophyletidae and Nanophyinae, not the 'closed' one of Curculionidae, and although the socket of the scapes is more narrowly encircling the base of the pedicel than it is in other Mesophyletidae, it is distinctly obliquely apically positioned, not ventrally, and the socket is clearly visible in apical view. Also contradicting a placement in Curculionidae are the antennal clubs of *Burmorhinus*, which are loosely 3(4)-segmented in *B. setosus* and also so indicated in *B. georgei*, the holotype preserving only segment 1 of one club but this with a distinctly setose apical surface and narrow articulation stem, sufficient evidence that its clubs are loose as well, not tightly compact as in Curculionidae. The third synapomorphy of Curculionidae, a pair of radial sclerites in the hindwing, cannot be assessed in *Burmorhinus* as the hindwings are not visible. The original placement of *Burmorhinus* in the curculionid subfamily Erirhininae [56] is also untenable, as two of the four characters on which it was based (scrobes directed towards the eye, tibiae with two apical spurs) occur widely in Curculionidae and also in Mesophyletidae and the other two (apex of the rostrum "with setae", tibial uncus "displaced" onto the inner apical angle) are misinterpretations. The rostrum of neither *B. georgei* nor *B. setosus* has an apicolateral setiferous groove as is characteristic of erirhinines [39], only a line of sparse setae as occurs in this position in many Mesophyletidae, and their tibiae do not have an uncus (only a small mucro in *B. georgei*). The purported affiliation of *Burmorhinus* with the erirhinine tribe Arthrostenini is also invalid, as its procoxae are in fact contiguous and it lacks scale-like setae, as occur in many genera related to *Arthrostenus* Schoenherr. Its strongly carinate and grooved, apically flattened rostrum and its long, flattened mandibles also do not accord with Arthrostenini.

Burmorhinus unequivocally agrees with Mesophyletidae in nearly all its characters except for the weak prosternal channel, but it shares the latter with *Rhadinomycter* (and also with *Palaeocryptorhynchus* if this belongs in the same family). *Burmorhinus* and *Rhadinomycter* are also unique in Mesophyletidae in having flattened, horizontal, non-exodont and horizontally articulating mandibles. *Burmorhinus* differs from *Rhadinomycter* in having two tibial spurs (one in *Rhadinomycter*), 3-segmented clubs and long setae on the body (Figure 51). From *Palaeocryptorhynchus* it is easily distinguishable by the lack of scales and of a receptacle on the mesoventrite for receiving the apex of the rostrum.

***Burmorhinus georgei* Legalov, 2018**

Burmorhinus georgei Legalov, 2018: 14 [56]

Redescription. Size. Length 2.94 mm, width 1.0 mm. Body elongate, slender; integument uniformly black, vestiture pale, brown and whitish, setae slender, acuminate. **Head** short, globular, dorsal outline in lateral view from base of rostrum continuing evenly between eyes; dorsally bulging somewhat over eyes; densely setose, setae fine, directed anteriad, coarser and longer on rostrum; forehead in middle with oval impression. **Eyes** large, rounded, flattened, positioned ventrolaterally (not visible in dorsal view), dorsally separated by basal width of rostrum anteriorly, further separated posteriorly. **Rostrum** densely setose, setae long, distinctly more curved and thicker than on head; from base to antennal insertions slightly compressed but gradually increasing in width, dorsally with strong median, paramedian and dorsolateral carinate ridges and deep intervening setose grooves; from antennal insertions to apex depressed; median and paramedian grooves and carinae abruptly ending at elongate subtrapezoidal smooth plateau (corresponding to the frons) in front of antennal insertions, this connected distally to distinct epistome, marked at each anterior corner by closely spaced pair of thick curved macrosetae; dorsolateral grooves reaching slightly beyond antennal insertions; laterally with indistinct setose scrobe-like groove reaching front of eye; antennal insertions in apical quarter, behind them with true lateroventral scrobes, separated from lateral setose groove by carina, reaching below eyes; in front of (and below) them with narrow groove reaching mandibular articulations (these lateral, horizontal), with 2 thick curved macrosetae near apex, at least one other smaller seta further back and 2 elongate macrosetae at antennal insertions; ventrally on each side with very long, thick curved macroseta behind hypostomal area. **Antennae.** Scapes apically oblique, with numerous sparse setae, articulation socket with pedicel very narrow; funicles with segment 1 bent at base, elongate, only slightly wider than others, 2–4 similar, elongate-cylindrical, 2 ca. 0.67 × as long as 1, others progressively shorter towards club, 5–7 subequal, subglobular; clubs with segment 1 obconical, ca. 2.0 × broader than last funicle segment (other club segments missing). **Mouthparts.** Mandibles symmetrical, long and narrow, apically strongly curved, outside with small rounded subflat projection with notch in front, inside with very large tooth at about middle, on left bicuspid, on right simple. Maxillary palps not clearly visible (only apical ovoid segment). Labium distinct, apicomedially with long narrow projection with group of ca. 4 setae at apex; palps 3-segmented, segments progressively narrower distad, segment 1 longer than 2 and with long thick macroseta, 2 short, globular, 3 slender, slightly longer and ca. half as wide as 2. **Thorax.** Pronotum elongate, slightly narrower than elytra; densely coarsely punctate, with edges of rostral channel terminating posteriorly in a rounded tooth. Pronotum sparsely setose, setae multidirectional; sides substraight, weakly curved; base broadly convex, without marginal ridge; posterior corners rounded and obsolete. Scutellum with whitish setae. Metaventrite flatly concave, coarsely sparsely punctate, somewhat prominent posterolaterally but not raised into weals. **Elytra** narrowly elongate, sparsely setose, setae mainly brown; striae and interstriae subequal in width; interstriae raised, punctorugose, more densely setose than striae; bases weakly concave; humeri narrowly rounded, tightly fitting with pronotal corners; sides straight, parallel-sided, marginal groove indistinct, very narrow, subequal for entire length, with broad shallow anterior notch. **Legs.** Procoxae strongly projecting, subconical; mesocoxae prominent; metacoxae only weakly transverse, not prominent. Femora constricted preapically, outside at apex shallowly (profemora) to deeply truncately emarginate, without distinct comb of curved setae, inside excavate in apical half (profemora) or quarter (meso- and metafemora) to receive tibiae. Tibiae abruptly curved basally, subcompressed, substraight, outside edge subcostate, with dense long curved thick whitish setae, other setae slender, straight; apically with distinct, well developed apical flanges lined with coarse fringing setae, on meso- and metatibiae these extending shortly along distal outer margin, with short dentate mucro on inner apical side and 2 slender spurs, increasing in length from pro- to metatibiae. Tarsi progressively longer from fore- to hindlegs, nearly as long as tibiae, densely setose ventrally, less so dorsally, seemingly progressively narrower from fore- to hindlegs; tarsites 1–2 apically subtruncate to slightly rounded, 1 elongate, widening apically, 2 similar, shorter, ca. half as long as 1, very slightly wider, 3 with short digitate slightly compressed lobes broadly connected basally and not concave along inner edge, cryptotarsite globular, prominent, 5 about as long as or shorter than 1 + 2; claws divergent-divaricate. **Abdomen**. Ventrites 1 and 2 large, flatly aligned, coarsely deeply sparsely punctate, elongate, subequal in length, each slightly longer than 3 + 4; 3–5 stepped, at slightly higher level, 1 with very broadly convex intercoxal projection; 3 and 4 subequal in length, sparsely setose, 5 slightly longer than 4, very densely setose, apically broadly rounded.

Material examined. Holotype (ISEA no. MA 2017/1): excellently preserved and well visible specimen, with diffuse whitish coating over most of surface but surface still well visible, left eye distorted, globular (right eye flattened), missing club and last five funicle segments of left antenna, last 2 or 3 club segments of right antenna, claw segment of left middle leg and claws of left metatarsus; re-prepared from much larger discoidal amber piece from which the original description was prepared, now in elongate rectangular cuboid, 3.9 × 1.2 × 1.1–2.1 mm; amber clear yellow with diffuse gritty particles and few larger masses of organic material near right side of specimen, without fractures or other obscurities.

Figure 51. *Burmorhinus setosus* sp. n., holotype. Habitus, dorsolateral (**a**); habitus, right lateral (**b**); habitus, left lateral (**c**); head, antennae, and prothorax, right lateral (**d**); same, left lateral (**e**); forelegs, left (**f**); mesotibia (**g**); left metatibia (**h**); forelegs (**i**). Scale bars: 0.5 mm (a–c).

Remarks. This distinctive species is readily distinguishable from *B. setosus* by its long narrow form, the multidirectional setae on the pronotum, the long dense setae on the interstriae and tarsites 5 being shorter than 1–3, as well as the inner apical tooth on the tibiae. In details, particularly of the rostrum, it displays numerous differences from *B. setosus*, most notably the deep grooves on the rostrum with long curved setae and thick intervening carinae. The holotype is likely to be a male due to its preapically inserted antennae.

Burmorhinus setosus Clarke & Oberprieler, **sp. n.** (Figure 51)

Description. Size. Length 2.45 mm, width 1.05 mm. Body black, concolorous, rostrum slightly paler; with whitish golden vestiture on body and legs, setae generally tapering at both ends. **Head** short, dorsal outline continuing evenly from base of rostrum to between eyes; forehead flat, densely setose, setae fine, directed anteriad, on rostrum and head subequal; ventrally bulging. **Eyes** lateral, large, subglobular, subhemispherical, slightly longer than deep; dorsally separated by about half basal rostral width anteriorly, narrowing between eyes and further separated posteriorly. **Rostrum** sparsely setose, setae short, curved, subequal to setae on head; stout, subcylindrical to subcompressed from base to antennal insertions, dorsally with median ridge with minute impunctate asetose groove in middle, with 2 other low paramedian carinae, these effaced or confused distad from antennal insertions; grooves between carinae shallow, lined with setae directed anteromesad (basally) or mesad (further from head); from antennal insertions to apex slightly depressed, slightly expanding apicad; median and paramedian grooves and carinae abruptly ending at narrow ogival plateau in front of antennal insertions, this connected distally to distinct epistome, marked by 2 elongate macrosetae along oblique edges; dorsolateral grooves reaching slightly beyond antennal insertions; antennal insertions in apical quarter, marked dorsally by distinct tubercles, behind them with deep lateroventral scrobes reaching eyes, in front of (and below) them with narrow groove reaching mandibular articulations (these lateral, horizontal), with 2 thick curved macrosetae near apex and at least 2 other smaller setae at antennal insertions; ventrally on each side with very long thick curved macroseta behind hypostomal area. **Antennae** dark brown, concolorous; scapes apically oblique, asetose, articulation socket with pedical very narrow; funicles with segment 1 bent at base, elongate, subequal in length to 2 + 3, distinctly broadened in apical half and ca. 2.0 × wider than others, 2–5 similar, elongate-cylindrical, increasingly obconical and progressively shorter towards club, 6–7 subequal, subglobular; clubs with segments 1 and 2 subequal in length and width, 1 distinctly obconical, ca. 2.5 × broader than last funicle segment, 2 subcylindrical, both apically with dense ring of short modified setae, 3 + 4 subfusiform, elongate, slightly longer and narrower than 1 + 2, indistinctly separated, 4 acute. **Mouthparts.** Mandibles symmetrical, small, broad at base, apically narrowly pointed, outside with weak notch just before base, inside on each side with at least one acute tooth Maxillary palps short, 3-segmented, apical segment apically rounded, subequal in length to 2, slightly narrower. Labial palps very slender, widely separated (not properly visible). **Thorax.** Prothorax not or slightly proclinate (area unclear), with anterior lateral margins seemingly drawn out into broad, weak ocular lobe, about as long as broad, distinctly narrower than elytra at base, distinctly punctosetose. Pronotum sparsely evenly setose, setae directed anteromesad, overall forming wavy pattern; sides broadly rounded; base distinctly sinuate, closely fitting to elytra, posterior angles acute, slightly extended. Scutellar shield tiny, prominent, longer than wide, with whitish setae. Metaventrite more finely punctosetose than dorsal side, raised into transverse weals. **Elytra** broadly elongate, sparsely setose, setae golden brown; punctostriate, striae asetose, punctures large; interstriae ca. 2.0 × broader than striae, broadly convex, interstriae 7 and 8 densely setose, basally confluent forming flatly produced humeri; bases broadly obtuse, weakly sinuate, humeri flatly produced, then sharply inflexed; sides slightly sinuate, marginal groove distinct, densely setose, subequal in width for entire length, with narrow anterior marginal notch; apices with small lobe before suture; sutural flanges narrow, equal. Hindwings fully developed. **Legs** robust. Procoxae globular, moderately projecting. Femora constricted preapically, at outside at most weakly emarginate, with comb of curved setae; inside excavate in apical one third or more to receive tibiae. Tibiae abruptly curved basally, subcylindrical, outside rounded, with short sparse recurved setae, inside with longer, denser setae in distal third; apically with well-developed flanges lined with coarse fringing setae, on meso- and metatibae these extending shortly along distal outer margin and associated with patch of longer oblique setae on outer side, without mucro on inner apical side; with 2 short narrow acute spurs. Tarsi progressively longer from fore- to hindlegs, of foreleg slightly shorter, of middle and hindlegs slightly longer than tibiae, densely setose ventrally, less so dorsally; tarsite 1 elongate, broad, subtriangular, apicolaterally roundly lobate, 2 shorter, ca. half as long as and much narrower than 1, apically excised, 3 deeply lobate, lobes elongate-ovoid, subparallel-sided, slightly depressed, broadly connected basally and distinctly concave along inner edge, cryptotarsite narrow, indistinct, recessed into base of 3, 5 elongate, much longer than 1 + 2; claws divergent. **Abdomen.** Ventrites 1 and 2 large, subflatly aligned, finely sparsely punctosetose, elongate, subequal in length; each slightly longer than 3 + 4, 3–5 stepped, at slightly higher level, 1 with very broadly convex intercoxal projection; 3 and 4 subequal in length, sparsely setose, 5 longer than 4 (not properly visible), apically broadly rounded.

Material examined. Holotype (NIGP154223): very well preserved, intact and well visible specimen, dorsal surface unobstructed, ventral surface less visible due to gritty haze near specimen, also with some distortion of (mainly ventral) head and prothorax (with crack or wrinkle on left side of pronotum), with left wing partly extended; in irregular 6-sided prism, 4.0 × 3.9 × 3.5 mm, with dorsum of weevil parallel to large curved face; amber hazy yellow-brown with darker halo around specimen, with diffuse microscopic particles, without fractures, bubbles, or debris.

Derivation of name. The species is named for its conspicuous covering of setae, the name being an adjective.

Remarks. This species is similar to *B. georgei* but distinguishable from the latter by its broader body, uniformly short whitish-golden setae, lack of a tibial mucro, small tibial spurs and the different dimensions of the tarsites, 1 being long and broad, 2 much shorter and narrower and 5 as long as 1–3. The condition and visibility of the prosternum is unfortunately poor due to it being depressed and obscured by some fine debris. The sides of the rostral channel have been somewhat flattened (pushed in), but the anterior lateral margin of the prothorax on both sides is traceable as a continuous edge from the front to the coxae. Although it is clear that the specimen has a prosternal channel, it is less certain whether it also has ocular lobes as occur in *B. georgei* and *Rhadinomycter perplexus*, because the flattening of the channel has obscured the curvature of the correct area (and also because these lobes were likely broadly rounded as they are in *R. perplexus*). Although the eyes seem somewhat bulging (they are, however, more dorsally positioned than in *B. georgei*, with a narrow flat space between them), the head is also distorted ventrally, likely accentuating the protrusion of the eyes. The holotype seemingly also represents a male.

Genus ***Rhadinomycter*** Clarke & Oberprieler, **gen. n.**

Type species: *Rhadinomycter perplexus* Clarke & Oberprieler, sp. n.

Description. Size. Length 2.45 mm, width 0.78 mm. **Head** short, subglobular-compressed, strongly convex dorsally and ventrally, not constricted behind eyes. **Eyes** large, weakly protruding, coarsely facetted, dorsally separated by about width of rostrum basally, anteriorly further separated posteriorly; forehead flatly convex, without paired tubercles between anterior margin of eyes. **Rostrum** slightly longer than pronotum, slender, strongly compressed to antennal insertions (partly a distortion), broader, slightly depressed to apex, weakly downcurved; antennal insertions lateral, in apical quarter to third (possibly a male trait; see also Remarks for *R. perplexus*), with scrobes behind them extending toward eye, in front of them laterally with a few long erect setae. Gular suture not discernible (ventral side of head not visible). **Antennae** geniculate, long; scapes elongate, cylindrical, apically slightly evenly expanded, about as long as funicles, not reaching eyes, articulation socket of funicle segment 1 apicoventral; funicles 7-segmented, segments 1–5 progressively shorter towards club, 6–7 subglobular; clubs short, loose, 4-segmented. **Mouthparts**. Labrum absent. Mandibles long (but partly concealed in dorsal view), flat, horizontal, non-exodont, articulation plane horizontal. **Thorax**. Prothorax not proclinate, with anterior lateral margins sinuate in lateral view, drawn out into broadly rounded ocular lobe. Pronotum strongly convex, laterally roundly subvertical, without tooth, posterior corners rounded and obsolete, tightly fitting onto elytra; surface coarsely densely punctate, sparsely setose, setae short, diverse, not forming dense coloured patches; notosternal sutures apparently absent. Prosternum with channel for reception of rostrum; procoxal cavities medially confluent, closer to hind margin of hypomeron. Scutellar shield densely setose. Mesocoxal cavities closed laterally. Metanepisterna not discernible (obstructed). Mesoventrite short, anteriorly sloping, with spiniform projection between coxae. Metaventrite about 3 times longer than mesoventrite, flatly concave. **Elytra** elongate, narrow, basally extended over base of pronotum, with narrowly rounded humeri, posteriorly weakly declivous, lateral margin substraight, apically conjointly rounded, not exposing pygidium; sutural flanges not visible; surface very coarsely punctostriate, without scutellary striole, punctures large, open, interstriae flat, narrower than striae, sparsely setose, setae short, diverse, indistinct, not forming dense coloured patches. **Legs**. Procoxae large, prominent, medially contiguous; mesocoxae elongate, narrowly globular, widely separated; metacoxae elongately globular, widely separated by broadly rounded process of ventrite 1. Trochanters short, oblique. Femora long, subcylindrical, outside rounded, inside excavate in distal half (profemora) to quarter (meso- and metafemora), receiving tibiae in repose, walls of groove at apex flatly, roundly extended, shearing against basal part of tibia. Tibiae substraight, compressed, distally expanded, outer edge subcostate, smooth, sparsely setose, with longer stiff setae in distal half, apex obliquely truncate, with 1 fixed spur. Tarsi shorter than tibiae, narrow, subflat; tarsite 1 apically truncate, 2 shorter, subquadrate, apically truncate, 3 shortly bilobed, lobes digitate, narrow, 5 slender, apically expanded; claws divergent-divaricate, simple, with ventrobasal seta. **Abdomen** with ventrites 1 and 2 very elongate, fused (suture less distinct than others), subflatly aligned, each about as long as 3–5, 3 slightly longer than 4, 5 subequal to 3 + 4, apically broadly rounded.

Derivation of name. The genus is named for its slender shape, the name derived from the Greek adjective *rhadinos* (slender) and noun *mykter* (G: *mykteros*) (nose, beak) and being masculine in gender.

Remarks. This genus is most similar to *Burmorhinus*, sharing the characters of a prosternal channel for reception of the rostrum and postocular lobes, but it can be readily distinguished by the distinctly 4-segmented clubs and single spur on the tibiae. Other differences from *Burmorhinus* include the rostrum being slightly longer than the pronotum, the scrobes extending towards the eyes, the shorter and more distinctly 4-segmented clubs, the flat narrow interstriae, the short, diverse, indistinct elytral setae and the more elongate, narrowly globular mesocoxae. *Rhadinomycter* is unique among Burmese amber weevils as known in being the only species in which the articulation socket of the scape with the pedicel is apicoventrally located (but still of the 'open' type of geniculation) and in having a long acute process on the mesoventrite projecting forwards from the base of the intercoxal process.

Rhadinomycter perplexus Clarke & Oberprieler, **sp. n.** (Figure 52)

Decription. Size. Length 2.45 mm, width 0.78 mm. Narrow, slightly depressed, coarsely punctate. **Head** short, globular, subcompressed (partly distorted); dorsal outline in lateral view forming sinus at base of rostrum, dorsally convex but not bulging over eyes; forehead strongly narrowly convex; sparsely finely setose, setae directed anteriad; ventral outline of head in lateral view forming strong angle at base of rostrum. **Eyes** ventrolateral, subcircular in outline. **Rostrum** sparsely, shortly setose, setae indistinct dorsally, with median, paramedian and dorsolateral carinae and deep intervening setose grooves; from antennal insertions to apex somewhat depressed, median and paramedian grooves and carinae ending at elongate trapezoidal smooth plateau (frons) in front of antennal insertions, this connected to a distinctly narrower, long and apically subemarginate epistome with close pair of curved macrosetae at each anterior corner, junction of frons and epistome laterally marked by pair of long thick curved macrosetae, further back with additional erect macroseta set in groove; laterally in front of antennal insertions with narrow groove reaching mandibular articulation (these lateral, horizontal), with single thick curved macroseta in near apex, 2 smaller setae further back (one near antennal insertions); ventrally on each side with very long thick curved macroseta behind hypostomal area. **Antennae.** Scapes slender, apically rounded in dorsal view, articulation socket with funicle segment 1 very narrow, not visible in dorsal view, slightly visible in apical view; funicles with segment 1 bent at base, elongate-obconical, slightly wider than others, 2–5 obconical, narrower, progressively shorter towards club, 2 ca. 0.67 × as long as 1, 6 and 7 subequal, subglobular, slightly broader than 5; clubs with segments 1–2 subequal in length, apically subangulately rounded, 1 obconical, longer than wide, ca. 2.0 × broader than 4, apical width ca. 3.0 × basal width, 2 subquadrate, basal width ca. 2.0 × broader than that of 1, 3 much narrower than 2, more broadly inserted into it than 2 is into 1, 4 ca.

half as long as 3, distinctly narrower, apically subsubtruncate. **Mouthparts.** Mandibles symmetrical, elongate, apically strongly curved, ca. half their length or more concealed by elongate epistome (mandibles not well visible on outside and inside; appearing very short, curved). **Thorax.** Prothorax elongate, ca 2.0 × longer than wide, with larger prothoracic volume, evenly curved from pronotal disc to sides; densely, coarsely punctate, punctures deep, irregularly distributed; with edges of rostral channel not terminating posteriorly in a tooth. Pronotum sparsely setose, setae stout, erect (some thickened, club-like); sides substraight, subparallel, gently curved anteriad; base broadly convex, without marginal ridge; posterior corners rounded, indistinct. Scutellar shield recessed a little beneath elytral surface, setae pale. Metaventrite broadly concave, seemingly wider posteriorly, narrowing anteriorly towards mesocoxae; coarsely punctate. **Elytra** narrowly elongate, subflat, seemingly fused at suture and to thorax and abdomen; sparsely setose, some setae almost clubbed, others curved or erect; striae distinct, much wider than interstriae anteriorly, punctures very large, deep, decreasing in size caudad, intervals between punctures flat, much less than puncture diameter; interstriae not raised (including sutural interstria), less distinct than striae anteriorly (broader, more distinct in posterior third); elytral bases strongly concave; sides straight, subparallel for most of length, marginal groove and anterior marginal notch absent. **Legs.** Procoxae prominent, subflatly conical, meso- and metacoxae weakly prominent, slightly elongate, metacoxae not transverse. Femora constricted preapically, outside at apex weakly emarginate, without distinct comb of curved setae; inside excavate in apical quarter to receive tibiae. Tibiae abruptly curved basally, subcompressed, sparsely setose, with erect setae; outside edge subcostate, with sparse curved pale setae, other setae slender, straight; apically with tarsal articulation surfaces suboblique (pro- and mesotibiae) or distinctly oblique (metatibiae), with distinct, well developed apical flanges lined with coarse fringing setae; with single slender tibial spur, shorter on protibiae, much longer on meso- and metatibiae; mucro absent; metatibiae slender basally, unevenly expanded distad, outer edge upturned in distal half, apically expanded, with edge of outer apical flange elongate, this and smaller inner flange lined with coarse fringing setae. Tarsi progressively longer and narrower from fore- to hindlegs, nearly as long as tibiae, densely setose ventrally, less so dorsally; tarsites 1 obconical, ca. 1.5 × longer than 2, apically truncate, 2 constricted basally, then subquadrate, wider than 1, apically truncate, 3 elongately lobate, not pedunculate, wider than 2, of anterior and middle legs subequal in length, of hindlegs distinctly shorter than these, lobes of 3 very short, less half as long as 5, 5 slender, gradually expanding distad, distinctly curved, longer than 1–2. **Abdomen.** Ventrites flat, coarsely punctate, 1 and 2 very elongate, 2 slightly longer than 1, 3–5 extremely short, 3 about third as long as 2, 4 slightly shorter than 3; sutures between 2 and 5 very deep, straight or subarcuate.

Material examined. Holotype (NIGP154224): well preserved, mostly undistorted and largely intact specimen, with head, rostrum and anterior prothorax somewhat compressed, rostrum partly obscured by fractures and compression, legs well visible, club, most of pedicel and funicle segments 2–7 of right antenna cut away with amber; in flat 7-sided block, 5.3 × 2.5 × 1.2 mm; amber clear yellow with murky patches, bubbles, and debris variously obscuring right-ventral views and apex of abdomen.

Derivation of name. The species is named for its perplexing characters, in particular the prosternal channel and the long, retractable legs (tibiae).

Remarks. This distinctive, possibly flightless species is characterised by a coarse, dense punctation, a dorsal sinus separating the rostrum from the head and a compressed prothorax. The apex of the rostrum has an extended epistome that conceals much longer mandibles than can be seen apically in dorsal view. In lateral view the articulation sockets are set much further back from the dorsal apex, such that much of the length of the mandibles is concealed in dorsal view. In the holotype the head and rostrum have been stretched asymmetrically so that the eye and antennal insertion on the left side are positioned further anteriorly than those on the right side.

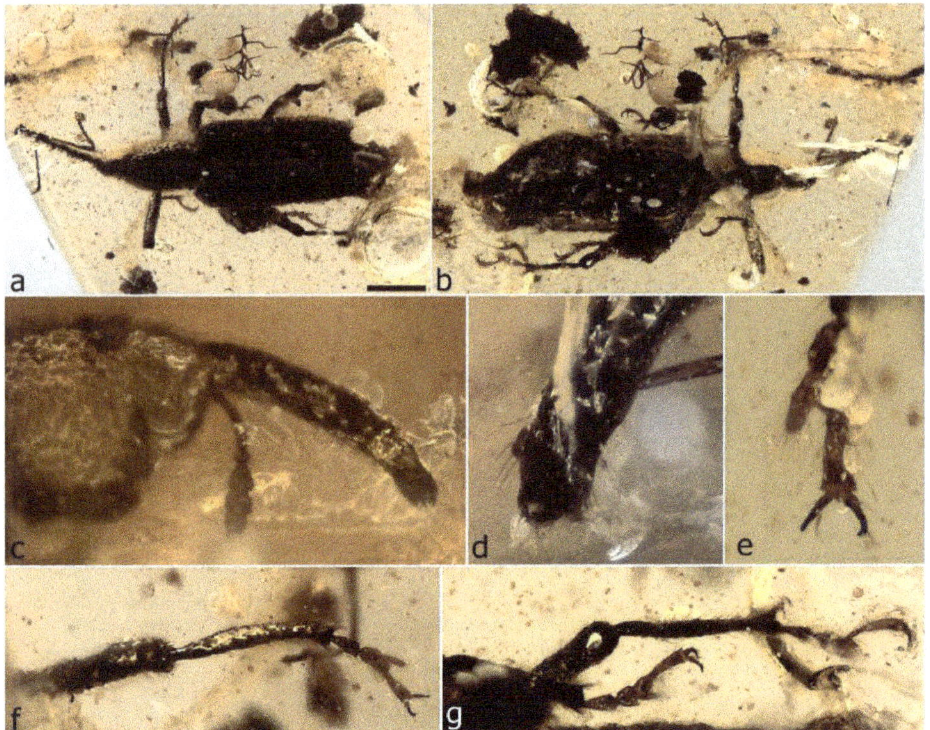

Figure 52. *Rhadinomycter perplexus* sp. n., holotype. Habitus, dorsal (rostrum appearing longer in this image due to a visual artifact caused by the angle of the block surface) (**a**); habitus, ventral (**b**); head and prothorax, right lateral (**c**); apex of rostrum and mandibles, dorsolateral (**d**); protarsus, dorsal (**e**); front leg (**f**); left meso-and metatibia (**g**). Scale bars: 1.0 mm (a).

Genus ***Gnomus*** Clarke & Oberprieler, **gen. n.**

Type species: *Gnomus brevis* Clarke & Oberprieler, sp. n.

Description. Size. Length 2.1–2.7 mm, width 0.9 mm or more. **Body** short, compact; sparsely setose. **Head** shortly conical, not constricted behind eyes. **Eyes** elongate, flat, not strongly protruding, coarsely facetted, dorsally separated by width of rostrum, without tubercles between them, forehead flat. **Rostrum** about as long as pronotum, stout, slightly to strongly curved; antennal insertions lateral, behind middle of rostral length. Gular suture not discernible. **Antennae** geniculate; scapes straight, cylindrical, apically slightly inflated; funicles slightly longer than scape, 7-segmented; clubs elongate, broad, loosely articulated, 4-segmented, segment 4 about as long as 3, broad to narrow. **Mouthparts**. Labrum absent. **Thorax**. Prothorax not proclinate, with anterior lateral margins vertical in lateral view. Pronotum short, convex, laterally rounded, without tooth, posterior corners rounded, not fitting closely onto elytra; notosternal sutures closed; procoxal cavities medially confluent. Scutellar shield transverse. Mesocoxal cavities laterally closed (by meso- and metaventrite). Metanepisternal sutures distinct. **Elytra** shortly elongate, with broadly rounded humeri, posteriorly strongly declivous, apically individually rounded, not exposing pygidium (last tergite extruded, triangular, slightly convex, but fitting under elytra); sutural flanges unknown; surface punctostriate, without scutellary striole. **Legs**. Procoxae short, prominent; mesocoxae subglobular, slightly projecting, metacoxae transverse. Trochanters short, oblique. Femora short, flatly subcylindrical, inflated in distal half, unarmed, outside rounded. Tibiae short, robust, outer edge rounded, distally setose, apically with 2 spurs. Tarsi almost as long as tibiae; tarsite 1 elongate, apically truncate, 2 shorter, triangular, apically truncate, 3 bilobed, lobes narrow to digitate, 5 slightly shorter than 1 + 2; claws divaricate, basally swollen, with ventrobasal seta. **Abdomen** with ventrites subequal, 1-4 progressively slightly shorter, 5 about as long as 4.

Derivation of name. The genus is named for its small, gnome-like appearance, from the Neo-Latin noun *gnomus*, meaning a dwarf; the gender of the name is masculine.

Remarks. *Gnomus* is largely a form genus, aggregating three species of similarly short and compact shape with non-dentate tarsal claws. The single specimens of all three species are poorly preserved (visible) and thus cannot be definitely assessed as being congeneric, but there are no sufficient discernible character differences between them to justify the erection of different genera for them. We selected the best-preserved specimen as the type species (Figure 53). We submitted the holotype of *G. spinipes* for CT scanning, but the density difference between the specimen and the surrounding amber was too small to allow the generation of suitable CT images. If further similar specimens are discovered, it may be possible to describe this genus more accurately and assess its proper species composition. *Gnomus* is similar in shape to the smaller *Cretocar luzzii* Gratshev & Zherikhin, 2000, described from the slightly younger New Jersey amber, and a careful comparison of these genera is warranted.

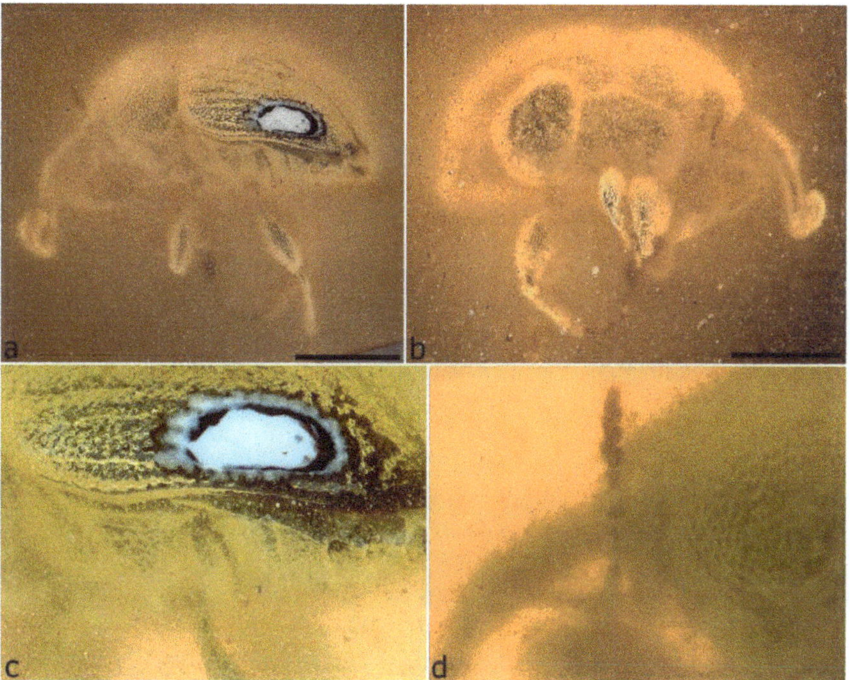

Figure 53. *Gnomus brevis* sp. n., holotype. Habitus, left lateral (**a**); habitus, right lateral (**b**); left elytron, left lateral (**c**); head, antenna and eye, left lateral (**d**). Scale bars: 1.0 mm (a,b).

Gnomus brevis Clarke & Oberprieler, **sp. n.** (Figure 53)

Description. Size. Length 2.72 mm, width ca. 1.4 mm. **Antennae.** Funicles with segments short, transverse but not clearly discernible; clubs flattened, apical segment bluntly triangular. **Thorax.** Pronotum with surface shallowly tuberculate, probably sparsely setose but setae indiscernible. **Elytra** posteriorly strongly evenly declivous, lateral margins bisinuate, emarginate over metacoxae; striae narrow, probably sparsely setose. **Legs**. Tibiae slightly flattened, spurs small (1 visible on metatibiae, possibly 2 on all tibiae). Tarsi with lobes of tarsite 3 narrow but not digitate. **Abdomen** with last tergite extruded, triangular, slightly convex, fitting under elytra (not espoused as pygidium), ventites subequal, 1–4 progressively slightly shorter, 5 about as long as 4, sides flat, high, fitting under elytra.

Material examined. Holotype (NIGP154225): well preserved, intact specimen, not compressed or distorted but poorly visible due to cloudiness of amber, with large flat bubble on right side and smaller one at tip of rostrum, thin silvery layer of air over part of surface, left elytron cut into at side; in centre of thin rectangular cuboid ca. 6.0 × 43 × 1.8 mm; amber densely packed with many minute impurities, giving it an opaque look.

Derivation of name. The genus is named for its short body, *brevis* being a Latin adjective.

Remarks. This is the best-preserved of the three species assigned to *Gnomus*. It differs from *G. spinipes* mainly in its more strongly curved rostrum and from *Gnomus* sp. in having normal tibial spurs.

Gnomus spinipes Clarke & Oberprieler, **sp. n.** (Figure 54a–e)

Description. **Size.** Length 2.1 mm, width not measurable. **Head.** Rostrum long, slender, slightly downcurved; antennal insertions median. **Antennae.** Scapes elongate slender, apically weakly expanded; funicles with segment 1 broader than others, 2–7 slightly progressively shorter towards clubs, 7 distinctly broader than 6; clubs with segments 1–3 subequal, obconical, progressively slightly longer apicad, 4 broadly inserted into 3, flattened, shorter than 3. **Mouthparts.** Not visible. **Thorax.** Prothorax not properly visible. Scutellum short, transverse. **Elytra** sparsely setose, setae short recurved, pale; strongly declivous, with very narrow marginal ridge subequal in width for entire length. **Legs.** Tibiae with stiff, suberect setae in distal half, spurs normal, paired, on metatibiae unequal and with long thicker fringing seta adjacent to inner spur. Tarsi with tarsite 1 apically subtruncate, 2 weakly excised, 3 with short lobes, ca. half as long as 5, digitate; claws angulate. **Abdomen.** Ventrites shortly, sparsely, finely setose; setae pale. Ventrite 5 broadly rounded apically.

Material examined. Holotype (NIGP154226): intact but poorly preserved and poorly visible specimen, compressed and pulled away from amber, leaving an impressed surface around most of specimen; in cuboid 5.5 × 5.2 × 2.5 mm; amber clear yellow with diffuse small debris particles and bubbles.

Derivation of name. The species is named for its spiny legs, especially the tibiae; the name is a noun in apposition.

Remarks. This species is only very poorly characterisable, because the single specimen has retracted from the surface of the amber, resulting in a large silvery halo around the body and rostrum that obscures most of its features. Its assignment to *Gnomus* is therefore only tentative. It differs from *Gnomus* sp. in having two normal, distadly directed spurs and the tibiae with suberect, stiff setae. From *G. brevis* it is distinguishable by the digitate (finger-like) lobes of tarsites 3.

***Gnomus* sp.** (Figure 54f–h)

Material examined. One specimen (NIGP156990); length 2.08 mm, width not measurable: intact but very poorly preserved, heavily distorted, partly decomposed and poorly visible, compressed and crumpled, covered with irregular bubbles; in slab with one rounded face, 5.3 × 4.5 × 1.7 mm; amber clear yellow, with large ovoid fracture near head, without any debris obscuring specimen.

Remarks. This specimen is the only one we have seen other than the Nemonychidae and *Palaeocryptorhynchus burmanus* that may not belong in Mesophyletidae. However, because of its poor preservation (strongly crumpled and partially obscured by bubbles) we do not name and describe a species for it here. Its critical observable characters include subbasally inserted antennae with evidently 5- or possibly 6-segmented funicles, short scapes (not determinable whether the antennae are non- or subgeniculate) and 4-segmented clubs and seemingly subvertical, exodont but apically pointed mandibles with 2 outer recurved teeth and no inner teeth (a unique mandible structure among known Burmese amber weevils). Thoracic details are well discernible; the notosternal sutures are open and the mesocoxal cavities laterally open as well. Ventrites 1 and 2 are fused and the others seemingly free, and a pygidium is indicated but the area is too crumpled to clearly assess this (the apex may just be distended).

Figure 54. *Gnomus spinipes* sp. n., holotype (**a**–**e**) and *Gnomus* sp. (**f**–**h**). Habitus, left lateral (**a**); habitus, right lateral (**b**); right hindleg (**c**); protibia and tarsi (**d**); right antenna (**e**). Habitus, right caudolateral (**f**); head and rostrum, left lateral (**g**); habitus, left lateral (**h**). Scale bars: 1.0 mm (a,b,f).

The legs are relatively well preserved and the tarsi are distinctive, with digitate lobes of tarsites 3 and angulate claws. The claws, however, are unusual in that the triangular inner tooth is situated at the middle of the elongate claw rather than at the base, and the claws lack a ventrobasal seta. The

4-segmented clubs, fused ventrites 1 and 2, laterally open pro- and mesocoxal cavities and subbasally inserted antennae are characters of Mesophyletidae, in particular of some genera of Aepyceratinae. However, the other antennal characters and the lack of ventrobasal claw setae do not agree with most Mesophyletidae (although ventrobasal setae are apparently absent in some species). Although the specimen is thus distinctive from other Mesophyletidae, we retain it in *Gnomus* for now, pending better-preserved material becoming available. It was submitted for CT scanning, but this produced no useful images to permit a better assessment of its characters.

This species is distinctive among Burmese amber weevils in having a pale reddish-brown cuticle (not so apparent under LED lighting) with a colourful metallic sheen (on elytra, seemingly other regions) and tibiae with divergent spurs, one of which is normal and directed distad and the other one is fixed and perpendicular to the tibial axis. Divergent spurs are so far otherwise known from one or two species related to *Compsopsarus*, of which we received specimens from the AMNH but which require further preparation to adequately describe and classify.

Genus ***Hukawngius*** Clarke & Oberprieler, **gen. n.**

Type species: *Hukawngius crassipes* Clarke & Oberprieler, sp. n.

Description. Size. Length 3.8 mm, width 5.0 mm. **Head** short, conical. **Eyes** lateral, slightly protruding, coarsely facetted, dorsally narrowly separated, by less than width of rostrum at base, forehead without tubercles. **Rostrum** slightly longer than pronotum, stout, cylindrical; antennal insertions lateral, behind them with faint scrobes extending to eye, in front of them without lateral row of setae. Long single gular suture indicated. **Antennae** geniculate; scapes cylindrical apically inflated; funicles slightly longer than scape, 7-segmented; clubs loosely articulated, 4-segmented **Mouthparts**. Labrum absent. Mandibles exodont, with 2 teeth on outside, anterior one forming short flat Y with sharp apical tooth, articulation plane horizontal. **Thorax**. Prothorax proclinate, with anterior lateral margins oblique in lateral view. Pronotum laterally rounded, without tooth, posterior corners angulate but not produced to fit closely onto elytra; surface sparsely setose; notosternal sutures closed. Prosternum moderately long; procoxal cavities medially confluent, close to posterior margin of hypomeron. Scutellar shield transverse. Mesocoxal cavities laterally closed (by meso- and metaventrite). Metanepisternal sutures distinct. Mesoventrite very short, anteriorly strongly sloping. Metaventrite longer, convex, without transverse weals before metacoxae. **Elytra** with well-developed, broadly rounded humeri, posteriorly steeply declivous, apically individually rounded, not exposing pygidium; sutural flanges not visible; surface punctostriate, without scutellary striole, sparsely irregularly setose. **Legs**. Procoxae elongate, prominent, medially contiguous; mesocoxae subglobular, narrowly separated; metacoxae flat, transversely elongate. Trochanters short, oblique. Femora subcylindrical, strongly inflated in distal half, unarmed, outside rounded. Tibiae straight, subcylindrical, as long as femora, outside rounded, apex obliquely truncate in pro- and mesotibiae, truncate in metatibiae, with 2 spurs. Tarsi robust, 0.67 × as long as tibia; tarsite 1 apically truncate, 2 apically slightly excised, 3 deeply bilobed, 5 as long as 1 + 2; claws divaricate, strongly dentate with ventrobasal seta at apex of tooth. **Abdomen** with ventrites 1 to 2 broad and long, subequal in length, 3 and 4 each slightly shorter than 2, 5 as long as 4, wide.

Derivation of name. The genus is named after the Hukawng Valley in Myanmar, in which the amber mines are located; the gender of the name is masculine.

Remarks. *Hukwangius* is distinguishable from all other mesophyletid genera with dentate claws by the lack of any ridge or crest on the outside of the tibiae (with or without serrulation/crenulation). From *Anchineus* (which may or may not have serrulate tibiae; the legs in the holotype are distorted) it differs by its much larger size, its longer, thinner antennae, exodont horizontal mandibles, cylindrical tibiae and broadly lobed tarsites 3 (Figure 55). In the shape of its eyes it is similar to *Aphelonyssus latus*, but the latter is much smaller and has the tibiae and also the femora finely crenulate on the outside.

Figure 55. *Hukawngius crassipes* sp. n., holotype. Habitus, left lateral (**a**); habitus, right lateral (**b**); habitus, dorsal (**c**); head and antenna, left lateral (**d**); legs (**e**); eyes, frontal (**f**); apex of rostrum and mandibles, dorsal (**g**); head, frontal (**h**); tarsi, ventral (**i**). Scale bars: 1.0 mm (a–c,e); 0.2 mm (g,i).

Hukawngius crassipes Clarke & Oberprieler, **sp. n.** (Figure 55)

Description. Size. Length 3.79 mm, width 5.0 mm. **Head** slightly constricted behind. **Eyes** large, hemispherical, dorsally close together. **Rostrum** strongly curved, dorsally forming slight sinus with head in profile; antennal insertions in basal third of rostral length. **Antennae** long; scapes slightly curved, thin proximally but gradually inflated towards apex; funicles with segment 1 spindle-shaped, 2 slightly shorter, others progressively shorter towards club; clubs long, apical segment about as long as 3, slightly flattened. **Thorax**. Pronotum broadly parabolic; convex, surface sculpture not discernible (decomposed, with large white spots), setae pale, thin, acute, dorsally directed anteriad, laterally directed anteromesad; notosternal sutures with upright stem, then abruptly bent and curved anteriad. Scutellar shield slightly rounded, short, setose. **Elytra** short, broad; striae indistinct, broad, setae pale, thin, acute, recurved caudad. **Legs**. Femora short, robust, faintly sinuate, thickest in about distal third. Tibiae stout, outside with a few small, low denticles visible in lateral view but not in a row in dorsal view, with dense setae in apical half, spurs small, pale, flat (best visible on left mesotibia in apical view from right side, also on protibia). Tarsi with tarsite 1 moderately elongate, apically broadening, 2 broader and shorter, 3 with lobes shortly pedunculate, 5 basally thin, strongly broadening apicad; claws very wide, ventral tooth almost right-angled. **Abdomen** as for genus.

Material examined. Holotype (NIGP156991): relatively well preserved, intact specimen, not compressed or distorted, with surface of pronotum and part of left elytron somewhat damaged; in centre of irregular rounded block ca. 7.5 × 5.0 × 3.5 mm; amber clear with few impurities but with flat vertical film along left side of specimen and horizontal crack around specimen.

Derivation of name. The species is named for its stout legs, from the Latin adjective *crassus* (stout) and noun *pes* (a foot); the name is a noun in apposition.

Remarks. This species is seemingly isolated in the group with dentate tarsal claws and geniculate antennae in the form of the legs, with cylindrical (rounded) tibiae and low tubercles not arranged in a row on the outside. Uniquely for this group, it also has the antennae inserted in the basal third of the rostral length and large hemispherical eyes only very narrowly separated dorsally.

Genus ***Mekorhamphus*** Poinar, Brown & Legalov, 2016

Mekorhamphus Poinar, Brown & Legalov, 2016: 158 [54] (type species, by original designation: *Mekorhamphus gyralommus* Poinar, Brown & Legalov, 2016)

Description. Size. Length 2.84–3.18 mm, width 0.95–1.81 mm. Body and appendages black. **Head** short to shortly elongate, strongly constricted and dorsoventrally bulging behind eyes. **Eyes** large, strongly protruding, coarsely facetted, dorsally separated by about basal width of rostrum anteriorly but further separated posteriorly; forehead with a pair of elongate carinate tubercles between eyes. **Rostrum** longer than pronotum to nearly as long as body, slender, cylindrical, strongly curved; antennal insertions lateral, with scrobes behind them reaching slightly below eye, in front of them laterally with a few long erect setae in apical quarter or more. Gular suture not clearly discernible. **Antennae** geniculate, long; scapes long, slender, cylindrical, apically clavate, about as long as funicle, reaching below eyes; funicles 7-segmented, segments 2–5 subequal in shape, long, slender, apically slightly widened, 6–7 stouter, shorter; clubs large, much broader than funicles, loosely articulated, 4-segmented, segment 4 acute, subflattened, about as long as 3. **Mouthparts**. Labrum absent. Mandibles small, exodont, articulation horizontal to possibly oblique. **Thorax**. Prothorax proclinate, often strongly so, with anterior lateral margins oblique in lateral view. Pronotum convex, laterally rounded, with or without tooth, posterior corners distinct, angulate, fitting closely onto elytra; surface coarsely punctorugose, densely setose, setae reclinate, directed anteromesad, mixed brown and coloured; notosternal sutures closed, vertical then abruptly curved anteriad, this part distinctly sulciform (a narrow groove). Prosternum moderately long; procoxal cavities medially confluent, usually closer to posterior margin of hypomeron. Scutellar shield densely setose. Mesocoxal cavities laterally closed. Metanepisterna distinct, densely setose. Mesoventrite short, anteriorly sloping. Metaventrite longer, convex, raised into distinct transverse weals. **Elytra** broad, basally shortly broadly lobed over pronotum, with rounded humeri, lateral margin strongly sinuate to roundly emarginate in middle;

posteriorly declivous, apically individually to nearly conjointly rounded, not exposing pygidium; sutural flanges narrow, subequal; surface punctostriate, without scutellary striole, interstriae flat, low, finely rugose, densely setose, setae mainly short (sometimes distinctly longer), mixed brown and coloured, thin, reclinate, directed caudad, interstriae without dense patches of coloured setae. **Legs**. Procoxae large, prominent, medially contiguous; mesocoxae subglobular, narrowly separated; metacoxae flat, transversely elongate. Trochanters short, oblique. Femora long, subcylindrical, inflated in distal half, outside rounded (profemora) or crenulate in distal half to third (meso-, metafemora). Tibiae compressed, outer edge carinate (protibiae; crenulate in *M. gracilipes* and *M. tenuicornis*) or crenulate (meso- and metatibiae), with longer denser setae in distal half, apex obliquely truncate, with 2 spurs; metatibiae dorsally notched at apex (except *M. beatae*). Tarsi almost as long as tibiae; flattened, densely setose; tarsite 1 elongate-triangular, apically excised, 2 shorter, triangular, apically deeply excised, 3 deeply bilobed, lobes pedunculate, inner margin sometimes concave at base, 5 long, slender, apically slightly expanded; claws divaricate, dentate with ventrobasal seta. **Abdomen** with ventrites 1 and 2 slightly longer than 3, 3 and 4 subequal, sutures between 2 and 5 distinctly grooved.

Derivation of name. The name of the genus is improperly formed and latinised; formed for a "prolonged rostrum" as stated it should have been *Mecinorhamphus*, as the Greek adjective for prolonged is *mekynos* (*mekos* is a noun and means length) and the Greek letter and sound *k* should have been latinised to *c* (as the Greek ending *-os* of *rhamphos* has been to the Latin *-us*). An unfortunate concoction but nonetheless nomenclaturally valid.

Remarks. *Mekorhamphus* was placed in its own tribe, Mekorhamphini, based on its horizontally moving exodont mandibles, 3-segmented maxillary palps, postmedially inserted antennae, elongate prosternum, contiguous procoxal cavities, distinct elytral striae, free ventrites, trochanters separating femora from coxae and first tarsites narrow and weakly extended. Only a few of these characters are correct as given (palps, procoxal cavities, striae); others cannot be compared in the other tribes from which Mekorhamphini were distinguished, because the characters cannot be properly seen in the other specimens, as we could observe, or are imprecisely defined (elongate prosternum, first tarsites). The genus was erected for a single species, *M. gyralommus*, and diagnosed by several characters, most of these being characteristic of all the species now known. *Mekorhamphus* is a character-rich genus distinguishable from all other Burmese amber weevils by the combination of the carinate tubercles between the eyes, crenulate meso- and metatibiae (sometimes also protibiae, e.g., *M. gracilipes*), broad, excised tarsites 1 and 2 (the latter more deeply so), strongly pedunculate lobes of tarsites 3, curved ventrites (one or more) and deeply grooved sutures between ventrites 2 and 5. The carinate tubercles between the eyes were not described for *M. gyralommus*, but they are present in this species too, just not distinctly visible because they are obscured by setal tufts and only discernible in direct dorsal or frontal views (e.g., Figure 56b,c,f–h and Figure 57f). *Mekorhamphus* is the only genus of Mesophyletidae that lacks an externally distinct gular suture as we could observe, this seemingly obliterated by the coarse punctorugose sculpturing of the venter. The deeply sulciform horizontal portion of the notosternal sutures and the coarse, dense, deep punctation of particularly the ventral side of the thorax are also characteristic of the genus. The lateral pronotal teeth (one on each side), as described for *M. gyralommus*, and the long fifth ventrite do not occur in all species of the genus but also in two *Myanmarus* species (though much more pronounced in *Myanmarus dentifer*). In most species, including *M. gyralommus*, the metatibiae (sometimes also the mesotibiae) are distinctly apically excised or notched on the outer side, whereas in other species it appears that the apex of the metatibia is instead strongly obliquely truncate, although compression may obscure the visibility of this character.

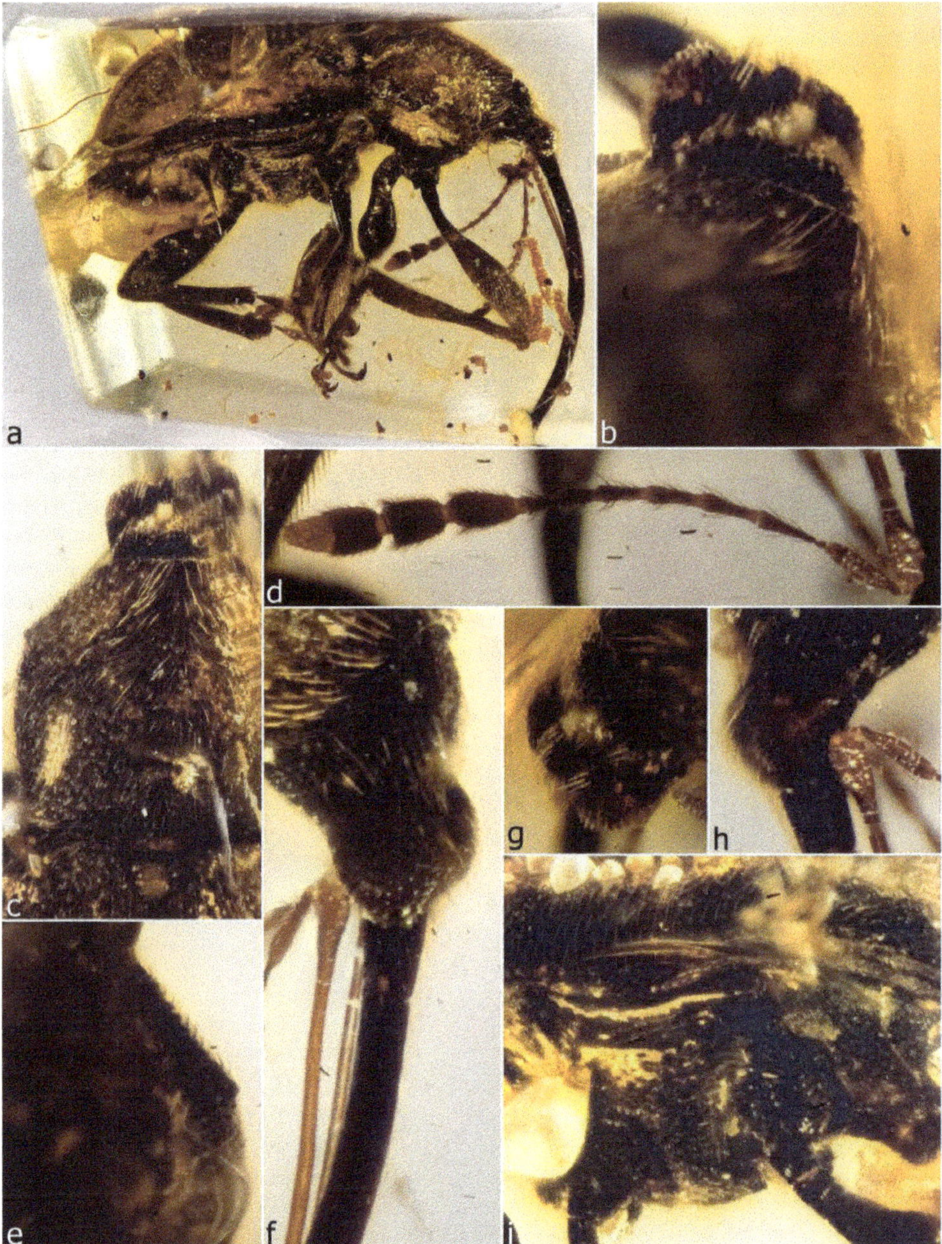

Figure 56. *Mekorhamphus gyralommus.* Habitus, right lateral (**a**); head showing tubercles and setal tufts between eyes, dorsoposterior (**b**); pronotum showing lateral denticle, dorsal (**c**); left funicle and club, left lateral (**d**); detail of pronotal denticle, dorsal (**e**); head, right dorsolateral (**f**); head, showing detail of tubercles and setal tufts, dorsolateral (**g**); eyes, left lateral (**h**); metathorax, left lateral (**i**).

Figure 57. *Mekorhamphus gyralommus*. Left hindleg, outer side (**a**); detail of left metatibia, outer side (**b**); tarsi, dorsal (**c**); detail of tarsus, dorsal (**d**); right protibia, inner side (**e**); *Mekorhamphus gyralommus*, holotype, head, dorsal view (**f**).

We recognise five species in *Mekorhamphus*, four of them here described. We include one other specimen in the genus, but we do not describe and name the species as the specimen is so poorly preserved that it is virtually impossible to adequately compare it with or distinguish it from other species, both the ones here described and possible future ones. Another congeneric specimen, received from the FMNH but too late to be prepared for inclusion in this paper, has a shorter, straighter rostrum and may represent another undescribed species or the male of one of the described ones, in which case substantial sexual dimorphism in rostrum length is indicated to occur in *Mekorhamphus*.

Mekorhamphus gyralommus Poinar, Brown & Legalov, 2016 (Figures 56 and 57)

Mekorhamphus gyralommus Poinar, Brown & Legalov, 2016: 158 [54]

Redescription. Size. Length 2.84–3.06 mm, width 1.42–1.54 mm. Body coarsely punctate, densely setose; setae short, subrecurved; sparser ventrally. **Head** shortly elongate, narrow; coarsely densely punctorugose, setose; with tubercles tracking inner margin of eyes, somewhat concealed by tufts of dense long curved whitish setae converging at base of rostrum and continuing along rostrum in grooves and thinning toward antennal insertions; forehead (between tubercles) grooved; ventrally coarsely punctorugose, no gular suture discernible. **Eyes** without interfacettal setae; slightly elongate. **Rostrum** about as long as elytra, strongly downcurved; dorsally with median, paramedian and dorsolateral irregular carinae extending to apex, and with deep intervening setose grooves in basal half reaching antennal insertions; antennal insertions slightly postmedian, in front of them with 5 short widely spaced setae in lateral groove. **Antennae.** Scapes apically distinctly oblique; funicles with segment 1 elongate, widening distad, slightly wider than rest of funicle, 2 slightly obconical, slightly shorter, 3–6 subcylindrical, slightly broader apically, 7 obconical; clubs about as long as funicle segments 3–7, segments 1 and 2 obconical, 1 apically rounded, 2 slightly broader, apically flattened, shorter than 1, 3 shorter, 4 subequal in length to 3, slightly paler, narrowly rounded, broadly inserted into 3. **Mouthparts.** Mandibles small, with 3 teeth on outside, 2 short triangular teeth basally, at least 1 large tooth on inside, apex forming V with apical-most inner and outer teeth, articulation oblique. **Thorax.** Prothorax proclinate, slightly narrower than elytra, coarsely punctate. Pronotum laterally with distinct but small denticle in apical third, base sinuate, closely fitting to elytra, posterior corners acutely angulate; with vestiture of mixed whitish and brownish setae. Prosternum short, ca. half as long as procoxae, anteriorly sloping, anterior margin substraight, prosternal process short, pointed; procoxal cavities slightly closer to posterior margin of hypomeron, this deeply, narrowly emarginate medially and with broad triangular hypomeral process. Scutellar shield elongately trapezoidal, covered with short whitish setae. **Elytra** with striae and interstriae subequal in width at least basally; strial punctures deep but narrow; interstriae prominent, more so basally, 1 (sutural) distinctly raised, lined with brownish setae, 8 more prominent than others, subcarinate, forming lateral margin visible in dorsal view, 9 subcarinate in basal third of length, becoming obsolete before about middle; elytral bases strongly sinuate; sides with distinct setose marginal groove, widening anteriad, with distinct anterior marginal notch; apices when fully closed weakly triangularly emarginate, nearly conjointly rounded; vestiture of largely whitish setae but mixed with some brownish setae. **Legs** long, slender; densely setose, covered with short subappressed golden or whitish setae, densest on femora; forelegs longer than middle legs, these slightly longer than hindlegs. Procoxae subconical. Protibiae swollen distally, outer edge carinate, broadly curved in distal half, inner edge straight with row of short oblique setae in distal half, apically without fringing setae, with 2 short spurs; mesotibiae apically abruptly curved inwards; metatibiae straight, shorter than mesotibiae, meso- and metatibae with outer edge finely carinate-crenulate, straight, notched at apex, inner and outer apical flanges with long coarse fringing setae, with 2 spurs, outer one on metatibiae robust, curved, inner one tiny, less than half length of larger spur. Tarsi at least half as long as tibiae (protarsi half, mesotarsi ca. 0.75 ×), mesotarsi shorter and narrower than protarsi, ventrally with dense setal pads; tarsite 1 ca. twice longer than 2, 3 strongly pedunculate, lobes elongate, ca. 0.5–0.75 × as long as 5, 5 slender, gradually expanding apicad. **Abdomen.** Ventrites subflatly aligned; densely setose, setae whitish; 1 and 2 subequal in length, 3 and 4 ca. half as long, subequal in length; sutures between 1 and 2 less distinctly grooved than between 2 and 5, sutures broadly curved, then kinked just before margin.

Material examined. Holotype (ISEA no. MA 2015/1), length 2.84 mm, width 1.42 mm: well preserved specimen, visible from all sides, slightly depressed, with eyes partly compressed, elytra and thoracic venter with fragmented debris coating, right mesotarsus missing, right protarsus, right metatarsus and left mesotarsus missing tarsites 3–5 (latter claw segment in amber), left metatarsus missing lobes of tarsite 3 and claws; re-prepared from slightly larger cuboid amber piece from which original description was prepared, now at one end of cuboid 7.8 × 4.3 × 3.3 mm; amber clear yellow with few large impurities, with two large discoidal fracture planes around specimen. Other material. One (probably female) specimen (GPIH no. 4987; coll. Gröhn no. 11145), length 3.06 mm, width 1.32 mm: near-perfectly preserved, intact and well-visible; re-prepared from ca. 20.0 × 14.0 mm amber cabochon, now in irregular wedge, 5.5 × 2.7 × 2.5 mm, with one curved face; amber clear yellow with few organic impurities, some larger scattered masses to left of specimen, with large bubble emanating from and obscuring apical ventrites and small bubble at apex of rostrum partly cut away and exposing mandibles (both infilled with resin; see Section 2.2), with numerous other smaller bubbles emanating from specimen and several small fractures on left and dorsal sides of specimen.

Remarks. This species is distinguishable from all other *Mekorhamphus* species by the tufts of long setae between the eyes and the consequently indistinct tubercles, in concert with the slightly elongate head and eyes. It was described as having slightly curved protibiae with a small mucro; the protibiae are in fact distinctly curved ('machete'-shaped) and no mucro is evident in either of the specimens we examined. In the holotype there is a slight notch at the outer angle of the right protibia, with a coarser seta, but this is not matched on the other protibia and there is no mucro on the inner angle. We examined a second, very similar but slightly larger, evidently conspecific specimen that is much better preserved than the holotype and differs mainly in having a longer rostrum and some details that appear related to distortion in the holotype. Of particular importance in matching the two specimens is the form of the head and tubercles, the straight inner protibial edge lined with oblique setae and the curved outer edge of the protibiae. *Mekorhamphus poinari* has similarly modified but differently shaped protibiae and also anterolateral pronotal teeth, but its eyes are rounder, the tubercles between the eyes more conical and the metatibial spurs diminished and projecting distad (possibly fixed, the inner one much smaller).

Mekorhamphus beatae Clarke & Oberprieler, **sp. n.** (Figures 58 and 59)

Description. Size. Length 3.01–3.31 mm, width 1.59 mm or more. Body coarsely, densely punctate; densely setose. **Head** grooved around lower sides of eyes. **Eyes** large, globular, facing forward, dorsally separated by about basal width of rostrum or less; forehead raised into 2 short, flat and blunt, keel-like longitudinal tubercles between eyes, distinctly grooved between them. **Rostrum** ca. 2.4× longer than pronotum; antennal insertions in ca. middle of rostral length, behind them with scrobes reaching base of eye, in front of them with lateral row of ca. 5 long, erect setae. **Antennae** very long; scapes slightly longer than funicle, terete, thin, apically oblique; funicles with segment 1 elongate, cylindrical, 2 subequal in length but slightly narrower proximally, 3–5 similar but progressively shorter, 6 and 7 slightly thicker, shorter; clubs slightly flattened, especially segment 4, much broader than funicles, segments 1–3 obconical, 2 shorter and wider, 4 distinct but fitting closely onto 3, flat. **Mouthparts.** Mandibles narrow, strongly exodont, outer edge with 3 large blunt teeth, middle one directed obliquely down, apical one forming broad sharp V with opposing inner one, the latter double (2 above each other), inner edge with 2 or 3 teeth, basal one(s) large. Labial palps long, 2-segmented **Thorax.** Pronotum broadly ogival, strongly convex, laterally rounded, without lateral tooth, posterior corners extended, fitting closely onto elytra, densely coarsely punctate-tuberculate, setae arising from punctures raised on small low tubercles, pale. Prosternum moderately long, strongly declivous, anterior margin broadly emarginate; procoxal cavities in about middle of prothorax. Scutellar shield small, prominent. **Elytra** shortly elongate, posteriorly gently declivous, punctostriate but striae indistinct; densely setose, setae directed mesocaudad to caudad; sides without carinate interstria 8, with setose marginal groove gradually widening anteriad, anterior marginal notch present; bases weakly sinuate, slightly extended over pronotum, apically individually rounded but closely fitting. **Legs** long, slender. Procoxae subconical; mesocoxae globular, subflat, narrowly separated. Femora inflated, slightly compressed, almost straight, outside with crenulate carina in distal 40 % of length, weaker on profemora. Tibiae substraight, slightly curved inwards at apex, outer side carinate (protibiae) or carinate-crenulate (meso- and metatibiae); apically truncate, with tarsal articulation surfaces oblique, with long coarse fringing setae, with 1 (protibiae) or 2 long stout spurs (meso- and metatibiae, on latter unequal, inner one larger). Tarsi ventrally with dense setal pads; tarsite 1 slightly longer than 2, 3 with pedunculate lobes ca. half as long as 5, 5 as long as 1 + 2, broadening distad. **Abdomen.** Ventrites stepped, densely setose, 3–4 with setae ca. 2.0× longer than those on 1–2, 1–2 subflatly aligned, subequal in length, 3–5 at slightly higher levels, 3–4 very short, subequal in length, ca. half as long as 2, 4 shorter than 3, 5 seemingly shorter than 4, retracted; suture between 1 and 2 slightly curved, between 2 and 3 slightly sinuate, between 3 and 4 distinctly sinuate, between 4 and 5 seemingly weakly sinuate, between 2 and 5 forming deep groove, with broad gap between ventrites 2 and 3 and 3 and 4 near middle.

Figure 58. *Mekorhamphus beatae* sp. n., holotype. Habitus, right lateral (**a**); head and prothorax, right lateral (**b**); apex of rostrum and mandibles, dorsal (**c**); ventrites, right lateral (**d**); prothorax and anterior pterothorax, right lateral (**e**); tarsites 2–5, dorsal (**f**); left antenna, left lateral (**g**); elytra, left lateral (**h**); right metatibia (outer side) showing crenulation (**i**); tarsi (**j**). Scale bars: 1.0 mm (a).

Material examined. Holotype (NIGP156992; Figure 58), length 3.01 mm, width 1.59 mm: exceptionally well preserved, nearly intact specimen, missing only claws of right hindleg, well visible (including mandibles) except from left side, surface details generally obscured by mostly even film of whitish debris; near edge of subovoid slab, 5.8 × 3.2 × 3.0 mm; amber clear yellow without impurities, with numerous bubbles and few small fractures on left and partly obscuring that side of specimen. Paratype (in private collection of Wei Ma, China; Figure 59), length 3.31 mm, width not measurable due to curvature of amber block: very well preserved, intact specimen, not compressed or distorted, well visible from both left and right but not from other sides due to shape of amber block; at an angle in centre of lenticular block with flat surface on left side of specimen, 16.1 × 10.3 × 6.3 mm; amber clear but with layer of sparse bubbles along right side of specimen, also other impurities but not obscuring specimen (which will be almost perfectly and totally visible if amber is cut down).

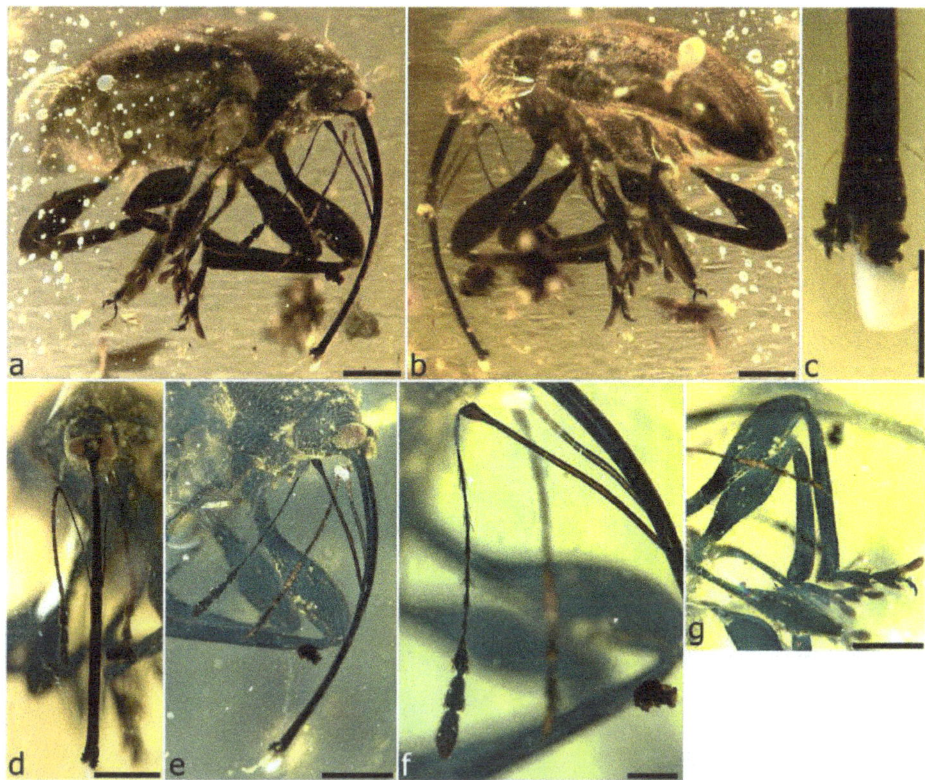

Figure 59. *Mekorhamphus beatae* sp. n., paratype. Habitus, right lateral (**a**); habitus, left lateral (**b**); apex of rostrum and mandibles (**c**); head and antennae, frontal (**d**); head and prothorax, right lateral (**e**); right antenna, lateral (**f**); anterior legs (**g**). Scale bars: 0.5 mm (a,b,d,e,g); 0.2 mm (c,f).

Derivation of name. This striking species is cordially named after Beate Oberprieler, wife of the senior author, in grateful recognition of her sacrifice of so many family hours and use of the family dinner table for our study of these fossils. The name also recalls the beauty and exceptional preservation of the specimens, the Latin adjective *beatus* meaning happy, fortunate, blessed.

Remarks. This species differs from *M. gyralommus* and *M. poinari* in having a laterally smooth prothorax, without any dentiform processes, and in a much longer and thinner rostrum. It differs from *M. gracilipes* in its shorter, stouter tarsi, dorsally carinate protibiae, less strongly curved ventrites and apically truncate metatibiae (not or weakly notched), and from *M. tenuicornis* it is distinguishable by its uniformly short elytral vestiture and dorsally carinate protibiae. The two specimens are virtually identical except that the paratype is slightly larger, and the very long, thin rostrum indicates both to be females. The holotype was submitted for CT scanning, but the contrast between the specimen and the amber was too low to permit any meaningful visualisation of the specimen.

Mekorhamphus gracilipes Clarke & Oberprieler, **sp. n.** (Figures 60 and 61)

Description. Size. Length 2.85–3.2 mm, width 1.5–1.8 mm. Entire body coarsely and densely punctate. **Eyes** large, globular, lateral, separated dorsally by about a basal rostral width. **Rostrum** ca. 2 × longer than pronotum (cut away apically); antennal insertions median, behind them with scrobes reaching eye, in apical quarter with lateral row of ca. 5 long, erect setae. **Antennae** very long, thin; scapes long, slightly longer than funicles, thin, apically oblique, reaching below eye; funicles with segment 1 slender, cylindrical, 2 subequal in length but slightly narrower proximally, 3–5 similar but progressively shorter, 6 and 7 slightly thicker, shorter; clubs slightly flattened, especially segment 4, segments 1–3 obconical, subequal in length; 4 broadly inserted into and slightly shorter than 3, apically narrowly rounded. **Mouthparts** (not preserved, cut away with amber). **Thorax.** Prothorax strongly proclinate. Pronotum broadly ogival, densely and coarsely punctate, laterally without tooth, sparsely setose, posterior angles acute, fitting closely to elytra. Prosternum long, anterior margin broadly arcuate, prosternal process short, pointed; procoxal cavities medially confluent, closer to posterior hypomeral margin. Scutellar shield narrowly roundedly triangular, raised, densely setose. **Elytra** short and broad, with well-developed, broadly rounded humeri, posteriorly gently declivous, apically individually rounded, surface indistinctly punctostriate, densely covered with long, recumbent setae; sides lacking any keel, margin with anterior marginal notch. **Legs** long, slender. Procoxae subconical; mesocoxae narrowly separated. Femora long, slightly compressed. Tibiae long and slender, pro- and metatibiae longer than mesotibiae; nearly straight, slightly widening distad, pro- and mesotibiae not bent inward apically, outer side crenulate, outer edge of metatibiae (and possibly mesotibiae, apices obscured or slightly distorted) distinctly emarginate apically; apex of pro- and mesotibiae obliquely truncate, of metatibiae squarely truncate, all with 2 long, sharp, slightly curved, equal spurs; apical edges with coarse fringing setae, tarsal articulation surfaces oblique. Tarsi shorter than tibiae; tarsite 1 longer and narrower than 2, 3 with lobes pedunculate, each ca. half as long as 5, 5 subequal to or slightly shorter than 1 + 2 on pro- and mesotarsi, longer than 1 + 2 on metatarsi. **Abdomen.** Ventrites subflatly aligned, curved (especially 2–4), 1 and 2 subequal in length, 3 half as long as 2, 4 slightly shorter than 3, 2–5 with long erect setae, 5 broad, slightly shorter than 4; sutures distinctly curved, between ventrites 1 and 2 very slightly curved at middle but laterally kinked (deflected posterodorsally), between 2 and 5 strongly arcuate, forming broad gap at middle (gap narrow between ventrites 1 and 2).

Material examined. Holotype (NIGP156993; Figure 60), length 3.18 mm, width 1.81 mm: exceptionally well preserved, nearly intact and well visible specimen, slightly depressed and distorted (prothorax, and right protibia bent inwards apically), with apical part of rostrum cut away with amber, with fragmented coating of whitish debris; in cuboid 7.2 × 4.3 × 3.5 mm, with one rounded face; amber clear yellow, with faint flow bands, few impurities, with several smaller subcircular fractures around and partly obscuring specimen. Paratype (NIGP156994; Figure 61), length 2.85 mm, width ca. 1.52 mm: reasonably well preserved specimen, not distorted but legs somewhat compressed, head, tip of rostrum and outside of left elytron cut off; in left side of irregular block with rectangular front surface 2.7 × 2.3 mm and tapering towards back, 4.0 mm long; amber clear but with curved layer of flow lines with bubbles along right side and small horizontal crack at back of specimen, also large bubble below right profemur.

Derivation of name. The species is named for its long, slender legs, especially the tarsi.

Remarks. *Mekorhamphus gracilipes* also differs from *M. gyralommus* and *M. poinari* in having a laterally smooth prothorax, without any dentiform processes. It is very similar to *M. beatae*, differing in the dorsally rounded profemora, longer thinner tarsi, finer tibial crenulation, distinctly apically notched metatibiae, flatly aligned and distinctly curved ventrites (especially 3 and 4) and short but distinctly visible ventrite 5. It is similar to *M. tenuicornis* in having long slender legs and distinctly apically notched metatibiae (also with long thick spurs), but it differs in having the pronotal setae not inserted on small tubercles and dense long recumbent setae on the elytra, as well as apically less strongly excised tarsites 2. The two specimens available for study are virtually identical, differing in no significant characters.

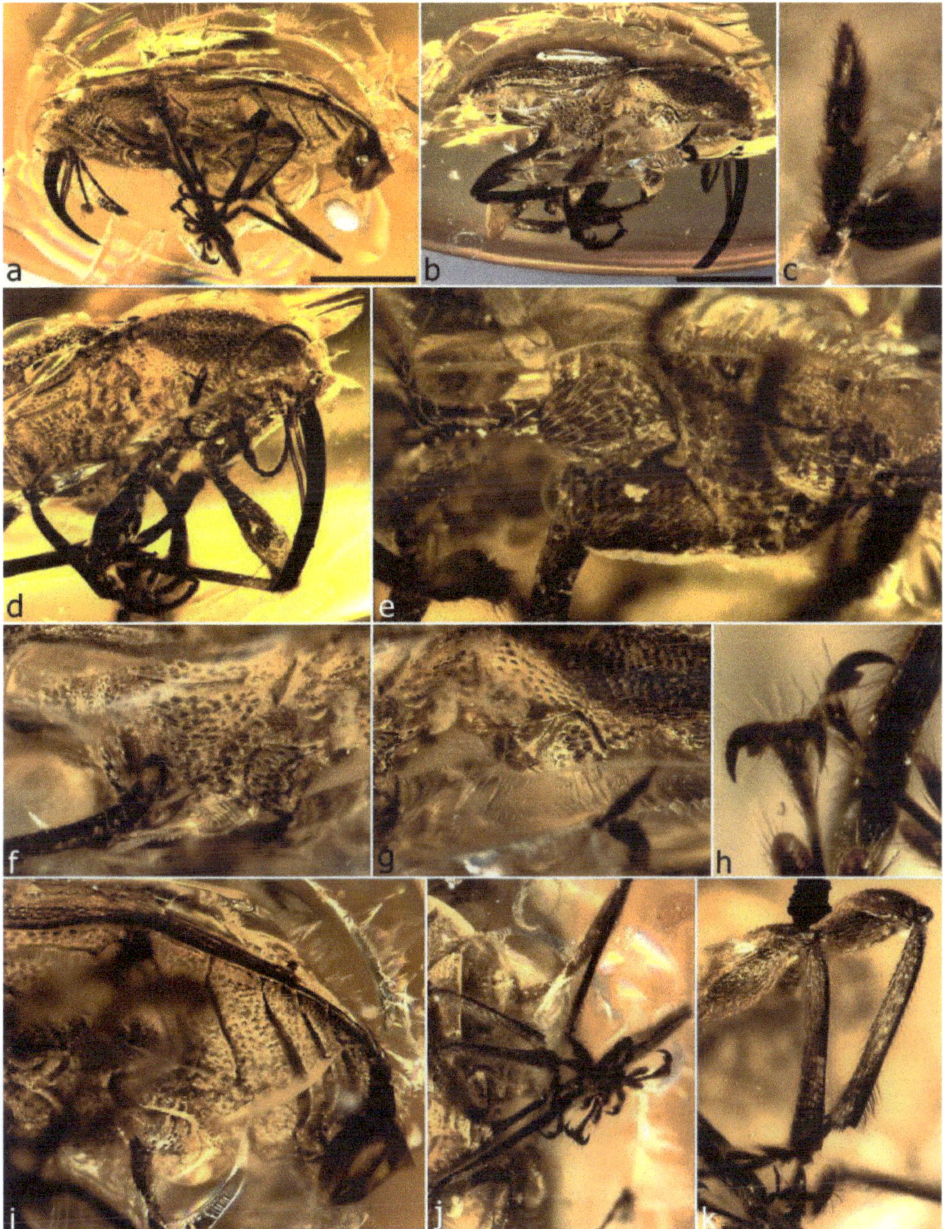

Figure 60. *Mekorhamphus gracilipes* sp. n., holotype. Habitus, left ventrolateral (**a**); habitus, right lateral (**b**); detail of antennal club (**c**); head and thorax, oblique frontal (**d**); prothorax and coxae, ventrolateral (**e**); metathorax, lateral (**f**); prothorax showing notosternal suture (**g**); tarsal claws (**h**); ventrites, ventrolateral (**i**); legs (**j**); meso- and metatibiae, outer sides (**k**). Scale bars: 1.0 mm (a,b).

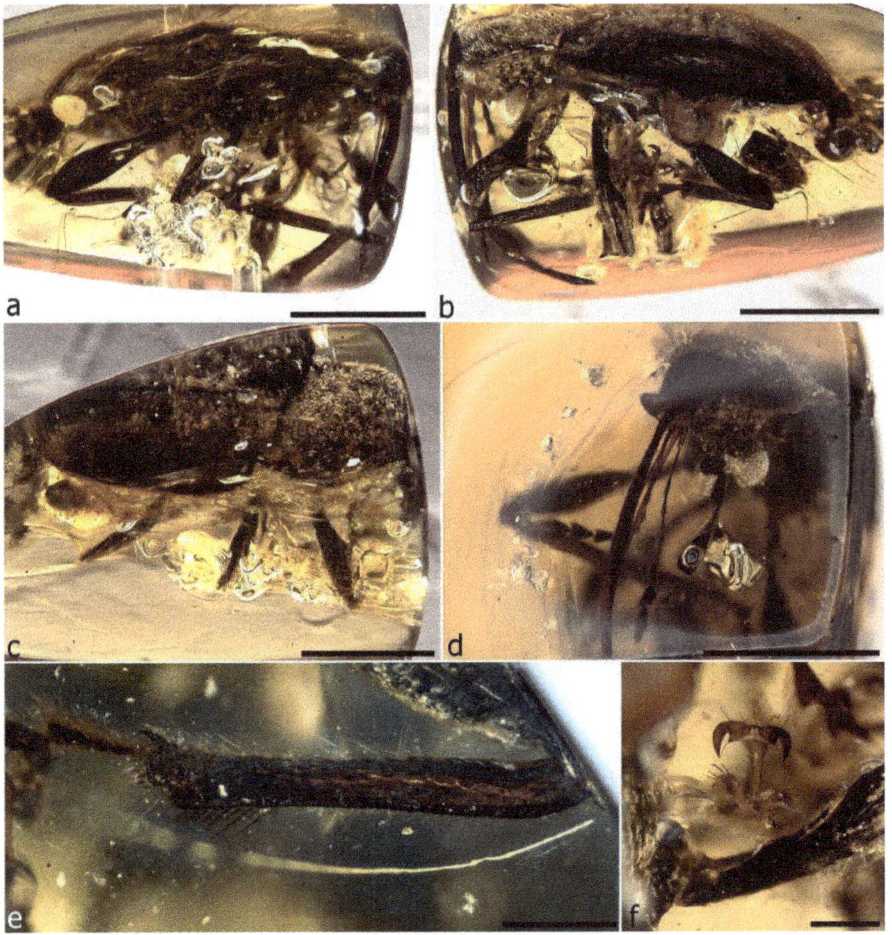

Figure 61. *Mekorhamphus gracillipes* sp. n., paratype. Habitus, right lateral (**a**); habitus, left lateral (**b**); habitus, dorsal (**c**); head and antennae, frontolateral (**d**); right metatibia showing subdued crenulation (**e**); claw, apical (**f**). Scale bars: 1.0 mm (**a–d**); 0.2 mm (e,f).

Mekorhamphus tenuicornis Clarke & Oberprieler, **sp. n.** (Figure 62)

Description. Size. Length 2.85 mm, width 0.95 mm. **Head** subglobular. **Eyes** dorsally narrowly separated. **Rostrum** longer than pronotum; antennal insertions in middle of rostral length, behind them with long narrow scrobes extending to eye, in apical quarter with lateral row of few long, erect setae. No gular suture visible. **Antennae.** Scapes apically abruptly clavate, strongly oblique; funicles slightly longer than scape, segments elongate, thin, 6 and 7 somewhat shorter and broader; clubs flattened, with segment 4 distinct. **Mouthparts**. Mandibles with 3 triangular teeth on outside and narrow apical point. **Thorax**. Prothorax strongly proclinate. Pronotum elongate, laterally without tooth, posterior angles acute, fitting onto elytra; sparsely setose, setae inserted on small tubercles, moderately long, thin, dark brown. Prosternum with position of procoxal cavities not discernible. Scutellar shield small, transverse (obscured). Metaventrite convex. **Elytra** elongate, with well-developed, broadly rounded humeri, anterior bases lobed, enclosing scutellar shield, posteriorly gently declivous, apically individually rounded; surface coarsely punctostriate but striae faint, interstriae narrow, roundly subcarinate, with sparsely vestiture of long, thin, dark brown, suberect setae directed caudad. **Legs**. Procoxae elongate, somewhat flattened, narrowly separated. Femora long, straight, slightly compressed, outside rounded (profemora) or crenulate in distal quarter of length (meso-, probably metafemora) Tibiae long, straight, outer edge crenulate, apex obliquely truncate, with 2 long, sharp spurs; metatibiae apically distinctly emarginate on outside. Tarsi nearly as long as tibiae; tarsites with long erect stiff setae, 1 longer than 2, 2 apically bilobed (deeply excised), 3 deeply bilobed, lobes pedunculate, 5 as long as 1 + 2. **Abdomen** with ventrites 1 to 2 subequal in length, 3 and 4 each half as long as 2, 5 not clearly discernible, apparently shortly triangular.

Figure 62. *Mekorhamphus tenuicornis* sp. n., holotype. Habitus, right lateral (**a**); habitus, left lateral (**b**); legs, right lateral (**c**); legs, left lateral (**d**); habitus, dorsal (**e**). Scale bars: 1.0 mm (a–c,e).

Material examined. Holotype (NIGP156995): reasonably well-preserved specimen, slightly distorted in dorsal view, with legs and rostrum compressed, pronotum dorsally slightly cut into; in middle but close to dorsal surface of rectangular cuboid 4.3 × 3.0 × 2.0 mm; amber clear but with many subparallel vertical flow lines on right and vertical layer of dense bubbles and impurities of varying sizes on left side of specimen, largely obscuring it.

Derivation of name. This species is named for its narrow antennae, the name being a Latin adjective.

Remarks. The species differs from *M. gyralommus* and *M. poinari* in having no lateral pronotal denticles and from *M. gracilipes* (both specimens) and *M. beatae* in the elytra being sparsely covered in long, thin, dark, suberect setae directed caudad and the bilobed (very deeply excised) tarsites 2. It is further distinct from *M. beatae* in the apically emarginate metatibiae with long thick spurs and from *M. gracilipes* in the distinctly crenulate tibiae, much slenderer funicles, with segments 6 and 7 slightly broader, and the differently shaped ventrites.

Mekorhamphus poinari Clarke & Oberprieler, **sp. n.** (Figure 63)

Description. Size. Length 2.94 mm, width 1.45 mm. Entire body coarsely and densely punctate. **Head.** Forehead with pair of flat and blunt, slightly oblique, keel-like tubercles between eyes. **Eyes** large, convex, semicircular, separated dorsally by about basal rostral width, very slightly further separated posteriorly. **Rostrum** ca. 1.5 × longer than pronotum; moderately curved; antennal insertions median, behind them with scrobes reaching base of eye, in front of them with long fine erect setae. **Antenna.** Scapes elongate, longer than funicle, reaching to below eye, apically oblique; funicles with segment 1 subequal in length to 2, 2 slightly shorter than 1, longer than 3, 3 longer than 4; clubs with segment 4 broadly inserted onto 3, narrowly rounded apically (other funicle segments and rest of club not visible). **Mouthparts.** Mandibles on outer side with 3 teeth, on inside with at least 2, articulation plane oblique. Maxillary palps projecting slightly obliquely forwards, 3-segmented, basal segments shorter, broader, than elongate, tapering apical segment. **Thorax.** Prothorax strongly proclinate. Pronotum broadly roundly subquadrate, narrowed anterolaterally, with small dentiform process at anterior half, with slight but weakly concave anteromedial collar; densely setose, setae mostly whitish mixed with fewer blackish setae; posterior corners distinctly angulate; notosternal sutures vertical, then abruptly angulated anteriad. Prosternum short, about half as long as procoxae, anterior margin emarginate, prosternal process short, acute; hypomeron ca. half as long as prosternum. Scutellar shield prominent, covered in light setae. Metaventrite short, ca. half as long as mesocoxa. **Elytra** punctostriate, densely setose, setae largely whitish; bases subsinuate, humeri broadly rounded; apices individually rounded, meeting almost evenly at suture. Legs long, slender, densely setose. Procoxae conical; mesocoxae narrowly separated, very large, subglobular, apparently longer than wide, longer than metaventrite. Femora with outer edges obscured. Tibiae long, slender, on outer edge carinate-crenulate; apically with tarsal articulation surfaces suboblique, with coarse fringing setae, few short dorso-apical setae and 2 spurs; protibiae somewhat slightly abruptly expanded apicad, outer edge in apical half slightly convexly produced, with matching slight broad concavity lined with dense oblique setae; mesotibiae straight; metatibiae slightly expanded apically, covered with long setae, outer side apically emarginate (also on mesotibiae) and with fine fringing setae lining apical edges; with 2 spurs, outer one more distinct. Tarsi with tarsite 1 apically excised, 2 shorter and wider than 1, deeply excised, 3 pedunculate, ca. 0.5–0.67 × as long as 5, 5 about as long at 1 + 2, longer on hindlegs. **Abdomen**. Ventrites setose, seemingly partly curved; 1 and 2 subequal, 1 somewhat inflated, 2–5 subflatly aligned, 3 ca. half as long as 2, 4 shorter than 3, 5 subequal in length to 4, thin and broadly rounded apically; sutures between 3 and 5 distinctly grooved.

Material examined. Holotype (NIGP157009), probably male: well preserved, moderately well visible specimen, with fragmented coating of debris over particularly dorsal side, missing claw segment of right protarsus and tarsites 3–5 of right mesotarsus, right fore- and middle legs partly cut away with amber at femorotibial joints, antennal clubs and most of funicles not visible; in angulate cuboid, 4.2 × 3.3 × 2.5 mm; amber clear yellow, somewhat hazy, with many impurities and bubbles obscuring clear view of specimen, one bubble near head and few on right side exposed and later infilled with resin (see Section 2.2), with large fracture running transversely–obliquely through block, intersecting specimen at pronotal–elytral juncture.

Derivation of name. This species is named after George Poinar Jr., in recognition of his wide-reaching contributions to the study of amber fossils, especially the biota preserved in Burmese amber, and for facilitating the study of type material housed in his collection (PACO), which greatly enhanced and enriched our work.

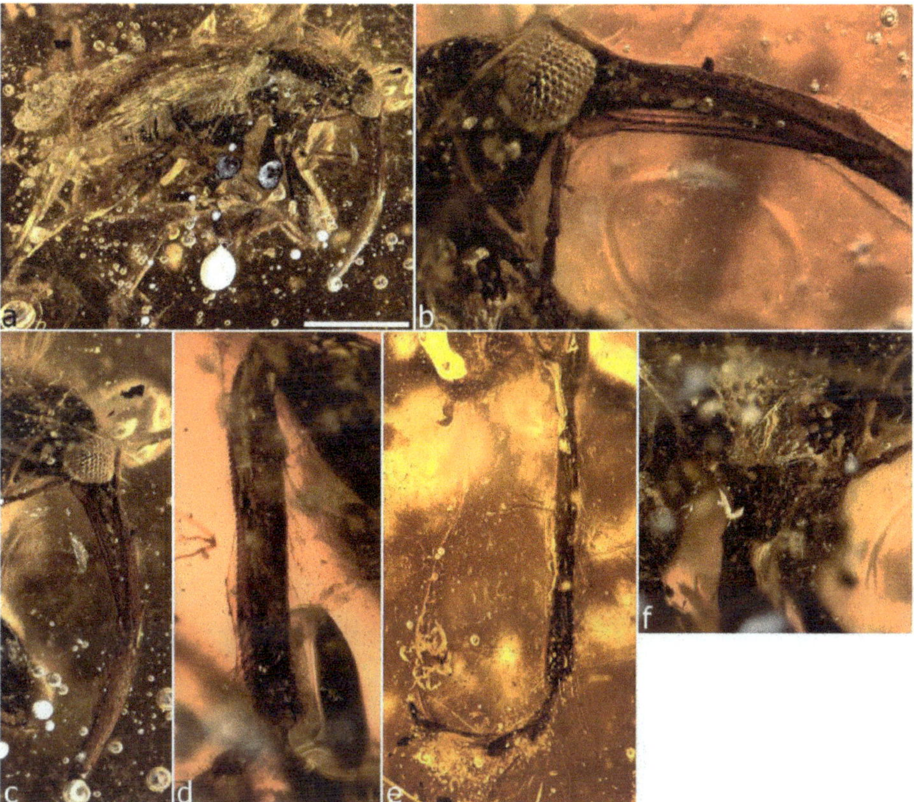

Figure 63. *Mekorhamphus poinari* sp. n., holotype. Habitus, right lateral (**a**); detail of head, right lateral (**b**); head, right lateral (**c**); right metatibia (**d**); left mesotibia and -tarsus, dorsal (**e**); prothorax showing notosternal suture, right lateral (**f**). Scale bar: 1.0 mm (a).

Remarks. This is the only other *Mekorhamphus* species in our sample with a small, flat, dentiform process on each side of the pronotum, as it occurs in *M. gyralommus*. It differs from this species in its rounder, hemispherical eyes with prominent tubercles between them, the less convex (hunched) elytra, the apically flatly enlarged protibiae curved shortly inwards at the apex, the large mesocoxae, leaving the disc of the metaventrite between them and the metacoxae very short, and the very short ventrite 5 (about as long as 4). Based on the length and thickness of the rostrum, the holotype appears to be a male.

Mekorhamphus **sp.** (Figure 64)

Material examined. One specimen (NIGP157010): badly distorted and compressed, black with milky greenish film around appendages; in centre of subcubic block of ca. 3 mm side length with rounded edges and convex dorsal side; amber clear without bubbles or impurities.

Remarks. This specimen appears similar to *M. gyralommus* in shape and the slightly sinuate and apparently distally widened protibiae, but the tubercles on its forehead are larger (though strongly compressed). It may represent a different species but is so poorly preserved that it cannot be properly diagnosed and compared with other *Mekorhamphus* species, and we therefore do not describe and name it.

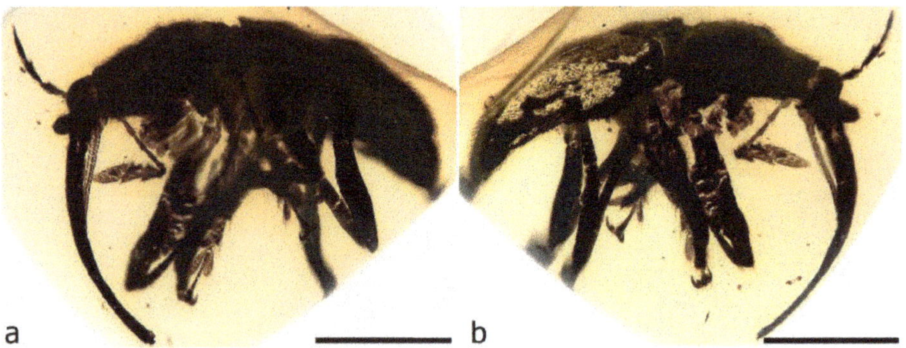

Figure 64. *Mekorhamphus* sp. Habitus, left lateral (**a**); habitus, right lateral (**b**). Scale bars: 1.0 mm (a,b).

Genus ***Compsopsarus*** Clarke & Oberprieler, **gen. n.**

Type species: *Compsopsarus reneae* Clarke & Oberprieler, sp. n.

Description. Size. Length 2.30–2.39 mm, width 0.62–1.05 mm. **Head** short, subspherical, dorsally and ventrally bulging. **Eyes** large, strongly and vertically flatly protruding, coarsely facetted, dorsally separated by ca. basal width of rostrum; forehead with pair of transverse, thick, carinate tubercles between eyes, without any tufts or patches of coloured setae above eyes. **Rostrum** about as long as pronotum, short, subcylindrical, curved; antennal insertions lateral, with deep scrobes behind and in front of them, in front of them laterally with a few long erect setae. **Antennae** geniculate, long; scapes long, subcylindrical, apically gradually inflated, slightly shorter than funicle; funicles 7-segmented, segments 2 and 3 slender obconical, others subglobular, broader; clubs seemingly 4-segmented, loosely articulated but segments close, segment 4 hardly distinguishable from 3, marked mainly by ring of setae. **Mouthparts**. Labrum absent. Mandibles small, scoop-like, vertical, non-exodont, endodont, articulation horizontal. Maxillary and labial palps long, slender, 3-segmented. **Thorax**. Prothorax proclinate, with anterior lateral margins oblique in lateral view. Pronotum strongly convex, collared anteriorly, laterally rounded, without tooth, posterior corners angulate, fitting closely onto elytra; surface finely densely punctate, densely setose, setae reclinate, directed anteromesad, disc with 3 pairs of patches of dense whitish setae, a small one at middle on either side of midline, another anterolaterally between midline and sides and one near middle of sides, anteriorly in shallow collared groove with most setae whitish; notosternal sutures closed, horizontal portion not visible. Prosternum short; procoxal cavities medially confluent, closer to anterior margin of prosternum. Scutellar shield prominently raised above elytral surface, densely setose. Mesoventrite short, anteriorly sloping. Mesocoxal cavities laterally closed by processes of meso- and metaventrite. Metaventrite longer, slightly convex, raised into weak transverse weals. Metanepisterna distinct, densely setose. **Elytra** elongate, basally extended over pronotum, with indistinctly rounded humeri, posteriorly declivous, lateral margin weakly sinuate; apically individually rounded, exposing pygidium; surface punctostriate, without scutellary striole, interstriae subflat, finely punctate, densely setose, setae short, thin, subrecurved, directed caudad, with at least 16 patches of whitish setae on interstriae (Figure 65). **Legs**. Procoxae large and conical, prominent, medially contiguous; mesocoxae globular, less prominent, narrowly separated; metacoxae flat, transversely elongate. Trochanters short, oblique. Femora long, strongly inflated in distal half, outside rounded. Tibiae compressed, apically expanded, outer edge costate, with dense long stiff setae in distal half, apex obliquely truncate, with 2 spurs. Tarsi shorter than tibiae, very slender; tarsite 1 apically excised, 2 shorter, broader, triangular, apically excised, 3 deeply bilobed, lobes pedunculate, 5 about as long as 1 + 2, apically expanded; claws divaricate, dentate (or possibly bifid; tooth distinctly curved inwards), ventrobasal seta not visible. **Abdomen** with ventrite 1 longer than 2, 3–4 subequal, 5 slightly shorter than 3 + 4, apically subtruncate; apical tergite exposed as vertical pygidium, densely setose.

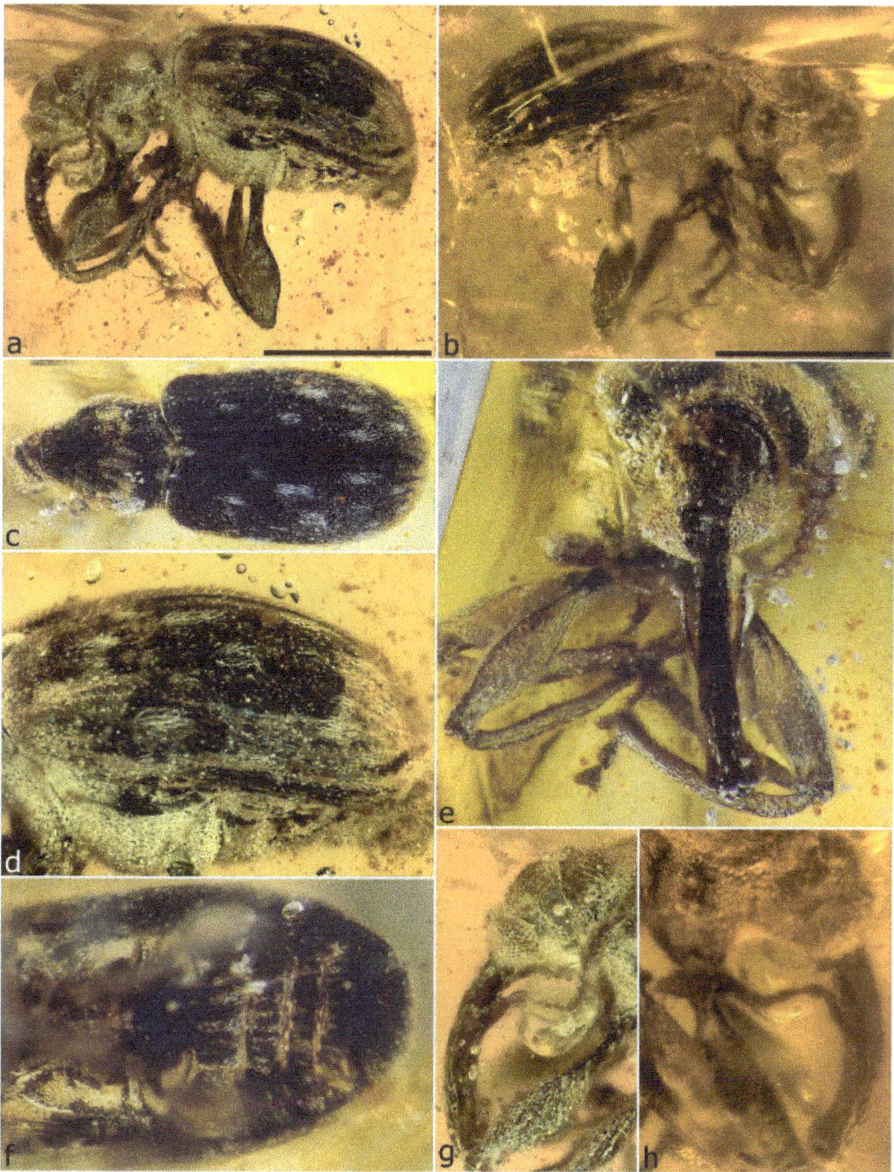

Figure 65. *Compsopsarus reneae* sp. n., holotype. Habitus, left lateral (**a**); habitus, right lateral (**b**); habitus, dorsal (**c**); elytra, left lateral (**d**); head, dorsal (**e**); ventrites, ventral (**f**); head and antenna, left lateral (**g**); same, right lateral (**h**). Scale bars: 1.0 mm (a,b).

Derivation of name. The name of the genus is formed from the Greek adjectives *kompsos* (elegant, pretty; Latin: *compsus*) and *psaros* (speckled, dappled, like a starling), in reference to the pleasing pattern of spots on the elytra of the type species, and it is masculine in gender.

Remarks. This extraordinary genus differs from all other Burmese amber weevils with geniculate antennae in having a pair of tubercles between the eyes, non-exodont, scoop-like vertical mandibles moving horizontally and an exposed flat (not cupulate) pygidium. It agrees with *Mekorhamphus* in the pair of tubercles between the eyes, but the form of the tubercles is different. It also differs from this genus in the lack of crenulation on the outside edge of the tibiae, the antennal scapes not reaching the eyes (separated by slightly more than length of funicle segment 1) and the slender tarsites 1 and 2. It agrees superficially with *Nugatorhinus* in having patches of whitish setae on the pronotum and elytra, but the setae in the patches are sparser and less distinct from other vestiture setae. *Nugatorhinus* also has tubercle-like structures between the eyes, but these are again different from those of *Compsopsarus*. A particularly unusual feature of this genus is the form of the claws, which seem intermediate between the dentate and bifid conditions, with the tooth flattened but curved inwards and situated slightly on the inner edge, similar to the structure of the claws of *Calyptocis*. We are aware of one other undescribed species of this genus, which lacks the setal patches and has more strongly punctostriate elytra, as well as two other similar specimens for which further study (block preparation) is required to ascertain their affinity.

Compsopsarus reneae Clarke & Oberprieler, **sp. n.** (Figures 65 and 66)

Description. Size. Length 2.30 mm, width 1.05 mm. Body small, compact, convex. **Head** seemingly weekly constricted behind eyes (eyes prominent, cuticle not constricted); with carinate tubercles between eyes converging posteriorly; distinctly, densely punctate. **Eyes** bulbous, somewhat protruding laterally, vertically subcompressed, slightly slanting backwards, with dense long interfacettal setae on top. **Rostrum** gradually widening and deepening distad, at mandibular articulations ca. wider than at base (possibly compression artefact); dorso-apically with oblique grooves delimiting narrow epistome; setose, setae small, appressed; antennal insertions slightly in front of middle. Scrobes strongly delimited, not quite reaching eyes. Mandibular sockets lateral, deep and thick. **Antennae** concolorous, club slightly darker; scapes elongate, not reaching eye (scape plus funicle segment 1 reaching eye), apically truncate; funicles with segment 1 elongate-ovoid, about 0.2 $\times$ legth of scape, 2 narrower, elongate obconical, 3–4 similar, progressively shorter, 5–7 subglobular, subequal; clubs densely, shortly setose, with whorls of numerous elongate slender setae; articulations between segments very broad, segments 1–2 subequal in length, 1 obconical, apically flat, 2 subconcave apically, 3 and 4 subfusiform, narrowing distad, 4 indistinct from 3 in most views. **Mouthparts.** Mandibles asymmetrical, right-superior; right mandible with 3 teeth on inside, basal tooth small, rounded, apical tooth curved inwards, with subequal subapical tooth; left mandible somewhat smaller, with apical teeth smaller, closer. **Thorax.** Prothorax only slightly narrower than elytra at humeri. Pronotum distinctly impressed anteriorly; sparsely setose, setae pale brown, with 3 denser patches of whitish setae, one paramedian in front of base, one at middle of side and one anterolaterally; base sinuate, closely fitting with elytra; posterior angles tightly fitting onto humeri. Mesepimera flush dorsally with posterolateral prothoracic margin and elytral margin. Metaventrite coarsely, densely punctosetose; setae densest posterolaterally; longer than mesocoxae. **Elytra** broadly transversely convex; finely, shallowly punctate; weakly punctostriate, striae indistinct from flat interstriae; humeri rounded, not projecting; sides weakly sinuate, flush with ventrites, with indistinct minutely punctosetose marginal groove, without anterior marginal notch. **Legs**. Tibiae distinctly (protibiae) or slightly (mesotibiae) curved or straight (metatibiae); apically with long coarse fringing setae interrupted dorsally, much thicker on metatibiae, with several elongate slender dorso-apical setae; on outer side of meso- and metatibiae with 2 rows of coarse setae in distal third. Protarsi distinctly longer than mesotarsi; tarsites 1 and 2 apically expanded, excised, protarsite 1 ca. 2.0 $\times$ longer than meso- and metatarsites 1, 2 shorter than 1, lobes of 3 ca. 0.67 $\times$ as long as 5; claws with inner tooth distinctly curved inward. **Abdomen**. Apical (exposed) tergite darkly pigmented, vertical. Ventrites distinctly stepped, each progressively higher, 1–2 more flatly aligned than others, densely setose; sutures straight/arcuate, intersegmental membranes visible between all ventrites, more protruding between ventrites 2 and 4.

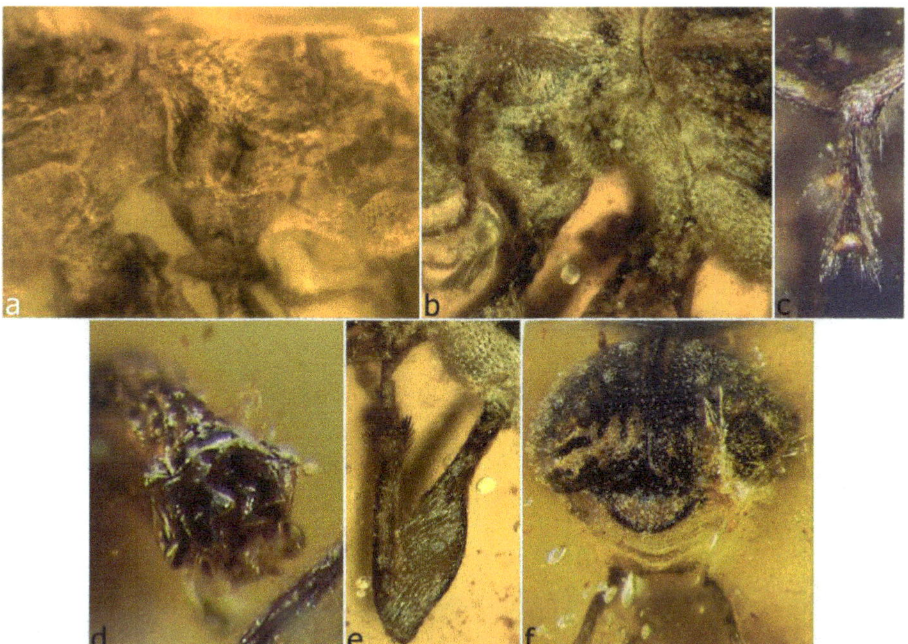

Figure 66. *Compsopsarus reneae* sp. n., holotype. Thorax, right lateral (**a**); thorax, left lateral (**b**); tarsites 1 and 2 (**c**); mandibles, apical (**d**); left hind leg (**e**); abdomen, showing pygidium, caudal.

Material examined. Holotype (NIGP157011): very well preserved, minimally distorted specimen, with left pro- and mesotarsi missing tarsites 3–5, other appendages intact, body surface largely visible but with fragmented surface coating of whitish material and/or minute bubbles; in irregular cuboid 3.0 × 2.2 × 1.5 mm; amber clear yellow with longitudinal fractures along right elytron and over right side of prothorax and head of specimen, partly obscuring that side, otherwise clear with minimal impurities.

Derivation of name. This attractive species is named after Renee Berentsen, partner of the first author, in recognition of all her tireless support and patience during our study of these fossils.

Remarks. This species can be readily distinguished from all other Burmese amber weevils by the generally dense coat of setae obscuring the elytral striae, the diffuse setal patches on the elytra and distinct dense vestiture on the ventrites. It also has long, dense interfacettal setae on the top of the eyes, a feature unique in our sample, and long dense setae in the apical half of the tibiae forming two subaligned rows on the outer side, as well as elongate coarse apical fringing setae. The holotype was submitted for CT scanning, but the contrast between the specimen and the amber was too low to permit any meaningful visualisation of the specimen.

***Compsopsarus* sp.** (Figure 67)

Material examined. One specimen (NIGP157012): very poorly preserved, compressed and distorted but intact; in middle of flat rectangular cuboid 4.7 × 4.6 × 1.5 mm; amber clear with many sparsely distributed pieces of debris but no cracks or flow lines.

Remarks. This heavily distorted specimen is not definitely assignable to any genus as here described. It agrees with *Compsopsarus reneae* in having a short rostrum, non-crenulate tibiae and seemingly an exposed pygidium, but it differs in not having interfacettal setae or setal patches on the elytra and its tarsites 1 and 2 being shorter and thicker. From *Mekorhamphus* it differs mainly by its shorter rostrum, short and broad funicles and clubs, lack of tibial crenulations and narrow tarsi with tarsites 1 apically truncate and 2 only weakly excised. We tentatively place this specimen in *Compsopsarus* as it does not reveal sufficient characters on which to base a different genus. Better-preserved specimens are needed to assess its taxonomic affinities. It is too poorly preserved to be properly diagnosed and compared with other similar species, and we therefore do not describe or name it.

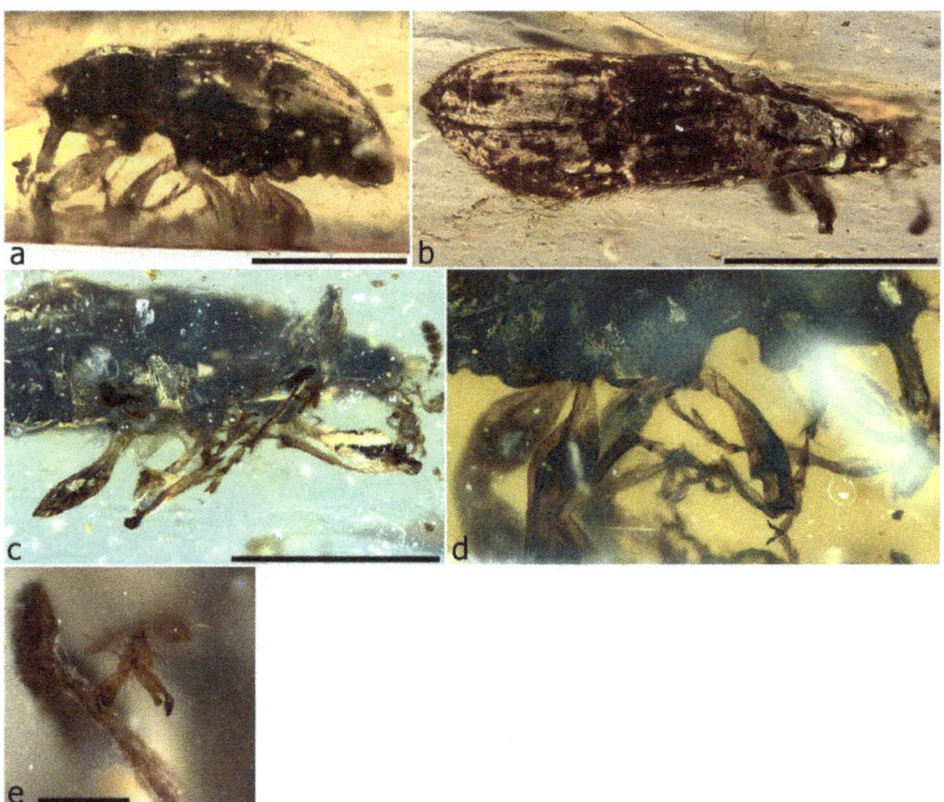

Figure 67. *Compsopsarus* sp. Habitus, left lateral (**a**); habitus, dorsal (**b**); legs and ventrites, ventral (**c**); same, right lateral (**d**); claw, apical (**e**). Scale bars: 1.0 mm (a–c); 0.1 mm (e).

Genus ***Myanmarus*** Clarke & Oberprieler, **gen. n.**

Type species: *Myanmarus caviventris* Clarke & Oberprieler, sp. n.

Description. Size. Length 2.5–2.75 mm, width 1.2–1.4 mm. Body black, appendages distinctly paler. **Head** long, strongly constricted and dorsoventrally bulging behind eyes. **Eyes** large, strongly laterally protruding, coarsely facetted, dorsally separated by about half basal width of rostrum in anterior half, much further separated posteriorly; forehead thinly linearly grooved, without pair of tubercles between anterior margin of eyes. **Rostrum** about 1.25 × longer than pronotum, slender, subcylindrical, moderately curved; antennal insertions lateral, behind them with scrobes reaching eye, in front of them lateroventrally with or without a few long erect setae. Single long gular suture present. **Antennae** geniculate, long; scapes elongate, cylindrical, apically clavate, about 4 × longer than funicle segment 1; funicles 7-segmented, much longer than scape, segments 2–7 progressively slightly decreasing in length; clubs large, loosely articulated, 4-segmented, segment 4 acute, about as long as 3. **Mouthparts**. Labrum absent. Mandibles small, exodont, articulation horizontal. **Thorax**. Prothorax proclinate, with anterior lateral margins oblique in lateral view. Pronotum slightly convex, laterally rounded, with or without tooth, posterior corners angulate, fitting closely onto elytra; surface coarsely rugose, densely setose, setae reclinate, directed anteromesad, some scattered whitish setae; notosternal sutures closed, vertical, bent or curved anteriad. Prosternum moderately short; procoxal cavities in middle of prothorax or closer to front of prosternum. Scutellar shield densely setose, anterior margin flush with basal elytral margins. Mesocoxal cavities laterally closed (by meso- and metaventrite). Metanepisterna distinct, densely finely setose. Mesoventrite short, anteriorly sloping. Metaventrite longer, slightly convex. **Elytra** elongate, basally extended over pronotum, with broadly rounded humeri, posteriorly declivous, lateral margin strongly sinuate to roundly emarginate in middle; apically individually rounded, not exposing pygidium; sutural flanges not visible; surface seemingly not or very weakly punctostriate, without scutellary striole, interstriae indistinct from striae, surface rugose, setose, setae long, thin, reclinate, directed caudad, with scattered whitish setae seemingly not arranged into a pattern. **Legs**. Procoxae subconical, large, prominent, medially contiguous; mesocoxae large, subglobular, narrowly separated; metacoxae flat, transversely elongate. Trochanters short, oblique. Femora long, subcylindrical, inflated in distal half, outside rounded (profemora) or subcrenulate along distal third (meso- and metafemora), unarmed. Tibiae straight, compressed, narrowing in apical quarter, outer edge carinate-crenulate to apex (protibiae) or attenuating before apex (meso- and metatibiae) and dorsal edge continued as small apical lobe; with dense long stiff setae in distal half, apices obliquely truncate, with 2 spurs. Tarsi almost as long as tibiae; tarsite 1 elongate, apically expanded, 2 shorter, subtriangular, apically subacute, 3 deeply bilobed, lobes pedunculate or not, 5 long, apically expanded; claws divaricate, dentate with ventrobasal seta at apex of tooth. **Abdomen** mostly with ventrites variously impressed to grooved, 1 and 2 fused, at about same level, with suture less distinct than others, subequal in length, 3–5 more distinctly stepped, sutures straight, 4 shorter than 3, 5 longer than 4, with distinct medial concavity.

Derivation of name. The genus is named after Myanmar, the literary name of the country of origin of the Burmese amber, in which the specimen is preserved; the gender of the name is masculine.

Remarks. *Myanmarus* is characterised by its darkly pigmented body with pale to very pale legs and antennae (most distinct in *M. caviventris* and *M. dentifer*), indistinct elytral striae, crenulate-serrulate tibiae and concave ventrites (one or more). Among the genera with crenulate tibiae it differs from *Mekorhamphus* in the lack of tubercles between the eyes and the structure of the tarsi (only weakly excised tarsites 2 and broad but not pedunculate lobes of tarsites 3) and agrees with this genus in the scapes reaching below the eyes. From *Elwoodius* it is most easily distinguished by the protruding but not conical eyes, the form of the tibial crenulation (pectinate in the metatibiae of *Eldoodius*) and the non-modified tibial spurs. With *Leptopezus* it agrees well in the form of the head and eyes and particularly in the structure of the metatibiae, but in *Myanmarus* the eyes are not depressed and the tarsi much broader and *Leptopezus* has unmodified ventrites. From other genera of Mesophyletiniae with dentate tarsal claws it can be distinguished by its crenulate tibiae and impressed ventrites. It differs from *Habropezus* also in its narrower body, crenulate tibiae and more strongly constricted head and from *Mesophyletis* also in having two subequal spurs on the tibiae, no pygidium and a strongly constricted head. It is also similar to *Bowangius* but larger and hairier and with the tibiae crenulate, not serrulate. The genus is represented by four species.

Myanmarus caviventris Clarke & Oberprieler, **sp. n.** (Figures 68 and 69, Video S6)

Description. Size. Length 2.74 mm, width 1.20 mm. Body black, densely setose, setae mainly black but with areas of whitish setae on head, pronotum and elytra; legs and antennae distinctly paler than body (and translucent). **Head** more highly convex dorsally than ventrally, coarsely punctate; shortly setose, some setae whitish; ventrally more evenly continuous with base of rostrum. **Eyes** slanting backwards, in dorsal view subtriangular; without interfacettal setae. **Rostrum** dorsally with weak median, paramedian and dorsolateral carinae and grooves reaching antennal insertions, these median, without scrobes in front of them, apically effaced (rostrum smoothly cylindrical in distal half); with few long erect setae on ventrolateral side. **Mouthparts**. Mandibles with 2 small pointed teeth on outer side, apex truncate, forming T. **Antennae**. Scapes long, slender, reaching behind front margin of eyes, apically truncate; funicles with segments with regular whorls of long setae, otherwise sparsely setose, segment 1 about as long as 2 + 3, 2–5 similarly elongately obconical, progressively shorter towards club, 6–7 similarly ovoid, 6 longer than 5, 7 shorter and narrower than 6; clubs with segments 1–2 apically oblique, 1 elongate, obconical, 2 slightly broader and longer than 1, 3 and 4 subequal in length to 2, 4 narrowly inserted into 3, distinctly articulating with 3. **Thorax.** Prothorax strongly proclinate. Pronotum laterally without teeth; with scattered whitish setae; rugose. Prosternum very short, forming thin strip in front of procoxae, prosternal process small, pointed; procoxal cavities closer to prosternal margin. Scutellar shield transverse, not raised above elytral bases, anterior margin continuous with elytral basal margin. Meso- and metathorax with scattered whitish setae; metaventrite with posterior edge on either side with row of sparse, stiff whitish setae (appearing reddish under some lighting) reaching over metacoxae. **Elytra** punctostriate, punctures deep but indistinct, striae and interstriae indistinct; lateral margin with marginal groove subequal in width for entire length, with anterior marginal notch; surface coarsely sculptured, sparsely setose, setae dark, fine, some diffuse whitish thicker longer setae on humeri and sides, denser on posterior third. **Legs** pale translucent yellowish. Femora on outer side rounded (profemora) or subcrenulate along distal third (meso- and metafemora); profemora longer than others. Tibiae flattened, substraight (metatibiae) or abruptly slightly curved inwards in apical quarter (pro- and mesotibiae), apically with small (indistinct) inner and outer flanges; protibiae ventrally with long elongate setae along distal half, increasing in density in slight apical emargination, tarsal articulation surfaces subtruncate; meso- and metatibiae with apical edges lined with coarse fringing setae, dorso-apically with several very long setae, tarsal articulation surfaces suboblique. Tarsi ca. 0.67 × as long as tibiae, ventrally densely setose, mesotarsi slightly shorter than protarsi; tarsite 1 elongate, apically expanded, subtruncate, on meso- and metatarsi slightly shorter than on protarsi, 2 ca. 0.67 × as long as 1, subtriangular, apically excised, with very long, thick dorso-apical setae, 3 deeply bilobed, lobes pedunculate, ca. half as long as 5, 5 elongate, slender. **Abdomen.** Apical tergites seemingly horizontally exposed (extruded). Ventrites slightly stepped; setose, with scattered white setae intermixed with darker setae; 1–3 subequal in length, 4 shorter than 3, 5 longer than 4, with distinct medial concavity, setose on either side of concavity but not inside it; sutures straight. Ovipositor with long hemisternites, each bearing elongate spically setose stylus.

Material examined. Holotype (NIGP157013), female: exceptionally well preserved, intact specimen with ovipositor extruded (connected to abdomen), largely undistorted, not compressed but surface details somewhat obscured by fine film of debris; in wedge 6.1 × 3.2 × 2.2 mm, with one large curved and large flat face and smaller edge faces exposing dorsal side and head; amber clear yellow, with large fracture plane obscuring left side and other inclusions partly obscuring rostral apex.

Derivation of name. The species is named for the peculiar large cavity on ventritre 5; the name being an adjective.

Remarks. This species has distinctive setal patterns on the elytra, with whitish setae intermixed with darker ones, as well as a line of thick whitish setae on the posterior edge of the metaventrite. In appearance it resembles *M. dentifer* but is distinguishable from the latter by the lack of the anterolateral tooth on the sides of the pronotum and the normal short spurs. It can be distinguished from *M. robustus*, also lacking pronotal teeth, in the concavity of ventrite 5, the longer slender tibiae and the metatibiae being emarginate on the outside before the apex. The holotype was submitted for CT scanning (Figure 69, Video S6), with good results, Figure 69 in particular revealing more distinct striae on the elytra and providing a clear view of ventral structures.

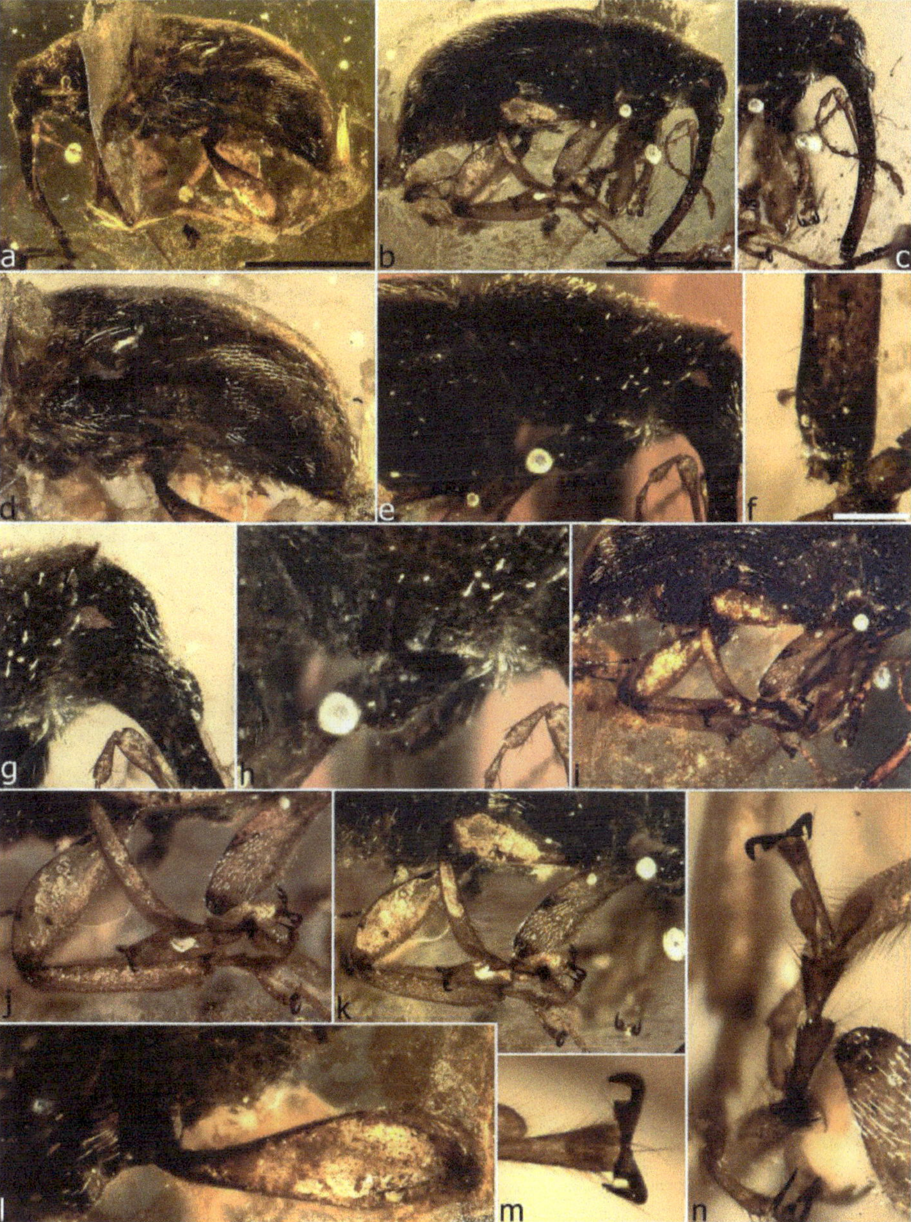

Figure 68. *Myanmarus caviventris* sp. n., holotype. Habitus, left lateral (**a**); habitus, right lateral (**b**); head and antennae, right lateral (**c**); elytra, left lateral (**d**); head and prothorax, right lateral (**e**); apex of rostrum and right mandible, dorsal (**f**); head, right lateral (**g**); prothorax showing notosternal suture, right lateral (**h**); abdomen showing ventrites, and legs, right lateral (**i**); legs, right lateral (**j**); same (**k**); left metacoxa and -femur, showing dense line of lighter setae on posterior metaventral margin (**l**); tarsal claw, ventral (**m**); protarsus, ventral (**n**). Scale bars: 1.0 mm (a,b); 0.1 mm (f).

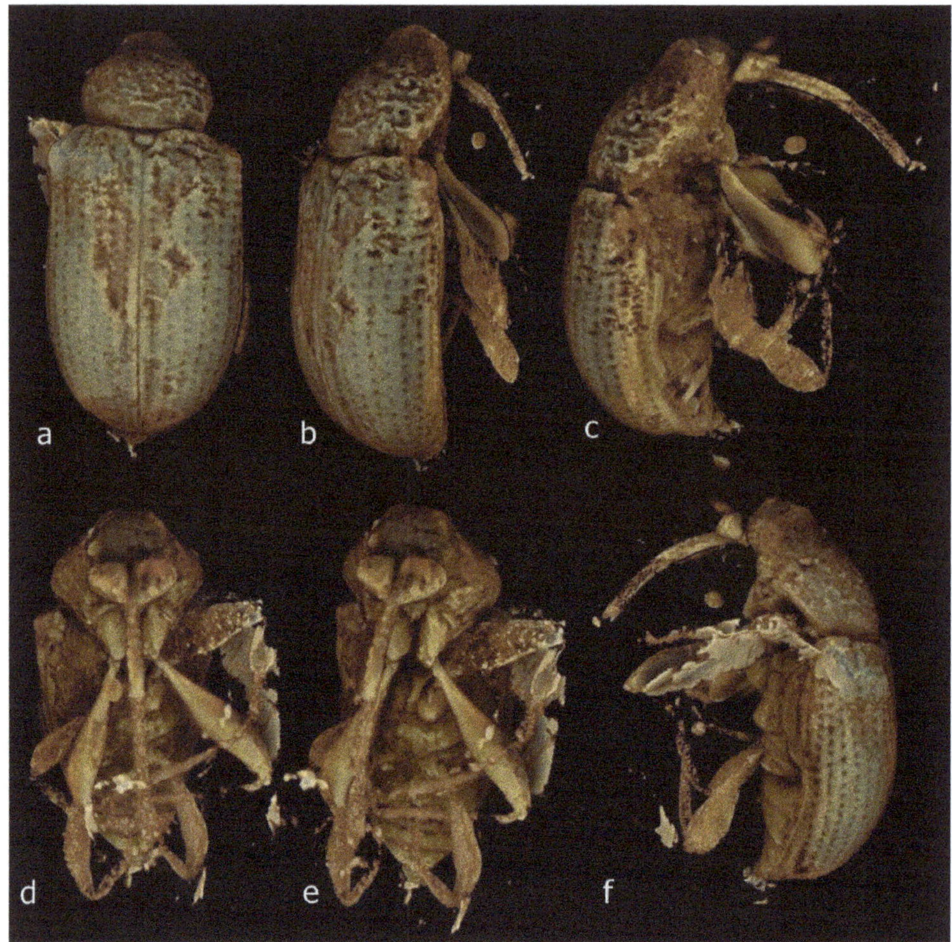

Figure 69. *Myanmarus caviventris* sp. n., holotype. Habitus images extracted from a micro-CT scanning reconstruction (see also Video S6). Dorsal, revealing punctate stria not clearly visible under light microscopy (**a**); dorsal oblique (**b**); right lateral (**c**); ventral (**d**); ventral (**e**); left lateral (**f**).

Myanmarus robustus Clarke & Oberprieler, **sp. n.** (Figure 70)

Description. Size. Length 2.5 mm, width 1.2 mm. **Head** shortly subconical, strongly constricted behind eyes. **Eyes** large, compressed, protruding, triangular in dorsal view, coarsely facetted, dorsally separated by less than width of rostrum at base. **Rostrum** slightly longer than pronotum, thin, cylindrical, slightly downcurved; antennal insertions in middle of rostral length, behind them with scrobes indicated, extending to eye, no lateral row of setae in front of them visible. No gular suture visible. **Antennae** Scapes long, straight, thin, apically abruptly clavate; funicles slightly longer than scape, segment 1 elongate, not inflated, 2 slightly shorter than 1, 3–5 elongate, thin, 6 and 7 shorter and slightly thicker; clubs long, flattened, segment 4 distinct, long, acute. **Mouthparts**. Mandibles flatly scoop-shaped with seemingly 3 large inner teeth (2 dorsally, 1 apically) and 2 low outer ones, apical one forming weak oblique T with inner one, articulation plane oblique. Maxillary palps apparently 3-segmented. **Thorax**. Pronotum elongate, laterally rounded, without tooth, strongly convex, surface moderately densely setose, setae short, thin, acute, directed anteromesad. Prosternum moderately long; procoxal cavities in about middle of prothorax. Scutellar shield indiscernible. **Elytra** elongate, with weak, broadly rounded humeri, posteriorly strongly evenly declivous; surface punctostriate but striae indistinct, densely setose, setae moderately long, thin, acute, directed caudad, odd interstriae slightly elevated (artefact?). **Legs**. Procoxae elongate. Tibiae short, slightly flattened, outer edge faintly crenulate/serrulate (visible on mesotibia in ventral view, also on protibia proximally), apex obliquely truncate, with 2 spurs (visible on protibiae). Tarsi stout, slightly shorter than tibiae; tarsite 1 long, narrow, apically slightly excised, 2 apically excised, 3 shortly bilobed, lobes not pedunculate, 5 as long as 1; claws divaricate, dentate with ventrobasal seta at apex of tooth. **Abdomen** with ventrites flat but seemingly not impressed, 1 to 2 subequal in length, about 2 × longer than 3 and 4, 5 apparently as long as 4.

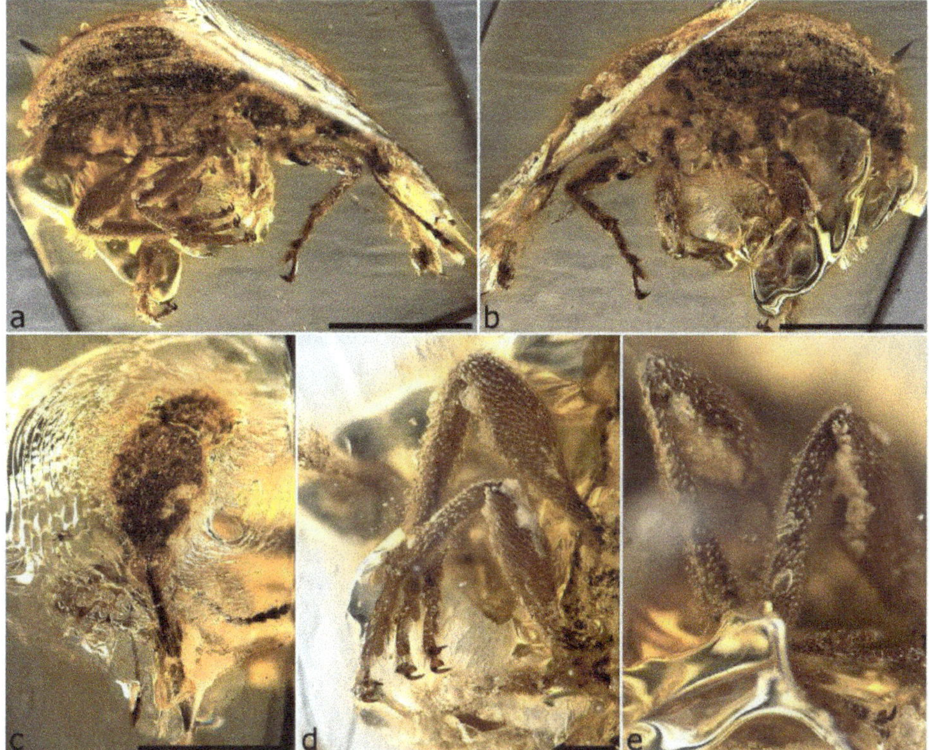

Figure 70. *Myanmarus robustus* sp. n., holotype. Habitus, right lateral (**a**); habitus, left lateral (**b**); head and prothorax, dorsal (**c**); legs, ventrolateral (**d**); tibiae (**e**). Scale bars: 1.0 mm (a–c); 0.2 mm (d).

Material examined. Holotype (NIGP157014): well preserved, intact specimen, not compressed or distorted, mostly well visible except for left side of rostrum, pronotum, and legs posteriorly; obliquely transverse in rectangular cuboid 4.6 × 3.3 × 2.2 mm; amber clear except for large reflective thin film of flow lines over rostrum, pronotum and basal part of elytra, with shallow crack along side of right elytron, transverse crack across elytral declivity and large, irregular, clear bubble beneath left side of crack enveloping middle and hindlegs, few other impurities.

Derivation of name. The species is named for its robust appearance, the name being an adjective.

Remarks. This species is similar to *M. caviventris* in lacking a distinct lateral prothoracic tooth, but it differs from that species in having whitish elytral setae, ventrite 5 without a concavity, the outer side of the meso- and metatibiae straight, not emarginate before the apex, and the outer side of the metafemora carinate (not crenulate) in the distal third. From *M. dentifer* and *M. diversiunguis* it is most readily distinguishable in not having a lateral tooth on the sides of the prothorax and from the latter also in having normal, dentate protarsal claws.

Myanmarus dentifer Clarke & Oberprieler, **sp. n.** (Figure 71)

Description. Size. Length 2.7 mm, width 1.03 mm. **Head** subglobular, small, narrow, strongly constricted behind eyes. **Eyes** large, subglobular, protruding, slanting towards back; dorsally broadly triangularly separated. **Rostrum** longer than pronotum, robust, slightly curved; antennal insertions in middle of rostral length, behind them with distinct narrow scrobes extending to eye, in front of them without lateral row of setae. **Antennae.** Scapes long, straight, thin, apically slightly clavate, not quite reaching eye in repose; funicles slightly longer than scape, segment 1 slightly inflated in middle, spindle-shaped, 2 as long as 1 but narrower, others narrower but progressively shorter towards club; clubs long, loosely articulated, flattened, segment 4 distinct, as long as 3. **Mouthparts**. Mandibles (widely opened) scoop-shaped with 3 large inner teeth (2 dorsally, 1 apically) and 2 low outer ones, apical one forming weak oblique T with inner one, articulation plane horizontal. **Thorax**. Prothorax strongly proclinate. Pronotum elongate, laterally with sharp, black, flat tooth just anteriorly of middle; densely setose, setae pale, long, narrow, directed anteromesad, sculpture not discernible; notosternal sutures sharply bent anteriad. Prosternum moderately long; procoxal cavities in about middle of prothorax. Scutellar shield weakly visible, round, slightly concave. Mesocoxal cavities laterally closed (by meso- and metaventrite). Metanepisternal sutures distinct. Mesoventrite short, anteriorly strongly sloping. Metaventrite longer, raised into oblique weals before metacoxae. **Elytra** short and broad, with weakly developed, broadly rounded humeri, posteriorly strongly declivous; surface indistinctly punctostriate, sparsely covered with thin, sharp setae directed caudad. **Legs**. Procoxae elongate, prominent. Femora straight, subcylindrical, metafemora more flattened (possibly an artefact). Tibiae long, flattened, outer edge crenulate (appearing serrulate in lateral view), apex obliquely truncate, with 2 slightly flattened spurs, the inner one larger on the metatibiae. Tarsi with tarsites 1 and 2 narrow, elongate, subtruncate, 5 shorter than 1 + 2; claws divaricate, strongly dentate. **Abdomen** with ventrites 1 to 2 subequal in length, 3, 4 and 5 about half as long, 1–3 with broad, deep median groove, its posterior angles on each ventrite strongly elevated into blunt, setose tubercles.

Material examined. Holotype (NIGP157015): eell preserved, intact specimen with slightly compressed tibiae, well visible from all sides, right metatarsus cut off except claws; at an angle in left side of irregular block 4.0 × 4.0 × 2.7; amber clear with several small dispersed impurities and larger irregular brown film below front legs and tip of rostrum.

Derivation of name. The species is named for the distinct teeth on the sides of the prothorax; the name is a noun in apposition.

Remarks. This species is distinctive in its medially conspicuously grooved ventrites 1 to 3, which apparently is not an artefact as all three ventrites are posteriorly extended into blunt tubercles next to the groove. It also has a distinctly laterally toothed prothorax, which otherwise only occurs in *M. diversiunguis* (but the tooth smaller), from which it also differs in having normally dentate protarsal claws (not bifid plus dentate). The lateral prothoracic teeth readily distinguish *M. dentifer* from *M. caviventris* and *M. robustus*.

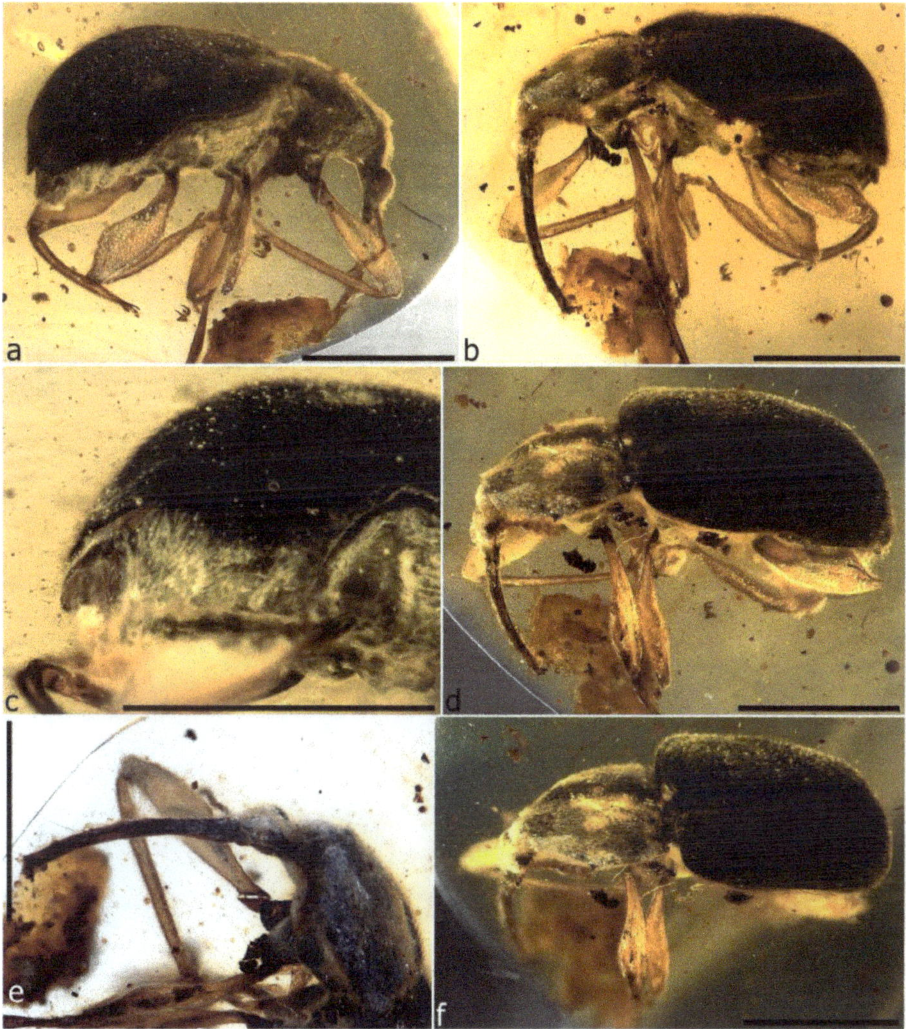

Figure 71. *Myanmarus dentifer* sp. n., holotype. Habitus, right lateral (**a**); habitus, left lateral (**b**); ventrites, ventral oblique (**c**); habitus, lateral oblique (**d**); head and prothorax, left lateral (**e**); habitus, dorsal oblique (**f**). Scale bars: 1.0 mm (a–f).

Myanmarus diversiunguis Clarke & Oberprieler, **sp. n.** (Figure 72)

Description. Size. Length 2.75 mm, width 1.38 mm. **Head** short, strongly constricted behind eyes. **Eyes** large, slightly elongate, protruding, dorsally narrowly separated along entire length, forehead very narrow, lower than dorsal margin of eyes, without tubercles. **Rostrum** slightly longer than pronotum, stout, slightly curved, subcylindrical; antennal insertions median, behind them with short scrobes extending to eye, in front of them with lateral row of ca. 5 long, erect setae in apical quarter of rostrum. No gular suture visible. **Antennae.** Scapes long, straight, thin, apically abruptly clavate; funicles slightly longer than scape, segment 1 inflated, others much narrower, elongate, progressively shorter towards club; clubs long, loosely articulated, flattened, segment 4 distinct, almost as long as 3. **Mouthparts**. Mandibles flatly scoop-shaped with 2 (possibly 3) large inner teeth

(1–2 dorsally, 1 apically) and 2 low outer ones, apical one forming weak oblique T with inner one, articulation plane oblique. Maxillae and labium indiscernible. **Thorax**. Prothorax strongly proclinate. Pronotum broadly roundly trapezoidal, laterally with small tooth; sparsely setose, setae pale, long, thin, directed anteromesad; notosternal sutures closed. Prosternum and hypomeron both short, about equal in length; procoxal cavities medially confluent, in middle of prothorax. Scutellar shield shortly transverse, rounded, faintly setose. Mesocoxal cavities laterally closed (by meso- and metaventrite). Metanepisternal sutures distinct. Mesoventrite short, anteriorly strongly sloping. Metaventrite longer, convex, with thick transverse weal before metacoxae. **Elytra** short and broad, with well-developed, broadly rounded humeri, posteriorly gently declivous, apically individually rounded, not exposing pygidium; sutural flanges not visible; surface coarsely punctostriate, without scutellary striole, sparsely covered with longish setae. **Legs**. Procoxae elongate, prominent, medially contiguous; mesocoxae subglobular, narrowly separated; metacoxae flat, transversely elongate. Trochanters short, oblique. Femora short, slightly compressed, pro- and mesofemora slightly inflated, metafemora strongly so, unarmed. Tibiae long, straight, compressed, outer edge crenulate, apex obliquely truncate, with 2 strong sharp spurs. Tarsi ca. half as long as tibiae; tarsite 1 elongate, on protarsi longer, on meso- and metatarsi shorter than 2 + 3, apically slightly excised, 2 shorter than 1, narrowly triangular, shallowly excised, 3 deeply bilobed, lobes short, 5 as long as 1, apically broadened; claws divaricate, on protarsi deeply bifid as well as dentate, with ventrobasal seta, on other tarsi only dentate. **Abdomen** with ventrites 1–3 broadly shallowly longitudinally impressed, posterior angles of impression on each ventrite elevated into blunt tubercle, 1 to 2 subequal in length, 3 and 4 slightly shorter, 5 not discernible.

Figure 72. *Myanmarus diversiunguis* sp. n., holotype. Habitus, right lateral (**a**); left lateral (**b**); dorsal (**c**); protibia and tarsus (**d**); mandibles (**e**); protarsal claws (**f–i**). Scale bars: 1.0 mm (a–d); 0.1 mm (e–i).

Material examined. Holotype (NIGP157016): not too well preserved but intact, slightly crumpled, dark specimen, with left front leg stretched out ventrad, surface of elytra with some pale spots, apex of left elytron slightly cut, otherwise well visible; in upper left part of rectangular cuboid 4.3 × 3.0 × 3.0–3.2 mm; amber clear with very few impurities, small crack and brown film next to left elytron.

Derivation of name. The species is named for its diverse tarsal claws, being normally dentate on the meso- and metatarsi but extraordinarily dentate as well as bifid on the protarsi. The name is an adjective.

Remarks. This species is one the most extraordinary of all Burmese amber weevils (as of weevils in general) in having the claws of its front tarsi distinctly bifid (with an inner long secondary claw) as well as dentate (with a shorter basal tooth, carrying a ventrobasal seta). This aptly demonstrates that the bifid and dentate conditions of the tarsal claws are not homologous and need to be distinguished. The prothorax of the holotype is somewhat distorted and the lateral teeth are not equal, the left one exaggerated due to a linear oblique depression in front of it, but as there is a small tooth in the same position on the undistorted right side of the prothorax, these teeth appear to be a real character of the species. The species also differs from all other *Myanmarus* species in its broadly impressed ventrites.

Genus ***Mesophyletis*** Poinar, 2006

Mesophyletis Poinar, 2006: 879 [29] (type species, subsequent designation (Poinar, 2008): *Mesophyletis calhouni* Poinar, 2006)

Redescription. Size. Length 2.12–2.82 mm, width 1.15–1.56 mm. **Head** short, transverse, constricted behind eyes, strongly domed dorsally, less convex ventrally. **Eyes** large, subelongate, protruding, coarsely facetted, dorsally separated by about basal width of rostrum; forehead grooved, without tubercles above eyes. **Rostrum** about as long as pronotum, subcylindrical, weakly downcurved; antennal insertions median, with scrobes behind them reaching eye, in front of them laterally with row of several of long erect fine setae. Single long gular suture present. **Antennae** geniculate, long; scapes elongate, cylindrical, apically inflated, shorter than funicle; funicles 7-segmented, segment 1 slightly longer and much broader than 2, 2–5 subequal, elongate-obconical, 6–7 similarly shortly ovoid, broader than 5, progressively shorter; clubs long, loosely articulated, 4-segmented, segment 4 broadly rounded, about as long as 3. **Mouthparts**. Labrum absent. Mandibles small, flat, exodont, articulation oblique. Maxillary and labial palps 3-segmented, both projecting obliquely forwards. **Thorax**. Prothorax strongly proclinate, with anterior lateral margins oblique in lateral view. Pronotum convex, laterally rounded, without tooth, posterior corners angulate, fitting closely onto elytra; surface coarsely tuberculate, sparsely finely setose, setae arising from minute punctures on top of tubercles, reclinate, directed anteromesad; notosternal sutures closed, vertical then abruptly deflected anteriorly. Prosternum short; procoxal cavities medially confluent, closer to anterior margin of prosternum. Scutellar shield asetose (or very sparsely setose), raised above elytral surface. Mesocoxal cavities laterally closed. Metanepisterna distinct, sparsely setose. Mesoventrite short, anteriorly sloping. Metaventrite longer, raised into transverse weals. **Elytra** elongate, basally extended over pronotum, with broadly rounded humeri, posteriorly declivous, lateral margin strongly sinuate to roundly emarginate in middle, apically individually rounded; sutural flanges not visible (elytra closed); surface weakly to distinctly punctostriate, without scutellary striole, interstriae tuberculate, setose, setae short, thin, reclinate, directed caudad. **Legs**. Procoxae large and conical, prominent, medially contiguous; mesocoxae subglobular, narrowly separated; metacoxae flat, transversely elongate. Trochanters short, oblique. Femora long, subcylindrical, inflated in distal half, unarmed, outer side serrulate in apical third to half (distinctly so on meso- and metafemora). Tibiae elongate straight, slender, compressed; outer edges serrulate, with dense long stiff fine setae in distal half of meso- and metatibiae, apically emarginate; apex subtruncate, with single strong apparently fixed inner spine and smaller free spur on pro- and mesotibiae, with 2 strong spurs on metatibiae, inner one fixed, broadened and flattened or slender and conical-rounded. Tarsi (as in *M. calhouni*; all missing from undescribed sp.) with tarsite 1 apically weakly excised, 2 shorter, subtriangular, apically excised, 3 deeply bilobed, lobes strongly pedunculate, 5 long, very slender, apically expanded; claws divaricate, dentate, probably with ventrobasal seta. **Abdomen**. Ventrites subflatly aligned, sutures straight; ventrites 1 and 2 subequal in length, 3 and 4 subequal, shorter than 2, 5 subequal in length to 3 + 4, broadly rounded.

Remarks. As the single type specimen of *Mesophyletis calhouni* is unavailable for study (see Section 2.1 above), this redescription is partly based on another evidently congeneric specimen we received from the FMNH but too late to fully include in this paper. It unfortunately lacks all its tarsi, and therefore the structure of the claws of the genus cannot be confirmed. In *Mesophyletis calhouni* the tarsi are very slender and the lobes of tarsites 3 are pedunculate. In the original description [29] the claws are reported as being bifid on the pro- and mesotarsi and "appendiculate (laminate)" on the metatarsi [29], which would be a very unusual condition in Mesophyletidae (different claws are otherwise present only in *Myanmarus diversiunguis*). The metatarsal claws as illustrated by Poinar [29] appear typically dentate (assuming the teeth carry a ventrobasal seta), but whether those of the mesotarsi are truly bifid (the secondary tooth attached on the inside rather than the ventral side of the primary one) is not clear in the photo (Figure 4 in [29]). In a later diagnosis of a tribe Mesophyletini [53], the claws are described as "widely divergent, with large tooth at base", which would accord with all claws being dentate. Confusion also surrounds the structure of the tibial spurs as reported in the literature [29,53]. In the original description [29] the metatibiae were described as "serrulate with 2 broad symmetrical spines at apex", with Figure 2 clearly showing the serrulation but only a single thick, curved and pointed spine (it is unclear from Figure 2 whether this is homologous with a spur). However, in the later paper [53] the apex of the metatibiae was described as "with two apical spurs" and so illustrated in Figure 10 (one spur very short, appearing to be broken off). From Figure 1 of the original description [29] and Figure 9 of the later paper [53] it is evident that both Figures 2 and 10 do depict metatibiae but that Figure 2 shows the left one (the inner side) and Figure 10 the right one (the outer side). As the spurs or spines must be identical on the two metatibiae, neither description in the literature can be correct. The only possible way to resolve this discrepancy is to assume that the metatibial spurs of *Mesophyletis calhouni* are constructed as they are in *Elwoodius conicops*, i.e., the inner one broad, flat and apparently fixed (as visible in Figure 2 in [29], obscuring the smaller outer one) and the outer one normal, slender and pointed (as visible in Figure 10 in [53], the broad inner one either broken off or only its apex visible due to the viewing angle). This metatibial spur configuration is also largely in agreement with our *Mesophyletis* specimen, which has two shorter robust but also unequal spurs on the metatibiae, with the inner one also fixed (but narrowly conical, not flattened). The pro- and mesotibiae of our specimen, however, have a small, slender free outer spur and an enlarged, apparently fixed but elongate, straight and narrowly conical inner spur that is directed distad, thus not agreeing with the pro- and mesotibial spurs of *E. conicops* (see under that species) but consistent with the original description of *M. calhouni*, which describes the pro- and mestotibiae as "bearing well-developed spine at apex". Other key observable characters of our specimen in agreement with *M. calhouni* are the large protruding eyes (recorded and drawn as round for *M. calhouni* but not so appearing in the published photo), relatively short curved rostrum with strongly exodont mandibles opening into a vertical position, serrate tibiae (recorded as "denticles absent" on the protibiae, although these also are easily overlooked—the denticles are less than half the size of the mesotibial ones in our specimen), sublevelled ventrites and likely the exposed tergites VII and VIII. The structure of the exposed abdominal apex is not clear in the photograph of *M. calhouni* but may be similar to that of the exposed tergite VIII in some *Bowangius* specimens. *Mesophyletis* is distinguishable from all other Burmese amber weevils by the above-listed characters, except that it apparently shares its metatibial spur structure with *Elwoodius conicops*, in which the inner, larger, fixed spur is flattened and asymmetrical on the metatibiae (Figure 91g). The large apical setiferous spike of *Echogomphus viridescens* (Figure 45l) is probably not homologous with a spur; see under this species. Fixed inner spurs of the metatibiae (and sometimes mesotibiae) also occur in some *Bowangius* species, but these are not much larger than the outer spurs but also apically slightly bent. *Mesophyletis* is easily distinguishable from *Elwoodius* by its normal eyes, these not being elongate-conical, and by the different spur configuration on the pro- and mesotibiae.

Mesophyletis was originally placed in a monotypic subfamily (Mesophyletinae, in a family 'Eccoptarthridae') and later as a tribe Mesophyletini in Carinae of a family 'Ithyceridae' [53], but several of its alleged diagnostic characters (e.g., absence of striae, exposed pygidium, swollen trochanters, bifid pro- and mesotarsal claws, procoxae separated by a keel-like prosternum) are probably not valid as described and require re-examination. In Burmese amber weevils it is sometimes easy to confuse basal breaks or cracks in the femora as the trochanterofemoral joint, perhaps giving the appearance of an enlarged trochanter; in all Mesophyletidae we examined (except *Aepyceratus*, in which the trochanters are sublobate) the trochanters are small and strongly oblique. Separation of the procoxal cavities in our sample of Mesophyletinae only occurs in some genera with simple or basally angulate claws (and not by a "keel-like" prosternum); all those in the group with dentate claws (in which *Mesophyletis* belongs) have the procoxae contiguous at least at the base and their cavities confluent. *Mesophyletis* is quite similar to various other mesophyletine genera, in particular *Bowangius* (and *Anchineus*), of which most species seem to have an exposed apical tergite (which may be a sexual trait) and several have a fused inner spur on the metatibiae, but it differs from *Bowangius* in having a subequal pair of spurs on each leg, subflatly aligned ventrites, more typically broader mandibles and slender tarsi.

Mesophyletis calhouni Poinar, 2006

Mesophyletis calhouni Poinar, 2006: 880 [29]

Material (not examined). Holotype, "body" length 2.8 mm (seemingly excluding rostrum, but not SL), 2.20 mm, width 1.21 mm; not studied, originally in amber collection of Ron Buckley, Sumter Ridge, Florence, KY, but sold to Deniz Eren in Turkey, where it is evidently not available for scientific study.

Remarks. As we were unable to study the holotype of this important species, despite repeated requests to its current owner (Deniz Eren, Turkey) to borrow the specimen, we are unable to provide a revised description or diagnosis of the species. According to the original description [29], the species is distinctive in having bicoloured elytra with areas of "castaneous colored squamae" and an "elliptical castaneous spot", also in having no striae, no serration on the protibiae and two broad symmetrical spurs on the metatibiae. Among our sample of Burmese amber weevils, several species have vestiture patterns generally involving patches or defined areas of paler setae, but all have the body cuticle unicolorous (or at least none have different regions of distinctly different colours as described for *M. calhouni*). In some specimens (e.g., *Compsopsarus reneae*), however, a distinct paler 'squamous' type pattern can be observed, created by a localised thin air layer seemingly divided into irregular cells forming a pattern. In our undescribed species of *Mesophyletis* the body cuticle is entirely black, the elytra are distinctly striate, the protibiae finely serrulate and the metatibial spurs unequal. According to Figure 2 in the original description [29], another distinctive feature of *M. calhouni* is the apical structure of the metatibiae (also indicated to be the same on the other tibia in Figure 1 in [29]), in which the inner edge appears to form a low rounded carina. Among our sample of specimens this character is otherwise present only in *Bowangius cyclops* (also on both metatibiae; see under that species), which also has an exposed tergite VIII (appearing similar to the apex of the abdomen in *M. calhouni* and other *Bowangius* species) and round hemispherical eyes. *Bowangius cyclops* can be distinguished from *M. calhouni* by the distinctly striate elytra, more robust tarsi with broad lobes of tarsites 3 and shorter tarsites 5 (only the metatarsal claw is preserved in *B. cyclops*, similar to that of *M. calhouni* in being dentate) and the distinctly thicker segment 4 of the funicles.

Genus ***Euryepomus*** Clarke & Oberprieler, **gen. n.**

Type species: *Euryepomus lophomerus* Clarke & Oberprieler, sp. n.

Description. Size. Length 2.8 mm, width ca. 1.4 mm. **Head** short, subconical. **Eyes** small, hemispherical, coarsely facetted, without tubercles between them. **Rostrum** about as long as pronotum, stout, subcylindrical, curved; antennal insertions postmedian, behind them with scrobes not quite extending to eye, in front of them without lateral row of setae discernible. No gular suture discernible.

Antennae geniculate, long; scapes relatively short, cylindrical, apically well clavate; funicles slightly longer than scape, 7-segmented, segment 1 elongate, 2 shorter, thinner, 3–7 very shorter; clubs loosely articulated, 4-segmented. **Mouthparts**. Labrum absent. Mandibles flat, exodont, curved down, articulation oblique. Maxillae and labium not discernible. **Thorax**. Prothorax strongly proclinate, with anterior lateral margins oblique in lateral view. Pronotum elongate, slightly convex, laterally rounded, without tooth, posterior corners rounded, fitting closely onto elytra; surface densely setose; notosternal sutures closed, upright then curved anteriad. Prosternum moderately long; procoxal cavities medially confluent. Mesocoxal cavities laterally closed (by meso- and metaventrite). Metanepisternal sutures distinct. Mesoventrite short, anteriorly strongly sloping. Metaventrite longer, slightly convex. **Elytra** broad, with prominent, broadly rounded, flattened humeri, apex obscured, probably individually rounded, not exposing pygidium; sutural flanges not visible; surface punctostriate, without scutellary striole, densely setose. **Legs**. Procoxae large, prominent, medially contiguous; mesocoxae subglobular, narrowly separated; metacoxae flat, transversely elongate. Trochanters short, oblique. Femora short, subcylindrical to slightly flattened, distally inflated, unarmed. Tibiae long, flattened, outer edge carinate, not crenulate, apex obliquely truncate, with 2 small spurs. Tarsi with tarsite 1 long, narrow, 2 shorter, triangular, 3 deeply bilobed, 5 as long as 1 + 2; claws divaricate, dentate with ventrobasal seta at apex of tooth. **Abdomen** with ventrites 1 and 2 fused, elongate.

Derivation of name. The genus is named for its broad, sharp shoulders, the name derived from the Greek adjective *eurys* (broad) and noun *epomis* (G: *epomidos*), the point of the shoulder, but being masculine in gender.

Remarks. *Euryepomus* belongs in the group of genera with carinate, crenulate or serrulate tibiae and subglobular eyes and is therefore most similar to *Mesophyletis*, *Mekorhamphus* and *Myanmarus*. From *Mekorhamphus* it is distinguishable by the lack of dorsal tubercles between the eyes, from *Myanmarus* by not having the ventrites impressed and from *Mesophyletis* by having carinate tibiae, free spurs and tarsites 3 with short, digitate lobes, the last having serrulate tibiae, one spur fixed and enlarged and tarsites 3 with long, pedunculate lobes.

Euryepomus lophomerus **Clarke & Oberprieler, sp. n.** (Figure 73)

Description. Size. Length 2.85 mm, width ca. 1.4 mm. **Head** strongly constricted behind eyes. **Eyes** dorsally separated anteriorly by less that width of rostrum but posteriorly by more (forehead triangular). **Rostrum** slightly longer than pronotum, dorsoventrally flattened (possibly a compression artefact); antennal insertions just behind middle of rostral length. **Antennae.** Scapes thin, not reaching eye in repose, apical club about twice as thick as shaft; funicles with segment 1 spindle-shaped, 2 shorter, thinner, 3–7 submoniliform; clubs long, slightly flattened, apical segment distinct, as long as segment 3, flattened. **Mouthparts**. Mandibles with 3 inner (dorsal) teeth, basal one broad and blunt, median one long and acute, apical one forming short, narrow, downcurved T at apex with corresponding inner one, outer (ventral) side with at least one small median tooth. **Thorax**. Pronotum slightly longer than broad; setae fine, long, sharp, suberect, directed anteriad. Prosternum moderately long; procoxal cavities in middle of prothorax. Scutellar shield small, transverse, slightly raised. **Elytra** with humeri laterally flattened, slightly explanate with short thick black carina; posteriorly gently declivous, lateral margin sinuate; punctures indistinct, setae long, straight, suberect, directed caudad. **Legs**. Profemora with outer edge rounded, non-carinate, meso- and metafemora with distinct black carina in distal half. Tibiae somewhat flattened, densely setose, spurs slender, small. Tarsi slightly longer than half of tibial length; tarsite 1 apically truncate, 2 apically slightly excised, 3 with lobes digitate, 5 very narrow between lobes of 3 but broadening distad. **Abdomen** with ventrites 1 and 2 each almost twice as longas 3 + 4, 5 as long as 4.

Material examined. Holotype (NIGP157017): well preserved, intact specimen, with rostrum slightly distorted, surface of pronotum party obscured by thin layer of small bubbles, otherwise well visible; slightly diagonally placed in centre of rectangular cuboid 4.4 × 3.2 × 2.05–2.25 mm; amber clear but with many impurities, especially on left side and around back of specimen, at back and below specimen with dense layer of flow lines obscuring caudal view, along left side also with narrow horizontal crack.

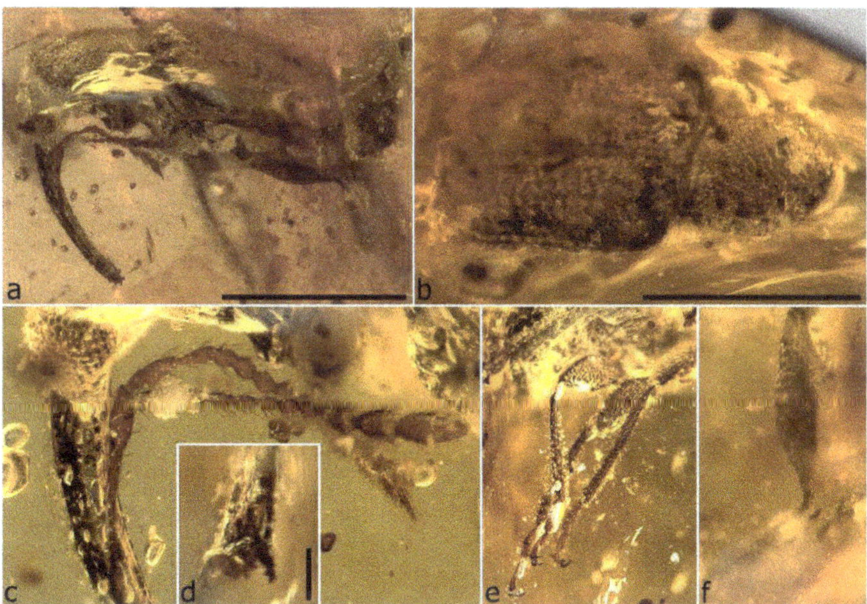

Figure 73. *Euryepomus lophomerus* sp. n., holotype. Habitus, left lateral (**a**); habitus, dorsal (**b**); eyes and antennae, left lateral (**c**); apex of rostrum and mandibles, apical (**d**); legs (**e**); left hindleg (**f**). Scale bars: 1.0 mm (a,b); 0.1 mm (d).

Derivation of name. The name of the species is a latinised adjective formed from the Greek nouns *lophos* (G: *lophou*), a crest or ridge, and *meros* (G: *mereos*), a thigh, for the distinct black ridge on the meso- and metafemora of the species.

Remarks. The single specimen of this species known thus far resembles that of *Mesophyletis calhouni* as illustrated by Poinar [29] in size and shape, but it differs mainly in its legs, the tibiae not being crenulate on the outside (only carinate) and with very small, normal spurs, the meso- and metafemora having a distinct, black carina in the distal half of the outer (dorsal) side, the lobes of tarsites 3 being narrow, digitate, not pedunculate, (all) the tarsal claws dentate and the ventrites slightly stepped (not flatly aligned).

Genus ***Periosocerus*** Clarke & Oberprieler, **gen. n.**

Type species: *Periosocerus deplanatus* Clarke & Oberprieler, sp. n.

Description. Size. Length 2.6–2.7 mm, width 1.35–1.42 mm. **Head** porrect, short, subconical, constricted behind eyes. **Eyes** large, strongly protruding laterally, facing forward or evenly semicircular in outline, coarsely facetted, without interfacettal setae, separated dorsally by about width of rostrum, without tubercles between them. **Rostrum** stout, subcylindrical but compressed; antennal insertions lateral, in apical third of rostral length, behind them with scrobes extending to eye, in front of them scrobes shortly extending, ventral margin with row of 6–7 long, erect setae. Gular suture single, long. **Antennae** geniculate, long, robust; scapes long, thick, cylindrical, apically gradually clavate; funicles as long as scape, 7-segmented, segment 1 long, 2–3 subequal but thinner, 4–5 shorter, 6 and 7 asymmetrical, all sparsely stiffly setose; clubs long, robust, loosely articulated, 4-segmented. **Mouthparts**. Labrum absent. Mandibles bluntly exodont, with 2 blunt upper teeth, articulation oblique. Maxillae exposed, palpiger about as wide as basal palp segment, palps 3-segmented, basal segment short, ca. 0.67

× as long and about as wide as 2, 3 fusiform, narrower than 2, about as long as 1 and 2, apically hyaline. Labium with palps inserted apically, porrect, 3-segmented, segment 1 apically oblique, with large seta, subequal in length to 3, 2 about 0.25 × as long as 3, 3 fusiform, apically hyaline (or with sensilla). **Thorax**. Prothorax not proclinate, with anterior lateral margins vertical in lateral view; laterally strongly inflexed. Pronotum longer than broad, slightly convex, laterally rounded, without tooth, posterior corners shortly extended to fit closely onto elytra; surface setose; notosternal sutures closed, vertical at base, then bent anteriad onto prothoracic margin. Prosternum moderately long; procoxal cavities medially confluent, situated about their length away from anterior margin of prosternum. Mesocoxal cavities laterally closed (by meso- and metaventrite). Metanepisterna setose, metanepisternal suture distinct. Mesoventrite short, anteriorly strongly sloping. Metaventrite longer, about as long as metacoxa, slightly convex. **Elytra** elongate, with broadly rounded humeri, apically individually rounded, not exposing pygidium; sutural flanges invisible; surface punctostriate, without scutellary striole, densely setose, setae long, fine, sharply pointed, directed caudad. **Legs**. Procoxae large, prominent, medially contiguous; mesocoxae subglobular, narrowly separated; metacoxae flat, transversely elongate. Trochanters short, oblique, recessed into coxae. Femora long, subcylindrical, strongly inflated in distal half, unarmed. Tibiae long, flattened, outer edge rounded or crenulate, apex truncate, with 2 spurs. Tarsi robust; tarsite 1 broadly triangular, 2 triangular, 3 deeply bilobed but short, 5 as long as 1 + 2, widening apicad; claws divaricate, dentate with ventrobasal seta at apex of tooth. **Abdomen** with ventrites 1 and 2 fused, slightly longer than 3.

Derivation of name. The genus is named for its large, thick antennae, the name formed from the Greek adjective *periosus* (immense) and noun *keras* (G: *keratos*), a horn, and being masculine in gender.

Remarks. This genus is similar to *Habropezus* in its elongate, depressed eyes and elytral sculpture and vestiture but differs mainly in its much larger, thicker antennae. It also has a flatter body and a longer prosternum (the procoxae placed further back). From *Mekorhamphus* it is readily distinguishable by lacking tubercles between the eyes. It is represented by two species.

Periosocerus deplanatus Clarke & Oberprieler, **sp. n.** (Figure 74)

Description. Size. Length 2.60 mm, width 1.35 mm. **Head** strongly constricted behind eyes, densely setose between eyes. **Eyes** elongate, compressed, in dorsal view strongly oblique, anteriorly separated by width of rostrum, forehead slightly impressed. **Rostrum** as long as pronotum, of even width throughout length, very slightly curved; dorsally sparsely setose, setae suberect, curved anteriad. **Antennae.** Funicles with segment 1 spindle-shaped, 2–3 almost as long but thinner, 4–5 shorter, 6–7 slightly longer than 4–5 but thicker; clubs slightly flattened, segments 1–3 asymmetrical, 4 bluntly triangular, about as long and almost as broad as 3. **Mouthparts**. Mandibles flat, with 2 blunt outer teeth, apical one forming square with inner apical one. **Thorax**. Pronotum slightly narrower than elytra, anteriorly not constricted, sides arcuate, widest across anterior third, basal margin arcuate, bent up, beaded, apical margin straight; surface shallowly rugose, densely setose, setae long, thin, sharp, directed mesad to anteromesad. Scutellar shield roundly subquadratic, flat, densely squamose. **Elytra** narrow, posteriorly sharply declivous, striae broad, shallow, punctures large, interstriae weakly convex, lateral margin faintly sinuate, with epipleural groove, anteriorly slight notched over head of metanepisternum. **Legs**. Procoxae in about middle of prosternal length. Femora rounded on outside. Tibiae straight, protibiae apically bent inwards, outer edge rounded, not crenulate, with long straight suberect setae in distal third. Tarsi almost as long as tibiae; tarsite 1 apically truncate, 2 apically strongly excised, 3 with lobes not pedunculate, 5 narrow. **Abdomen** with ventrites 1 and 2 slightly longer than 3, 4 slightly shorter than, 5 long, subtriangular.

Material examined. Holotype (NIGP157018): well preserved, intact specimen but strongly depressed, especially head and rostrum, left hindwing protruding at back; in centre of rectangular cuboid 5.4 × 3.8 × 2.75 mm; amber very clear with almost no impurities except small bubble well above pronotum, narrow slanting film of minute bubbles along side of right elytron and large elongate milky bubble beneath side of left elytron.

Derivation of name. The species is named for its flattened shape. Even though this is partly due to a compression artefact, the species appears to have been very flat in nature.

Figure 74. *Periosocerus deplanatus* sp. n., holotype. Habitus, right lateral (**a**); habitus, dorsal (**b**); habitus, left lateral (**c**); head, dorsal (**d**); left antenna (**e**); apical part of rostrum and mandibles (**f**); distal part of hindwing (**g**); left femora and tarsal claws (**h**). Scale bars: 1.0 mm (a–c); 0.2 mm (e,g,h).

Remarks. The species differs from *P. crenulatus* mainly in having the tibiae not crenulate on the outside, as well as by having longer tarsi but shorter antennal clubs. The single known specimen is strongly depressed, especially the head and rostrum, and does not allow a proper determination of all critical characters.

Periosocerus crenulatus Clarke & Oberprieler, **sp. n.** (Figures 75 and 76, Video S7)

Description. Size. Length 2.67 mm, width 1.42 mm. **Head** weakly constricted behind eyes, densely setose, finely punctorugose; ventrally moderately bulging. **Eyes** in dorsal view evenly semicircular, dorsally separated by basal width of rostrum anteriorly but further separated posteriorly. **Rostrum** ca. 0.67 × as long as pronotum, dorsally with short recurved setae from base to antennal insertions; with median, paramedian and dorsolateral carinae extending to antennal insertions, all but dorsolateral ones partly confused beyond antennal insertions; dorso-apically with pair of stout curved setae on each side; ventrally on either side of postmentum with pair of thick setae, postmentum distinct, projecting somewhat obliquely, narrowing toward antennal insertions, apically (suture with prementum) truncate. Scrobes narrowly extending anteriad from antennal insertions to mandibular articulations, distal 2 lateral setae distinctly thicker than others. **Antennae.** Scapes about as long as funicle; funicles with segment 1 fusiform, 2.0 × longer than 2, 2–4 elongate-obconical, subequal, 5 similar, ca. half as long as 4, 6–7 broader, distinctly obconical, apically oblique, seemingly more densely setose than 1–5 (like club segments); clubs with segments 1–2 obconical, 2 ca. 0.67 × as long as 1, apices of 1–3 distinctly oblique, 3 1.5 × longer than 4, 4 acute, narrowly inserted into 3. **Mouthparts.** Mandibles flat, outside and inside with 2 teeth, basal ones larger, apical ones smaller, those from opposite sides together forming apical Y. Maxillae with coarse seta on palpiger. **Thorax.** Pronotum slightly narrower than elytra, anteriorly slightly constricted; sides sinuate, straight in basal third, broadly convex anteriorly, in dorsal view seemingly tuberculate, almost toothed; base broadly convex in median half, slightly curved outwards to posterior angles, closely fitting onto elytra; coarsely punctorugose, setae directed anteriad and anteromesad. Prosternum elongate, about as long as procoxae, prosternal process short, pointed; hypomeron ca. 0.25 × as long as prosternum. Scutellar shield distinct, level with elytra, densely setose. **Elytra** with interstriae weakly prominent, punctorugose, striae and interstriae more prominent laterally; bases weakly sinuate, closely fitting over pronotum; lateral margin sinuate with distinct marginal groove, slightly widening anteriorly. **Legs.** Procoxae close to posterior hypomeral margin. Femora on outside rounded (profemora) or crenulate (meso- and metafemora); mesofemora strongly notched before apex, with curved, blunt, posteriadly projecting tooth on anterior edge of inner side. Tibiae with long suberect setae, denser longer setae in apical half; outer edge sparsely serrulate (protibiae) or crenulate with curved, pointed and closely spaced teeth (meso- and metatibae); apically somewhat expanded, apex with short narrow flange lined with coarse fringing setae, dorso-apically with stronger, coarser, spur-like seta (less distinct from other fringing setae on metatibiae) and slender elongate setae; with 2 elongate unequal spurs. Tarsi ca. half as long as tibiae; tarsite 1 apically subtruncate, 2 ca. half as long as 1, apically strongly excised, 3 with lobes subpedunculate, ca. half as long as 5, somewhat recessed into apex of 2, 5 long, very slender. **Abdomen.** Ventrites densely setose, setae much finer than dorsal setae; sutures substraight or arcuate, 1 and 2 subflatly alligned, lower than others, subequal in length, 3–4 progressively shorter, 5 longer than 4.

Material examined. Holotype (NIGP157019): well preserved specimen, not distorted or compressed, well visible (especially ventrally), with fragmented coating of whitish debris, with femorotibial joints of right legs and left hindleg, distal half of left mesotibia and right elytral apex and half of ventrite 5 cut away with amber, other structures intact with all mouthpart structures clearly visible, tarsi bunched beneath specimen (poorly visible); in triangular prism 5.1 × 3.2 × 2.5 mm with one long edge rounded off; amber clear yellow, with one large bubble above pronotum and several other smaller bubbles and few tiny fractures close to dorsal surface of specimen.

Derivation of name. The species is named for its crenulate tibiae, the name being a Latin adjective.

Remarks. *Periosocerus crenulatus* differs from *P. deplanatus* mainly in having the tibiae serrulate to crenulate on the outside and by having shorter tarsi but longer antennal clubs. An exceptional feature of the single known specimen is the astonishing preservation and unobstructed view of the mouthparts. The specimen was submitted for CT scanning, but the resulting images are not too clear (Figure 76, Video S7).

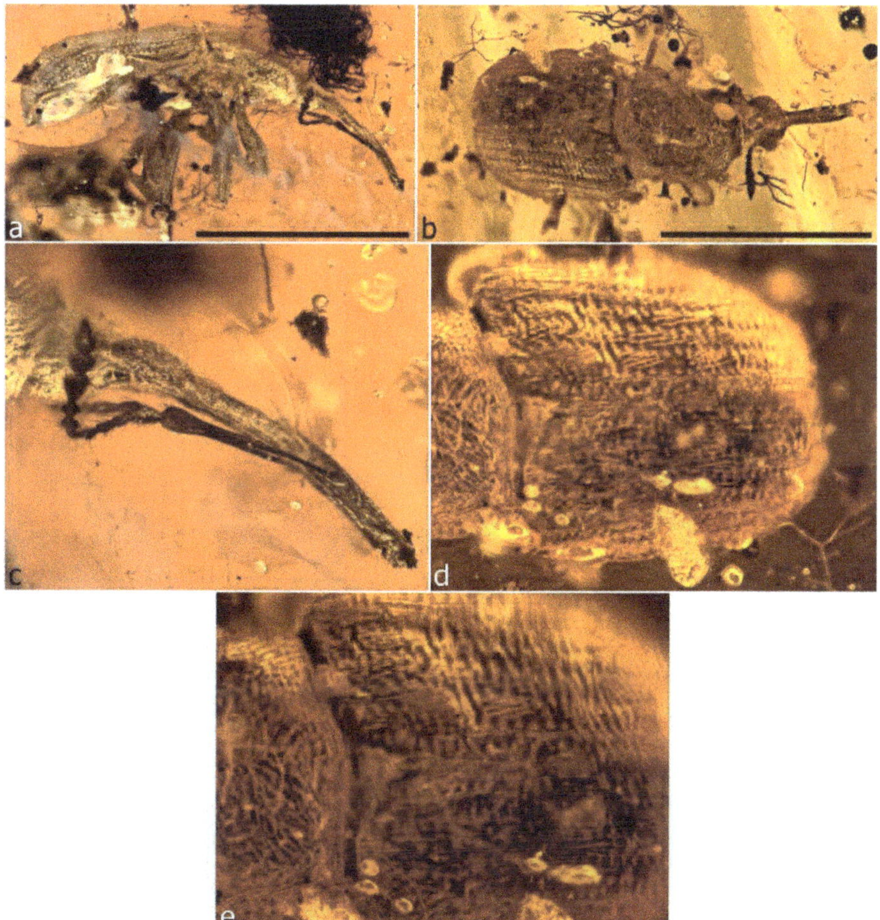

Figure 75. *Periosocerus crenulatus* sp. n., holotype. Habitus, right lateral (**a**); dorsal oblique (**b**); head and antenna, right lateral (**c**); elytra, dorsal oblique (**d**); same, detail (**e**). Scale bars: 1.0 mm (a,b).

Genus ***Habropezus*** Poinar, Brown & Legalov, 2016

Habropezus Poinar, Brown & Legalov, 2016: 160 [54] (type species, by original designation: *Habropezus plaisiommus* Poinar, Brown & Legalov, 2016)

Redescription. Size. Length 1.74–3.0 mm, width 0.68–1.2 mm. **Head** short, subconical, moderately constricted behind eyes, bulging dorsally, more strongly ventrally. **Eyes** large, elongate-compressed, strongly protruding, moderately coarsely facetted, without interfacettal setae, dorsally separated by about width of rostrum anteriorly but further separated posteriorly; forehead flat, without tubercles above eyes. **Rostrum** about 1.5 × longer than pronotum, subcylindrical, moderately curved; antennal insertions lateral, with scrobes behind them reaching eye, in front of them with or without (*H. kimpulleni*) row of sparse erect setae. Gular suture single, from base of head to underside of rostrum. **Antennae** geniculate, long; scapes elongate, cylindrical, apically only slightly inflated, almost as long as funicle; funicles 7-segmented, segment 1 subequal in length to and wider than 2, others thinner and progressively shorter towards club; clubs elongate, loosely articulated, 4-segmented, segment 4 acute, flattened, about as long as 3, broadly inserted into 3 in dorsal view, much more narrowly in lateral view (at least in *H. tenuicornis*). **Mouthparts**. Labrum absent. Mandibles small, exodont, articulation oblique. Maxillae and labium not clearly visible. **Thorax**. Prothorax slightly to strongly proclinate, with anterior lateral margins oblique in lateral view. Pronotum convex, laterally rounded, without tooth, posterior corners subacute, fitting closely onto elytra; surface coarsely punctorugose to closely tuberculate, sparsely to densely finely setose, setae reclinate, directed anteromesad; notosternal sutures closed, vertical then abruptly deflected anteriorly. Prosternum short; procoxal cavities medially confluent, closer to anterior prosternal margin. Scutellar shield densely setose, raised above elytral surface. Mesocoxal cavities laterally closed (by meso- and metaventrite). Metanepisterna distinct, setose. Mesoventrite short, anteriorly sloping. Metaventrite longer, raised into transverse weals before metacoxae. **Elytra** elongate, basally concave to receive basal pronotal margin, with broadly rounded humeri, posteriorly declivous, lateral margin strongly sinuate to roundly emarginate in middle, apically weakly individually to conjointly rounded, not exposing pygidium; sutural flanges narrow, equal; surface punctostriate, without scutellary striole, interstriae coarsely tuberculate, sparsely setose, setae long, thin, reclinate, directed caudad. **Legs**. Procoxae large, prominent, medially contiguous; mesocoxae subglobular to conical, projecting, narrowly separated; metacoxae flat, transversely elongate. Trochanters short, oblique. Femora long, subcylindrical, inflated in distal half, unarmed, meso- and metafemora on outside carinate to serrulate in distal third. Tibiae elongate, straight, slender, compressed, outer edges serrulate, with dense long stiff setae in distal half, apex subtruncate, with 2 spurs. Tarsi more than half length of tibiae; tarsite 1 apically subtruncate, 2 shorter, subtriangular, apically subtruncate, 3 deeply bilobed, 5 long, slender, apically expanded; claws divaricate, dentate with ventrobasal seta on proximal face of tooth. **Abdomen.** Tergite VII flat, apically sharply rimmed, fitting tightly onto caudal edge of ventrite 5. Ventrites 1 and 2 longer than 3, 1–4 progressively shorter, 5 shorter or slightly longer than 4, apically broadly rounded, flat.

Remarks. Along with *Mekorhamphus*, this genus was placed in a tribe Mekorhamphini based on several alleged characters, including its horizontally moving exodont mandibles, an elongate procoxal prosternum, contiguous procoxal cavities and swollen trochanters, and distinguished from the tribes Anchineini and Mesophyletini based on combinations of these. Our examination of the remarkably well preserved and largely unobstructed holotype of *H. plaisiommus* revealed that the original description is somewhat unclear on the ventral prothoracic details, describing the prosternum as elongate and almost equal in length to the procoxal cavities, whereas the hypomeron ("postcoxal portion") is stated as being short, this seemingly indicating a placement of the procoxal cavities closer to the posterior than the anterior prothoracic margin. However, the prosternum is in fact shorter than the hypomeron, and thus the procoxal cavities are more anteriorly positioned. The original description also mentions a fused labrum, mandibles probably with teeth on the external margins and moving horizontally, antennal scrobes directed toward the eyes and free ventrites. We found no evidence of a labrum (fused or otherwise; the apical part of the rostrum forms a truncate margin), the first two ventrites to be fused, only the last three free (as in other Mesophyletidae), but the mandibular dentition to be correct as stated. The rostrum in the holotype of *H. plaisiommus* is depressed, resulting in the antennal insertions being pushed to the ventral side (an artefact), obscuring the scrobes and leaving the mandibular articulation plane unverifiable. The antennal insertions would have naturally been in a lateral position, as they are in all other Mesophyletidae, and the mandibular articulation plane could have been horizontal or oblique given the clearly visible, typically flat structure and overlapping configuration of the mandibles in their repose position. The displacement of the ventrites (sunken) prevents an assessment of the exposure and structure of tergite VIII, which in several species of the very similar genus *Bowangius* is carinate along the upper edge and apically sharply flanged. This structure is seemingly absent or different in most *Habropezus* species (in *H. kimpulleni* tergite VII abuts directly onto ventrite 5), but more material is required to confirm this, especially if the condition present in our *Bowangius* specimens is a character of the male. Most of the above-mentioned details were only visible at magnifications greater than 100×, as is the serrulation of the tibiae, a critical feature not mentioned in the original description.

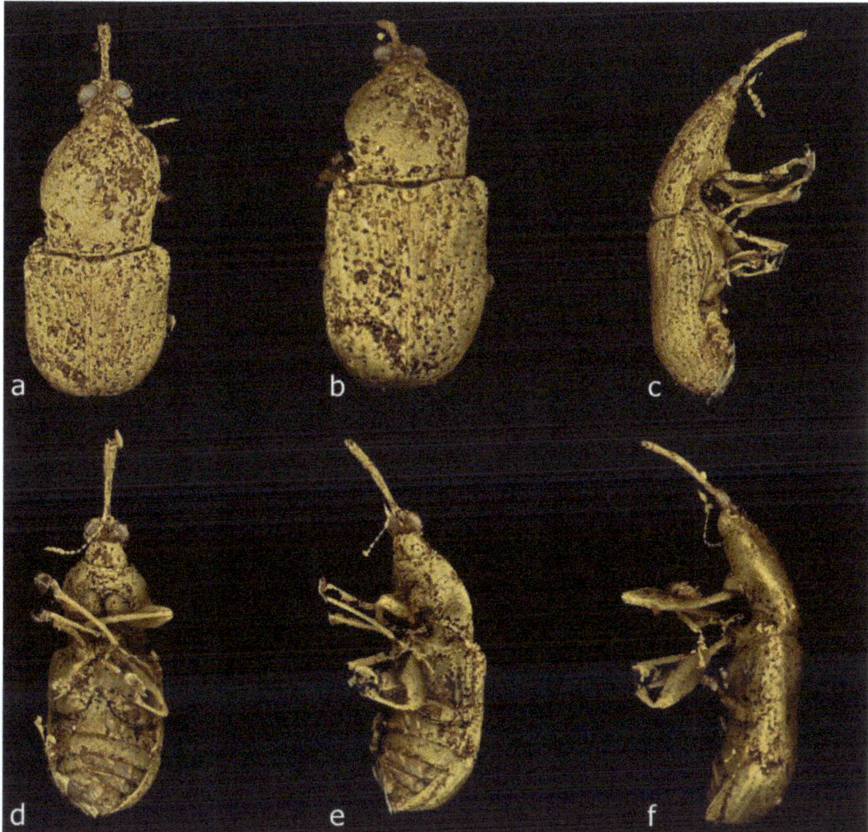

Figure 76. *Periosocerus crenulatus* sp. n., holotype. Habitus images extracted from a micro-CT scanning reconstruction (see also Video S7). Dorsal oblique (**a**); dorsal (**b**); right lateral (**c**); ventral (**d**); ventral oblique (**e**); left lateral (**f**).

Habropezus is difficult to distinguish from several other genera, seemingly forming a group of closely related taxa including *Bowangius*, *Euryepomus*, *Leptopezus*, *Mesophyletis* and *Periosocerus*. Among its most important diagnostic characters are the compressed eyes and posteriorly widening, flattened forehead, which distinguish it from all other genera with serrulate or crenulate tibiae except *Periosocerus*, from which it differs in its serrulate tibiae, distinctly punctostriate elytra with raised interstriae, distinct humeri, higher body with shorter prosternum and the proclinate prothorax. *Habropezus* currently contains four species, a pair of larger ones without lateral setae on the rostrum (*H. kimpulleni*) or with only fine, indistinct ones (*H. plaisiommus*) and a pair of smaller ones with long and very distinct such setae (*H. tenuicornis* and *H. incoxatirostris*).

Habropezus plaisiommus Poinar, Brown & Legalov, 2016

Habropezus plaisiommus Poinar, Brown & Legalov, 2016: 160 [54]

Redescription. Size. Length 3.0 mm (excl. rostrum, not SL). Small, uniformly black but with legs paler, dark brown; coarsely punctate, sparsely setose. **Head** setose. **Eyes** strongly, flatly protuberant. **Rostrum** apically slightly expanded, dorso-apically grooved, with at least 2 long setae at sides; antennal insertions slightly antemedian. **Antennae.** Scapes extending to near front margin of eye; funicles with segments 2–7 subequal in width, apically expanded; clubs ca. 2.0 × wider than funicle, segment 1 slightly longer and about as wide as 2, obconical, basal width less than that of 2, 3 slightly shorter than 4. **Mouthparts.** Mandibles with 2 large teeth on inner and outer edges; apically forming broad V, slightly emarginate, with apical outer tooth much larger than middle outer one but middle inner tooth much larger than apical inner tooth; articulation probably oblique (rostrum depressed). **Thorax.** Prothorax strongly proclinate. Pronotum sparsely, coarsely punctate, punctures distinct, setae fine, recurved; sides weakly curved; base slightly sinuate, fitting closely onto elytra. Prosternum short, anterior margin broadly emarginate, prosternal process short, pointed; hypomeron longer than prosternum. Scutellar shield subcircular, slightly convex. Mesocoxal cavities narrowly separated. **Elytra** evenly rugosely sculptured between punctures; punctures on interstriae finer than in striae; interstriae prominent, raised above striae, setose, setae arising from prominences; bases weakly sinuate, extended over pronotum; humeri slightly flatly projecting; sides weakly sinuate, without any lateral carina, margin pre-apically with row of 3–4 small teeth along edge; with marginal groove subequal in width for entire length, with anterior marginal notch. Hindwings present (wing visible beneath slightly parted elytra). **Legs** long, very slender, similarly setose. Femora very slender basally, inflated in distal half, with subappressed fine recurved whitish setae; outside slightly (profemora) to more distinctly (meso- and metafemora) serrulate in distal third. Tibiae with outer edges (protibiae) carinate in proximal third to half and sparsely serrulate in distal half (gap of 2–4 teeth between teeth), or (meso- and metatibiae) closely serrulate for entire length, with teeth curved, pointed; apically with shortly and broadly rounded outer flanges and slender fringing setae; dorso-apically with several very long fine setae; ventrally with 2 long dark spurs; protibiae ca. 1.25 × longer than others; metatibiae without inner apical flanges. Tarsi distinctly elongate-slender, progressively shorter from fore- to hindlegs; tarsite 1 elongate, ca. 2.0 × longer than 2, slightly expanded apically, 2 slightly wider, of metatarsi with dorso-apical setae denser and longer than others, 3 with lobes slightly wider apically than basally, slightly less than half length of claw segment, with inner edges of lobes distinctly continuous basally, of metatarsi narrower than other legs; claws with strong triangularly acute basal tooth. **Abdomen.** Ventrites sparsely setose, distinctly sparsely punctate (punctures similar to dorsal punctures, finer), rugosely microsculptured; 1–2 subflatly aligned, 2–5 stepped, 5 slightly longer than 4, apically rounded; sutures deeply grooved, straight.

Material examined. Holotype (PACO, with curatorial #Bu-C-48A): excellently preserved, intact, well visible and minimally distorted specimen (rostrum depressed), with several bubbles emanating from body, including large one from abdominal apex and from rupture of ventral side of right profemur; situated close to large curved face of irregular cuboid; amber clear yellow-brown with flow band above top left side of specimen and few impurities, but numerous large bubbles (see also [54]). Paratype (PACO, with curatorial #Bu-C-48B): heavily distorted and decomposed, poorly visible specimen, with head and prothorax crumpled and ruptured, most legs missing or not properly visible (left legs) or missing parts, antennae largely missing; in irregular cuboid with one end rounded; amber clear yellow, largely free of impurities, with large mass (interpreted a braconid wasp cocoon [54]) adjacent to left side. This specimen is not conspecific with *Habropezus plaisiommus* (see Genus and species *incertae sedis* below).

Remarks. This species is most similar to *H. kimpulleni*, agreeing in mandibular structure and dentition, apparently not having an apicolateral row of erect setae on the rostrum, which occurs in the other two species, and in having a similar tarsal configuration. It differs from *H. kimpulleni* in its long, slender, straight, serrulate tibiae (apically notched and crenulate in *H. kimpulleni*). From *H. tenuicornis* and *H. incoxatirostris* it differs in its rostrum not having apicolateral setae and in its regularly and closely serrulate tibiae.

Habropezus kimpulleni Clarke & Oberprieler, **sp. n.** (Figure 77)

Description. Size. Length 2.78 mm, width 1.2 mm. **Head** short, subconical, strongly constricted behind eyes. **Eyes** large, forehead between them triangular, flat. **Rostrum** relatively thick, slightly curved; antennal insertions in basal 2/5 of rostral length, in front of them without lateral row of setae. **Antennae** long; scapes long, thin; funicles longer than scape, segment 1 inflated (as thick as club of scape), 2–7 thinner, also long but gradually shortening towards club; clubs almost as long as funicle, slightly flattened, segment 1 as long and broad as 2, both obconical, segment 4 as long as 3, broadly inserted into 3, flat, acute. **Mouthparts**. Mandibles with 3 teeth on inner (dorsal) side, basal 2 large and acute, apical one smaller (and double on right mandible, with another tooth projecting obliquely ventrad from it), with 2 teeth on outer (ventral) side, basal one also larger but blunt, inner and outer apical teeth together forming broad sharp T; articulation oblique. **Thorax**. Prothorax strongly proclinate. Pronotum elongate, slightly convex, laterally strongly rounded, posterior margin bisinuate, distinctly rimmed; surface densely bluntly tuberculate, sparsely setose, setae long, thin, recumbent. Scutellar shield short, broad, covered with long setae directed caudad. Mesepimera large. Metanepisternal sutures distinct. **Elytra** with almost angular humeri, posteriorly gently declivous, apically weakly individually rounded; striae indistinct. **Legs**. Femora subcylindrical, profemora on outside rounded but meso- and metafemora finely crenulate in distal third. Tibiae flattened, outer edge distinctly crenulate, meso- and metatibiae apically excised on outside, apex obliquely truncate, with 2 large spurs. Tarsi almost as long as tibiae; tarsite 1 elongate, narrow, 2 shorter, narrowly triangular, 3 with lobes not pedunculate, 5 as long as 1 + 2. **Abdomen**. Ttergite VII strongly sclerotised, finely punctate. Ventrites 1 and 2 slightly longer than 3 + 4, 3 slightly longer than 4, 5 short, apically broadly rounded.

Material examined. Holotype (NIGP157020): extremely well preserved, intact specimen, not compressed or distorted, very well visible from all sides; in centre of rectangular cuboid 5.5 × 3.1 × 2.6 mm; amber very clear with almost no impurities except large brown film diagonally above head pronotum, white bubble below prosternum, large, irregular clear bubble below right side of abdomen, and clear bubble above elytral declivity.

Derivation of name. The species is cordially named after our colleague Kimberi Pullen for his longstanding and able assistance with weevil research and curation of the ANIC weevil collection, especially during the compilation of the catalogue of Australian weevils [58].

Remarks. This species is most similar to *H. plaisiommus*, the rostrum also without a distal row of erect lateral setae and the mandibles with a similar dentition, but it differs from the latter in its broader antennal clubs and several leg characters, the tibiae crenulate throughout their length (not carinate or serrulate), the meso- and metatibiae apically notched on the outside, tarsites 1 shorter (less than 2.0 × longer than 2) and the lobes of tarsites 3 less than half as long as tarsites 5. From *H. tenuicornis* and *H. incoxatirostris* it is most readily distinguishable by its rostrum not having an apicolateral row of sparse erect setae and its crenulate tibiae and longer tarsi.

Habropezus tenuicornis **Clarke & Oberprieler, sp. n.** (Figure 78)

Description. Size. Length 1.83 mm, width 0.9 mm. **Head** porrect, weakly constricted behind eyes. **Eyes** elongate, in dorsal view hemispherical, laterally protruding. **Rostrum** dorsally with 2 subparallel grooves along entire length; antennal insertions median, in front of them with ca. 6 long curved setae in groove reaching apex of rostrum; dorso-apically with 2 long setae at sides. **Antennae.** Scapes not quite reaching front margin of eye, shorter than funicles (ca. as long as first 5 funicle segments), apically truncate; funicles ca. 0.33 × as long as scapes, segments 1–4 subequal in width, 1 oval, longer than 2, 2–6 similarly narrow basally, apically expanded and rounded, 2–6 subequal in length, 6 slightly narrower than 5, 7 ca. 0.67 × as long as and slightly narrower than 6; clubs very long (almost as long as funicles), thin, segments 1–3 obconical, progressively shorter distally, 4 acute, tapering to a point, shorter than 3. **Mouthparts.** Mandibles seemingly exodont (not entirely visible), articulation oblique. **Thorax**. Prothorax slightly proclinate, narrower than elytra. Pronotum weakly collared anteriorly, widest at about middle; punctate, sparsely setose; base weakly sinuate and posterior corners distinctly angulate, fitting closely to elytra. **Elytra** densely setose; apically almost conjointly rounded. Hindwings present (partially extended). Scutellar shield slightly prominent. **Legs** and setae pale, spurs and fringing setae of tibiae dark. Procoxae conical, strongly projecting, others not visible. Femora narrow basally, moderately inflated apically; outside rounded (profemora) or serrulate (metafemora; mesofemora not properly visible). Tibiae long, slender, flattened, sparsely serrulate on outside, setose with stiff sparse longer setae in distal half; apically with tarsal articulation surfaces suboblique, with short narrow flanges lined with coarse fringing setae, with spurs large, distinct, subequal. Tarsi ca. 0.67 × as long as tibiae, slender; tarsite 1 narrower and ca. 2.0 × longer than 2, 3 lobate, not pedunculate, lobes short, finger-like, broad at base, slightly wider distally, ca. half as long as 5. **Abdomen** not visible.

Figure 77. *Habropezus kimpulleni* sp. n., holotype. Habitus, right lateral (**a**); habitus, left lateral (**b**); habitus, dorsal (**c**); head, ventral (**d**); elytra and apical tergite, apical (**e**); antennae (**f**); rostrum apex and mandibles, dorsolateral (**g**); same, right lateral (**h**); tarsi and left mesotibia (**i**); left metatibia (**j**). Scale bars: 1.0 mm (**a–c**); 0.2 mm (d,f,j); 0.1 mm (g–i).

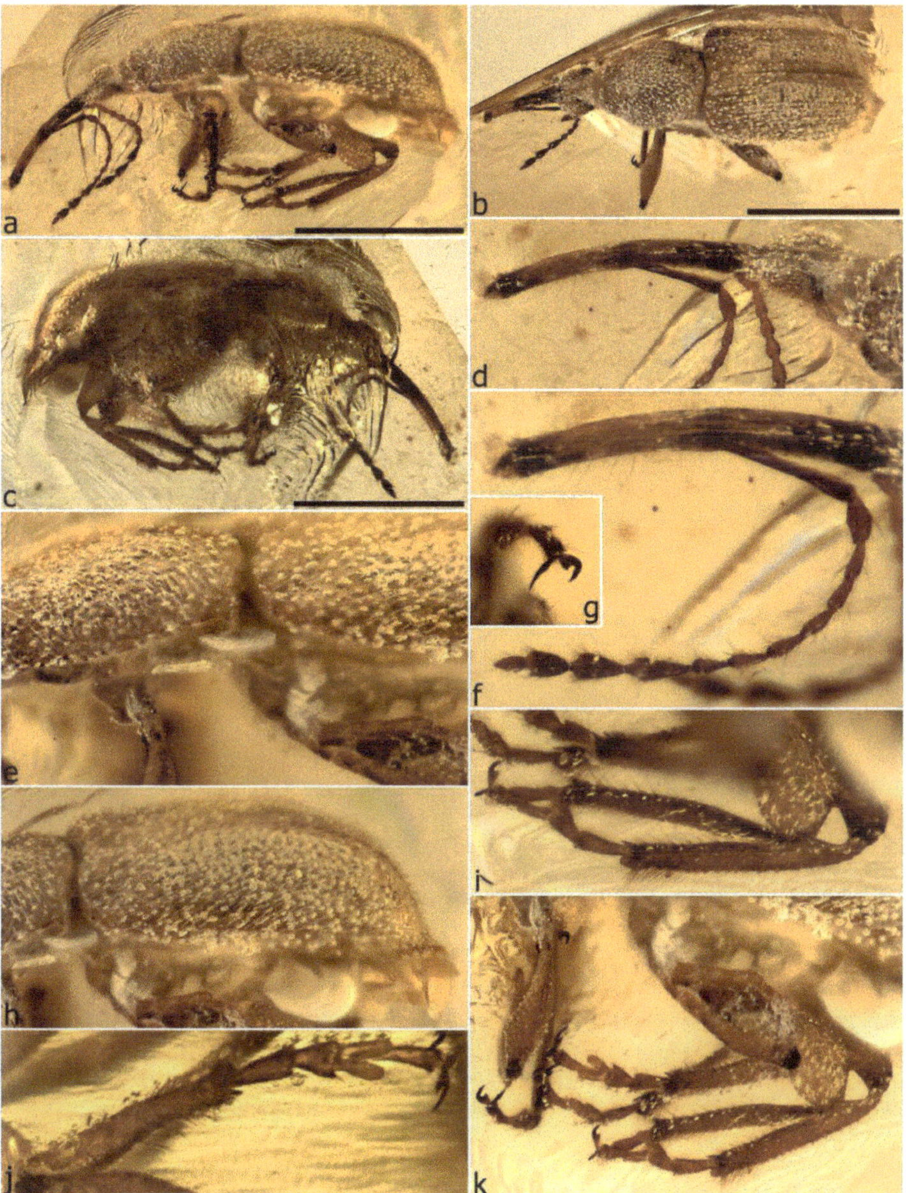

Figure 78. *Habropezus tenuicornis* sp. n., holotype. Habitus, left lateral (**a**); habitus, dorsal (**b**); habitus, right lateral, showing fracture plan obscuring right side (**c**); head (**d**); prothorax and humeral region of elytra (**e**); rostrum and antenna (**f**); left protarsal claw (**g**); elytra, left lateral (**h**); metatibiae (**i**); tibia and tarsus, ventral (**j**); legs (**k**). Scale bars: 1.0 mm (a–c).

Material examined. Holotype (NIGP157021), male: well preserved, intact but poorly visible specimen, not distorted or compressed but body nearly entirely covered by thin cloudy whitish substance and tiny bubbles, right side of prothorax and elytra concealed by fracture and flow band, head, rostrum and appendages well visible; in irregular triangular prismoid (with sixth face) 7.5 × 5.1 × 3.0 mm; amber clear yellow, without debris, other fractures or bubbles.

Derivation of name. The species is named for its very slender antennae, from the Latin adjective *tenuis* (slender) and noun *cornu* (G: *cornus*), the name being an adjective.

Remarks. This species differs from *H. plaisiommus* and *H. kimpulleni* in its smaller size, slightly shorter rostrum, thicker tibiae with indistinct serration, thicker tarsi with broader lobes of tarsites 3 and in having, like *H. incoxatirostris*, an apicolateral row of sparse erect setae on the rostrum and the outside of the meso- and metatibiae adorned with a row of smaller, widely separated denticles. From the last species it differs mainly in its evenly curved rostrum and serrulate tibiae. The holotype was submitted for CT scanning as much of its surface is obscured, but the contrast between the specimen and the amber was too low to permit any meaningful visualisation of the specimen.

Habropezus ncoxatirostris Clarke & Oberprieler, **sp. n.** (Figure 79)

Description. Size. Length 1.73 mm, width 0.68 mm. Body setae long, suberect. **Head** indicated to be constricted behind eyes and moderately bulging, impunctate. **Eyes** coarsely facetted. **Rostrum** distinctly kinked (bent at ca. 40°) at antennal insertions (probably partly a distortion); dorsally with grooves and carinae indicated; antennal insertions median, in front of them with lateral row of ca. 10 elongate slender setae in groove. **Antennae**. Scapes slender, shorter than funicles, gradually expanding apicad, not quite reaching eye, apex slightly oblique; funicles longer than scapes, segment 1 ca. 2.0 × longer than 2, broader than others, 2–6 subequal, 7 ca. 0.67 × as long as 6; clubs long, distinctly broader than funicles, segments 1–2 subequal, obconical with rounded apices, 3 ca. 0.67 × as long as 2, 4 broadly inserted, elongate-acute, slightly narrower and distinct from 3. **Mouthparts.** Mandibles with 2 teeth on outside and inside edges (at apices forming medially notched 'T'), basal inner tooth much larger, triangular. Maxillary palps visible but not clear. **Thorax.** Prothorax slightly proclinate. Pronotum narrower than elytra, punctorugose, sparsely setose, setae long, at apices of prominences; widest in front of middle, sides strongly narrowing anteriad and posteriad, laterally margined (but not carinate). Prosternum very short, ca. 0.33 × as long as procoxae, prosternal process short, angulate; procoxal cavities closer to prosternal margin; hypomeron longer than prosternum. Scutellar shield recessed into elytral bases, anterior margin concavely continuous with elytral basal margins. **Elytra** coarsely rugose, punctostriate; striae much broader than interstriae, punctures indistinct; interstriae prominent, subcarinate in anterior third, less distinct from striae posteriorly, seemingly with line of setae in striae and on interstriae, setae situated on cuticular prominences; sides with 2 macrosetae in basal third, with distinct punctosetose marginal groove subequal in width for entire length, with anterior marginal notch; bases weakly sinuate; humeri subflat; apically individually rounded, outer edges lined with minute teeth. Hindwings present. **Legs.** Procoxae conical; mesocoxae subglobular, weakly projecting, separated by unconnected processes of meso- and metaventrite. Femora subcylindrical, inflated in distal half, outside seemingly rounded. Tibiae with outer side costate, without teeth (protibiae) or subserrulate, with sparse fine denticles only (meso- and metatibiae); apically with short flanges lined with coarse fringing setae, dorso-apically with elongate slender setae; spurs small, indistinct, unequal, inner one slightly smaller on pro- and metatibiae but distinctly on metatibiae. Tarsi slender, nearly as long as tibiae, tarsites 1 and 2 gradually expanding apicad, 1 ca. 1.5 × longer than 2, 2 slightly wider, 3 strongly lobate, lobes elongate-oval, ca. half as long as 5 (shorter on protarsi), with inner margins concave, 5 extremely slender, clearly articulating with cryptotarsite; claws with basal tooth small, with wide arcuate gap between outer claw and inner edge of tooth. **Abdomen,** Ventrites seemingly stepped, sparsely setose, with scattered longer setae; ventrite 1 slightly longer than 2, 3 half as long as 2, 4 longer than 3, 5 short, ca. half as long as 4; sutures between ventrites deeply grooved, subarcuate.

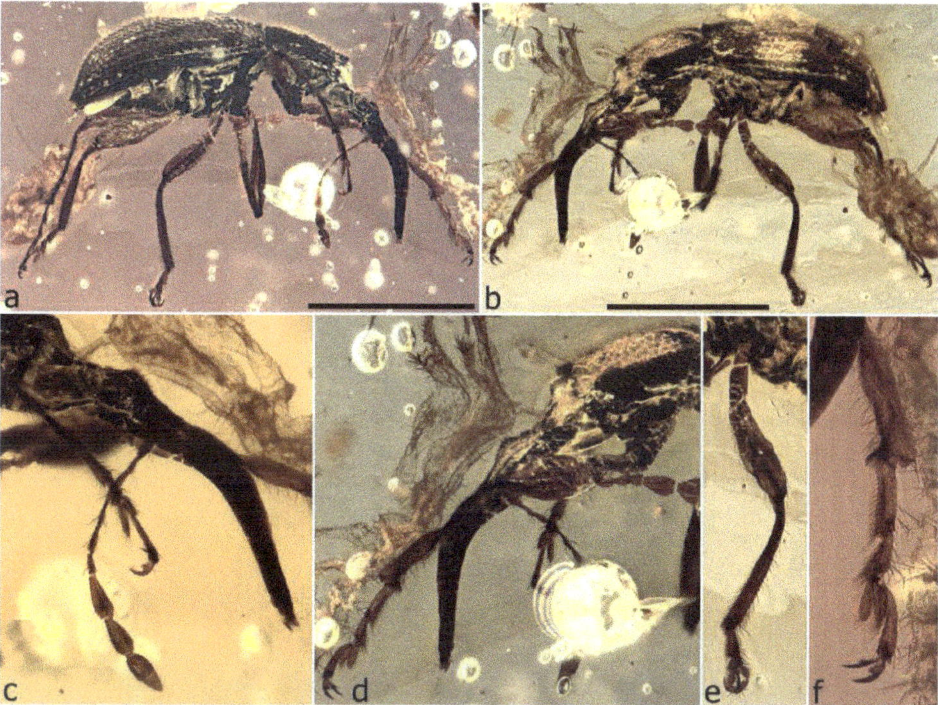

Figure 79. *Habropezus incoxatirostris* sp. n., holotype. Habitus, right lateral (**a**); habitus, left lateral (**b**); head, antenna and protibia and tarsus, right lateral (**c**); same, left lateral (**d**); left middle leg (**e**); left protibia and tarsus (**f**). Scale bars: 1.0 mm (a,b).

Material examined. Holotype (NIGP157022): poorly preserved, somewhat decomposed (mainly legs) and heavily distorted (crumpled) but largely intact specimen, with left protarsal claw and most of tarsite 3 cut away with amber and right mesotibia severed (prior to preservation) and missing, both hindwings partly extended, head, rostrum and eyes, pro- and pterothorax, legs, and ventrites heavily distorted (compressed, crumpled); in subcuboidal block, 5.3 × 2.6 × 1.9 mm, with one large curved side; amber clear yellow, with minor debris particles and another arthropod inclusion in front of head and rostrum, with large white spherule anteriorly and large mite posteriorly on left side obscuring left aspect of metatibiae.

Derivation of name. The species is named for its kinked rostrum, from the Latin adjective *incoxatus* (bent down).

Remarks. This species agrees with *H. tenuicornis* in having an apicolateral row of sparse erect setae on the rostrum and the outside of the meso- and metatibiae adorned with a row or smaller, widely separated denticles, which are apparently formed by the edges of setal sockets rather than being distinct curved teeth. From *H. tenuicornis* it differs in having broader antennal clubs (much broader than the funicles) and longer tarsi. It is also distinctive in its coarse pronotal and elytral sculpturing and the distinct striae, as well as by the rostrum being bent in the middle, at least some of these features possibly accentuated by distortion.

Genus *Leptopezus* Clarke & Oberprieler, **gen. n.**

Type species: *Leptopezus rastellipes* Clarke & Oberprieler, sp. n.

Description. Size. Length 2.12–2.82 mm, width 1.15–1.56 mm. **Head** short, transverse, constricted behind eyes, weakly convex dorsally, more strongly bulging ventrally. **Eyes** large, elongate, slightly depressed and protruding, coarsely facetted, without interfacettal setae, dorsally separated by about width of rostrum; forehead grooved, without tubercles above eyes. **Rostrum** ca. 1.25 × longer than pronotum, subcylindrical, downcurved; antennal insertions lateral, with scrobes behind them reaching eye, in front of them laterally with row of long erect setae. Single long gular suture present. **Antennae** geniculate, long; scapes elongate, cylindrical, apically inflated, almost as long as funicle; funicles 7-segmented, segment 1 subequal in length to and wider than 2, 2–3 subequal, elongate, subcylindrical, 4–7 progressively shorter towards club; clubs short, loosely articulated, 4-segmented, segment 4 broadly rounded, subequal in length to 3. **Mouthparts**. Labrum absent. Mandibles small, flat, exodont, articulation oblique. Maxillary and labial palps 3-segmented, both projecting obliquely forwards. **Thorax**. Prothorax proclinate, with anterior lateral margins oblique in lateral view. Pronotum convex, laterally rounded and without tooth, posterior corners produced to fit closely onto elytra; surface coarsely punctorugose, sparsely to densely finely setose, setae reclinate, directed anteromesad; notosternal sutures closed, vertical then abruptly bent anteriad. Prosternum moderately long; procoxal cavities medially confluent, in middle of prothorax. Scutellar shield densely setose, raised above elytral surface. Mesocoxal cavities laterally closed. Metanepisterna distinct, setose. Mesoventrite short, anteriorly sloping. Metaventrite longer, raised into transverse weals in front of metaxcoxae. **Elytra** elongate, basally extended over pronotum, with broadly rounded humeri, posteriorly declivous, lateral margin strongly sinuate to roundly emarginate in middle, apically individually rounded, not exposing pygidium; sutural flanges narrow; surface punctostriate, without scutellary striole, interstriae rugose, setose, setae short, thin, reclinate, directed caudad. **Legs**. Procoxae large and conical, prominent, medially contiguous; mesocoxae subglobular, narrowly separated; metacoxae flat, transversely elongate. Trochanters short, oblique. Femora long, subcylindrical, inflated in distal half, unarmed, profemora on outside rounded, meso- and metafemora serrulate in apical third. Tibiae elongate, straight, slender, compressed, outer edges serrulate (protibiae) or carinate-crenulate (meso- and metatibiae), with dense long stiff setae in distal half, of meso- and metatibiae apically emarginate; apex subtruncate, with 2 strong spurs. Tarsi more than half as long as tibiae; tarsite 1 apically subtruncate, 2 shorter, subtriangular, apically excised, 3 deeply bilobed (not or subpedunculate), 5 long, very slender, apically expanded; claws divaricate, dentate with ventrobasal seta on proximal face of tooth. **Abdomen** with ventrites 1 and 2 longer than 3, 1–4 progressively shorter, 5 not discernible.

Derivation of name. The genus is named for its very slender tarsi, especially the long, thin onychia, the name composed from the Greek adjective *leptos*, thin, and noun *peza*, a foot, and being masculine in gender.

Remarks. This genus is similar to *Habropezus* but differs in the more robust body form, more strongly globular and protruding eyes, different elytral sculpturing and the strongly crenulate and distinctly apically emarginate outsides of the meso- and metatibae. It is also similar to *Myanmarus* but has unmodified ventrites, and from *Mekorhamphus* it can be distinguished by the lack of tubercles between the eyes. It is represented by two species.

Leptopezus rastellipes Clarke & Oberprieler, **sp. n.** (Figure 80)

Description. Size. Length 2.82 mm, width 1.56 mm. Robust, short; weakly sclerotised, legs paler, rostrum darker. **Head** widening posteriorly; dorsally setose, setae whitish, directed anteromesad. **Eyes** strongly protuberant, roundly triangular, slanting backwards. **Rostrum** very long and thin, ca. 1.25 × longer than pronotum, strongly downcurved, subcylindrical, gradually narrowing before slightly expanding at apex; dorsally with median, paramedian and dorsolateral carinae extending to antennal insertions, all but dorsolateral ones somewhat obsolete or confused beyond antennal insertions; dorso-apically with 2 erect setae in short oblique sulcus bordering epistome; antennal insertions median, in dorsal view marked by weak protuberances, behind them with deep scrobes reaching eye, in front of them with ca. 10 widely spaced erect setae placed in groove extending to mandibular articulation. **Antennae.** Scapes extending to front margin of eye, apically slightly inflated; funicles with segments 1–3 subequal in length, segment 1 ovoid, ca. 2.0 × wider than 2, 2–3 subcylindrical, subequal, 4–7 progressively shorter, 4–6 similar to 3, more obconical, 7 subglobular; clubs short, about as long as funicle segments 3–7, segments 1–3 successively slightly shorter, 2 slightly wider than 1, 4 flat, broadly inserted into 3. **Mouthparts.** Mandibles with 4 teeth on outside, 2 very small basal denticles, larger ones at middle and apex, inside edge with 3 large teeth, apical teeth on each side long, slender, curved, forming short apical Y. Maxillae with galea subequal in length to palp, with apical brush; apical palp segment roundly conical, apically setose (or hyaline), penultimate segment subequal, basal segment (?) slightly shorter and narrower than penultimate one. Labial palps with last segment slender, cylindrical, twice longer than wide, apically hyaline (or microsetose). **Thorax.** Pronotum narrower than elytra; weakly collared anterolaterally, sides curved, basally slightly constricted; basal margin broadly curved, corners quadrately rounded; punctures strongly delimited. Prosternum short, prosternal process short, pointed. Scutellar shield slightly longer than wide. Metaventrite and metanepisterna coarsely punctate, punctures ca. 2.0 × diameter of elytral punctures. **Elytra** sparsely setose, setae very fine; interstriae flat, finely microsculptured; sides with distinct marginal groove, slightly thicker in anterior half, setose, with anterior marginal notch; bases nearly straight. Hindwings present, fully developed. **Legs** sparsely setose. Femora on outside rounded (profemora) or subcrenulate in distal third (meso- and metafemora). Tibae sparsely shortly setose, outer edge serrulate (protibiae) or sharply carinate-crenulate (meso- and metatibiae), ventrally in apical half with longer oblique setae; apically with short, narrow outer flanges lined with coarse long fringing setae, dorso-apically with several elongate slender setae, with 2 subequal apical spurs; protibiae very slender, much longer than mesotibiae, teeth along outer edge oblique, broadly spaced (about one tooth apart); meso- and metatibiae dorsally carinate-crenulate, teeth overlapping (difficult to distinguish), more densely setose in outer preapical emargination. Tarsi very long and slender, dorsally densely setose, ventrally with short dense setae; tarsite 1 elongate, ca. 2.0 × longer than 2, slightly widening apicad, 2 ca. half as long as 1, very slightly wider, dorso-apically with distinctly longer setae, 3 strongly lobate, lobes subpedunculate, short and flimsy, ca. half as long as 5, 5 elongate, longer than 1–2, extremely slender, gradually widening apicad; claws long and sharp, basal tooth strongly pointed. **Abdomen.** Ventrites finely punctate finely setose; 2–5 slightly stepped, sutures straight.

Material examined. Holotype (NIGP157023), probably female: well preserved (including mouthparts and rostrum) teneral specimen (weakly pigmented, internal structures visible), not distorted or compressed but partially damaged, with terminal ventrites and most of left elytron severed in amber block (but missing) and left eye and pro- and mesofemora ruptured, onychium and part of tarsite 3 lobes of right metatarsus cut away with amber, otherwise intact and well visible from all sides (except ventral prothorax) with surface unobstructed by debris, right wing partially extended; in irregular cuboid 8.0 × 3.0 × 3.2 mm with four flat sides, one large flat face and large curved flat face; amber clear yellow with several flow bands with diffuse particles and three small fractures near specimen, with line of small bubbles emanating from near left humerus.

Derivation of name. The species is named for its large, sharp, rake-like tarsal claws, from the Latin nouns *rastellus* (G: *rastelli*), a little rake, and *pes* (G: *pedis*), a foot; the name is a noun in apposition.

Remarks. This distinctive species is characterised by its long thin legs with strongly serrulate protibiae but distinctly crenulate meso- and metatibiae. It is most easily distinguishable from *L. barbatus* by the slender inconspicuous setae in front of the antennal insertions and the broad interstriae with distinct but thin and shallow striae. The holotype is a very teneral specimen, probably a female because of its long slender rostrum. The metendosternite is visible, distinctly Y-shaped and indicated to have either a short stalk or no stalk at all and anterior arms.

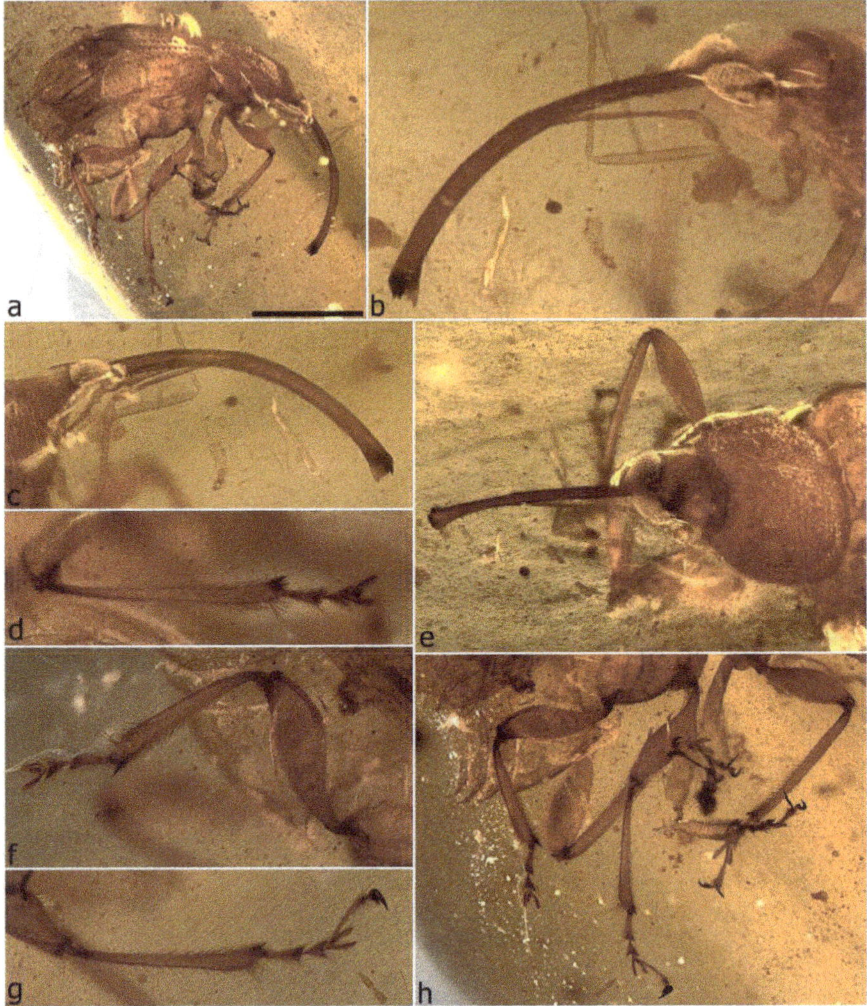

Figure 80. *Leptopezus rastellipes* sp. n., holotype. Habitus, right lateral (**a**); head and antenna, left lateral (**b**); same, right lateral (**c**); right metatibia and tarsus (**d**); head and pronotum, dorsal (**e**); right hindleg, outer side (**f**); right middle leg (**g**); legs (**h**). Scale bars: 1.0 mm (a).

Leptopezus barbatus Clarke & Oberprieler, **sp. n.** (Figure 81)

Description. Size. Length 2.12 mm, width 1.15 mm. **Eyes** slightly protruding, slanting backwards, forehead triangular. **Rostrum** slightly longer than pronotum, downcurved, thick at base, gradually narrowing and widening distad; antennal insertions in middle of rostral length, in front of them with lateral row of 7–8 long, erect setae curved anteriad. **Antennae.** Scapes straight, narrowly cylindrical, apically strongly clavate, reaching below eye; funicles not clearly visible; clubs long, segment l longer than 2, 1 and 2 widening apicad, 4 as long and broad as 3, flattened. **Mouthparts.** Mandibles not clearly discernible. **Thorax**. Pronotum elongate, laterally weakly curved; convex, densely setose, setae pale, long directed anteromesad, sculpture not discernible. Scutellar shield small, transverse, slightly rounded. Meso- and metaventrite distorted. **Elytra** short, broad, at base slightly sinuate, closely fitting onto prothorax, with broadly rounded, flattened humeri, posteriorly strongly declivous, apically individually rounded, not exposing pygidium; sutural flanges not visible; surface indistinctly punctostriate, densely setose, setae short, sharp, suberect, directed caudad. **Legs**. Coxae largely obscured, procoxae apparently contiguous, close to posterior hypomeral margin. Femora long, subcompressed, slightly inflated in middle, profemora on outside slightly carinate, meso- and metafemora distinctly carinate in distal half or more. Tibiae long, straight, compressed, outer edge faintly crenulate in meso- and metatibiae (in protibia not properly visible), metatibiae abruptly narrowed at apex; mesotibiae bent inwards at apex; tibiotarsal articulation surfaces (i.e., at apex of tibia) subtruncate to at most weakly oblique, spurs equal, protibiae apparently with single spur. Tarsi almost as long as tibiae; tarsite 1 elongate, narrow, ca. 2.0 × longer than 2, 2 shorter, apically deeply excised in protarsi, truncate in meso- and metatarsi, 3 deeply bilobed, lobes subpedunculate, 5 slightly longer than 1 + 2. **Abdomen**. Tergite VII domed, apical margin truncate-excised, slightly exposing VIII, this tightly fitting onto apical margin of ventrite 5. Ventrites not clear, 1 and 2 indicated to be longer than others.

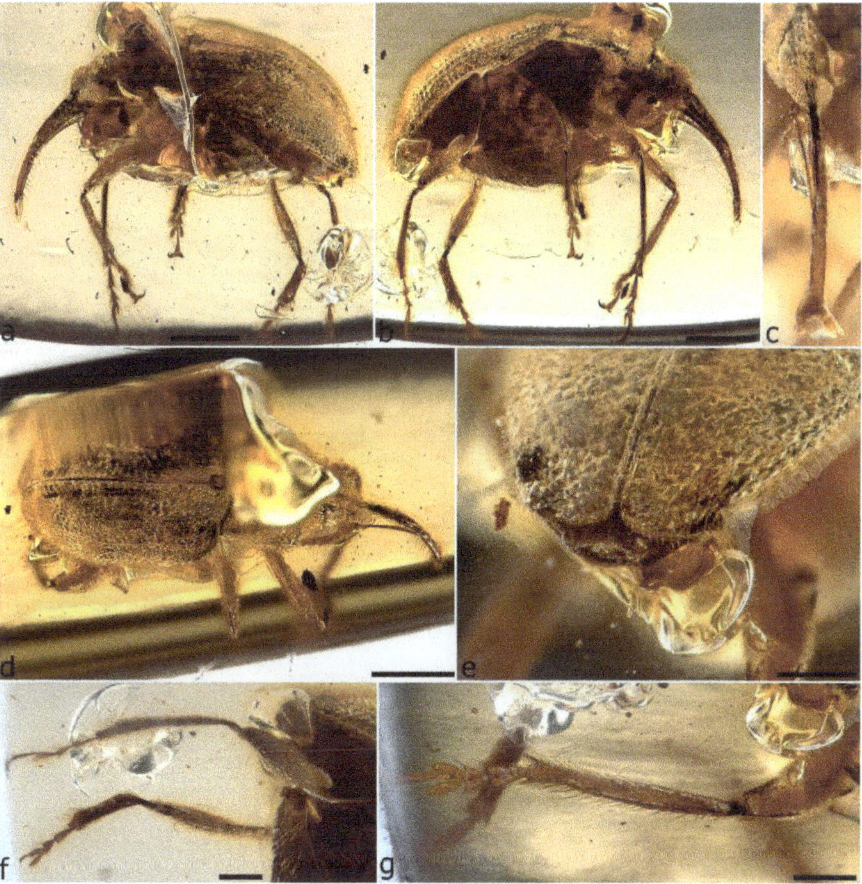

Figure 81. *Leptopezus barbatus* sp. n., holotype. Habitus, left lateral (**a**); habitus, right lateral (**b**); head, dorsal (**c**); habitus, dorsal (**d**); apices of elytra showing exposed tergite (**e**); hindlegs, lateral (**f**); right hindleg, posterior (**g**). Scale bars: 0.5 mm (a,b,d); 0.2 mm (e–g).

Material examined. Holotype (NIGP157024): not well preserved but intact specimen, body and rostrum depressed, legs compressed, mostly well visible except for underside (including head) and left side; at slight angle in centre of rectangular cuboid 3.9 × 3.53 × 1.7 mm; amber very clear with no impurities except large clear bubble between hindlegs, but large, irregular, dark bubble on pronotum, extending dorsad to edge of amber block and ventrad down left side of specimen, also with irregular dark layer of air on underside between legs, with small cracks between femora.

Derivation of name. The species is named for the distinct lateral rows of sparse, stiff, erect setae on the rostrum in front of the antennal insertions, giving it a bearded (*barbatus* in Latin) appearance; the name is a Latin adjective.

Remarks. This species is readily characterised by the conspicuous lateral row of long, erect, curved setae in the apical half of the rostrum. Whereas this row of setae occurs in many mesophyletids, it it usually fairly indistinct (as it also is in *L. rastellipes*). From the latter species *L. barbatus* further differs in having very indistinct interstriae (properly discernible only laterally), the shorter rostrum and the carinate (not crenulate) outer edge of the metafemora.

Genus *Anchineus* Poinar & Brown, 2009

Anchineus Poinar & Brown, 2009: 264 [59] (type species, by original designation: *Anchineus dolichobothris* Poinar & Brown, 2009)

Redescription. Size. Length ca. 1.6 mm, width 0.85 mm. **Head** convex dorsally, more so ventrally. **Eyes** small, elongate, protruding, facing forwards, coarsely facetted, dorsally separated by ca. width of rostrum anteriorly; forehead flat, without tubercles. **Rostrum** slightly longer than pronotum, cylindrical, moderately curved; antennal insertions lateral, slightly postmedian, with scrobes behind them not reaching eye, also with scrobes in front of them indicated, laterally without long erect setae. Single long gular suture indicated (not clearly discernible). **Antennae** geniculate, long; scape, funicle segments 1 and 2 and club segments 3 and 4 distinctly paler than rest of antenna; scapes elongate, slender, subcylindrical, apically only slightly inflated, shorter than funicle; funicles 7-segmented, segment 1 about as long as 2, much broader, others narrower, 2 elongate slender, cylindrical, ca. 2.0× longer than 3, narrower, 3–7 progressively slightly shorter towards club; clubs large, loosely articulated, 4-segmented, segment 4 acute, about as long as and broadly inserted into 3. **Mouthparts.** Labrum absent. Mandibles small, flat, vertical, exodont, articulation apparently vertical. Maxillae with galea pointed, densely setose apically and along inner side; palps 3-segmented, projecting obliquely from rostral apex, apical segment ca. half as long as penultimate segment. Labial palps slender, 3-segmented, directed forward. **Thorax.** Prothorax transverse, strongly proclinate, with anterior lateral margins oblique in lateral view. Pronotum slightly narrowed anteriorly, nearly as wide as elytra at base, without lateral tooth, posterior corners not extended, fitting closely onto elytra; surface sparsely setose, setae reclinate, directed anteromesad; notosternal sutures closed. Prosternum moderately long; procoxal cavities probably medially confluent. Mesoventrite short, anteriorly sloping. Metaventrite longer. **Elytra** elongate, basally lobed over base of pronotum, with indistinct rounded humeri, posteriorly declivous, lateral margin strongly sinuate to roundly emarginate in middle, apically conjointly rounded, probably not exposing pygidium; sutural flanges narrow, equal; surface punctostriate, without scutellary striole, interstriae flat, finely granulate interspersed with tubercles, sparsely setose, setae short, thin, reclinate, directed caudad. **Legs.** Procoxae conical, prominent, probably medially contiguous; mesocoxae subglobular, moderately projecting, narrowly separated; metacoxae flat, transversely elongate. Trochanters short, oblique. Femora long, subcylindrical, inflated in distal half, narrowed before apex, outside rounded. Tibiae substraight, compressed, slender, outer edges carinate, serrulate on protibiae (probably others also), with denser longer setae in distal half, apex obliquely truncate, with 2 spurs. Tarsi almost as long as tibiae; tarsite 1 apically subtruncate, 2 shorter, subtriangular, apically weakly excised, 3 deeply bilobed, lobes digitate, 5 long, apically expanded; claws divaricate, dentate with ventrobasal seta. **Abdomen.** Tergite VIII with curved line where elytral apices fit, with apex fitting evenly into concave margin of ventrite 5. Ventrite 1 longer than 2, 2–4 subequal, 5 slightly longer than 3, apically broadly rounded.

Remarks. This genus was originally placed in Curculionidae based on its geniculate antennae, well-developed scrobes, an upper position of the rostrum and small trochanters [59]. Although these characters are correct as given, the antennal geniculation in *Anchineus* is of the open type as in all other Mesophyletidae and the genus does not have compact clubs, as occur in "the great majority of taxa" in Curculionidae. Several unusual differentiating characters as given in the original description are seemingly unique to *Anchineus*, but some of them are misinterpretations that have not been corrected in subsequent publications, as follows.

Scrobes extending in front of the antennal insertions. These were described as "foveiform scrobes" with depressions for the reception of the club, and our examination of the holotype revealed them to be real. Although many other genera of Burmese amber weevils also have the scrobes seemingly extending in front of the antennal insertions (though rarely as distinctly as behind them), none of these have such foveiform impressions as evident in *Anchineus*.

Vertical mandibles. Our examination of the holotype also revealed that the original description of the mandibles as "laterally flattened and would have moved vertically" is correct. Assuming a flattened form, the vertical position of the mandibles is relatively clear from the photographs of the rostrum apex in lateral view (Figures 6 and 7 in [59]), but these do not resolve the distinction between rostrum and mandible nor show the position of the articulation sockets. We can corroborate both the flattened (though slightly scoop-shaped) form and the vertical position (in repose!) of the mandibles as originally figured and also that the movement plane at a minimum would have been very strongly oblique (subvertical), as the articulation sockets are not visible in lateral view and the rostrum is not distorted at the apex. However, the authors compared the vertical position of the mandibles of *Anchineus* with that of *Mesophyletis*. In the figures of the latter genus (Figures 1 and 3 in [29]), the right mandible is shown in the vertical *open* position typical of most Mesophyletinae, which is not the same as the position when the mandibles are closed. The mandibles of *Anchineus* and *Mesophyletis* are thus not comparable, as the repose position is vertical in the former but horizontal in the latter and the articulation sockets are vertical in the former but oblique in the latter.

Tibiae with three apical spurs. We could discern in the holotype the tibial part described as "bearing 3 apical spurs" [59] and illustrated in Figure 6 of [53] (incorrectly labelled as the mesotibia) for the only tibia on which there appear to be three spurs (the right metatibia); the other tibiae cannot be viewed from a comparable angle, e.g., Figures 5 and 7 [53], and show only one or two spurs. Under high magnification (ca. 120 × with a Leica M165C microscope) we could determine that the supposed third spur on the right metatibia is actually a narrow piece of debris, texturally different from the two real spurs. *Anchineus* therefore has only two tibial spurs, like all other Burmese amber weevils (in fact all weevils, unless one or both are lost).

Apex of tarsites 5 with "unguitractor plate". The tarsus of the holotype that shows the supposed "unguitractor plate" (Figure 9 in [59]) with an apically lobed fifth segment is the only tarsus so appearing in the specimen, and this appearance is a simple artefact of depression of the segment; the lobes identified in the figure are actually the apices of the fifth segment appearing thicker and drawn out due to the compression. We have seen a similar artefact in some other specimens. Furthermore, an unguitractor plate does not occur in any extant weevils.

Procoxae separated by narrow prosternal process. We could not achieve a clear view of the ventral prothorax in the holotype because this area is heavily distorted (crumpled), but the procoxal cavities are indicated to be confluent and without a narrow process (at least without one that contacts the hypomeral process). All comparable Mesophyletidae we examined also have the procoxal cavities confluent.

Pygidium "probubly slightly exposed". Our examination of the holotype did not reveal a true pygidium, i.e., a permanently exposed, vertically orientated tergite VII (or even VIII) when the elytra are closed and the abdomen is in its normal resting position (as is present in *Calyptocis*, *Acalyptopygus*, *Echogomphus* and *Compsopsarus*), but in *Anchineus* the apex of the abdomen may have been exposed in a way similar as that in genera such as *Bowangius*, in which tergite VIII is carinate and in repose orientated subvertically.

Anchineus is similar to *Bowangius* but differs in having the mandibles articulating in a vertical position and a trapezoidal pronotum (wider at the base, although some distortion apparent). It also differs in having digitate lobes of tarsites 3, postmedially inserted antennae and the scapes in repose terminating well in front of the eyes (possibly a male trait). Although much of the head of the holotype of *A. dolichotbothris* is obscured from view, the somewhat forward-facing eyes and flat forehead are also distinctly different from those of *Bowangius*. Apparently its protibiae but not its meso- and metatibiae are serrulate; if the latter indeed lack this serrulation (not due to decomposition or wear in life), this would be another critical difference from *Bowangius*.

Anchineus dolichobothris Poinar & Brown, 2009

Anchineus dolichobothris Poinar & Brown, 2009: 266 [59]

Redescription. Size. Length 1.6 mm ("body", seemingly excluding rostrum, not SL). **Head** probably short. **Eyes** black. **Rostrum** moderately curved; scrobes extending somewhat dorsolaterally to just before anterior margin of eye; distally of insertions laterally weakly concave, seemingly with weak depressions (Figure 1 inset [59]); with few short setae dorsally near apex and ventrally, without elongate setae. **Antennae.** Scapes not reaching eye, articulation with funicle segment 1 suboblique; funicles with segment 1 with bend just distally of condyle, expanding apicad, 3 slightly longer than 4, 4–7 subequal; clubs with segments 1–2 shortly obconical, subequal in length, 1 slightly narrower at base and apex; 3 and 4 subequal in length, weakly differentiated. **Mouthparts.** Mandibles without subapical teeth along external edge, with a rounded blunt tooth at outer apex, a rounded apical tooth at inner apex (these ones projecting forwards) and larger rounded basal tooth along inner edge (Figures 6 and 7 [59]). **Thorax.** Prothorax strongly proclinate, with anterior lateral margins oblique in lateral view. Pronotum slightly narrowed anteriorly, basal margin broadly sinuate. Scutellar shield triangular. **Elytra** metallic black, shiny, seemingly glabrous but sparsely and shortly setose (Figure 5 [59]), more densely so apically; sides with distinct marginal groove, subequal in width along entire length, with anterior marginal notch before humeri; striae narrow, linear, with small indistinct punctures; interstriae broad, more than 3.0 × wider than striae, micropunctate. **Legs.** Metacoxae strongly narrowed laterad. Protibiae long, slender, slightly thicker in middle, shortly sparsely setose, longer and denser ventrally in apical third, outer edge carinate, sparsely serrulate from about middle to just before apex, teeth widely irregularly spaced (width of several teeth), slightly elongately concave on inner edge before apex, lined with short oblique setae, apically weakly flanged, with sparse coarse fringing setae; mesotibiae widening at about basal third, then subequal, outer edge with irregularly longer setae in distal half, longer than short oblique setae along inside, inner edge in distal third curved slightly inwards, apex with larger flanges, lined with coarse fringing setae; metatibiae with setae in apical third longer than those of mesotibiae, forming loose patch; apically with long coarse fringing setae. Tarsi short, protarsi slightly more than half as long as protibia, mesotarsi only slightly shorter than protarsi; tarsite 1 ca. 1.5 × longer than 2, expanding apicad, 3 distinctly bilobed, lobes digitate (tubular; rounded apically), slightly widening apicad, ca. 0.67 × as long as 5, with inner sides distinctly concave in basal half, 5 expanding distad, curved; claws dentate, with basal tooth large and flat. **Abdomen.** Tergites VII and VIII probably partly exposed, subequal in length, VII with pale area on either side of midline (possibly wing binding patch). Ventrite 5 apically strongly concavely curved upwards beneath elytral apices.

Material examined. Holotype (PACO, 'accession' #B-C-41): poorly preserved, heavily distorted and somewhat decomposed specimen, surface details largely unobstructed, head retracted into prothorax, this crumpled, pterothorax and abdomen sunken into body cavity, elytra pushed inwards with left partially overlapping right, exposing abdominal apex, tarsites 3–5 of left metatarsus missing; at one end of irregular rectangular cuboid; amber clear brown, with several small impurities and bubbles; with another coleopteran inclusion at opposite end of block, partially cut away during preparation (see also [59]).

Remarks. This species is distinguishable from all other known Burmese amber weevils by the combination of its very small size, glabrous-appearing and slightly metallic black elytra with deep narrow striae and the digitate lobes of tarsites 3. From *Bowangius glabratus*, also with metallic and seemingly glabrous elytra, it most readily differs by its smaller size and narrow lobes of tarsites 3.

Genus ***Bowangius*** Clarke & Oberprieler, **gen. n.**

Type species: *Bowangius cyclops* Clarke & Oberprieler, sp. n.

Description. Size. Length 1.86–2.87 mm, width 0.59–1.3 mm. Body small, elongate, sparsely setose, **Head** elongate, convex dorsally, more strongly bulging ventrally, constricted behind eyes. **Eyes** large, round to elongate, strongly protruding, coarsely facetted, dorsally separated by nearly width of rostrum throughout; forehead flat, without tubercles between eyes. **Rostrum** as long as or slightly longer than pronotum, subcylindrical, weakly downcurved; antennal insertions lateral, with scrobes behind them and in front, in front of them laterally with a few long erect (usually indistinct) setae (absent in *B. glabratus*). Single long gular suture present. **Antennae** geniculate, long; scapes elongate, cylindrical, apically inflated; funicles 7-segmented, segment 1 broad, 2–5 narrower, progressively shorter towards club, 6–7 broader, subglobular; clubs large, loosely articulated, 4-segmented, segment 4 acute, about as long as 3. **Mouthparts**. Labrum absent. Mandibles small, exodont, articulation oblique. **Thorax**. Prothorax weakly to strongly proclinate, with anterior lateral margins oblique in lateral view. Pronotum slightly convex, laterally rounded, without tooth, posterior corners not extended, fitting closely onto elytra; surface coarsely punctate, sparsely setose, setae reclinate, directed anteromesad; notosternal sutures closed, vertical then curved anteriad. Prosternum moderately long; procoxal cavities medially confluent, in middle of prothorax. Scutellar shield densely setose. Mesocoxal cavities closed by meso- and metaventrites. Mesoventrite short, anteriorly sloping. Metaventrite longer, raised into transverse weals in front of metacoxae. Metanepisterna distinct, sparsely setose. **Elytra** elongate, basally closely fitting to pronotum, with broadly rounded flatly produced humeri, posteriorly declivous, lateral margin strongly sinuate to roundly emarginate in middle, apically individually rounded, not exposing pygidium; sutural flanges narrow, equal; surface punctostriate, without scutellary striole, interstriae convex, finely tuberculate, setose, setae long, thin, reclinate, directed caudad. **Legs**. Procoxae large, prominent, medially contiguous; mesocoxae subglobular, narrowly separated; metacoxae very large, flat, transversely elongate. Trochanters short, oblique. Femora long, subcylindrical, inflated in distal half, outside rounded or carinate in distal half to third. Tibiae straight, compressed, outer edge serrulate, teeth sharp, pointed, separate at their bases, with sparse long stiff setae in distal half, apex obliquely truncate, with 2 spurs. Tarsi shorter than tibiae; tarsite 1 apically excised to subtruncate, 2 shorter, triangular, apically excised or subtruncate, 3 deeply bilobed, lobes not or subpedunculate, 5 long, apically expanded; claws divaricate, dentate with ventrobasal seta at apex of tooth. **Abdomen.** Tergite VIII sometimes (apparently in males) exposed, with tranverse carinate ridge along upper edge and sharp apical flange. Ventrites 1 and 2 slightly longer than 3, 3 slightly longer than 4, 5 as long as 3, apically broadly rounded.

Derivation of name. The genus is cordially named for Professor Bo Wang, of the Nanjing Institute of Geology and Palaeontology, for making these many interesting amber fossils available to us for study. The gender of the name is masculine.

Remarks. This genus is difficult to distinguish from *Habropezus*, being generally similar in shape and sculpture and also having serrulate tibiae, but it has different, more globular eyes (strongly so, subhemispherical in *B. cyclops* and *Bowangius* sp. 1) and a narrower grooved forehead, shorter stouter antennae and different, seemingly more specialised mandibles with a slenderer apical 'T'. The eyes of *B. tanaops* and *B. zhenuai* are somewhat flattened in comparison with those of *B. cyclops* and in this sense are more in agreement with *Habropezus* than with *B. cyclops*; however, the forehead is broad between the eyes and posteriorly widening in *Habropezus*, whereas in *Bowangius* it is very narrow, at most as broad as the base of the rostrum, and not widening posteriorly. Another distinctive character of *Bowangius* is the exposed and carinate tergite VIII in some specimens (apparently males; they also have a shorter rostrum), although this character may also be present in *Anchineus*. From this genus

Bowangius can be distinguished by its more distinctly exodont mandibles and their horizontal repose position. *Bowangius* is represented by up to eight species as known, although the specimens of two of the species that we do not describe are not preserved adequately enough to ascertain whether they represent different species (and even belong in *Bowangius*).

Bowangius cyclops Clarke & Oberprieler, **sp. n.** (Figure 82)

Description. Size. Length 2.05 mm, width 0.76 mm. **Head.** Head punctate, punctures finer than on prothorax; densely setose behind and between eyes. **Eyes** large, subglobular, dorsally separated by ca. half width of rostrum. **Rostrum** ca. 1.25 × longer than pronotum, weakly downcurved; antennal insertions in middle of rostral length, behind them with scrobes extending to front of eye; apex produced into narrowly rounded lobe. **Antennae.** Scapes about as long as funicles, reaching to just below eye; funicles with segment 1 slightly longer than 2, 3 elongate-obconical, shorter than 2, 4 shorter, distinct, ca. 2.0 × broader than long, 5 similar to 3, shorter, 6–7 subglobular, 7 slightly longer, broader than 6; clubs with segments 1–2 subequal, obconical, 3 slightly shorter, 4 subequal in length to 3, narrowly acute apically. **Mouthparts.** Mandibles slender, abruptly curved from base, outside with small bluntly rounded tooth in middle and smaller pointed tooth at apex, inside with larger sharper curved tooth (ca. 3.0 × longer than outer middle tooth) in middle and smaller pointed tooth at apex; apically subemarginate, with inner and outer angles formed by apical teeth. **Thorax.** Prothorax strongly proclinate. Pronotum densely punctate. Prosternum short, subequal in length to hypomeron. Scutellar shield elongate-quadratic, densely setose, slightly lower than elytra. **Elytra** with bases sinuate, formed by swellings at base of interstriae 2–5, sharply carinate and extended over base of pronotum; humeri flatly concavely extended; sides lacking any keel (carinate interstriae), with marginal groove irregularly lined with setae, with anteromarginal notch; apices rounded; striae and interstriae irregularly setose. **Legs** long, slender. Procoxae subconical, contiguous; mesocoxae bulbous, prominent, possibly slightly elongate; metacoxae prominent. Protibiae long and slender, nearly straight, not or only weakly widening distally, apically roundly truncate, spurs slender; mesotibiae long and slender, nearly straight, only slightly widening apicad, outer edge narrowing before apex, spurs short, robust; metatibiae with rounded flange in apical third, with one large flat apparently fixed spur and one normal narrower spur; tarsal articulation surfaces oblique, apically with coarse fringing setae. Tarsi at least half as long as tibiae; tarsite 1 large, subtriangular, apically excised, 2 wider than and ca. half as long as 1, apically excised, 3–4 not preserved, 5 shorter than 1 + 2, slightly curved (basal teeth possibly asymmetrical on metatarsal claws). **Abdomen.** Tergite VIII exposed, concave, forming small apical setose flange, delimited on upperside by arcuate carina. Ventrites setose; 1–2 subequal in length, subflatly aligned, 3–5 stepped, successively higher, 3–4 successively shorter, 5 very broadly rounded, subequal in length to 3; sutures straight, not forming broad gaps.

Material examined. Holotype (NIGP157025): very well preserved, minimally distorted and compressed specimen with anterolateral thoracic and elytral structures somewhat obscured (most significantly on left side), body covered by fragmented coating of whitish debris or secretions, otherwise well visible, with rostrum, antennae and left legs slightly compressed/decomposed, mandibles well-preserved, fully opened and clearly visible, left metatarsus, tarsites 3–5 of left pro- and mesotarsi and right fore-, middle and hindlegs missing (inner lobe of tarsite 3 of right protarsus still attached and claw segment of evidently the right metatarsus in vicinity of this leg); in narrow elongate cuboid, 2.7 × 2.0 × 1.6 mm); amber clear, with minimal impurities; with several dark flat flow lines and shearing planes parallel to specimen.

Derivation of name. The species is named for its large round eyes, from the Greek nouns *kyklos* (G: *kyklou*), meaning circle, and *ops* (G: *opos*), meaning eye.

Remarks. This distinctive species is distinguished from all other *Bowangius* species by its large globular (almost hemispherical) eyes, flatly concavely extended humeri, short inner flange on the apical third of the metatibiae and more elongate and slightly depressed body. It shares with *B. tanaops* and *B. zhenuai* an exposed tergite VIII, with a broad carina at the base and a thick apical rim. *Bowangius cyclops* is also similar to *A. dolichobothris* in its antennae, the scape being relatively short and both it and funicle segments 1 and 2 being pale, but it differs in the more elongate pronotum, coarsely rugose and densely setose elytra and strongly exodont mandibles.

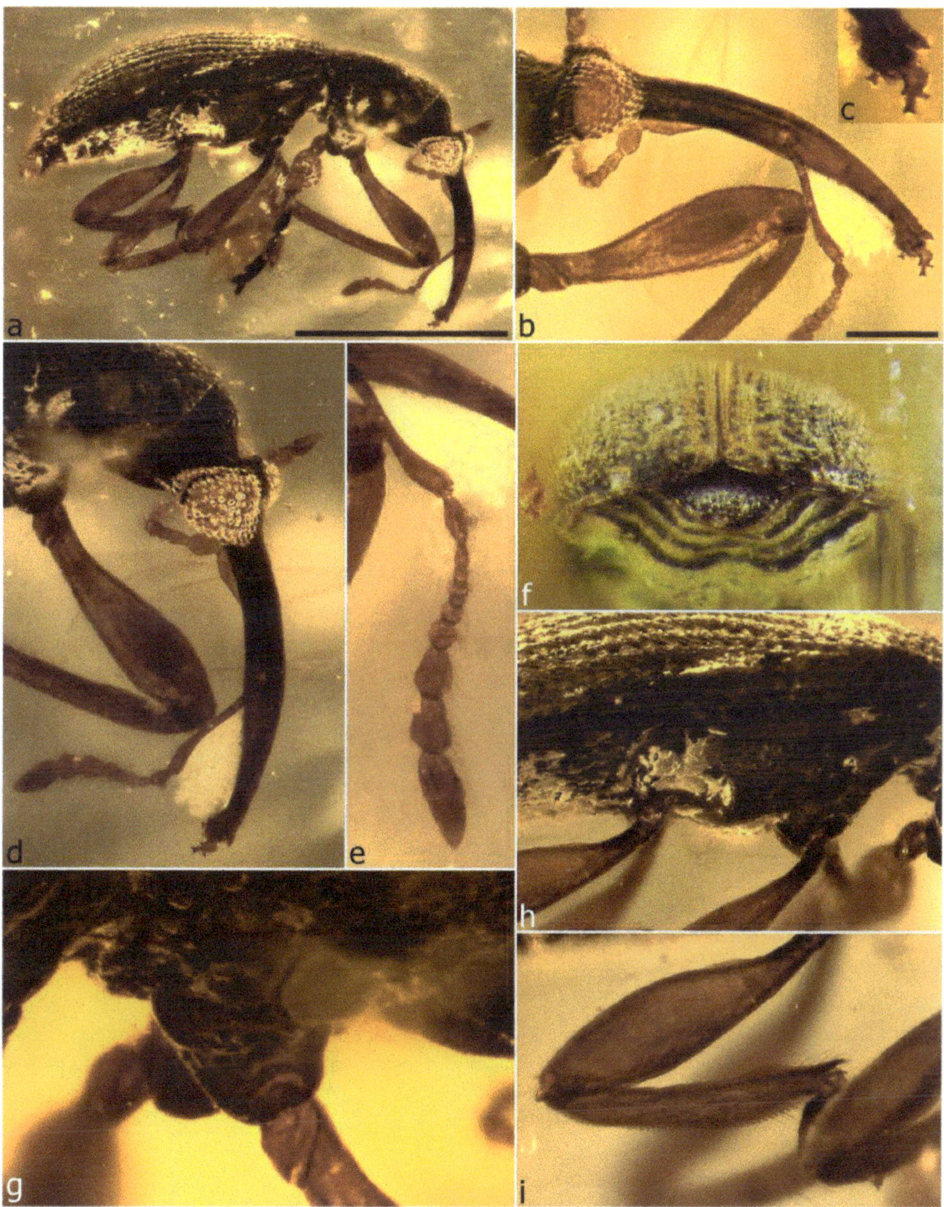

Figure 82. *Bowangius cyclops* sp. n., holotype. Habitus, right lateral (**a**); head and rostrum, right lateral (**b**); mandibles, right lateral (**c**); prothorax and head, right lateral (**d**); right antenna, right lateral (**e**); abdomen and elytra, caudal (**f**); detail of notosternal suture, right lateral (**g**); elytra, pterothorax and ventrite 1, right lateral (**h**); right hindleg (**i**). Scale bars: 1.0 mm (a); 0.2 mm (b).

***Bowangius* sp. 1** (Figure 83)

Material examined. One specimen (in private collection of Mr. Huijun Huang, China), 2.12 mm long: intact, pale yellowish-brown (teneral), extremely well preserved, not distorted, well visible from both lateral sides but not from dorsal and ventral sides due to shape of amber block; in left corner of large flat ellipsoid 20.2 × 15 × 5.7 mm; amber clear but with many impurities away from specimen, also containing a large leaf, probably of a fern.

Remarks. This specimen is exceptionally well preserved and shows many structures and characters in extreme detail and clarity, but due to its position and orientation in the flat lenticular amber piece its dorsal and right side are partly obscured, and other aspects are distorted due to the curvature of the amber surface. It would be perfectly and totally visible if it were cut out of the amber block in a cuboid aligned with its sides. It is very similar to *B. cyclops* in size and nearly all discernible characters, differing only in having a longer, thinner rostrum, the outside of the femora finely carinate in the distal half, tarsites 2 apically truncate (excised in *B. cyclops*) and tergite VIII not exposed beyond VII. The longer rostrum and hidden tergite VIII suggest that the specimen is a female, and if the carination of the femora and the shape of the tarsites are also subject to sexual dimorphism, this specimen may well be the female of *B. cyclops*. Because of this and because the specimen is housed in a private collection and its accessibility to science therefore not assured, we prefer not to describe it as another species.

***Bowangius* sp. 2** (Figure 84)

Material examined. One specimen (in private collection of Mr. Huijun Huang, China), 2.8 mm long, ca. 1.3 mm wide: extremely well preserved, intact with extended ovipositor, not distorted, superbly visible from left side and largely from right and ventral side but not from dorsal side due to shape of amber block; slightly off centre to left in irregular flat ellipsoid 15 × 11 × 7.1 mm; amber very clear but with many impurities, in particular a layer of sparse small bubbles along left side of specimen, also a large, elongate, clear bubble beneath rostrum and dark structure (possibly a flower) between legs on right side.

Remarks. This specimen is very well preserved and shows a large suite of characters, including an extended ovipositor, but for proper scientific study it needs to be cut out of the large, cabochon-shaped amber block so that its critical structures can be properly assessed. It is similar to *B. cyclops* in having large, subglobular eyes, but it is slightly larger, has a higher body, thinner rostrum, more strongly clavate scapes, longer funicular segments, the outer side of the meso- and metafemora distally carinate (not rounded) and tarsites 2 apically truncate (not excised). Its most valuable feature in evidence is the extruded ovipositor, showing long, slender, weak gonocoxites with thin dorsal and ventral sclerotised baculi (rods) and a long subapical stylus with an apical tuft of setae. We do not describe this species as the specimen is housed in a private collection, in which accessibility of a holotype to science cannot be guaranteed.

***Bowangius* sp. 3** (Figure 85).

Material examined. One specimen (NIGP157026), 1.89 mm long, 0.59 mm wide: poorly preserved but intact, body and especially rostrum and legs compressed and slightly distorted, reasonably well visible from all sides; in centre of rectangular cuboid 4.45 × 3.86 × 1.62 mm; amber clear but with many small impurities not obstructing view of specimen significantly.

Remarks. This distorted and inadequately preserved specimen appears to belong in *Bowangius* mainly due to its globular eyes (similar to those of *B. cyclops*) and serrulate meso- and metatibiae (the protibiae being carinate only), but it displays too few unequivocal characters for a definite assignment to this genus and a comparison with its species. We therefore do not describe and name it as another species here.

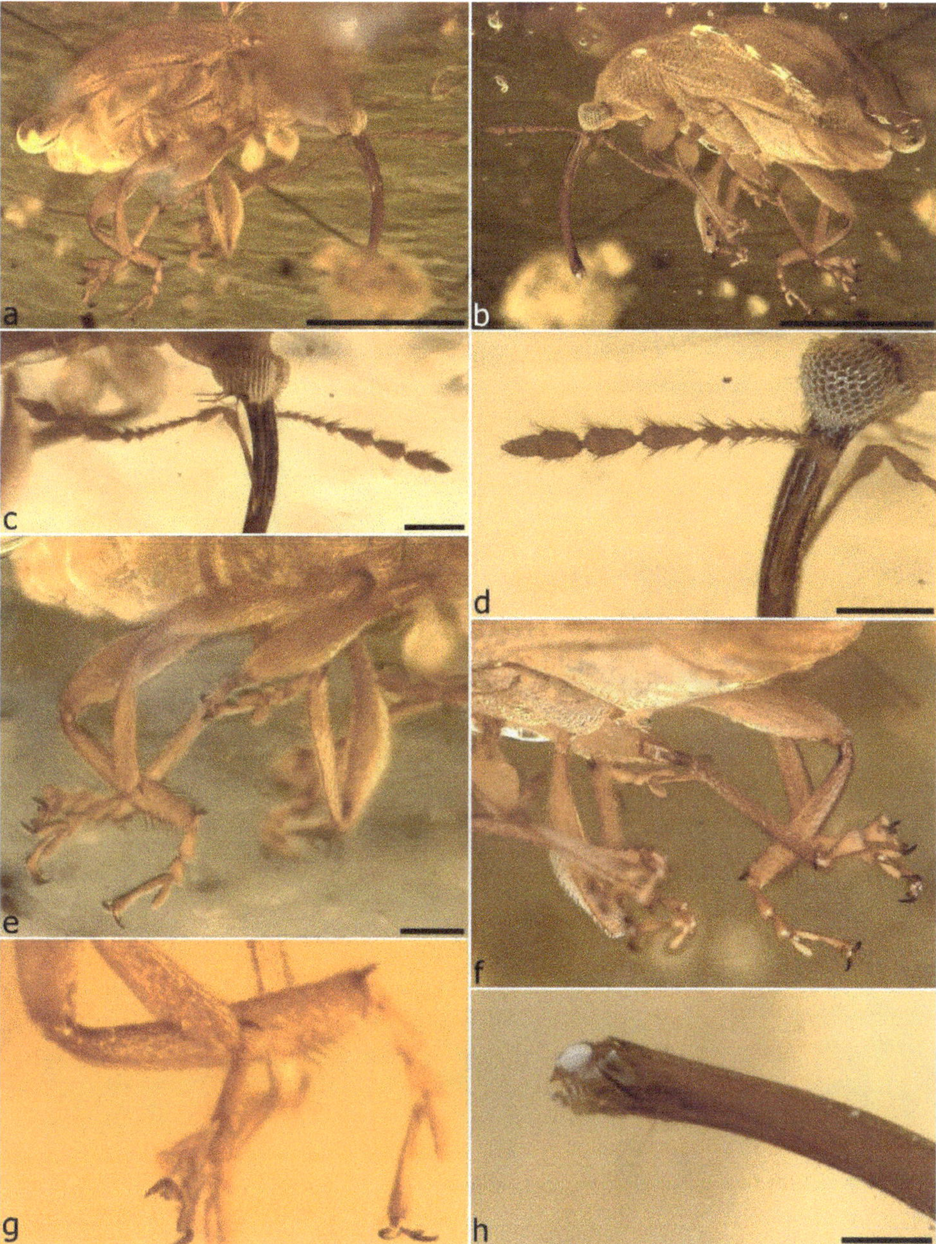

Figure 83. *Bowangius* sp. 1, Habitus, right lateral (**a**); habitus, left lateral (**b**); head and antennae, right lateral (**c**); eye, antenna and base of rostrum, left lateral (**d**); legs, right lateral (**e**); legs, left lateral (**f**); detail of metatibia and tarsi, right lateral (**g**); detail of rostrum apex showing mandibular articulation socket (**h**). Scale bars: 1.0 mm (a,b); 0.2 mm (c–e); 0.1 mm (h).

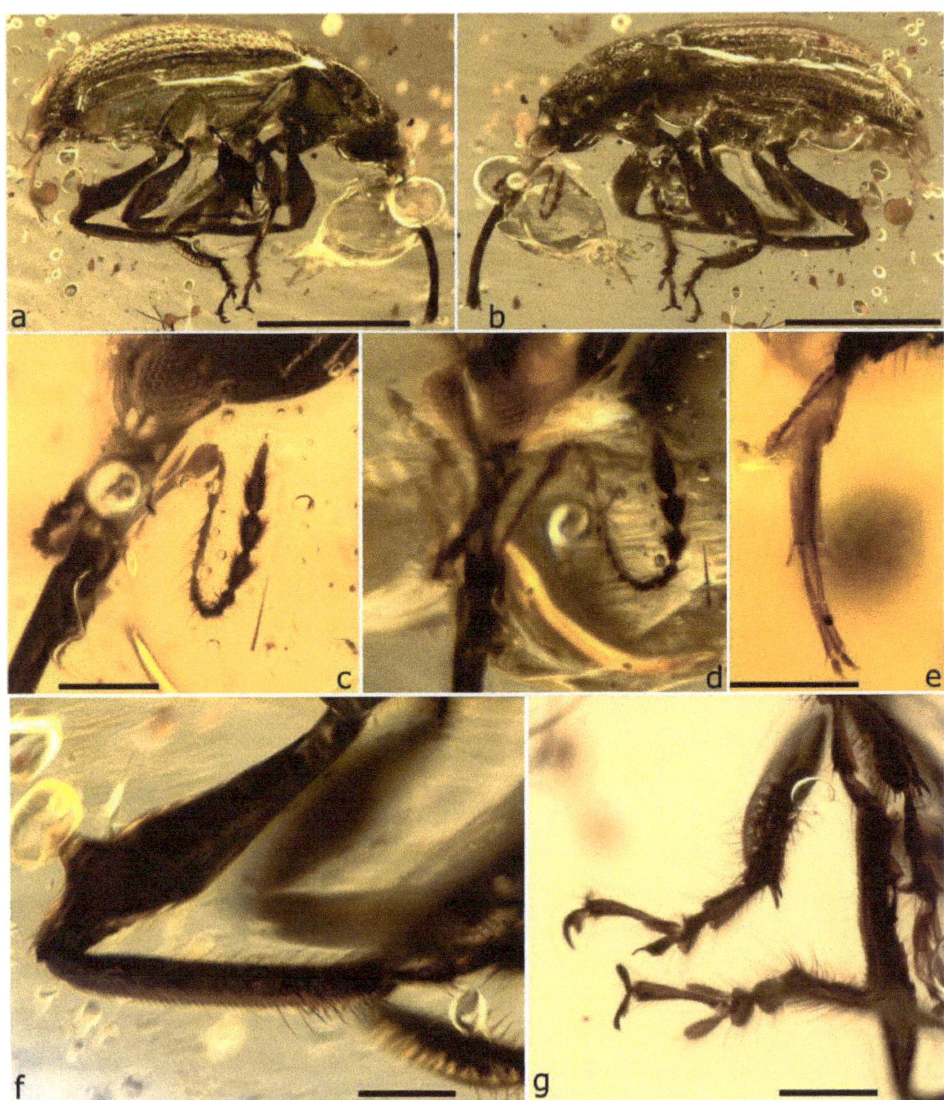

Figure 84. *Bowangius* sp. 2. Habitus, right lateral (**a**); left lateral (**b**); head and antenna, left lateral (**c**); same (**d**); ovipositor (**e**); left hindleg, posterior (**f**); tarsi (**g**). Scale bars: 1.0 mm (a,b); 0.2 mm (c,e–g).

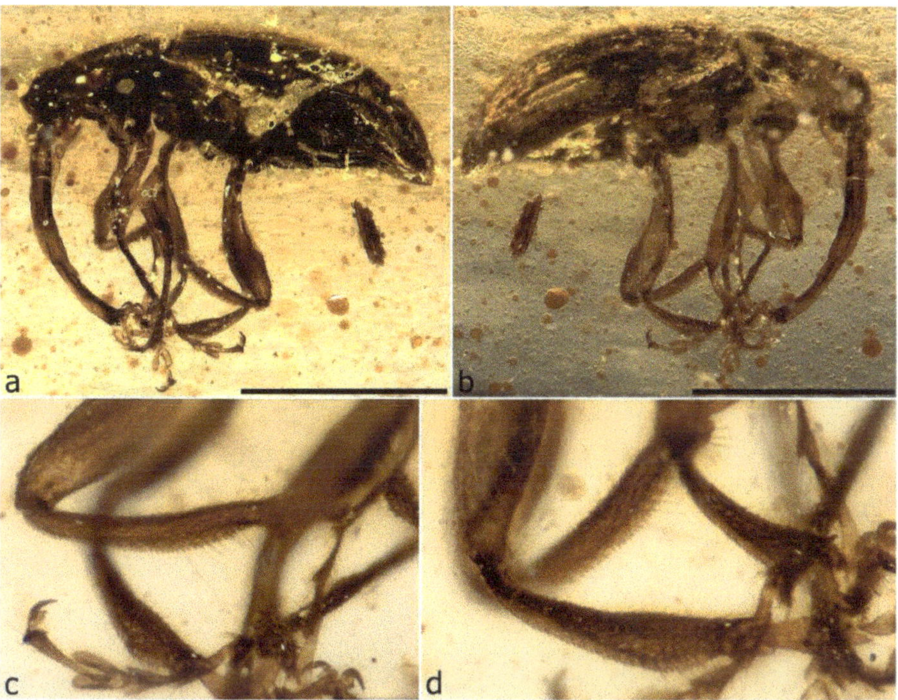

Figure 85. *Bowangius* sp. 3. Habitus, left lateral (**a**); habitus, right lateral (**b**); legs (**c**); legs, showing detail of metatibial serrulation (**d**). Scale bars: 1.0 mm (a,b).

Bowangius tanaops Clarke & Oberprieler, **sp. n.** (Figure 86)

Description. Size. Length 1.93 mm, width 0.7 mm. **Head** weakly constricted behind eyes; forehead flat. **Eyes** slightly depressed, somewhat elongate, dorsally separated by about basal width of rostrum. **Rostrum** ca. 1.25 × longer than pronotum; weakly curved; dorsally with distinct median, paramedian and dorsolateral carinae and intervening grooves from base to antennal insertions, grooves linearly setose at least to antennal insertions, carinae and grooves confusedly continued to near apex; dorso-apically subtruncate, with few setae, epistome distinct, V-shaped, delimited by grooves; antennal insertions median, behind them with deep scrobes reaching front of eye, in front of them continuing narrowly to mandibular articulation; rostrum basally on ventral side without paramedian tubercles. Mandibular articulations oblique, angled upward. **Antennae** paler than rostrum; scapes apically oblique, reaching just below front margin of eye; funicles with segment 1 broader than others, 2 elongate, fusiform, shorter than 1, 2–7 progressively shorter towards club, 6–7 subglobular, subequal, broader than 2–5; clubs with segments 1–2 subequal in length, apically oblique, anterior sides somewhat produced to rounded point, 3 narrower, slightly shorter than 2, 4 slightly more than half length of 3, seemingly flattened. **Mouthparts**. Mandibles on outside with 2 subequal teeth. Maxillary palps obliquely projecting from apex of rostrum, 3-segmented; terminal segment narrower than penultimate, hyalinae, apically with slender seta. Labial palps projecting forwards. **Thorax.** Prothorax weakly proclinate; moderately roundly inflexed laterally, lateroventrally less punctorugose than pronotum. Pronotum elongate, longer than wide, narrowing anteriad from about distal third, slightly emarginate behind middle, base broadly convex in middle, tightly fitting to elytra, posterior angles acutely angulate; surface evenly contoured, punctorugose, setose, setae uniformly whitish, indistinct, short, largely directed mesad to anteromesad. Prosternum short, less than half as long as procoxae, anterior margin broadly arcuate, prosternal process short, narrow, pointed; procoxal cavities close to posterior hypomeral margin, hypomeron ca. half as long as prosternum. Scutellar shield subeven with elytra, with anterior edge subcontinuous with elytral bases. Metaventrite coarsely punctate. **Elytra** with bases bisinuate, weakly extended over pronotum, humeri broadly rounded, slightly flatly or concavely produced, formed by confluent interstriae 6 and 7, tightly fitting onto pronotum, outer apical edges smooth (not toothed), lateral margins with punctosetose marginal groove gradually increasing in width anteriad, with distinct anterior marginal notch; surface rugose, sparsely setose, punctures restricted to striae, striae and interstriae subequal in width, stria 5 somewhat deeper basally than others, interstriae prominent, somewhat raised, distinct also on declivity. Hindwings present. **Legs** darkly coloured as body. Procoxae conical, contiguous; mesocoxae roundly conical, prominent; metacoxae large, broadly transverse. Femora rounded on outside, without carina or teeth. Tibiae on outer side carinate-serrulate along entire length, teeth distinctly curved, separated, less distinct on protibiae; protibiae slightly longer than meso- and metatibiae, with more prominent inner apical flange, meso- and metatibiae slightly curved inwardly near apex; apically indistinctly flanged, with coarse, sparse fringing setae and elongate dorso-apical setae, with 2 apical spurs, one fixed and larger on meso- and metatibiae. Tarsi ca. 0.75 × as long as tibiae; tarsites 1–2 apically expanded, 1 ca. 2.0 × longer than 2, very slender basally, apically subtruncate, 2 more shortly triangular, apically excised, 3 with lobes weakly pedunculate, subovoid, very narrow basally, weakly concave along inner margins, ca. half as long as 5, cryptotarsites distinct, 5 extremely slender basally, slightly flattened. **Abdomen.** Tergite VII strongly sclerotised, distinctly vertically margined laterally (but not apically), moderately densely punctosetose, laterally with distinct (wing binding?) patches on either side of base; tergite VIII divided by arcuate carina into basal and apical parts, with apical part subconcave, vertical, densely punctosetose, apically delimited by marginal ridge, then inflexed (appearing as sixth ventrite as seen). Ventrites sparsely setose; 1–2 subequal in length, subflatly aligned, with narrow suture, 3–5 slightly stepped, separated by more distinct, straight to arcuate sutures, 3 slightly shorter than 2, 4 slightly shorter than 3, 5 about as long as 3, apically broadly rounded, with distinct lip.

Material examined. Holotype (NIGP157027): exceptionally well preserved specimen, well visible from all sides, missing tarsus of right hindleg, otherwise intact, with abdominal apex (tergites VII and VIII) distended, completely exposed; in triangular slab with longest edge face curved, 3.9 × 3.1 × 1.7 mm; amber clear yellow with several transverse fractures over dorsal side of specimen, one large one above pronotum, with several large bubbles in vicinity of specimen, partly obscuring ventral pro- and mesothorax.

Derivation of name. The species is named for its elongate eyes, from the Greek adjective *tanaos*, meaning elongate, and the Greek noun *ops* (G: *opos*), meaning eye.

Remarks. This species is distinguishable from *B. cyclops* by its more elongate, slightly depressed eyes, its ventrally strongly bulging head, at most slightly flattened humeri, lack of flanges along the inner edge of the metatibiae and the basally narrower elytra (across humeri). Although the holotype appears relatively slender compared with congeners, it is slightly distorted (compressed), as particularly evident in the preservation of the ventrites and possibly also in the extended apical tergites (VII and VIII), which are protruding and well visible. From *B. zhenuai* it is distinguishable by the narrower body, curved rostrum, more posteriorly positioned procoxae and more distinct and anteromesadly directed pronotal setae (this last feature shared with *B. cyclops*). The holotype was submitted for CT scanning, but the contrast between the specimen and the amber was too low to permit any meaningful visualisation of the specimen.

Bowangius zhenuai Clarke & Oberprieler, **sp. n.** (Figure 87)

Description. Size. Length 1.86 mm, width 0.78 mm. **Head** slightly constricted behind eyes; interocular area narrow, dorsally maximally slightly less than basal width of rostrum, forehead flat. **Eyes** somewhat depressed, elongate. **Rostrum** slightly longer than head + pronotum; substraight; dorsally with distinct median, paramedian and dorsolateral carinae and intervening setose grooves from base to antennal insertions, carinae and grooves confusedly continued to near apex; antennal insertions median, behind them with deep scrobes extending to front of eye, in front of them with similar groove and with row of short slender indistinct setae; rostrum basally on ventral side in front of eyes with raised paramedian line of curved tubercles with a short seta between tubercles. **Antennae.** Scapes apically oblique, reaching well below front margin of eye, shorter than funicles; funicles with segment 1 broader than others, 2–5 elongate, slender, progressively shorter towards club, 6–7 broader, 6 expanded apically, slightly broader than 7, 7 subglobular; clubs with segments 1–2 subequal, distinctly obconical, apically suboblique, 2 basally wider than 1, 3 ca. half as long as 2, 4 broadly inserted, flattened. **Mouthparts.** Mandibles on outside with 2 small subequal denticles, on inside with 3 (or 4) teeth, mesal 2 larger, largest with subemarginate apex (sub-bicuspid), apical part elongate, slender, apex narrowly T-shaped; articulations oblique, also upwardly oblique. **Thorax.** Prothorax strongly proclinate. Pronotum widest at about middle, gradually narrowing anteriad and posteriad, base weakly convex in middle, tightly fitting onto elytra, posterior angles roundly angulate (not acute); surface punctorugose, sparsely shortly setose, setae pale whitish, indistinct, short, largely directed anteriad. Prosternum very short, ca. 0.2 × as long as procoxae, anterior margin distinctly narrowly emarginated, prosternal process short, broadly triangular; procoxal cavities closer to anterior prosternal margin, hypomeron ca. 3.0 × longer than prosternum. Scutellar shield narrower transverse, densely setose. Metaventrite finely, sparsely punctate. **Elytra** with bases bisinuate, closely fitting, shortly extended over pronotum, humeri broadly rounded, somewhat flatly produced, formed by confluent interstriae 6 and 7, outer apical edges smooth, not toothed, lateral margins with punctosetose marginal groove increasing in width anteriad, with distinct anterior marginal notch; surface with strial punctures very small, fine, striae as broad as or broader than interstriae, interstriae prominent, interstria 6 obsolete before humeri, 6 and 7 confluent near humeri, interstriae much less distinct on declivity, sparsely shortly setose, setae pale, whitish. Hindwings present. **Legs.** Legs much paler than body. Procoxae subconical; mesocoxae nearly as prominent as procoxae; metacoxae large, strongly transverse, ovoid. Femora on outside rounded, without carina or teeth. Tibiae on outer side carinate-serrulate along entire length, teeth distinctly curved, approximate on mesotibiae (appearing crenulate due to viewing angle), less distinct on protibiae; protibiae longer than meso- and metatibiae, the latter straight (not distinctly curved inwardly near apex); apically with coarse elongate fringing setae and elongate dorso-apical setae, with 2 subequal apical spurs, strong and curved on meso- and metatibiae, inner spur on metatibiae shorter, apparently fixed, more curved than outer spur. Tarsi ca. half (protarsi) to 0.75 × (meso- and metatarsi) as long as tibiae, protarsi slightly longer than others; tarsites 1–2 apically expanded, 1 ca. 2.0 × longer than 2, very slender basally, apically weakly excised, 2 more triangular, apically excised, dorso-apically with several elongate black setae, 3 with lobes weakly pedunculate, subovoid, distinctly concave along inner margins, slightly more than half as long as 5, cryptotarsite distinct, 5 extremely slender basally, slightly flattened. **Abdomen.** Tergite VII setose at least apically, distinctly edged laterally; tergite VIII flattened, divided by concavely arcuate carina into basal and apical parts, with apical part flatly concave, densely, finely setose. Ventrites 1–2 subequal in length, subflatly aligned with narrow suture, 3–5 slightly stepped, separated by more distinct, straight to arcuate sutures, 3 shorter than 2, 4 shorter than 3, 5 subequal in length to 3, apically broadly rounded, with distinct lip.

Material examined. Holotype (NIGP169513): exceptionally well preserved, intact specimen, not distorted or compressed, well visible from all sides, with right hindwing partly extended, abdominal apex (tergites VII and VIII) partly distended and visible; in trapezoidal block 4.0 × 2.0 × 1.4 mm with all faces flat; amber clear yellow with darker flow bands, with diffuse minute particles and with few minor cracks around side of specimen.

Derivation of name. The species is named after Zhenua Liu, currently a Ph.D. student at the ANIC, for his assistance with obtaining specimens for this study and for his stimulating discussions about and his work on the beetle fauna preserved in Burmese amber.

Remarks. This species is also readily distinguishable from *B. cyclops* by its elongate, slightly depressed eyes and the lack of diagnostic characters of that species. It is most similar to *B. tanaops*, from which it differs in its straighter rostrum, row of flat tubercles on either side of the gular suture and distinctly anteriadly directed, fine pronotal setae. It is also much broader, with distinct and broadly rounded humeri and more coarsely punctostriate elytra, the procoxae situated closer to the anterior prosternal margin and the mesotibiae straight, with both spurs free.

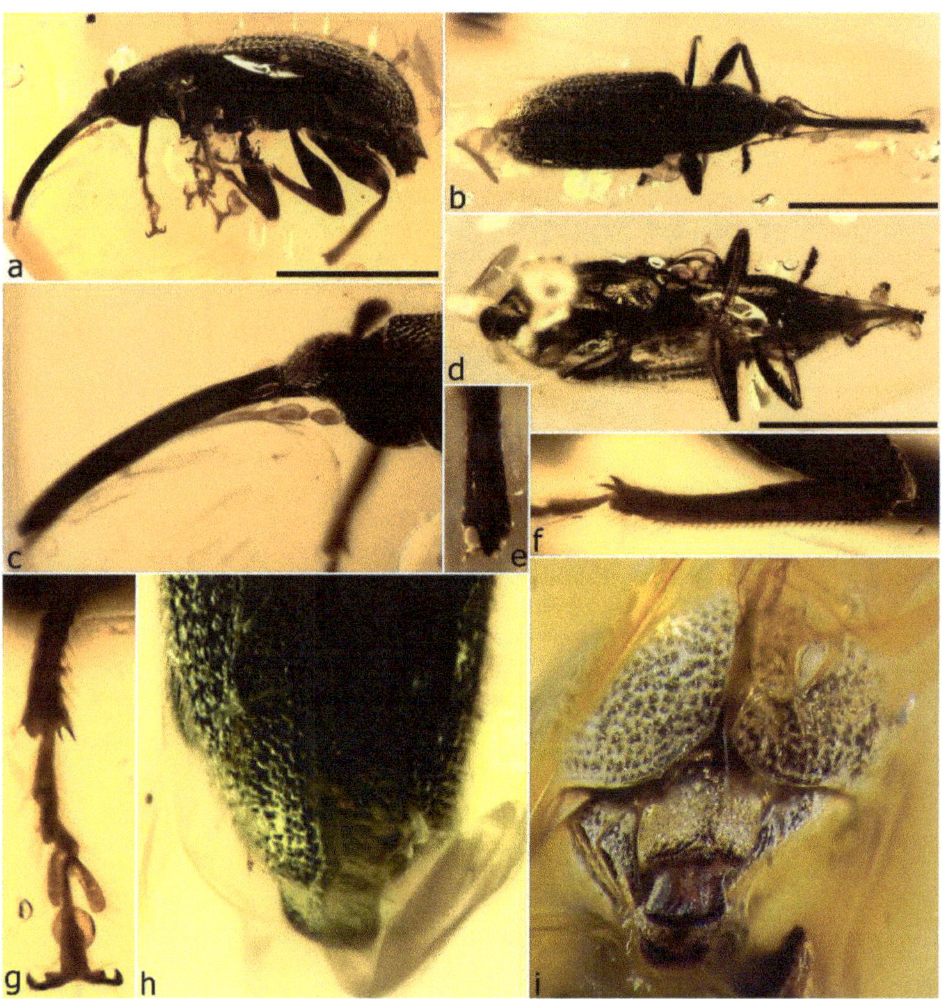

Figure 86. *Bowangius tanaops* sp. n., holotype. Habitus, left lateral (**a**); habitus, dorsal (**b**); head, left lateral (**c**); habitus, ventral (**d**); apex of rostrum and mouthparts, dorsal (**e**); left metatibia (**f**); left protarsus (**g**); apical half of elytra and exposed tergite VII, dorsal (**h**); apex of abdomen and elytra, caudal (**i**). Scale bars: 1.0 mm (a,b,d).

Figure 87. *Bowangius zhenuai* sp. n., holotype. Habitus, left (**a**); habitus, right (**b**); habitus, dorsal (**c**); habitus, ventral (**d**); head and pronotum, dorsal (**e**); apex of elytra and abdomen, caudal (**f**); metatibiae (**g**); apex of rostrum and mandibles, left (**h**); head and rostrum, oblique right (**i**); tarsus (**j**); tarsal claws (**k**). Scale bars: 1.0 mm (a–d).

Bowangius glabratus Clarke & Oberprieler, **sp. n.** (Figure 88)

Description. Size. Length 2.86 mm, width unmeasurable. Body seemingly robust, probably broad, integument uniformly metallic black, vestiture short, pale. **Head** short, probably globular, dorsally somewhat bulging; densely setose, setae fine, directed anteriad, not continuing onto rostrum; forehead in middle narrow, setose. Eyes large, probably slightly elongate, dorsally separated by about basal width of rostrum or less. Rostrum asetose, dorsally with median, paramedian and dorsolateral ridges and intervening grooves; antennal insertions in middle of rostrum length, behind them with scrobes extending to eye, in front with narrow grooves reaching mandibular articulations, without row of setae in groove. **Antennae.** Scapes elongate, apically strongly clavate, oblique; apical articulation socket very narrow; funicles with segment 1 elongate, only slightly wider than others; 2–5 similar, elongate-cylindrical, slender; 6–7 shorter, thicker, 7 shorter than 6; clubs with segment 1 obconical, shorter, narrower than 2, 2–3 subequal, 3 somewhat rounder, 4 conical, broadly inserted onto 3, apically acute. **Mouthparts.** Mandibles symmetrical, short and broad, apically forming strongly notched V, outside with 2 small pointed teeth, one at base and one in middle, inside with very large tooth in about middle, adjacent to apical inner tooth of V. Maxillary palps short, 3-segmented, narrowing apicad. **Thorax.** Prothorax probably proclinate. Pronotum with posterior corners angulate, closely fitting onto elytra; sparsely setose, setae directed anteromesad, fine, short and pale. **Elytra** with humeri narrowly rounded, tightly fitting to pronotal corners, marginal groove distinct, narrow, subequal for entire length, with broad anterior marginal notch; surface nearly smooth, sparsely setose, interstriae much broader than striae, flatly convex, setae short, fine. **Legs.** Femora on outside rounded (profemora) or serrulate in apical third (meso- and metafemora). Tibiae slightly curved, compressed; protibiae longer than others, carinate and sparsely setose on outside, meso- and metatibiae on outside serrulate, with dense long setae in apical quarter; apically with distinct well-developed flanges lined with coarse fringing setae, with 2 short slender spurs. Tarsi with tarsite 1 apically slightly excised, 2 excised, about as wide as 1, 3 with short ovoid lobes, broadly connected basally and concave along inner edges; 5 slightly shorter than 1 + 2; claws divaricate, with triangular basal tooth. **Abdomen.** Ventrites flatly aligned, finely sparsely punctate; 1–2 fused, 3–4 progressively shorter, 5 about as long as 3, apically broadly rounded.

Material examined. Holotype (NIGP169514): reasonably well preserved, largely intact but heavily distorted specimen (compressed and crumpled), well visible from all sides (including mandibles and appendages), appendages severed as follows: scape of right antenna broken in several places and funicle between segments 2 and 3, right profemora from trochanter and protarsites 1 and 2 and 4 and 5 and outer claw, left protarsites and right mesotarsites 4 and 5, right metatarsites 1 and 2 and left metatarsus from tibia, also missing claws; in subtrapezoidal slab 5.6 × 3.3 × 0.9–1.1 mm, with three flat edges and rounded edge in front of rostrum; amber clear brownish-yellow, hazy, with diffuse microscopic particles, without any bubbles or cracks.

Derivation of name. This species is named for its glabrous, black appearance, the name being a Latin adjective.

Remarks. This species can be distinguished from all other known *Bowangius* species by the combination of its larger size, glossy elytra without sculpture and with extremely fine short vestiture (the elytra appearing nearly glabrous) and its carinate protibiae. The combination of carinate protibiae (without teeth) and serrulate meso- and metatibiae is rare, otherwise occurring in our sample only in *Louwiocis megalops*, *Habropezus incoxatirostris* and *Bowangius* sp. 3, possibly also in *Mesophyletis calhouni* (described as smaller denticles on the mesotibiae and none on the protibiae [29]). Despite the distorted condition of the holotype, *B. glabratus* is not or hardly decomposed and perfectly visible from all sides, allowing confirmation of its uniqueness among our sample of Burmese amber weevils. We tentatively assign this large specimen to *Bowangius*, but additional better-preserved specimens are needed to confirm this placement.

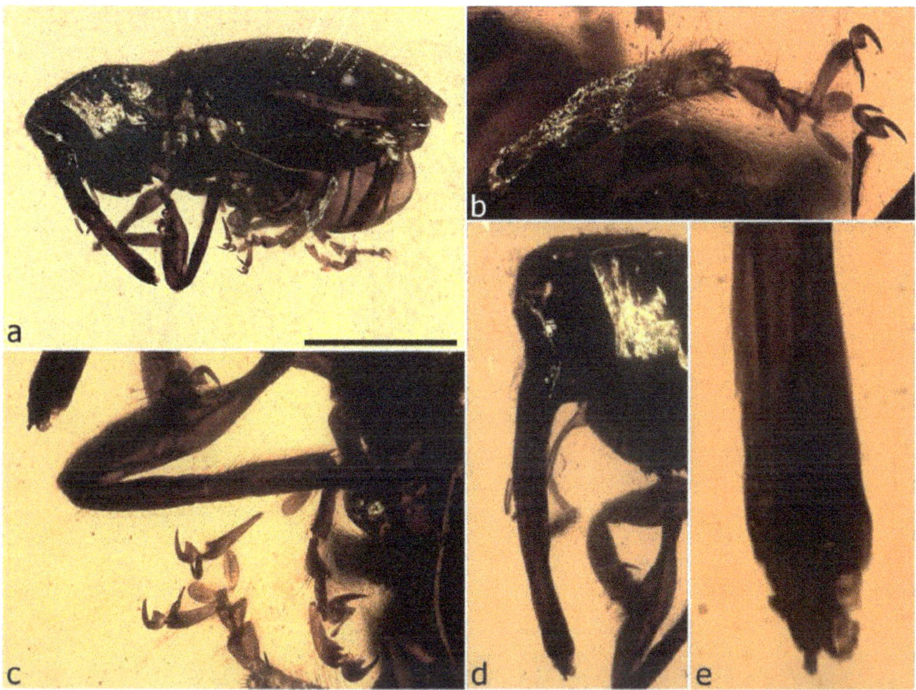

Figure 88. *Bowangius glabratus* sp. n., holotype. Habitus, left lateral (**a**); left metatibia (**b**); protibia and tarsi (**c**); head and rostrum, left lateral (**d**); apical part of rostrum, left lateral (**e**). Scale bar: 1.0 mm (a).

Bowangius sp. 4 (Figure 89)

Material examined. One specimen (NIGP169515), female; length 2.62 mm, width not measurable: reasonably well preserved, intact but heavily distorted (compressed and crumpled), well visible from all sides (including mandibles and appendages); in cuboid 3.1 × 2.8 × 1.9 mm; amber clear yellowish-green, without any bubbles or cracks or debris.

Remarks. While well visible, the specimen had decayed somewhat and is crumpled and collapsed, with key areas of the thorax crushed and associated characters rendered invisible or uninterpretable. Important among these are the basal leg segments, abdominal ventrites and pterothoracic sclerites. However, the mouthparts and many other important details of the legs are visible. On the characters that can be assessed reliably, it appears to belong in *Bowangius*. Its most charactertistic features are the sparse, semi-appressed short body setae and the crenulate-serrulate tibiae. The extruded ovipositor reveals elongate, apically obliquely pointed gonocoxites, each subapically with a tubular truncate stylus with few short setae.

Genus ***Louwiocis*** Clarke & Oberprieler, **gen. n.**

Type species: *Louwiocis megalops* Clarke & Oberprieler, sp. n.

Description. Size. Length 2.2 mm, width 1.2 mm. **Head** large, bulbous, faintly constricted behind eyes. **Eyes** large, vertically elongate, anterior and posterior margins almost straight; hardly prominent, coarsely facetted; dorsally separated by basal width of rostrum, without tubercles between them. **Rostrum** as long as pronotum, stout, cylindrical, inserted in middle of head and dorsally forming deep acute sinus with head; antennal insertions lateral, behind them with scrobes extending to eye, in front of them with lateral row of setae. Gular suture seemingly single, long. **Antennae** geniculate; scapes relatively short, straight, cylindrical, apically clavate; funicles slightly longer than scape, 7-segmented, segment 1 elongate, 2 subequal but thinner, rest progressively shorter towards club; clubs loosely articulated, 4-segmented, apical segment distinct. **Mouthparts**. Labrum absent. Mandibles flat, exodont, vertical in open position, inner (dorsal) side with 2 large teeth and smaller apical one forming short thick V with opposing outer tooth, outer (ventral) side with 2 other much smaller and lower teeth; articulation plane oblique. Maxillary palps 3-segmented. **Thorax**. Prothorax proclinate, with anterior lateral margins oblique in lateral view. Pronotum roundly trapezoidal, laterally rounded, without tooth, posterior corners produced to fit closely onto elytra; surface sparsely setose, setae directed anteromesad; notosternal sutures closed, upright above procoxal cavity, then curved anteriad. Prosternum seemingly short, reduced (not clearly visible); procoxal cavities medially confluent, close to anterior border of prosternum. Scutellar shield distinct. Mesocoxal cavities laterally closed (by meso- and metaventrite). Metanepisternal sutures distinct. Mesoventrite short, anteriorly strongly sloping. Metaventrite longer, flat with slight transverse weals before metacoxae. **Elytra** elongate, with well-developed, broadly rounded humeri, posteriorly gently declivous, apically individually rounded but almost subtruncate together, not exposing pygidium; sutural flanges not visible; surface punctostriate, without scutellary striole, sparsely setose. **Legs**. Procoxae elongate, prominent, medially contiguous; mesocoxae subglobular, narrowly separated; metacoxae flat, transversely elongate. Trochanters short, oblique. Femora subcylindrical, inflated in distal half, unarmed, meso- and metafemora on outside with crenulate carina in distal third to half. Tibiae long, straight, outer edge carinate or serrulate, apex with 1 (protibiae) or 2 (meso- and metatibiae) spurs; meso- and metatibiae with row of sparse, long erect setae on inside in distal half. Tarsi longer than half length of tibiae; tarsite 1 elongate, 2 triangular, 3 deeply bilobed, 5 as long as 1; claws divaricate, dentate with ventrobasal seta at apex of tooth. **Abdomen** with ventrites 1–2 fused, subequal in length, 3–4 each slightly shorter than 2, 5 as long as 4.

Derivation of name. The genus is cordially named after the late Schalk Louw (1952–2018) in recognition of his work on the weevil fauna of southern Africa; the name is composed of his surname and the Greek noun *kis* (G: *kios*) (weevil or beetle) and is masculine in gender.

Remarks. On its dentate claws, exodont mandibles and serrulate tibial edges, *Louwiocis* relates to the group of genera including *Mekorhamphus, Mesophyletis, Myanmarus, Habropezus, Bowangius, Anchineus* and *Elwoodius.* It differs from these in its vertically elongate, hardly protruding eyes. From *Aphelonyssus* it differs additionally in its much larger eyes, more distinct elytral striae longer body setae, apically bent mesotibiae and longer first and truncate second tarsites. It includes a single species among the material available for study.

Louwiocis megalops Clarke & Oberprieler, **sp. n.** (Figure 90)

Description. Size. Length 2.18 mm, width 1.2 mm. **Eyes** large, vertically subrectangular (anterior and posterior margins straight), hardly protruding. **Rostrum** slightly curved; antennal insertions in middle of rostral length, in front of them with lateral row of ca. 5 long, erect setae. **Antennae** long; scapes thin, apically slightly clavate; funicles with segment 1 elongate, slightly inflated, 2 as long as 1 but narrower, others also narrow, progressively becoming shorter towards club; clubs long, flattened, segment 4 distinct, shorter and narrower than 3, flat. **Mouthparts**. Mandibles flat, with 2 large recurved teeth on inner (dorsal) side, smaller one at apex forming short V with similar one on outer (ventral) side, this with 2 much smaller teeth. **Thorax**. Pronotum strongly convex, surface rugulose, sparsely setose, setae fine, pale. Scutellar shield elongate, posteriorly rounded, slightly raised, setose. **Elytra** with surface sparsely setose, setae short, yellowish, suberect, slanting caudad. **Legs**. Femora long, medially inflated; profemora on outside rounded, meso- and metafemora with crenulate carina in distal half. Tibiae long, flattened, outer edge finely serrulate (faintly carinate on protibiae), apex obliquely truncate, extended shortly inwards, mesotibiae bent inwards at apex; protibial spur short, cylindrical, meso- and metatibial spurs equal, free, slightly curved and flattened. Tarsi with tarsite 1 long, narrow, apically truncate, 2 half as long, apically truncate, 3 with lobes not pedunculate. **Abdomen** as for genus.

Material examined. Holotype (NIGP169516): well preserved, intact specimen, not compressed or distorted, right middle leg outstretched ventrad, right front leg broken off at trochanter, well visible from all sides except posterior right; in dorsal half of flat rectangular cuboid 4.2 × 3.65 × 1.2 mm; amber very clear without any impurities except two small elongate clear bubbles over right metathorax and larger flat clear bubble over posterior side of right elytron.

Figure 89. *Bowangius* sp. Habitus, right lateral (**a**); habitus, left lateral (**b**); ovipositor (**c**); head and antenna, right lateral (**d**); right elytral surface, right lateral (**e**); legs and ventrites (**f**); rostral apex and mandibles, right lateral (**g**); legs, left lateral (**h**); same, showing crenulation of metatibia and spurs (**i**); tarsus (**j**); metatibia (**k**); tarsi and claws (**l**). Scale bars: 1.0 mm (a,b).

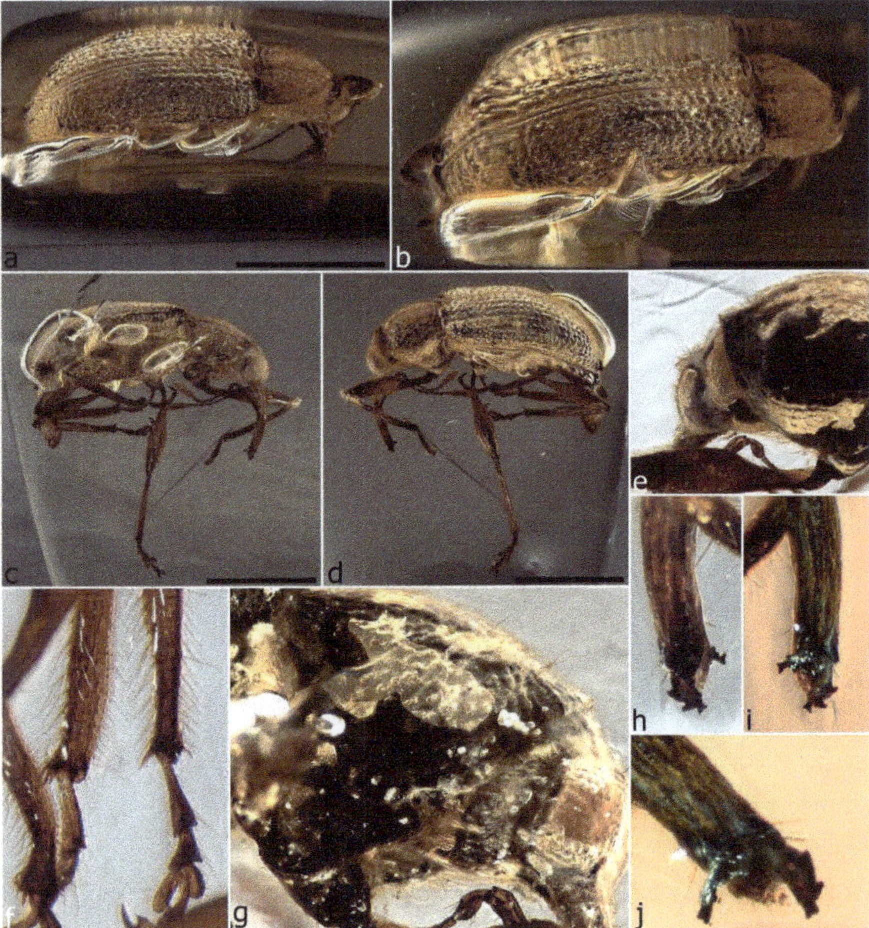

Figure 90. *Louwiocis megalops* sp. n., holotype. Habitus, dorsal oblique (**a**); detail of elytra, dorsal oblique (**b**); habitus, right lateral (**c**); habitus, left lateral (**d**); head and prothorax, left lateral (**e**); tibiae (**f**); head and thorax, right lateral (**g**); apex of rostrum and mandibles, left lateral (**h**); same, right oblique (**i**); same, frontal (**j**). Scale bars: 1.0 mm (a–d).

Derivation of name. The species is named for its large, subrectangular, hardly protruding eyes, the name formed from the Greek adjective *megas* (large) and noun *ops* (eye) and being a noun in apposition.

Remarks. The vertically elongate, almost straight-sided eyes distinguish this species from all other Burmese amber weevils as studied. In the holotype the mandibles are very well preserved and visible in their vertical, open position.

Genus ***Elwoodius*** Clarke & Oberprieler, **gen. n.**

Type species: *Elwoodius conicops* Clarke & Oberprieler, sp. n.

Description. Size. Length 3.8 mm, width 2 mm. **Head** short, subconical. **Eyes** large, conical, strongly protruding, coarsely facetted, dorsally separated by basal width of rostrum, without tubercles between them, forehead slightly impressed. **Rostrum** relatively short, as long as pronotum, stout, cylindrical; antennal insertions lateral, behind them with scrobes extending to eye, in front of them with lateral row of erect setae. Gular suture single, from base of head to underside of rostrum. **Antennae** geniculate; scapes long, cylindrical, apically clavate; funicles slightly longer than scape, 7-segmented, segment 1 elongate, thick, rest thinner, subequal in length but progressively shortened; clubs long, loosely articulated, 4-segmented. **Mouthparts**. Labrum absent. Mandibles small, very narrow, flat, exodont, outside with 3 teeth, basal and median one large, triangular, apical one smaller, narrower, forming small T with similar inner tooth; articulation oblique. Maxillary palps long, 3-segmented. **Thorax**. Prothorax strongly proclinate, with anterior lateral margins oblique in lateral view. Pronotum broader than long, laterally rounded, without tooth, posterior corners angled, fitting closely onto elytra, surface sparsely setose; notosternal sutures closed. Prosternum short; procoxal cavities medially confluent, close to anterior margin of prosternum. Scutellar shield short, broad, slightly convex. Mesocoxal cavities laterally closed (by meso- and metaventrite). Metanepisternal sutures distinct. Mesoventrite short, anteriorly strongly sloping. Metaventrite longer, raised into narrow transverse weals before metacoxae. **Elytra** elongate, with weakly, broadly rounded humeri, posteriorly gently declivous, apically individually angled, not exposing pygidium; sutural flanges narrow, equal; surface finely punctostriate, without scutellary striole, very sparsely irregularly setose, setae directed caudad. Hindwings present. **Legs**. Procoxae large, prominent, close to anterior margin of prosternum, medially contiguous; mesocoxae subglobular, narrowly separated; metacoxae flat, transversely elongate. Trochanters short, oblique. Femora subcylindrical, inflated in distal half, unarmed, on outside with crenulate carina. Tibiae flattened, outer edge serrulate (protibiae) to pectinate (meso- and metatibiae), apex with 2 spurs, inner one on meso- and metatibiae fixed and modified. Tarsi almost as long as tibiae; tarsite 1 elongate, triangular, 2 triangular, 3 deeply bilobed; claws divaricate, dentate with ventrobasal seta at apex of tooth. **Abdomen.** Last tergite (seemingly VII) apically with sharply rimmed setiferous cavity. Ventrites 1–2 fused, equal in length, each slightly longer than 3–4, 5 long, subtriangular, all with long erect setae.

Derivation of name. The genus is respectfully named after the late Elwood C. Zimmerman (1912–2004), in recognition of his huge contributions to the study of the weevil faunas of the Pacific region and Australia; the name is masculine in gender.

Remarks. *Elwoodius* is readily distinguishable from all other mesophyletid genera by its extraordinary narrowly conical, laterally protruding eyes and the flattened meso- and metatibaie with their pectinate outer edges and modified inner spurs. Another unusual feature is the setose apical cavity of the last tergite, which appears unique not only among the Burmese amber weevils but among weevils in general. Only a single species is included in *Elwoodius* thus far.

Elwoodius conicops Clarke & Oberprieler, **sp. n.** (Figure 91)

Description. Size. Length 3.79 mm, width 1.9 mm. **Head** strongly constricted behind eyes, vertex abruptly higher than forehead. **Eyes** strongly elongate–conical, about 2 × wider than long, protruding laterad as well as above forehead. **Rostrum** slightly curved, basally extending smoothly onto forehead; antennal insertions in apical third of rostral length, in front of them with lateral row of 6–7 long, erect setae. **Antennae** long; scapes thin, apically abruptly shortly clavate; funicles with segment 1 slightly spindle-shaped, 2–5 slightly shorter, much thinner but thickening apicad, 6–7 shorter and thicker; clubs long, slightly flattened, terminal segment basally as broad as segment 3, slightly shorter, bluntly triangular. **Mouthparts**. Mandibles narrow, curved, outer teeth low but acute, apex narrow, acute. **Thorax**. Pronotum broadly roundly trapezoidal, slightly convex; surface very sparsely setose, setae (discernible laterally) very fine, pale, sharp, directed mesad to anteromesad. Prosternum deeply roundly emarginate (allowing head and rostrum to flex down onto procoxae). **Elytra** with fine, moderately long, sharply pointed white setae, arising both from strial punctures and on interstriae. **Legs**. Procoxae long, contiguous throughout their length, anteriorly slightly compressed and angled apart (seemingly to receive rostrum when flexed down). Femora long, strongly inflated distally, thickest in about distal quarter, outside with crenulate carina in distal 0.4 of length. Tibiae slightly curved inwards at apex, outer edge serrulate on protibiae, pectinate on meso- and metatibiae; apex truncate, spurs on protibiae small, thin, narrow equal, on meso- and metatibiae unequal, the inner one fixed (fused), on mesotibiae subcylindrical and curved inwards, on metatibiae broad and flat. Tarsi with tarsite 1 narrowly triangular, apically excised, 2 shorter, triangular with thin base, apically excised, 3 deeply bilobed but short, lobes not pedunculate, 5 slightly shorter than 1 + 2. **Abdomen** with ventrites in about middle of length with irregular transverse band of long, erect silvery white setae of varying lengths.

Material examined. Holotype (ANIC 25-073777): very well preserved, intact specimen, not compressed or distorted, both hindwings extruded posteriorly, right elytron slightly cut laterally, well visible from all sides; in centre of rectangular cuboid 5.15 × 4.1–3.7 × 3.2 mm with convex dorsal surface; amber very clear with few scattered impurities and a number of small clear bubbles over pronotum, a sinuate layer of flow lines above eyes and faint fault layer on inside of left legs to edge of amber block, also fine cracks above dorsum of specimen.

Derivation of name. The species is named for its pronounced conical eyes, from the Greek nouns *konos*, meaning cone, and *ops*, meaning eye; the name is a noun in apposition.

Remarks. This is one of the most peculiar species of Buremese amber weevils, immediately recognisable by its large conical eyes and its pectinate meso- and metatibiae with enlarged, fixed inner spurs. Although clearly a mesophyletid (one of the few specimens with the long single gular suture of the group well visible), it approaches the habitus of the family Curculionidae in having its anterior prosternal margin deeply emarginate and its procoxae anteriorly narrowed and splayed, evidently allowing the head and rostrum to flex down and rest on the procoxae. This condition may be interpreted as a morphological precursor to the short prosternal channel of *Burmorhinus*.

Genus ***Aphelonyssus*** Clarke & Oberprieler, **gen. n.**

Type species: *Aphelonyssus latus* Clarke & Oberprieler, sp. n.

Description. Size. Length 2.4 mm, width 1.2 mm. **Head** short, subglobular, very faintly constricted behind eyes. **Eyes** relatively small, flatly hemispherical, coarsely facetted, dorsally separated by width of rostrum, without tubercles between them, placed anteriorly on head (Figure 92e,f). **Rostrum** shorter than pronotum, stout, cylindrical, only very slightly curved; antennal insertions lateral, behind them with deep scrobes extending to eye, in front of them with lateral row of few long, erect setae. Gular suture single, long but indistinct posteriorly. **Antennae** geniculate; scapes short, straight, cylindrical, apically clavate; funicles slightly longer than scape, 7-segmented, segment 1 elongate, 3–7 thinner and gradually shorter towards club; clubs loosely articulated, 4-segmented but segment 4 not well distinct from 3. **Mouthparts**. Labrum, mandibles, maxillae and labium unknown (apex of rostrum of single specimen cut off). **Thorax**. Prothorax strongly proclinate, with anterior lateral margins oblique in lateral view (Figure 92a). Pronotum roundly ogival, laterally rounded, slightly explanate, without tooth, posterior corners angled, fitting closely onto elytra; notosternal sutures closed. Prosternum long, strongly declivous, anteriorly deeply roundly emarginate, surface moderately densely setose, setae suberect, directed anteriad; procoxal cavities medially confluent, close to posterior margin of hypomeron. Scutellar shield short, broad, rounded. Mesocoxal cavities laterally closed (by meso- and metaventrite). Metanepisternal sutures distinct. Mesoventrite short, anteriorly sloping. Metaventrite longer, very weakly convex, with distinct discrimen. **Elytra** shortly elongate, with weakly, broadly rounded humeri, posteriorly gently declivous, apically individually rounded, not exposing pygidium; sutural flanges narrow, equal; surface coarsely punctostriate, without scutellary striole; sparsely setose, setae short, suberect, directed caudad. **Legs**. Procoxae short, subglobular, slightly prominent, medially contiguous; mesocoxae globular, narrowly separated; metacoxae flat, transversely elongate. Trochanters short, oblique. Femora short, strongly inflated in distal half, subcylindrical to subcompressed, unarmed, meso- and metafemora with outer edge faintly serrulate in distal third. Tibiae short, straight, meso- and metatibiae with outer edge finely serrulate, apex with 2 spurs (at least on metatibiae). Tarsi slightly shorter than tibiae; tarsites 1 and 2 triangular, apically excised, 3 deeply bilobed, 5 as long as than 1 + 2; claws divaricate, dentate with ventrobasal seta at apex of tooth. **Abdomen** with ventrites 1 and 2 fused, subequal, 2 slightly longer than 3, 3 slightly longer than 4, 5 as long as 4, crescentic.

Derivation of name. The genus is named for its simple, smooth surface and stout, almost straight rostrum, the name derived from the Greek adjective *aphelos* (even, smooth) and verb *nysso* (to pierce) and being masculine in gender.

Figure 91. *Elwoodius conicops* sp. n., holotype. Habitus, right lateral (**a**); habitus, dorsal oblique (**b**); head and rostrum (**c**); habitus, left lateral (**d**); head (**e**); apex of rostrum and mandibles, dorsal (**f**); left metatibia (**g**); head and thorax, ventral (**h**); mesotibiae and tarsus (**i**); mesotibiae showing spurs (**j**); mesotibiae, detail of dorsal pectination (**k**); tergite VII, dorsal oblique (**l**); tergite VII, dorsal (**m**); mesotibial spurs (**n**). Scale bars: 1.0 mm (a,b,d); 0.2 mm (c,e,f,m).

Figure 92. *Aphelonyssus latus* sp. n., holotype. Habitus, right lateral (**a**); habitus, dorsal (**b**); habitus, left lateral (**c**); habitus, ventral oblique (**d**); head and eyes, dorsal (**e**); prothorax and head, right lateral (**f**); pronotum, dorsal (**g**); ventrites (**h**); ventrites, left lateral (**i**); meso- and metafemora and metatibia (**j**); right antenna (**k**); right metafemur and tibia (**l**). Scale bars: 0.5 mm (a–d); 0.2 mm (f).

Remarks. Due to its serrulate femora and tibiae, this genus belongs in the *Mekorhamphus-Mesophyletis-Habropezus-Bowangius* group, but its non-dentate head and prothorax and round, anteriorly placed (forward-facing) eyes set it apart from all these genera. Its broad, smooth shape (Figure 92) is also atypical of Mesophyletidae, although shared by *Elwoodius*, which has very different (conical) eyes and legs. Thus far it appears to be an isolated genus in the family, containing a single species.

Aphelonyssus latus Clarke & Oberprieler, **sp. n.** (Figure 92)

Description. Size. Length 2.35 mm, width 1.2 mm. **Eyes** large, circular in outline, flat, only slightly protruding, facing forwards, forehead impressed. **Rostrum** slightly curved; antennal insertions in middle of rostral length. **Antennae** long; scapes apically abruptly clavate; funicles with segment 1 as long as 2 + 3, thinly spindle-shaped, 2–3 thinner, 4–7 progressively thickening; clubs short, about 2 × thicker than funicle segment 7, segments laterally rounded, segment 4 short, not well distinct from 3, flattened, apically broadly rounded. **Thorax**. Pronotum slightly broader than long, slightly convex, base straight; surface shallowly coarsely punctate, setae fine, short, stiff, suberect; procoxal cavities placed about their length from anterior margin. **Elytra** broad; surface with strial punctures large, shallow, setae short, fine, acute, brown. **Legs**. Profemora subcylindrical, outer edge rounded, meso- and metafemora compressed, outer edge serrulate in distal third (visible in caudal view). Tibiae slightly flattened, broadening distad, metatibiae apically bent inwards, apex truncate, spurs thin but quite long (only discernible on metatibiae). Tarsi with tarsite 1 apically excised, corners rounded, 2 shorter, apically deeply excised with corners rounded, appearing almost bilobed, 3 with lobes subpedunculate. **Abdomen** as for genus.

Material examined. Holotype (NIGP169517): very well preserved, intact specimen, not compressed or distorted, with tip of rostrum cut off, left elytron and part of pronotum with patches of silvery film, well visible from all sides except middle of left side; slightly angled in centre of rectangular cuboid 4.4 × 3.8 × 1.6 mm; amber very clear without impurities but with four large interconnected clear bubbles with brown walls on left anterior side of specimen, extending into a diagonal brown flow line on underside across left front leg and similar subparallel one further back (not obscuring specimen).

Derivation of name. The species is named for its broad shape, especially of the elytra, the name being the Latin adjective *latus*.

Remarks. This species has a characteristic look due to its broad, flattish, smooth shape and small, flat, forward-facing eyes, appearing more curculionid- than mesophyletid-like. However, its loosely articulated antennal clubs, serrulate meso- and metafemora and -tibiae and indicated long single gular suture show that it belongs in the latter family. Unfortunately the rostrum of the holotype is cut away at the apex and the mandibles are not preserved.

Incertae sedis

Tribe Palaeocryptorhynchini Poinar & Legalov, 2015

Paleocryptorhynchini [*sic*] Poinar & Legalov, 2015: 562 [53] (type genus, by original designation: *Palaeocryptorhynchus* Poinar, 2009 (misspelled as *Paleocryptorhynchus*))

Remarks. The original spelling of the name, Paleocryptorhynchini, is incorrect as the original spelling of its type genus is *Palaeocryptorhynchus*, not *Paleocryptorhynchus* as cited when the tribe was described [53]. This mistake has been perpetuated in later literature. Given the uncertainty regarding the relationships and the family-group assignment of the single genus and species placed in this tribe (see below), it is a rather redundant taxon.

Genus ***Palaeocryptorhynchus*** Poinar, 2009

Palaeocryptorhynus Poinar, 2009: 587 [60] (multiple original spelling; rejected by Legalov, 2010) [61]

Palaeocryptorhynchus Poinar, 2009: 588 [60] (multiple original spelling; accepted by Legalov, 2010 [61], though misspelled as *Paleocryptorhynchus*) (type species, by original designation: *Palaeocryptorhynchus burmanus* Poinar, 2009)

Redescription. Size. Length 6.7 mm, width ca. 2.1 mm. **Head** short, subspherical, dorsally strongly domed, distinctly sculptured and squamose. **Eyes** ventrally positioned, subflat, dorsally (near base of rostrum) separated by about half basal width of rostrum; forehead convex between eyes. **Rostrum** slightly longer than pronotum, slender, weakly curved, broad basally, covered with suberect scales directed posteriad, in repose retracted in prosternal channel and mesothoracic receptacle. **Antennae**, antennal insertions and mouthparts not visible. **Thorax**. Prothorax tightly fitting onto elytra and pterothorax; proclinate, anterior lateral margins drawn out into strong narrowly rounded ocular lobes partly covering eyes when rostrum in repose; sides weakly convex, abruptly constricted in anterior third, more strongly rounded just before base. Pronotum slightly narrower than elytra, widest basally; distinctly collared in anterior third, extending over head; densely squamose, scales directed mesad and anteromesad, appressed to integument, differently coloured; basal margin strongly bisinuate, medially broadly subangularly rounded, fitting into broad emargination in base of elytra, concavely curved laterad to sides, posterior corners rounded, not acute; notosternal sutures obsolete or obscured by rounded, flattened scales. Prosternum forming deep channel for reception of rostrum, with carinate edges terminating posteriorly as small denticles between procoxae, these marking posterolateral corners of a concave plate partly extending between procoxae. Scutellar shield slightly raised above elytral surface. Mesoventrite with curved receptacle for apex of rostrum in repose, situated between procoxae and partly between mesocoxae. **Elytra** elongate, basally lobed over base of pronotum, with broadly rounded humeri, posteriorly declivous, lateral margin substraight to weakly emarginate in middle, apically strongly pointed, subconjointly rounded, not exposing pygidium; surface distinctly punctostriate, without scutellary striole, strial punctures large, seemingly without setae, squamose; interstriae 7–9 confluent at humerus. **Legs** long, slender. Pro- and mesocoxae conical, strongly protruding, broadly separated. Trochanters short, oblique. Femora long, subcylindrical, outside rounded, armed, inside with large flattened triangular tooth near middle, inside excavate in distal half, receiving tibiae in repose, walls of groove at apex flatly and roundly extended. Tibiae sinuate, subcompressed, outer edge rounded, apically extended into long curved uncus-like tooth; apex obliquely truncate, without spurs. Tarsi very long, almost as long as tibiae; tarsite 1 subcylindrical, apically rounded, ventrally densely setose, 2 shorter, subtriangular, apically subtruncate, 3 strongly lobate, 4 (cryptotarsite) distinct; claws divergent, simple, without distinct basal angle or tooth. **Abdomen** not visible.

Remarks. The name of the genus was spelled in two ways in the original paper, as "*Palaeocryptorhynus* gen. nov." in the genus heading on page 587 and as *Palaeocryptorhynchus* in the title, in the discussion (page 588) and in combination with the species name [60]. Legalov [61] selected the latter spelling as the valid one, though inadvertently and also misspelling it as *Paleocryptorhynchus* (as he did in later papers).

The taxonomic position of this genus remains enigmatic and uncertain. Its assignment to the family Curculionidae [60] appears cogent based on characters such as the deflexed rostrum retracting into a channel between the proxocae, the vestiture of scales and the long tibial unci. However, a prosternal rostral channel also occurs in the mesophyletid genera *Burmorhinus* and *Rhadinomycter* (albeit not posteriorly closed by a receptacle), and as the antennae of *Palaeocryptorhynchus* are hidden, it is unknown whether their clubs are also compact like those of Curculionidae. Furthermore, its subfamilial and tribal placement in Curculionidae is problematic. It was originally assigned to the erstwhile subfamily Cryptorhynchinae (*sensu lato*), based on its deep narrow rostral channel, sharply demarcated mesosternal receptacle and long tibial unci [60], but it was subsequently placed in the subfamily Erirhininae based on its allegedly subapically inserted antennae and relatively prominent eyes [53] (given that a mesosternal receptacle also occurs in the erirhinine genus *Aonychus* Schoenherr). This placement is, however, unconvincing for several reasons. Firstly, our examination of the type specimen of *P. burmanus* did not reveal any evidence of the position of the antennal insertions; the structure indicated as such by Legalov and Poinar (Figure 11 in [53]) is the left posterior angle of the prosternal channel (also visible in ventral view in Figure 6 in [60]), and all extant erirhinine genera with a rostral channel (*Afghanocryptus* Voss, *Aonychus*, *Arthrostenus*, *Desmidophorus* Dejean, *Ocladius* Schoenherr, *Tadius* Pascoe) have medially inserted antennae (which is also the common position in Cryptorhynchini). Secondly, the only such erirhinine genus with prominent eyes is *Arthrostenus*, but this has a very weak rostral channel without any posterior receptacle. Thirdly, the tibiae of *P. burmanus* have a fairly long, curved uncus (directed more or less along the tibial axis), whereas the tibiae of all these erirhinine genera have a mucro (inserted at the inner angle of the tibial apex and directed perpendicular to the tibial axis). There is therefore no character in evidence to support the placement of *Palaeocryptorhynchus* in Erirhininae. Its mesosternal receptacle, the long unci and also its general habitus (the tightly retractable rostrum and legs) are in much greater agreement with Cryptorhynchini, as originally concluded by its author [60]. An assignment to this tribe is, however, also problematic, because this is indicated to be much too young (50–60 Ma), and even the larger CCCMS clade, in which the tribe belongs, is estimated to be only ca. 75 million years old [6]. Additionally, the erirhinine taxa as compared above are seemingly too young to be in contention as possible relatives of *Palaeocryptorhynchus*. This, together with the fact that no further such specimen has been reported from Burmese amber, places doubt on its origin and age. The amber piece is said to have been obtained from the Noije Bum mine in the Hukawng Valley in Myanmar [60], and in our exposure to UV light it fluoresced in the same way as the other Burmese amber weevil specimens we studied in PACO, suggesting that it is of the same origin and age as these. If this is true, the specimen may not, after all, be a representative of the family Curculionidae. Its identity can probably only be clarified if more specimens are found and reveal critical structures such as the antennae and mouthparts. For the time being, this taxon should be treated as *incertae sedis* and not be used to infer ages and other evolutionary scenarios in Curculionoidea.

Palaeocryptorhynchus burmanus Poinar, 2009

Palaeocryptorhynchus burmanus Poinar, 2009: 588 [60]

Redescription. Size. Length 6.7 mm, width ca. 2.1 mm. **Head** strongly directed posteroventrad, bulging slightly over eyes, with thin shiny groove separating eyes from head. **Eyes** weakly flatly protruding, separated by about half basal width of rostrum. **Rostrum** narrowing to about middle, then expanding to apex; antennal insertions not visible. **Antennae** not visible. **Thorax.** Pronotum as long as wide, covered in flat scales, mixed circular and ovoid in shape, paler near midline, distinctly darker either side of midline forming oblique separated broad bands in posterior two-thirds; with low indistinct median ridge extending from base to about posterior two-thirds, slightly narrowly beaded at base and contacting scutellar shield; prothoracic ocular lobes not densely setose. Scutellar shield oblong, (probably) setose. **Elytra** with 10 strongly carinate interstriae; interstria 3 strongly protuberant in basal fifth, forming elongate prominence raised higher than level of other interstriae. **Legs.** Profemora seemingly with long flange in basal third of inner edge fitting into groove on outer face of mesofemora (possibly an artefact of leg distortion); tibiae slightly shorter than femora. Tarsi with tarsite 1 more than twice longer than 2, not apicolaterally lobed, 2 subtriangular, narrow basally, apically expanded, subtruncate, 3 with lobes short, semicircularly expanded (not pedunculate), 5 elongate, more than twice longer than tarsite 3 lobes, slender at base, somewhat expanding apically, sparsely setose. **Abdomen** not visible.

Material examined. Holotype (PACO, with curatorial #B-C-42): exceptionally well-preserved, seemingly intact specimen, not compressed or distorted except for collapsed ventrites (not visible), well visible from all except right side, with mite attached to left side of first ventrite; in irregular block with two flat sides, one flat edge, and two rounded sides, orientated with sides parallel to flat faces; amber clear dark brown, with flow band adjacent to and largely obscuring right side of specimen, with another insect inclusion near pronotum and few bubbles emanating from abdomen.

Remarks. Our redescription of this species (and genus) is based largely on the original description [60], updated where applicable to include other characters (or terminology) consistently evaluated and used in our study. The antennal insertions, if subapical as suggested [53], would seem more likely to be positioned further distally on the rostrum than indicated in the figure, more in the vicinity of the maximal width of the spatulate apical part. Some other details mentioned in previous descriptions are also at odds with our examination of the specimen and need further clarification, such as the supposed two groups of setae and the uncus/mucro on the tibiae, the structure of the ventrites and the presence (or absence) of deciduous mandibular cusps. We could not see any of these features, although it is possible that examination of the specimen immersed in a liquid could render some of them visible. The original description mentions black and whitish scales in *P. burmanus*; we add here that these scales are also iridescent greenish. The peculiar flange of the profemora, seemingly locking onto the mesofemora when the legs are retracted, is also a most unusual character, seemingly not reported for any extant weevil. From our examination of the specimen it is not clear whether this is a real feature or a distortion artefact. Unfortunately the right side of the specimen is not visible to further assess this. The study of this enigmatic specimen would benefit from trimming down the amber block as much as possible, so as to allow a clearer view of all its features, and CT scanning may also be able to reveal further taxonomically critical characters (e.g., the antennae or the presence of sclerolepidia).

Genus and species *incertae sedis*

Diagnosis. Head. Rostrum longer than pronotum. **Antennae.** Scapes elongate, slender, apically swollen, not reaching anterior margin of eye. **Mouthparts.** Mandibles small, exodont (single tooth visible). **Thorax.** Pronotal vestiture comprising dense, mixed brown and whitish setae, forming loose clumps (but not seemingly part of a pattern). **Elytra** punctostriate; marginal groove present; vestiture as on pronotum; apically steeply declivous. **Legs** very slender; protibiae not serrulate along distal edge; tarsi narrow, tarsite 1 of meso- and metatarsi seemingly shorter than 2 (left hindleg), tarsite 3 lobes elongate, subpedunculate; claws divaricate, basally angulate with ventrobasal seta. **Abdomen.** Ventrites slightly stepped, 1–3 progressively shorter, 3 and 4 subequal, 5 subequal to 4.

Material Examined. See Material Examined section under *Habropezus plaisiommus* above.

Remarks. The type series of *Habropezus plaisiommus* includes a single paratype specimen. Our examination of it revealed that it is not conspecific with the holotype and belongs in a different genus. While enough details could be discerned in the specimen to conclude this much, it is so poorly preserved (distorted, quite decomposed and missing critical structures) and insufficiently visible that it is impossible to confidently place it in one of the genera treated here. Most of the legs are either missing, not visible or missing parts, the antennae are largely missing, the head and prothorax are crumpled and the ventral side and elytra are poorly visible. Nevertheless, the combination of observable characters suggests that this specimen may represent a different, as yet unknown genus. Though unlikely, further trimming of the amber block may still reveal further details, enable a generic assignment and clarify some details seemingly unusual as we observed (e.g., narrow tarsi with tarsites 1 shorter than 2).

Taxa excluded from Curculionoidea or from Burmese amber

Subfamily Palaeotylinae Poinar, Vega & Legalov, 2018

Palaeotylinae Poinar, Vega & Legalov, 2018: 2 [62] (type genus: *Palaeotylus* Poinar, Vega & Legalov, 2018)

Genus ***Palaeotylus*** Poinar, Vega & Legalov, 2018

Palaeotylus Poinar, Vega & Legalov, 2018: 2 [62] (type species, by original designation: *Palaeotylus femoralis* Poinar, Vega & Legalov, 2018)

Palaeotylus femoralis Poinar, Vega & Legalov, 2018

Palaeotylus femoralis Poinar, Vega & Legalov, 2018: 2 [62]

Remarks. This genus and species was very recently described and assigned to a new family taxon of Platypodinae, described as a subfamily Palaeotylinae (treating the platypodines as a family distinct from Curculionidae, contrary to the common current classification). The placement of the genus in Platypodinae was based on it having femoral "mycangia", elongate tarsi, a broad head (as broad as the pronotum) and tibiae without outer denticles and spurs, although the authors noted that it differs from platypodines in several critical characters, having six-segmented funicles, loosely three-segmented clubs, coarsely facetted eyes, the tibiae with apical teeth, tarsites 1 shorter than 2–4 combined, the meso- and metatarsi with tarsites 2 and 3 bilobed and the pronotum without mycangia [62]. In Platypodinae the funicles are usually four-segmented (rarely two-, three- or five-segmented), the clubs are lenticular one-segmented, the tarsites are all cylindrical, tarsites 1 are longer than 2–5 together and mycangia occur on the pronotum, never on the femora. In addition, some of the structures of *Palaeotylus femoralis* seem to be misinterpreted in the description: The very loose, stalked club segments appear natural in the photograph in Figure 2, not damaged and in reality compact as in the "reconstruction" presented in Figure 3, and the tarsi appear to be four-segmented (Figures 6 and 8) rather than five-segmented as drawn (Figures 7 and 9). Further, weevils (including Platypodinae) have no "cerci" on the abdomen nor a segmented, cylindrical spiculum ventrale projecting from it. On the basis of the characters presented in the description of *Palaeotylus femoralis*, the specimen cannot be accepted as belonging in Platypodinae (or in Curculionoidea). Careful re-examination, and probably trimming of its amber block, are required to assess its real taxonomic affinities and classification.

Representatives of Platypodinae (of "tesserocerine affinities") have been reported from Burmese amber in the literature before [63], but no such specimens have been described and this record remains unconfirmed. The photographs of one of them as we have seen (courtesy of Anthony Cognato) reveal that it is not a platypodine either but one of the curious bostrichoids that have been known from Burmese amber for some time and are currently under study. Very similar (possibly conspecific) specimens examined by us have a large free labrum, flabellate three-segmented antennal clubs, large posteriorly open procoxal cavities, flattened protruding metacoxae, large unequal tibial spurs (the larger one serrate), pentamerous tarsi with cylindrical segments on all legs (formula 5-5-5) and five free ventrites, all characters that are incompatible with Curculionoidea including Platypodinae. A similar species was recently described in Bostrichidae as *Poinarius burmaensis* Legalov [64].

The occurrence of Platypodinae in Burmese amber is also highly unlikely because no authentic specimens have been reported among the several thousand beetles known from this amber by now and because of the indicated age of the subfamily (ca. 75 Ma, [6]), which is too young for Burmese amber.

Subfamily Scolytinae

Two inclusions in Burmese amber have been placed in this subfamily. The first, named *Cryphalites rugosissimus*, was the first beetle described from this amber [28], but it was later recognised as being misidentified and tentatively assigned to the family Colydiidae [23], now classified as a subfamily of Zopheridae. A colour photograph of the specimen published [23] clearly shows that it is not a scolytine. The second scolytine described from Burmese amber is *Microborus inertus*, as discussed below.

Genus ***Microborus*** Blandford, 1897

Microborus Blandford, 1897: 879 [65] (type species, by monotypy: *Microborus boops* Blandford, 1897)

Remarks. This small genus is extant mainly in the Neotropical region, with eight species, but it has recently also been confirmed from Africa, with two species in Cameroon and one in Madagascar [66]. It is currently placed in the tribe Hexacolini but appears to occupy a more isolated position in Scolytinae [66]. It was recorded from Burmese amber with the species *B. inertus* Cognato & Grimaldi, forcing the conclusion that it was once more widely distributed and has remained morphologically conserved (unchanged) for 100 million years [67]). However, recent dated estimates of weevil phylogeny have found much younger ages for the subfamily Scolytinae, between 82 Ma [14] and an early Cenozoic age [6], making this scenario unlikely and casting doubt on the origin and age of the *M. inertus* specimen (see below).

***Microborus inertus* Cognato & Grimaldi**, 2009

Microborus inertus Cognato & Grimaldi, 2009: 95 [67]

Material examined. Holotype (AMNH JZC-Bu228): well preserved and mostly visible, intact specimen, with ventral side partially obscured by whitish emulsion-like substance; at one corner of subquadrangular slab 7.0 × 9.0 × 1.0 mm; amber clear-yellowish with several fracture planes near specimen and with minor flow bands and impurities.

Remarks. This specimen was obtained from a large amount of unprocessed amber sent to the AMNH from Myanmar by a Canadian company, the locality given as the Tanai village in Kachin State [67], so not the mine at Noije Bum from which the date of Burmese amber has been obtained [20].

As we were able to borrow the specimen, we checked it under UV light and found the amber not to fluoresce like typical Burmese amber does (Figure 1m), confirming that it is not from the Noije Bum mine and evidently a different (younger) amber. The specimen also has a white, partially opaque layer between its pronotum and elytra and along its venter, similar to the characteristic *Verlumung* of Baltic amber (Figure 3d). This milky emulsion of minute bubbles, apparently caused by decomposition of gases and moisture, is typical of Baltic amber [68,69] and does not evidently occur in Burmese amber (also not present in any of the specimens studied by us, although a pale translucent film occurs on the underside of the *Habropezus tenuicornis* holotype, Figure 78a,e,h). The nature and provenance of the amber enclosing the *Microborus inertus* specimen should be investigated using infrared light or nuclear magnetic resonance; Baltic amber has a characteristic infrared spectrum [69] and Burmese amber a characteristic NMR spectrum [70]. Pyrolysis gas chromatography can also be used to confirm Burmese amber [71]. The conclusion that the *Microborus inertus* specimen is not embedded in Burmese amber is further supported by the fact that, among the many tens of thousands of beetles that have now been recovered from Burmese amber, no further scolytines have been found. Age estimates of Scolytinae cannot therefore be based on this specimen and its alleged age. Further, the issue of compromised dates of Burmese amber inclusions is evidently not limited to the *Microborus inertus* specimen, as we have also found undescribed staphylinid inclusions in amber from the same source to glow in the same dull yellowish-green colour under UV light, not to fluoresce blue like Burmese amber does.

4. Discussion

Before this study the knowledge of the Burmese amber weevil fauna was based on isolated descriptions of only a few specimens (and the taxa mostly misclassified), which did not allow much interpretation of the peculiarities of the fauna. From our more comprehensive study of a much larger set of specimens and taxa, it is now possible to draw some first conclusions about the salient features and adaptations of the fauna and put it into an ecological and evolutionary context. Based on the additional specimens we have seen but are not describing here, there is no doubt that many more specimens and more taxa (genera and species) of this extraordinary fauna will be discovered in due course, which should allow refinement of the preliminary extrapolations presented here.

4.1. Diversity of Burmese Amber Weevils

By now about 110 weevil specimens appear known from Burmese amber. This includes the 12 specimens so far described, the 64 presented in this paper, another 18 we received late from the AMNH and the FMNH and another 14 we have seen on photographs in books [25,55] and on various internet sites (some no longer available there). Undoubtedly there are more specimens in private collections that have not been documented anywhere, and with more and more Burmese amber inclusions becoming available for sale, there is little doubt that this number will increase substantially. Even at this stage, the weevil fauna in Burmese amber emerges as one of the richest Mesozoic faunas known of this beetle group, rivalling that of the Upper Jurassic Karatau site in Kazakhstan, of which about the same number of specimens has been recorded in the literature but more apparently exist. It certainly far exceeds the latter fauna in terms of the preservation of morphological details, however.

As recognised in this study, the Burmese amber weevil fauna currently comprises 70 species classified in 36 genera and two families (Nemonychidae and Mesophyletidae) (see Appendix A). Only three species are assignable to the former family, the remainder all representing an evidently extinct family that appears to be as diverse, in taxa as in forms, as several of the smaller extant ones. The large number of species and the dearth of conspecific specimens is somewhat surprising. In consideration of preservation artefacts and of sexual differences as exhibited by extant weevils, we initially regarded several inclusions as conspecific with others, but after trimming the amber blocks and carefully studying the specimens under high magnification, it became apparent that they differ in characters that are species-specific in extant weevils and thus represent different species. It generally proved easier to recognise different species among these amber weevils than different genera.

In size the Burmese amber weevil fauna ranges between 10 and 1.6 mm in body length (excluding the rostrum) as measured. A specimen pictured by Xia et al. [55], p. 115, is given as being 18 mm long, but this probably includes the rostrum, the body then being about 12 mm long, comparable with the largest ones we have examined. Most of the species range between 2 and 5 mm in size, but a few are larger (*Calyptocis* 5.6 mm, *Burmocorynus* 6.15–6.7, *Aepyceratus* 6.9 mm, *Petalotarsus* 4.1–7.25 mm, *Cetionyx* 8.2–10 mm) and a few are smaller (*Bowangius* 2.8–1.9 mm, *Gnomus* 2.8–2.1 mm, *Guillermorhinus* 1.8 mm, *Anchineus* 1.6 mm).

In shape the Burmese amber weevils are relatively uniform. Most species have a high, loose body form with long legs and long to very long tarsi, but the smaller ones (*Anchineus*, *Bowangius* and *Habropezus*) have somewhat lower, narrower bodies. The only significant deviations from this body plan occur in *Petalotarsus*, and to some degree in *Burmocorynus*, whose species have flatter, more robust bodies with shorter and stouter legs, especially shorter and separated procoxae, and in *Palaeocryptorhynchus*, which has the withdrawn ('cryptic') body and appendages of certain extant Curculionidae. Nearly all species have a porrect head and a long, slender rostrum, which often reaches two-thirds of the body length (up to 0.82 % in *Mekorhamphus beatae*), translating into 1.7–2.7 mm in length in *Mekorhamphus* and 5.3–7.2 mm in *Cetionyx*. The antennae of species with such long rostrums are invariably geniculate, often very long and slender as well. Shorter rostrums (about as long as the pronotum) occur in relatively few species, nearly always in conjunction with non-geniculate antennae. Except for *Palaeocryptorhynchus*, the vestiture of all species consists only of setae, never of scales. The eyes are mostly large to very large, strongly protruding and coarsely facetted.

4.2. Specialised Features

The most noticeable feature of the Burmese amber weevil fauna is the preponderance of geniculate antennae. Of the 70 species recognised here, only 12 have antennae that are not distinctly geniculate (although then subgeniculate, with an elongate scape, in all but the three species of Nemonychidae). This makes the Mesophyletidae the third group of weevils with geniculate antennae, the others being the extant Curculionidae and Nanophyinae (Brentidae). As shown in the chapter on morphology above, the geniculate antenna of Mesophyletidae is of the 'open' type as it is in Nanophyinae, not of the 'closed' one as occurs in Curculionidae. The geniculate antenna is concomitant with the long rostrum in these weevils, the scape folding back into a long narrow scrobe along the sides of the rostrum, behind the antennal insertions. This modification of the antenna enables especially females to insert their entire rostrum into plant tissues, past the antennal insertions, and reach deep-lying plant organs for oviposition [2]. The larger mesophyletids such as *Cetionyx* would thus have been able to drill oviposition holes to about 7 mm in depth. The mesophyletid ovipositors that are exposed (in *Bowangius* sp. 2 and sp. 4, *Debbia gracilirostris*) are correspondingly thin and elongate, evidently likewise adapted to be inserted into long and narrow holes for laying eggs. This adaptation indicates that these mesophyletids were highly specialised herbivores, likely ovule or seed predators, on a par with extant curculionid tribes such as Anthonomini, Curculionini and others.

Another striking feature of Mesophyletidae are the mostly strongly exodont, flat mandibles. These resemble those of Rhynchitinae but differ foremost in having two large inner (dorsal) teeth in addition to the outer ones and the apex narrowly T- or Y-shaped (see chapter on morphology). They also have a very different, oblique articulation plane and appear to have opened from a more or less horizontal position into a vertical one, then projecting forwards from the rostrum (not sideways as in Rhynchitinae). This motion would have given them a sawing or rasping rather than a biting action and enabled them to cut a ∧- or ∩-shaped hole into plant tissues, of the same width as the rostrum diameter. By analogy with the oblique mandibles of *Ergania* (Figure 4j,n) and *Pimelata* Pascoe (Curculionini), the Allocorynini (Belidae) and the seed-feeding species of *Antliarhinus* (Brentidae) and the vertical mandibles of other Curculionini, it appears that such mandibles are more effective and efficient in drilling deep holes with long thin rostrums, especially into harder or tougher plant tissues (such as nuts in several Curculionini and cycad sporophylls and seeds in Allocorynini and *Antliarhinus*), as the hole does not have to be larger than the diameter of the rostrum. Together with the long rostrum, these unique exodont mandibles suggest that many Mesophyletidae were similarly adapted to pierce inner plant organs, such as ovules and seeds, and that they were in fact more specialised piercers than extant Rhynchitinae, which mostly pierce or cut terminal shoots and flower or fruit buds [52].

The third outstanding characteristic of the Burmese amber weevils are the long legs and especially the long tarsi with their sharp claws. In the Mesophyletidae the tibiae are often flattened and externally strengthened by a crenulate or serrulate carina, the spurs are generally large and sometimes fused to the apex of the tibia, the tarsi are large, flat and very flexible (tarsites 1 and 2 apically deeply excised to bilobed) and the claws are widely divaricate and mostly strongly and sharply dentate. These features suggest that the weevils were well adapted to an arboreal life, able to tightly grip onto smooth plant surfaces such as leaves or climb among flimsy plant parts during feeding and oviposition. Their loose body form, with the elytra not tightly locked together at the apex (individually rounded, the sutural flanges narrow and equal) and to the pygidium (which is sometimes exposed), indicates that they were able fliers and probably quick to take to flight. Somewhat in contrast to this is the more robust and stockier shape of *Petalotarsus*. Its more compact, flatter body, with a shorter thicker rostrum, shorter, separated procoxae, shorter legs and broader tarsi, suggests a more sedentary lifestyle and, at least in *P. oxycorynoides*, an ability to squeeze into tight spaces, such as between leaf sheaths or under bark. Other body shapes, such as the short compact one of *Gnomus*, indicate yet different lifestyles.

Also outstanding among the characters of the Mesophyletidae are the large, protruding and coarsely facetted eyes. In comparison, similar-sized extant Rhynchitinae have much less protruding and more finely facetted eyes, as have various curculionid tribes of similar size and body shape (e.g., Acalyptini, Ceutorhynchni, Curculionini, Eugnomini, Storeini). Whereas the large size and subglobular to conical shape of the mesophyletid eyes indicate that they were able to see well, the large facets suggest that the eyes may have been adapted to conditions of low light, such as in dense canopy or lower vegetation strata in thick forests. Similarly coarse eyes occur among extant weevils in many Nemonychidae, in some Caridae (*Carodes* and an undescribed genus) and in particular in the curculionine tribes Ochyromerini and Rhamphini, all of which are arboreal groups. Similar but much smaller eyes occur in several weevil taxa inhabiting leaf litter, especially Phrynixini, which also appear to be adapted to see in relative darkness. As Nemonychidae, Caridae, Ochyromerini and Rhamphini are generally active during the day, as far as known, the large and coarsely facetted eyes of Mesophyletidae do not necessarily suggest a nocturnal life style, although this cannot be excluded.

4.3. Host Associations

The resin that formed the Burmese amber has been attributed to araucarioid trees similar to the extant genus *Agathis* [70], but no specific araucariaceous taxon has yet been described from Burmese amber. Conifers identified in the amber thus far are two fragments of leafy shoots assigned to the extant monotypic genus *Metasequioa* (Cupressaceae) [24], some wood fragments with araucarioid pitting [70], and some araucarioid or podocarpoid leaves and a cone [25]. Other plant groups identified in Burmese amber include marchantiophytes (liverworts), bryophytes (mosses), pteridophytes (ferns) and twelve species of angiosperms, most of them flowers that cannot or only tentatively be placed in extant families; for a list see [26]. It is the last group that is of particular interest for the weevils preserved in Burmese amber.

The three nemonychid species thus far found in Burmese amber, *Burmonyx zigrasi*, *Burmomacer kirejtshuki* and *Guillermorhinus longitarsis*, all assignable to the now predominantly southern-hemisphere subfamily Rhinorhynchinae, are likely to have lived in association with conifers and particularly with Araucariaceae, as the extant Rhinorhynchinae in Australia and South America do [42,45]. No extant nemonychid is known to live on Cupressaceae [44]. The larvae of *Burmonyx*, *Burmomacer* and *Guillermorhinus* can be assumed to have also developed in the microstrobili of their hosts, feeding on pollen, as most extant Nemonychidae do (except the small angiosperm-associated tribe Rhynchitomacrini and genus *Nemonyx*). In view of the large amount of Burmese amber that has been mined and the tens of thousands of beetles that have been found in it, it is surprising that only three nemonychids have been discovered in this amber so far.

The other Burmese amber weevils, the Mesophyletidae, were most probably associated with the early angiosperms that appear to have been quite diverse in Burmese amber. Among the twelve angiosperm species described, the two grasses, *Programinis burmitis* and *P. laminatus* [72], are unlikely to have served as mesophyletid hosts, as grass-adapted taxa among modern Curculionidae are usually elongate and conspicuously striped or spotted and not shaped like mesophyletids, although some mesophyletids may have inhabited the inflorescences of *Programinis* (which were, however, very small [72]). Also by analogy with modern curculionids, the staminate flowers of *Palaeoanthella huangii* and *Cascolaurus burmitis* were probably less attractive to mesophyletids due to the more ephemeral and less nutritious nature of such flowers. The pistillate and bisexual flowers, however, containing nutritious ovules, would have been much more suitable to sustain developing mesophyletid larvae, and the long thin rostrums with exodont mandibles of most adult weevils appear eminently suited to piercing the calyx and ovules of these flowers. The pistillate flowers of *Lachnociona terriae* and the bisexual ones of *Antiquifloris latifibris*, *Tropidogyne pikei* and *T. pentaptera* were about 5 mm long and wide and their ovules thus in easy reach of the rostrums of the larger mesophyletids such as *Mekorhamphus*, whose rostrums measure 1.7–2.7 mm in length, whereas the smaller (0.8–2.1 mm diameter) bisexual flowers and calyces of *Eoëpigynia burmensis*, *Jamesrosea burmensis*, *Micropetasos burmensis* and *Endobeuthos paleosum* could have served as brood sites for the larvae of the smaller mesophyletids such as *Anchineus* and *Bowangius* (ca. 2 mm body length). Especially the deep calyces and hypanthia of the *Eoëpigynia*, *Lachnociona* and *Tropidogyne* flowers appear like ideal brood sites for weevils, and it is very unlikely that they would not have been utilised as such by mesophyletids. If the largest known mesophyletids (*Burmocorynus* and *Cetionyx*) also developed in flowers or fruits, as seems likely given their long rostrums, even larger flowers can be expected to have occurred among the angiosperms of the Burmese amber flora.

The angiosperms described from Burmese amber all appear to represent basal family group taxa, a few tentatively assignable to extant families but several not. Four have been interpreted as magnoliids, *Antiquifloris latifibris* as an extinct lineage [73] and the other three as Laurales, *Jamesrosea burmensis* as the sister taxon of Atherospermataceae and Gomortegaceae [74], *Palaeoanthella huangii* with possible affinities to Monimiaceae [75] and *Cascolaurus burmitis* as a basal lineage of Lauraceae resembling the extant genus *Litsea* [76]. The other taxa have been assigned to the core eudicot clade [77], *Micropetasos burmensis* with no affinity to any modern family [78] but *Lachnociona terriae* near the rosid families Brunelliaceae and Cunoniaceae [79], *Tropidogyne pikei* and *T. pentaptera* as related to Cunoniaceae, with similarity to the genus *Ceratopetalum* [80,81], *Eoëpigynia burmensis* in the asteroid family Cornaceae *sensu lato*, with similarities to the genus *Cornus* [82], and *Endobeuthos paleosum* with similarities to Dilleniaceae [83]. *Programinis* has been identified as a grass-like monocot of an early bambusoid type, originally only assigned to the order Poales [72] but subsequently to Poaceae [84].

4.4. Evolutionary Significance

The dominant component of the Burmese amber weevil fauna is represented by a diverse but distinct family-group taxon, whose characters do not accord with those of any of the eight extant families (after [6]) and which thus needs to be treated as a separate, ninth family of weevils, the Mesophyletidae. In its characters it displays strong affinities to the 'middle group' of weevil families (Belidae, Attelabidae, Caridae), being quite different from the more basal families Cimberididae, Nemonychidae and Anthribidae and also from the more derived Brentidae and Curculionidae. It appears related to Attelabidae, sharing the characteristic long single gular suture of this family, but it differs in many other characters (absence of scutellary strioles, typically geniculate antennae, flat exodont oblique mandibles, dentate tarsal claws and covered pygidium) and cannot be accommodated in Attelabidae without a significant widening of the concept of the latter. The antennae, mandibles and tarsal claws also differentiate the Mesophyletidae from the small southern family Caridae, which furthermore has ventrally inserted antennae and a short gular suture.

The long thin rostrum, the exodont oblique mandibles, the large eyes and the broad dentate tarsal claws indicate that most mesophyletids were specialised phytophages, evidently adapted to an arboreal life as flower or seed predators of the early angiosperms that began to diversify in the conifer forests of the middle Cretaceous. As many of the flowers thus far discovered in Burmese amber are assigned to the basal magnoliid clade and as extant magnoliids are largely pollinated by weevils [85], it seems likely that mesophyletid weevils may have also played a role in the pollination of the early, Cretaceous representatives of this plant group. The weevil pollinators of extant magnoliids belong to the curculionid tribes Ochyromerini and Storeini *sensu lato*, the former pollinating members of the Annonaceae and seemingly Magnoliaceae and Schizandraceae and the latter pollinating species of Eupomatiaceae, Myristicaceae and Winteraceae [85]. Possibly nursery pollination mutualisms as exist between extant Storeini and their hosts species of *Eupomatia* and *Myristica* may also have already developed between mesophyletids and their hosts in the middle Cretaceous.

Poinar recently presented evidence that the flora preserved in Burmese amber has strong connections to Gondwana [86], both regarding the araucariaceous trees that may have produced the resin that formed the amber and several of the angiosperms whose flowers are preserved in it (affiliated with Gondwanan families such as Cunoniaceae, Monimiaceae and Dilleniaceae). Also some insects appear to show such connections to the extant Gondwanan fauna. The underlying reason for this connection given is the geological origin of the West Burma Block, which was originally attached to eastern Gondwana (north-western Australia) and rafted to its current location in South-East Asia later, possibly during the Upper Jurassic, although the angiosperm fossils with proposed Gondwanan affinities in Burmese amber suggest that it could not have occurred before the Lower Cretaceous [86]. However, recent interpretations of tectonic evidence indicate that the West Burma block was already attached to Asia by the Upper Triassic [87–89]. The weevils in Burmese amber provide no specific support for a Gondwanan origin of the biota preserved in this amber. The three species of Nemonychidae are readily assignable to the subfamily Rhinorhynchinae, which presently occurs in Australia and the Southern Pacific region (New Caledonia and New Zealand) as well as in South America [45]. Fossils assignable to this subfamily have been recorded from the Upper-Jurassic Talbragar Fish Bed in Australia [90], indicating that it has been present in Eastern Gondwana for at least 150 Ma. It is mainly associated with Araucariaceae, on whose pollen the larvae feed. This plant family is today largely restricted to the Southern Hemisphere but had an almost worldwide distribution in the Mesozoic [91–93], and Rhinorhynchinae can therefore also be expected to have occurred quite widely in the Northern Hemisphere in the past, including in South-East Asia, without any specific Gondwanan affinity. Furthermore, results from pyrolysis gas-chromatography and mass spectrometry suggest that Burmese amber may be derived from conifers of the northern family Pinaceae rather than Araucariaceae [94], which would make a Gondwana connection for Rhinorhynchinae even less likely and also account for their scarcity in this amber. The biogeographical affinities of the family Mesophyletidae are more difficult to assess, as no extant representatives remain. Its closest affinities are with the extant families Attelabidae and Caridae. Attelabidae are represented in Australia with only three genera [58], in New Caledonia with only one species [95], in New Zealand with none and in South America with only the genus *Minurus* Waterhouse outside of the tropical areas, and the family is therefore evidently not of Gondwanan origin. By contrast, the small family Caridae is restricted to Australia, New Guinea and southern South America and seemingly of Gondwanan origin as, despite various claims in the literature, no authentic fossils of it are known from the Northern Hemisphere [16]. The relationships of Mesophyletidae to these two families are in need of further study, but if inclusions in other Cretaceous ambers (in Europe and North America) can be assigned to Mesophyletidae, this family may have a Laurasian or Pangaean rather than Gondwanan origin.

Given the taxic and morphological diversity of Mesophyletidae as currently evident, this family appears to represent the first (earliest) diversification of weevils on angiosperms. Its extent over time and space is still unclear, but it may be represented in other Cretaceous ambers and sedimentary *Lagerstätten* as well and have prevailed for some time during the Upper Cretaceous. When it disappeared is also unknown, but it seems to have become extinct by the Cenozoic era, its ecological niche on later angiosperms filled by later diversifications of 'flower weevils' of Brentidae and Curculionidae and of rhynchitine Attelabidae. Along with the numerous other curious beetles and other insects preserved in Burmese amber, the mesophyletids evidently represent the unique weevil fauna of an extinct ecosystem. Judging by the frequency with which new specimens are discovered in Burmese amber, its known diversity stands to increase and its relationships and evolutionary history will become increasingly better understood over time.

5. Conclusions

Our study has revealed the existence of a remarkable diversity of weevils in the middle Cretaceous, both in numbers of genera and species and in specialised morphological structures that rival, and in some instances exceed, those present in extant weevils. These specialisations indicate that the weevils were as highly adapted to their hosts (evidently early angiosperms, as preserved in Burmese amber) as modern weevils are to theirs and that they likely occupied similar niches as flower or seed predators, perhaps as wood-borers and with other lifestyles as well. Noticeably absent from Burmese amber are short-snouted leaf-feeders and trunk-borers, as represented by the subfamilies Entiminae and Scolytinae among extant weevils, suggesting that these niches were either not present in this mid-Cretaceous ecosystem or not exploitable by mesophyletids. Their diversity nonetheless shows that weevils had evolved manifold shapes and sizes and morphological adaptations 100 million years ago, when the angiosperm-dominated ecosystems of today started to emerge [96]. With this diversity the Burmese amber weevils fill a large and important lacuna in the fossil history of weevils, given the severe paucity of weevil fossils known from the Cretaceous period.

Our study also highlights the difficulties encountered in the study of weevil fossils, in which critical structures and characters are often not preserved or are distorted or obscured by other inclusions or amber impurities. Three approaches have to be taken to mitigate or surmount this problem: specimen preparation, examination of larger numbers of specimens and careful comparison with extant weevils and their characters. Study of specimens in large and unsuitably shaped amber pieces is generally a futile exercise, as critical characters (even well-preserved ones) can usually not be observed accurately and are easily misinterpreted. For scientific study it is imperative to cut the amber block as close to the specimen as possible and into a cuboid or similar shape that allows undistorted views of the characters, and the size, shape and characteristics of the block should also be provided in descriptions so as to aid interpretation of the specimen and of structures that are obscured or distorted by the amber. The description of taxa based on single specimens is an equally precarious undertaking, as no fossil has all structures preserved adequately, and interpretations of apparent characters are not infrequently proved wrong by their comparison in other specimens. As an example, the failure to discern a gular suture in a specimen does not necessarily mean that it is absent, as another similar specimen may show it up very clearly. Accurate interpretation of characters of fossils requires careful comparison of equivalent features in extant weevils and a general familiarity with extant weevils and their salient characters. As we have shown, characters such as geniculate antennae, exodont mandibles, crenulate tibiae and 'toothed' claws are not identical (homologous) in extant weevils, and classifications of fossils based on crude similarities in such features are usually mistaken.

As a result of these shortcomings, the identification and classification of Burmese amber weevils in the literature is nearly always incorrect or at least severely compromised. The eleven species so far described (not counting *Microborus inertus*) have been classified in four families, Nemonychidae, Belidae, Caridae (as 'Ithyceridae') and Curculionidae. Only two of these assignments, of *Burmonyx* and *Burmomacer* to Nemonychidae, proved to be correct. Although some misclassification of fossils is to be expected given their imperfect preservation, the lack of proper peer review evidently also plays a role in this. Many papers describing Burmese amber weevils have clearly not been scrutinised by competent reviewers, and we are also aware of cases of reviewer critiques having been disregarded by journal editors. As a consequence, the descriptions and classifications of Burmese amber weevils (as of other weevil fossils) are largely unreliable and should not be accepted at face value in the construction of phylogenetic and evolutionary scenarios unless or until they have been vetted by suitable peers. This is all the more important for fossils to be used in calibrating phylogenetic estimates and inferring lineage ages; these have to be rigorously assessed for both their identity and their age, using specific criteria such as those set out by Parham et al. [97]. If either their provenance or their identity is not unequivocally demonstrated in the description and has not been vetted by peer review, they should not be used to calibrate phylogenetic trees and infer lineage ages, much less to construct evolutionary scenarios.

Supplementary Materials: The following are available online at https://zenodo.org/record/2526793#.XCNDQ8QRWUI. Video S1: Video clip showing a 3D micro-CT scanning reconstruction of the holotype of *Calyptocis brevirostris;* Video S2: Video clip showing a 3D micro-CT scanning reconstruction of the holotype of *Cetionyx batiatus;* Video S3: Video clip showing a 3D micro-CT scanning reconstruction of the holotype of *Petalotarsus oxycorynoides;* Video S4: Video clip showing a 3D micro-CT scanning reconstruction of the holotype of *Opeatorhynchus comans;* Video S5: Video clip showing a 3D micro-CT scanning reconstruction of the holotype of *Debbia gracilirostris;* Video S6: Video clip showing a 3D micro-CT scanning reconstruction of the holotype of *Myanmarus caviventris;* Video S7: Video clip showing a 3D micro-CT scanning reconstruction of the holotype of *Periosocerus crenulatus.*

Author Contributions: D.J.C. and R.G.O. conceived and carried out the study and wrote the paper (Conceptualization; Writing—original draft); A.L. composed the CT images (Methodology) and D.D.M. contributed to writing the paper and acquired financial support for the project leading to this publication (Writing—original draft; funding acquisition).

funding: D.J.C.'s research was supported by the US National Science Foundation grant #DEB1355169 (to Duane McKenna), and the Strategic Priority Research Program (B) of the Chinese Academy of Sciences (grant XDB26000000 to Bo Wang) supported the acquisition of specimens.

Acknowledgments: We are greatly indebted to Bo Wang (NIGP) for kindly making the vast majority of the specimens available for this study for a considerable period of time. We sincerely thank Steve Davis (AMNH), Carsten Gröhn (GPIH) and Marek Wanat (University of Wroclav), Dong Ren (CNUB) Shuhei Yamamoto (FMNH), and Yu-Lingzi Zhou (ANIC) for contributing other specimens, Andrei Legalov (ISEA) for loaning type specimens in his care, David Grimaldi for loaning the types of *Burmonyx zigrasi* and *Microborus inertus* and George Poinar Jr. for allowing access to type specimens of described species maintained at Oregon State University. We are grateful to Debbie Jennings (ANIC) for the majority of the photography used in our study, to Chenyang Cai (NIGP) for photographs of *Nugatorhinus chenyangi* and for assisting in the description of this species, to Steve Davis and Chris Marshall for hosting D.J.C. during visits to AMNH and Oregon State University, respectively, to Corrie Morreau (FMNH) for access to her microscope for photography of specimens, to Michael Turner (ANU) for CT scanning work, to Cristian Beza-Beza (University of Memphis) for help with UV photography and to Adam Ślipiński, Yu-Lingzi Zhou and Zhenua Liu (ANIC) for stimulating discussions about the beetle fauna preserved in Burmese amber.

Conflicts of Interest: The authors declare no conflict of interest.

Appendix Checklist and Classification of Burmese Amber Weevils

Family Nemonychidae

Subfamily Rhinorhynchinae

Genus ***Burmonyx*** Davis & Engel, 2014
    Burmonyx zigrasi Davis & Engel, 2014

GenusGenus ***Guillermorhinus*** Clarke & Oberprieler, **gen. n.**
    Guillermorhinus longitarsis Clarke & Oberprieler, **sp. n.**

Genus ***Burmomacer*** Legalov, 2018
    Burmomacer kirejtshuki Legalov, 2018

Family Mesophyletidae stat. n.

Subfamily Aepyceratinae

Genus ***Aepyceratus*** Poinar, Brown & Legalov, 2017
Aepyceratus hyperochus Poinar, Brown & Legalov, 2017

Genus ***Platychirus*** Clarke & Oberprieler, **gen. n.**
Platychirus beloides Clarke & Oberprieler, sp. n.

Genus ***Rhynchitomimus*** Clarke & Oberprieler, **gen. n.**
Rhynchitomimus chalybeus Clarke & Oberprieler, sp. n.

Genus ***Nugatorhinus*** Clarke & Oberprieler, **gen. n.**
Nugatorhinus albomaculatus Clarke & Oberprieler, sp. n.
Nugatorhinus chenyangi Clarke & Oberprieler, sp. n.

Genus ***Calyptocis*** Clarke & Oberprieler, **gen. n.**
Calyptocis brevirostris Clarke & Oberprieler, sp. n.

Genus ***Acalyptopygus*** Clarke & Oberprieler, **gen. n.**
Acalyptopygus astriatus Clarke & Oberprieler, sp. n.
Acalyptopygus brevicornis Clarke & Oberprieler, sp. n.
Acalyptopygus elongatus Clarke & Oberprieler, sp. n.
Acalyptopygus lingziae Clarke & Oberprieler, sp. n.

Subfamily Mesophyletinae

Genus ***Cetionyx*** Clarke & Oberprieler, **gen. n.**
Cetionyx batiatus Clarke & Oberprieler, sp. n.
Cetionyx terebrans Clarke & Oberprieler, sp. n.
Cetionyx ursinus Clarke & Oberprieler, sp. n.

Genus ***Burmocorynus*** Legalov, 2018
Burmocorynus jarzembowskii Legalov, 2018
Burmocorynus longus Clarke & Oberprieler, sp. n.

Genus ***Petalotarsus*** Clarke & Oberprieler, **gen. n.**
Petalotarsus curculionoides Clarke & Oberprieler, sp. n.
Petalotarsus cylindricus Clarke & Oberprieler, sp. n.
Petalotarsus oxycorynoides Clarke & Oberprieler, sp. n.
***Petalotarsus* sp.**

Genus ***Opeatorhynchus*** Clarke & Oberprieler, **gen. n.**
Opeatorhynchus comans Clarke & Oberprieler, sp. n.

Genus ***Echogomphus*** Clarke & Oberprieler, **gen. n.**
Echogomphus viridescens Clarke & Oberprieler, sp. n.
Genus *Cyrtocis* Clarke & Oberprieler, **gen. n.**
Cyrtocis gibbus Clarke & Oberprieler, sp. n.
Genus *Ocriocis* Clarke & Oberprieler, **gen. n.**
Ocriocis binodosus Clarke & Oberprieler, sp. n.

Genus ***Electrocis*** Clarke & Oberprieler, **gen. n.**
Electrocis dentitibialis Clarke & Oberprieler, sp. n.

Genus ***Debbia*** Clarke & Oberprieler, **gen. n.**
Debbia gracilirostris Clarke & Oberprieler, sp. n.

Genus ***Burmorhinus*** Legalov, 2018
Burmorhinus georgei Legalov, 2018
Burmorhinus setosus Clarke & Oberprieler, sp. n.

Genus ***Rhadinomycter*** Clarke & Oberprieler, **gen. n.**
Rhadinomycter perplexus Clarke & Oberprieler, sp. n.

Genus ***Gnomus*** Clarke & Oberprieler, **gen. n.**
Gnomus brevis Clarke & Oberprieler, sp. n.
Gnomus sp.
Gnomus spinipes Clarke & Oberprieler, sp. n.

Genus ***Hukawngius*** Clarke & Oberprieler, **gen. n.**
Hukawngius crassipes Clarke & Oberprieler, sp. n.

Genus ***Mekorhamphus*** Poinar, Brown & Legalov, 2016
Mekorhamphus beatae Clarke & Oberprieler, sp. n.
Mekorhamphus gracilipes Clarke & Oberprieler, sp. n.
Mekorhamphus gyralommus Poinar, Brown & Legalov, 2016
Mekorhamphus poinari Clarke & Oberprieler, sp. n.
Mekorhamphus tenuicornis Clarke & Oberprieler, sp. n.
Mekorhamphus sp.

Genus ***Compsopsarus*** Clarke & Oberprieler, **gen. n.**
Compsopsarus reneae Clarke & Oberprieler, sp. n.
Compsopsarus sp.

Genus ***Myanmarus*** Clarke & Oberprieler, **gen. n.**
Myanmarus caviventris Clarke & Oberprieler, sp. n.
Myanmarus dentifer Clarke & Oberprieler, sp. n.
Myanmarus diversiunguis Clarke & Oberprieler, sp. n.
Myanmarus robustus Clarke & Oberprieler, sp. n.

Genus ***Mesophyletis*** Poinar, 2006
Mesophyletis calhouni Poinar, 2006

Genus ***Euryepomus*** Clarke & Oberprieler, **gen. n.**
Euryepomus lophomerus Clarke & Oberprieler, sp. n.

Genus ***Periosocerus*** Clarke & Oberprieler, **gen. n.**
Periosocerus crenulatus Clarke & Oberprieler, sp. n.
Periosocerus deplanatus Clarke & Oberprieler, sp. n.

Genus ***Habropezus*** Poinar, Brown & Legalov, 2016
Habropezus incoxatirostris Clarke & Oberprieler, sp. n.
Habropezus kimpulleni Clarke & Oberprieler, sp. n.
Habropezus plaisiommus Poinar, Brown & Legalov, 2016
Habropezus tenuicornis Clarke & Oberprieler, sp. n.

Genus ***Leptopezus*** Clarke & Oberprieler, **gen. n.**
Leptopezus barbatus Clarke & Oberprieler, sp. n.
Leptopezus rastellipes Clarke & Oberprieler, sp. n.

Genus ***Anchineus*** Poinar & Brown, 2009
Anchineus dolichobothris Poinar & Brown, 2009

Genus ***Bowangius*** Clarke & Oberprieler, **gen. n.**
Bowangius cyclops Clarke & Oberprieler, sp. n.
Bowangius glabratus Clarke & Oberprieler, sp. n.
Bowangius tanaops Clarke & Oberprieler, sp. n.
Bowangius zhenuai Clarke & Oberprieler, sp. n.
Bowangius sp. 1
Bowangius sp. 2
Bowangius sp. 3
Bowangius sp. 4

Genus ***Louwiocis*** Clarke & Oberprieler, **gen. n.**
Louwiocis megalops Clarke & Oberprieler, sp. n.

Genus ***Elwoodius*** Clarke & Oberprieler, **gen. n.**
Elwoodius conicops Clarke & Oberprieler, sp. n.

Genus ***Aphelonyssus*** Clarke & Oberprieler, **gen. n.**
Aphelonyssus latus Clarke & Oberprieler, sp. n.
Incertae sedis

Genus ***Palaeocryptorhynchus*** Poinar, 2009
Palaeocryptorhynchus burmanus Poinar, 2009
Genus and species *incertae sedis*
Taxa excluded from Curculionoidea or from Burmese amber

Genus ***Palaeotylus*** Poinar, Vega & Legalov, 2018
Palaeotylus femoralis Poinar, Vega & Legalov, 2018

Genus ***Microborus*** Blandford, 1897
Microborus inertus Cognato & Grimaldi, 2009

References

1. Kuschel, G. A phylogenetic classification of Curculionoidea to families and subfamilies. *Mem. Entomol. Soc. Wash.* **1995**, *14*, 5–33.
2. Oberprieler, R.G.; Marvaldi, A.E.; Anderson, R.S. Weevils, weevils, weevils everywhere. *Zootaxa* **2007**, *1668*, 491–520.
3. Holloway, B.A. Anthribidae (Insecta: Coleoptera). *Fauna N. Z.* **1982**, *3*, 1–264.
4. Zimmerman, E.C. *Australian Weevils (Coleoptera: Curculionoidea). Vol. 1. Orthoceri. Anthribidae to Attelabidae. The Primitive Weevils*; CSIRO: Melbourne, Australia, 1994.
5. Marvaldi, A.E.; Sequeira, A.S.; O'Brien, C.W.; Farrell, B.D. Molecular and morphological phylogenetics of weevils (Coleoptera, Curculionoidea): do niche shifts accompany diversification? *Syst. Biol.* **2002**, *51*, 761–785. [CrossRef] [PubMed]
6. Shin, S.; Clarke, D.J.; Lemmon, A.R.; Moriarty-Lemmon, E.; Aitken, A.L.; Haddad, S.; Farrell, B.D.; Marvaldi, A.E.; Oberprieler, R.G.; McKenna, D.D. Phylogenomic data yield new and robust insights into the phylogeny and evolution of weevils. *Mol. Biol. Evol.* **2017**, *35*, 823–836. [CrossRef] [PubMed]
7. Anderson, R.S. Weevils and plants: phylogenetic versus ecological mediation of evolution of host plant associations in Curculionidae (Curculioninae). *Mem. Entomol. Soc. Can.* **1993**, *165*, 197–232. [CrossRef]
8. Thompson, R.T. Observations on the morphology and classification of weevils (Coleoptera: Curculionoidea) with a key to the major groups. *J. Nat. Hist.* **1992**, *26*, 835–891. [CrossRef]
9. Alonso-Zarazaga, M.A.; Lyal, C.H.C. *A World Catalogue of Families and Genera of Curculionoidea (Insecta: Coleoptera) (excepting Scolytidae and Platypodidae)*; Entomopraxis: Barcelona, Spain, 1999; 315p, ISBN 84-605-9994-9.
10. Alonso-Zarazaga, M.A.; Lyal, C.H.C. A catalogue of family and genus group names in Scolytinae and Platypodinae with nomenclatural remarks (Coleoptera: Curculionidae). *Zootaxa* **2009**, *2258*, 1–134.

11. Bouchard, P.; Bousquet, Y.; Davies, A.E.; Alonso-Zarazaga, M.A.; Lawrence, J.F.; Lyal, C.H.C.; Newton, A.F.; Reid, C.A.M.; Schmitt, M.; Ślipiński, S.A.; Smith, A.B.T. Family-group names in Coleoptera (Insecta). *ZooKeys* **2011**, *88*, 1–972. [CrossRef]
12. Marvaldi, A.E.; Morrone, J.J. Phylogenetic systematics of weevils (Coleoptera: Curculionoidea): a reappraisal based on larval and adult morphology. *Insect Syst. Evol.* **2000**, *31*, 43–58.
13. McKenna, D.M.; Sequeira, A.S.; Marvaldi, A.E.; Farrell, B.D. Temporal lags and overlap in the diversification of weevils and flowering plants. *Proc. Nat. Acad. Sci. USA* **2009**, *106*, 7083–7088. [CrossRef] [PubMed]
14. Gunter, N.L.; Oberprieler, R.G.; Cameron, S.L. Molecular phylogenetics of Australian weevils (Coleoptera: Curculionoidea): exploring relationships in a hyperdiverse lineage through comparison of independent analyses. *Aust. Entomol.* **2016**, *55*, 217–233. [CrossRef]
15. McKenna, D.D. Temporal lags and overlap in the diversification of weevils and flowering plants: recent advances and prospects for additional resolution. *Am. Entomol.* **2011**, *57*, 54–55. [CrossRef]
16. Oberprieler, R.G.; Anderson, R.S.; Marvaldi, A.E. 3 Curculionoidea Latreille, 1820: Introduction, Phylogeny. In *Handbook of Zoology. Arthropoda: Insecta. Coleoptera, Beetles. Volume 3: Morphology and Systematics (Phytophaga)*; Leschen, R.A.B., Beutel, R.G., Eds.; Walter de Gruyter: Berlin, Germany; Boston, MA, USA, 2014; pp. 285–300.
17. Franz, N.M.; Engel, M.S. Can higher-level phylogenies of weevils explain their evolutionary success? A critical review. *Syst. Entomol.* **2010**, *35*, 597–606. [CrossRef]
18. Johnson, A.J.; McKenna, D.D.; Jordal, B.H.; Cognato, A.I.; Smith, S.M.; Lemmon, A.R.; Lemmon, E.L.M.; Hulcr, J. Phylogenomics clarifies repeated evolutionary origins of inbreeding and fungus farming in bark beetles (Curculionidae, Scolytinae). *Mol. Phylogenet. Evol.* **2018**, *127*, 229–238. [CrossRef] [PubMed]
19. McKenna, D.D.; Clarke, D.J.; Anderson, R.; Astrin, J.J.; Brown, S.; Chamorro, L.; Davis, S.R.; de Medeiros, B.; del Rio, M.G.; Haran, J.; et al. Morphological and molecular perspectives on the phylogeny, evolution and classification of weevils (Coleoptera: Curculionoidea): Proceedings from the 2016 International Weevil Meeting. *Diversity* **2018**, *10*, 64. [CrossRef]
20. Shi, G.; Grimaldi, D.A.; Harlow, G.E.; Wang, J.; Wang, J.; Yang, M.; Lei, W.; Li, Q.; Li, X. Age constraint on Burmese amber based on U-Pb dating of zircons. *Cretac. Res.* **2012**, *37*, 155–163. [CrossRef]
21. Chhibber, H.L. *The Mineral Resources of Burma*; Macmillan & Co.: London, UK, 1934; 320p.
22. Sun, T.T.; Kleismantas, A.; Nyunt, T.T.; Zheng, M.R.; Krishnaswamy, M.; Ying, L.H. Burmese amber from Hti Lin. *J. Gemmol.* **2015**, *34*, 606–615. [CrossRef]
23. Ross, A.J.; York, P.V. A list of types and figured specimens of insects and other inclusions in Burmese amber. *Bull. Nat. Hist. Mus. Geol.* **2000**, *56*, 11–20.
24. Grimaldi, D.A.; Engel, M.S.; Nascimbene, P.C. Fossiliferous Cretaceous amber from Myanmar (Burma): Its rediscovery, biotic diversity, and paleontological significance. *Am. Mus. Novit.* **2002**, *3361*, 1–71. [CrossRef]
25. Zhang, W. *Frozen Dimensions. The Fossil Insects and Other Invertebrates in Amber*; Chongqin University Press: Chongqin, China, 2017; 698p.
26. Ross, A.J. *Burmese (Myanmar) Amber Taxa, On-Line Checklist v. 2018.2*. 2018. Available online: http://www.nms.ac.uk/explore/stories/natural-world/burmese-amber (accessed on 3 November 2018).
27. Ross, A.; Mellish, C.; York, P.; Crighton, B. Chapter 12. Burmese Amber. In *Biodiversity of Fossils in Amber from the Major World Deposits*; Penney, D., Ed.; Siri Scientific Press: Castleton, UK, 2010; pp. 208–235.
28. Cockerell, T.D.A. Arthropods in Burmese amber. *Am. J. Sci.* **1917**, *44*, 360–368. [CrossRef]
29. Poinar, G., Jr. *Mesophyletis calhouni* (Mesophyletinae), a new genus, species and subfamily of Early Cretaceous weevils (Coleoptera: Curculionoidea: Eccoptarthridae) in Burmese amber. *Proc. Entomol. Soc. Wash.* **2006**, *108*, 878–884.
30. McKenna, D.D.; Wild, A.L.; Kanda, K.; Bellamy, C.L.; Beutel, R.G.; Caterino, M.S.; Farnum, C.W.; Hawks, D.C.; Ivie, M.A.; Jameson, M.L.; et al. The beetle tree of life reveals that Coleoptera survived end-Permian mass extinction to diversify during the Cretaceous terrestrial revolution. *Syst. Entomol.* **2015**, *40*, 835–880. [CrossRef]
31. Jordal, B.H.; Cognato, A.I. Molecular phylogeny of bark and ambrosia beetles reveals multiple origins of fungus farming during periods of global warming. *BMC Evol. Biol.* **2012**, *12*, 1–10. [CrossRef] [PubMed]
32. Gohli, J.; Kirkendall, L.R.; Smith, S.M.; Cognato, A.I.; Hulcr, J.; Jordal, B.H. Biological factors contributing to bark and ambrosia beetle species diversification. *Evolution* **2017**, *71*, 1258–1272. [CrossRef] [PubMed]

33. Xing, L.; McKellar, R.C.; Xu, X.; Li, G.; Bai, M.; Persons, W.S.; Miyashita, T.; Benton, M.J.; Zhang, J.; Wolfe, A.P.; et al. A feathered dinosaur tail with primitive plumage trapped in mid-Cretaceous amber. *Curr. Biol.* **2016**, *26*, 3352–3360. [CrossRef] [PubMed]
34. Xing, L.; Sames, B.; McKellar, R.C.; Xi, D.; Bai, M.; & Wan, X. A gigantic marine ostracod (Crustacea: Myodocopa) trapped in mid-Cretaceous Burmese amber. *Sci. Rep.* **2018**, *8*, 1365. [CrossRef]
35. Bisulca, C.; Nascimbene, P.C.; Elkin, L.; Grimaldi, D.A. Variation in the deterioration of fossil resins and implications for the conservation of fossils in amber. *Am. Mus. Novit.* **2012**, *3734*, 1–19. [CrossRef]
36. Nascimbene, P.; Silverstein, H. The preparation of fragile Cretaceous ambers for conservation and study of organismal inclusions. In *Studies on Fossils in Amber, with Particular Reference to the Cretaceous of New Jersey*; Grimaldi, D., Ed.; Backhuis Publishers: Leiden, The Netherlands, 2000; pp. 93–102.
37. Azar, D.; Perrichot, V.; Néraudeau, D.; Nel, A. New psychodids from the Cretaceous ambers of Lebanon and France, with a discussion of *Eophlebotomus connectens* Cockerell, 1920 (Diptera, Psychodidae). *Ann. Entomol. Soc. Am.* **2003**, *96*, 117–126. [CrossRef]
38. Limaye, A. Drishti: a volume exploration and presentation tool. In *Developments in X-Ray Tomography VIII*; Stock, S.R., Ed.; SPIE: Bellingham, WA, USA, 2012; Volume 8506. [CrossRef]
39. Oberprieler, R.G. 3.7.1 Brachycerinae Billberg, 1820. In *Handbook of Zoology. Arthropoda: Insecta. Coleoptera, Beetles. Volume 3: Morphology and Systematics (Phytophaga)*; Leschen, R.A.B., Beutel, R.G., Eds.; Walter de Gruyter: Berlin, Germany; Boston, MA, USA, 2014; pp. 424–451.
40. Riedel, A.; dos Santos Rolo, T.; Cecilia, A.; van de Kamp, T. Sayrevilleinae Legalov, a newly recognised subfamily of fossil weevils (Coleoptera, Curculionoidea, Attelabidae) and the use of synchrotron microtomography to examine inclusions in amber. *Zool. J. Linn. Soc.* **2012**, *165*, 773–794. [CrossRef]
41. Oberprieler, R.G.; Thompson, R.T.; Peterson, M. Darwin's forgotten weevil. *Zootaxa* **2010**, *2675*, 33–46.
42. Kuschel, G. Nemonychidae of Australia, New Guinea and New Caledonia. In *Australian Weevils (Coleoptera: Curculionoidea). Vol. 1. Orthoceri: Anthribidae to Attelabidae: The Primitive Weevils*; Zimmerman, E.C., Ed.; CSIRO: Melbourne, Australia, 1994; pp. 563–637.
43. Oberprieler, R.G.; Scholtz, C.H. The genus *Urodontidius* Louw (Anthribidae: Urodontinae) rediscovered and its biological secrets revealed: a tribute to Schalk Louw. *Diversity* **2018**, *10*, 92. [CrossRef]
44. Anderson, R.S.; Oberprieler, R.G.; Marvaldi, A.E. 3.1 Nemonychidae Bedel, 1882. In *Handbook of Zoology. Arthropoda: Insecta. Coleoptera, Beetles. Volume 3: Morphology and Systematics (Phytophaga)*; Leschen, R.A.B., Beutel, R.G., Eds.; Walter de Gruyter: Berlin, Germany; Boston, MA, USA, 2014; pp. 301–309.
45. Kuschel, G.; Leschen, R. Phylogeny and taxonomy of the Rhinorhynchinae (Coleoptera: Nemonychidae). *Invertebr. Syst.* **2011**, *24*, 573–615.
46. Legalov, A.A. Two new weevil tribes (Coleoptera: Curculionidae) from Burmese amber. *Hist. Biol.* **2018**. [CrossRef]
47. Legalov, A.A.; Azar, D.; Kirejtshuk, A.G. A new weevil (Coleoptera; Nemonychidae; Oropsini trib. nov.) from Lower Cretaceous Lebanese amber. *Cretac. Res.* **2017**, *70*, 111–116. [CrossRef]
48. Peris, D.; Davis, S.R.; Engel, M.S.; Delclòs, X. An evolutionary history embedded in amber: Reflection of the Mesozoic shift in weevil-dominated (Coleoptera: Curculionoidea) faunas. *Zool. J. Linn. Soc.* **2014**, *171*, 534–553. [CrossRef]
49. Davis, S.R.; Engel, M.S. A new nemonychid weevil from Burmese amber (Coleoptera: Curculionoidea). *ZooKeys* **2014**, *405*, 127–138. [CrossRef]
50. Poinar, G., Jr. Type genus for Mesophyletinae, a subfamily of Early Cretaceous weevils (Coleoptera: Curculionoidea: Eccoptarthridae) in Burmese amber. *Proc. Entomol. Soc. Wash.* **2008**, *110*, 262. [CrossRef]
51. Poinar, G.O., Jr.; Bown, A.E.; Legalov, A.A. A new weevil. *Aepyceratus hyperochus* gen. et sp. nov., Aepyceratinae subfam. nov., (Coleoptera: Nemonychidae) in Burmese amber. *Cretac. Res.* **2017**, *77*, 75–78. [CrossRef]
52. Riedel, A. 3.4 Attelabidae Billberg, 1820. In *Handbook of Zoology. Arthropoda: Insecta. Coleoptera, Beetles. Volume 3: Morphology and Systematics (Phytophaga)*; Leschen, R.A.B., Beutel, R.G., Eds.; Walter de Gruyter: Berlin, Germany; Boston, MA, USA, 2014; pp. 328–355.
53. Legalov, A.A.; Poinar, G., Jr. New tribes of the superfamily Curculionoidea (Coleoptera) in Burmese amber. *Hist. Biol.* **2015**, *27*, 558–564. [CrossRef]
54. Poinar, G.O., Jr.; Bown, A.E.; Legalov, A.A. A new weevil tribe, Mekorhamphini trib. nov. (Coleoptera, Ithyceridae) with two new genera in Burmese amber. *Ukrainian J. Ecol.* **2016**, *6*, 157–164. [CrossRef]

55. Xia, F.; Yang, G.; Zhang, Q.; Shi, G.; Wang, B. *Burmese Amber: Lives through Time and Space*; Science Press: Beijing, China, 2015; 197p.
56. Legalov, A.A. A new weevil, *Burmorhinus georgei* gen. et sp. nov. (Coleoptera; Curculionidae) from the Cretaceous Burmese amber. *Cretac. Res.* **2018**, *84*, 13–17. [CrossRef]
57. Oberprieler, R.G. 3.7 Curculionidae Latreille, 1802. In *Handbook of Zoology. Arthropoda: Insecta. Coleoptera, Beetles. Volume 3: Morphology and Systematics (Phytophaga)*; Leschen, R.A.B., Beutel, R.G., Eds.; Walter de Gruyter: Berlin, Germany; Boston, MA, USA, 2014; pp. 423–424.
58. Pullen, K.R.; Jennings, D.; Oberprieler, R.G. Annotated catalogue of Australian weevils (Coleoptera: Curculionoidea). *Zootaxa Monogr.* **2014**, 1–481.
59. Poinar, G., Jr.; Bown, A.E. *Anchineus dolichobothris*, a new genus, species and family of Early Cretaceous weevils (Curculionidae: Coleoptera) in Burmese amber. *Proc. Entomol. Soc. Wash.* **2009**, *111*, 263–270. [CrossRef]
60. Poinar, G., Jr. *Palaeocryptorhynchus burmanus*, a new genus and species of Early Cretaceous weevils (Coleoptera: Curculionidae) in Burmese amber. *Cretac. Res.* **2009**, *30*, 587–591. [CrossRef]
61. Legalov, A.A. Checklist of Mesozoic Curculionoidea (Coleoptera) with description of new taxa. *Balt. J. Coleopterol.* **2010**, *10*, 71–101.
62. Poinar, G.O., Jr.; Vega, F.E.; Legalov, A.A. New subfamily of ambrosia beetles (Coleoptera: Platypodidae) from mid-Cretaceous Burmese amber. *Hist. Biol.* **2018**, 1–6. [CrossRef]
63. Jordal, B.H. Molecular phylogeny and biogeography of the weevil subfamily Platypodinae reveals evolutionarily conserved range patterns. *Mol. Phylogenet. Evol.* **2015**, *92*, 294–307. [CrossRef] [PubMed]
64. Legalov, A.A. New auger beetle (Coleoptera; Bostrichidae) from mid-Cretaceous Burmese amber. *Cretac. Res.* **2018**, *92*, 210–213. [CrossRef]
65. Blandford, W.F.H. Insecta. Coleoptera. Rhynchophora. Scolytidae. [Cont.]. *Biol. Centrali-Am.* **1897**, *4*, 169–176.
66. Jordal, B. Ancient diversity of Afrotropical *Microborus*: three endemic species—not one widespread. *ZooKeys* **2017**, *710*, 33–42. [CrossRef] [PubMed]
67. Cognato, A.I.; Grimaldi, D. 100 million years of morphological conservation in bark beetles (Coleoptera: Curculionidae: Scolytinae). *Syst. Entomol.* **2009**, *34*, 93–100. [CrossRef]
68. Hoffeins, C. On Baltic amber inclusions treated in an autoclave. *Polish J. Entomol.* **2012**, *81*, 165–183. [CrossRef]
69. Weitschat, W.; Wichard, W. *Atlas of Plants and Animals in Baltic Amber*; Verlag Dr. Friedrich Pfeil: München, Germany, 2002; 256p.
70. Poinar, G., Jr.; Lambert, J.B.; Wu, Y. Araucarian source of fossiliferous Burmese amber: spectroscopic and anatomical evidence. *J. Bot. Res. Inst. Tex.* **2007**, *1*, 449–455.
71. Zherikhin, V.V.; Ross, A.J. A review of the history, geology and age of Burmese amber (Burmite). *Bull. Nat. Hist. Mus. Geol.* **2000**, *56*, 3–10.
72. Poinar, G.O., Jr. *Programinis burmitis* gen. et sp. nov. and *P. laminatus* sp. nov., Early Cretaceous grass-like monocots in Burmese amber. *Aust. Syst. Bot.* **2004**, *17*, 497–504. [CrossRef]
73. Poinar, G.O., Jr.; Buckley, R.; Chen, H. A primitive mid-Cretaceous angiosperm flower, *Antiquifloris latifibris* gen. & sp. nov., in Myanmar amber. *J. Bot. Res. Inst. Tex.* **2016**, *10*, 155–162.
74. Crepet, W.L.; Nixon, K.C.; Grimaldi, D.; Riccio, M. A mosaic Lauralean flower from the Early Cretaceous of Myanmar. *Am. J. Bot.* **2016**, *103*, 290–297. [CrossRef]
75. Poinar, G.O., Jr.; Chambers, K.L. *Palaeoanthella huangii* gen. and sp. nov., an Early Cretaceous flower (Angiospermae) in Burmese amber. *SIDA Contrib. Bot.* **2005**, *21*, 2087–2092.
76. Poinar, G.O., Jr. A mid-Cretaceous Lauraceae flower, *Cascolaurus burmitis* gen. et sp. nov., in Myanmar amber. *Cretac. Res.* **2017**, *71*, 96–101. [CrossRef]
77. The Angiosperm Phylogeny Group. An update of the Angiosperm Phylogeny Group classification for the orders and families of flowering plants: APG IV. *Bot. J. Linn. Soc.* **2016**, *181*, 1–20. [CrossRef]
78. Poinar, G.O., Jr.; Chambers, K.L.; Wunderlich, J. *Micropetasos*, a new genus of Angiosperms from mid-Cretaceous Burmese amber. *J. Bot. Res. Inst. Tex.* **2013**, *7*, 745–750.
79. Poinar, G.O., Jr.; Chambers, K.; Buckley, R. An Early Cretaceous angiosperm fossil of possible significance in rosid floral diversification. *J. Bot. Res. Inst. Tex.* **2008**, *2*, 1183–1192.
80. Chambers, K.L.; Poinar, G.O., Jr.; Buckley, R.T. *Tropidogyne*, a new genus of Early Cretaceous eudicots (Angiospermae) from Burmese amber. *Novon* **2010**, *20*, 23–29. [CrossRef]

81. Poinar, G.O., Jr.; Chambers, K.L. *Tropidogyne pentaptera*, sp. nov., a new mid-Cretaceous fossil angiosperm flower in Burmese amber. *Palaeodiversity* **2017**, *10*, 135–140. [CrossRef]
82. Poinar, G.O., Jr.; Chambers, K.L.; Buckley, R. *Eoëpigynia burmensis* gen. and sp. nov., an Early Cretaceous eudicot flower (Angiospermae) in Burmese amber. *J. Bot. Res. Inst. Tex.* **2007**, *1*, 91–96.
83. Poinar, G.O., Jr.; Chambers, K.L. *Endobeuthos paleosum* gen. et sp. nov., fossil flowers of uncertain affinity from mid-Cretaceous Myanmar amber. *J. Bot. Res. Inst. Tex.* **2018**, *12*, 133–139.
84. Poinar, G., Jr. Silica bodies in the Early Cretaceous *Programinis laminatus* (Angiospermae: Poales). *Palaeodiversity* **2011**, *4*, 1–6.
85. Caldara, R.; Franz, N.M.; Oberprieler, R.G. 3.7.10 Curculioninae Latreille, 1802. In *Handbook of Zoology. Arthropoda: Insecta. Coleoptera, Beetles. Volume 3: Morphology and Systematics (Phytophaga)*; Leschen, R.A.B., Beutel, R.G., Eds.; Walter de Gruyter: Berlin, Germany; Boston, MA, USA, 2014; pp. 589–628.
86. Poinar, G.O., Jr. Burmese amber: evidence of Gondwanan origin and Cretaceous dispersion. *Hist. Biol.* **2018**. [CrossRef]
87. Metcalf, I. Gondwana dispersion and Asian accretion: tectonic and palaeogeographic evolution of eastern Tethys. *J. Asian Earth Sci.* **2013**, *66*, 1–33. [CrossRef]
88. Metcalfe, I. Tectonic evolution of Sundaland. *Bull. Geol. Soc. Malaysia* **2017**, *63*, 27–60.
89. Sevastjanova, I.; Hall, R.; Rittner, M.; Paw, S.M.T.L.; Naing, T.T.; Alderton, D.H.; Comfort, D. Myanmar and Asia united, Australia left behind long ago. *Gondwana Res.* **2016**, *32*, 24–40. [CrossRef]
90. Oberprieler, R.G.; Oberprieler, S.K. *Talbragarus averyi* gen. et sp. n., the first Jurassic weevil from the southern hemisphere (Coleoptera: Curculionoidea: Nemonychidae). *Zootaxa* **2012**, *3478*, 256–266.
91. Stockey, R.A. Mesozoic Araucariaceae: morphology and systematic relationships. *J. Plant Res.* **1994**, *107*, 493–502. [CrossRef]
92. Kershaw, P.; Wagstaff, B. The southern conifer family Araucariaceae: history, status and value for paleoenvironmental reconstruction. *Annu. Rev. Ecol. Syst.* **2001**, *32*, 397–414. [CrossRef]
93. Sequeira, A.S.; Farrell, B.D. Evolutionary origins of Gondwanan interactions: how old are Araucaria beetle herbivores? *Biol. J. Linn. Soc.* **2001**, *74*, 459–474. [CrossRef]
94. Dutta, S.; Mallick, M.; Kumar, K.; Mann, U.; Greenwood, P.F. Terpenoid composition and botanical affinity of Cretaceous resins from India and Myanmar. *Int. J. Coal Geol.* **2011**, *85*, 49–55. [CrossRef]
95. Kuschel, G. Curculionoidea (weevils) of New Caledonia and Vanuatu: basal families and some Curculionidae. In *Zoologia Neocaledonica 6. Biodiversity Studies in New Caledonia*; Grandcolas, P., Ed.; Mémoires du Muséum National d'Histoire Naturelle: Paris, France, 2008; Volume 197, pp. 99–249.
96. Magallón, S.; Gómez-Acevedo, S.; Sánchez-Reyes, L.L.; Hernández-Hernández, T. A metacalibrated time-tree documents the early rise of flowering plant phylogenetic diversity. *New Phytol.* **2015**, *207*, 437–453. [CrossRef]
97. Parham, J.F.; Donoghue, P.C.J.; Bell, C.J.; Calway, T.D.; Head, J.J.; Holroyd, P.A.; Inoue, J.G.; Irmis, R.B.; Joyce, W.G.; Ksepka, D.T.; et al. Best practices for justifying fossil calibrations. *Syst. Biol.* **2012**, *61*, 346–359. [CrossRef] [PubMed]

Communication

Replacement Names for *Elwoodius* Clarke & Oberprieler and *Platychirus* Clarke & Oberprieler (Coleoptera: Curculionoidea: Mesophyletidae)

Dave J. Clarke [1,*] and Rolf G. Oberprieler [2,*]

[1] Department of Biological Sciences, University of Memphis, 3700 Walker Ave, Memphis, TN 38152, USA
[2] CSIRO, Australian National Insect Collection, G.P.O. Box 1700, Canberra, ACT 2601, Australia
* Correspondence: djclarke@memphis.edu (D.J.C.); rolf.oberprieler@csiro.au (R.G.O.); Tel.: +1-773-573-2000 (D.J.C.)

http://zoobank.org/urn:lsid:zoobank.org:pub:urn:lsid:zoobank.org:pub:91D0E440-F2F0-4E3C-BA20-F74D99960BD1
Received: 14 January 2019; Accepted: 19 January 2019; Published: 22 January 2019

Abstract: In a recent paper we published on the weevil fauna preserved in Burmese amber, two newly proposed generic names were subsequently identified as preoccupied names (*Elwoodius* Clarke & Oberprieler and *Platychirus* Clarke & Oberprieler). We propose the name *Zimmiorhinus* as a replacement name for *Elwoodius* Clarke & Oberprieler and *Burmophyletis* as a replacement name for *Platychirus* Clarke & Oberprieler.

Keywords: homonym; taxonomy; Curculionoidea; Mesophyletidae; Cretaceous

---

1. Introduction

The recent paper by Clarke et al. published in *Diversity* 11(1) [1] proposed the new name of *Elwoodius* Clarke & Oberprieler, with type species *Elwoodius conicops* Clarke & Oberprieler, and *Platychirus* Clarke & Oberprieler, with type species *Platychirus beloides* Clarke & Oberprieler, for two weevil species assigned to the extinct family Mesophyletidae. Each of these species were represented by unique holotypes preserved in Burmese amber. The name *Elwoodius* Clarke & Oberprieler is preoccupied by the name of an extant weevil genus, *Elwoodius* Colonnelli, 2014, with type species *Elwoodius barbatus* (Curculionidae) from Socotra Is., Yemen [2]. The name *Platychirus* Clarke & Oberprieler is preoccupied by the synonymic name of an extant fly genus, *Platychirus* Agassiz, 1846 (Syrphidae) [3], being an intentional though unjustified emendation [4,5] of the valid name *Platycheirus* Lepeletier & Serville, 1828 [6], but an available name. Both of these available names had escaped our searches for homonyms of all generic names proposed in our paper.

Neither of these two homonymic names proposed by Clarke & Oberprieler [1] have any synonyms; therefore, for each of them a substitute name must be proposed with its own author and date, in accordance with Article 60.3 of the International Code of Zoological Nomenclature [7]. The purpose of this *Communication* is to propose these replacement names.

2. Replacement Names

Genus ***Zimmiorhinus*** Clarke & Oberprieler, **new name**
urn:lsid:zoobank.org:act:D99B830B-A66F-4D0C-80DA-CE8FA72805AF

Elwoodius Clarke & Oberprieler, 2018, non *Elwoodius* Colonnelli, 2014 [2]. Type species: *Elwoodius conicops* Clarke & Oberprieler, 2018.

Remarks/Derivation of name. Like the preoccupied name *Elwoodius* Clarke & Oberprieler, this replacement name is proposed in honor of the late Elwood C. Zimmerman, affectionately known as 'Zimmie'. Its gender is masculine. The single species included in the genus is *Zimmiorhinus conicops* (Clarke & Oberprieler, 2018), **comb.n.**

Genus ***Burmophyletis*** Clarke & Oberprieler, **new name**
urn:lsid:zoobank.org:act:1E6017AF-77D8-4B3D-8B1F-738AE63B2910

Platychirus Clarke & Oberprieler, 2018, non *Platychirus* Agassiz, 1846 [3]. Type species: *Platychirus beloides* Clarke & Oberprieler, 2018.

Remarks/Derivation of name. This replacement name for *Platychirus* Clarke & Oberprieler is derived from Burma, the old name for Myanmar, where the specimen was found, and the Greek word *phyle*, tribe or race, in reference to the genus *Mesophyletis*, the type genus of Mesophyletidae. Its gender is feminine. The single species included in the genus is *Burmophyletis beloides* (Clarke & Oberprieler, 2018), **comb.n.**

Acknowledgments: We thank Enzo Colonnelli for alerting us to the existence of the name *Elwoodius* Colonnelli, which also facilitated the discovery of the name *Platychirus* Agassiz.

Conflicts of Interest: The authors declare no conflict of interest.

References

1. Clarke, D.J.; Limaye, A.; McKenna, D.D.; Oberprieler, R.G. The Weevil Fauna Preserved in Burmese Amber—Snapshot of a Unique, Extinct Lineage (Coleoptera: Curculionoidea). *Diversity* **2018**, *11*, 1. [CrossRef]
2. Colonnelli, E. Apionidae, Nanophyidae, Brachyceridae and Curculionidae except Scolytinae (Coleoptera) from Socotra Island. *Acta Entomol. Musei Natl. Prague* **2014**, *54*, 295–422.
3. Agassiz, L. *Nomenclatoris Zoologici Index Universalis, Continens Nomina Systematica Classium, Ordinum, Familiarum et Generum Animalium Omnium, Tam Viventium Quam Fossilium, Secundum Ordinem Alphabeticum Unicum Disposita, Adjectis Homonymiis Plantarum, Nec Non Variis Adnotationibus Et Emendationibus*; Jent et Gassmann: Soloduri, 1846.
4. Evenhuis, N.L. Family Syrphidae. Catalog of the Fossil Flies of the World (Insecta: Diptera) Website. Available online: http://hbs.bishopmuseum.org/fossilcat/fosssyrph.html (accessed on 9 January 2019).
5. Young, A.D.; Marshall, S.A.; Skevington, J.H. Revision of *Platycheirus* Lepeletier and Serville (Diptera: Syrphidae) in the Nearctic north of Mexico. *Zootaxa* **2016**, *4082*, 1–317. [CrossRef] [PubMed]
6. Latreille, P.A.; Lepeletier, A.L.M.; Serville, J.G.A.; Guérin, F.E. Entomologie, ou histoire naturelle des crustacés, des arachnides et des insectes. In *Encylopédie Méthodique. Histoire Naturelle*; Tome Dixième: Paris, France, 1828; pp. 345–833.
7. International Commission on Zoological Nomenclature. INTERNATIONAL CODE OF ZOOLOGICAL NOMENCLATURE Fourth Edition [Incorporating Declaration 44, Amendments of Article 74.7.3, with Effect from 31 December 1999 and the Amendment on E-Publication, Amendments to Articles 8, 9, 10, 21 and 78, with Effect from 1 January 2012]. 2012. Available online: http://www.iczn.org/iczn/index.jsp (accessed on 14 January 2019).

Article

An Illustrated Synoptic Key and Comparative Morphology of the Larvae of Dryophthorinae (Coleoptera, Curculionidae) Genera with Emphasis on the Mouthparts

M. Lourdes Chamorro

Systematic Entomology Laboratory, Agricultural Research Service, United States Department of Agriculture, c/o National Museum of Natural History, Smithsonian Institution, P.O. Box 37012, MRC-168, Washington, DC 20013, USA; lourdes.chamorro@ars.usda.gov; Tel.: +1-202-633-1019

Received: 1 October 2018; Accepted: 10 December 2018; Published: 2 January 2019

Abstract: This study provides an illustrated synoptic key and comparative morphology to the 38 known larvae of dryophthorine genera representing seven subtribes in four of the five tribes: *Cactophagus* LeConte, *Cosmopolites* Chevrolat, *Cyrtotrachelus* Schoenherr, *Diathetes* Pascoe, *Diocalandra* Faust, *Dryophthoroides* Roelofs, *Dryophthorus* Germar, *Dynamis* Chevrolat, *Eucalandra* Faust, *Eugnoristus* Schoenherr, *Foveolus* Vaurie, *Mesocordylus* Lacordaire, *Metamasius* Horn, *Metamasius* (=*Paramasius* Kuschel), *Myocalandra* Faust, *Nassophasis* Waterhouse, *Nephius* Pascoe, *Odoiporus* Chevrolat, *Phacecorynes* Schoenherr, *Polytus* Faust, *Poteriophorus* Schoenherr, *Rhabdoscelus* Marshall, *Rhinostomus* Rafinesque, *Rhodobaenus* LeConte, *Rhynchophorus* Herbst, *Scyphophorus* Schoenherr, *Sipalinus* Marshall, *Sitophilus* Schoenherr, *Sparganobasis* Marshall, *Sphenophorus* Schoenherr, *Stenommatus* Wollaston, *Temnoschoita* Chevrolat, *Trigonotarsus* Guerin-Meneville, *Trochorhopalus* Kirsch, *Tryphetus* Faust, *Xerodermus* Lacordaire, and *Yuccaborus* LeConte. Only *Prodioctes* Pascoe was not included due to lack of specimens to examine. Seven genera are reported here for the first time. Detailed line drawings of the mouthparts of 37 genera are provided. The synoptic key is a multi-entry key, different from a traditional, single entry dichotomous key, which allows the user to identify dryophthorine larvae using any combination of characters (couplets). A total of 52 characters are included. This study provides support for the retention of Stromboscerini in the subfamily.

Keywords: weevil larvae; palm weevils; invasive species; comparative morphology

---

1. Introduction

Some of the most devastating weevil pests that have plagued human agriculture globally have been dryophthorines (e.g., granary weevil (*Sitophilus granarius* (Linnaeus)) and palm weevils (*Rhynchophorus ferrugineus* (Olivier), *R. vulneratus* (Panzer), and *R. palmarum* Linnaeus)). Dryophthorines are major pests of various important agricultural commodities such as banana (e.g., *Cosmopolites sordidus* (Germar), *Odoiporus longicollis* (Olivier), and *Polytus mellerborgii* (Boheman)), agave (the sisal weevil *Scyphophorus acupunctatus* Gyllenhal), coconut palm (*Rhinostomus barbirostris* (Fabricius), *Rhabdoscelus obscurus* (Boisduval), and *R. palmarum*), sugar cane (*Metamasius hemipterus* (Linnaeus) and *Rhabdoscelus obscurus*), oil and sago palm (*Rhabdoscleus obscurus, Rhynchophorus palmarum* and *R. ferrugineus*), and date palm (*Rhynchophorus palmarum*) among other monocotyledons.

Worldwide, Dryophthorinae include approximately 1200 species in 153 genera and five tribes (Cryptodermatini, Dryophthorini, Orthognathini, Rhynchophorini and Stromboscerini;) with most species associated with woody monocots [1,2]. Cryptodermatini include the Oriental genus *Cryptoderma* Ritsema; Dryophthorini include the cosmopolitan *Dryophthorus* Germar and *Stenommatus*

Wollaston and the Oriental/Australasian *Psilodryophthorus* Wollaston. Orthognathini are subdivided into two subtribes, Rhinostomina, which include *Rhinostomus* Rafinesque and *Yuccaborus* LeConte, and Orthognathina containing *Mesocordylus* Lacordaire, *Orthognathus* Schoenherr, and *Sipalinus* Marshall. The two most diverse tribes are Stromboscerini with 12 genera and Rhynchophorini, which include 125 genera in six subtribes [3].

The monophyly of Dryophthorinae has been supported by a number of higher-level phylogenic studies of the Curculionoidea [4–6]. However, the molecular-based phylogeny presented by McKenna et al. [7], which includes a taxon sampling of nine dryophthorines, does not recover a monophyletic Stromboscerini + Dryophthorinae. The phylogeny of Dryophthorinae is currently under study based on combined analyses of molecular and morphological data. The current comparative study of the morphological features of the immature stages, here presented in the form of a synoptic key, will serve as a valuable source of data towards inferring the phylogeny of the subfamily.

In an attempt to provide flexibility and usability of diagnostic tools, and circumvent the inherent limitations of single entry keys, a synoptic key structure is here adopted to provide the user with multiple start options when attempting to identify larval dryophthorines. In addition, an updated and overarching tool to identify Dryophthorinae larvae is sorely needed given the agricultural importance of this group and the limitations of currently available diagnostic tools [8]. Presently, to attempt an identification of dryophthorine larvae, a combination of several publications must be consulted, of which Anderson's [9] and May's [10] work stand out. Anderson's [9] dichotomous keys work best with a perfectly preserved, prepared, and mounted specimen, which may not always be available. These keys also do not allow for the potential use of obvious characters of the head, thorax, and abdomen as a primary way to quickly eliminate distinct taxa. All currently known larvae representing 37 dryophthorine genera, with the exception of *Prodioctes* Pascoe (known as the rhizome weevil in India) are included in this key and incorporates seven (eight if *Paramasius* is granted generic status) new ones (Table 1). Three genera, *Dynamis* Chevrolat, *Dryophthoroides* Roelofs, and *Xerodermus* Lacordaire, were coded based on published records [11–13]. Detailed line drawings of the mouthparts of 37 genera, including reproductions of Gardner's [11,12] illustrations of *Dryophthoroides* and *Xerodermus*, are provided in alphabetical order at the end of the key (Figures 56–92).

Table 1. Genera of known dryophthorinae larvae arranged based on current classification, biogeographic range, and depository (Australian National Insect Collection, ANIC; British Museum of Natural History, BMNH; Instituto Nacional de Pesquisas de Amazonias, INPA; Museu Zoologico de Sao Paulo, MZSP; National Taiwan Normal University, Taipei, NTNU; United States National Museum, USNM). *Paramasius* Kuschel is here included as a separate genus from *Metamasius* Horn; however, this was done in an attempt to uncover potential characters that may distinguish this once valid genus from *Metamasius*. This table references the figure plates of each genus.

	Tribe	Subtribe	Figure	Genus (Described)	Genus (Previously Undescribed)	Region	Depository/Source
Dryophthorinae	Orthognathini	Rhinostomina	77	*Rhinostomus* Rafinesque		Circumtropical	USNM
			92	*Yuccaborus* LeConte		New World	USNM
		Orthognathina	3, 6, 66		*Mesocordylus* Lacordaire	Neotropical	USNM
			4, 81, 94	*Sipalinus* Marshall		Oriental	NTNU, USNM
	Rhynchophorini	Rhynchophorina	4, 7, 58	*Cyrtotrachelus* Schoenherr		Oriental	USNM
				Dynamis Chevrolat		Neotropical	[13]
			7, 11, 79	*Rhynchophorus* Herbst		Circumtropical	USNM
		Polytina	7, 74	*Polytus* Faust		Cosmopolitan	USNM
		Sphenophorina	56	*Cactophagus* LeConte		New World	USNM
			3, 7, 11, 57	*Cosmopolites* Chevrolat		Cosmopolitan	USNM
			4, 6, 59	*Diathetes* Pascoe		Australasian	ANIC
			1, 7, 63	*Eucalandra* Faust		Neotropical	USNM
			3, 6, 65		*Foveolus* Vaurie	Neotropical	MZSP
			3, 11, 67	*Metamasius* Horn		New World	USNM
			3, 7, 69		*Nassophasis* Waterhouse	Oriental	BMNH, USNM
			3, 6, 71	*Odoiporus* Chevrolat		Oriental	USNM
			3, 6, 72		*Paramasius* Kuschel	Neotropical	INPA
			2, 6, 11, 73, 93	*Phacecorynes* Schoenherr		Afrotropical	USNM
			4, 7, 75		*Poteriophorus* Schoenherr	Oriental	USNM, NTNU
				Prodioctes Pascoe		Oriental	N/A
			76	*Rhabdoscelus* Marshall		Oriental	USNM
			7, 11, 78	*Rhodobaenus* LeConte		New World	USNM
			2, 80	*Scyphophorus* Schoenherr		New World	USNM
			4, 6, 83		*Sparganobasis* Marshall	Australasian	ANIC
			2, 7, 84, 85	*Sphenophorus* Schoenherr		Nearctic/Holarctic	USNM
			2, 87	*Temnoschoita* Chevrolat		Afrotropical	BMNH, USNM
			4, 6, 88, 94	*Trigonotarsus* Guerin-Meneville		Australasian	ANIC
			3, 7, 11, 89	*Trochorhopalus* Kirsch		Afrotropical	USNM
		Litosomina	3, 6, 64		*Eugnoristus* Schoenherr	Afrotropical	BMNH
			1, 82	*Sitophilus* Schoenherr		Cosmopolitan	USNM
			1, 90		*Tryphetus* Faust	Oriental	USNM
		Diocalandrina	1, 7, 60	*Diocalandra* Faust		Oriental	USNM
			7, 68	*Myocalandra* Faust		Oriental	USNM
	Dryophthorini		7, 62	*Dryophthorus* Germar		Circumtropical	USNM
			1, 11, 86	*Stenommatus* Wollaston		New World	USNM
	Stromboscerini		61	*Dryophthoroides* Roelofs		Oriental	[12]
			1, 7, 70	*Nephius* (=*Anius*) Pascoe		Oriental	BMNH
			91	*Xerodermus* Lacordaire		Oriental	[11,12]

2. Materials and Methods

2.1. Specimens

Specimens included in this study are deposited in the following institutions, as indicated in Tables 1 and 2:

Australian National Insect Collection, ANIC;
British Museum of Natural History, BMNH;
Instituto Nacional de Pesquisas de Amazonias, INPA;
Museu Zoologico de São Paulo, MZSP;
National Taiwan Normal University, Taipei, NTNU; and
United States National Museum, USNM.

2.2. Slides/Associations

Associations and identifications were made by competent authorities (listed in Table 2 of the material examined) based on larva-pupa-adult associations, collected from known host plant, and/or through rearing (Figures 93 and 94). Therefore, ambiguity or uncertainty are indicated with a pound symbol (#) next to the taxon name or remains uncategorized and is listed at the bottom of the character (couplet). Taxon names with an asterisk (*) indicate polymorphism. When multiple species of a genus were available for study, these were examined and in the case of *Sitophilus* and *Sphenophorus*, also illustrated and coded. The identity of *Mesocordylus* is based on identification of the pupa and therefore it is considered tentative in this study. However, the larva is identified as belonging in the subtribe Orthognathina. *Paramasius* is here included as a separate genus from *Metamasius*; however, this was done in an attempt to uncover potential characters that may confirm synonymy or possibly resurrection from *Metamasius*.

All of Cotton's and Anderson's slides housed at the USNM were examined. In the interest of time, not all were given a unique identifier or included in Table 2. Incomplete taxa for which specimens were not available for examination are *Dynamis*, *Dryophthoroides*, *Rhinostomus*, and *Xerodermus*. Their inclusion in the key largely reflects literature sources and and therefore were not included in all couplets. *Prodioctes* is not included and information on the larva of this taxon remains elusive. The only specimen available for study of *Rhinostomus* was of a prepared slide containing only the labrum/epipharynx; consequently the head, remained of the mouthparts, thorax, and abdomen were not examined. Finally, variation across instars or populations was not significantly studied.

2.3. Slide Preparation

Slide preparation of the mouthparts follows Chamorro et al. [14] with the additional step of clearing the mouthparts with 10% KOH prior to placing them on the mounting medium (PVA #6371A, Bioquip Products). The reader should consult May [10,15] for a detailed description of slide preparation procedures. Her work is a fundamental resource for the study of the immature stages of weevils.

2.4. Drawings and Imaging

Pencil drawings were rendered by the author with the aid of a drawing tube mounted on a Zeiss Axiophot compound microscope, for the mouthparts, and a Zeiss Discovery v8 stereomicroscope for the head, thorax, and abdomen of large specimens. The pencil drawings were scanned and imported as templates into Adobe Illustrator (Adobe Systems) to render digital vector graphics. Plates were arranged and labeled with Adobe Illustrator and Adobe InDesign (for color photographs).

Images were taken with an Olympus PEN5 camera mounted on a Zeiss Discovery v8 stereomicroscope. Individual images were taken at different focal planes and combined to create a single image with Zerene Stacker (Zerene Systems, LLC, Richland, WA, USA).

2.5. Terminology

Terminology follows Oberprieler et al. [16] and Chamorro et al. [17] (Figure 74). Initially, character and character state management and coding were facilitated with the use of the program vSysLab: a virtual Systematics Laboratory [18] and LUCID (lucidcentral.org).

2.6. Key: Structure and User Instructions

The synoptic key follows Holloway [19] and this publication should be consulted for relevant references and for a detailed description and example on how to use a synoptic key. Briefly, a synoptic key can best be described as a static (paper) LUCID key, which allows the user to identify a specimen using any combination of characters (couplets). A synoptic key is multi-entry, in contrast to a traditional, single entry dichotomous key. A LUCID key will be generated as a follow-up to this study.

This study focuses largely on the mouthparts, however some characters of the head, chaetotaxy, and overall body characters, such as thoracic and abdominal spiracles and shape of posterior segments (VIII and IX), are also included. The user is encouraged to first eliminate taxa by referencing the included characters of the head, general body form, spiracles, caudal processes, and, following dissection of the mouthparts, the chaetotaxy of the labrum such as *als* and *ams* of the epipharynx (indicated in grey). The key is arranged into the following nine categories, each containing a variable number of couplets (characters): habitus (3 couplets), head (3), thorax (1), abdomen (4), labrum/epipharynx (20), mandibles (4), hypopharynx (2), maxillae (10), and labium (5). The taxon chosen to illustrate a given character is underlined for each couplet. The taxon or taxa chosen to illustrate a given character state (lead) is underlined. Taxa not included in a given couplet are listed at the end of each couplet. Information on possible host commodity/host plant is included in Table 2 for most specimens as part of the material examined and this information, as well as generalized distribution data included in Table 2, may prove helpful in the identification process. Finally, as suggested by Marshall [20] (p. 395): "If the key below is used to identify specimens involved in published research (print or online), please cite this work and include the full citation in the list of references."

To begin, select any character (couplet) (but preferably ones indicated in grey, which tend to be diagnostic at higher levels or do not require dissections of the mouthparts) and identify the character state (lead) that best describes the feature observed on the specimen. Each state is illustrated to avoid uncertainty. List the taxa under the selected character state (lead). To narrow down the list, choose a subsequent character (couplet), determine which state (lead) best describes your specimen and delete the name(s) that do not appear on the list under that state (lead). Continue this elimination process until a single taxon name remains.

Table 2. Material examined including locality data and depository.

Figure	USNMENT Code	Species	Determiner	Host	Location	Depository
56	USNMENT01448507	*Cactophagus validus* (LeConte)	R.T. Cotton		Leng 335	USNM
56	USNMENT01119907	*Cactophagus validus* (LeConte)	D.M.Anderson		California, Garden Grove, A.C. Davis	USNM
56	USNMENT01119908	*Cactophagus validus* (LeConte)	R.T. Cotton		Leng 335	USNM
57	USNMENT01448170	*Cosmopolites sordidus* (Germar)	D.M.Anderson	in banana stalk	Puerto Rico, Bayamon, 11.vi.1943, San Juan #8597, 43 9200	USNM
57	USNMENT01119912	*Cosmopolites sordidus* (Germar)	D.M.Anderson	in trunks-banana	Florida, Forth Myers, 15.ii.'44, 44 2934	USNM
58	USNMENT01119896	*Cyrtotrachelus thompsoni* Alonso-Zarazaga & Lyal	Mr. Lee	Bamboo	Taiwan: E120.63108, N23.84535, 26.viii.2015	USNM
59	USNMENT01160268	*Diathetes morio* Pascoe	R. Oberprieler	*Pandanus*	Australia: 16°04′07″S, 145°27′49″E, QLD, Daintree NP, Pilgrim Sands, 19.Feb. 1998, R. Oberprieler, RO507	ANIC
60	USNMENT01448168	*Diocalandra taitensis* (Guerin-Meneville)	DMAnderson	petiole of coconut	USA: HI, Honolulu, 2-20-29, O.H.S.	USNM
62	USNMENT01448160	*Dryophthorus americanus* Bedel	DMAnderson	in rotting pine	USA: MD, near Plummers Island, 25.vii.1913, Schwarz & Barber	USNM
62	USNMENT01448161	*Dryophthorus americanus* Bedel	DMAnderson	Old, rotten chestnut	USA: VA, Vienna, 19.iii.1942, J.C. Bridwell	USNM
63	USNMENT01448167	*Eucalandra setulosa* (Gyllenhal)	D.M.Anderson	in stem, solid bamboo	Venezuela, Yaracuy, 23.x.'43, McClure, #24	USNM
64	USNMENT01070971	*Eugnoristus braueri* Kolbe	N/A	attacking coconut	Seychelles, 19	BMNH
65	USNMENT01160269	*Foveolus c.f. maculatus* O'Brien	de Madeiros	Euterpe oleracea	Brazil:1-1.497, -48.461, CU069 Pará, Belem, Ilha do Combu, 29.v.2012, B. de Madeiros	MZSP
66	USNMENT01448502	*Mesocordylus* (Prob.) sp. (larva & pupa)	Chamorro	Ex: *Papyrus*	Costa Rica: Port intercepted, 25.ii.2017	USNM
67	USNMENT01448165	*Metamasius* (*ritchiei* Marshall?)	R.T. Cotton	from pineapple plants	Panama, Canal Zone, Summit	USNM
67	USNMENT01448166	*Metamasius hemipterus* (Linnaeus)	R.T. Cotton		Leng 335	USNM
67	USNMENT01448164	*Metamasius ritchiei* Marshall	R.T. Cotton		Jamaica	USNM
67	USNMENT01448162	*Metamasius ritchiei* Marshall	D.M.Anderson	Pineapple	Jamaica, nr. Kingston, Sept. 16, 1917, A.H. Ritchie colr., A-500	USNM
67	USNMENT01119914	*Metamasius ritchiei* Marshall	D.M.Anderson	Pineapple	Jamaica, nr. Kingston, Sept. 16, 1917, A.H. Ritchie colr., A-500	USNM

Table 2. *Cont.*

Figure	USNMENT Code	Species	Determiner	Host	Location	Depository
67	USNMENT01119913	*Metamasius ritchiei* Marshall	R.T. Cotton	in pineapple	Leng 335, Jamaica, W.I.	USNM
67	USNMENT01448163	*Metamasius ritchiei* Marshall	R.T. Cotton	in pineapple	Leng 335, Jamaica, W.I.	USNM
68	USNMENT01448173	*Myocalandra* sp.	W.H. Anderson	dried bamboo sticks	Japan, 28.i.1941, 41 1683	USNM
68	USNMENT01448172	*Myocalandra* sp.	W.H. Anderson	dried bamboo sticks	Japan, 28.i.1941, 41 1683	USNM
69	USNMENT01070977	*Nassophasis* sp.	N/A	*Calanthe cardioglossa*	Vietnam: intercepted Miami CBP; 6/16/2003, #246766	USNM
70	USNMENT01070967	*Nephius* (=Anius) *pauperatus* (Pascoe)	N/A	Bark *Alaeocarpus*	INDIA: Samsing, Kalimpong Bengels, N.G. Chatterjee +Mite, x.1933; BMNH 1949-505, T.2194J.77(S.E. 1631)	BMNH
71	USNMENT01160265	*Odoiporus longicollis* (Linnaeus)	M.L. Chamorro	In *Musa*	Taiwan, Sun Moon Lake, 23.vii.2015	USNM
72	USNMENT01160257	*Paramasius cristulatus* Vannin	B. de Madeiros	*Clusia* c.f. *nemorosa* fruit	Brazil: Amazonas, Manaus, Reserva Duche, -2.93024, -59.97397, 20.v.2012, F. Cabral	INPA
73	USNMENT01448191	*Phacecorynes zamiae* (Gyllenhal)	D.M.Anderson	*Encephalartos caffra*	Africa, Port Elizabeth, 21.v.36, EQA 36291	USNM
74	USNMENT01448169	*Polytus mellerborgii* (Boheman)	DMAnderson	banana corm	USA: HI, Honolulu, 3-27-35, O.H.Swezey	USNM
75	USNMENT01119878	*Poteriophorus uhlemanni* (Schultze)	Chamorro		Taiwan, Lanyu	USNM
75	USNMENT01070986	*Poteriophorus uhlemanni* (Schultze)	Chamorro		Taiwan, Lanyu	USNM
75	USNMENT01070985	*Poteriophorus uhlemanni* (Schultze)	Chamorro		Taiwan, Lanyu	USNM
76	USNMENT01448505	*Rhabdoscelus obscurus* (Boisduval)?	R.T. Cotton		Tahiti	USNM
77	USNMENT01448185	*Rhinostomus barbirostris* (Fabricius)	W.H. Anderson		Brazil, sent by G. Bondar, rec'd 16.x.1942	USNM
78	USNMENT01448193	*Rhodobaenus tredecimpunctatus* (Illiger)	D.M. Anderson	Stems of *Eupatotium*	Aug. 5, 1917. Hopk. U.S. Forest insect colln.	USNM
78	USNMENT01448194	*Rhodobaenus tredecimpunctatus* (Illiger)	D.M. Anderson	Stems of *Eupatotium*	Aug. 5, 1917. Hopk. U.S. Forest insect colln.	USNM
78	USNMENT01448184	*Rhynchophorus cruentatus* (Fab.)	R.T. Cotton		Leng 335	USNM
79	USNMENT01448183	*Rhynchophorus palmarum* (L.) (Prob.)	W.H. Anderson	in pineapple fruit	Mexico, Loma Bonita, Oaxaca 17.vi.'47, El Paso #49108, 47 8789	USNM
80	USNMENT01075012	*Scyphophorus acupunctatus* Gyllenhal			Mex: Magdalena, Sonora	USNM

Table 2. *Cont.*

Figure	USNMENT Code	Species	Determiner	Host	Location	Depository
80	USNMENT01075013	*Scyphophorus acupunctatus* Gyllenhal	R.T. Cotton		Arizona Leng 335	USNM
80	USNMENT01075007	*Scyphophorus acupunctatus* Gyllenhal	M.L. Chamorro		Arizona, Fort Grant, Hobbard #869	USNM
80	USNMENT01075013	*Scyphophorus acupunctatus* Gyllenhal	R.T. Cotton		Arizona Leng 335	USNM
80	USNMENT01119877	*Scyphophorus yuccae* Horn	R.T. Cotton		Arizona Leng 335	USNM
80	USNMENT01448506	*Scyphophorus yuccae* Horn	D.M. Anderson		California, Arrowhead, Apr. '93, U.S.D.A. No. 4083	USNM
81	USNMENT01070976	*Sipalinus gigas* (Fabricius)	D.M. Anderson	"Acrylic sheets"	Japan:intercepted 04/23/75, USDA 75-7265, New York, NY CBP	USNM
82	USNMENT01160334	*Sitophilus granarius* (Linnaeus)	D.M. Anderson	Valonea nuts	Greece, 23.vi.1937, NY 72028	USNM
82	USNMENT01160255	*Sitophilus linearis* (Herbst)	D.M. Anderson	Tamarind bean pods	Trinidad, 22.iii.1942, 42 3894	USNM
82	USNMENT01448500	*Sitophilus oryzae* (Linnaeus)	D.M. Anderson	In Chestnut	Italy?, 31.i.1944, 44 2571	USNM
83	USNMENT01160266	*Sparganobasis subcruciata* Marshall	R. Oberprieler	Oil palm plantation	Papua New Guinea, 8°51′14S, 147°41′43E,28.IX.2010, C.F. DEWHURST	ANIC
84, 85	USNMENT01119910	*Sphenophorus* [*Calendra*] sp.		In Tule, in roots	Louisiana, Morse, 23.viii.1943, Spec. Surv. #1581, 43 13469	USNM
84, 85	USNMENT01448181	*Sphenophorus aequalis* Gyllenhal	R.T. Cotton		Leng 336, Texas, R.T. Cotton	USNM
84, 85	USNMENT01119911	*Sphenophorus cariosa* (Olivier)	D.M. Anderson	*Cyperus exaltatus*	Texas, Pierce, Dec. 3, 1907, J.D. Mitchell	USNM
84, 85	USNMENT01448182	*Sphenophorus discolor* (Mannerheim)	R.T. Cotton		Leng 336, R.T. Cotton	USNM
84, 85	USNMENT01448179	*Sphenophorus maidis* Chittenden	D.M. Anderson		USA. South Carolina., Lee Co., Meredith, O.L. Cartwright, 1928	USNM
84, 85	USNMENT01119906	*Sphenophorus maidis* Chittenden	D.M. Anderson	in corn	USA., South Carolina, U.S.D.A. 2809, Box 13/13	USNM
84, 85	USNMENT01448180	*Sphenophorus pertinax* (Olivier)	R.T. Cotton		Leng 336, R.T. Cotton	USNM
84, 85	USNMENT01448177	*Sphenophorus pontederiae* Chittenden	D.M. Anderson	in rootstocks Pontederia cordata	Massachusetts, Stoughton, aug. 1924, D.H. Blake collr.	USNM
84, 85	USNMENT01119909	*Sphenophorus pontederiae* Chittenden	D.M. Anderson	in rootstocks Pontederia cordata	Massachusetts, Stoughton, aug. 1924, D.H. Blake collr.	USNM
84, 85	USNMENT01448178	*Sphenophorus pontederiae* Chittenden	R.T. Cotton		Leng 336	USNM
86	USNMENT01448176	*Stenommatus musae* Marshall	Anderson/Swezey	ex. Banana corm	USA, Honolulu, 27.iii.1935, O.H. Swezey	USNM

Table 2. *Cont.*

Figure	USNMENT Code	Species	Determiner	Host	Location	Depository
86	USNMENT01448175	*Stenommatus musae* Marshall	Anderson/Swezey	ex. Banana corm	USA, Honolulu, 27.iii.1935, O.H. Swezey	USNM
86	USNMENT01448174	*Stenommatus musae* Marshall	Anderson/Swezey	ex. Banana corm	USA, Honolulu, 27.iii.1935, O.H. Swezey	USNM
87	USNMENT00677022	*Temnoschoita* sp.	D.M. Anderson	Coconut palm stems	Togo, intercepted New York, NY CBP, 05/21/80 #013366	USNM
88	USNMENT01160267	*Trigonotarsus calandroides* Guérin-Méneville	R. Oberprieler	*Xanthorrhoea* stump	Australia: ACT, Uriarra Rd. 7.vii.1976, D.P. Crane	ANIC
89	USNMENT01448190	*Trochorhopalus strangulatus* Gyllenhal	D.M. Anderson	in sugar cane	Victorias, Occ. Negras, Nov. 10, 1929, Pierce coll., # CC.80	USNM
89	USNMENT01448120	*Trochorhopalus strangulatus* Gyllenhal	D.M. Anderson	in sugar cane	Victorias, Occ. Negras, Nov. 10, 1929, Pierce coll., # CC.80	USNM
89	USNMENT01448192	*Trochorhopalus strangulatus* Gyllenhal	D.M. Anderson	in sugar cane	Victorias, Occ. Negras, Nov. 10, 1929, Pierce coll., # CC.80	USNM
90	USNMENT01119895	*Tryphetus incarnatus* Gyllenhal	M.L. Chamorro	Fabaceae seed	Thailand, intercepted in Los Angeles CBP, 3.xi.2015	USNM
92	USNMENT01448501	*Yuccaborus lentiginosus* Casey	D.M. Anderson		USA, Texas, Yucca Ridge, 10 mi N.E. Brownsvile, 4.vi.1904, H.S. Barber	USNM

3. Results

The following diagnostic characters were listed by May [15] (p. 663) to distinguish Dryophthorinae larvae. Characters 1, 2, 11, and 12 are among the most salient features.

> "Larva, 1. Body shape typically expanded between Abd IV and Abd V, narrowing abruptly (seed inhabiting forms such as *Sitophilus* Schoenherr are subspherical). 2. Abd VIII/IX together forming a depressed dorsal disk. 3. Dorso- and ventropleural abdominal areas subdivided into 2, 3, or 4 superimposed lobes. 4. Anus subterminal or ventral. 5. Head usually free, longer than wide, entire behind; postoccipital condyles obsolete. 6. Labrum rounded, rarely distinctly trilobate. 7. Tormae well developed, subparallel, or convergent, often joined by a basal bridge. 8. Maxilla with dorsal malal setae usually branched; some setae of other mouthparts may also be branched. 9. Hypopharynx often densely pubescent. 10. Hyphopharyngeal bracon clear (*Sitophilus granarius* and *S. zeamais* are exceptions). 11. Spiracles ovate-fringed with an outer skin fold narrowly pigmented, or subtriangular with unpigmented skin fold; aligned vertically with airtubes dorsal. 12. Spiracles of Abd VIII positioned on dorsal disk, caudad."

Known larvae of some Molytinae (Anderson's [9] Cholini and most Hylobiini) resemble dryophhtorine larvae in that they possess almost identical abdominal spiracles [9]. Anderson [9] stressed the importance of the orientation of abdominal spiracles on segment VIII as the key feature distinguishing dryophthorines from these groups. In Dryophthorinae, the spiracular airtubes of segment VIII are directed caudad and not dorsad. The larvae of at least three molytine genera, *Sternechus* Schoenherr, *Anchonus* Schoenherr, and *Heilipus* Germar, possess abdominal spiracles with airtubes directed nearly posterad [9].

The following key presents illustrated characters (Figures 1–55) useful to distinguish known dryophthorine larvae. In addition, detailed illustrations of the mouthparts of the following 36 genera are provided at the end of the key (Figures 56–92): *Cactophagus* LeConte (Figure 56), *Cosmopolites* Chevrolat (Figure 57), *Cyrtotrachelus* Schoenherr (Figure 58), *Diathetes* Pascoe (Figure 59), *Diocalandra* Faust (Figure 60), *Dryophthoroides* Roelofs (Figure 61), *Dryophthorus* Germar (Figure 62), *Eucalandra* Faust (Figure 63), *Eugnoristus* Schoenherr (Figure 64), *Foveolus* Vaurie (Figure 65), *Mesocordylus* Lacordaire (Figure 66), *Metamasius* Horn (Figure 67), *Myocalandra* Faust (Figure 68), *Nassophasis* Waterhouse (Figure 69), *Nephius* Pascoe (Figure 70), *Odoiporus* Chevrolat (Figure 71), *Paramasius* Kuschel (currently a junio synonym of *Metamasius*) (Figure 72), *Phacecorynes* Schoenherr (Figure 73), *Polytus* Faust (Figure 74), *Poteriophorus* Schoenherr (Figure 75), *Rhabdoscelus* Marshall (Figure 76), *Rhinostomus* Rafinesque (Figure 77, only labrum/epipharynx), *Rhodobaenus* LeConte (Figure 78), *Rhynchophorus* Herbst (Figure 79), *Scyphophorus* Schoenherr (Figure 80), *Sipalinus* Marshall (Figure 81), *Sitophilus* Schoenherr (Figure 82), *Sparganobasis* Marshall (Figure 83), *Sphenophorus* Schoenherr (Figures 84 and 85), *Stenommatus* Wollaston (Figure 86), *Temnoschoita* Chevrolat (Figure 87), *Trigonotarsus* Guerin-Meneville (Figure 88), *Trochorhopalus* Kirsch (Figure 89), *Tryphetus* Faust (Figure 90), *Xerodermus* Lacordaire (Figure 91), and *Yuccaborus* LeConte (Figure 92). *Dynamis* Chevrolat is not illustred. Among characters considered important to distinguish among the tribes and genera are: the number of frontal setae of the head, the shape of the thoracic and abdominal spiracles, the shape of abdominal segments VIII and IX, the chaetotaxy of the labrum and epipharynx, and the chaetotaxy and overall shape of the mala. Photographs of live adult, larva, and pupa *in situ* in their hosts are included for *Phacecorynes*, *Sipalinus*, and *Trigonotarsus* (Figures 93 and 94).

3.1. Synoptic Key to Larval Dryophthorinae

3.1.1. Habitus

1. Body length of mature larva

a. **Small (~2–8 mm) (Figure 1)**
Diocalandra, Dryophthoroides, Dryophthorus, Eucalandra*, Myocalandra*, Nephius* (=*Anius*), *Polytus, Stenommatus, Tryphetus, Xerodermus*

b. **Medium (~8–16 mm) (Figures 2 and 3)**
Cactophagus, Cosmopolites, Diathetes, Diocalandra*, Eucalandra*, Eugnoristus, Foveolus, Mesocordylus* (probably), *Metamasius, Myocalandra*, Nassophasis, Odoiporus, Paramasius, Phacecorynes, Poteriophorus*, Rhabdoscelus, Rhodobaenus, Scyphophorus, Sparganobasis*, Sphenophorus, Temnoschoita, Trochorhopalus*

c. **Large (~16–40 mm) (Figure 4)**
Cyrtotrachelus, Diathetes, Dynamis, Poteriophorus*, Rhinostomus, Rhynchophorus, Sipalinus, Sitophilus, Sparganobasis*, Trigonotarsus, Yuccaborus*

2. Expansion of abdomen ventrolaterally from segments II, III, or IV to V or VI, abruptly narrowed posterad

a. **Present**
Cactophagus, Cosmopolites, Cyrtotrachelus, Diathetes, Diocalandra, Dryophthoroides [12], *Dryophthorus, Dynamis* [13], *Eucalandra, Eugnoristus, Foveolus, Mesocordylus* (probably), *Metamasius, Myocalandra, Nassophasis, Nephius* (=*Anius*), *Odoiporus, Paramasius, Phacecorynes, Polytus, Poteriophorus, Rhabdoscelus, Rhodobaenus, Rhynchophorus, Scyphophorus, Sipalinus, Sparganobasis, Sphenophorus, Stenommatus, Temnoschoita, Trigonotarsus, Trochorhopalus, Xerodermus* [11], *Yuccaborus*

b. **Absent**
Sitophilus, Tryphetus

Figure 1. Larva, lateral view: (**a**) *Sitophilus*; (**b**) *Stenommatus*; (**c**) *Diocalandra*; (**d**) *Tryphetus*; (**e**) *Eucalandra*; and (**f**) *Nephius*.

Figure 2. Larva, lateral view: (**a**) *Phacecorynes*; (**b**) *Sphenophorus*; (**c**) *Scyphophorus*; and (**d**) *Temnoschoita*.

Figure 3. Larva, lateral view: (**a**) *Cosmopolites*; (**b**) *Mesocordylus* (probably); (**c**) *Eugnoristus*; (**d**) *Foveolus*; (**e**) *Metamasius*; (**f**) *Nassophasis*; (**g**) *Odoiporus*; (**h**) *Paramasius*; and (**i**) *Trochorhopalus*.

Figure 4. Larvae, lateral view: (**a**) *Cyrtotrachelus*; (**b**) *Diathetes*; (**c**) *Poteriophorus*; (**d**) *Sipalinus*; (**e**) *Sparganobasis*; and (**f**) *Trigonotarsus*.

3. *Location of anus (=position of segment X) (Figure 5)*

a. **Ventral, subapical**
Cactophagus, *Cosmopolites*, *Cyrtotrachelus*, *Diathetes*, *Diocalandra*, *Dryophthoroides* [12], *Dryophthorus*, *Dynamis* [13], *Eucalandra*, *Foveolus*, *Mesocordylus* (probably), *Metamasius*, *Myocalandra*, *Nassophasis*, *Nephius* (=*Anius*), *Odoiporus*, *Paramasius*, *Phacecorynes*, *Polytus*, *Poteriophorus*, *Rhabdoscelus*, *Rhinostomus*, *Rhodobaenus*, *Rhynchophorus*, *Scyphophorus*, *Sipalinus*, *Sparganobasis*, *Sphenophorus*, *Stenommatus*, *Temnoschoita*, *Trigonotarsus*, *Trochorhopalus*, *Xerodermus* [11], *Yuccaborus*

b. **Terminal**
Eugnoristus, *Sitophilus*, *Tryphetus*

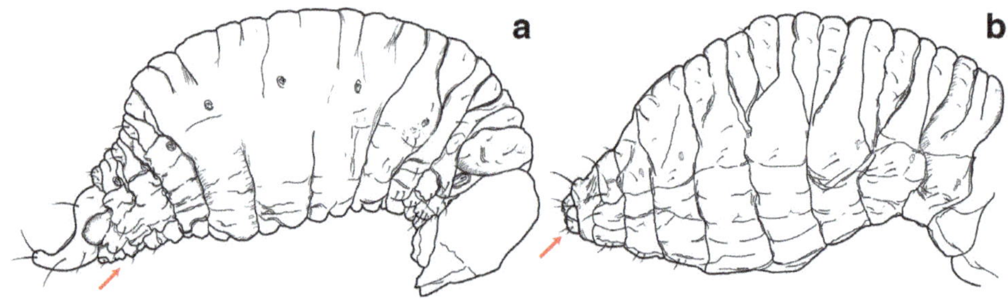

Figure 5. Thorax and abdomen, larva, lateral view: (**a**) *Phacecorynes zamiae*; and (**b**) *Eugnoristus braueri*.

3.1.2. Head

1. Number of setae on frons (fs) (Figures 6 and 7)

a. 8 (4 pairs) (*fs1* absent)
Cyrtotrachelus#

b. 10 (5 pairs)
Cactophagus, Cosmopolites, Diathetes, Diocalandra (fs1 minute), Dynamis [13], *Eucalandra, Eugnoristus, Foveolus, Mesocordylus* (probably), *Metamasius, Myocalandra, Nassophasis, Nephius* (=*Anius*), *Odoiporus, Paramasius, Phacecorynes, Polytus, Poteriophorus, Rhabdoscelus, Rhinostomus, Rhodobaenus, Rhynchophorus, Scyphophorus, Sipalinus, Sparganobasis, Sphenophorus, Sitophilus, Temnoschoita, Trigonotarsus, Trochorhopalus, Tryphetus, Yuccaborus*

c. 12 (6 pairs)
Dryophthorus, Stenommatus

Not included: *Dryophthoroides, Xerodermus*

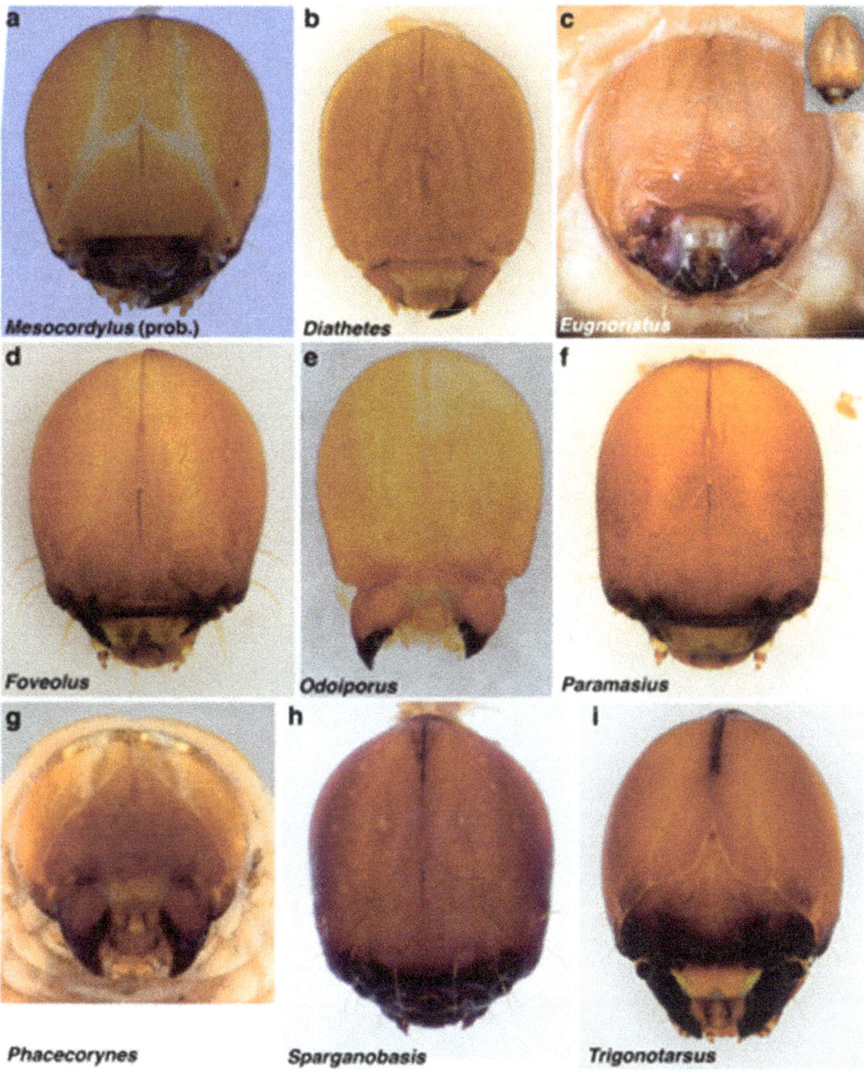

Figure 6. Head, larva, dorsal view: (**a**) *Mesocordylus*; (**b**) *Diathetes*; (**c**) *Eugnoristus*; (**d**) *Foveolus*; (**e**) *Odoiporus*; (**f**) *Paramasius*; (**g**) *Phacecorynes*; (**h**) *Sparganobasis*; and (**i**) *Trigonotarsus*.

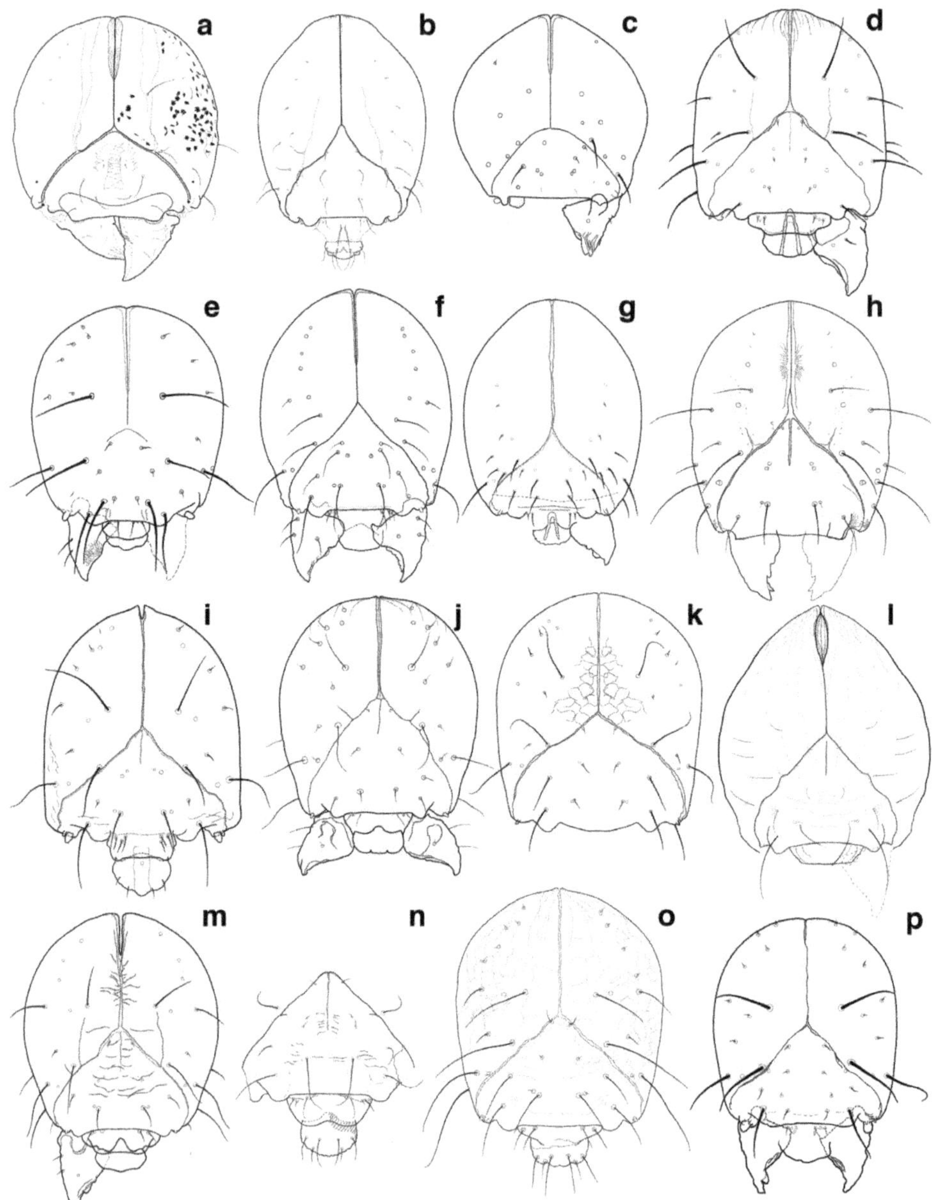

Figure 7. Head, larva, dorsal view: (**a**) *Cyrtotrachelus*; (**b**) *Eucalandra*; (**c**) *Nephius* (=*Anius*); (**d**) *Cosmopolites*; (**e**) *Dryophthorus*; (**f**) *Diocalandra*; (**g**) *Myocalandra*; (**h**) *Nassophasis*; (**i**) *Polytus*; (**j**) *Rhodobaenus*; (**k**) *Rhynchophorus*; (**l**) *Poteriophorus*; (**m**) *Sphenophorus maidis*; (**n**) *Sphenophorus pontederiae*; and (**o**) *Trochorhophalus*.

2. *Minute, clumped asperites (leopard-like) surface (Figure 8)*

a. Prominent laterally and medially
<u>*Cyrtotrachelus*</u>

b. **Absent**
Cactophagus, *Cosmopolites*, *Diathetes*, <u>*Diocalandra*</u>, *Dryophthorus*, *Eucalandra*, *Eugnoristus*, *Foveolus*, *Mesocordylus* (probably), *Metamasius*, *Myocalandra*, *Nassophasis*, *Nephius* (=*Anius*), *Odoiporus*, *Paramasius*, *Phacecorynes*, *Polytus*, *Poteriophorus*, *Rhabdoscelus*, *Rhodobaenus*, *Rhynchophorus*, *Scyphophorus*, *Sipalinus*, *Sitophilus*, *Sparganobasis*, *Sphenophorus*, *Stenommatus*, *Temnoschoita*, *Trigonotarsus*, *Trochorhopalus*, *Tryphetus*, *Yuccaborus*

Not included: *Dryophthoroides*, *Xerodermus*

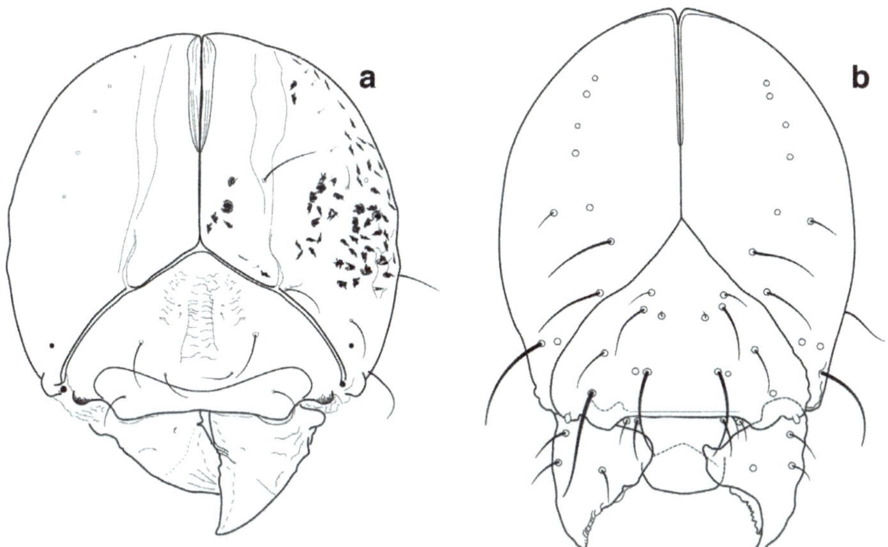

Figure 8. Head, larva, dorsal view: (**a**) *Cyrtotrachelus*; and (**b**) *Diocalandra*.

3. Sensorium of antenna (Figure 9)

a. **Apical, convex sensorium moderately sized in relation to cushion-like, basal segment**
Cactophagus, *Cosmopolites*, *Cyrtotrachelus*, *Diathetes*, *Dynamis* [13], *Eucalandra*, *Eugnoristus*, *Foveolus*, *Mesocordylus* (probably), *Metamasius*, *Nassophasis*, *Odoiporus*, *Paramasius*, *Phacecorynes*, *Polytus#**, *Poteriophorus*, *Rhabdoscelus*, *Rhodobaenus*, *Rhynchophorus*, *Scyphophorus*, *Sipalinus*, *Sparganobasis*, *Sphenophorus*, *Temnoschoita*, *Trigonotarsus*, *Trochorhopalus*, *Yuccaborus* (seemingly 2 sensoria)

b. **Conical, sensorium relatively enlarged**
Diocalandra, *Dryophthorus*, *Myocalandra#*, *Nephius* (=*Anius*), *Polytus#**, *Sitophilus*, *Stenommatus*, *Xerodermus* [11], *Tryphetus*

Not included: *Dryophthoroides* (but possibly b), *Rhinostomus*

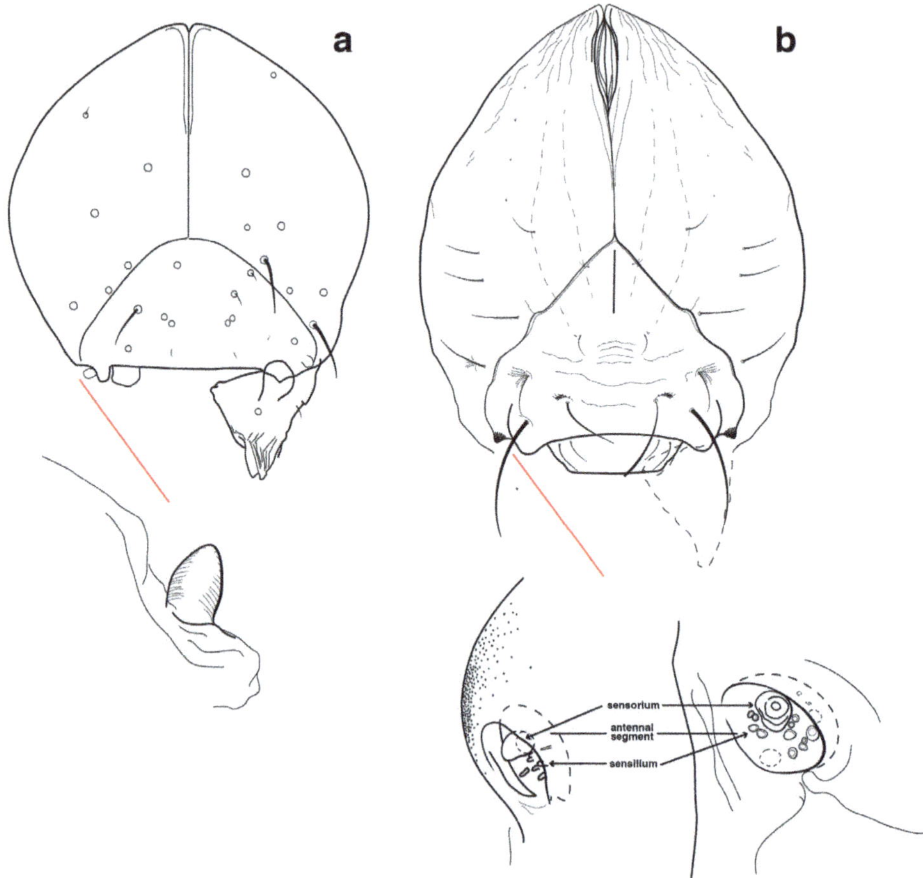

Figure 9. Head, dorsal view and detail of antenna: (**a**) *Nephius* (=*Anius*); and (**b**) *Poteriophorus*.

3.1.3. Thorax

1. Relative length of orifice and airtube of thoracic spiracle (Figure 10)

a. **Orifice 4 times longer than air tube**
*Cosmopolites, Temnoschoita, Yuccaborus#**

b. **Orifice 3–4 times longer than air tube (elongate)**
Cactophagus, Metamasius ritchei, Odoiporus, Rhabdoscelus*, Sparganobasis, Sphenophorus*, Trigonotarsus*

c. **Orifice 2.5–2 times longer than air tube length**
Diocalandra, Eugnoristus, Foveolus, Paramasius, Phacecorynes, Rhabdoscelus, Sitophilus granarius*, Sphenophorus pontederiae, Tryphetus*

d. **Orifice as long as air tube**
Diathetes, Nephius (=Anius), Nassophasis, Scyphophorus, Sitophilus granarius, Sphenophorus maidis, Sphenophorus cariosus, Xerodermus* [11]

e. **Orifice shorter than length of air tube (less than 3/4 length)**
Drophthorus, Metamasius ritchiei? (USNMENT01448165), Mesocordylus#, Myocalandra, Polytus, Poteriophorus, Rhodobaenus, Stenommatus, Trochorhopalus*

f. **Lacks air tubes**
Cyrtotrachelus, Dynamis [13], *Rhynchophorus, Sipalinus#** see [12], *Yuccaborus** (perhaps a separate state is warranted to accommodate the uniquely shaped spiracles (rounded, non-slit like orifice with small, non-scalloped airtubes) present in *Yuccaborus* and perhaps *Sipalinus* and *Cyrtotrachelus*.)

g. **Spiracles functionally absent, vestigial**
*Eucalandra#, Sipalinus#**

Not included: *Dryophthoroides, Rhinostomus*

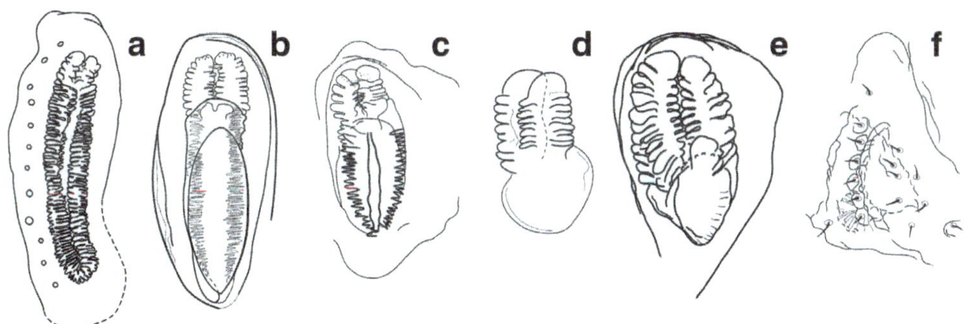

Figure 10. Thoracic spiracles, larva: (**a**) *Cosmopolites sordidus;* (**b**) *Metamasius ritchie;* (**c**) *Phacecorynes zamiae;* (**d**) *Nephius (=Anius) pauperatus;* (**e**) *Trochorhopalus strangulatus;* and (**f**) *Rhynchophorus palmarum.*

3.1.4. Abdomen

1. Functional spiracles of segments I-VII (Figure 11)

a. **Prominent and all sub-equal in size**
Cactophagus, Diathetes, Diocalandra, Eucalandra, Eugnoristus, Foveolus, Mesocordylus (probably), *Metamasius, Nassophasis, Odoiporus, Paramasius, Poteriophorus, Rhabdoscelus, Phacecorynes, Rhodobaenus, Scyphophorus, Sitophilus, Sparganobasis, Sphenophorus, Temnoschoita, Trigonotarsus, Tryphetus*

b. **Very small**
Cosmopolites, Myocalandra*, Xerodermus** [11], *Sipalinus* (Gardner [12] states these are present on all segments, however, spiracles are usually very difficult to observe and may be misinterpreted as "e".) # [12]

c. **Present only on VII**
Dryophthoroides [12], *Nephius (=Anius)*

d. **Present only on I and VII**
Stenommatus

e. **Present, but II, III, and IV reduced in size by about half**
Trochorhopalus

f. **None discernable, absent**
Cyrtotrachelus, Cosmopolites, Dryophthorus, Polytus, Dynamis, Myocalandra** (may be interpreted as being absent), *Rhynchophorus, Xerodermus* [12], *Yuccaborus*

Not included: *Rhinostomus*

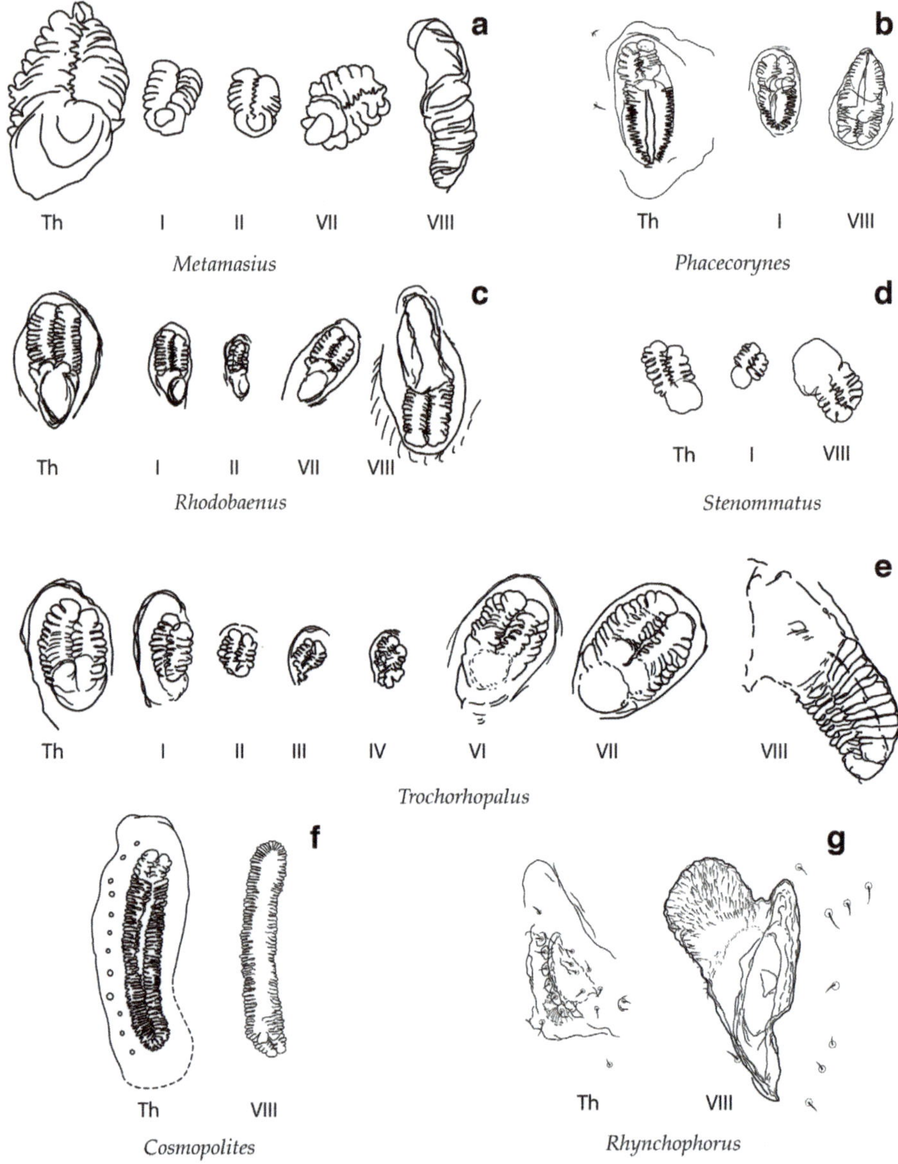

Figure 11. Thoracic and abdominal spiracles, larva: (**a**) *Metamasius*; (**b**) *Phacecorynes*; (**c**) *Rhodobaenus*; (**d**) *Stenommatus*; (**e**) *Trochorhopalus*; (**f**) *Cosmopolites*; and (**g**) *Rhynchophorus*.

2. *Abdominal segment VIII (Figure 12)*

a. With a pair of digitate projections

Dryophthoroides [12], *Dryophthorus* (reduced), *Mesocordylus* (probably) (reduced), *Nephius (=Anius)*, *Sipalinus*, *Xerodermus* [11,12], *Yuccaborus*

b. Without a pair of digitate projections

Cactophagus, *Cosmopolites*, *Cyrtotrachelus*, *Diathetes*, *Diocalandra*, *Dynamis* [13], *Eucalandra*,

Eugnoristus, Foveolus, Metamasius, Myocalandra, Nassophasis, Odoiporus, Paramasius, Phacecorynes, Polytus, Poteriophorus, Rhabdoscelus, Rhodobaenus, Rhynchophorus, Scyphophorus, Sitophilus, Sparganobasis, Sphenophorus, Stenommatus, Temnoschoita, Trigonotarsus, Trochorhopalus, Tryphetus

Not included: *Rhinostomus*

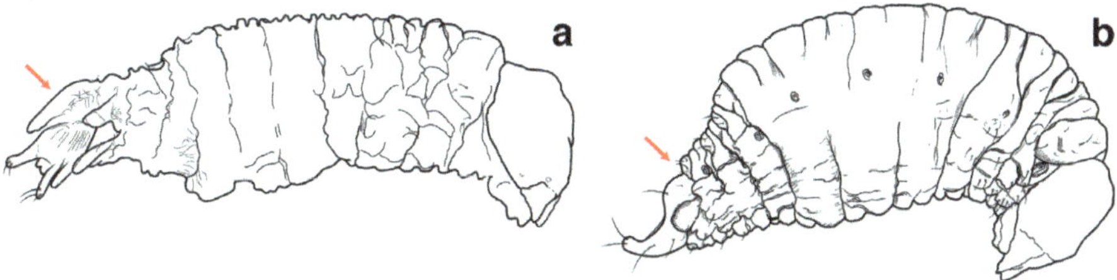

Figure 12. Larvae, lateral view: (**a**) *Nephius* (=*Anius*); (**b**) *Phacecorynes*.

3. Posterior margin of abdominal segment IX (Figure 13)

a. **Without a pair of projections, may be broad and truncate**
Cactophagus, Cosmopolites, Cyrtotrachelus, Diocalandra, Eucalandra, Eugnoristus, Foveolus, Metamasius, Myocalandra, Nassophasis, Odoiporus, Paramasius, Polytus, Poteriophorus, Rhabdoscelus, Rhynchophorus, Sitophilus, Sparganobasis, Sphenophorus, Temnoschoita, Trochorhopalus, Tryphetus

b. **With 1 pair of digitate projections**
Diathetes, Dryophthorus, Dryophthoroides [12], *Phacecorynes, Rhodobaenus, Scyphophorus, Stenommatus, Trigonotarsus, Xerodermus* [11,12], *Yuccaborus*

c. **With 2 pairs of digitate projections**
Mesocordylus (probably), *Nephius (=Anius), Sipalinus*

Not included: *Dynamis, Rhinostomus*

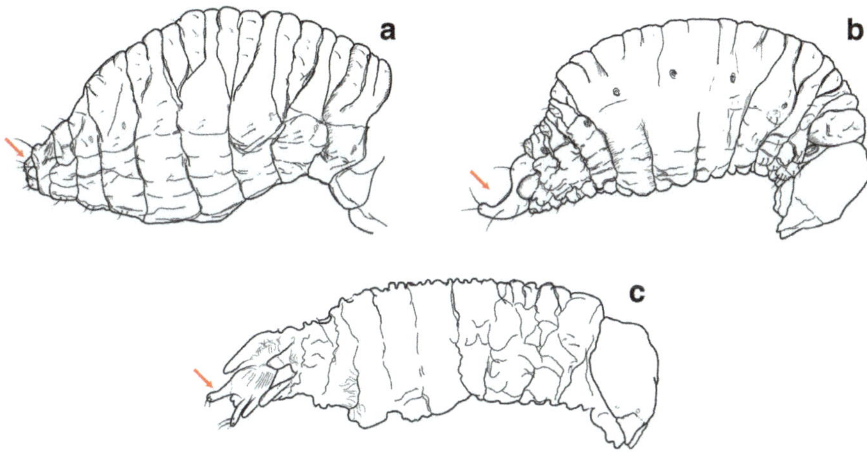

Figure 13. Larvae, lateral view: (**a**) *Eugnoristus* (abdomen only); (**b**) *Phacecorynes*; and (**c**) *Nephius* (=*Anius*).

4. Shape of posterior abdominal digitate projections of segment IX (Figure 14)

a. **Not applicable**
Cactophagus, *Cosmopolites*, *Cyrtotrachelus*, *Diathetes** (may be misinterpreted as this), *Diocalandra*, *Eucalandra*, *Eugnoristus*, *Foveolus*, *Metamasius*, *Myocalandra*, *Nassophasis*, *Odoiporus*, *Paramasius*, *Polytus*, *Poteriophorus*, *Rhabdoscelus*, *Rhynchophorus*, *Sitophilus*, *Sparganobasis*, *Sphenophorus*, *Temnoschoita*, *Trochorhopalus*, *Tryphetus*

b. **Short and bluntly conical**
*Diathetes**, *Rhodobaenus*, *Trigonotarsus*

c. **1.5 to 2× longer than wide (long and broad; reference widest section)**
Dryophthorus, *Dryophthoroides* [12], *Mesocordylus* (probably), *Nephius* (=*Anius*), *Phacecorynes*, *Scyphophorus*, *Sipalinus*, *Stenommatus*, *Xerodermus* [11,12], *Yuccaborus*

Not included: *Dynamis*, *Rhinostomus*

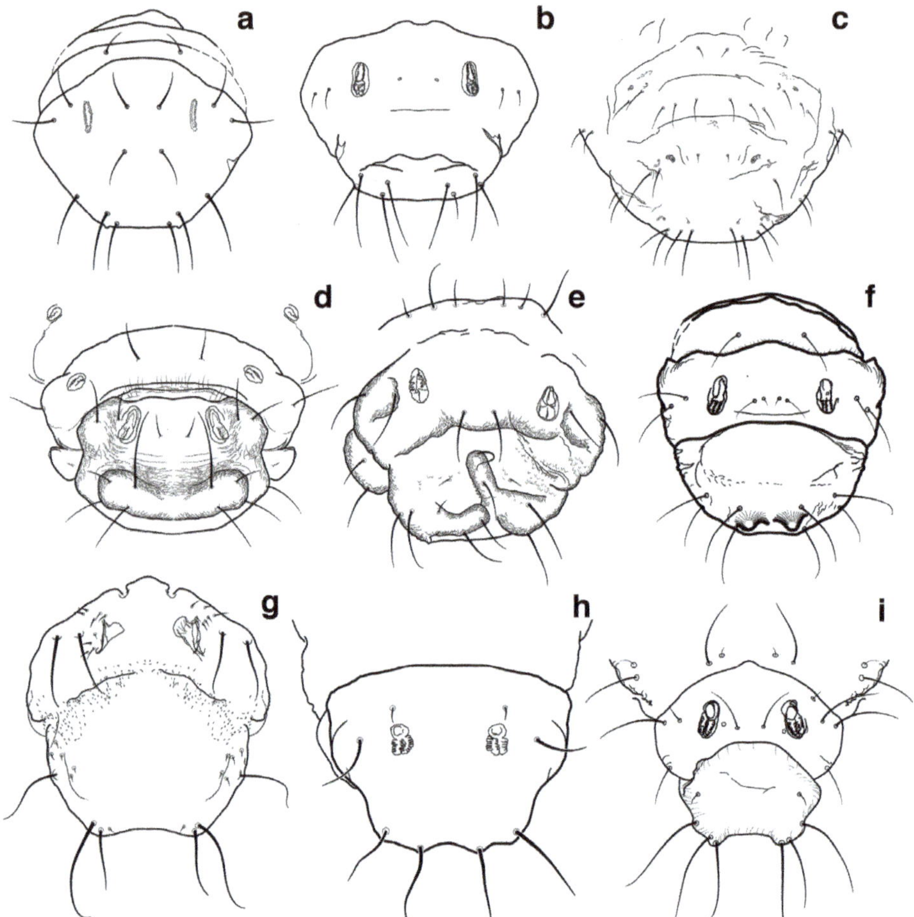

Figure 14. Segments VIII and IX, larva, caudal view: (**a**) *Cosmopolites*; (**b**) *Metamasius*; (**c**) *Myocalandra*; (**d**) *Nassophasis*; (**e**) *Phacecorynes*; (**f**) *Rhodobaenus*; (**g**) *Rhynchophorus*; (**h**) *Stenommatus*; and (**i**) *Trochorhopalus*.

3.1.5. Labrum/Eipharynx

1. *Shape of labral seta 2 (lms2) (Figure 15)*

a. **Entire**
Cactophagus, Cyrtotrachelus, Diathetes, Diocalandra, Dryophthorus, Eugnoristus, Mesocordylus (probably), Metamasius, Myocalandra, Nassophasis, Nephius (=Anius), Paramasius Phacecorynes, Polytus, Poteriophorus, Rhabdoscelus, Rhodobaenus, Rhynchophorus, Scyphophorus, Sipalinus, Sitophilus, Sphenophorus, Stenommatus, Temnoschoita*, Trigonotarsus Trochorhopalus, Tryphetus, Yuccaborus*

a. **Branched (at least 1)**
Cosmopolites, Eucalandra, Foveolus, Metamasius hemipterus, Odoiporus, Temnoschoita**

Not included: *Dryophthoroides, Dynamis, Rhinostomus, Sparganobasis* (broken), *Xerodermus*

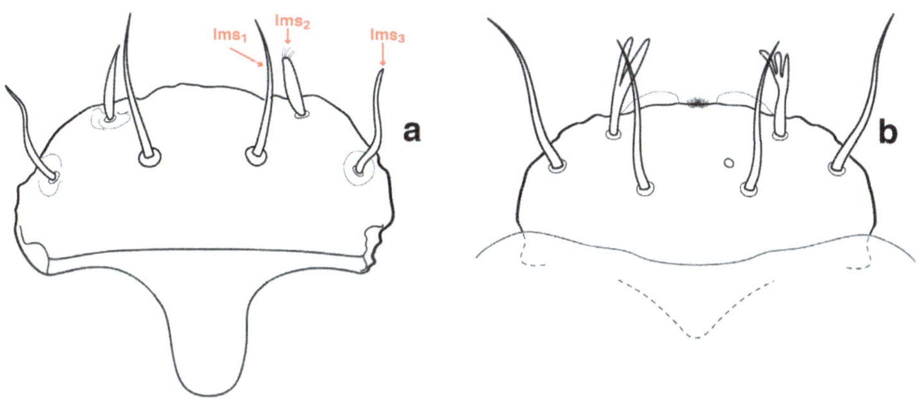

Figure 15. Labrum, larva: (**a**) *Poteriophorus* sp.; and (**b**) *Metamasius hemipterus; lms1–3*: labral seta.

2. *Peg-like pore basally on labrum (Figure 16)*

a. **Absent**
Cosmopolites, Dryophthorus, Nassophasis#, Sitophilus, Sphenophorus#*, Stenommatus*

b/c. **Single medially/submedially**
Cactophagus (Right), *Diocalandra* (Medially), *Eucalandra* (Left)#*, *Eugnoristus* (L), *Foveolus* (L), *Mesocordylus* (probably) (M), *Metamasius* (M), *Myocalandra* (M), *Nephius* (=*Anius*)#, *Paramasius* (R/L), *Phacecorynes* (R), *Polytus* (M), *Poteriophorus* (R), *Rhabdoscelus* (R), *Rhodobaenus* (R), *Sitophilus oryzae**(Medial/L), *Sparganobasis* (L), *Temnoschoita* (R), *Trochorhopalus* (M, slight R), *Tryphetus* (M)

d. **Two submedially**
Cyrtotrachelus, Diathetes, Eucalandra#, Odoiporus, Rhynchophorus, Scyphophorus, Sipalinus, Sphenophorus* maidis, Trigonotarsus, Yuccaborus*

Not included: *Dryophthoroides, Xerodermus*

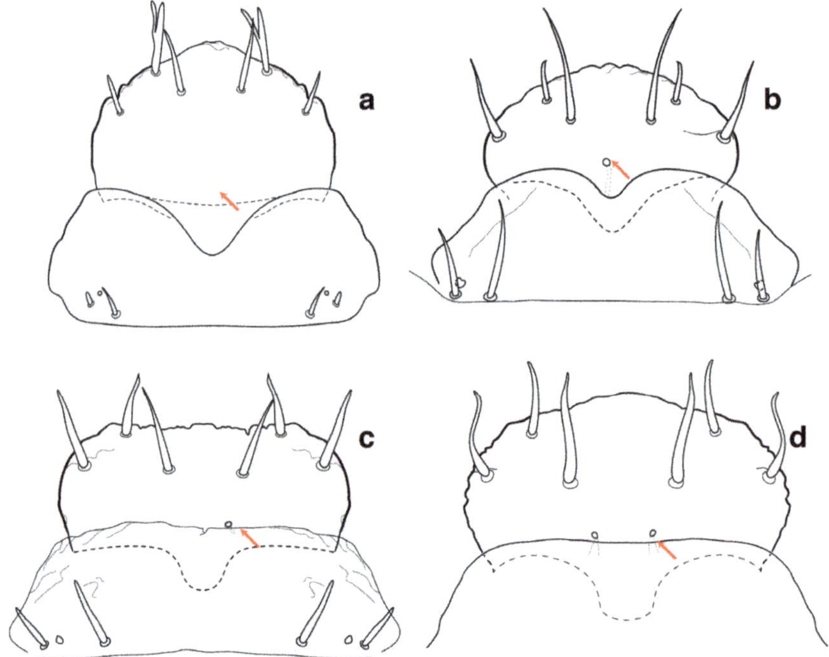

Figure 16. Labrum, larva: (**a**) *Cosmopolites;* (**b**) *Myocalandra;* (**c**) *Rhodobaenus;* and (**d**) *Sphenophorus maidis*.

3. *Shape of epipharynx (Figure 17)*

a. **Sub-rectangular, margin evenly rounded or moderately to weakly produced medially**
Cactophagus, Cyrtotrachelus, Diathetes, Diocalandra, Dryophthoroides [12], *Dryophthorus, Dynamis* [13], *Eucalandra, Foveolus, Mesocordylus* (probably), *Myocalandra, Nassophasis, Nephius* (=*Anius*), *Odoiporus, Paramasius, Phacecorynes, Poteriophorus, Rhabdoscelus#*, Rhinostomus, Rhynchophorus, Scyphophorus, Sipalinus, Sitophilus, Sparganobasis, Sphenophorus, Stenommatus, Trigonotarsus* (medially emarginated), *Trochorhopalus, Tryphetus, Xerodermus** ([11], p. 41) "Labrum strongly transverse, anterior margin trilobed, posterior margin extended medially."), *Yuccaborus*

b. **Sub-rectangular, margin truncate**
Eugnoristus, Metamasius, Rhabdoscelus#, Rhodobaenus, Temnoschoita*

c. **Sub-quadrate, markedly produced medially**
Cosmopolites, Polytus

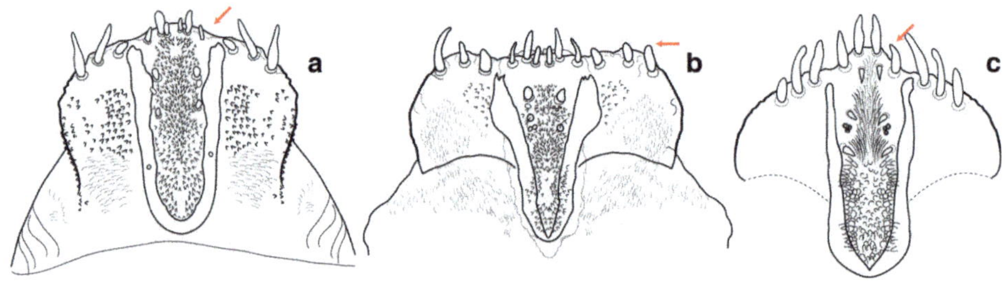

Figure 17. Epipharynx, larva: (**a**) *Phacecorynes;* (**b**) *Rhodobaenus;* and (**c**) *Polytus*.

4. *Lateral asperites/setae on epipharynx (Figure 18)*

a. **Absent**
Diocalandra, Dryophthoroides [12], *Dryophthorus, Myocalandra, Polytus, Sitophilus, Stenommatus, Tryphetus*

b. **Present**
Cactophagus, Cosmopolites, Cyrtotrachelus (present basolaterally), *Diathetes, Eucalandra, Eugnoristus, Foveolus, Mesocordylus* (probably) (present basolaterally), *Metamasius, Nassophasis, Nephius (=Anius), Odoiporus, Paramasius, Phacecorynes, Poteriophorus, Rhabdoscelus, Rhinostomus, Rhodobaenus, Rhynchophorus, Scyphophorus, Sipalinus, Sparganobasis, Sphenophorus, Temnoschoita, Trigonotarsus, Trochorhopalus, Yuccaborus*

Not included: *Dynamis, Xerodermus*

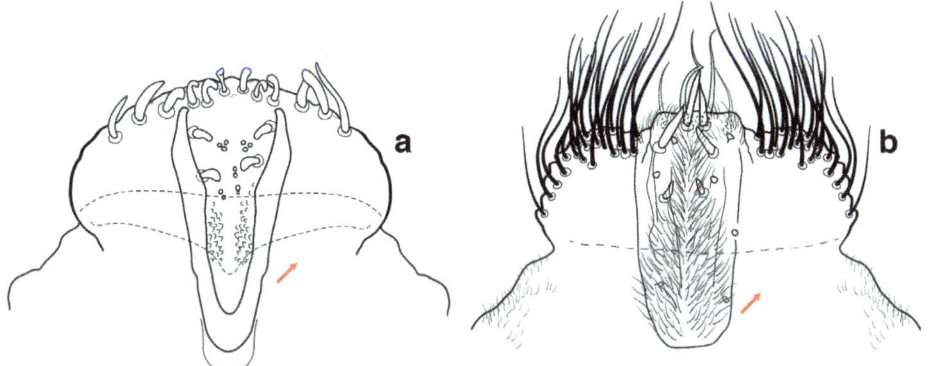

Figure 18. Epipharynx, larva: (**a**) *Myocalandra;* and (**b**) *Cyrtotrachelus.*

5. *Modification of lateral setae on epipharynx (Figure 19)*

a. **Setae absent**
Diocalandra, Dryophthoroides [12], *Dryophthorus, Myocalandra, Polytus, Sitophilus, Stenommatus, Tryphetus*

b. **Not modified (straight, tapering)**
*Cosmopolites**, *Cyrtotrachelus, Diathetes, Eucalandra, Mesocordylus* (probably), *Metamasius, Rhinostomus, Sipalinus, Sphenophorus**, *Trigonotarsus, Yuccaborus** (on one side)

c. **Twisted**
Nephius (=Anius)

d. **Curved mesally**
*Foveolus** (anteriorly), *Paramasius, Trochorhopalus* (with rounded asperites medially)

e. **Asperites throughout**
Eugnoristus, Poteriophorus, Rhabdoscelus

f. **Asprites distally or medially, setose proximally (and laterally)**
*Cactophagus, Cosmopolites**, *Foveolus**, *Nassophasis, Odoiporus, Phacecorynes, Rhodobaenus, Rhynchophorus, Scyphophorus, Sparganobasis, Sphenophorus**, *Temnoschoita, Yuccaborus**

Not included: *Dynamis, Xerodermus*

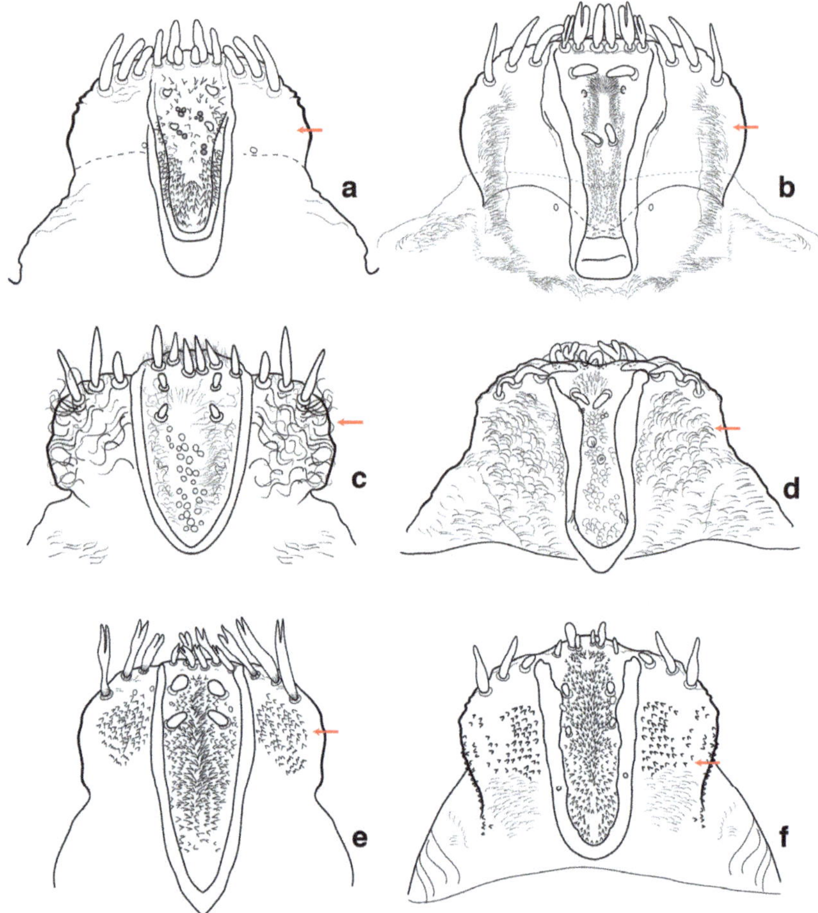

Figure 19. Epipharynx, larva: (**a**) *Diocalandra;* (**b**) *Metamasius;* (**c**) *Nephius* (=*Anius*); (**d**) *Trochorhopalus;* (**e**) *Eugnoristus;* and (**f**) *Phacecorynes.*

6. Distribution of lateral microsetae/asperites of epipharynx (Figure 20)

a. **Setae absent**
Diocalandra, Dryophthoroides [12], *Dryophthorus, Myocalandra, Polytus, Sitophilus, Stenommatus, Tryphetus*

b. **1/3–1/4 distal length of labral rods**
Eucalandra, Eugnoristus, Rhabdoscelus

c/d. **Variously patterned, entire length or between 1/2–2/3 distal length of labral rods**
Cactophagus, Cosmopolites, Foveolus, Metamasius, Nassophasis, Nephius (=*Anius*), *Odoiporus, Paramasius, Phacecorynes, Poteriophorus, Rhinostomus, Rhodobaenus, Rhynchophorus, Sparganobasis, Sphenophorus, Temnoschoita, Trochorhopalus*

e. **Throughout area 2/3 or less proximal length of labral rods**
Cyrtotrachelus, Diathetes, Mesocordylus (probably), *Scyphophorus, Sipalinus, Trigonotarsus#, Yuccaborus*

Not included: *Dynamis, Xerodermus*

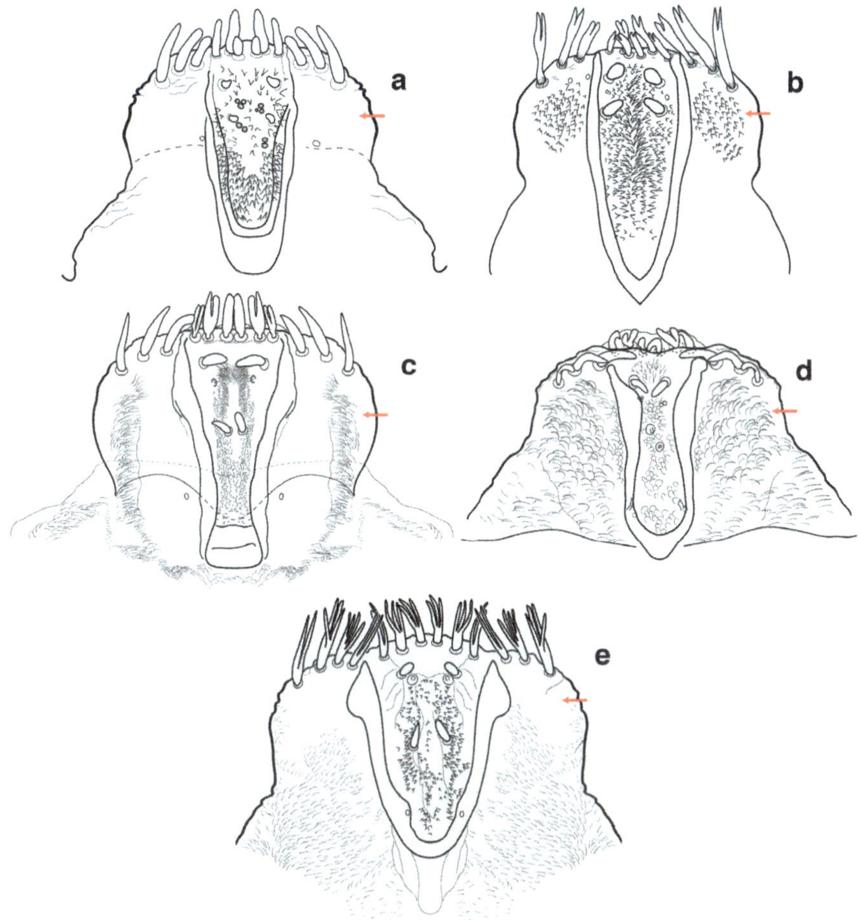

Figure 20. Epipharynx, larva: (**a**) *Diocalandra;* (**b**) *Eugnoristus;* (**c**) *Metamasius;* (**d**) *Trochorhopalus;* and (**e**) *Sipalinus.*

*7. Lateral microsetae/asperites of epipharynx (***Rhynchophorus** *and* **Cyrtotrachelus** *have large setae on the lateral margin) (Figure 21)*

a. **Setae absent**
Diocalandra, Dryophthoroides [12], *Dryophthorus, Myocalandra, Polytus, Sitophilus, Stenommatus, Temnoschoita, Tryphetus*

b. **Not reaching lateral margin of labrum**
Cactophagus, Cyrtotrachelus, Eucalandra, Eugnoristus, Foveolus, Mesocordylus (probably), *Metamasius, Paramasius, Poteriophorus, Rhabdoscelus Rhodobaenus, Rhynchophorus, Sphenophorus, Trochorhopalus, Yuccaborus*

c. **Reaching lateral margin of labrum**
Cosmopolites, Diathetes, Nassophasis, Nephius (=*Anius*), *Odoiporus, Phacecorynes, Rhinostomus, Scyphophorus, Sipalinus* (minimally), *Sparganobasis, Trigonotarsus*

Not included: *Dynamis, Xerodermus*

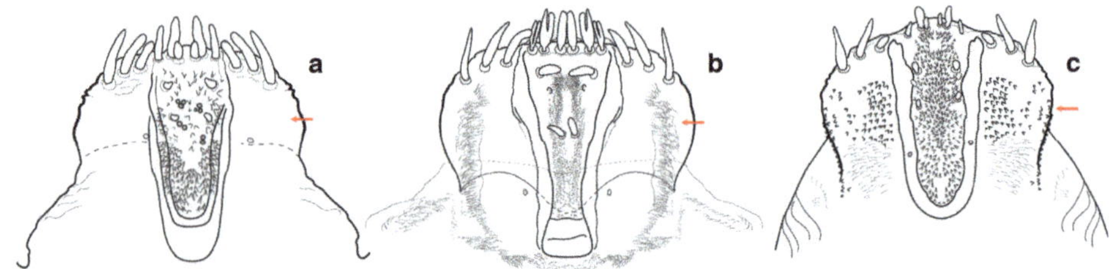

Figure 21. Epipharynx, larva: (**a**) *Diocalandra*; (**b**) *Metamasius*; and (**c**) *Phacecorynes*.

8. Mesal asperites on epipharynx (Figure 22)

a. **Absent**
Sitophilus, *Tryphetus*

b. **Present over labral rods**
Eucalandra

c. **Present medially**
Cactophagus, *Cosmopolites*, *Cyrtotrachelus*, *Diathetes*, *Diocalandra*, *Dryophthoroides* [12], *Dryophthorus*, *Eugnoristus*, *Foveolus*, *Mesocordylus* (probably), *Metamasius*, *Myocalandra*, *Nassophasis*, *Nephius* (=*Anius*), *Odoiporus*, *Paramasius*, *Phacecorynes*, *Polytus*, *Poteriophorus*, *Rhabdoscelus*, *Rhinostomus*, *Rhodobaenus*, *Rhynchophorus*, *Scyphophorus*, *Sipalinus*, *Sparganobasis*, *Sphenophorus*, *Stenommatus*, *Temnoschoita*, *Trigonotarsus*, *Trochorhopalus*, *Yuccaborus*

Not included: *Dynamis*, *Xerodermus*

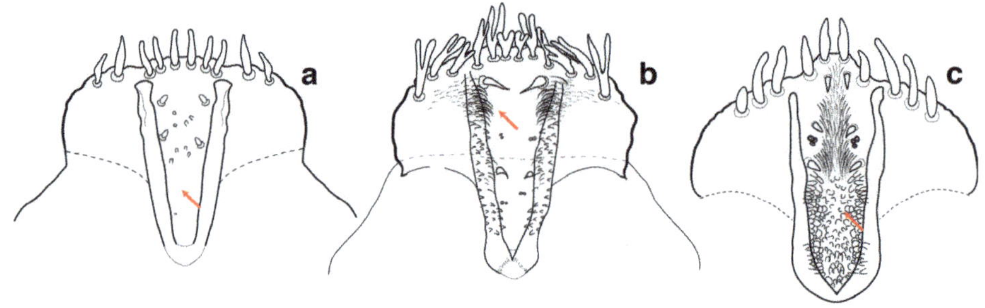

Figure 22. Epipharynx, larva: (**a**) *Sitophilus*; (**b**) *Eucalandra*; and (**c**) *Polytus*.

9. Modification of mesal asperites of epipharynx (Figure 23)

a. **Asperites absent**
Sitophilus, *Tryphetus*

b. **Setae only**
*Cosmopolites**, *Cyrtotrachelus*, *Sphenophorus*

c. **Asperites only**
Cactophagus, *Cosmopolites**, *Diathetes*, *Diocalandra*, *Dryophthorus*, *Eugnoristus*, *Mesocordylus* (probably), *Myocalandra*, *Odoiporus*, *Paramasius*, *Phacecorynes*, *Poteriophorus*, *Rhinostomus*, *Rhodobaenus*, *Sipalinus*, *Sparganobasis*, *Stenommatus**, *Trochorhopalus* (possibly a few setae)

d. **Setae and asperites**
Eucalandra, *Foveolus*, *Metamasius*, *Nassophasis*, *Nephius* (=*Anius*) (rounded asperites), *Polytus*, *Rhabdoscelus*, *Rhynchophorus*, *Scyphophorus*, *Stenommatus**, *Temnoschoita*, *Trigonotarsus*, *Yuccaborus*

Not included: *Dynamis, Dryophthoroides, Xerodermus* (([11], p. 41) " . . . between the posteriorly conjoined rods is a pair of short setae and numerous skin points.") (maybe "a")

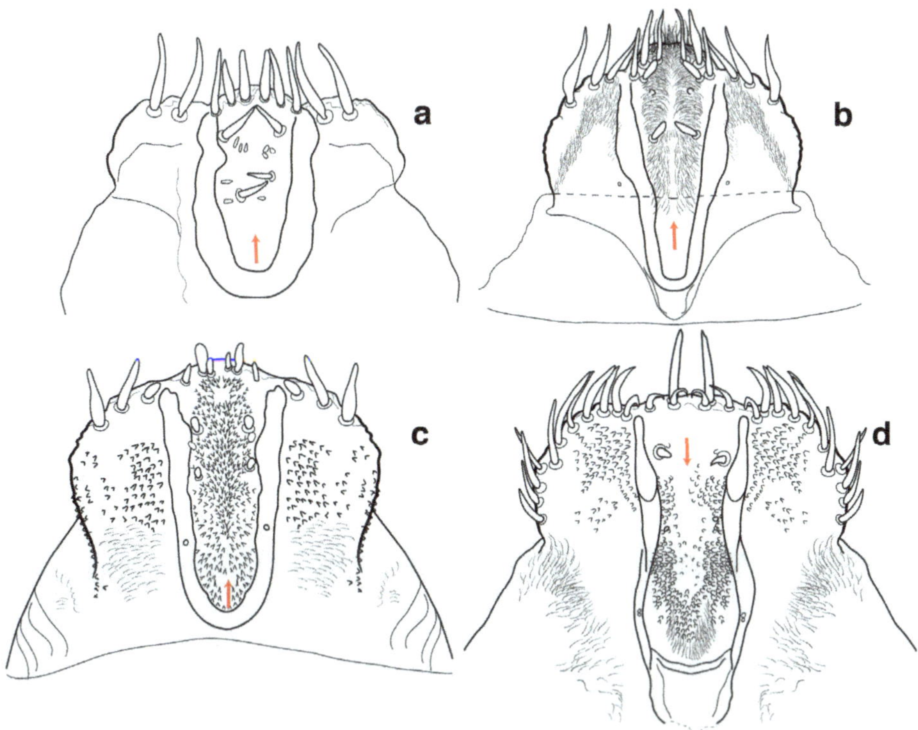

Figure 23. Epipharynx, larva: (**a**) *Tryphetus*; (**b**) *Sphenophorus*; (**c**) *Phacecorynes*; and (**d**) *Rhynchophorus*.

10. Orientation/placement of mesal asperites of epipharynx (Figure 24)

a. **Asperites absent or on labral rods**
Eucalandra, *Sitophilus, Tryphetus*

b. **Forming a well defined elongate, mesal glabrous area**
*Foveolus**, *Metamasius*, *Nassophasis, Rhinostomus, Rhodobaenus, Sphenophorus**

c. **Forming a well defined subrectangular, mesal glabrous area**
*Foveolus** (could be interpreted as "b"), *Paramasius*, *Temnoschoita*

d. **Glabrous mesal area, but loosely defined by asperites**
Cyrtotrachelus, Diocalandra, Myocalandra, Nephius (=*Anius*), *Phacecorynes, Rhynchophorus, Sipalinus, Sphenophorus**, *Stenommatus*

e. **Without mesal glabrous area (between asperites)**
Cactophagus, Cosmopolites, Diathetes, Dryophthoroides [12], *Dryophthorus, Eugnoristus, Mesocordylus* (probably), *Odoiporus, Polytus, Poteriophorus, Rhabdoscelus, Scyphophorus, Sparganobasis, Trigonotarsus*, *Trochorhopalus*, *Yuccaborus*

Not included: *Dynamis, Xerodermus*

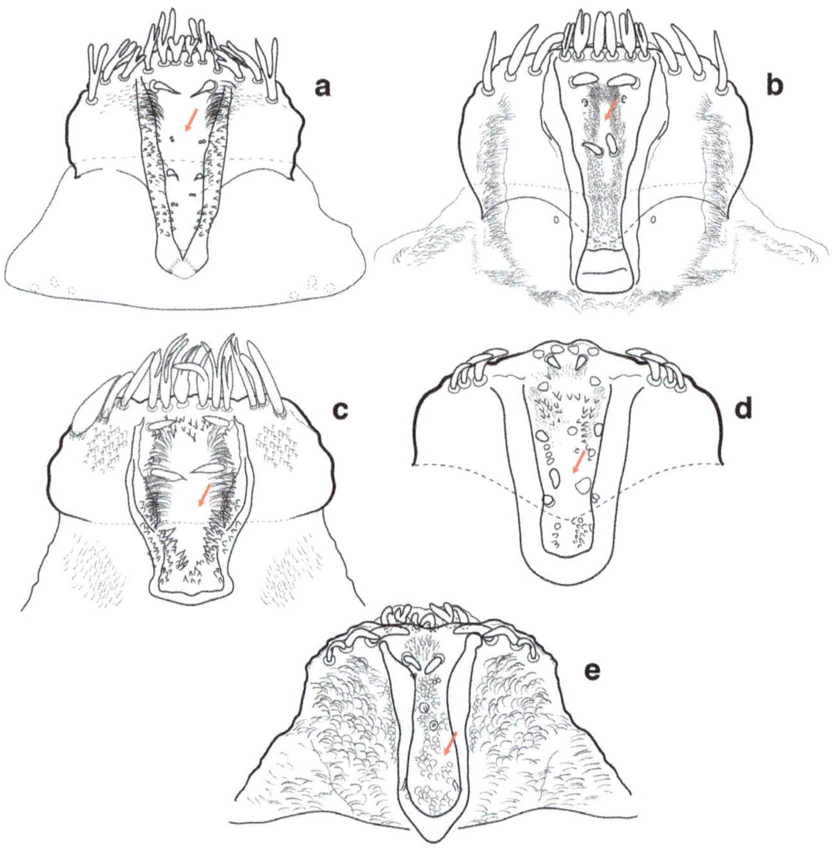

Figure 24. Epipharynx, larva: (**a**) *Eucalandra*; (**b**) *Metamasius*; (**c**) *Temnoschoita*; (**d**) *Stenommatus*; and (**e**) *Trochorhopalus*.

11. Number of anterolateral epipharyngeal setae (als) (Figure 25)

a. **2 on each side**
Sitophilus, *Tryphetus*

b. **3 on each side**
Cactophagus, *Cosmopolites*, *Diathetes*, *Diocalandra*, *Dryophthoroides* [12], *Dryophthorus*, *Eucalandra*, *Eugnoristus*, *F7veolus*, *Mesocordylus* (probably), *Metamasius*, *Myocalandra*, *Nassophasis*, *Nephius* (=*Anius*), *Odoiporus*, *Paramasius*, *Phacecorynes*, *Polytus*, *Poteriophorus*, *Rhabdoscelus*, *Rhodobaenus*, *Scyphophorus*, *Sipalinus** (on one side, but probably unusual), *Sparganobasis*, *Sphenophorus*, *Stenommatus*, *Temnoschoita*, *Trigonotarsus*, *Trochorhopalus*, *Xerodermus* [11]

c. **4 on each side**
*Sipalinus**, *Yuccaborus**

d. **5 on each side**
*Yuccaborus**

e. **10–11 on each side**
Dynamis [13], *Rhinostomus* (10), *Rhynchophorus* (11)

f. **At least 14 on each side**
Cyrtotrachelus

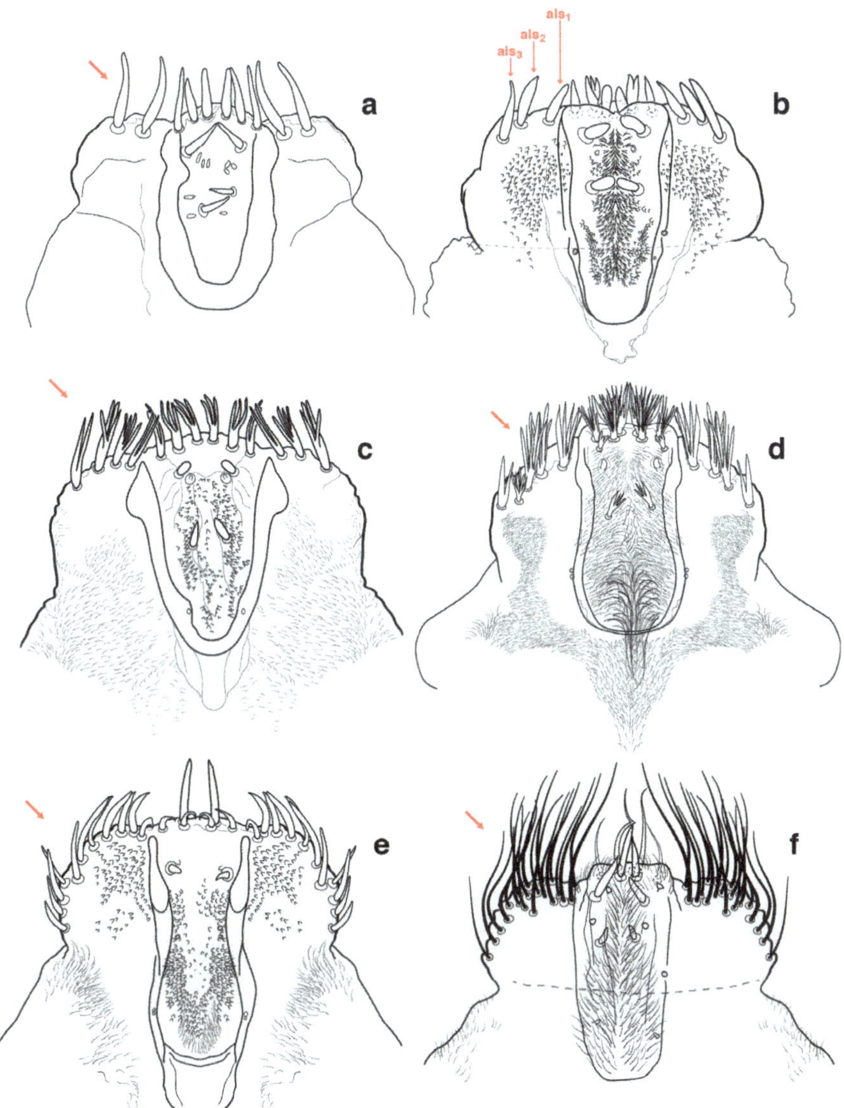

Figure 25. Epipharynx, larva: (**a**) *Tryphetus*; (**b**) *Poteriophorus*; (**c**) *Sipalinus*; (**d**) *Yuccaborus*; (**e**) *Rhynchophorus*; and (**f**) *Cyrtotrachelus*. *Als*1–3: anterolateral setae of epipharynx.

12. Shape of anterolateral epipharyngeal setae (als 1, 2, 3) (Figure 26)

a. **Tapering to apex**
Cactophagus, *Cyrtotrachelus*, *Diathetes*, *Diocalandra*, *Dryophthoroides* [12], *Dryophthorus*, *Mesocordylus* (probably), *Metamasius*, *Myocalandra*, *Nassophasis*, *Nephius* (=*Anius*), *Paramasius*, *Phacecorynes*,

Polytus, Poteriophorus, Rhabdoscelus, Rhodobaenus, Rhynchophorus, Scyphophorus (sometimes subapically expanded before tapering), Sitophilus, *Sphenophorus*, Stenommatus, Temnoschoita, Trigonotarsus, Trochorhopalus, Tryphetus, Xerodermus* [11]

b. **Bifurcate**
Cosmopolites, Eucalandra, Eugnoristus, Foveolus (*als1* bifurcate, *als2* trifurcate), *Odoiporus, Sphenophorus**

c. **Deeply trifurcate, multifurcate (tufted)**
Rhinostomus, Sipalinus, Yuccaborus

Not included: *Dynamis, Sparganobasis* (broken/ equally weathered?)

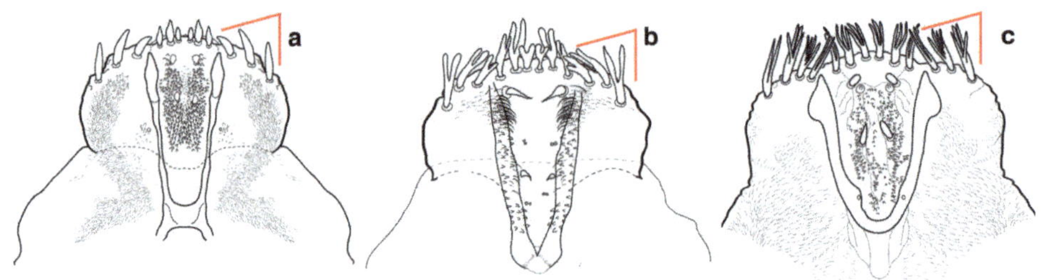

Figure 26. Epipharynx, larva: (**a**) *Cactophagus;* (**b**) *Eucalandra;* and (**c**) *Sipalnus.*

13. Shape of anteromedian epipharyngeal seta 1 (ams1) (Figure 27)

a. **Entire**
Cactophagus, Cyrtotrachelus, Diathetes, Diocalandra, Dryophthoroides [12], *Dryophthorus, Mesocordylus* (probably), *Myocalandra, Nephius* (=*Anius*), *Phacecorynes, Polytus, Rhodobaenus, Rhynchophorus, Sitophilus, Sphenophorus*, Stenommatus, Trigonotarsus#, Tryphetus, Xerodermus* [11]

a. **Bifurcate**
Cosmopolites, Eucalandra, Eugnoristus, Metamasius, Nassophasis, Paramasius, Poteriophorus*, Rhabdoscelus, Scyphophorus*, Sphenophorus*, Temnoschoita, Trochorhopalus*

c. **Trifurcate (largely)**
Cosmopolites, Foveolus** (could be interpreted as "c"), *Odoiporus, Poteriophorus*, Scyphophorus*, Sipalinus*

d. **Multifurcate, tufted**
Foveolus, Rhinostomus, Yuccaborus*

Not included: *Dynamis, Sparganobasis* (broken/weathered all equally?)

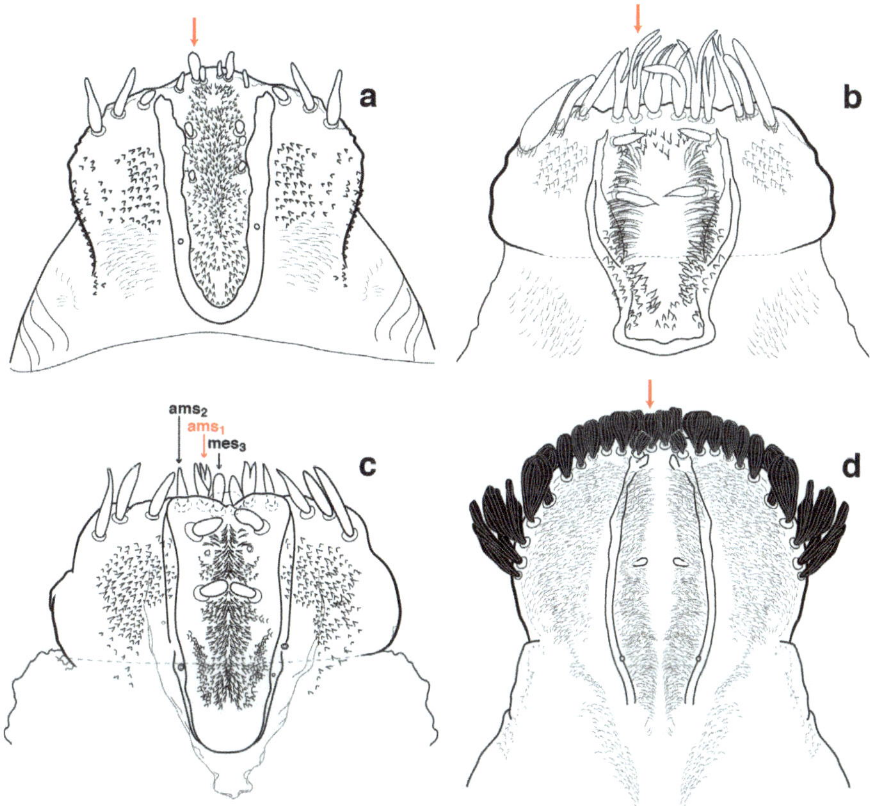

Figure 27. Epipharynx, larva: (**a**) *Phacecorynes*; (**b**) *Temnoschoita*; (**c**) *Poteriophorus*; and (**d**) *Rhinostomus*. *Mes3:* median epipharyngeal setae; *ams1, 2:* anteromedian setae of epipharynx.

14. Shape of anteromedian epipharyngeal seta 2 (ams2) (Figure 28)

a. **Entire**
Cactophagus, *Cosmopolites**, *Cyrtotrachelus*, *Diathetes*, *Diocalandra*, *Dryophthoroides* [12], *Dryophthorus*, *Mesocordylus* (probably), *Myocalandra*, *Nassophasis*, *Nephius* (=*Anius*), *Paramasius*, *Phacecorynes*, *Polytus*, <u>*Poteriophorus*</u>, *Rhabdoscelus*, *Rhodobaenus*, *Rhynchophorus*, *Scyphophorus*, *Sitophilus*, *Sphenophorus*, *Stenommatus*, *Trigonotarsus* (but could be broken), *Trochorhopalus*, *Tryphetus*, *Xerodermus* [11]

b. **Bifurcate**
*Cosmopolites**, *Eucalandra*, *Eugnoristus*, *Foveolus*, <u>*Metamasius*</u>, *Odoiporus*, *Sipalinus*, *Temnoschoita*

c. **Tufted (deeply furcate)**
Rhinostomus, <u>*Yuccaborus*</u>

Not included: *Dynamis*, *Sparganobasis* (broken/weathered all equally?)

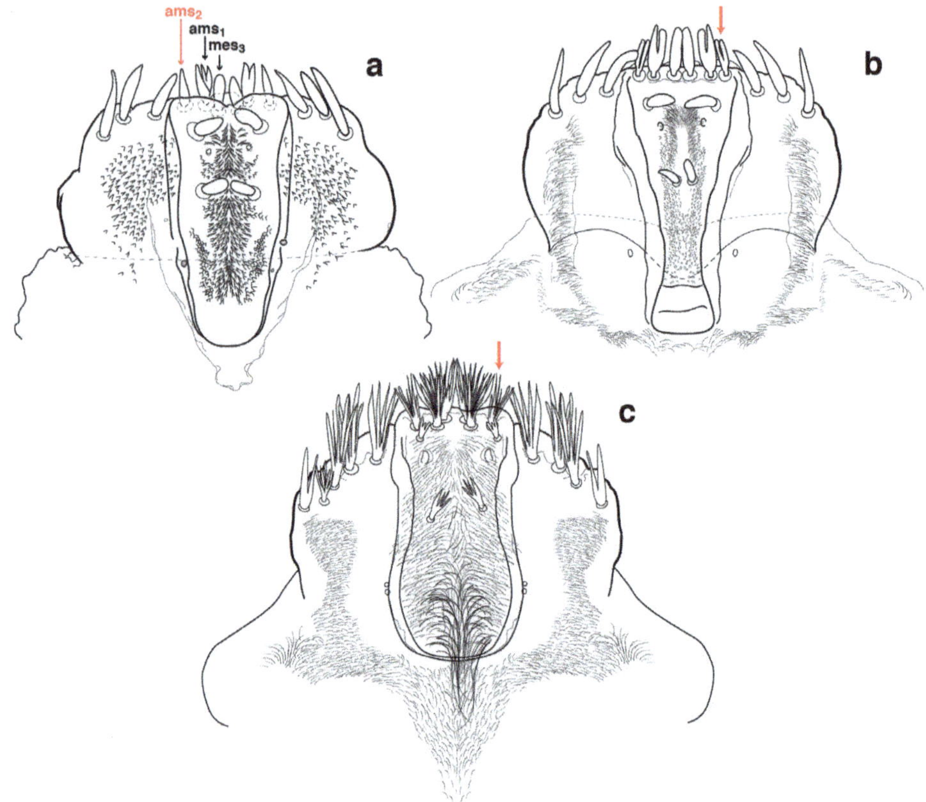

Figure 28. Epipharynx, larva: (**a**) *Poteriophorus*; (**b**) *Metamasius*; and (**c**) *Yuccaborus*. *Mes3:* median epipharyngeal setae; *ams1, 2:* anteromedian setae of epipharynx.

15. Relative length of ams1 and ams2 (Figure 29)

a. **Sub-equal**
Cactophagus, Diocalandra, Dryophthoroides [12], *Dryophthorus, Eucalandra, Eugnoristus*, Foveolus, Mesocordylus* (probably), *Myocalandra, Nassophasis, Nephius* (=*Anius*), *Odoiporus, Paramasius, Polytus, Poteriophorus, Rhabdoscelus, Rhinostomus, Scyphophorus*, Sipalinus, Sitophilus, Sphenophorus, Stenommatus#, Trochorhopalus, Tryphetus, Yuccaborus*

b. ***ams1* approximately twice as long as *ams2***
Cyrtotrachelus, Diathetes, Eugnoristus, Phacecorynes, Rhynchophorus, Scyphophorus**

c. ***ams1* approximately 1/3 longer than *ams2***
Cosmopolites, Metamasius, Rhodobaenus, Temnoschoita

Not included: *Sparganobasis* (broken/weathered), *Trigonotarsus* (broken/weathered; maybe b/c), *Xerodermus*

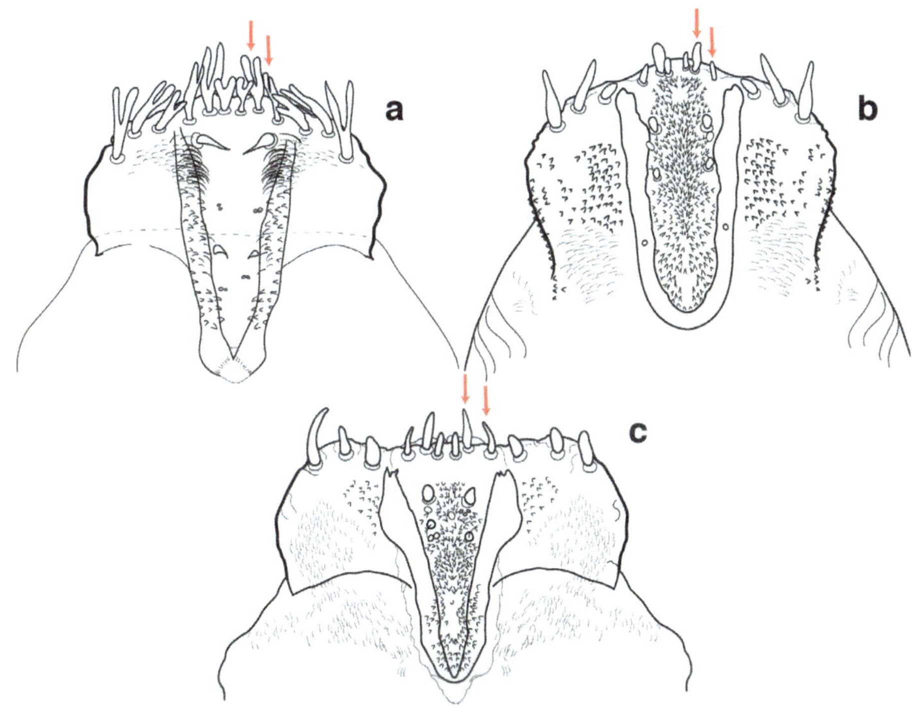

Figure 29. Epipharynx, larva: (**a**) *Eucalandra*; (**b**) *Phacecorynes*; and (**c**) *Rhodobaenus*.

16. Position of median epipharyngeal setae 3 (mes3) (Figure 30)

a. ***mes3* anterad on margin between 2 *ams* pairs; linear (this interpretation gives the appearance of 6 *ams*)**
Cactophagus, Eucalandra, Foveolus, Metamasius, Myocalandra, Nassophasis, Paramasius, Phacecorynes, Rhabdoscelus, Rhodobaenus, Sipalinus, Sitophilus, Sphenophorus, Temnoschoita, Trigonotarsus, Trochorhopalus, Tryphetus

b. ***mes3* proximal and almost directly in front of *ams1*; not linear**
Cosmopolites, Cyrtotrachelus, Diathetes, Diocalandra, Dryophthoroides [12], *Dryophthorus#, Eugnoristus, Mesocordylus* (probably), *Nephius* (=*Anius*), *Odoiporus, Polytus, Poteriophorus, Rhynchophorus, Scyphophorus, Sparganobasis, Stenommatus, Xerodermus* [12], *Yuccaborus*

c. ***mes3* proximal and almost directly in front of *ams2*; not linear**
Rhinostomus

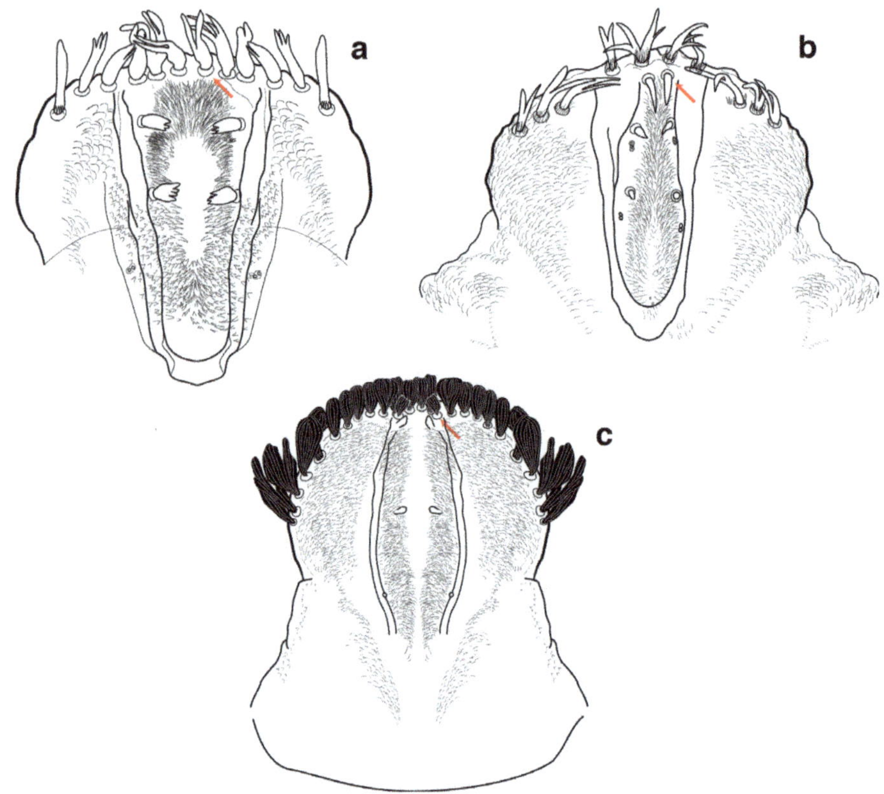

Figure 30. Epipharynx, larva: (**a**) *Foveolus;* (**b**) *Cosmopolites;* and (**c**) *Rhinostomus.*

17. Minute setae apically on labrum/epipharynx (Figure 31)

a. Absent

Cactophagus, Cosmopolites, Diathetes, Diocalandra, Dryophthoroides [12], *Dryophthorus, Eucalandra, Eugnoristus, Foveolus, Mesocordylus* (probably), *Metamasius*, Myocalandra, Nassophasis, Odoiporus, Phacecorynes, Polytus, Poteriophorus, Rhabdoscelus, Rhinostomus, Rhodobaenus, Rhynchophorus, Scyphophorus, Sipalinus, Sitophilus, Sparganobasis, Sphenophorus pontederiae*, Stenommatus, Temnoschoita, Trigonotarsus, Trochorhopalus, Tryphetus, Yuccaborus*

b. Present

Cyrtotrachelus, Metamasius, Nephius (=Anius), Paramasius, Sphenophorus**

Not included: *Dynamis, Xerodermus*

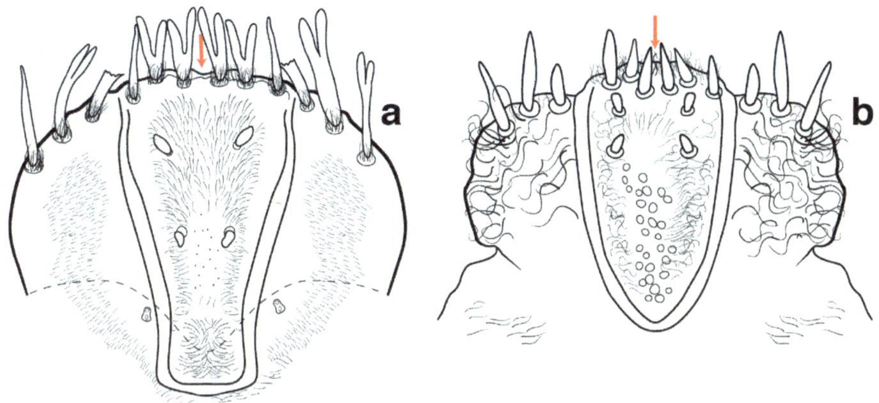

Figure 31. Epipharynx, larva: (**a**) *Sphenophorus pontederiae*; and (**b**) *Nephius* (=*Anius*).

18. Short setae at the base of ams1, ams2, and mes3 (Figure 32)

a. Absent

Cactophagus, Diathetes, Diocalandra, Dryophthoroides [12], *Dryophthorus, Eucalandra, Eugnoristus, Foveolus, Mesocordylus* (probably), *Metamasius, Myocalandra, Nephius* (=*Anius*), *Odoiporus, Phacecorynes, Polytus, Poteriophorus, Rhabdoscelus, Rhinostomus, Rhodobaenus, Rhynchophorus, Scyphophorus, Sipalinus, Sitophilus, Sparganobasis, Sphenophorus maidis*, Stenommatus, Temnoschoita, Trigonotarsus, Trochorhopalus, Tryphetus, Yuccaborus*

b. Present

*Cosmopolites, Cyrtotrachelus, Nassophasis, Paramasius, Sphenophorus pontederiae**

Not included: *Dynamis, Xerodermus*

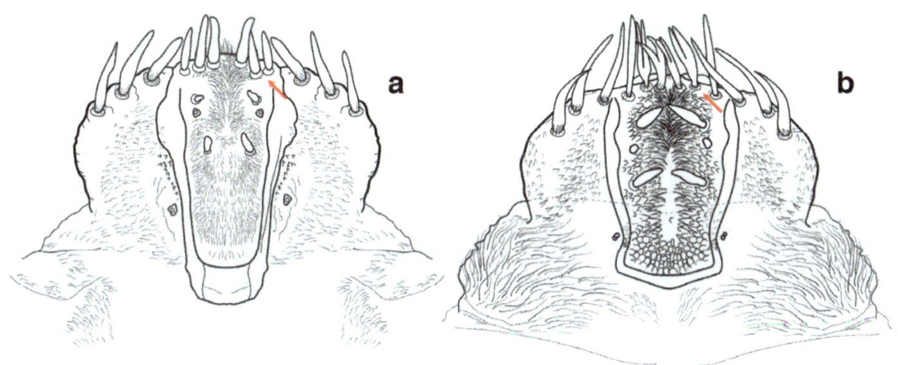

Figure 32. Epipharynx, larva: (**a**) *Sphenophorus maidis*; and (**b**) *Nassophasis*.

19. Shape of mes3 apically (Figure 33)

a. Entire

Cactophagus, Cosmopolites, Cyrtotrachelus, Diathetes, Diocalandra, Dryophthorus, Mesocordylus (probably), *Metamasius, Myocalandra, Nephius* (=*Anius*), *Paramasius, Phacecorynes, Polytus, Poteriophorus, Rhabdoscelus, Rhodobaenus, Rhynchophorus, Scyphophorus, Sitophilus, Sphenophorus*, Stenommatus, Trochorhopalus, Tryphetus, Xerodermus* [11]

b. **Bifurcate**
Eucalandra, *Eugnoristus*, *Foveolus*, *Nassophasis*, *Sphenophorus**, *Temnoschoita*

c. **Trifurcate**
Odoiporus, *Sipalinus*

d. **Tufted**
Rhinostomus, *Yuccaborus*

Not included: *Dynamis*, *Sparganobasis* (broken/weathered all equally?), *Trigonotarsus* (broken, weathered, maybe "a")

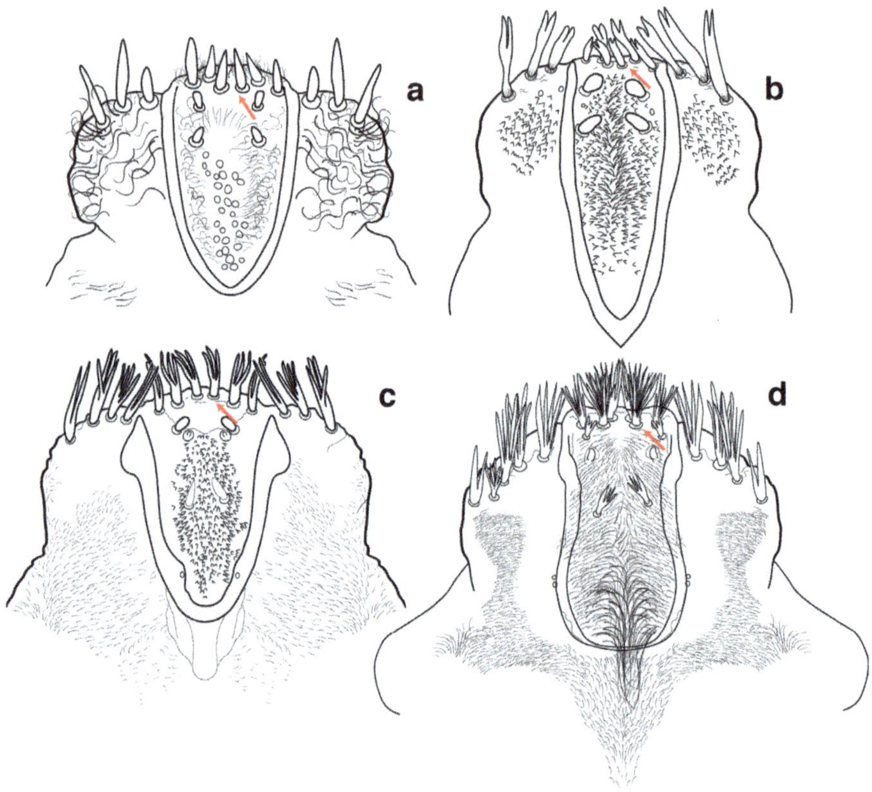

Figure 33. Epipharynx, larva: (**a**) *Nephius* (=*Anius*); (**b**) *Eugnoristus*; (**c**) *Sipalinus*; and (**d**) *Yuccaborus*.

20. Condition of mes1 (Figure 34)

a. **Present, unmodified**
Cactophagus, *Cosmopolites*, *Cyrtotrachelus*, *Diathetes*, *Diocalandra*, *Dryophthoroides* [12], *Dryophthorus*, *Eucalandra*, *Eugnoristus*, *Mesocordylus* (probably), *Metamasius*, *Myocalandra*, *Nassophasis*, *Nephius* (=*Anius*), *Odoiporus*, *Paramasius*, *Phacecorynes*, *Polytus*, *Poteriophorus*, *Rhabdoscelus*, *Rhinostomus*, *Rhodobaenus*, *Rhynchophorus**, *Scyphophorus*, *Sipalinus* (very lightly sclerotized), *Sitophilus*, *Sparganobasis*, *Sphenophorus*, *Stenommatus*, *Temnoschoita*, *Trigonotarsus*, *Trochorhopalus*, *Tryphetus*, *Xerodermus* [11]

b. **Present, minute (approximately 1/4 size of *mes2*)**
Odoiporus, *Rhynchophorus**

c. **Trifurcate (tufted)**
Foveolus (shallow), *Yuccaborus*

Not included: *Dynamis*

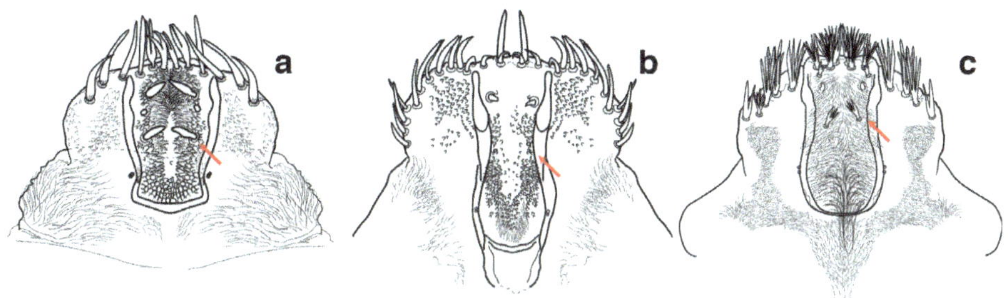

Figure 34. Epipharynx, larva: (**a**) *Nassophasis*; (**b**) *Rhynchophorus palmarum*; (**c**) *Yuccaborus*.

3.1.6. Mandibles (Orientation of Mandibles on a Slide Preparation May Obscure Some of the Characters)

1. Number of incisor cusps of mandible (Figure 35)

a. **Single**
Cactophagus, Eucalandra, Eugnoristus, Nephius (=Anius), Poteriophorus, Rhabdoscelus, Scyphophorus, Sipalinus#, Sitophilus granarius, Sitophilus linearis, Sparganobasis (with sub-apical lobe), *Sphenophorus, Trigonotarsus, Yuccaborus*

b. **Bidentate**
Cosmopolites, Cyrtotrachelus, Diathetes, Diocalandra, Dryophthoroides [12], *Foveolus, Mesocordylus* (probably), *Myocalandra, Nassophasis, Odoiporus, Paramasius, Phacecorynes, Rhodobaenus, Rhynchophorus, Stenommatus, Temnoschoita, Trochorhopalus#, Tryphetus, Xerodermus* [11]

c. **Tridentate**
Polytus

Not included: *Dryophthorus, Metamasius, Rhinostomus*

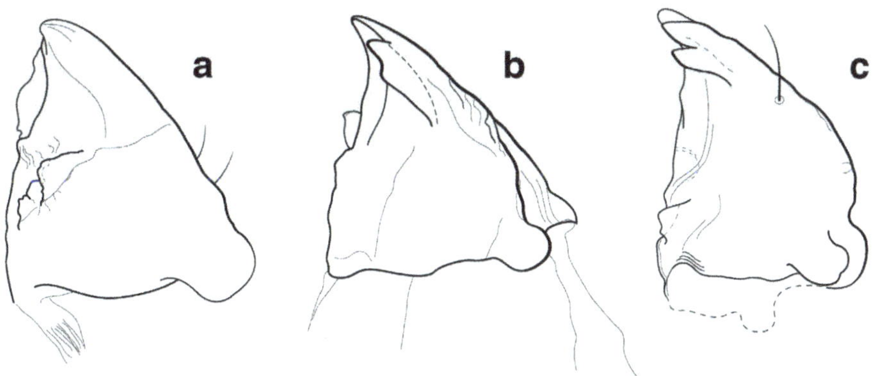

Figure 35. Mandible, larva: (**a**) *Eucalandra*; (**b**) *Nassophasis*; and (**c**) *Polytus*.

2. Internal edge with elongate concavity (Figure 36)

a. **Absent**
Cactophagus, Cyrtotrachelus, Dryophthorus, Eucalandra, Myocalandra, Nassophasis, Nephius (=*Anius*), *Poteriophorus, Rhynchophorus, Scyphophorus, Sitophilus granarius, Sitophilus linearis, Sparganobasis, Sphenophorus, Temnoschoita, Trigonotarsus, Trochorhopalus, Yuccaborus*

b. **Present**
Cosmopolites, Diathetes (apparently doble), *Diocalandra, Eugnoristus, Foveolus, Mesocordylus#, Odoiporus, Paramasius, Phacecorynes, Polytus, Rhabdoscelus, Rhodobaenus, Stenommatus*

Not included: *Dryophthoroides, Dynamis, Metamasius, Sipalinus, Rhinostomus, Xerodermus*

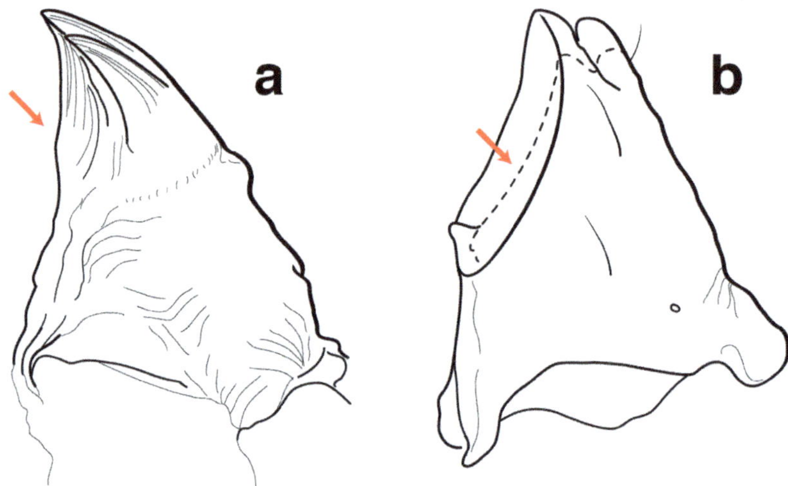

Figure 36. Mandible, larva: (**a**) *Cyrtotrachelus*; and (**b**) *Polytus*.

3. Pair of deltoid projections at base of concavity (Figure 37)

a. **Absent**
Cactophagus, Cyrtotrachelus, Diathetes, Diocalandra, Dryophthorus, Eucalandra, Eugnoristus, Mesocordylus (probably), *Metamasius, Myocalandra, Nassophasis, Nephius* (=*Anius*), *Paramasius, Phacecorynes, Poteriophorus, Rhabdoscelus, Rhodobaenus, Rhynchophorus, Scyphophorus, Sipalinus, Sitophilus, Sparganobasis, Sphenophorus, Stenommatus, Temnoschoita, Trigonotarsus, Trochorhopalus, Tryphetus, Yuccaborus*

b. **Present**
Cosmopolites, Foveolus, Odoiporus, Polytus

Not included: *Dryophthoroides, Dynamis, Rhinostomus, Xerodermus*

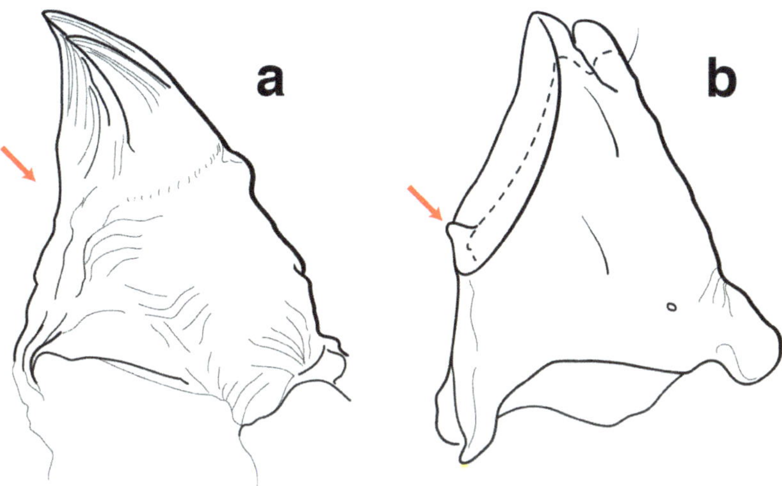

Figure 37. Mandible, larva: (**a**) *Cyrtotrachelus*; and (**b**) *Polytus*.

4. Granular area on subapical surface of mandible (Figure 38)

a. **Absent**
Cactophagus, Cosmopolites, Cyrtotrachelus, Diathetes, Diocalandra, Eucalandra, Eugnoristus, Foveolus, Mesocordylus (probably), *Metamasius, Myocalandra, Nassophasis, Nephius* (=*Anius*), *Odoiporus, Paramasius, Phacecorynes, Polytus, Poteriophorus, Rhabdoscelus, Rhodobaenus, Rhynchophorus, Scyphophorus, Sipalinus, Sitophilus**, *Sparganobasis, Sphenophorus, Stenommatus, Temnoschoita, Trigonotarsus, Trochorhopalus, Tryphetus*

b. **Present**
Dryophthorus, Sitophilus granarius, Yuccaborus* (basally, reduced)

Not included: *Dryophthoroides, Dynamis, Rhinostomus, Xerodermus*

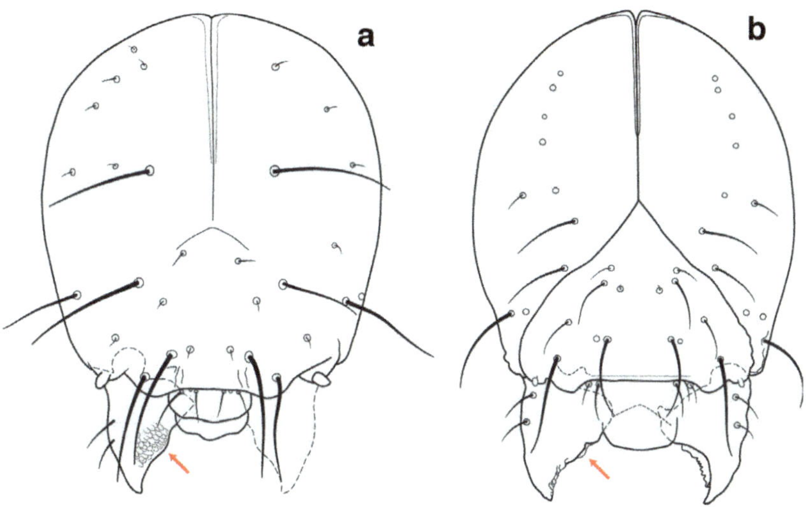

Figure 38. Head and mandibles, larva, anterior view: (**a**) *Dryophthorus*; and (**b**) *Diocalandra*.

3.1.7. Hypopharynx

1. *Setation of lateral lobes of hypopharynx (Figure 39)*

a. **Reduced setation**
Diocalandra, Dryophthorus, Eucalandra#, Eugnoristus#, Myocalandra, Sitophilus#, Stenommatus, Tryphetus#*

b. **Pubescent**
Cactophagus, Cosmopolites, Cyrtotrachelus, Diathetes, Diocalandra, Foveolus, Mesocordylus* (probably), *Metamasius, Nassophasis, Nephius* (=*Anius*), *Odoiporus, Paramasius, Phacecorynes, Polytus, Poteriophorus, Rhabdoscelus, Rhodobaenus, Rhynchophorus, Scyphophorus, Sipalinus, Sparganobasis, Sphenophorus, Temnoschoita, Trigonotarsus, Trochorhopalus, Yuccaborus*

Not included: *Dryophthoroides, Rhinostomus, Xerodermus*

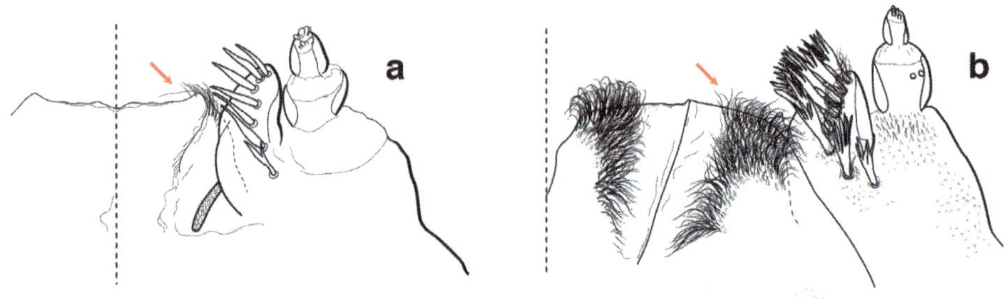

Figure 39. Hypopharynx, mala, and maxillary palp, larva, dorsal view: (**a**) *Myocalandra*; and (**b**) *Nassophasis*.

2. *Location of setae on hypopharynx (Figure 40)*

a. **Laterad, medially glabrous**
Cyrtotrachelus, Diocalandra, Dryophthorus, Mesocordylus (probably), *Myocalandra, Nassophasis, Nephius (=Anius)*, Phacecorynes, Rhodobaenus, Sitophilus, Sphenophorus*, Stenommatus, Tryphetus*

b. **Laterad with scattered medial setae**
Cosmopolites, Foveolus, Nephius (=Anius), Polytus*

c. **Covering anterior margin not extending beyond length of mala**
Eugnoristus#, Temnoschoita

d. **Covering anterior margin and extending posterad beyond mala, usually in a narrow pubescent lateral strip resulting in a glabrous medial area**
Cactophagus, Diathetes, Eucalandra, Metamasius, Odoiporus, Paramasius, Poteriophorus, Rhabdoscelus, Rhynchophorus, Scyphophorus, Sipalinus, Sparganobasis#, Sphenophorus, Trigonotarsus, Trochorhopalus, Yuccaborus*

Not included: *Dryophthoroides, Dynamis, Rhinostomus, Xerodermus*

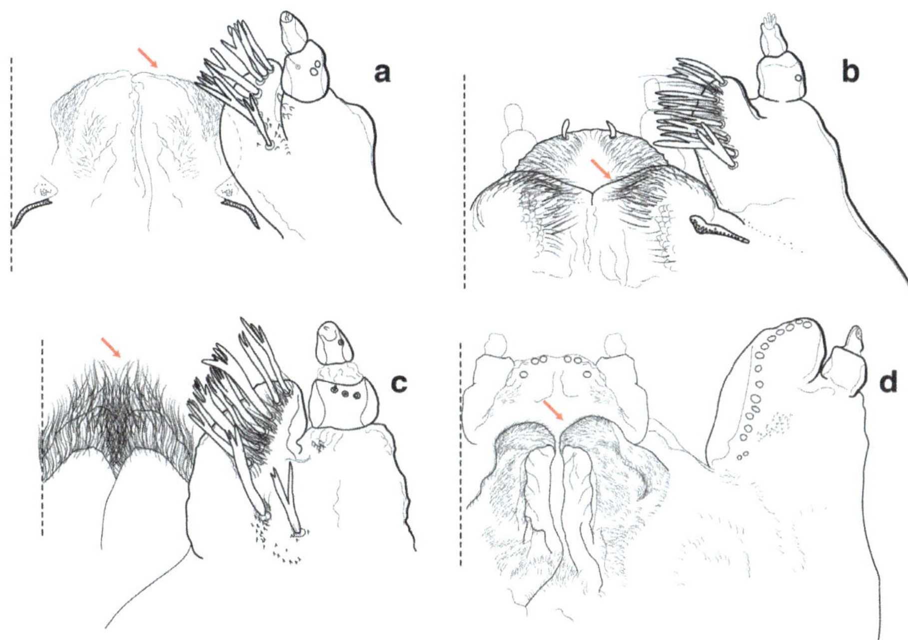

Figure 40. Hypopharynx, mala, and maxillary palp, larva, dorsal view: (**a**) *Diocalandra*; (**b**) *Polytus*; (**c**) *Temnoschoita*; and (**d**) *Sipalinus*.

3.1.8. Maxilla

1. Shape of mala (Figure 41)

a. Digitate (anterior margin rounded)
Cactophagus, *Cyrtotrachelus*, *Diathetes*, *Diocalandra*, *Dryophthoroides* [12], *Dryophthorus*, *Eucalandra*, *Mesocordylus* (probably), *Myocalandra*, *Nassophasis*, *Nephius* (=*Anius*), *Odoiporus*, *Paramasius*, *Phacecorynes*, *Poteriophorus*, *Rhabdoscelus*, *Scyphophorus*, *Sparganobasis*, *Sphenophorus** (but could be interpreted as "b" by some), *Temnoschoita*, *Trochorhopalus*, *Tryphetus*

b. Subquadrate (anterior margin angled but distinctly truncate)
Cosmopolites, *Eugnoristus*, *Metamasius*, *Polytus*, *Rhodobaenus*, *Sipalinus*, *Sitophilus*, *Sphenophorus** (but could be interpreted as "a" by some), *Yuccaborus*

c. Quadrate (anterior margin truncate)
Foveolus, *Rhynchophorus*, *Stenommatus*, *Trigonotarsus*

Not included: *Xerodermus*, *Rhinostomus*

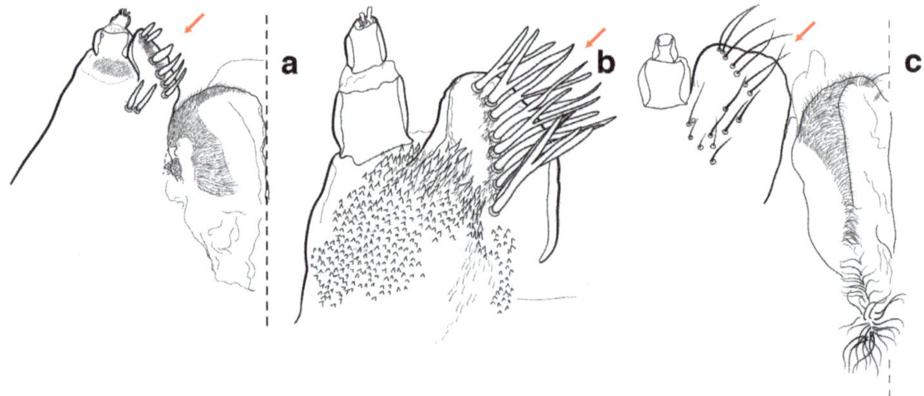

Figure 41. Mala and maxillary palp, larva, dorsal view: (**a**) *Cactophagus*; (**b**) *Cosmopolites*; and (**c**) *Rhynchophorus*.

2. Asperites/microsetae on dorsal mala (Figure 42)

a. **None discernable**
Cyrtotrachelus, *Dryophthoroides* [12], *Dryophthorus*, *Mesocordylus* (probably), *Myocalandra*, *Rhabdoscelus*, *Sitophilus*, *Stenommatus*

b. **Present but not between *dms***
Diocalandra, *Nassophasis*, *Phacecorynes*, *Rhodobaenus* *(may be misinterpreted as "c"), *Rhynchophorus*, *Sipalinus*, *Trochorhopalus*, *Tryphetus*

c. **Present at base of and between *dms***
Cactophagus, *Cosmopolites* (asperites mostly), *Diathetes (dms 3–8)*, *Eucalandra*, *Eugnoristus*, *Foveolus (dms 2–8)*, *Metamasius*, *Nephius* (=*Anius*), *Odoiporus* (copious amounts), *Paramasius (dms 1–8)*, *Polytus*, *Poteriophorus*, *Rhodobaenus** (but setae not asperites), *Scyphophorus*, *Sparganobasis* (only at base of *dms 2–8)*, *Sphenophorus*, *Temnoschoita*, *Trigonotarsus*, *Yuccaborus*

d. **Present between *dms 4–8* and as a distinct rounded area of microsetae (*dms 1–3*)**
Trigonotarsus

Not included: *Rhinostomus*, *Xerodermus*

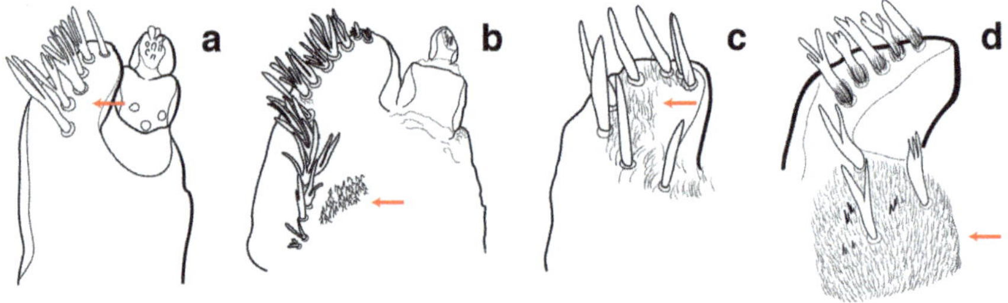

Figure 42. Mala and maxillary palp, larva, dorsal view: (**a**) *Dryophthorus*; (**b**) *Sipalinus*; (**c**) *Rhodobaenus*; and (**d**) *Trigonotarsus*.

3. Asperites/microsetae on stipes distally (base of maxillary palp), dorsal view (Figure 43)

a. **None discernable**
Dryophthoroides [12], *Dryophthorus*, *Foveolus* (but asperites at base of mala), *Metamasius*,

Myocalandra, Phacecorynes, Polytus, Rhabdoscelus, Sipalinus, Sitophilus, Sphenophorus, Stenommatus, Tryphetus*

b. **Asperites present**
Cactophagus, Cosmopolites, Diathetes, Diocalandra, Eucalandra, Eugnoristus, Mesocordylus (probably) (serrate), *Nassophasis, Odoiporus, Paramasius, Poteriophorus, Rhodobaenus* (very fine), *Rhynchophorus, Scyphophorus, Sparganobasis, Sphenophorus** (very fine), *Temnoschoita, Trigonotarsus, Trochorhopalus, Yuccaborus*

c. **Elongate and thin setae present**
Nephius (=Anius)

d. **Elongate and stout setae present**
Cyrtotrachelus

Not included: *Dynamis, Rhinostomus, Xerodermus*

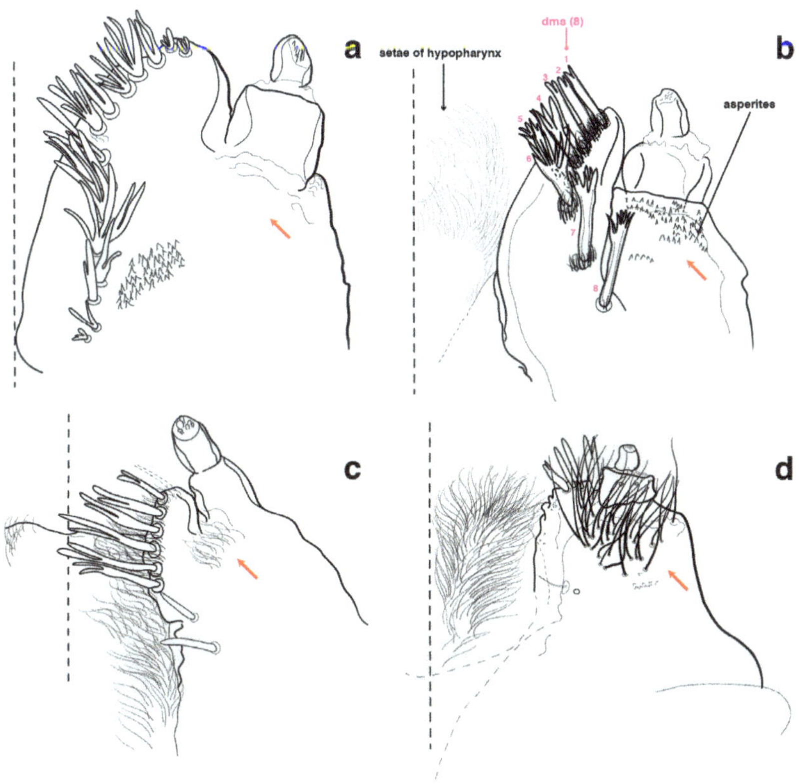

Figure 43. Hypopharynx, mala, and maxillary palp, larva, dorsal view: (**a**) *Sipalinus*; (**b**) *Poteriophorus*; (**c**) *Nephius (=Anius)*; and (**d**) *Cyrtotrachelus*. *Dms*: dorsal malar setae.

4. Arrangement of dorsal malar setae (dms) (Figure 44)

a. **Regularly aligned in single row (*dms 7,8* may sometimes be more lateral, but always in a single file)**
Cactophagus (*dms 7,8* laterad), *Cosmopolites, Diathetes, Diocalandra, Dryophthoroides* [12], *Dryophthorus, Eucalandra* (*dms 7,8* somewhat distant from *dms 3–8*), *Eugnoristus, Foveolus* (*dms*

7,8 somewhat distant from *dms 1–6*), *Mesocordylus* (probably), *Metamasius* (*dms 7,8* somewhat distant from *dms 3–8*), *Myocalandra*, *Nassophasis* (*dms 7,8* somewhat distant from *dms 1–6*), *Nephius* (=*Anius*), *Odoiporus*, *Paramasius* (*dms 7,8* somewhat distant from *dms 1–6*), *Phacecorynes*, *Polytus*, *Poteriophorus* (*dms 7,8* somewhat distant from *dms 1–6*), *Rhabodscelus* (*dms 7,8* somewhat distant from *dms 1–6*), *Rhodobaenus* (*dms 7,8* somewhat distant from *dms 1–6*), *Scyphophorus* (*dms 7,8* somewhat distant from *dms 1–6*), *Sipalinus*, *Sitophilus oryzae*#(some *Sitophilus* may have dms irregularly distributed), *Sparganobasis*, *Sphenophorus* (*dms 7,8* somewhat distant from *dms 1–6*), *Stenommatus*, *Temnoschoita* (*dms 7,8* somewhat distant from *dms 1–6*), *Trochorhopalus*, *Tryphetus*, *Xerodermus* [11], *Yuccaborus*

b. **Regularly distributed medially, proximally somewhat clumped**
Cyrtotrachelus, *Rhynchophorus*, *Trigonotarsus*#

Not included: *Dynamis*, *Rhinostomus*

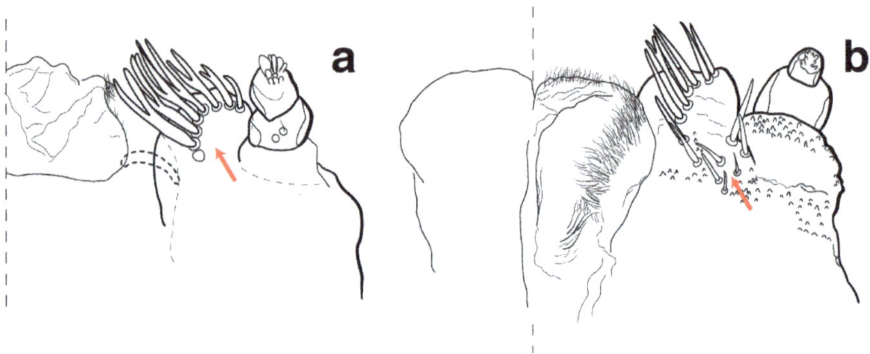

Figure 44. Hypopharynx, mala, and maxillary palp, larva, dorsal view: (**a**) *Stenommatus*; and (**b**) *Rhynchophorus*.

5. Number of dorsal malar setae (dms) (Figure 45)

a. **Less than five (5)**
Sitophilus (but unclear)

b. **Six (6)**
Dryophthoroides# ([12], (p. 253) " ... row of about six fissile setae and one or two simple setae."), *Dryophthorus*, *Eugnoristus** (but not entirely clear), *Tryphetus*#*, *Xerodermus* [11]

c. **Seven (7)**
Myocalandra

d. **Seven (7) (with a small pore)**
Stenommatus

e. **Eight (8)**
Cactophagus, *Cosmopolites*, *Diathetes*, *Diocalandra*, *Eucalandra*, *Eugnoristus**, *Foveolus*, *Mesocordylus* (probably), *Metamasius*, *Nassophasis*, *Nephius* (=*Anius*), *Paramasius*, *Phacecorynes*, *Polytus*, *Poteriophorus*, *Rhabdoscelus*, *Rhodobaenus*, *Scyphophorus*, *Sparganobasis*, *Sphenophorus*, *Temnoschoita*, *Trigonotarsus*, *Trochorhopalus*, *Tryphetus*#*

f. **Ten (10)**
Odoiporus (10–11), *Yuccaborus*

g. **Fourteen (14)**
Rhynchophorus

h. **Fifteen (15)**
Sipalinus

i. **More than 30**
Cyrtotrachelus

Not included: *Dynamis*, *Rhinostomus*

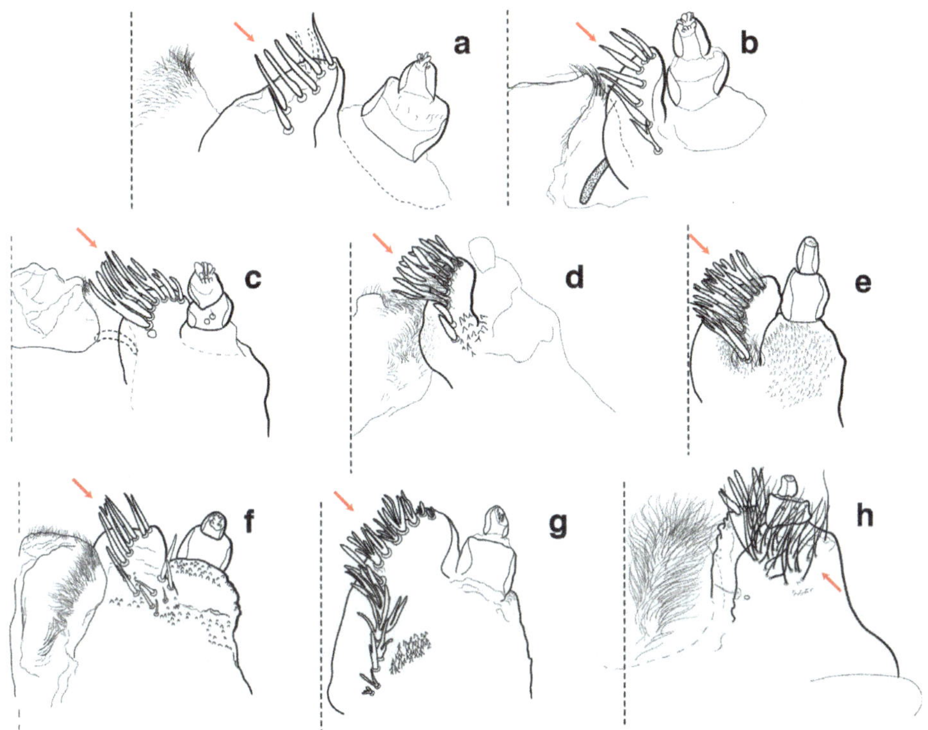

Figure 45. Hypopharynx, mala and maxillary palp, larva, dorsal view: (**a**) *Tryphetus;* (**b**) *Myocalandra;* (**c**) *Stenommatus* (**d**) *Eucalandra;* (**e**) *Odoiporus;* (**f**) *Rhynchophorus;* (**g**) *Sipalinus;* and (**h**) *Cyrtotrachelus.*

6. Shape of dorsal malar setae (dms) (if at least one is branching, select b, see next couplet) (Figure 46)

a. **Entire**
Cactophagus, *Cyrtotrachelus* (possibly some distally may be superficially branched), *Phacecorynes*, *Rhodobaenus*, *Rhynchophorus* (occasional bifurcate seta), *Sitophilus*, *Tryphetus*

b. **Branching**
Cosmopolites, *Diathetes*, *Diocalandra*, *Dryophthoroides* [12], *Dryophthorus*, *Eucalandra*, *Eugnoristus*, *Foveolus*, *Mesocordylus* (probably), *Metamasius*, *Myocalandra*, *Nassophasis*, *Nephius* (=*Anius*), *Odoiporus*, *Paramasius*, *Polytus*, *Poteriophorus*, *Rhabdoscelus*, *Scyphophorus*, *Sipalinus*, *Sparganobasis*, *Sphenophorus*, *Stenommatus*, *Temnoschoita*, *Trigonotarsus*, *Trochorhopalus*, *Xerodermus* [11], *Yuccaborus* (tufted)

Not included: *Dynamis*, *Rhinostomus*

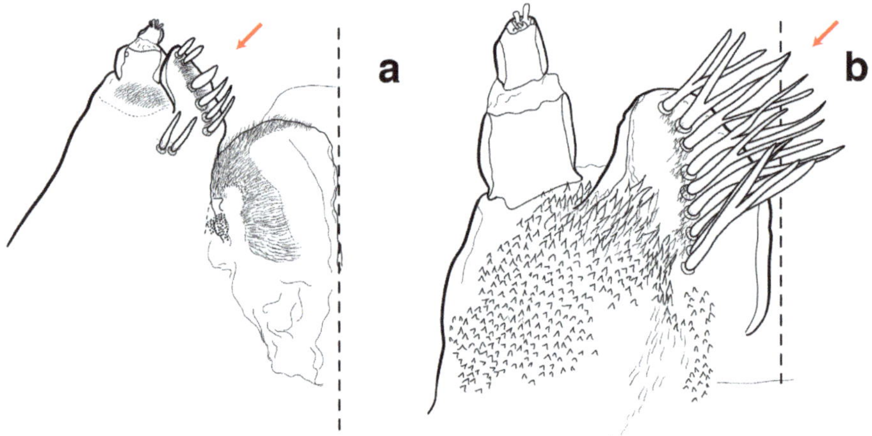

Figure 46. Hypopharynx, mala and maxillary palp, larva, dorsal view: (**a**) *Cactophagus*; and (**b**) *Cosmopolites*.

7. Progression of branching of dorsal malar setae (dms) (this is a more generalized version of character 8 below) (Figure 47)

a. Entire, not branching

Cactophagus, *Cyrtotrachelus* (possibly some distally may be superficially branched), *Phacecorynes*, *Rhodobaenus*, *Rhynchophorus* (occasional bifurcate seta), *Sitophilus*, *Tryphetus*

b. All *dms* equally branching

Cosmopolites, *Diocalandra*, *Eugnoristus*, *Foveolus*, *Nassophasis*, *Odoiporus*, *Paramasius**, *Polytus*, *Scyphophorus*, *Sipalinus*, *Temnoschoita*, *Yuccaborus*

c. Branching *dms* distally decreasing

Diathetes (*dms 3–8* branching), *Dryophthoroides* [12], *Dryophthorus*, *Mesocordylus* (probably), *Myocalandra*, *Poteriophorus*, *Rhabdoscelus*, *Stenommatus*, *Trigonotarsus*

d. Branching *dms* distally increased

Eucalandra, *Metamasius*, *Nephius* (=*Anius*), *Paramasius**, *Sparganobasis* (*dms 8* bifurcate, others trifurcate?), *Sphenophorus*, *Trochorhopalus*

Not included: *Dynamis*, *Rhinostomus*, *Xerodermus*

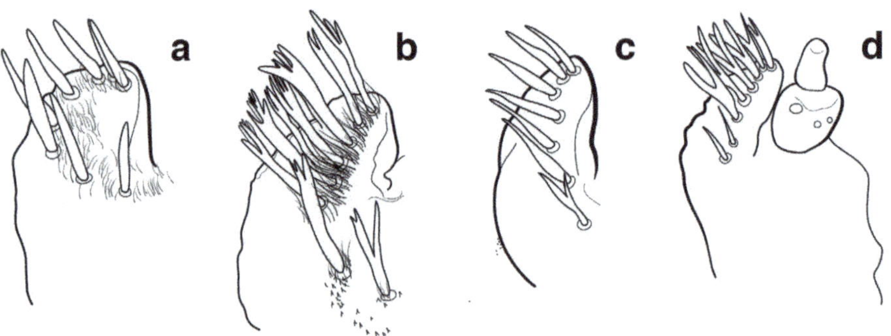

Figure 47. Mala and maxillary palp (**d** only), larva, dorsal view: (**a**) *Rhodobaenus*; (**b**) *Temnoschoita*; (**c**) *Myocalandra*; and (**d**) *Eucalandra*.

8. *Variation of branching of dms (slight plasticity may exist and should be considered; this couplet is a more complicated (less generalized) version of couplets #6, 7 above) (Figure 48)*

a. **Not branching**
Cactophagus, *Cyrtotrachelus* (distal setae may be superficially branched), *Phacecorynes*, *Rhodobaenus*, *Rhynchophorus* (occasional bifurcate seta), *Sitophilus*, *Tryphetus*

b. **All bifurcate**
Cosmopolites, *Diocalandra*, *Eugnoristus*, *Paramasius**, *Polytus*, *Sparganobasis*

c. **Some *dms* entire, others bifurcate**
Diathetes [*dms 1–2* entire, *dms 3–8* bifurcate (shallow)], *Dryophthorus* (*dms 2–6* bifurcate, *dms 1* entire), *Eucalandra* (*dms 1–6* bifurcate, *dms 7*, *dms 8* entire), *Mesocordylus* (probably) (*dms 1–5* entire, *dms 6–8* bifurcate), *Metamasius* (*dms 1–6* bifurcate, *dms 7*, *dms 8* entire; sometimes *dms 7* or *8* may be bifurcate), *Myocalandra* (*dms 1–6* entire, *dms 7* bifurcate), *Nephius* (=*Anius*) (*dms 1–6* bifurcate, *dms 7*, *dms 8* entire), *Paramasius** (*dms 1–7* bifurcate, *dms 8* enitre), *Sphenophorus* (*dms 1–6* bifurcate, *dms 7*, *dms 8* entire), *Stenommatus* (*dms 1* entire, *dms 2–7* bifurcate), *Trochorhopalus* (*dms 1–6* bifurcate, *dms 7*, *dms 8* entire)

d. **Some *dms* bifurcate, others multifurcate**
*Dryophthoroides#** (([12], p. 253) " ... row of about six fissile setae and one or two simple setae.") [12], *Poteriophorus* (*dms 1–3* bifurcate, *dms 4–8* tufted, shallow and multi-furcate), *Rhabdoscelus* (*dms 1–2, 8* bifurcate, *dms 3–7* multifurcate), *Scyphophorus* [*dms 1–6* bifurcate, *dms 7, 8* tufted (deeply and multi-furcate)], *Trigonotarsus* (*dms 1–5* bifurcate, *dms 6–8* multifurcate)

e. **All tufted (shallow and multi-furcate)**
*Dryophthoroides#** [12], *Foveolus* (with long setae at base), *Nassophasis*, *Odoiporus*, *Temnoschoita*

f. **All tufted (deeply and multi-furcate)**
Sipalinus, *Yuccaborus*

Not included: *Dynamis*, *Rhinostomus*, *Xerodermus* ([11], (p. 41) " ... mala with several branched setae, six of them in a longitudinal row not extending far from apex.")

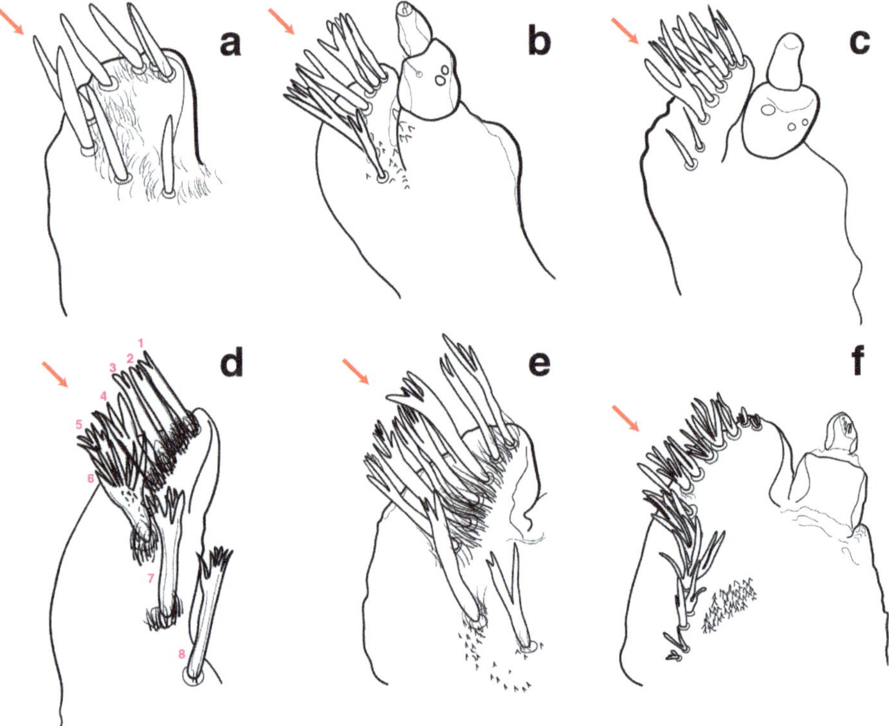

Figure 48. Mala and maxillary palp, larva, dorsal view: (**a**) *Rhodobaenus;* (**b**) *Diocalandra;* (**c**) *Eucalandra;* (**d**) *Poteriophorus;* (**e**) *Temnoschoita;* and (**f**) *Sipalinus*.

9. Ventral malar setae (vms) (Figure 49)

a. **Unbranched**
Cactopahgus, Cosmopolites, Cyrtotrachelus, Diathetes, Diocalandra, Dryophthorus, Eucalandra, Mesocordylus (probably), *Myocalandra, Nephius* (=*Anius*), *Paramasius, Phacecorynes, Rhodobaenus, Rhynchophorus, Scyphophorus, Sitophilus, Sparganobasis Sphenophorus, Stenommatus, Tryphetus*

b. **One (1) branched, remainder unbranched**
Polytus, Rhabdoscelus, Trigonotarsus#

c. **Two (2) branched**
Foveolus#, Eugnoristus, Metamasius, Sipalinus, Trochorhopalus*

d. **More than two branched**
Foveolus#, Nassophasis, Odoiporus* (3), *Poteriophorus, Temnoschoita, Yuccaborus* (at least 3)

Not included: *Dryophthoroides, Dynamis, Rhinostomus, Xerodermus*

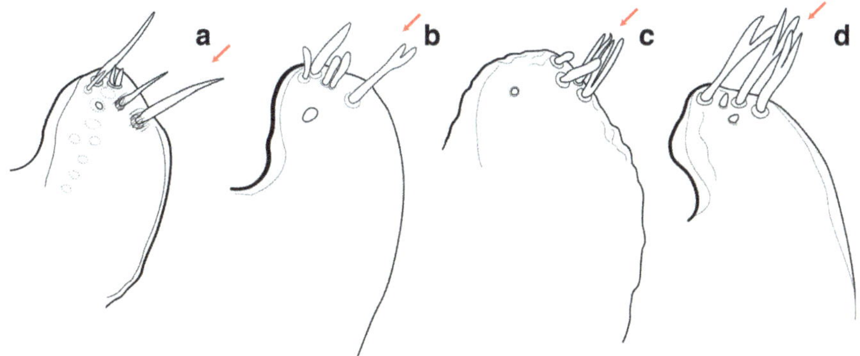

Figure 49. Mala, larva, dorsal view: (**a**) *Cosmopolites;* (**b**) *Polytus;* (**c**) *Sipalinus;* and (**d** *Nassophasis*.

10. Pair of short, contiguous vms4 and vms3 (bilobed vms3?) (Figure 50)

a. **Absent**
Cactopahgus, Diathetes, Diocalandra, Dryophthorus, Eugnoristus, Foveolus#, Metamasius, Myocalandra, Nassophasis, Nephius* (=*Anius*), *Odoiporus, Paramasius, Phacecorynes, Poteriophorus, Rhabdoscelus, Rhodobaenus, Rhynchophorus, Scyphophorus, Sipalinus, Sphenophorus, Stenommatus, Temnoschoita, Trigonotarsus* (apparently very small one present, but unclear), *Trochorhopalus, Tryphetus, Yuccaborus*

b. **Present**
Cosmopolites, Cyrtotrachelus, Eucalandra, Foveolus#, Mesocordylus* (probably), *Polytus, Sitophilus, Sparganobasis* (bilobed *vms3*?)

Not included: *Dryophthoroides, Dynamis, Rhinostomus, Xerodermus*

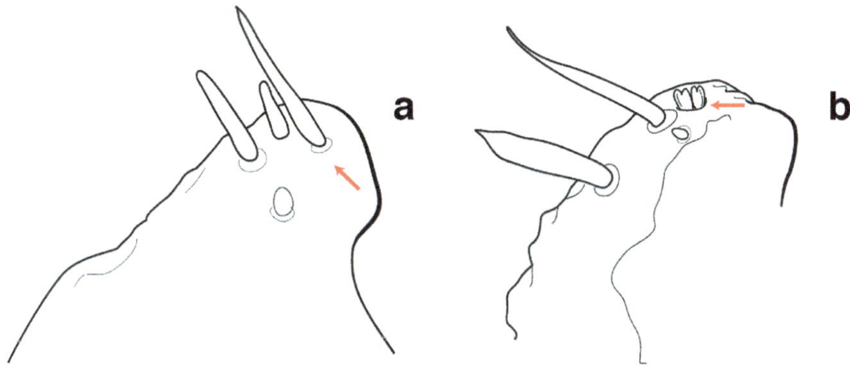

Figure 50. Mala, larva, ventral view: (**a**) *Tryphetus;* and (**b**) *Cyrtotrachelus*.

3.1.9. Labium

1. Ligula with copious apical setation (Figure 51)

a. **Absent**
Cactopahgus, Cosmopolites, Cyrtotrachelus, Diathetes, Eucalandra, Mesocordylus (probably), *Nassophasis, Odoiporus, Phacecorynes, Poteriophorus, Rhabdoscelus, Rhodobaenus, Rhynchophorus, Scyphophorus, Sipalinus, Sitophilus, Sparganobasis, Trigonotarsus, Tryphetus, Yuccaborus*

b. Present

Diocalandra, Dryophthorus (small amounts of setae), *Eugnoristus, Foveolus, Metamasius, Myocalandra, Nephius* (=*Anius*), *Paramasius, Polytus, Sphenophorus, Stenommatus* (reduced), *Temnoschoita, Trochorhopalus* (reduced), *Xerodermus* [11]

Not included: *Dryophthoroides, Dynamis, Rhinostomus*

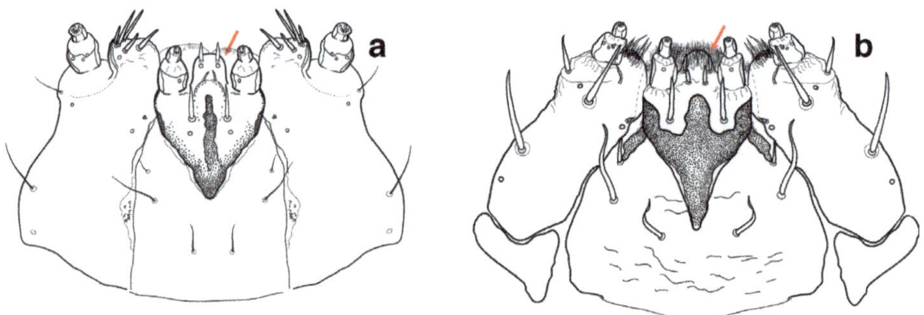

Figure 51. Labium, labio-maxillary complex: (**a**) *Rhynchophorus*; and (**b**) *Sphenophorus*.

2. *Ligular seta 1 (lgs1) (Figure 52)*

a. Entire

Cactopahgus, Cyrtotrachelus, Diathetes, Diocalandra, Dryophthorus, Eugnoristus, Foveolus, Mesocordylus (probably), *Metamasius, Myocalandra, Nassophasis, Nephius* (=*Anius*), *Paramasius, Poteriophorus, Phacecorynes, Polytus, Rhabdoscelus, Rhodobaenus, Rhynchophorus, Scyphophorus, Sitophilus, Sparganobasis* (broken/weathered), *Sphenophorus, Stenommatus, Temnoschoita, Trigonotarsus, Trochorhopalus, Tryphetus, Xerodermus* [11]

b. Branched

Cosmopolites, Eucalandra, Odoiporus

c. Tufted

Yuccaborus

Not included: *Dryophthoroides, Dynamis, Rhinostomus, Sipalinus*

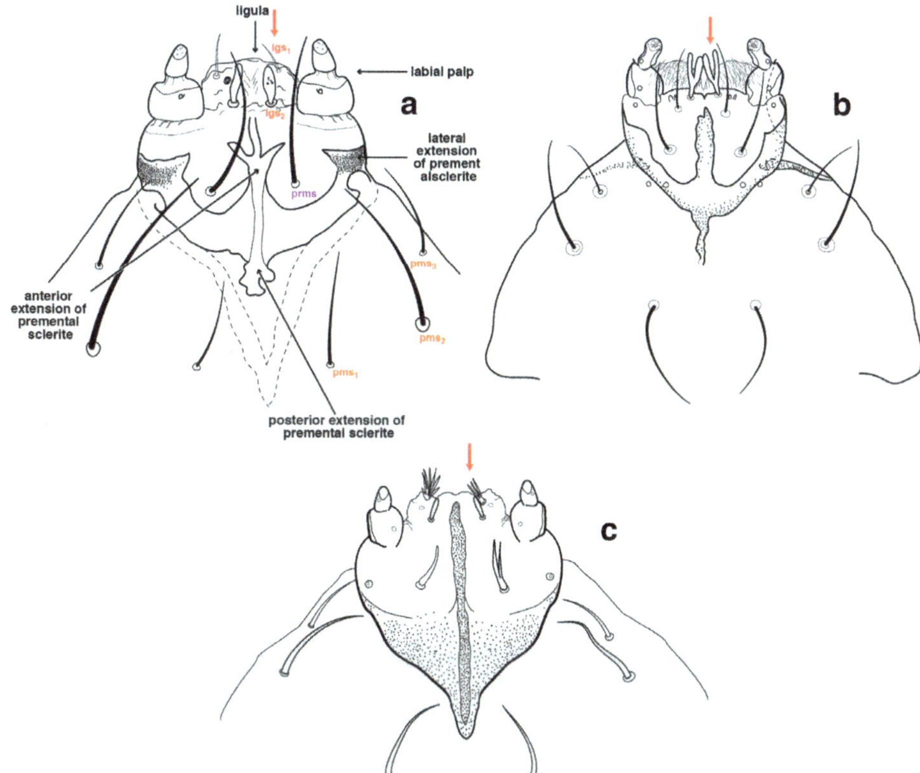

Figure 52. Labium, larva: (**a**) *Poteriophorus*; (**b**) *Eucalandra*; and (**c**) *Yuccaborus*. *Lgs1, 2*: ligular setae.

3. Position of postmental setae (pms1, 2, 3) (Figure 53)

a. Linear

Dryophthorus, *Mesocordylus* (probably), <u>*Stenommatus*</u>

b. Not linear

Cactophagus, *Cosmopolites*, *Cyrtotrachelus*, *Diathetes*, *Diocalandra*, *Eucalandra*, *Eugnoristus*, *Foveolus*, *Metamasius*, *Myocalandra*, *Nassophasis*, *Nephius* (=*Anius*), *Odoiporus*, *Paramasius*, *Phacecorynes*, <u>*Polytus*</u>, *Poteriophorus*, *Rhabdoscelus*, *Rhodobaenus*, *Rhynchophorus*, *Scyphophorus*, *Sipalinus*, *Sitophilus*, *Sparganobasis*, *Sphenophorus*, *Temnoschoita*, *Trigonotarsus*, *Trochorhopalus*, *Tryphetus*, *Yuccaborus*

Not included: *Dryophthoroides*, *Dynamis*, *Rhinostomus*, *Xerodermus*

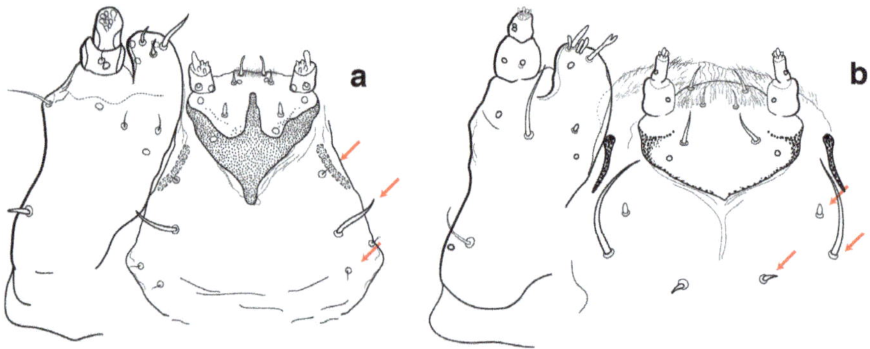

Figure 53. Labium, larva: (**a**) *Stenommatus*; and (**b**) *Polytus*.

4. Second labial palpomere (Figure 54)

a. **Absent/reduced**
Dryophthorus, *Stenommatus*

b. **Present, prominent**
Cactophagus, *Cosmopolites*, *Cyrtotrachelus*, *Diathetes*, *Diocalandra*, *Dryophthoroides* [12], *Eucalandra*, *Eugnoristus*, *Foveolus*, *Mesocordylus* (probably), *Metamasius*, *Myocalandra*, *Nassophasis*, *Nephius* (=*Anius*), *Odoiporus*, *Paramasius*, *Phacecorynes*, *Polytus*, *Poteriophorus*, *Rhabdoscelus*, *Rhodobaenus*, *Rhynchophorus*, *Scyphophorus*, *Sipalinus*, *Sitophilus*, *Sparganobasis*, *Sphenophorus*, *Temnoschoita*, *Trigonotarsus*, *Trochorhopalus*, *Tryphetus*, *Xerodermus* [11], *Yuccaborus*

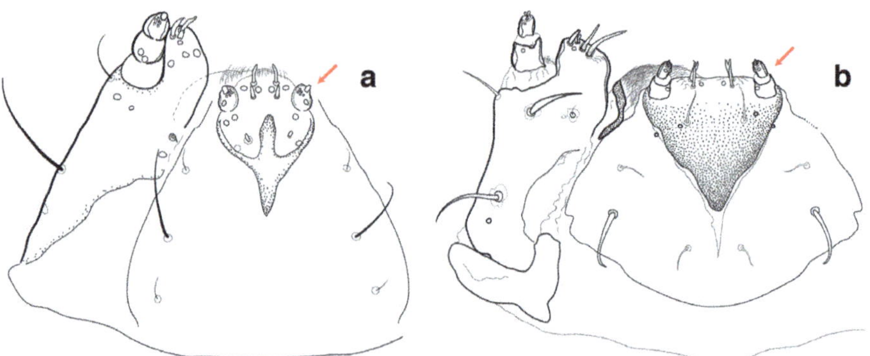

Figure 54. Labium, larva: (**a**) *Dryophthorus*; and (**b**) *Cosmopolites*.

5. Pigmentation of premental sclerite (Figure 55)

a. **Trident**
Cactophagus, *Cyrtotrachelus*, *Diocalandra*, *Dryophthorus*, *Dryophthoroides*# [12], *Eucalandra*, *Eugnoristus* (anteriorly reduced), *Foveolus*, *Mesocordylus* (probably), *Metamasius*, *Myocalandra*, *Nassophasis*, *Paramasius*, *Phacecorynes*, *Polytus**, *Poteriophorus*, *Rhabdoscelus*, *Rhynchophorus*, *Scyphophorus*, *Sitophilus*, *Sparganobasis*, *Sphenophorus*, *Stenommatus*, *Temnoschoita*, *Trigonotarsus*#*, *Trochorhopalus*, *Tryphetus*, *Xerodermus* [11]

b. **Trident, with prominent, narrow, elongate anterior extension (almost reaching apex of ligula)**
Sipalinus, *Trigonotarsus*#*, *Yuccaborus*

c. **Triangular**
Cosmopolites, *Polytus**, *Rhodobaenus*

d. **Effaced**
Diathetes, *Nephius (=Anius)*, *Odoiporus*

Not included: *Dynamis*, *Rhinostomus*

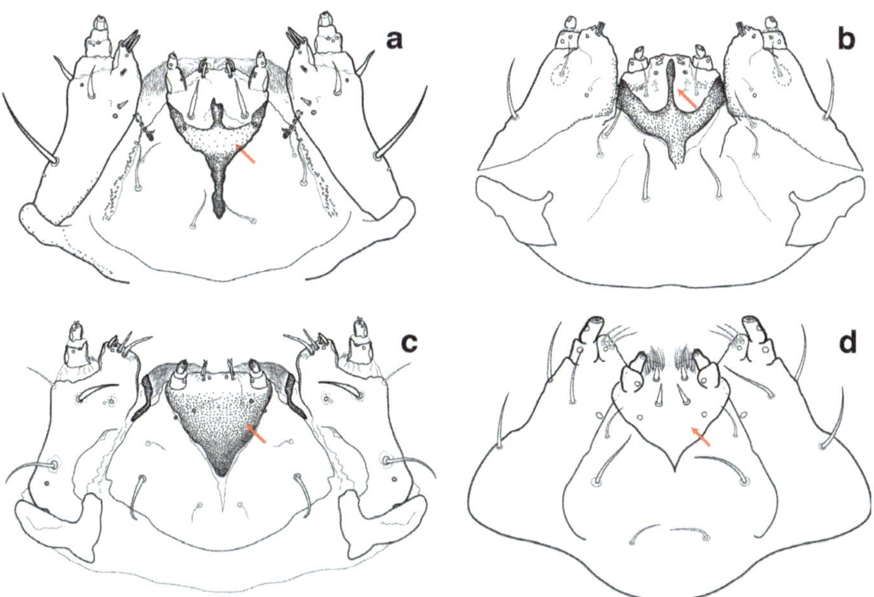

Figure 55. Labium, larva: (**a**) *Phacecorynes*; (**b**) *Sipalnus*; (**c**) *Cosmopolites*; and (**d**) *Nephius*.

PLATES OF THE MOUTHPARTS OF 36 DRYOPHTHORINE GENERA

(in alphabetical order)

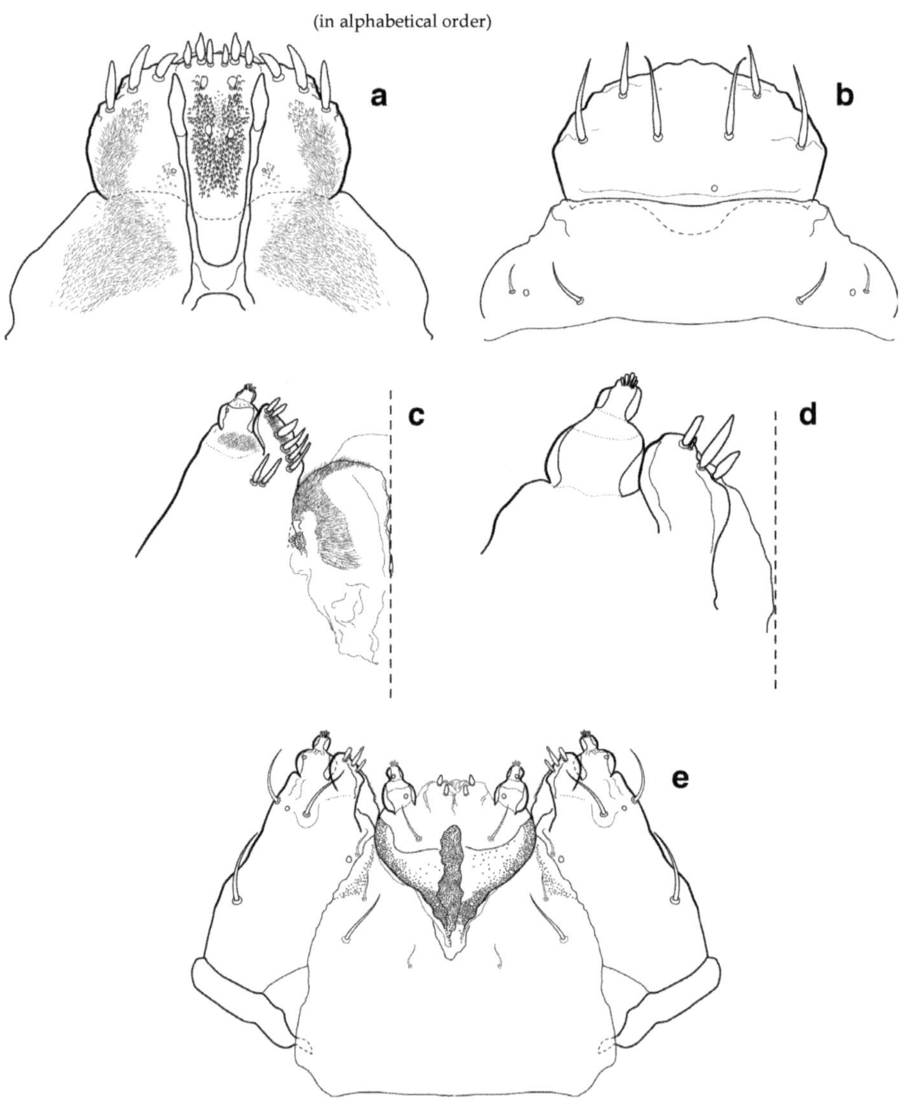

Figure 56. *Cactophagus validus* mouthparts: (**a**) epipharynx; (**b**) clypeus and labrum; (**c**) detail of hypopharynx, mala, and maxillary palp, dorsal view; (**d**) detail of mala and maxillary palp, ventral view; and (**e**) labio-maxillary complex, ventral view.

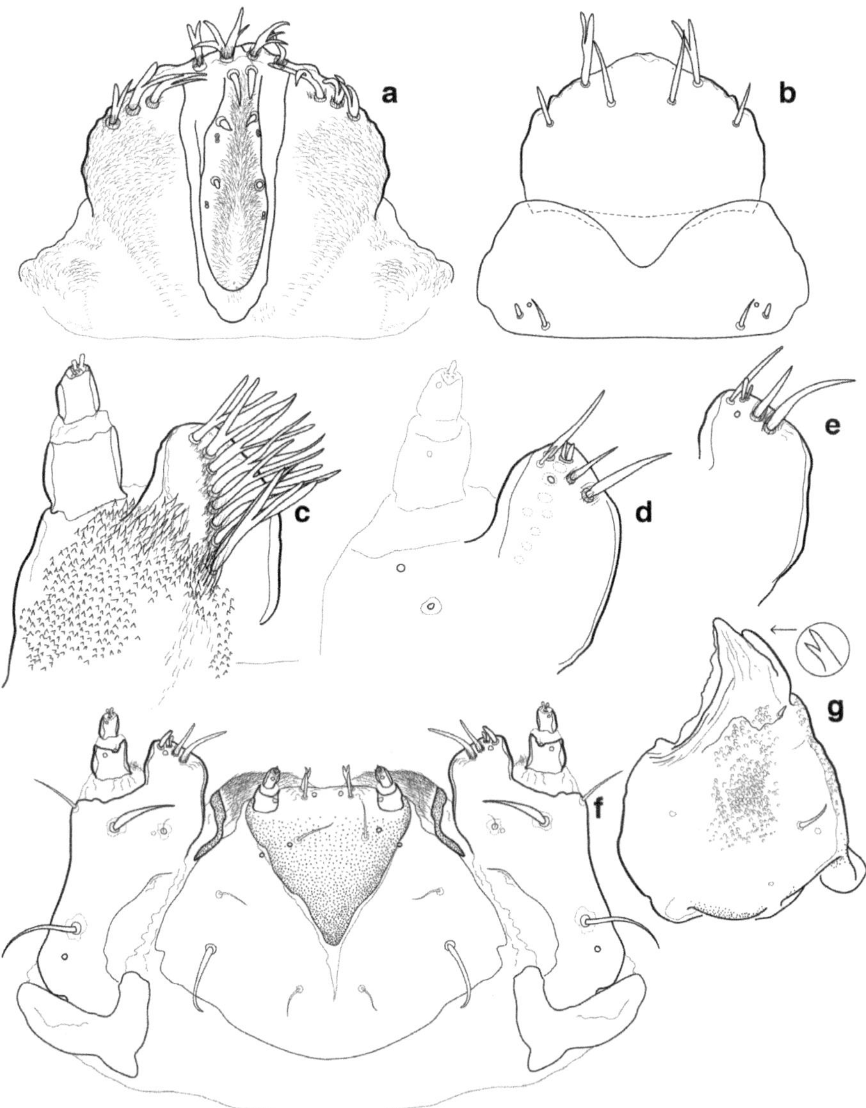

Figure 57. *Cosmopolites sordidus* mouthparts: (**a**), epipharynx; (**b**) clypeus and labrum; (**c**) detail of mala and maxillary palp, dorsal view; (**d**) detail of mala and maxillary palp, ventral view; (**e**) detail of mala and maxillary palp, opposite side, ventral view; (**f**) labio-maxillary complex, ventral view; and (**g**) mandible and detail of cusps.

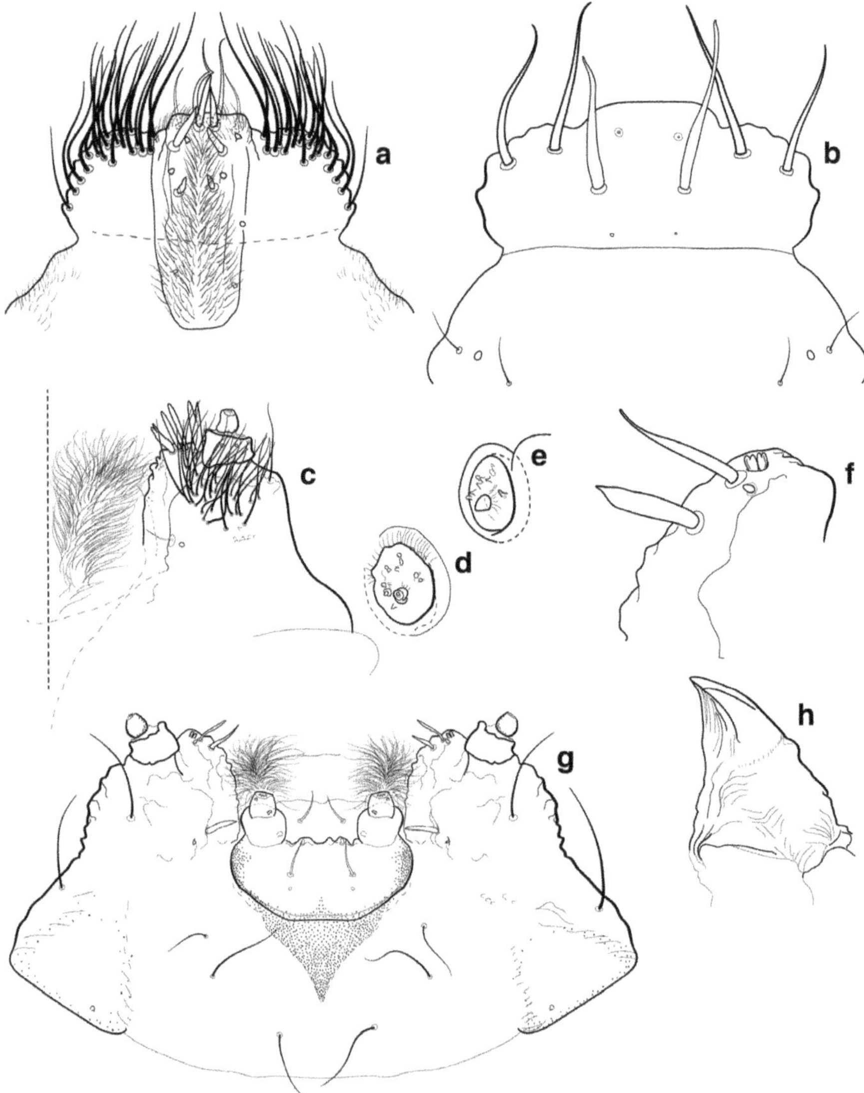

Figure 58. *Cyrtotrachelus thompsoni* mouthparts: (**a**) epipharynx; (**b**) clypeus and labrum; (**c**) detail of hypopharynx, mala, and maxillary palp, dorsal view; (**d**) antenna, anterior view; (**e**) antenna, oblique lateral view; (**f**) detail of mala, ventral view; (**g**) labio-maxillary complex, ventral view; and (**h**) mandible.

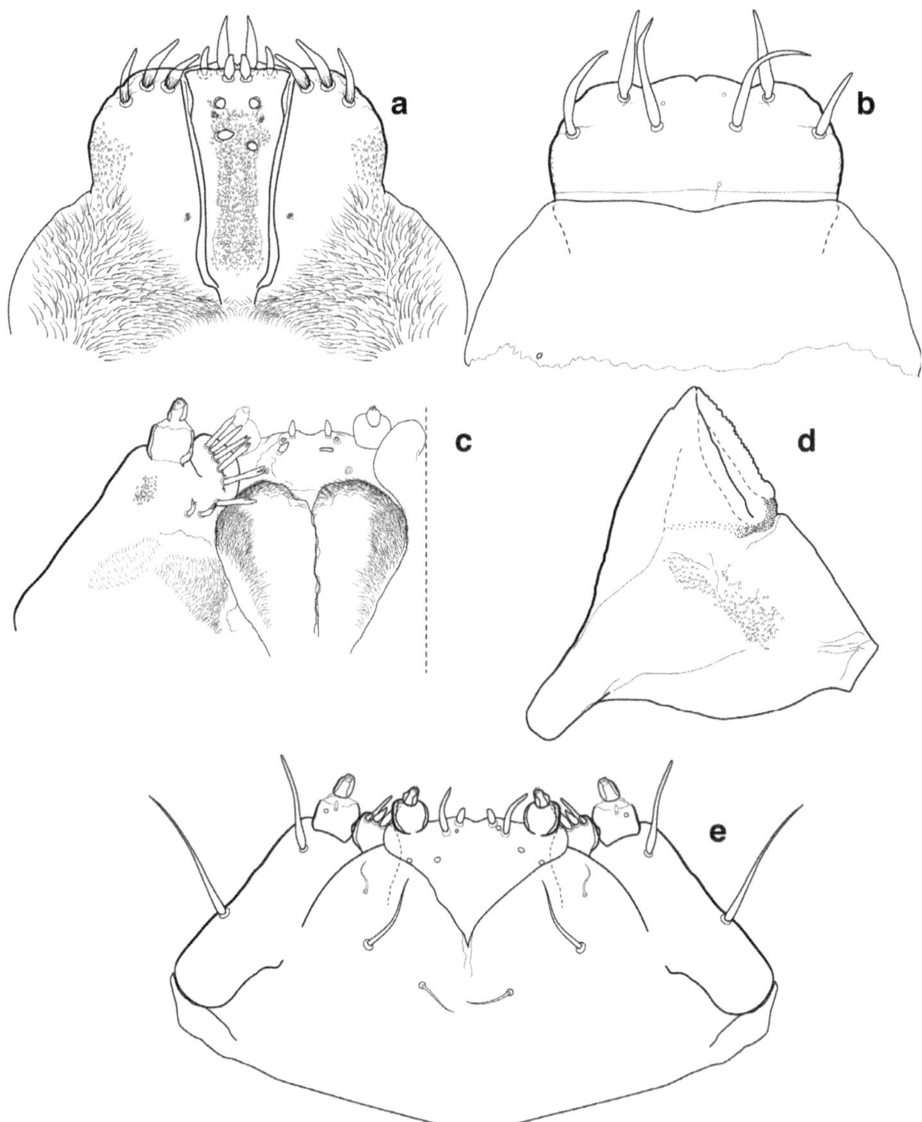

Figure 59. *Diaphetes morio* mouthparts: (**a**) epipharynx; (**b**) clypeus (incomplete) and labrum; (**c**) detail of hypopharynx, mala, and maxillary palp, dorsal view; (**d**) mandible; and (**e**) labio-maxillary complex, ventral view.

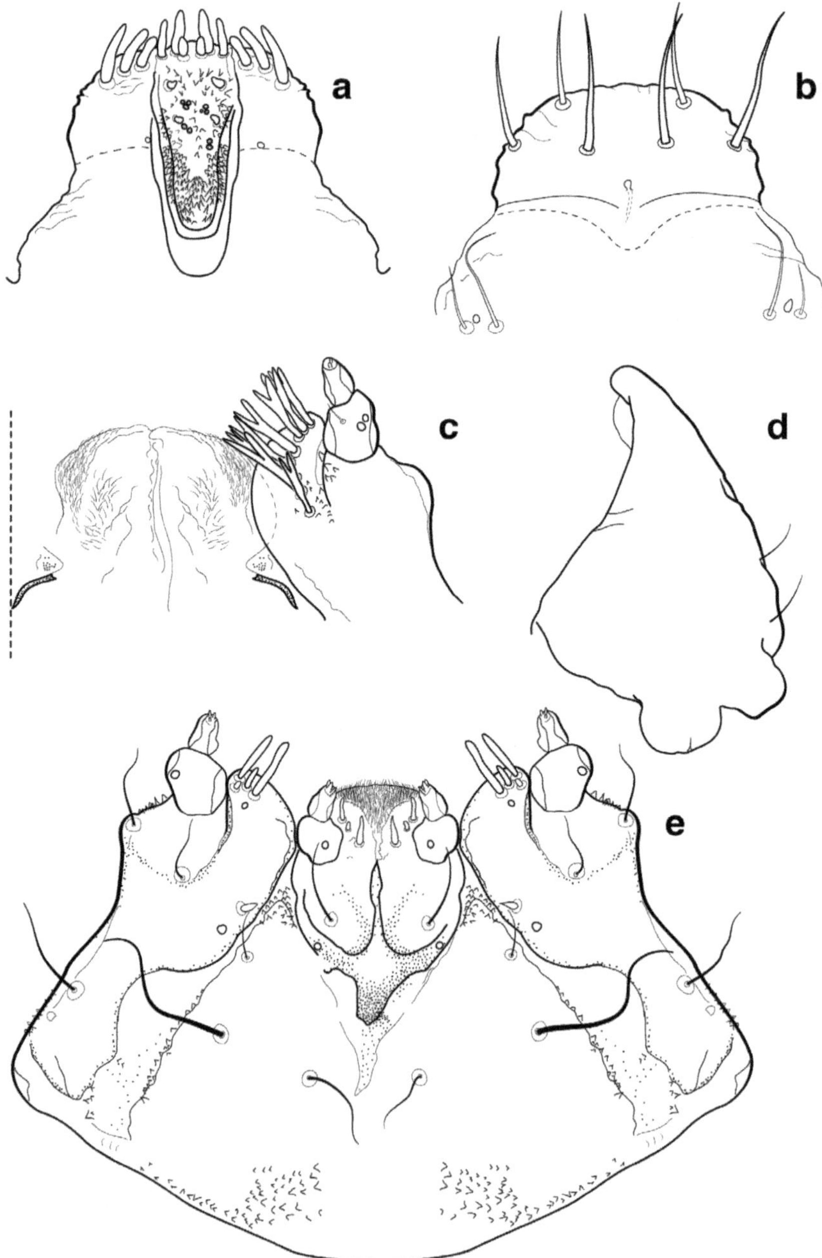

Figure 60. *Diocalandra frumenti* mouthparts: (**a**) epipharynx; (**b**) clypeus and labrum; (**c**) detail of hypopharynx, mala, and maxillary palp, dorsal view; (**d**) mandible; and (**e**) labio-maxillary complex, ventral view.

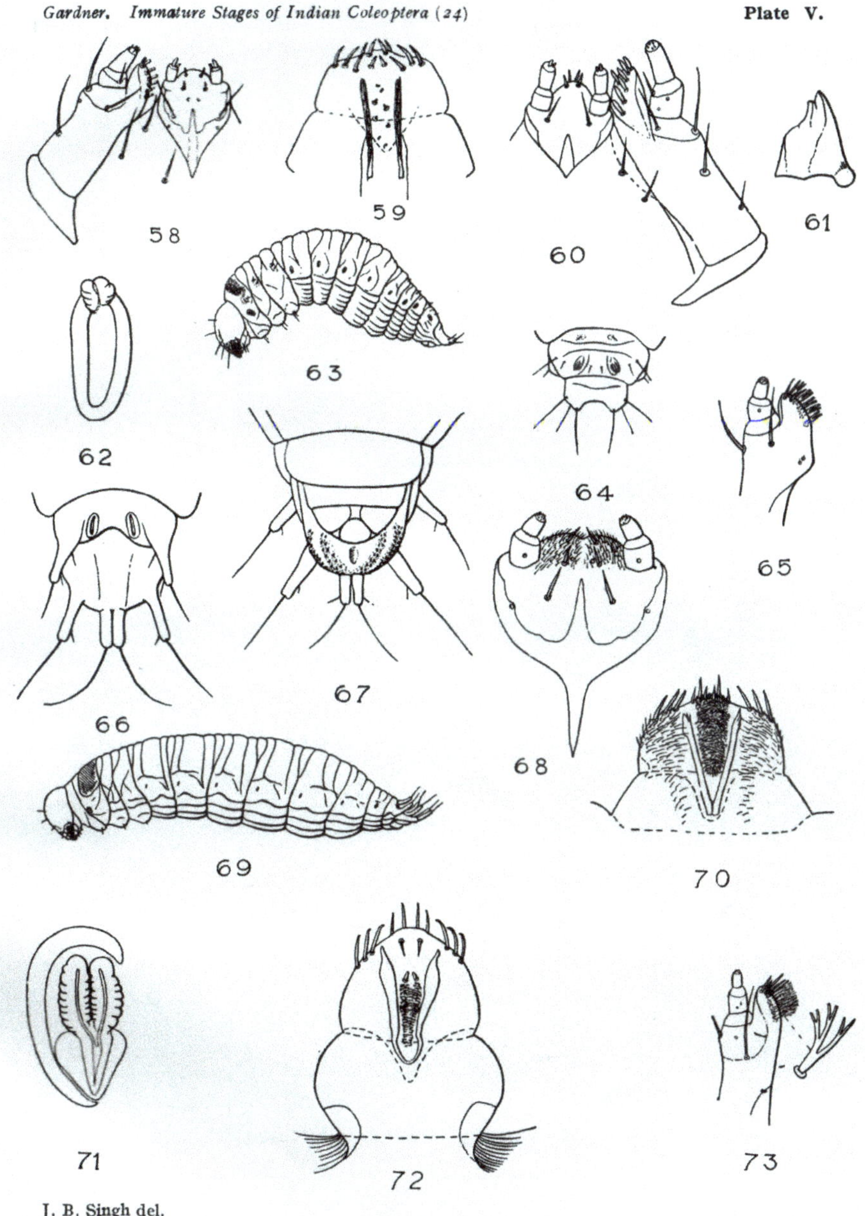

Figure 61. Reproduction of Gardner's ([12], p. 260) Plate V: "Larvae of Curculionidae. Figure 58. *Himatinum lalli*, ventral mouthparts. Figures 59 and 60. *Phloeophagosoma aesculi*. 59. Epipharynx. 60. Ventral mouthparts. Figures 61 and 62. *Macrorhyncholus ventilaginis*. 61. Mandible. 62. Spiracle. Figures 63 and 64. *Trochorrhopalus* [sic] *balwanti*. 63. Larva. 64. Caudal extremity, dorsal view. Figures 65–71. *Anius* [*Nephius*] *pauperatus*. 65. Maxillary mala and palp. 66. Caudal extremity, dorsal view. 67. Caudal extremity, ventral view. 68. Labium. 69. Larva. 70. Epipharynx. 71. Spiracle. Figures 72 and 73. *Dryophthoroides parvungulis*. 72. Epipharynx. 73. Maxillary mala and palp."

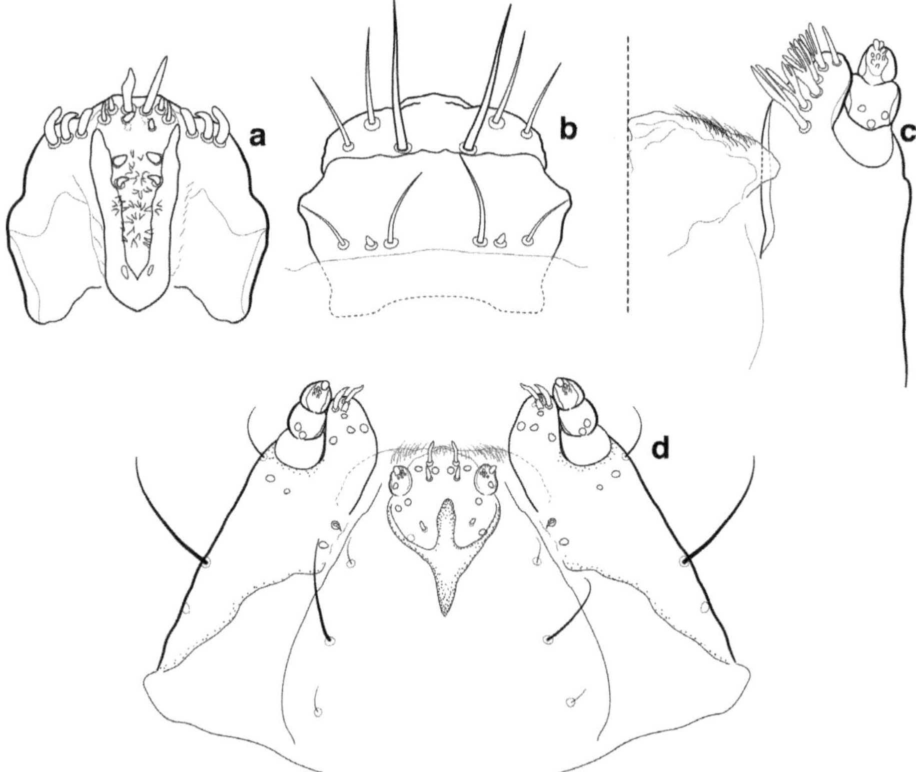

Figure 62. *Dryophthorus americanus* mouthparts: (**a**) epipharynx; (**b**) clypeus and labrum; (**c**) detail of hypopharynx, mala, and maxillary palp, dorsal view; and (**d**) labio-maxillary complex, ventral view.

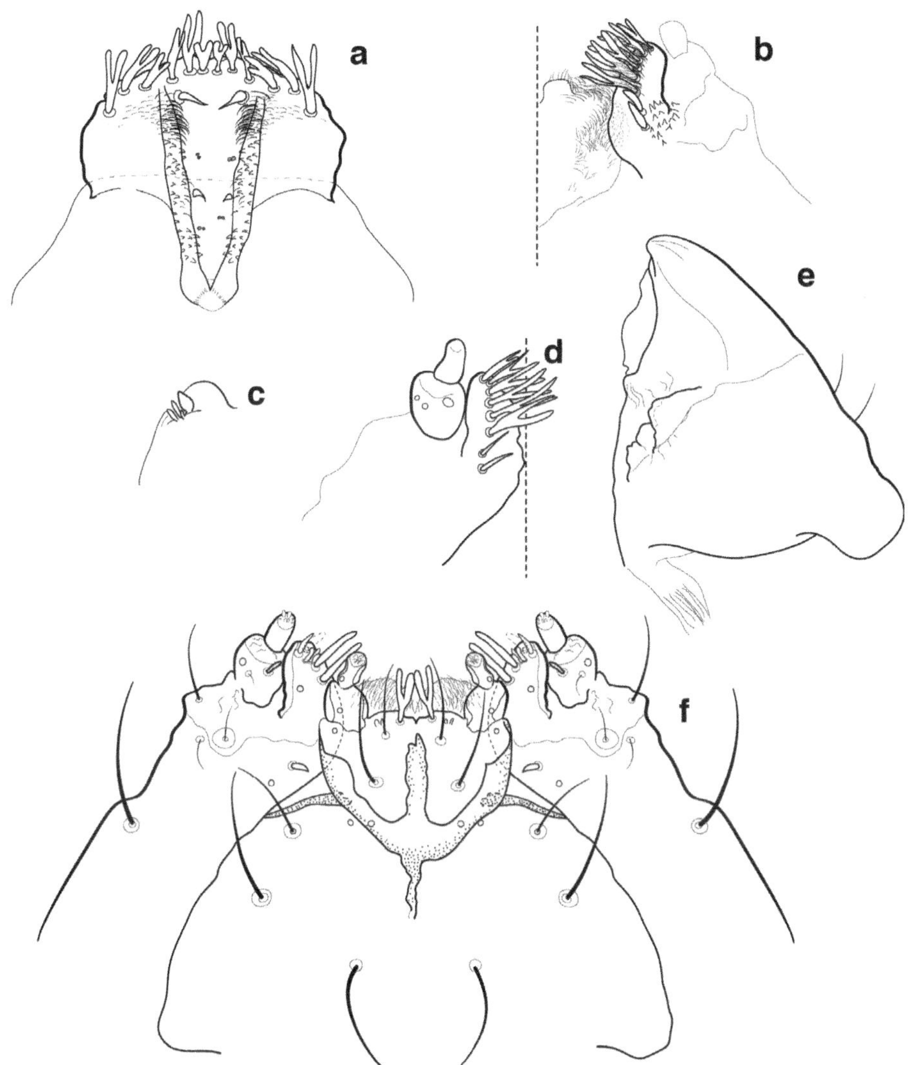

Figure 63. *Eucalandra setulosa* mouthparts: (**a**) epipharynx; (**b**) mala and hypopharynx, dorsal view; (**c**) antenna, lateral view; (**d**) detail of mala and maxillary palp, ventral view; (**e**) mandible; and (**f**) labio-maxillary complex, ventral view.

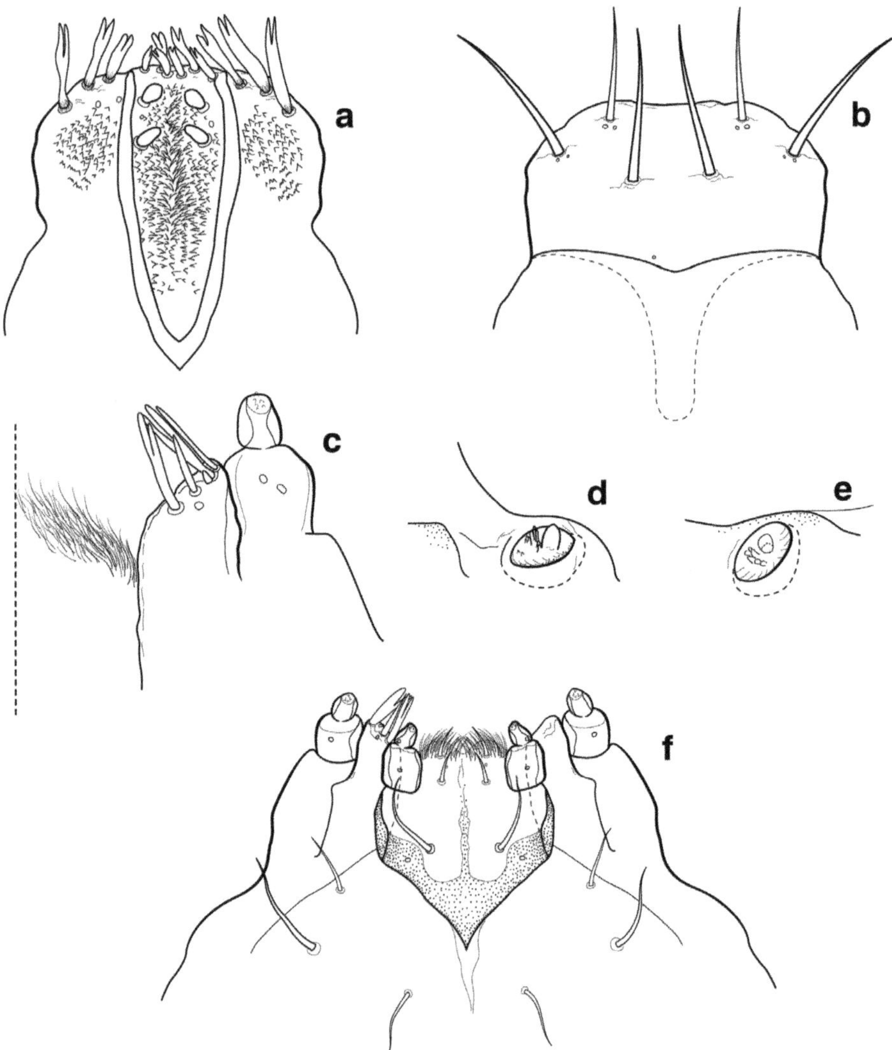

Figure 64. *Eugnoristus braueri* mouthparts: (**a**) epipharynx; (**b**) clypeus (incomplete) and labrum; (**c**) hypopharynx, mala, and maxillary palp, dorsal view; (**d**) antenna, oblique-lateral view; (**e**) antenna, antero-lateral view; and (**f**) labio-maxillary complex, ventral view.

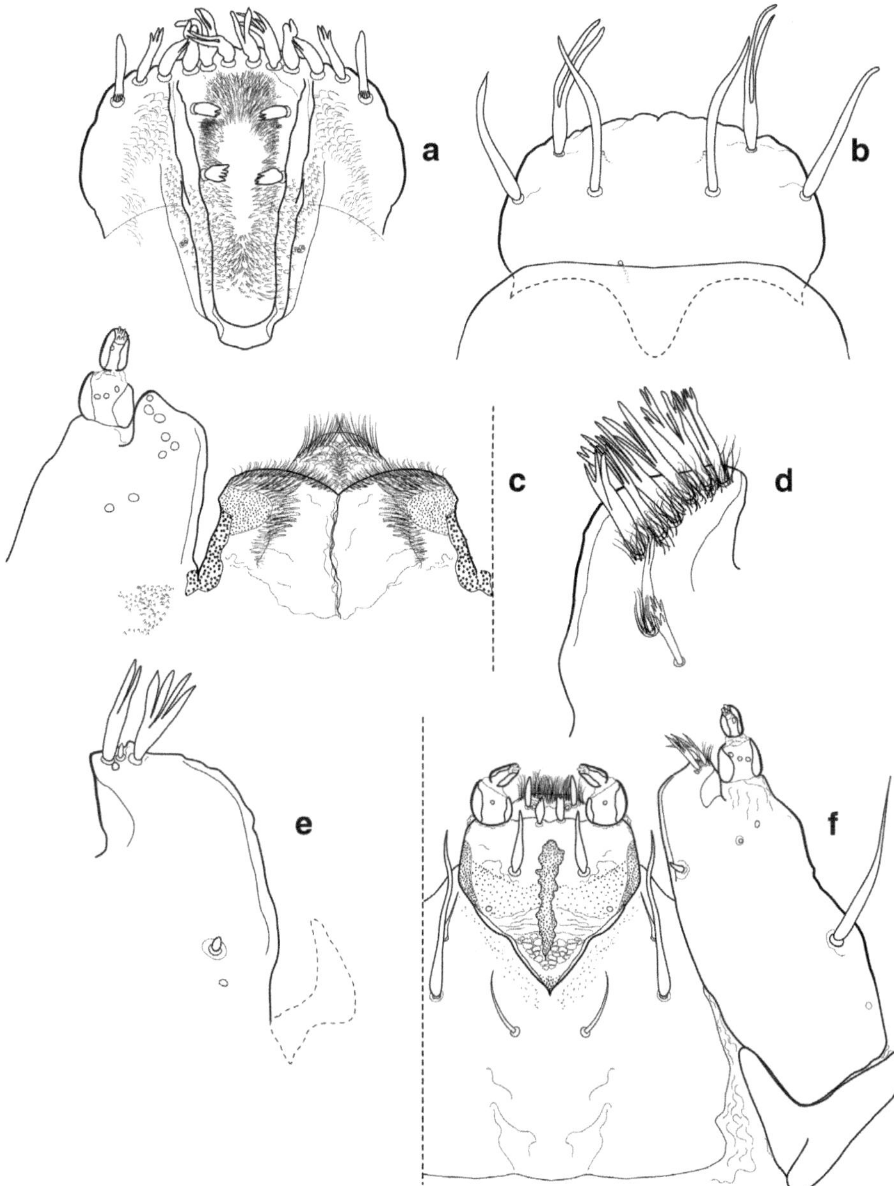

Figure 65. *Foveolus* c.f. *maculatus* mouthparts: (**a**) epipharynx; (**b**) clypeus (incomplete) labrum; (**c**) hypopharynx, mala, and maxillary palp, dorsal view; (**d** detail of mala and dorsal malar setae; (**e**) detail of mala and ventral malar setae; and (**f**) labio-maxillary complex, ventral view.

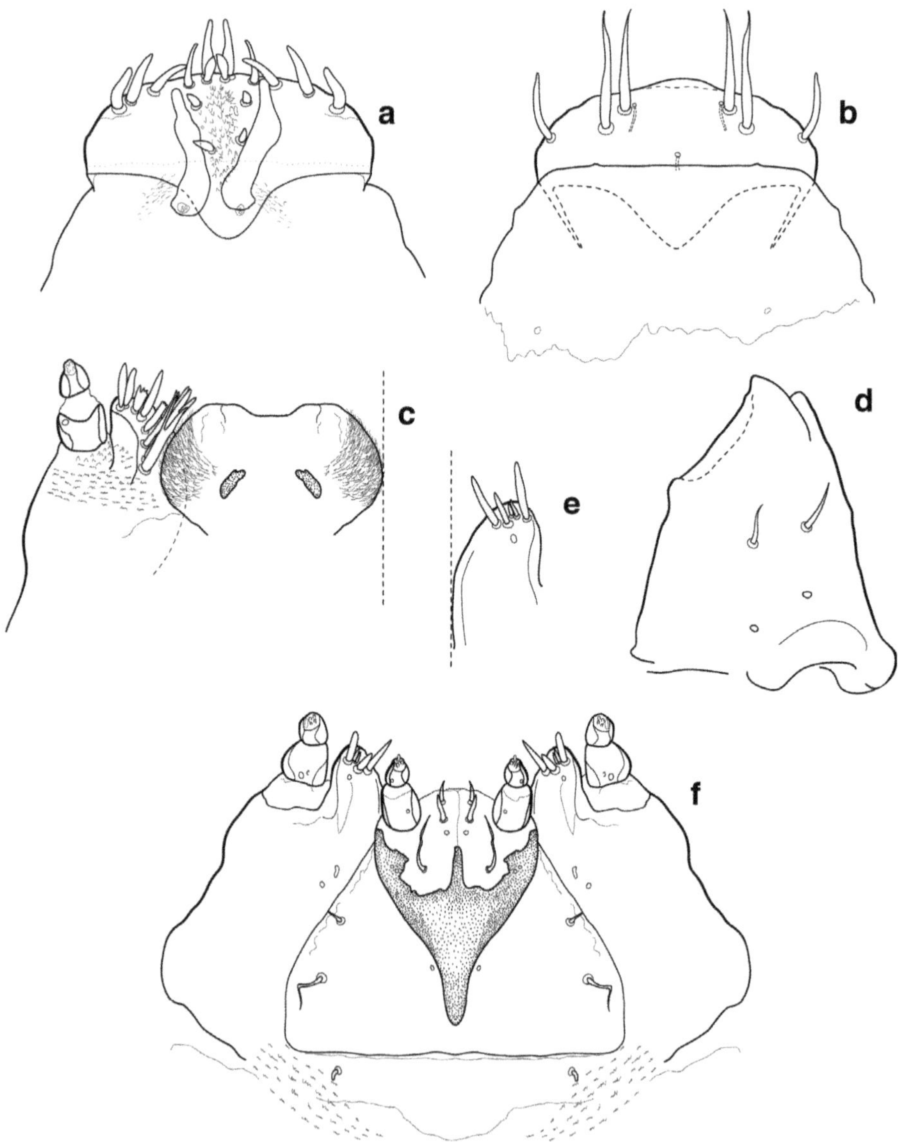

Figure 66. Probably *Mesocordylus* sp. mouthparts: (**a**) epipharynx; (**b**) clypeus (incomplete) and labrum; (**c**) detail of hypopharynx, mala and maxillary palp, dorsal view; (**d**) mandible; (**e**) detail of mala and ventral malar setae; and (**f**) labio-maxillary complex, ventral view.

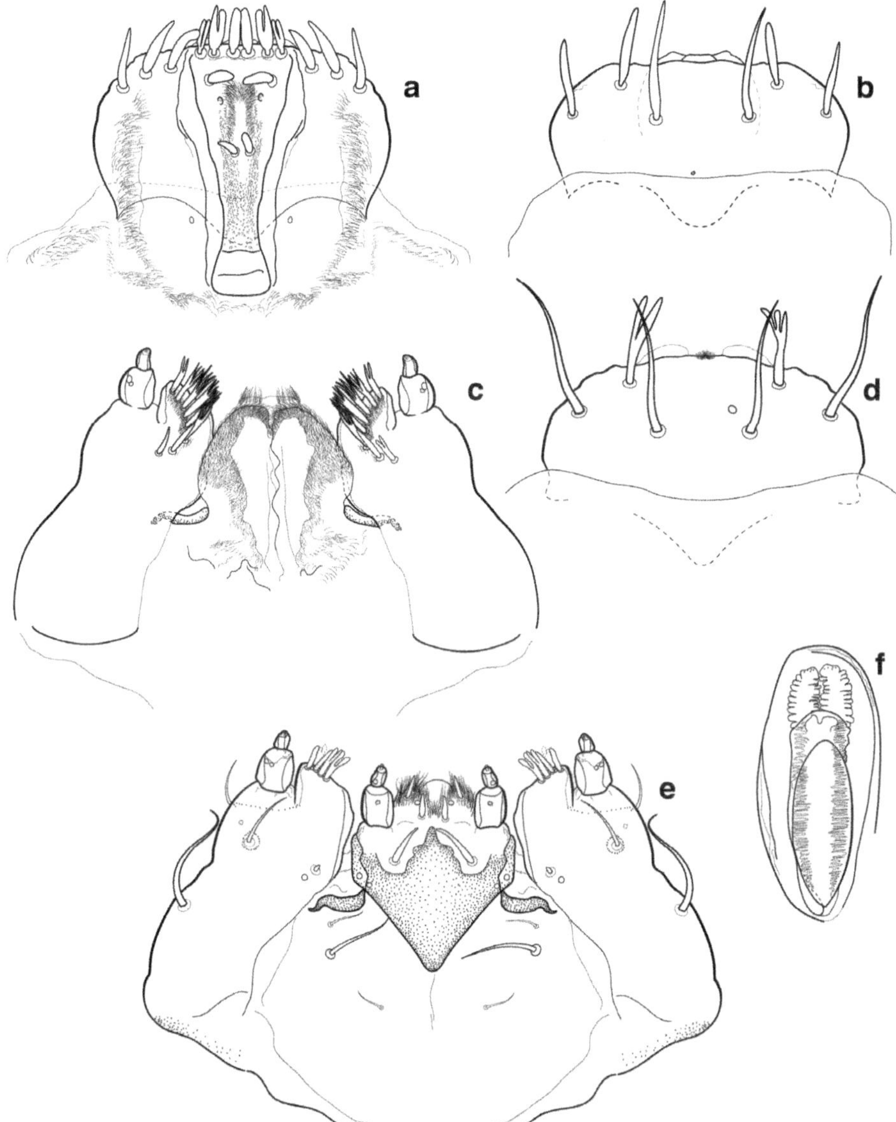

Figure 67. *Metamasius ritchiei* mouthparts: (**a**) epipharynx; (**b**) clypeus (incomplete) and labrum; (**c**) maxillae and hypopharynx, dorsal view; (**d**) *Metamasius hemipterus* clypeus (incomplete) and labrum; (**e**) labio-maxillary complex, ventral view; and (**f**) thoracic spiracle.

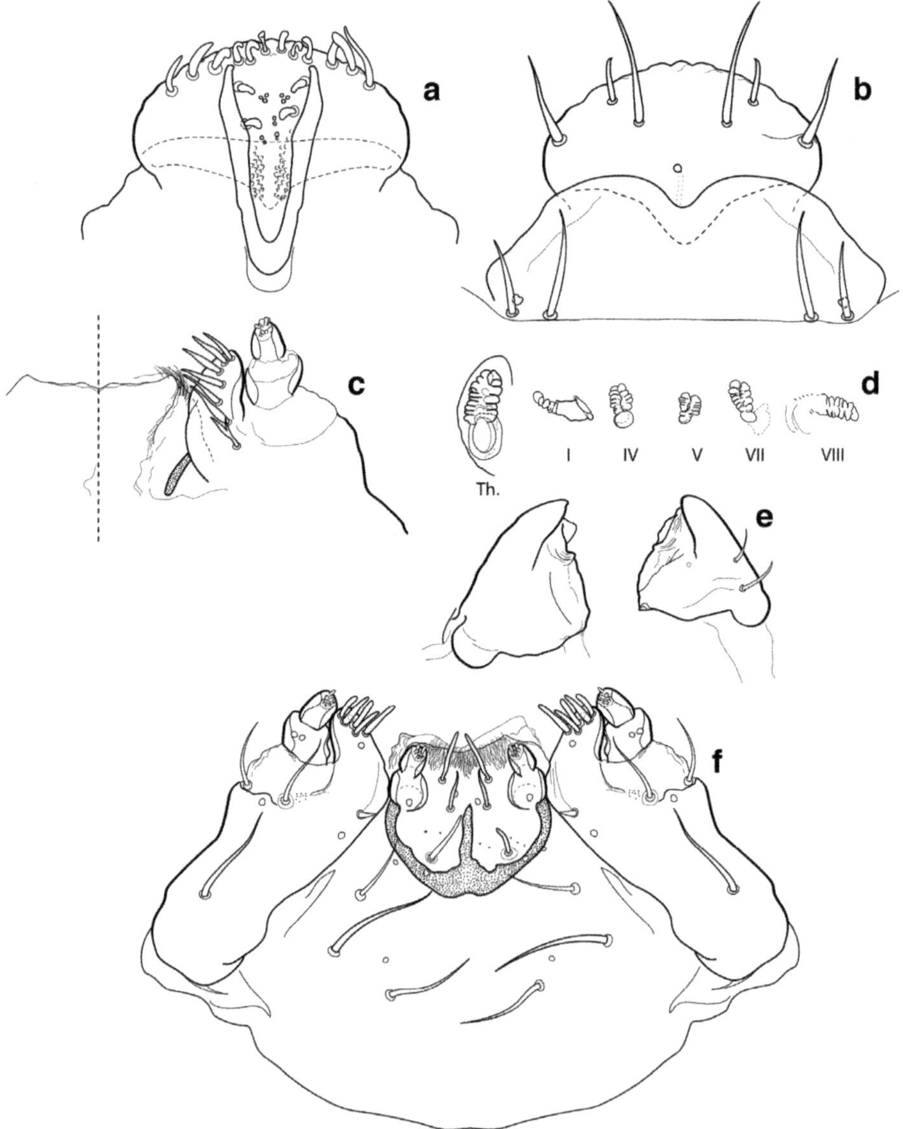

Figure 68. *Myocalandra* sp. Mouthparts: (**a**) epipharynx; (**b**) clypeus and labrum; (**c**) maxillae and hypopharynx, dorsal view; (**d**) spiracles; (**e**) mandibles; and (**f**) labio-maxillary complex, ventral view.

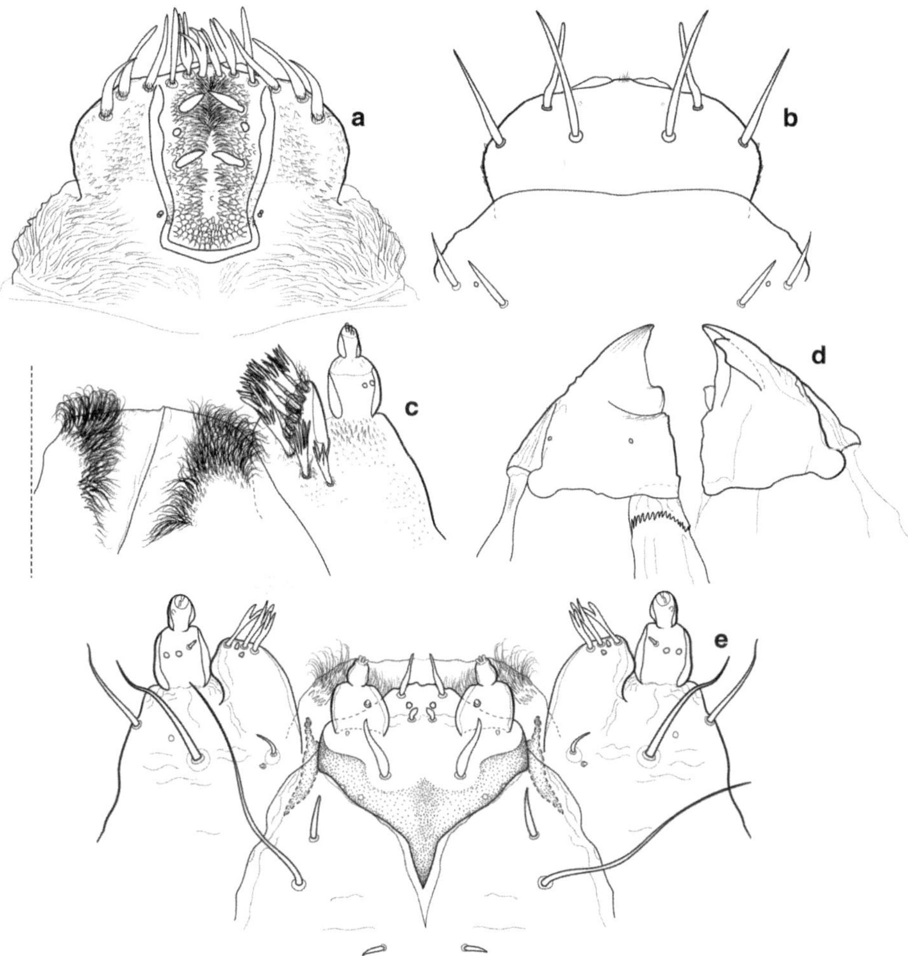

Figure 69. *Nassophasis* sp. Mouthparts: (**a**) epipharynx; (**b**) clypeus and labrum; (**c**) maxillae and hypopharynx, dorsal view; (**d**) spiracles; (**e**) mandibles; and (**f**) labio-maxillary complex, ventral view.

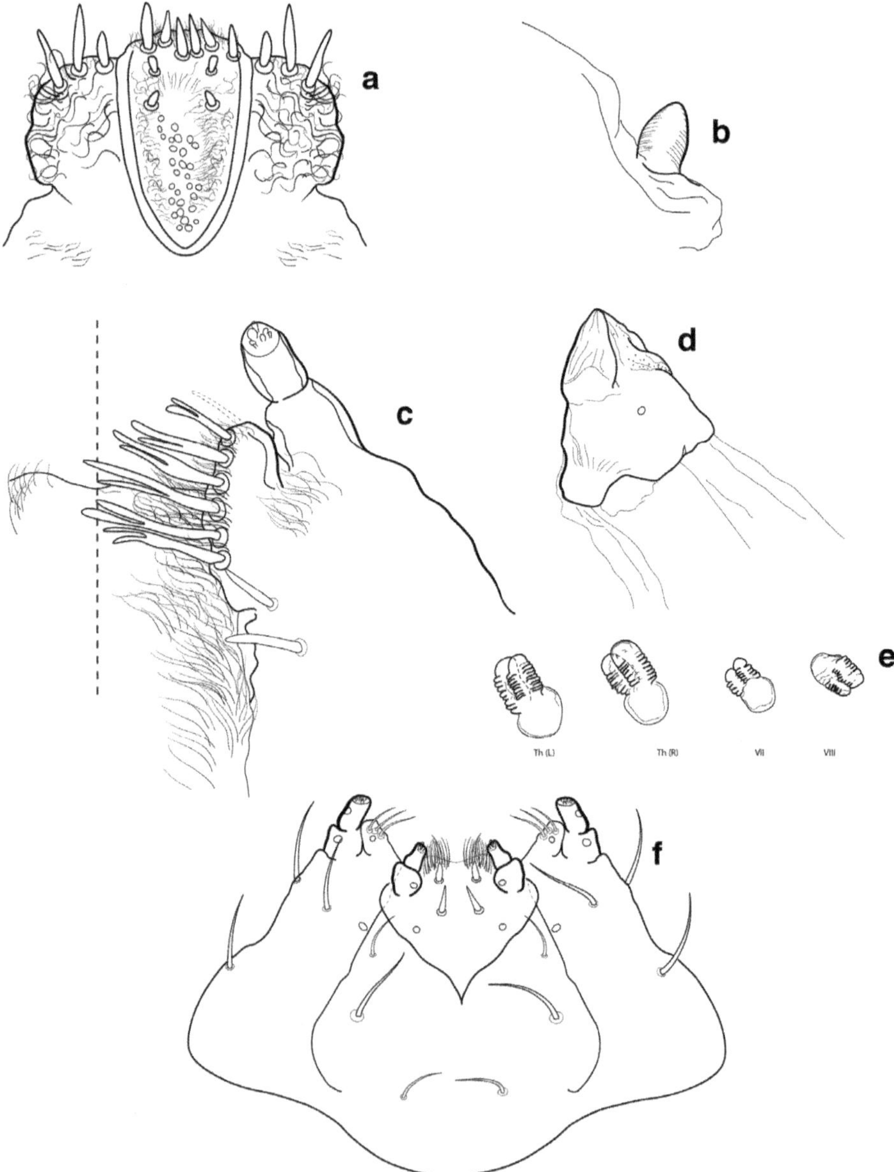

Figure 70. *Nephius* (=*Anius*) *pauperatus* mouthparts: (**a**) epipharynx; (**b**) antenna, lateral view; (**c**) maxillae and hypopharynx, dorsal view; (**d**) mandible; (**e**) spiracles; and (**f**) labio-maxillary complex, ventral view. Th (L): thorax, left side; Th (R): thorax, right side; VII: 7th abdominal segment; VIII: 8th abdominal segment.

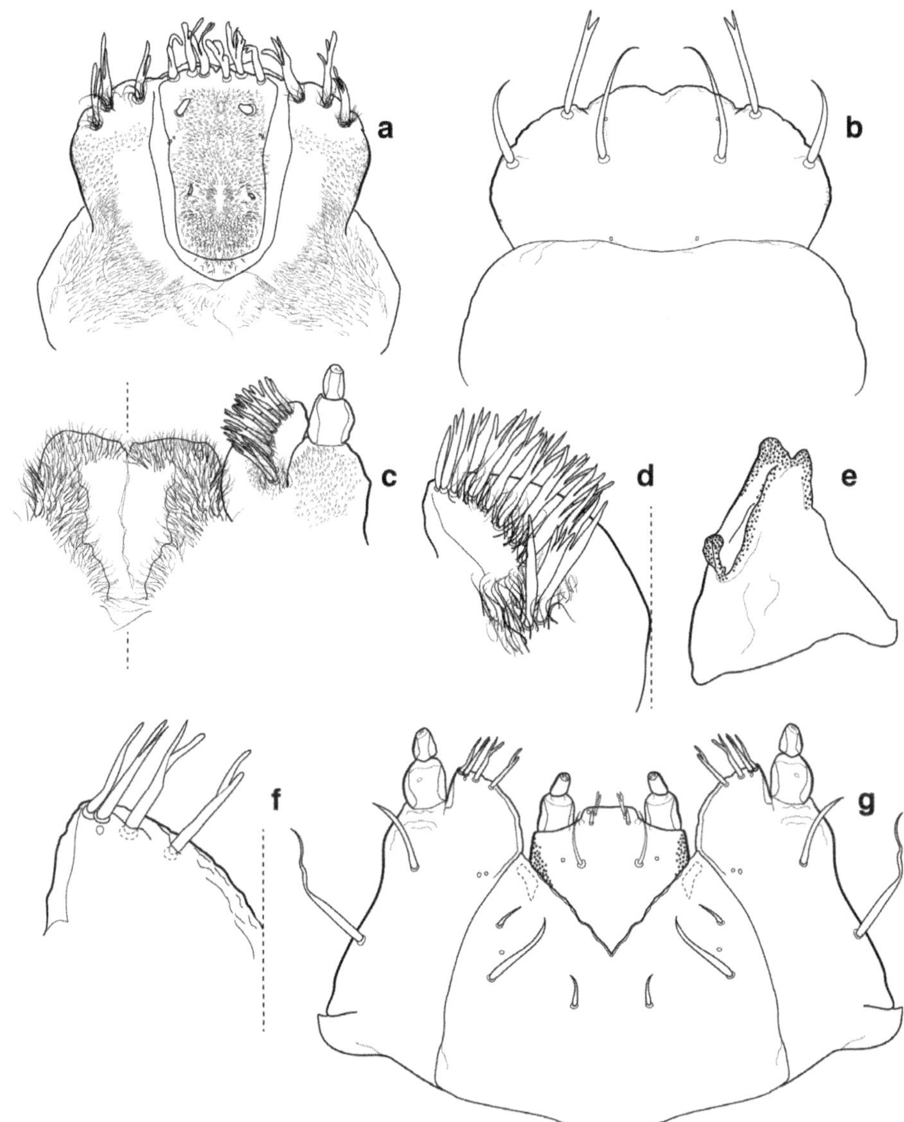

Figure 71. *Odoiporus longicollis* mouthparts: (**a**) epipharynx; (**b**) clypeus (incomplete) and labrum; (**c**) detail of hypopharynx, mala, and maxillary palp, dorsal view; (**d**) detail of mala and dorsal malar setae; (**e**) mandible; (**f**) detail of mala and ventral malar setae; and (**g**) labio-maxillary complex, ventral view.

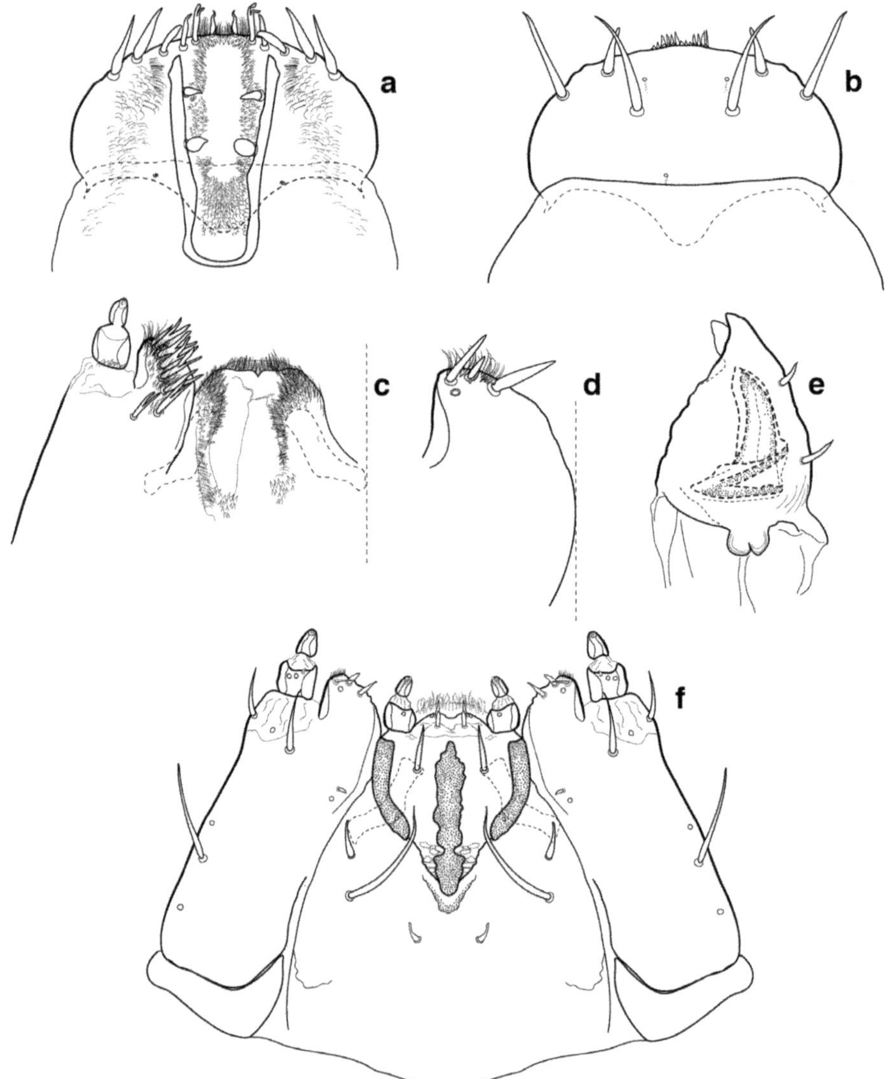

Figure 72. *Paramasius cristulatus* mouthparts: (**a**) epipharynx; (**b**) labrum; (**c**) maxillae and hypopharynx, dorsal view; (**d**) mala, ventral view; (**e**) mandible, internally with scerotized process; and (**f**) labio-maxillary complex, ventral view.

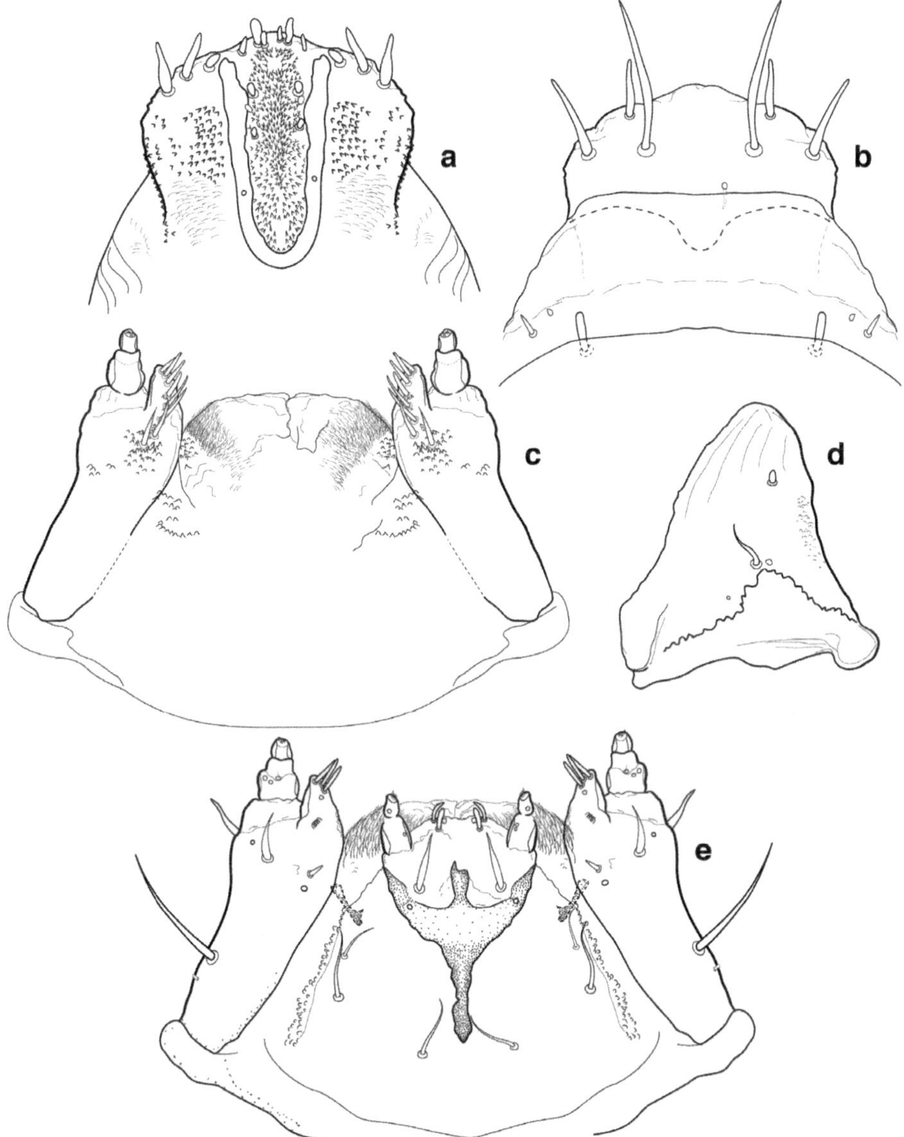

Figure 73. *Phacecorynes zamiae* mouthparts: (**a**) epipharynx; (**b**) clypeus and labrum; (**c**) maxillae and hypopharynx, dorsal view; (**d**) mandible; and (**e**) labio-maxillary complex, ventral view.

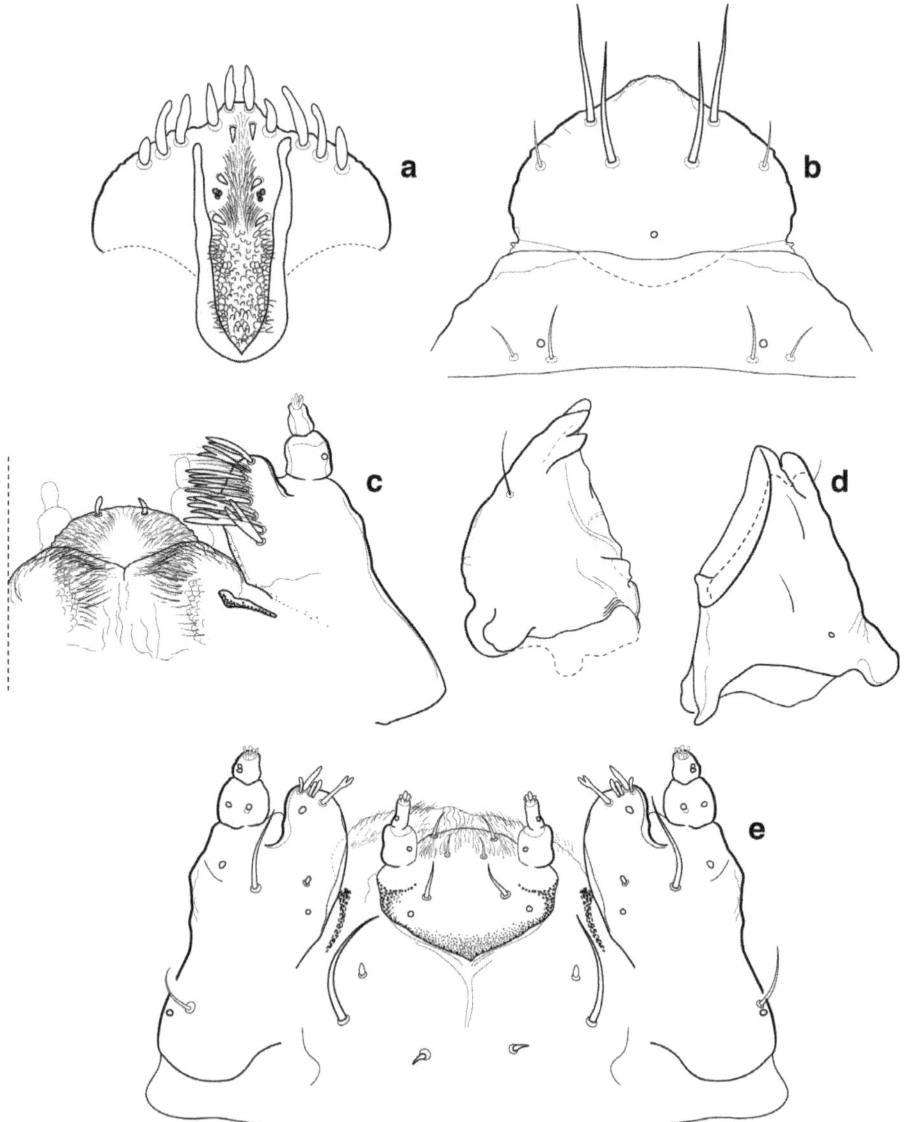

Figure 74. *Polytus mellerborgii* mouthparts: (**a**) epipharynx; (**b**) clypeus and labrum; (**c**) maxillae and hypopharynx, dorsal view; (**d**) mandibles; and (**e**) labio-maxillary complex, ventral view.

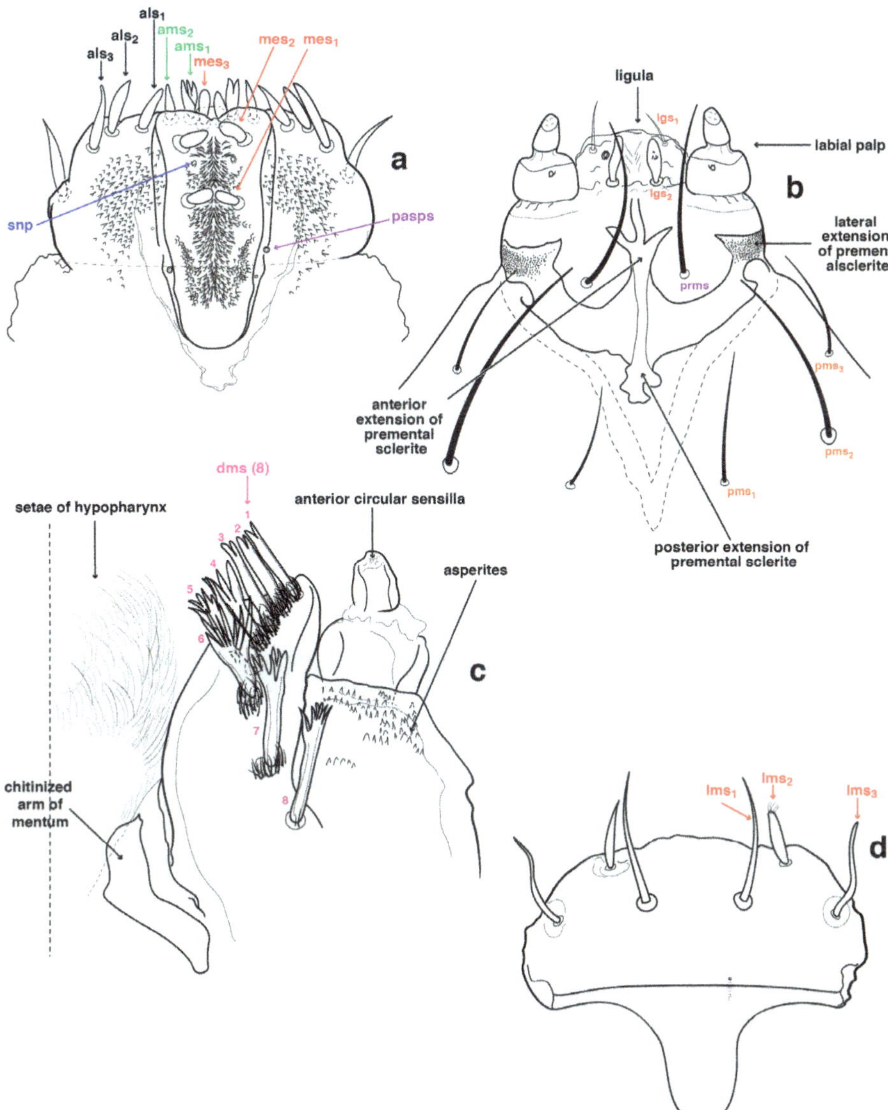

Figure 75. *Poteriophorus uhlemanni* mouthparts: (**a**) epipharynx; (**b**) labium; (**c**) detail of maxilla and hypopharynx, dorsal view; and (**d**) labrum. *Als*: anterolateral setael; *ams*: anteromedian setae; *dms*: dorsal malar setae; *lgs*: ligular setae; *lms*: labral setae; *mes*: median epipharyngeal setae; *pasps*: posterior accessory sensory pores; *pms*: postmental setae; *prms*: premental setae; *snp*: sensory pores of epipharynx.

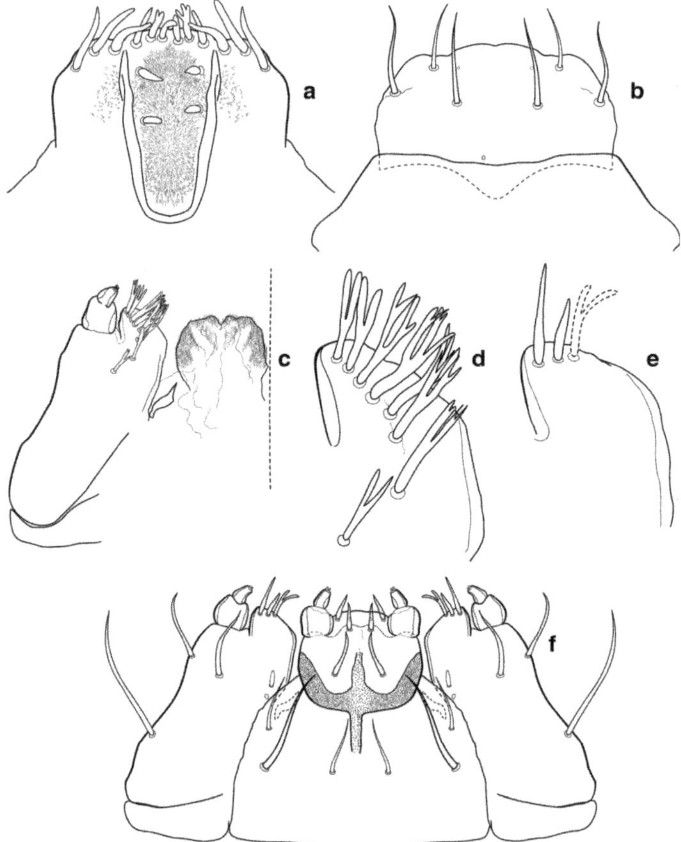

Figure 76. *Rhabdoscelus* prob. *obscurus* mouthparts: (**a**) epipharynx; (**b**) clypeus (incomplete) and labrum; (**c**) maxillae and hypopharynx, dorsal view; (**d**) detail of mala and dorsal malar setae; (**e**) detail of mala and ventral malar setae; and (**f**) labio-maxillary complex, ventral view.

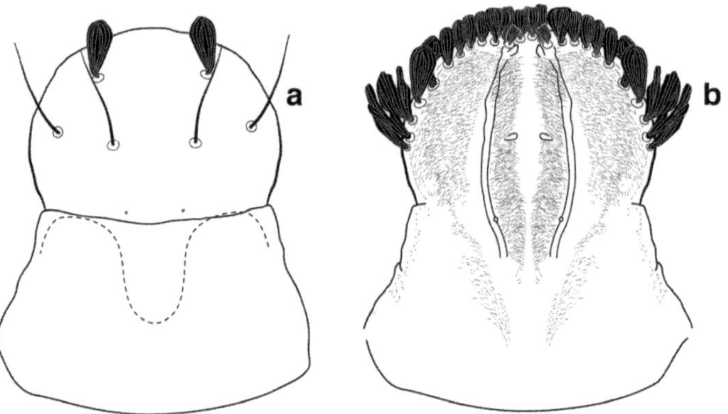

Figure 77. *Rhinostomus barbirostris* mouthparts: (**a**) clypeus (without setae) and labrum; and (**b**) epipharynx.

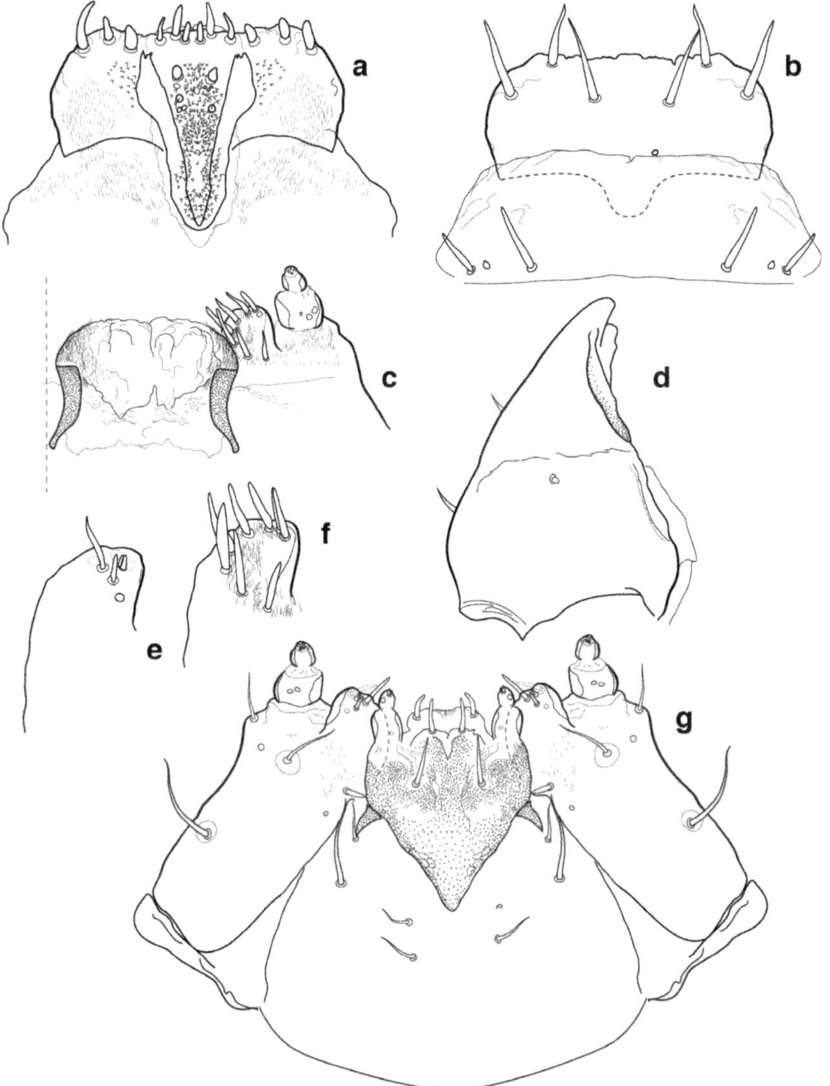

Figure 78. *Rhodobaenus tredecempunctatus* mouthparts: (**a**) epipharynx; (**b**) clypeus and labrum; (**c**) mala, maxillary palp, and hypopharynx, ventral view; (**d**) mandible; (**e**) detail, mala, ventral view; (**f**) detail, mala, dorsal view; and (**g**) labio-maxillary complex, ventral view.

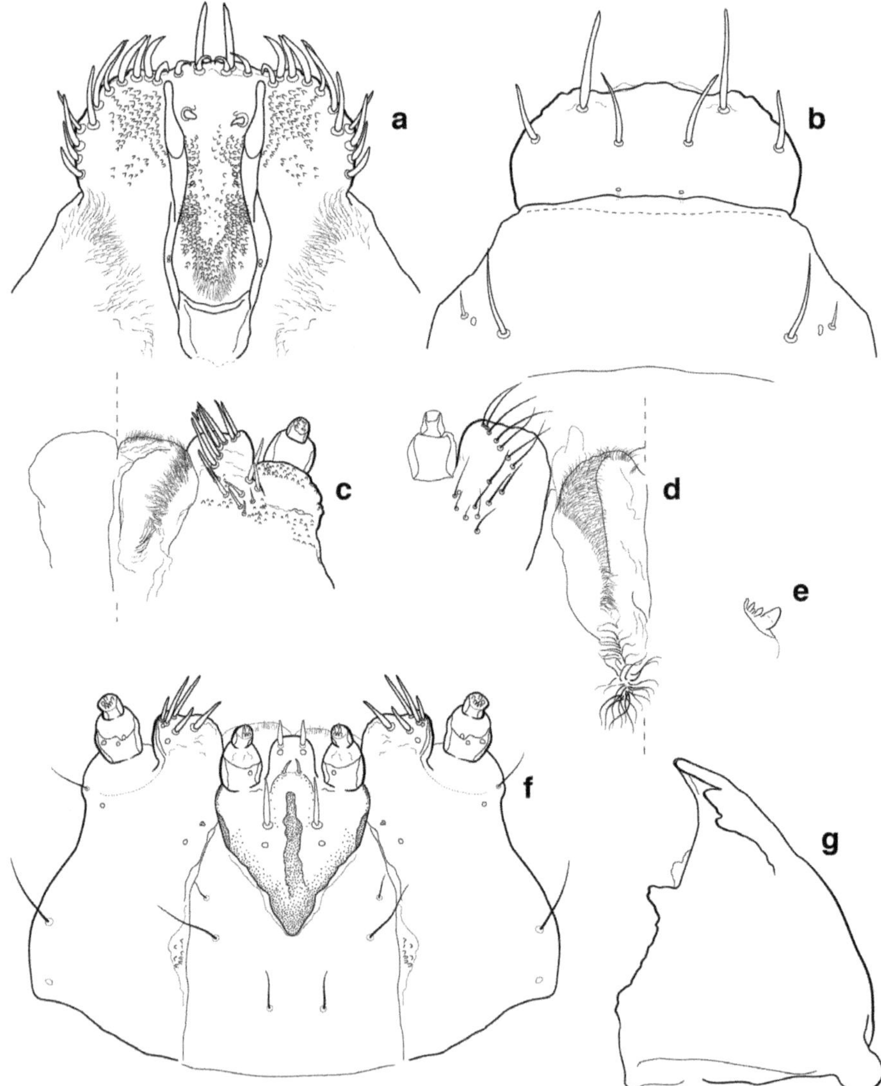

Figure 79. *Rhynchophorus palmarum* mouthparts: (**a**) epipharynx; (**b**) clypeus and labrum; (**c**) mala and maxillary palp, dorsal view; (**d**) detail, mala and hypopharynx, ventral view; (**e**) maxillae, hypopharynx, labium, dorsal view; and (**f**) labio-maxillary complex, ventral view.

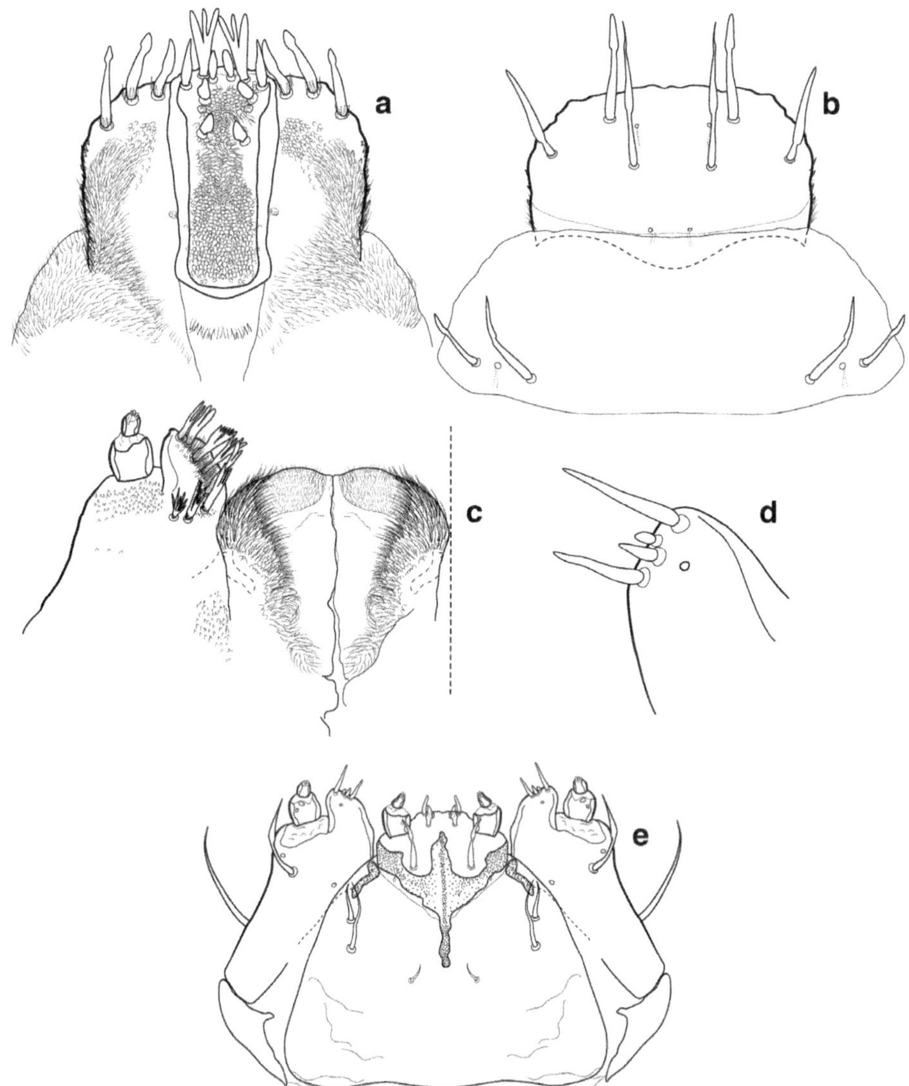

Figure 80. *Scyphophorus acupunctatus* mouthparts: (**a**) epipharynx; (**b**) clypeus and labrum; (**c**) mala, maxillary palp, and hypopharynx, dorsal view; (**d**) detail of mala and ventral malar setae; and (**e**) labio-maxillary complex, ventral view.

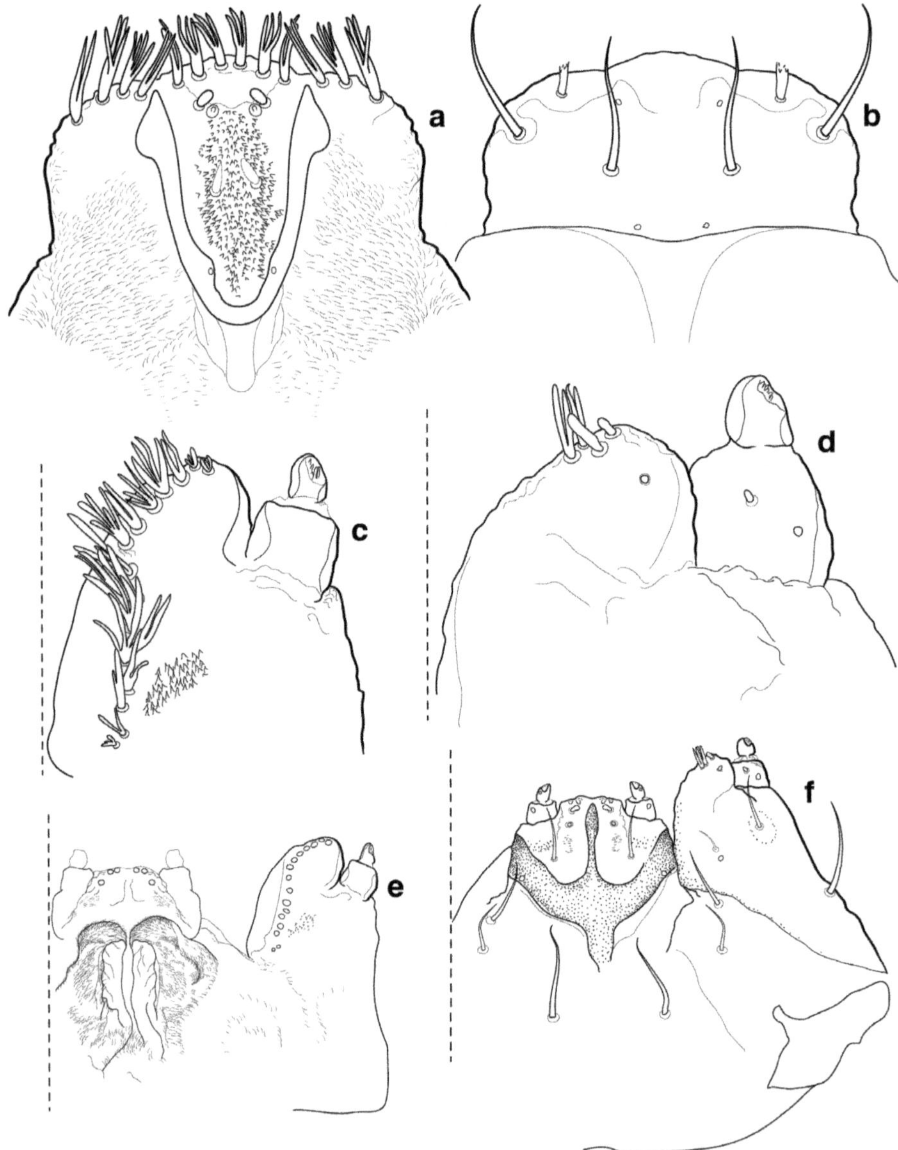

Figure 81. *Sipalinus gigas* mouthparts: (**a**) epipharynx; (**b**) clypeus (incomplete) and labrum; (**c**) detail, mala and maxillary palp, dorsal view; (**d**) mala and maxillary palp, ventral view; (**e**) labio-maxillary complex, dorsal view; and (**g**) labio-maxillary complex, ventral view (*lms2*: labial setae 2 broken).

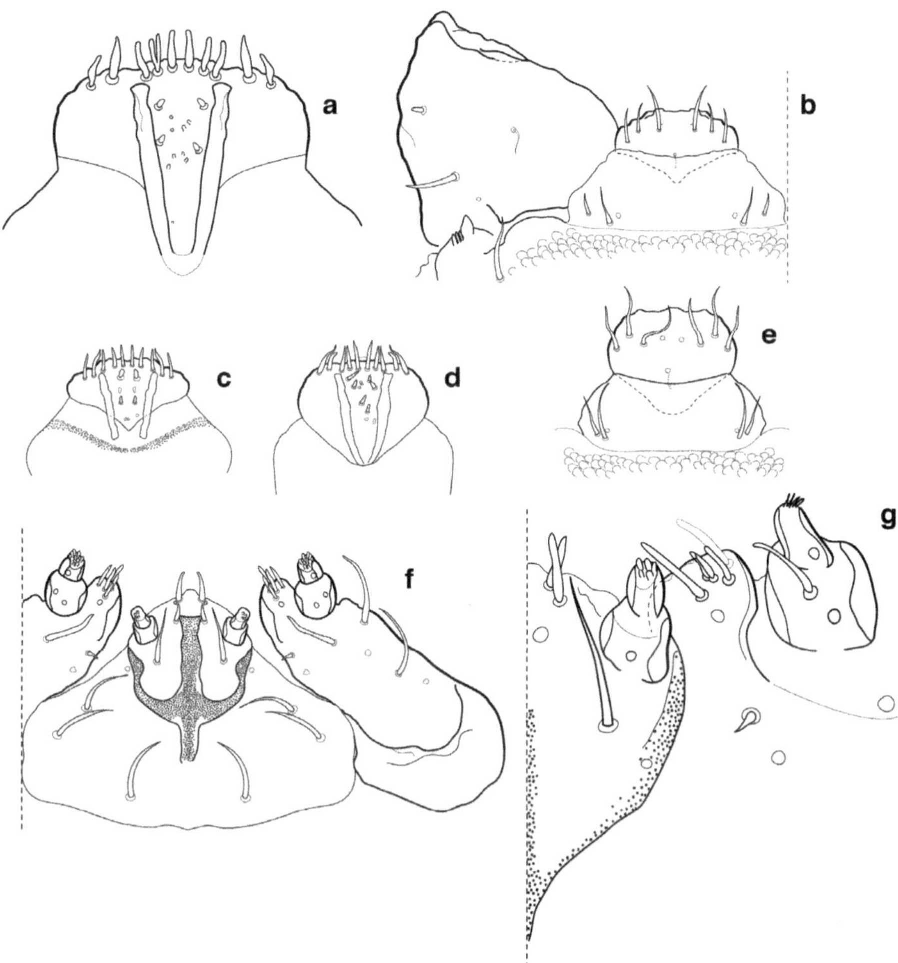

Figure 82. *Sitophilus* spp. mouthparts: (**a**) *Sitophilus granarius* epipharynx; (**b**) *Sitophilus linearis*, labrum, clypeus, mandible, antenna and anterior margin of frons; (**c**) *Sitophilus linearis*, epipharynx; (**d**) *Sitophilus oryzae*, epipharynx; (**e**) *Sitophilus oryzae*, labrum, clypeus, and anterior margin of frons; (**f**) *Sitophilus linearis*, labio-maxillary complex, ventral view; and (**g**) *Sitophilus granarius* detail, mala and labium, ventral view.

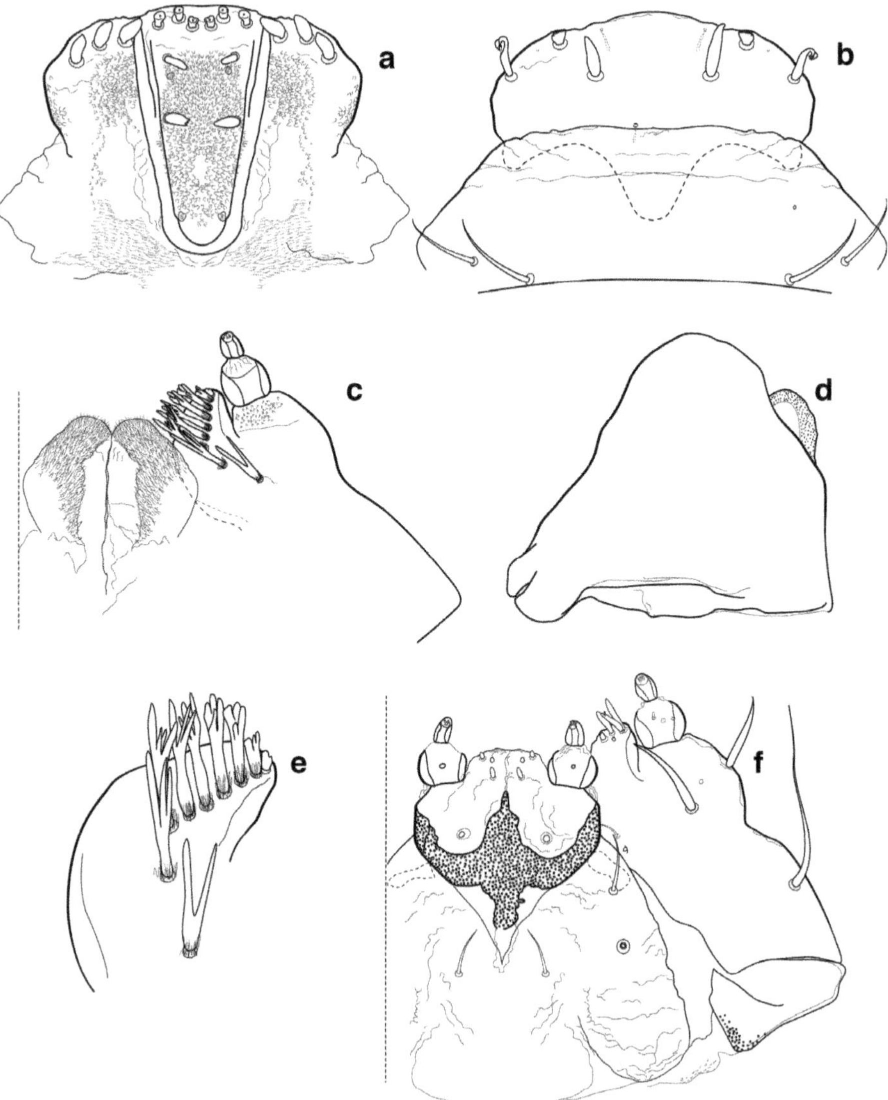

Figure 83. *Sparganobasis subcruciata* mouthparts: (**a**) epipharynx, setae broken/weathered; (**b**) clypeus and labrum, setae broken/weathered; (**c**) mala and hypopharynx, dorsal view; (**d**) mandible; (**e**) detail, mala and dorsal malar setae, some setae broken/weathered; and (**f**) labio-maxillary complex, ventral view, some setae broken/weathered.

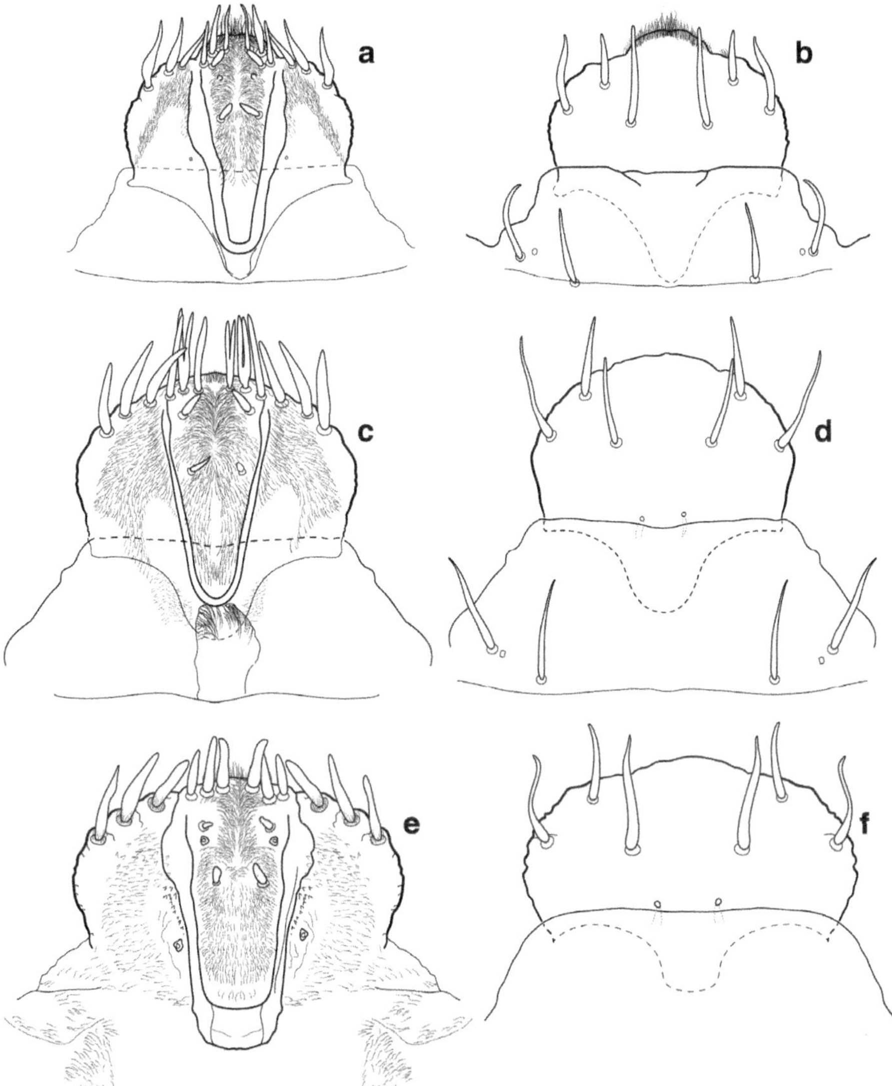

Figure 84. *Sphenophorus* spp. mouthparts: (**a**) *Sphenophorus discolor*, epipharynx; (**b**) *Sphenophorus discolor*; clypeus and labrum; (**c**) *Sphenophorus aequalis*, epipharynx; (**d**) *Sphenophorus aequalis*, clypeus and labrum; (**e**) *Sphenophorus maidis*, epipharynx; and (**f**) *Sphenophorus maidis*, clypeus (incomplete) and labrum.

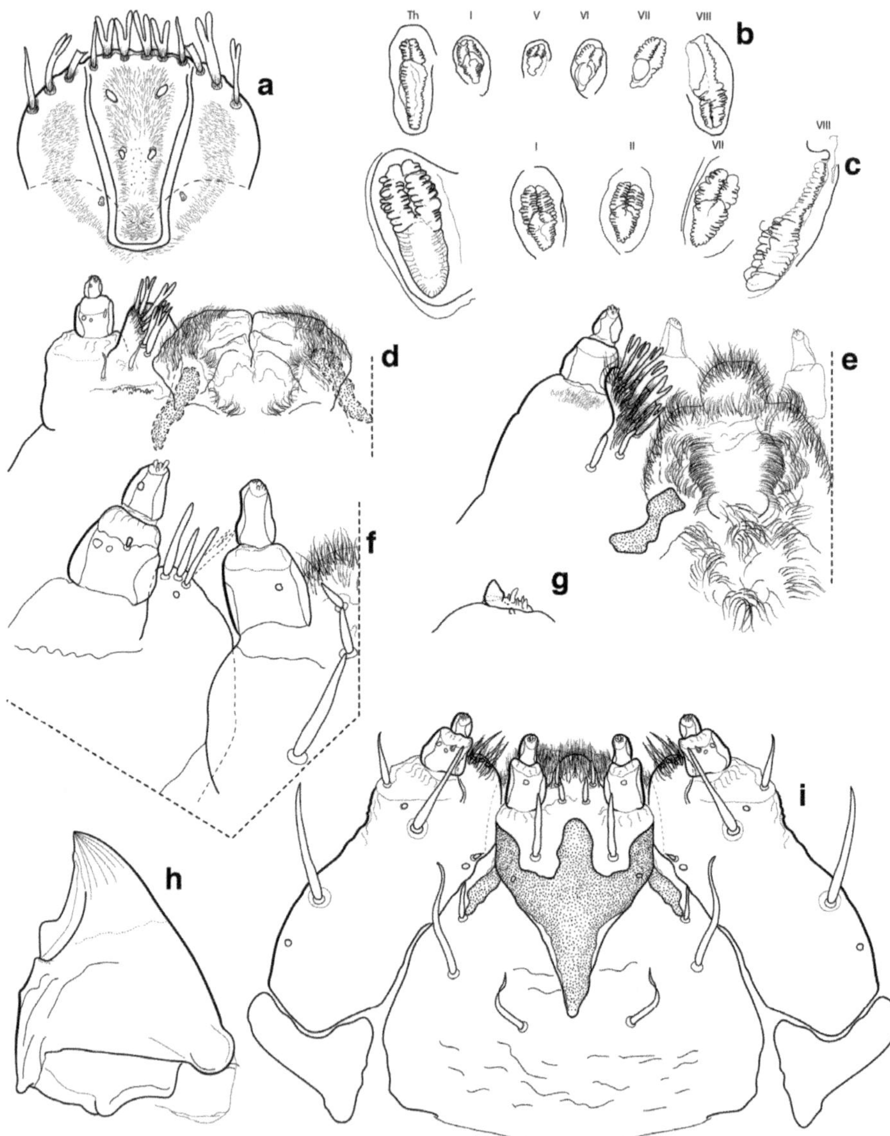

Figure 85. *Sphenophorus* spp. mouthparts and spiracles: (**a**) *Sphenophorus pontederiae*, epipharynx; (**b**) *Sphenophorus pontederiae*, spiracles; (**c**) *Sphenophorus maidis* spiracles; (**d**) *Sphenophorus maidis*, maxillae and hypopharynx, dorsal view; (**e**) *Sphenophorus maidis*, maxillae and hypopharynx, dorsal view; (**f**) *Sphenophorus maidis*, detail, mala and labium, ventral view; (**g**) *Sphenophorus maidis*, antenna; (**h**) *Sphenophorus maidis*, mandible; and (**i**) *Sphenophorus maidis*, labio-maxillary complex, ventral view.

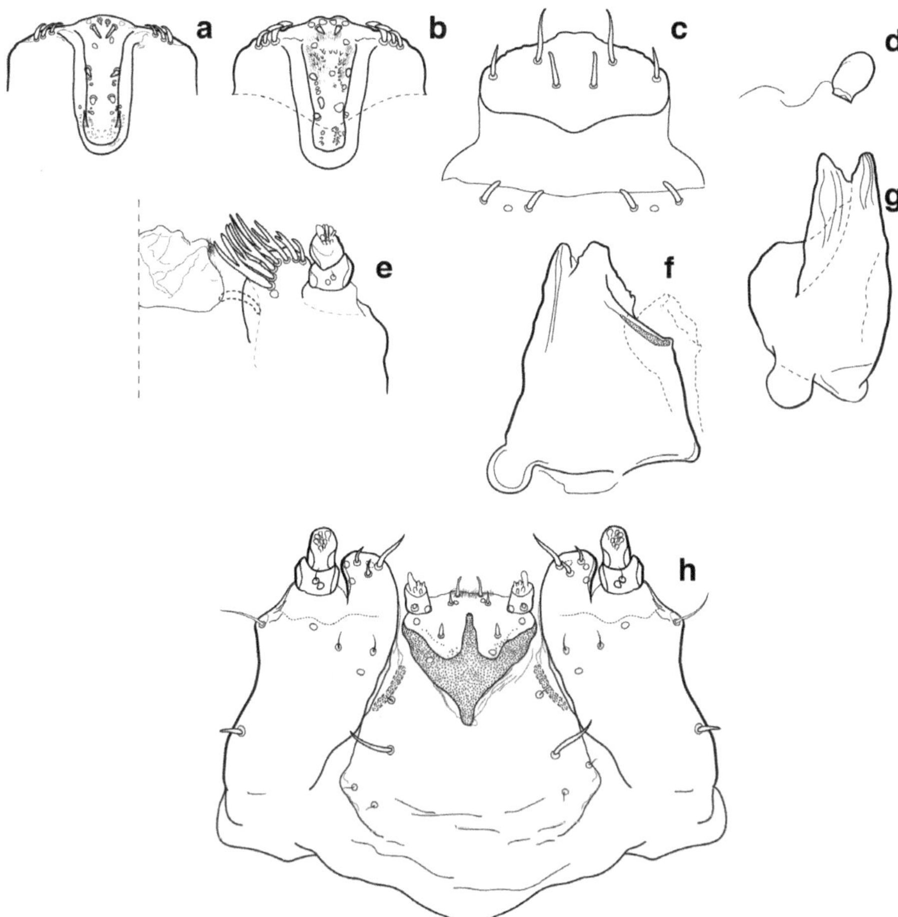

Figure 86. *Stenommatus musae* mouthparts: (**a**) epipharynx; (**b**) epipharynx, variation; (**c**) clypeus and labrum; (**d**) antennae; (**e**) mala, hypopharynx, maxillary palp, dorsal view; (**f**) mandible; (**g**) mandible; and (**h**) labio-maxillary complex, ventral view.

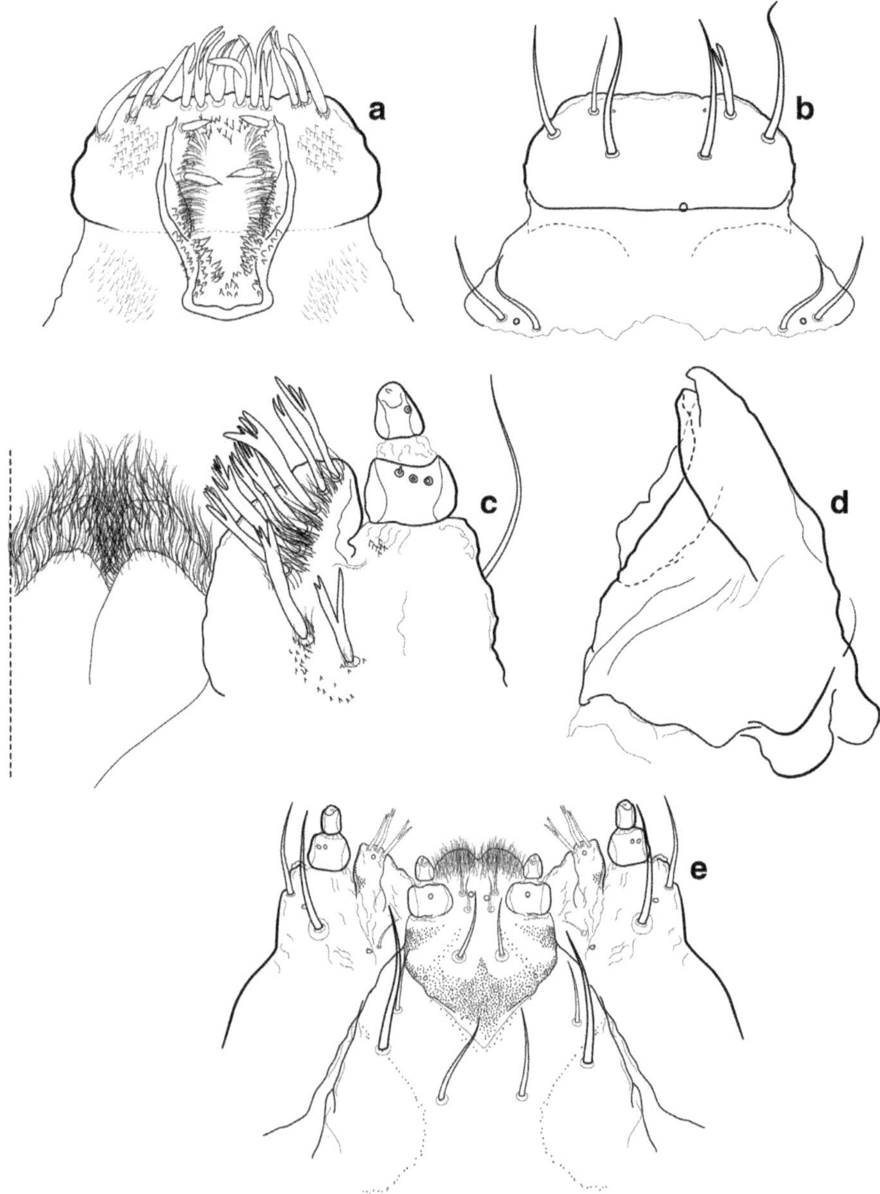

Figure 87. *Temnoschoita* sp. Mouthparts: (**a**) epipharynx; (**b**) clypeus and labrum; (**c**) detail of hypopharynx and maxillae, dorsal view; (**d**) mandible; and (**e**) labio-maxillary complex, ventral view.

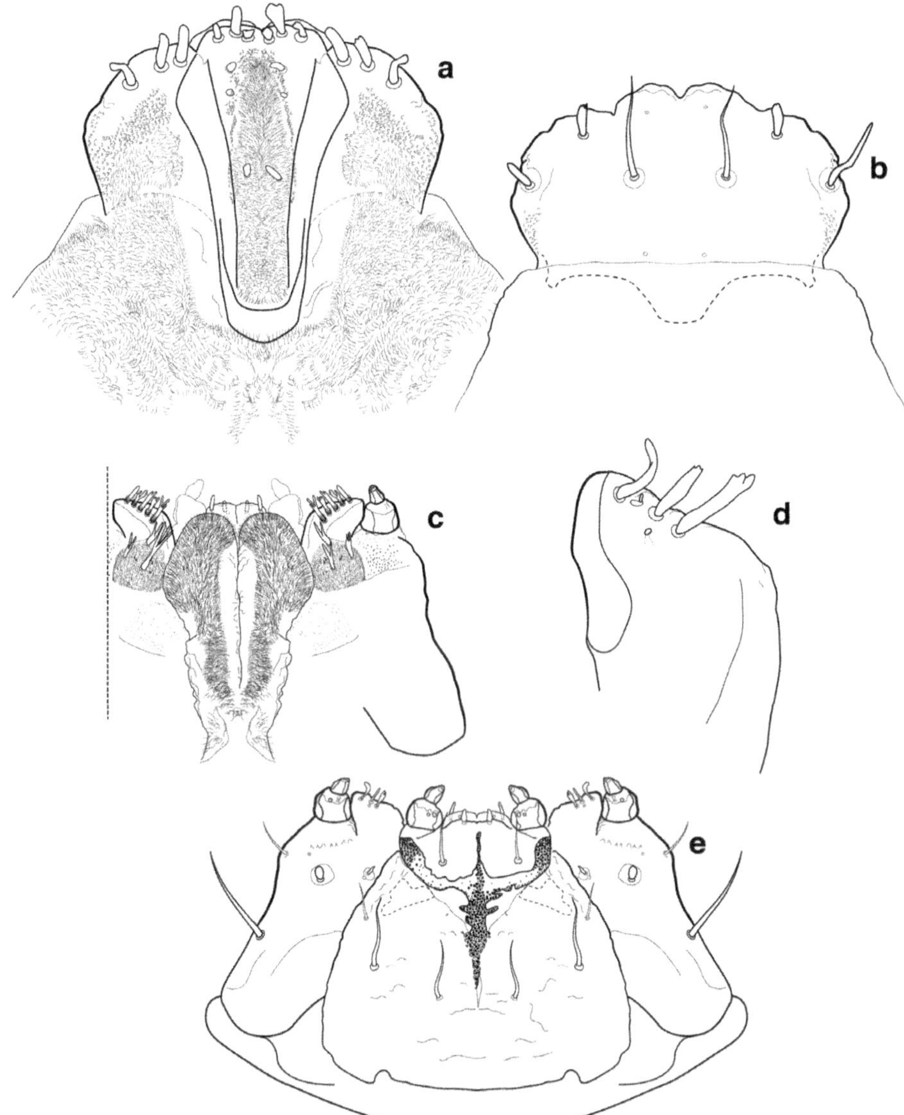

Figure 88. *Trigonotarsus calandroides* mouthparts: (**a**) epipharynx; (**b**) clypeus (incomplete) and labrum; (**c**) hypopharynx, and maxillary palp, dorsal view; (**d**) detail of mala and ventral malar setae; and (**e**) labio-maxillary complex, ventral view.

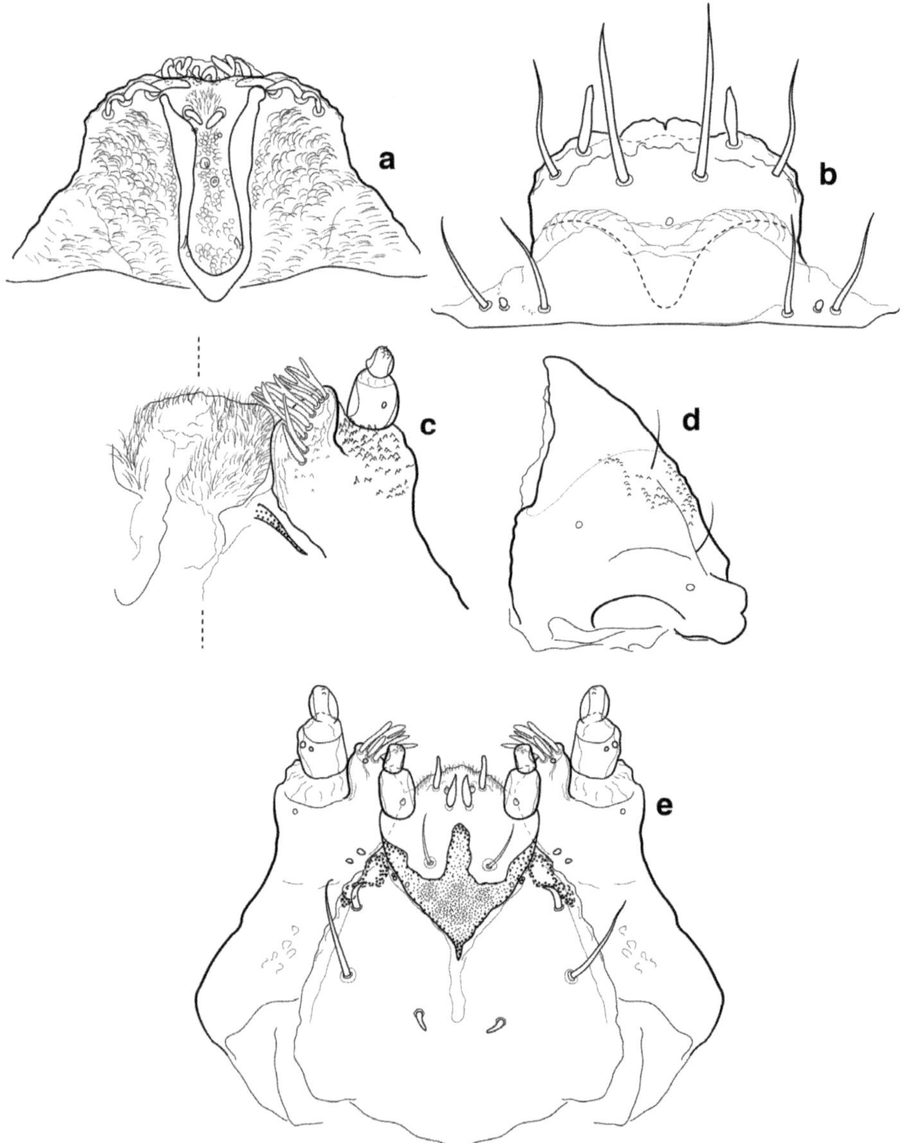

Figure 89. *Trochorhopalus strangulatus* mouthparts: (**a**) epipharynx; (**b**) clypeus and labrum; (**c**) detail of hypopharynx, mala, and maxillary palp, dorsal view; (**d**) mandible; and (**e**) labio-maxillary complex, ventral view.

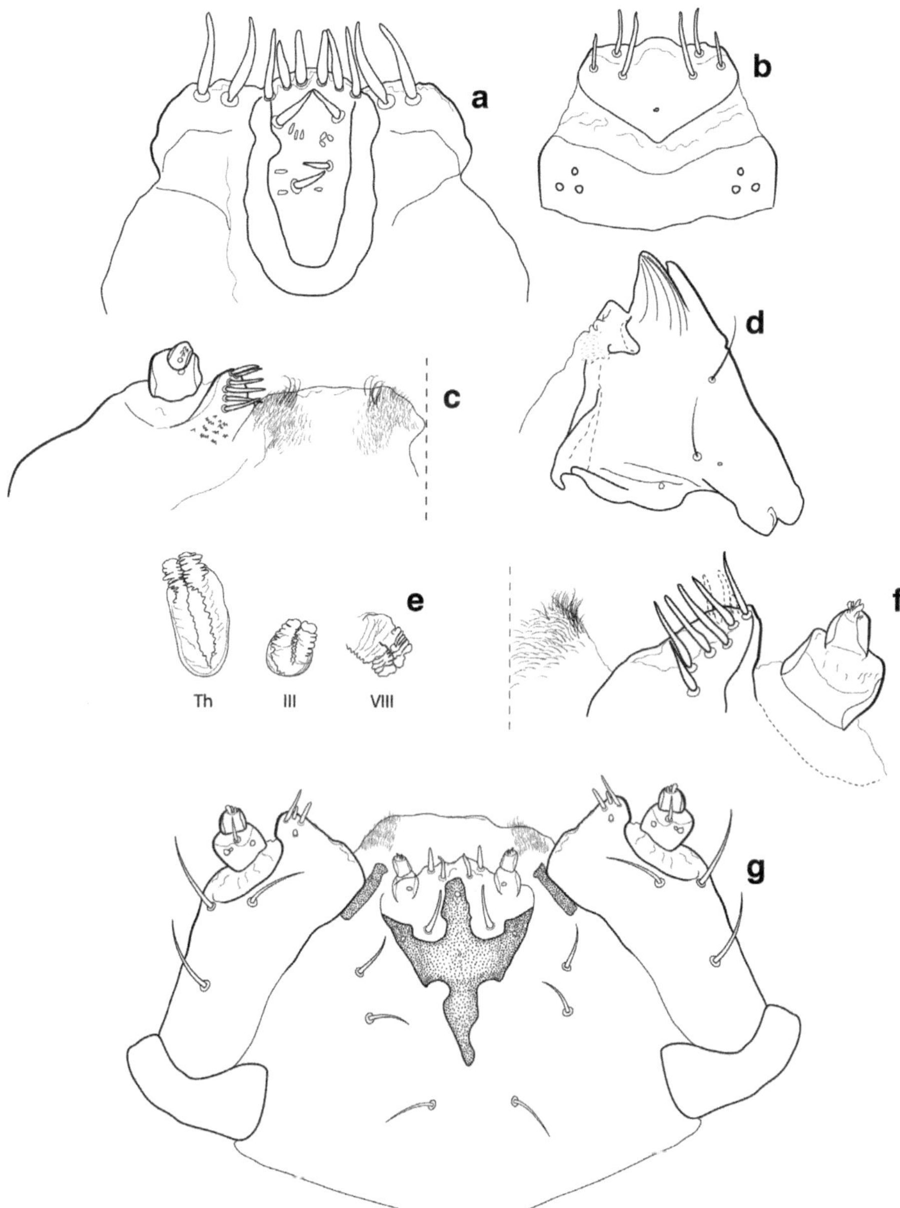

Figure 90. *Tryphetus incarnatus* mouthparts: (**a**) epipharynx; (**b**) labrum and clypeus; (**c**) detail of hypopharynx, mala, and maxillary palp, dorsal view; (**d**) mandible; (**e**) spiracles; (**f**) detail, mala, hypopharynx, maxillary palp, dorsal view; and (**g**) labio-maxillary complex, ventral view. Th: thorax; III: 3rd abdominal segment; VIII: 8th abdominal segment.

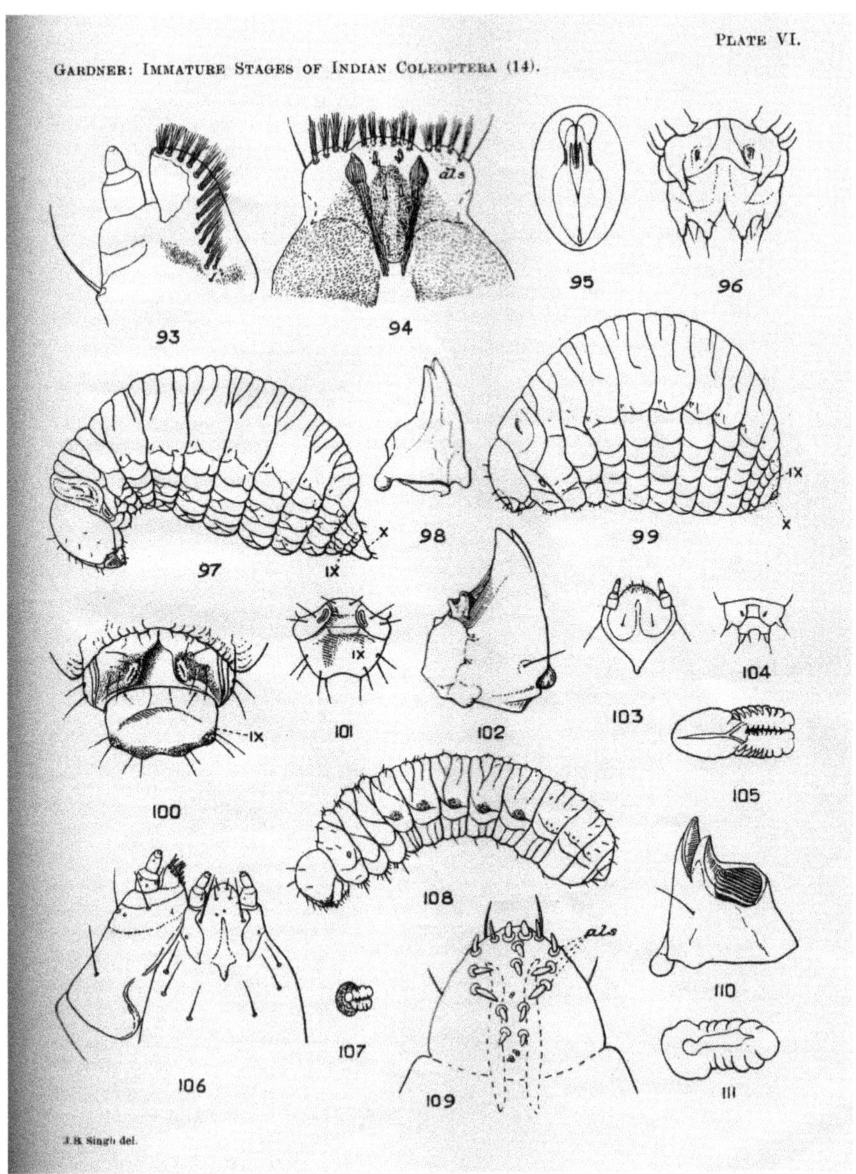

Figure 91. Reproduction of Gardner ([11], 48) Plate VI: "Larvae of Curculionidae. Figures 93–96. *Sipalus [Sipalinus] hypocrita* Boh. 93. Maxillary mala, dorsal view. 94. Epipharynx. 95. Abdominal spricale. 96. Posterior extremity, dorsal view. Figures 97 and 98. *Cyrtotrachelus longipes* F. 97. Lateral view. [1] ([1] In Figure 97 the numerals IX and X should have been interchanged). 98. Mandible. Figure 99. *Calandra [Sitophilus] rugicollis* Casey. Figure 100. *Rhynchophorus ferrugienus* Ol., posterior extremity in dorsal view. Figures 101 and 102. *Odoiporus longicollis* Ol. 101. Posterior extremity in dorsal view. 102. Mandible. Figures 103–105. *Xerodermus himalayanus* Mshl. 103. Labium. 104. Posterior extremity in dorsal view. 105. Spiracle. Figures 106–110. *Cossonus binodosus* Mshl. 106. Ventral mouthparts. 107. Abdominal spiracle. 108. Lateral view. 109. Epipharynx. 110. Mandible. Figure 111. *Brachydemnus himalayensis* Stebb., an abdominal spiracle."

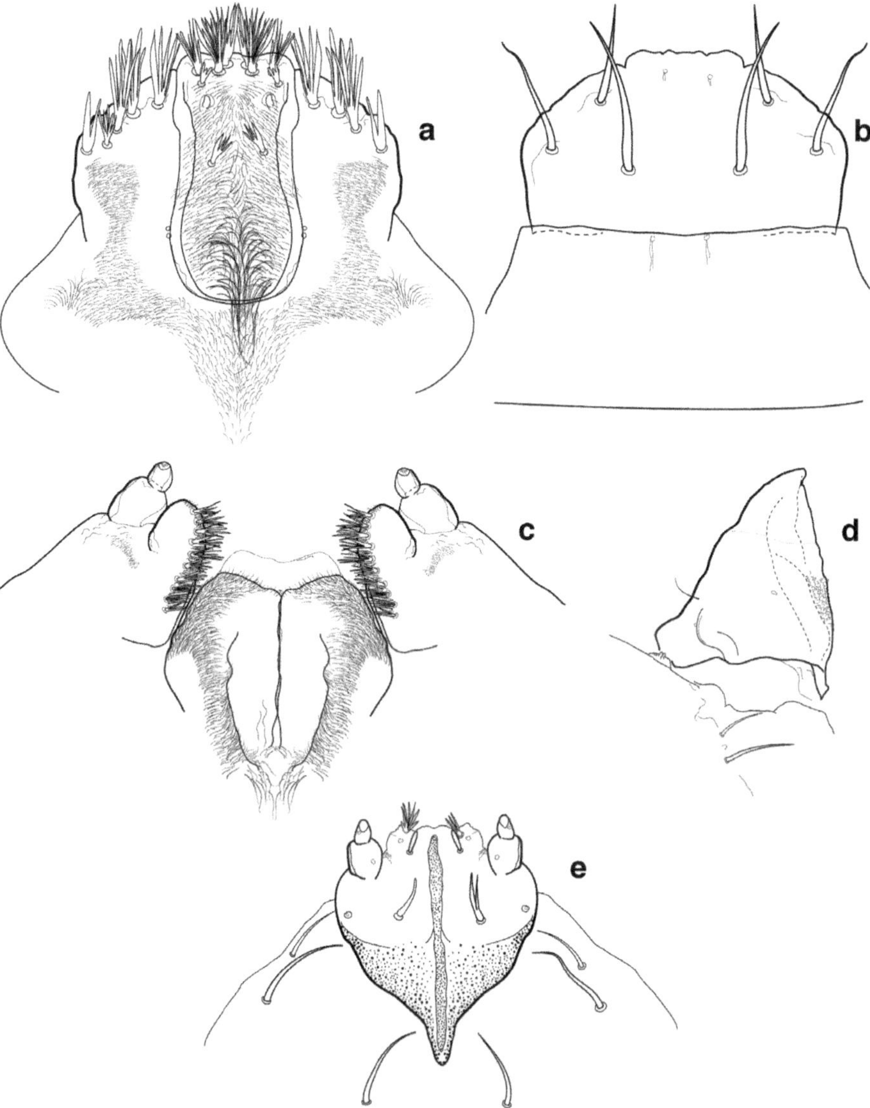

Figure 92. *Yuccaborus lentiginosus* mouthparts: (**a**) epipharynx; (**b**) clypeus (incomplete) and labrum; (**c**) detail of hypopharynx and maxillae, dorsal view; (**d**) mandible and antenna; and (**e**) labium, ventral view.

Figure 93. *Phacecorynes variegatus*: (**a**) adult; (**b**) variable form, adult; (**c**) larva, lateral view; (**d**) larva, lateral view partially concealled; (**e**) pupa, ventral view; *Phacecorynes sommeri*: (**f**) adult; (**g**) larva, lateral view; and (**h**) pupa, lateral view. Photographs courtesy of Rolf Oberprieler.

Figure 94. *Sipalinus gigas*.: (**a**) adult; (**b**) external damage; (**c**) larva, caudal view; (**d**) larva, lateral view; *Trigonotarsus calandroides* (**e**) adult, lateral view;: (**f**) larva, lateral view; (**g**) larva undergoing peristaltic behavior, lateral view; and (**h**) pupa, ventral view. Photographs courtesy of Rolf Oberprieler.

4. Discussion

Fifty-two (52) fully illustrated characters of dryophthorine larvae are provided to assist in the identification of this economically important group and difficult to identify life stage. The larvae of some dryophthorine genera, such as *Sipalinus*, *Scyphophorus*, and *Cosmopolites*, are readily recognizable based on a combination of salient characters such as the presence of caudal digitate processes and/or the absence of functional spiracles on the abdomen. This key attempts to provide the user with the freedom to initiate the identification process from any couplet in the key. This identification tool focuses rather heavily on characters of the mouthparts.

This study includes, for the first time, information on the larva of six previously unknown genera of Dryophthorinae: *Mesocordylus* (probably), *Foveolus*, *Nassophasis*, *Poteriophorus*, *Sparganobasis*, *Eugnoristus*, and *Tryphetus*. In addition, the larva of the once valid genus *Paramasius*, now treated as a junior synonym of *Metamasius*, is described for the first time. Data on the larva of *Paramasius* will be included in ongoing phylogenetic and comparative studies to determine whether synonymy of this genus is supported.

Chaetotaxy of the body, although historically relied upon by authors to segregate the major groups within Dryophthorinae [9], was not used here because of extensive amount of missing data due to rubbing or deterioration of specimens. Furthermore, characters previously used by May [15] and Marvaldi and Morrone [6], such as the number of abdominal superimposed lobes (2, 3, or 4) were not included. Additional features that presented challenges and were not included due to difficulty assessing homology are the presence and location of the posterior accessory sensory pores (*pasps*) and of the sensory pores (*snp*).

The current comparative study of dryophthorine larvae is intended to serve as a source of data for ongoing phylogenetic analyses of the subfamily, and preliminary results suggests the presence of significant signal for understanding relationships among higher groups within Dryophthorinae. Zimmerman [21] provided the most comprehensive overview of dryophthorines to date (his Rhynchophoridae) and discussed the placement of the now tribes Stromboscerini and Dryophthirini, which Anderson [9] placed collectively under Stromboscerinae. Grebennikov [22] hypothesized Dryophthorini and Stromboscerini to be sister taxa and suggested the possible exclusion of *Nephius* from Stromboscerini. Study of known stromboscerine larvae *Dryophthoroides*, *Xerodermus* [11,12], and *Nephius* (=*Anius*) suggests retention of the tribe in Dryophthorinae and supports its distinction from Dryophthorini, here represented by *Stenommatus* and *Dryophthorus*. The retention of *Nephius* in Stromboscerini remains uncertain. These are preliminary results and a morphological and molecular phylogenetic analysis is currently underway.

Funding: This research received no external funding.

Acknowledgments: The addition of the new larvae of dryophthorine genera would not have been possible without the generosity and assistance of Ramesha Barikkad (Kerala University), Beulah Gardner (UKNHM), Chia-Lung Huang (National Taiwan Normal University), Bruno de Madeiros (MCZ), and Rolf Oberprieler (ANIC). Elizabeth Roberts (USDA ARS SEL) completed the majority of the digital illustrations of the body and head and Lucrecia Rodriguez (USDA ARS SEL) generated several habitus images. Their contributions are greatly appreciated. This manuscript benefited greatly from the comments, edits and careful evaluation of three anonymous reviewers; their dedication and time is greatly appreciated. Matt Buffington and Petra Lacayo-Chamorro offered invaluable assistance during the course of this study and their help is greatly appreciated. The USDA is an equal opportunity provider and employer. Mention of trade names or commercial products in this publication is solely for the purpose of providing specific information and does not imply recommendation or endorsement by the USDA.

Conflicts of Interest: The author declares no conflict of interest.

References

1. Oberprieler, R.G.; Marvaldi, A.E.; Anderson, R.S. Weevils, weevils, weevils everywhere. *Zootaxa* **2007**, *1668*, 491–520. [CrossRef]
2. Anderson, R.S.; Marvaldi, A.E. 3.7.3 Dryophthorinae Schoenherr, 1825. In *Handbook of Zoology. Arthropoda: Insecta: Coleoptera*; Leschen, R.A.B., Beutel, R.G., Eds.; De Gruyter: Berlin, Germany; München, Germany; Boston, MA, USA, 2014; pp. 477–483.
3. Alonso-Zarazaga, M.A.; Lyal, C.H.C. *A World Catalogue of Families and Genera of Curculionoidea (Insecta: Coleoptera) (Excepting Scolytidae and Platypodidae)*; Entomopraxis: Barcelona, Spain, 1999.
4. Kuschel, G. A phylogenetic classification of Curculionoidea to families and subfamilies. *Mem. Entomol. Soc. Wash.* **1995**, *14*, 5–33.

5. Thompson, R.T. Observations on the morphology and classification of weevils (Coleoptera, Curculionoidea) with a key to major groups. *J. Nat. Hist.* **1992**, *26*, 835–891. [CrossRef]
6. Marvaldi, A.E.; Morrone, J.J. Phylogenetic systematics of weevils (Coleoptera: Curculionoidea): A reappraisal based on larval and adult morphology. *Insect Syst. Evol.* **2000**, *31*, 43–58. [CrossRef]
7. McKenna, D.D.; Sequeira, A.S.; Marvaldi, A.E.; Farrell, B.D. Temporal lags and overlap in the diversification of weevils and flowering plants. *Proc. Natl. Acad. Sci. USA* **2009**, *106*, 7083–7088. [CrossRef] [PubMed]
8. Chamorro, M.L.; Huang, C.L. First description of the Immature Stages of *Poteriophorus* Schoenherr: The larva, pupa and biology of *P. uhlemanni* (Curculionidae: Dryophthorinae) discovered through Dawu Traditional Ecological Knowledge. *Coleopt. Bull.* **2019**.
9. Anderson, W.H. Larvae of Some Genera of Calendrinae (=Rhynchophorinae) and Stromboscerinae (Coleoptera: Curculionidae). *Ann. Entomol. Soc. Am.* **1948**, *41*, 413–437. [CrossRef]
10. May, B.M. An Introduction to the Immature Stages of Australian Curculionoidea. In *Australian Weevils*; Csiro Publishing: Melbourne, Australia, 1994; Volume II, pp. 365–728.
11. Gardner, C.M. Immature stages of Indian Coleoptera (14) (Curculionidae). *Indian For. Rec. (Entomol. Ser.)* **1934**, *20*, 1–48, Plates I–VI.
12. Gardner, J.C.M. Immature stages of Indian Coleoptera (24, Curculionidae contd.). *Indian For. Rec. (New Ser.) Entomol.* **1938**, *3*, 227–260, Plates I–V.
13. Wattanapongsiri, A. A Revision of the Genera *Rhynchophorus* and *Dynamis* (Coleoptera: Curculionidae). Ph.D. Thesis, Oregon State University, Corvallis, OR, USA, 1966; p. 418.
14. Chamorro, M.L.; Volkovitsh, M.G.; Poland, T.M.; Haack, R.A.; Lingafelter, S.W. Preimaginal Stages of the Emerald Ash Borer, *Agrilus planipennis* Fairmaire (Coleoptera: Buprestidae): An Invasive Pest on Ash Trees (*Fraxinus*). *PLoS ONE* **2012**, *7*, e33185. [CrossRef] [PubMed]
15. May, B.M. Larvae of Curculionoidea (Insecta: Coleoptera): A systematic overview. *Fauna N. Z.* **1993**, *28*, 226p.
16. Oberprieler, R.G.; Anderson, R.S.; Marvaldi, A.E. 3 Curculionoidea Latreille, 1802: Introduction, Phylogeny. In *Handbook of Zoology. Arthropoda: Insecta: Coleoptera*; Leschen, R.A.B., Beutel, R.G., Eds.; De Gruyter: Berlin, Germany; München, Germany; Boston, MA, USA, 2016; pp. 285–300.
17. Chamorro, M.L.; Persson, J.; Torres-Santana, C.W.; Keularts, J.; Scheffer, S.J.; Lewis, M.L. Molecular and Morphological Tools to Distinguish *Scyphophorus acupunctatus* Gyllenhal, 1838 (Curculionidae: Dryophthorinae): A New Weevil Pest of the Endangered Century Plant, *Agave eggersiana* from St. Croix, U.S. Virgin Islands. *Proc. Entomol. Soc. Wash.* **2016**, *118*, 218–243. [CrossRef]
18. Jonhnson, N.F. Future taxonomy today: New tools applied to accelerate the taxonomic process. In *Systema Naturae 250: The Linnaean Ark*; Polaszek, A., Ed.; CRC Press Taylor & Francis Group: London, UK, 2010; pp. 137–147.
19. Holloway, B.A. Anthribidae (Insecta: Coleoptera). *Fauna N. Z.* **1982**, *3*, 272p.
20. Marshall, C.J. Two new species of rain beetle (Coleoptera: Pleocomidae: *Pleocoma* LeConte, 1856) in the Pacific Northwest of the United States of America. *Zootaxa* **2018**, *4471*, 387–395. [CrossRef] [PubMed]
21. Zimmerman, E. *Australian Weevils. Volume III. Nanophyidae, Rhynchophoridae, Erirhinidae, Curculionidae: Amycterinae, Literature Consulted*; CSIRO Publications: Melbourne, Australia, 1993; 854p.
22. Grebennikov, V.V. Dryophthorinae weevils (Coleoptera: Curculionidae) of the forest floor in Southeast Asia: Illustrated overview of nominal Stromboscerini genera. *Zootaxa* **2018**, *4418*, 1–15. [CrossRef] [PubMed]

Article

Description of Four New Species of the Afrotropical Weevil Genus *Afroryzophilus* (Coleoptera, Curculionidae)

Roberto Caldara [1,]* and Michael Košťál [2]

1 Via Lorenteggio 37, 20146 Milano, Italy
2 Střelecká 459, 50002 Hradec Králové, Czech Republic; michael.kostal@iol.cz
* Correspondence: roberto.caldara@gmail.com

Received: 2 April 2018; Accepted: 17 May 2018; Published: 23 May 2018

Abstract: Four new species belonging to the Afrotropical weevil genus *Afroryzophilus* Lyal, 1990 (Coleoptera, Curculionidae, Brachycerinae, Tanysphyrini) are described: *A. centrafricanus* n. sp. (Central African Republic), *A. congoanus* n. sp. (Democratic Republic of the Congo), *A. kuscheli* n. sp. (Senegal), and *A. somalicus* n. sp. (Somalia). Previously, this genus was monotypic, based only on *A. djibai* Lyal, 1990 from Senegal. The five species of this genus are very similar to each other in external morphology, varying only in the width of the forehead and that of the third tarsomere, the length of the fifth tarsomere and the pattern of dorsal seta-like scales. However, the male genitalia show clear interspecific differences.

Keywords: Coleoptera; Curculionidae; Brachycerinae; Tanysphyrini; *Afroryzophilus*; new species; Afrotropical region

1. Introduction

The genus *Afroryzophilus* was described by Lyal [1] from a single taxon, *A. djibai* Lyal, 1990, attacking rice in Senegal. Subsequently, no other species of this genus has been described and no other author dealt specifically with this genus. In recent years, we had an opportunity to study several African specimens of *Afroryzophilus* belonging to four closely related species. The comparison of six of the 69 paratypes of *A. djibai* revealed that they all belong to undescribed species.

The aim of this paper is to describe these new species and to redefine the genus in light of morphological characters of all its species.

2. Materials and Methods

2.1. Descriptions and Illustrations

For new species, holotypes were generally used for description whereas best-preserved paratypes were used for illustrations in some cases.

2.2. Measurements and Photographs

All measurements were made under a stereomicroscope (Wild, Heerbrugg, Switzerland) using an ocular micrometer. The body length is interpreted as the distance between the anterior eye margin and the elytral apex. Index Rl/Rw is interpreted as the ratio of rostrum length from its base to the apex without mandibles to the medial length of pronotum, index Ew/Pw as the ratio of the maximum elytral width in the humeral region to the maximum pronotal width.

Whole body photos were made by a high-resolution camera (Canon EOS 50D, Canon Inc., Tokyo, Japan) and a macro zoom lens (Canon MP-E 65 mm, Canon Inc.). Male genital structures were dissected and treated for several days in 10% KOH. Male genitalia were photographed in glycerol with the same camera under a laboratory microscope (Intraco Micro LMI T PC, Intraco Micro, Czech Republic). The multilayer pictures were processed using the software Combine ZP.

2.3. Diagnosis

A cluster of all characters were used to identify a particular species.

2.4. Terminology

We followed the online glossary of weevil characters proposed in the International Weevil Community Website (http://weevil.info/glossary-weevil-characters) (accessed on 18 March 2018).

2.5. Acronyms and Abbreviations

The materials studied are housed in various collections and are identified by the following acronyms:

BMNH Department of Entomology, The Natural History Museum, London, UK
CA Roberto Caldara, Milano, Italy
KO Michael Košťál, Hradec Králové, Czech Republic
MSNM Museo civico di Storia Naturale, Milano, Italy
RMCA Musée Royal de l'Afrique Centrale, Tervuren, Belgium

The following abbreviations were used: E: Elytra; P: Pronotum; R: Rostrum; l: length; w: width.

3. Taxonomy

3.1. The Genus *Afroryzophilus* Lyal

A very accurate description, and illustrations of this genus, including all structures of male and female genitalia, as we could ascertain by the study of six paratypes (BMNH) of *A. djibai*, were made by Lyal [1]. After the description of the following four new species, the original description remains almost unchanged. Therefore, we simply report a detailed diagnosis of the most important characters of *Afroryzophilus*.

With regard to genitalia, the new species show no substantial differences from those illustrated in detail by Lyal [1] except for the shape of the penis.

Diagnosis. Length 2.1–3.4 mm (rostrum excluded). Integument completely covered with appressed broad scales, almost rounded, mostly pitted on pronotum; and dorsally also with some recumbent to subrecumbent narrower seta-like scales. Rostrum long, in males slightly, in females more distinctly longer than pronotum, weakly curved in lateral view, tapering toward apex; densely squamose in basal half, almost glabrous in apical half (Figure 1f,g). Antennae densely squamose, inserted at middle of rostrum in males, just beyond middle of rostrum in females; scrobe parallel to rostrum length, not reaching eye. Scape not reaching eye, funicle longer than scape, 6-segmented. Mandibles with two exterior non-dehiscent teeth. Forehead (part of head between eyes) slightly narrower to distinctly broader than rostrum at base. Eyes large, not prominent.

Pronotum moderately wider than long, with slightly rounded sides, weakly convex. Prosternum with anterior margin concave, without longitudinal canal. Postocular lobes smoothly rounded. Scutellum small, oval. Elytra wider than pronotum, humeri prominent, sides parallel; interstriae broad, wider than striae, flat; striae with narrow deep punctures. Wings present. Metasternum anteriorly convex, posteriorly concave, especially in males. Femora clavate, unarmed. Tibiae weakly flexuose, mucronate, mucro stout, curved, abruptly directed ventrad. Tarsi with third tarsomere more or less large, distinctly bilobed, fifth tarsomere short, retracted to lobes of third tarsomere, claws small, separated from base.

Abdomen with ventrite 1 as long as ventrite 2, ventrites 3–5 shorter, not in the same plane as ventrites 1–2.

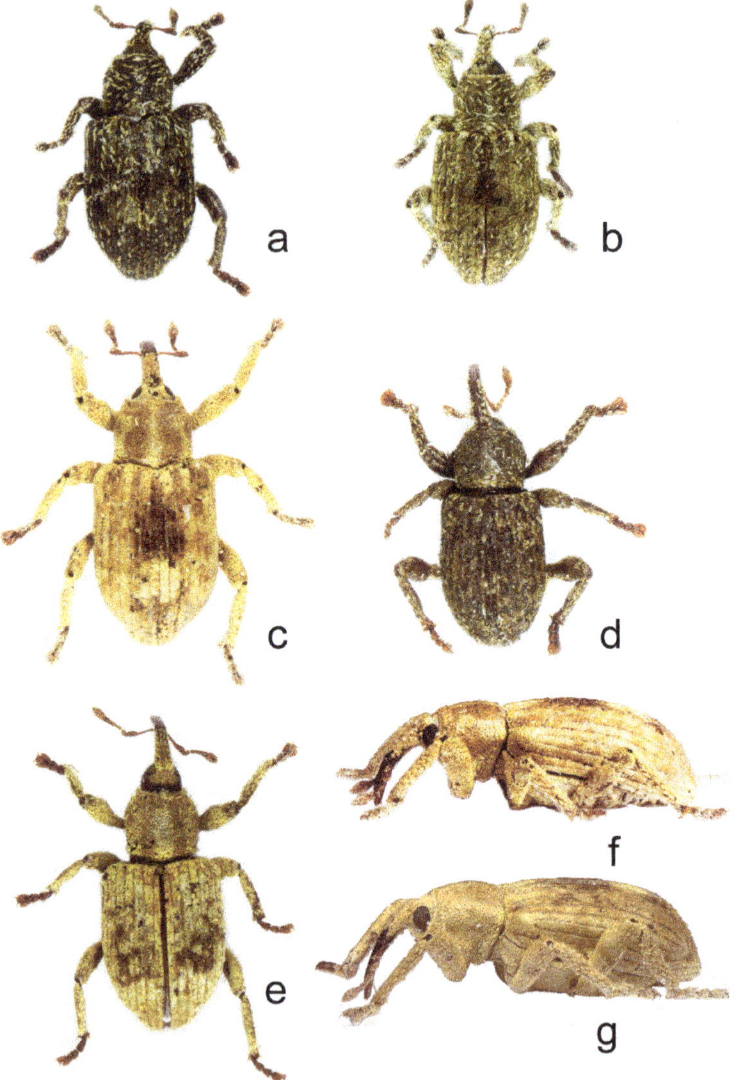

Figure 1. Habitus of (**a**) *Afroryzophilus djibai* ♂; (**b**) *A. centrafricanus* ♂; (**c**) *A. kuscheli* ♂; (**d**) *A. congoanus* ♂; (**e**) *A. somalicus* ♀; (**f**) *A. kuscheli* ♂in lateral view; (**g**) *A. kuscheli* ♀in lateral view. (Not at the same scale).

Male genitalia. Tergite VII with posterior margin medially emarginate. Tegmen with parameroid lobes weakly sclerotized, connate in their whole length, narrowed at base; then subparallel, tapered at apex, slightly shorter than body of penis. Body of penis more than half the length of apodemes. Tectum slender, united together with pedon with apodemes. Endophallus with small tooth-, comma- or crescent-like sclerite inside body of penis, with a flagellum more or less sclerotized, and as long as apodemes.

Female genitalia. Tergite VII quadrate. Tergite VIII quadrate, with posterior margin rounded, exposed at rest. Sternite VIII elongate, lateral sclerotizations of apical plate distinct for less than half the total length of sternite. Spermathecal duct arising part-way along bursa.

3.2. Treatment of the New Species

***Afroryzophilus centrafricanus* n. sp.** (Figures 1b, 2b, 3b, 4b and 5b)

Type series. Holotype, male "Central African Rep., Bamingui-Bangoran Pr. [Prefecture] 75 km SSW Ndele, 8–12 July 2011, 450 m, A. Kudrna Jr. Lgt". (MSNM). Paratypes: Same data as holotype (1 male and 4 females, RC, MK).

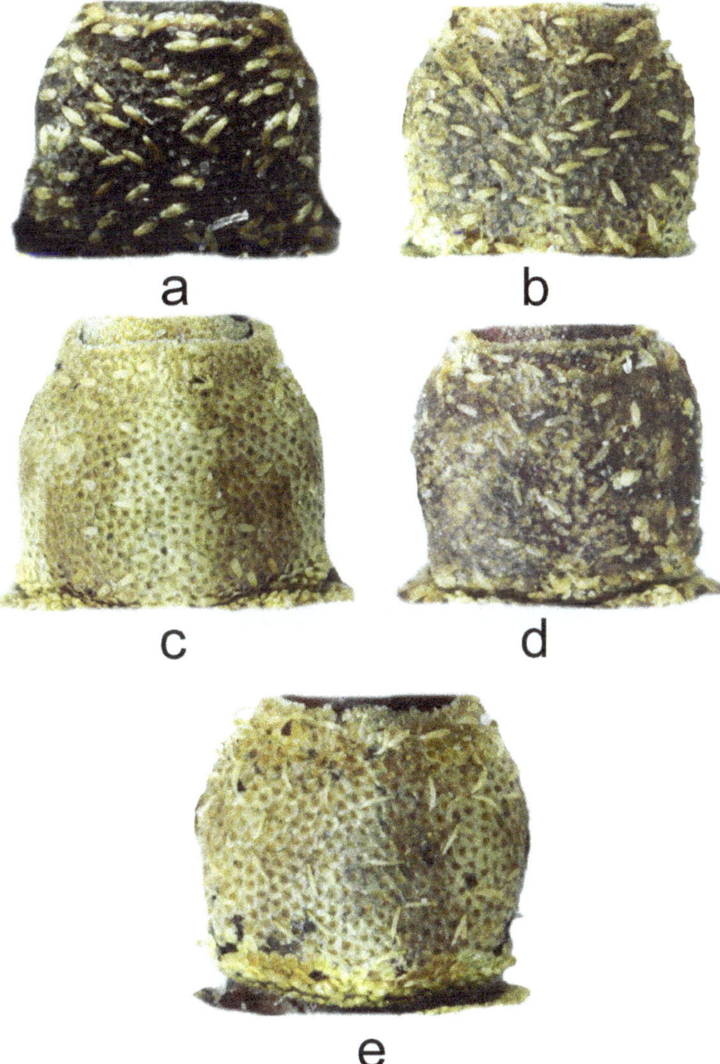

Figure 2. Scales on pronotum of (**a**) *Afroryzophilus djibai*; (**b**) *A. centrafricanus*; (**c**) *A. kuscheli*; (**d**) *A. congoanus*; (**e**) *A. somalicus*. (Not at the same scale).

Figure 3. Scales on elytra of (**a**) *Afroryzophilus djibai*; (**b**) *A. centrafricanus*; (**c**) *A. kuscheli*; (**d**) *A. congoanus*; (**e**) *A. somalicus*. (Not at the same scale).

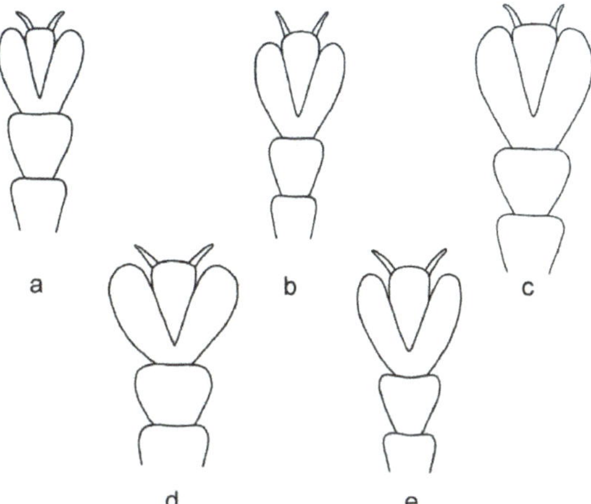

Figure 4. Protarsi of (**a**) *Afroryzophilus djibai;* (**b**) *A. centrafricanus;* (**c**) *A. kuscheli;* (**d**) *A. congoanus;* (**e**) *A. somalicus.* (Not at the same scale).

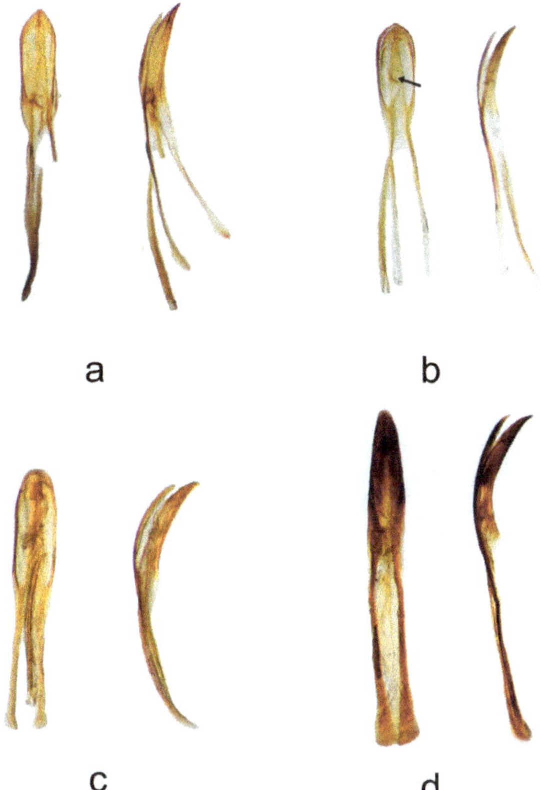

Figure 5. Penis in ventrodorsal (left) and lateral (right) view of (**a**) *Afroryzophilus djibai;* (**b**) *A. centrafricanus;* (**c**) *A. kuscheli;* (**d**) *A. congoanus.* Black arrow indicates a sclerite inside the endophallus. (Not at the same scale).

Description. Holotype. Male. Length 2.3 mm. Habitus as in Figure 1b. Integument completely, densely covered with decumbent and appressed scales, these are oval to polygonal, on pronotum moderately pitted at sides, mainly light brown; on basal half of interstria 1 also dark brown, and whitish on base of pronotum along middle and at side, base of odd elytral interstriae and third quarter of interstria 1, and with robust, subrecumbent, posteriorly distinctly recurved, lanceolate brown seta-like scales, these distinct and unevenly sparsely distributed on pronotum, arranged in single regular rows on each elytral interstria and numerous on rostrum, femora and tibiae. Rostrum slightly longer than pronotum (Rl/Pl 1.05), in lateral view distinctly, evenly curved, markedly tapered from base to apex; in dorsal view slightly narrowing from base to antennal insertion, then parallel-sided. Forehead 1.4 times broader than rostrum at base. Pronotum slightly wider than long (Pw/Pl 1.15), with weakly rounded sides, widest at middle, weakly convex. Elytra distinctly longer than wide (El/Ew 1.48), distinctly wider than pronotum (Ew/Pw 1.52), with parallel sides, weakly convex. Tarsi with second and third tarsomere moderately transverse, only slightly wider than long, fifth tarsomere slightly projecting beyond lobes of third tarsomere. Penis (Figure 5b) with body short (l/w 2.6), parallel-sided in dorsal view, distinctly thin in lateral view, with apodemes twice as long as body, only slightly enlarged at their extremities; endophallus with weakly sclerotised flagellum.

Variability. Length 2.2–2.6 mm. Females as male, except for rostrum distinctly longer (Rl/Rw 5.65) and in lateral view almost subparallel-sided from base to apex; in dorsal view moderately narrowing from base to antennal insertion, then parallel-sided to apex. Apart from sexual characters no significant differences between the holotype and the paratypes.

Etymology. This species name is a Latin adjective that refers to the country in which the type locality is situated.

Remarks. This species is the most closely related to *A. dijbai*, from which it differs externally only by slightly thinner seta-like scales of the elytral interstriae and the fifth tarsomere in dorsal view, slightly protruding from the third tarsomere (not protruding in *A. dijbai*) (see Figure 4b vs. Figure 4a). On the contrary, the shape of the body of the penis is distinctly different in these two species, being shorter and distinctly thinner in lateral view in *A. centrafricanus* (see Figure 5b vs. Figure 5a).

Distribution. Central African Republic.

***Afroryzophilus kuscheli* n.sp.** (Figure 1c,f–g, Figures 2c, 3c, 4c and 5c)

Type series. Holotype, male "Senegal-Kaolack, Norio du Rip, 24 July 2009, Moretto" (MSNM). Paratypes: Same data as holotype (1 male and 3 females, RC, MK).

Description. Holotype. Length 3.0 mm. Habitus as in Figure 1c. Integument completely, densely covered with decumbent and appressed scales, these are oval to polygonal, on pronotum moderately pitted at sides, light brown, with large macula at middle of first three elytral interstriae, also dark brown and with longitudinal vittae at midline and sides of pronotum, large macula at apical third of first three elytral interstriae and short whitish vittae on odd elytral interstriae, and with robust subrecumbent, posteriorly distinctly recurved, robust lanceolate brown seta-like scales, being distinct and unevenly sparsely distributed on pronotum, arranged in single regular rows on each elytral interstria and numerous on rostrum, femora and tibiae. Integument with vestiture as in *A. centrafricanus* with regard to colour and pattern, except for very sparse seta-like scales on pronotum; and thinner and sparse (except for on interstria 1) on elytra, especially on even interstriae, almost indistinct since completely flattened. Rostrum moderately long, slightly longer than pronotum (Rl/Rw 4.17, Rl/Pl 1.13); in lateral view distinctly and evenly curved, distinctly tapered from base to apex; in dorsal view slightly narrowing from base to antennal insertion, then parallel-sided. Forehead 1.3 times broader than rostrum at base. Pronotum slightly wider than long (Pw/Pl 1.10), with weakly rounded sides, widest at middle, weakly convex. Elytra distinctly longer than wide (El/Ew 1.39), markedly wider than pronotum (Ew/Pw 1.59), with parallel sides, weakly convex. Tarsi with second and third tarsomere moderately transverse, only slightly wider than long, fifth tarsomere not projecting beyond lobes of third tarsomere. Penis (Figure 5c) with body moderately long (l/w 3.2), parallel-sided in dorsal view, moderately robust in lateral view, with apodemes only slightly longer than body, somewhat enlarged at their extremities; endophallus with distinctly sclerotized flagellum.

Variability. Length 2.7–3.4 mm. Females as males, except for rostrum distinctly longer (Rl/Rw 5.67) and in lateral view almost subparallel-sided from base to apex; in dorsal view moderately narrowing from base to antennal insertion, then slightly widened to apex. Apart from sexual characters no significant differences between the holotype and the paratypes.

Etymology. This species is named in the memory of Willy Kuschel, one of the greatest experts on weevils, who particularly studied the erirhinines and gave them a completely new taxonomic arrangement.

Remarks. To date, this is the largest species in this genus. Apart from the shape of the penis (see Figure 5c vs. Figure 5a vs. Figure 5b), *A. kuscheli* differs from *A. djibai* and *A. centrafricanus* by smaller seta-like scales, which are very sparse on the pronotum and the elytra, and indistinct on the even interstriae, being completely flattened and not curved in the middle. Moreover, the dorsal pattern seems to be characterised by more contrasting colours, with distinct large dark brown and white maculae on the disc, and white vittae on the pronotum, and less distinctly on the odd elytral interstriae.

Distribution. Western Senegal.

***Afroryzophilus congoanus* n. sp.** (Figures 1d, 2d, 3d, 4d and 5d)

Type series. Holotype, male "Congo Belge: P.N.A. [Parc National Albert, currently Parc National des Virunga], 14-15-VIII-1952, P. Vanschuytbroeck & J. Kekenbosch, 767-70/Massif Ruwenzori, Kalonge, 2.210 m" (RMCA). Paratype: "Congo belge: P.N.A., Kanyabayongo (Kabasha), 1760 m, 6-xii-1934, G.F. de Witte: 870/R. Dét. uu 4470/Erirrhinide ou Tychiide" (1 female, RMCA).

Description. Holotype. Length 2.1 mm. Habitus as in Figure 1d. Integument with regards to colour and pattern as in *A. djibai*, vestiture with scales on pronotum and more robust elytra, more numerous and darker in colour. Rostrum moderately long, slightly longer than pronotum (Rl/Rw 4.10, Rl/Pl 1.15), in lateral view distinctly and evenly curved, distinctly tapered from base to apex; in dorsal view slightly narrowed from base to antennal insertion, then parallel-sided. Forehead slightly narrower than rostrum at base. Pronotum moderately wider than long (Pw/Pl 1.20), with weakly rounded sides, widest at middle, weakly convex. Elytra distinctly longer than wide (El/Ew 1.40), distinctly wider than pronotum at base (Ew/Pw 1.37), with parallel sides, weakly convex. Tarsi with second and third tarsomere distinctly transverse, with fifth tarsomere not projecting beyond lobes of third tarsomere. Penis (Figure 5d) with body long (l/w 4.3), gradually narrowed from base to apex in dorsal view, moderately robust in lateral view, with apodemes moderately longer than body and distinctly enlarged at their extremities; endophallus with distinctly sclerotized flagellum.

Variability. Female as male, except for rostrum distinctly longer and in lateral view almost subparallel-sided from base to apex; in dorsal view slightly narrowed from base to antennal insertion, then parallel-sided to apex; length 2.4 mm.

Etymology. The name is a Latin adjective that refers to the country in which the type locality is situated.

Remarks. Due to the forehead being only slightly narrower than the rostrum at base, this species seems to be close to *A. somalicus*. However, it differs from this species by the third tarsomere being broader and distinctly wider than the second tarsomere, and by the seta-like scales on pronotum and elytra being more robust, shorter and more numerous.

Distribution. Democratic Republic of the Congo (Eastern provinces).

***Afroryzophilus somalicus* n. sp.** (Figures 1e, 2e, 3e and 4e)

Type series. Holotype, female "Coll. Mus. Tervuren, Somalie: Afgoi [Afgooye], August 1977, Leg. Olmi" (RMCA).

Description. Female. Length 2.8 mm. Integument completely, densely covered with decumbent and appressed, oval to polygonal scales; scales on pronotum distinctly and deeply pitted, mainly light brown; on basal half of elytra, especially on perisitural interstriae, light brown scales intermixed with light and slightly darker scales, with distinct micaceous reflections; lanceolate, seta-like scales apically distinctly recurved, almost transparent with silvery reflections, recumbent to semi-erect on pronotum and elytra, moderately thin on rostrum, femora and tibiae, very sparse on pronotum, a little more numerous and barely visible on elytra, more numerous on basal half of rostrum and legs. Rostrum long (Rl/Rw 5.02, Rl/Pl 1.16), in lateral view almost subparallel-sided from base to apex; in dorsal view slightly narrowed from base to antennal insertion, then parallel-sided to apex. Forehead slightly narrower than rostrum at base. Pronotum moderately wider than long (Pw/Pl 1.17), with weakly rounded sides, widest at middle, weakly convex. Elytra distinctly longer than wide (El/Ew 1.46), markedly wider than pronotum at base (Ew/Pw 1.52), with parallel sides, weakly convex. Tarsi with second and third tarsomere moderately broad, almost as long as wide, fifth tarsomere slightly projecting beyond lobes of third tarsomere.

Etymology. The name is a Latin adjective that refers to the country in which the type locality is situated.

Remarks. This species differs from the others by the more deeply pitted scales on the pronotum. Moreover, the scales of the dorsal vestiture are light brown with distinct micaceous reflections. Probably also the pattern of the elytral vestiture is different from that of the other species; however, this difference needs to be confirmed by examination of further specimens. The seta-like scales are thinner and longer both on pronotum and elytra and more raised on the tibiae.

Distribution. South-western Somalia (Lower Shebelle Region).

Key to the species

1. Forehead 1.3–1.4 times broader than rostrum at base. 2
– Forehead slightly narrower than rostrum at base. 4
2. Seta-like scales of dorsal vestiture subrecumbent, more distinct, more numerous both on pronotum and all elytral interstriae (Figure 2a,b and Figure 3a,b). Elytral vestiture on disc at most with poorly distinct large dark brown macula (Figure 3a,b). 3
– Seta-like scales of dorsal vestiture recumbent, less visible, sparser especially on even elytral interstriae (Figures 2c and 3c). Elytral vestiture on disc with two distinct large maculae, one dark brown and one white (Figure 3c) .... *kuscheli* **n. sp.**
3. Fifth tarsomere in dorsal view not projecting beyond lobes of third tarsomere (Figure 4a). Seta-like scales of elytral interstriae robust. .... *dijbai* Lyal
– Fifth tarsomere in dorsal view slightly projecting beyond lobes of third tarsomere (Figure 4b). Seta-like scales of elytral interstriae slightly thinner. .... *centrafricanus* **n. sp.**
4. Tarsi shorter, second and third tarsomere wider than long (w/l 1.50 and 1.25 respectively) (Figure 4d). Broad scales of dorsal vestiture opaque, those covering pronotum shallowly pitted (Figure 2d,e); seta-like scales more robust and more numerous (Figures 2d and 3d). .... *congoanus* **n.sp.**
– Tarsi longer, second and third tarsomere as wide as long (Figure 4e). Broad scales of dorsal vestiture with distinct micaceous reflections, those covering pronotum deeply pitted (Figures 2e and 3e). Seta-like scales thinner and sparser, especially on pronotum (Figure 2e). .... *somalicus* **n. sp.**

4. Discussion

At the time of description, due to the structure of the genitalia, particularly the presence of a pedo-tectal penis, this genus was placed by Lyal [1] in Erirhininae, however, without specifying a tribe. Subsequently, Alonso-Zarazaga and Lyal [2] placed *Afroryzophilus* in Erirhininae, Erirhinini. Very recently, this genus was transferred by Oberprieler [3] to Tanysphyrini, which were considered a tribe of the Brachycerinae, this subfamily included all Erirhininae sensu Alonso-Zarazaga and Lyal [2].

We agree with this placement, as *Afroryzophilus* possesses all the main characters distinctive for this tribe [3]: Rostrum moderately long, without apico-lateral setiferous grooves but usually with one or a few long setae in the same position; dorsum of rostrum usually densely squamose from base to antennal insertion but conspicuously glabrous in front of it; funicle six-segmented, with basal segment enlarged and apically densely fringed with setae or tuft hairs, tibiae mucronate, without spurs, bevels or corbels (though sometimes with false corbels); tarsi usually flattened with short onychium and barely or not markedly, to completely withdrawn into lobes of segment 3, claws simple, divaricate; ventrite 5 usually with a pair of thin setal tufts or single long setae; tegmen laterally fused (not hinged), usually with complete but weakly sclerotized dorsal plate.

Lyal [1] already discussed the differences between *Afroryzophilus* and the other erirhinine genera, pointing out the characteristic shape of the mandibles, which are toothed externally in this genus, and between *Afroryzophilus* and other genera of Brachycerinae feeding on rice, i.e., *Echinocnemus* Schoenherr, 1843 (Erirhinini) and *Lissorhoptrus* LeConte, 1876 (Tanysphyrini), both with the onychium distinctly longer than the third tarsomere. However, only the former genus is distributed in Africa. In fact, the only other genus of Tanysphyrini known from the Afrotropical region is *Araxus* Marshall, 1955, whose biology is unknown although it was hypothesised that it might feed on mosses [3]. This genus, which is distributed in western and central Africa but also in Madagascar [4,5], clearly differs from *Afroryzophilus* by the seven-segmented funicle, the stout rostrum, poor sexual dimorphism, the very stout tibiae and the lack of pitted scales on the pronotum.

Concerning the five species of this genus, they differ from each other externally only by a few characters, mainly by the shape of the tarsi and the width of the forehead, the shape of the seta-like scales on pronotum and elytra and, to a less extent, by the pattern of the dorsal vestiture. Other characters, such as the shape of pronotum and elytra, are almost identical in all species. In contrast, the male genitalia are clearly different in each of the four species of which the male is known.

It was reported that *A. djibai* is a potential pest of rice in Senegal [1,6]. Unfortunately, we have no data on the host plants of the new species and do not know whether they might also affect rice cultivation. However, *A. centrafricanus*, *A. kuscheli*, and *A. somalicus* were collected in a plain, the latter two near the coast, and all near a river, and in areas where rice is possibly cultivated [7]. In contrast, the fourth species, *A. congoanus*, was collected at high altitude in a National Park, far from populated areas. Thus, it is evident that this species does not live on rice. It is highly probable that, as in *Lissorhoptrus* [3], the species of *Afroryzophilus* can also feed on different plants belonging to the family Poaceae.

Author Contributions: Both authors contributed equally to the design, analysis and writing of the paper.

Acknowledgments: We are thankful to Max Barclay (BMNH) for providing us with paratypes of *A. djibai*, which were essential to our study. We are also grateful to two anonymous referees who allowed us to improve our paper with interesting suggestions.

Conflicts of Interest: The authors declare no conflicts of interest.

References

1. Lyal, C.H.C. A new genus and species of rice weevil from the Sahel (Coleoptera: Curculionidae: Erirhininae). *Bull. Entomol. Res.* **1990**, *80*, 183–189. [CrossRef]
2. Alonso-Zarazaga, M.A.; Lyal, C.H.C. *A World Catalogue of Families and Genera of Curculionoidea (Insecta: Coleoptera) (Excepting Scolytidae and Platypodidae)*; Entomopraxis S.C.P: Barcelona, Spain, 1999; p. 315.
3. Oberprieler, R. Coleoptera, Beetles: Morphology and Systematics (Phytophaga). In *Handbook of Zoology: Arthropoda: Insecta*; Leschen, R.A.B., Beutel, R.G., Eds.; De Gruyter: Berlin/Heidelberg, Germany; Boston, MA, USA, 2014; Volume 3.
4. Caldara, R.; O'Brien, C.W. On the systematic position and nomenclature of some species of the genus Bagous Germar, 1817 (Coleoptera: Curculionidae). *G. it. Ent.* **1994**, *7*, 1–4.

5. Heinrichs, E.A.; Barrion, A.T. *Rice-Feeding Insects and Selected Natural Enemies in West Africa: Biology, Ecology, Identification*; International Rice Research Institute and Abidjan (Côte d'Ivoire), WARDA–The Africa Rice Center: Los Baños, Philippines, 2004; p. 243.
6. Marshall, G.A.K. On the Curculionidae (Coleoptera) of Angola. II. *Publ. Cult. Camp. Diam. Angola* **1958**, *38*, 111–154.
7. Somado, E.A.; Guei, R.G.; Nguyen, N. Overwiew: Rice in Africa. In *NERICA®: The New Rice for Africa—A Compendium*; Somado, E.A., Guei, R.G., Keya, S.O., Eds.; Africa Rice Center (WARDA): Cotonou, Benin; FAO: Rome, Italy; Sasakawa Africa Association: Tokyo, Japan, 2008; p. 9.

Communication

Validation of the Names of Four Weevil Species Described by Caldara & Košťál, Description of Four New Species of the Afrotropical Weevil Genus *Afroryzophilus* (Coleoptera, Curculionidae); *Diversity* 2018, *10*, 37

Roberto Caldara [1,*] and Michael Košťál [2]

[1] Via Lorenteggio 37, 20146 Milano, Italy
[2] Střelecká 459, 50002 Hradec Králové, Czech Republic; michael.kostal@iol.cz
* Correspondence: roberto.caldara@gmail.com

http://zoobank.org/urn:lsid:zoobank.org:pub:84D9D43F-47FB-40D4-9310-1D867BECDDDB
Received: 31 July 2018; Accepted: 6 August 2018; Published: 9 August 2018

Abstract: Four new species of the erirhinine genus *Afroryzophilus* Lyal, 1990 from Africa are described, *A. centrafricanus* **sp. n.**, *A. congoanus* **sp. n.**, *A. kuscheli* **sp. n.** and *A. somalicus* **sp. n.**, with bibliographic reference to fuller descriptions and illustrations in the recent paper by Caldara & Košťál (2018) published in the journal *Diversity* 10 (2), 37, in which the names were not made available under the rules of the International Code of Zoological Nomenclature dealing with electronic publication.

Keywords: Brachycerinae; Tanysphyrini; *Afroryzophilus*; new species; Afrotropical region

---

1. Introduction

The recent paper by Caldara & Košťál published in *Diversity* 10 (2) [1] was not in full compliance with the International Code of Zoological Nomenclature [2] regarding publication of online taxonomic papers. Article 8.5. states that, to be considered published [within the meaning of the Code], "a work issued and distributed electronically must be registered in the Official Register of Zoological Nomenclature (ZooBank) (see Article 78.2.4) and contain evidence in the work itself that such registration has occurred" (Article 8.5.3.). Because the paper by Caldara & Košťál (2018) was not registered in ZooBank prior to publication and therefore evidence of registration was not included in it, the new taxonomic names proposed in the paper are not available under the Code [3]. The purpose of this paper is to make those names available.

To fulfill the requirements of Article 8.5, this paper has been registered in ZooBank, under the LSID above, and the names of the species described below have also been registered, following recommendation 10B of the Code. Their LSIDs are given under each name. To meet the requirements of Article 13.1.2. of the Code, the names listed below are accompanied by a bibliographic reference to their full descriptions and are thereby made available from the publication of this paper. The wording of Article 13.1.2. is somewhat ambiguous as to the status of descriptions based on bibliographic reference, so to avoid any further problems we have added below a brief description differentiating each taxon and a holotype designation with the repository identified; these are repeated from the original paper [1].

2. New Nomenclatural Acts

Afroryzophilus centrafricanus Caldara & Košťál, **sp. n.**

Afroryzophilus centrafricanus Caldara & Košťál, 2018: 4 [1] (not available)

http://zoobank.org/urn:lsid:zoobank.org:act:024B33CD-3AC6-4077-BDC2-DEF596639721

Description. Integument completely, densely covered with decumbent and appressed scales, pronotum moderately pitted at sides, mainly pale brown, basal half of interstria 1, base of pronotum along middle and at sides also dark brown, base of odd elytral interstriae and third quarter of interstria 1 whitish and with robust, subrecumbent, posteriorly distinctly recurved, lanceolate brown seta-like scales, these distinct and unevenly sparsely distributed on pronotum, arranged in single regular rows on each elytral interstria. Forehead 1.4 times broader than rostrum at base. Tarsi with second and third tarsomere moderately transverse, only slightly wider than long, fifth tarsomere slightly projecting beyond lobes of third tarsomere. Penis with body short and parallel-sided in dorsal view, distinctly thin in lateral view, with apodemes twice as long as body, only slightly enlarged at their extremities; endophallus with weakly sclerotized flagellum. Length 2.2–2.6 mm. See Caldara & Košťál, 2018: 4, Figures 1b, 2b, 3b, 4b and 5b [1] for full description.

Holotype, ♂: "Central African Rep., Bamingui-Bangoran Pr. [Prefecture] 75 km SSW Ndele, 8–12 July 2011, 450 m, A. Kudrna Jr. Lgt". (Repository: Museo Civico di Storia Naturale, Milano, Italy). Paratypes listed in [1].

Distribution. Central African Republic.

Afroryzophilus kuscheli Caldara & Košťál, **sp. n.**

Afroryzophilus kuscheli Caldara & Košťál, 2018: 7 [1] (not available)

http://zoobank.org/urn:lsid:zoobank.org:act:0C0AE3AD-FFCC-43BE-AA31-728AC8056848

Description. Integument with vestiture as in *A. centrafricanus* with regard to colour and pattern, except for very sparse seta-like scales on pronotum and thinner and sparse (except on interstria 1) on elytra, especially on even interstriae, almost indistinct as completely flattened. Forehead 1.3 times broader than rostrum at base. Tarsi with second and third tarsomere moderately transverse, only slightly wider than long, fifth tarsomere not projecting beyond lobes of third tarsomere. Penis with body moderately long and parallel-sided in dorsal view, moderately robust in lateral view, with apodemes only slightly longer than body, somewhat enlarged at their extremities; endophallus with distinctly sclerotized flagellum. Length 2.7–3.4 mm. See Caldara & Košťál, 2018: 7–8, Figures 1c,f,g, 2c, 3c, 4c and 5c [1] for full description.

Holotype, ♂: "Senegal-Kaolack, Norio du Rip, 24 July 2009, Moretto" (Repository: Museo Civico di Storia Naturale, Milano, Italy). Paratypes listed in [1].

Distribution. Senegal.

Afroryzophilus congoanus Caldara & Košťál, **sp. n.**

Afroryzophilus congoanus Caldara & Košťál, 2018: 8 [1] (not available)

http://zoobank.org/urn:lsid:zoobank.org:act:178C0859-A8F9-4AA2-B889-0E3054950C66

Description. Integument with colour and pattern as in *A. djibai*, vestiture with scales on pronotum and elytra more robust, more numerous and darker in colour. Forehead slightly narrower than rostrum at base. Tarsi with second and third tarsomere distinctly transverse, with fifth tarsomere not projecting beyond lobes of third tarsomere. Penis in dorsal view with body long and gradually narrowed from base to apex, in lateral view moderately robust, with apodemes moderately longer than body and distinctly enlarged at their extremities; endophallus with distinctly sclerotized flagellum. Length 2.10–2.40 mm. See Caldara & Košťál, 2018: 8, Figures 1d, 2d, 3d, 4d and 5d [1] for full description.

Holotype, ♂: "Congo Belge: P.N.A. [Parc National Albert, currently Parc National des Virunga], 14-15-VIII-1952, P. Vanschuytbroeck & J. Kekenbosch, 767-70/Massif Ruwenzori, Kalonge, 2.210 m" (Repository: Musée Royal de l'Afrique Centrale, Tervuren, Belgium). Paratype listed in [1].

Distribution. Democratic Republic of Congo.

Afroryzophilus somalicus Caldara & Košťál, **sp. n.**

Afroryzophilus somalicus Caldara & Košťál, 2018: 8 [1] (not available)

http://zoobank.org/urn:lsid:zoobank.org:act:1B43833E-251D-4F0E-BC6E-1B998E73DF28

Description. Integument completely, densely covered with decumbent and appressed scales, pronotum distinctly and deeply pitted, mainly pale brown; on basal half of elytra, especially on perisutural interstriae, with pale brown scales intermixed with pale and slightly darker scales, with distinct micaceous reflections; lanceolate, seta-like scales apically distinctly recurved, almost transparent with silvery reflections, recumbent to semi-erect on pronotum and elytra, moderately thin on rostrum, femora and tibiae, very sparse on pronotum, a little more numerous and barely visible on elytra, more numerous on basal half of rostrum and legs. Forehead slightly narrower than rostrum at base. Tarsi with second and third tarsomere moderately broad, almost as long as wide, fifth tarsomere slightly projecting beyond lobes of third tarsomere. Length 2.80 mm. See Caldara & Košťál, 2018: 8–9, Figures 1e, 2e, 3e and 4e [1] for full description.

Holotype, ♀: "Coll. Mus. Tervuren, Somalie: Afgoi [Afgooye], August 1977, Leg. Olmi" (Repository: Musée Royal de l'Afrique Centrale, Tervuren, Belgium).

Distribution. Somalia.

Author Contributions: Both authors contributed equally to the design, analysis and writing of the paper.

Acknowledgments: We are very grateful to the editorial staff of *Diversity* and especially to Dr. Christopher Lyal (The Natural History Museum, London) and Dr. Rolf Oberprieler (CSIRO Australian National Insect Collection) for their assistance in resolving this matter and for ensuring compliance with the relevant articles of the Code.

Conflicts of Interest: The authors declare no conflict of interest.

References

1. Caldara, R.; Košťál, M. Description of four new species of the Afrotropical weevil genus *Afroryzophilus* (Coleoptera, Curculionidae). *Diversity* **2018**, *10*, 37. [CrossRef]
2. International Commission on Zoological Nomenclature. INTERNATIONAL CODE OF ZOOLOGICAL NOMENCLATURE Fourth Edition [Incorporating Declaration 44, Amendments of Article 74.7.3, with Effect from 31 December 1999 and the Amendment on E-Publication, Amendments to Articles 8, 9, 10, 21 and 78, with Effect from 1 January 2012]. 2012. Available online: http://www.iczn.org/iczn/index.jsp (accessed on 26 July 2018).
3. International Commission on Zoological Nomenclature. Amendment of Articles 8, 9, 10, 21 and 78 of the International Code of Zoological Nomenclature to expand and refine methods of publication. *Zookeys* **2012**, *219*, 1–10. [CrossRef]

MDPI
St. Alban-Anlage 66
4052 Basel
Switzerland
Tel. +41 61 683 77 34
Fax +41 61 302 89 18
www.mdpi.com

Diversity Editorial Office
E-mail: diversity@mdpi.com
www.mdpi.com/journal/diversity

www.ingramcontent.com/pod-product-compliance
Lightning Source LLC
Chambersburg PA
CBHW051702210326
41597CB00032B/5346